Signal Transduction

Signal Transduction

Bastien D. Gomperts
Division of Biosciences, University College London, UK

IJsbrand M. Kramer
European Institute of Chemistry and Biology, University of Bordeaux, France

Peter E.R. Tatham
Division of Biosciences, University College London, UK

ELSEVIER

AMSTERDAM • BOSTON • HEIDELBERG • LONDON • NEW YORK • OXFORD
PARIS • SAN DIEGO • SAN FRANCISCO • SINGAPORE • SYDNEY • TOKYO
Academic Press is an imprint of Elsevier

Academic Press is an imprint of Elsevier
30 Corporate Drive, Suite 400, Burlington, MA 01803, USA
32 Jamestown Road, London NW 1 7BY, UK
525 B Street, Suite 1900, San Diego, CA 92101-4495, USA

Library of Congress Cataloging-in-Publication Data
A catalog record for this book is available from the Library of Congress

British Library Cataloging in Publication Data
A catalogue record for this book is available from the British Library

ISBN: 978-0-12-369441-6

For information on all Academic Press publications
visit our website at www.elsevierdirect.com

Typeset by Macmillan Publishing Solutions
(www.macmillansolutions.com)

09 10 11   10 9 8 7 6 5 4 3 2 1

Contents

Contents

Contents

Contents

Contents

Contents

Contents

Contents

Preface

Introducing the first edition of *Signal Transduction*, we asked the question, for whom had this book been written, and the simplest response was that it had been written for ourselves. We added the pious hope that it might also be instructive and entertaining for students and professionals at many levels. So why a second edition? The main reasons must be that we have not only been spurred on by the enormous progress in the field but also encouraged by a number of very favourable reviews and letters of appreciation that have offered constructive criticism and new ideas for the development of the text.

In the intervening period (around seven years), the general territory covered by the expression 'signal transduction' has spread inexorably wider, but with the passing of time, not just one, but now two of the authors (BDG and PERT) have gone beyond the age at which it is normal, in British universities at least, to close up their labs and depart their desks. So at the outset, it is beholden on these two greybeards to acknowledge the major contribution of IJsbrand Kramer who undertook the drafting of most of the second half of this new edition. Without him, there would be no second edition. Yet, as before, the aim throughout has been to create a single text as if conceived by one mind, written by one hand. In this way, we hope that it avoids the worst excesses of the skimpily edited texts compiled from articles by multiple specialist authors.

Although we have touched the leading edges of the subject, we have also endeavoured to provide an elementary basis with some historical background to all the topics covered. There has been no attempt to be comprehensive and we are aware that many important topics that well qualify for inclusion are conspicuous only by their absence. However, the omissions in the new edition are not the same as those of the first. Gone now are the major sections on the cell cycle and apoptosis; instead we have new chapters covering nuclear receptors, development, and cancer therapy. As previously, we have been the main beneficiaries as students of our own subject. As we learned more, we were encouraged to challenge, or at least to re-examine, some of the well-established dogmas. We have also learned to respect the wisdom of our forebears, whose freedom of thought and sometimes serendipitous discoveries in the 19th and early 20th centuries led to the creation of the modern sciences of physiology, pharmacology, biochemistry, and cell biology, and related clinical fields, especially endocrinology and immunology.

The book conveniently divides in two parts. The first nine chapters provide the nuts and bolts of what might be termed 'classical' signal transduction.

They concentrate mainly on hormones, their receptors, and the generation and actions of second messengers, particularly cyclic nucleotides and calcium. It was the advances in this area, particularly the discovery of the G proteins, that originally gave rise to the expression 'signal transduction', although the word 'transduction' was stolen from elsewhere (see definitions, below). Chapter 10 extends the paradigm, dealing with the intracellular receptors that, in the nucleus, influence transcription. Then, in Chapters 11–23, attention is concentrated on processes set in action by growth factors and adhesion molecules, particularly through the covalent modification of proteins and inositol-containing lipids, for example by phosphorylation, dephosphorylation, cleavage or ubiquitylation. An important, though not exclusive, impetus to research in this area has been the quest to understand the cellular transformations underlying cancer, with the hope of devising effective therapeutic procedures. Finally, Chapter 24 reflects back on a theme that runs throughout the book. It deals with the concept of protein structural domains and illustrates their central importance in signalling mechanisms.

In preparing the book, we have had the benefit of advice and opinions from many friends and colleagues. These include John Blenis (Boston), Alex Bridges (Ann Arbor), Zhijan Chen (Dallas), Jean Dessolin (Bordeaux), Elisabeth Genot (Bordeaux), John Kuriyan (Dallas), Michel Laguerre (Bordeaux), Patrick Lemaire (Marseille), David Litchfield (London, Ontario), Alfonso Martinas-Arias (Cambridge, UK), Joan Massagué (New York), Juan Modolell (Madrid), Alexandra Newton (La Jolla), François Schweisguth (Paris), Pat Simpson (Cambridge, UK), Nick Tonks (Cold Spring Harbor), Colin Ward (Parkville), Xuewu Zhang (Dallas). Geoffrey Strachan advised on the translation of French and German texts into contemporary (19th century) English.

Many others, too numerous to name individually, have given us the benefit of their knowledge and understanding. In acknowledgement of their contribution we offer the following quotation by one of the pioneers of signal transduction:[1]

Of course, the authors of this paper would themselves never have recognized the expression 'signal transduction' and it would be a further 100 years before it made its appearance in the biological literature. The sensations brought about by pituri, an alkaloid that according to Ringer and Murrell induces some of the pharmacological effects of atropine (courage, infuriation, frustration, and headaches), are not dissimilar to those experienced by us in the writing of this book. Indeed, they will be familiar to any students and investigators in this and other fields of research. However, we should not take this too far. When Ringer tested the effects of the application of pituri on four men,[2] he noted that it also caused drowsiness, faintness, pallor, giddiness, hurried and superficial breathing, dilatation of the pupils, general weakness with convulsive twitchings and in large doses copiously increased salivary secretion.

ON PITURI. By SYDNEY RINGER, M.D., and WILLIAM MURRELL, M.R.C.P., *Lecturer on Practical Physiology at the Westminster School of Medicine, and Assistant Physician to the Royal Hospital for Diseases of the Chest.*

QUITE recently a student of University College, London, whose name we have unfortunately forgotten, gave us a small packet containing a few twigs and broken leaves of the powerful and interesting drug Pituri. These we placed in Mr Gerrard's hands, and he kindly made first an extract from which he obtained a minute quantity of an alkaloid, and with this he made a solution containing one part of the alkaloid to twenty of water.

Baron Mueller, from an examination of the leaves of pituri, is of opinion that it is derived from Duboisia Hopwoodii. Pituri is found growing in desert scrubs from the Darling River and Barcoote to West Australia. The natives, it is said to fortify themselves during their long foot marches, chew the leaves for the same purpose as Cocoa leaves are used in Bolivia. Dr G. Bennett in the New South Wales Medical Gazette, May, 1873, says Pituri is a stimulating narcotic and is used by the natives of New South Wales in like manner as the Betel of the East. It seems to be a substitute for tobacco.

It is generally met with in the form of dry leaves, usually so pulverized that their character cannot be made out.

The use of pituri is confined to the men of a tribe called Mallutha. Before any serious undertaking, they chew these dried leaves, using about a tea-spoonful. A few twigs are burnt and the ashes mixed with the leaves. After a slight mastication the bolus is placed behind the ear (to increase it is supposed its strength), to be again chewed from time to time, the whole being at last swallowed. The native after this process is in a sufficiently courageous state either to transact business or to fight. When indulged in to excess, it is said to induce a condition of infuriation. In persons not accustomed to its use pituri causes severe headache.

It also antagonized the action of muscarine on the heart. This has not been our experience, which leads one to wonder who, among their students, colleagues, and servants may have offered themselves up as willing – or perhaps less than willing – guinea pigs in the furtherance of scientific research. Ringer and his friends apparently eschewed membership of the very honourable brotherhood of self-experimenters of which the more famous members include Sir Humphry Davey, who breathed nitrous oxide as well as other more noxious gases; John Scott Haldane, who too inhaled lethal gases; and more recently Barry Marshall, who swallowed a culture of *Helicobacter pylori* to show that it caused stomach ulcers and who with Robin Warren was awarded the Nobel Prize in Physiology or Medicine in 2005. Another member of this fraternity, Charles Edouard Brown-Séquard, figures prominently in the first chapter.

References

1. Ringer S, Murrell W. On pituri. *J Physiol*. 1878:377–383.
2. Ringer S. On the action of pituri on man. *Lancet*. 1879:290–291.

From the Shorter Oxford English Dictionary (3rd edition, 1994, with corrections 1977, © Oxford University Press:

Transduction (trɒnsˌdʌ·kʃən). *rare.* 1656. [ad. L. *tra(ns)ductionem, tra(ns)ducere*; see TRADUCE.] The action of leading or bringing across.

Traduce (trădiū·s), *v.* 1533. [ad. L. *traducere* to lead across, etc.; also, to lead along as a spectacle, to bring into disgrace; f. *trans* across + *ducere* to lead.] †1. *trans.* To convey from one place to another; to transport –1678. †b. To translate, render; to alter, modify, reduce –1850. †c. To transfer from one use, sense, ownership, or employment to another –1640. †2. To transmit, esp. by generation –1733. †b. *transf.* To propagate –1711. †c. To derive, deduce, obtain *from* a source –1709. 3. To speak evil of, esp. (now always) falsely or maliciously; to defame, malign, slander, calumniate, misrepresent 1586. †b. To expose (to contempt); to dishonour, disgrace (*rare*) –1661. †4. To falsify, misrepresent, pervert –1674. 1. b. Milton has been traduced into French and overturned into Dutch SOUTHEY. a. Vertue is not traduced in propagation, nor learning bequeathed by our will, to our heires 1606. 3. The man that dares t., because he can With safety to himself, is not a man COWPER. b. By their own ignoble actions they t., that is, disgrace their ancestors 1661. 4. Who taking Texts .. traduced the Sense thereof 1648. Hence **Tradu·cement**, the, or an, action of traducing; defamation, calumny, slander. **Tradu·cingly** *adv.*

From the Oxford English Dictionary (2nd edition, 2008 © Oxford University Press) online:

transduction

(trɑːnsˈdʌkʃən, træns-) [ad. L. *transductiōn-em* (usually *trāductiōnem*), n. of action f. *tra(ns)dūcĕre*: see TRADUCE.]

1. The action of leading or bringing across. *rare.*

> **1656** BLOUNT *Glossogr.*, *Transduction*, a leading over, a removing from one place to another. *a***1816** BENTHAM *Offic. Apt. Maximized, Introd. View* (1830) 19 In lieu of *adduction*, as the purpose requires, will be subjoined *abduction, transduction,*..and so forth.

2. The action or process of transducing a signal.

> **1947** *Jrnl. Acoustical Soc. Amer.* XIX. 307/1 It is rather interesting..that the direct method of electronic transduction, instead of the indirect method of employing a conventional transducer and then amplifying the output with a vacuum tube, has not been developed. **1970** J. EARL *Tuners & Amplifiers* iv. 87 Low impedance pickup cartridges..using the moving-coil principle of transduction. **1975** *Nature* 17 Apr. 625/1 The transduction of light energy into neural signals is mediated in all known visual systems by a common type of visual pigment.

3. *Microbiology.* The transfer of genetic material from one cell to another by a virus or virus-like particle.

> **1952** ZINDER & LEDERBERG in *Jrnl. Bacteriol.* LXIV. 681 To help the further exposition of our experiments, we shall use the term transduction for genetically unilateral transfer in contrast to the union of equivalent elements in fertilization. **1960** [see F III. 1l]. **1971** *Nature* 18 June 466/1 It has been suggested that transduction of genes by viruses was an important mechanism in evolution for spreading useful mutations between organisms not formally related. **1977** *Lancet* 9 July 94/2 These were derived by selection of sensitive variants from gentamicin-resistant strains or by transduction of this resistance to sensitive strains.

Hence **transˈductional** *a.*, of or pertaining to (genetic) transduction.

> **1956** *Genetics* XLI. 845 (*heading*) Linear inheritance in transductional clones. **1980** *Jrnl. Gen. Microbiol.* CXIX. 51 Transductional analysis revealed that one of the four mutations carried by strain T-693 was responsible for constitutive synthesis of both isoleucine and threonine biosynthetic enzymes.

Notes

For protein structural data we have made use of:

- The Protein Data Bank (http://www.pdbr.org): H. M. Berman, J. Westbrook, Z. Feng et al., The protein data bank. Nucleic Acid Res 28, 235–242 (2000).
- For chemical structures we wish to acknowledge the use of the EPSRC's Chemical Database Service at Daresbury: D. A. Fletcher, R. F. McMeeking, and D. Parkin,The United Kingdom Chemical Database Service, J Chem Inf Comput Sci 36, 746–749 (1996).

Most of the protein structures were generated using the following software:

- PyMOL (Warren Delano, Delano Scientific LLC, Palo Alto, CA, USA; http://www.delanoscientific.com/).

Other structures were obtained using

- RasMol (Roger Sayle and E. James Milner-White, RasMol: Biomolecular graphics for all, Trends Biochem Sci 20, 374 (1995))
- CHIME (Eric Martz, University of Massachusetts, Amherst, MA, USA and MDL Information Systems, Inc., San Leandro, CA, USA.)

How to view stereo images of molecular structures

There is much more information in a stereoscopic image than in a conventional flat projection. Inexpensive viewers for looking at stereo pairs are available via the Internet, but with a little effort most people can learn how to view stereo images with unaided eyes, unless you are unfortunate enough to have an ophthalmological condition such as one very weak eye. Practice is the key.

There are two ways of seeing a 3D image by observing a stereo pair. You may either cross your eyes, so that the left eye views the right-hand image and vice versa, or you may allow your eyes to diverge, so that each eye looks at the image in front of it. Computer-generated images are usually for divergent viewing. To view the images in this book (and most printed images), you should view the left image with the left eye and the right image with the right eye. If the two do not readily fuse into a single 3D image, try the following.

First touch your nose to the page between and below the stereo pair. The two images will now be superimposed, but the picture will be very blurred (although you may notice some three-dimensionality at this stage). Now move the page slowly away from you, but take care not to rotate it. Concentrate on the 3D aspect and wait for your eyes to bring it into focus.

Note that if you attempt to view a divergent pair using the convergent strategy (crossed eyes) the image will be 3D but inverted in a confusing way.

References

We have tried to provide original text sources to nearly all the statements, experiments, and discoveries discussed. The main reason for this is that we ourselves have necessarily had to extend the treatment of nearly all the topics presented far beyond the areas of our own experience or expertise. Thus, the comprehensive lists are there to provide us with some sort of reassurance that what we have written has not simply been conjured out of the air. Also, because we have made a particular feature of presenting original historical source material by quotation, which necessarily required referencing, it seemed logical also to include literature references to modern sources as well. Thus we hope that this book may serve as a valuable resource, in the manner of a basic literature review, for anyone wanting to explore further.

Abbreviations

Historically, the names of many biologically active compounds and proteins (mainly enzymes) were related to effect or function and the abbreviations are well established and fairly obvious. More recently, as waves of new gene products have landed upon the literature, the matter of naming and abbreviation has become completely random. In order not to disrupt the text with a succession of clumsy definitions and translations, dedicated tables have been appended at the end of each chapter (from Chapter 12 onwards), linking the names to their abbreviations and their accession numbers in the SwissProt and OMIM databases.

Genes and gene products

According to convention, acronyms and abbreviations printed in lower case indicate genes (e.g. ras). When capitalized, they indicate their protein products (Ras). The genes of yeast are printed in upper case (RAS). The prefixes v- and c- indicate viral or cellular origin (v-Ras, C-Ras).

Amino acids

Generally the standard three-letter abbreviations are used. Where this is inconvenient or otherwise unsuitable, the single-letter codes have been used. Both sets of abbreviations are listed in the following table.

Amino acid	Three-letter code	Single-letter code
Alanine	Ala	A
Arginine	Arg	R
Asparagine	Asp	N
Aspartate	Asp	D
Cysteine	Cys	C
Glutamate	Glu	E
Glutamine	Gln	Q
Glycine	Gly	G
Histidine	His	H
Isoleucine	Ile	I
Leucine	Leu	L
Lysine	Lys	K
Methionine	Met	M
Phenylalanine	Phe	F
Proline	Pro	P
Serine	Ser	S
Threonine	Thr	T
Tryptophan	Trp	W
Tyrosine	Tyr	Y
Valine	Val	V

Prologue: Signal Transduction, Origins, and Ancestors

Transduction, the word and its meaning: one dictionary, different points of view

The expression *signal transduction* first made its mark in the biological literature in the 1970s[1] and appeared as a title word in 1979.[2–4] Physical scientists and electronic engineers had earlier used the term to describe the conversion of energy or information from one form into another. For example, a microphone transduces sound waves into electrical signals. The widespread use of the term in bio-speak was triggered by an important review by Martin Rodbell, published in 1980 (**Figure 1.1**).[5] He was the first to draw attention to the role of GTP and GTP-binding proteins in metabolic regulation and he deliberately borrowed the term to describe their role. By the year 2000, 12% of all papers using the term *cell* also employed the expression *signal transduction*.

Alfred G. Gilman and Martin Rodbell, awarded the Nobel Prize in 1994 'for their discovery of G-proteins and their role in signal transduction in cells'.

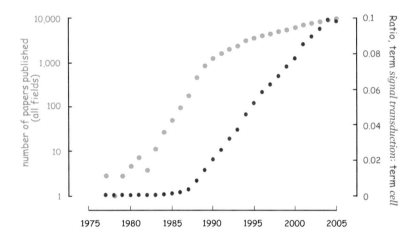

FIG 1.1 Occurrence of the term *signal transduction*. The left-hand axis records all papers using the term *signal transduction* traced through the PubMed database. The right-hand axis records the proportion of papers using the term *cell* that also use the term *signal transduction*.

Hormones, evolution, and history

These chemical messengers . . . or 'hormones' (from the Greek ὁρμων, meaning excite or arouse), as we may call them, have to be carried from the organ where they are produced to the organ which they affect, by means of the bloodstream, and the continually recurring physiological needs of the organism must determine their repeated production and circulation throughout the body.[6]

The plasma membrane barrier

In the main, when we consider signal transduction we are concerned about how external influences, particularly the presence of specific hormones, can determine what happens inside their target cells. There is a difficulty, since the hormones, being mostly hydrophilic (or lipophobic) substances, are unable to pass through membranes, so that their influence must somehow be exerted from outside. The membranes of cells, although very thin (3–6 nm) are effectively impermeable to ions and polar molecules. Although K^+ ions might achieve diffusional equilibrium over this distance in water in about 5 ms, they would take some 12 days (280 h) to equilibrate across a phospholipid bilayer (under similar conditions of temperature, etc.). Likewise, the permeability of membranes to polar molecules is low. Even for small molecules such as urea, membrane permeability is about 10^4 times lower than that of water. So for a hormone such as adrenaline (epinephrine), the rate of permeation is too low to measure. The evolution of receptors has accompanied the development of mechanisms which permit external chemical signalling molecules, the first messengers, to direct the activities of cells in a variety of ways with high specificity and precise control in terms of extent and duration. With some

Steroids, prostaglandins, etc. are hydrophobic and can therefore traverse cell membranes to interact with their intracellular receptors.

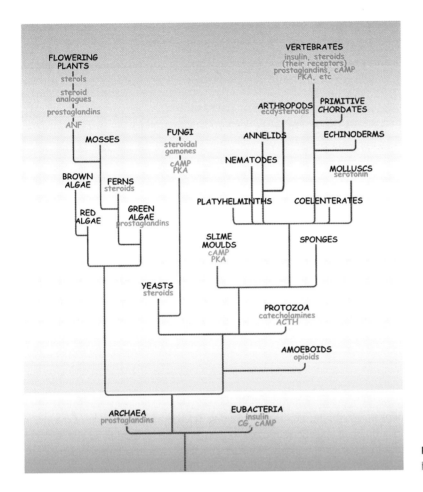

FLOWERING PLANTS
sterols
steroid analogues
prostaglandins
ANF

MOSSES

BROWN ALGAE

FERNS
steroids

RED ALGAE

GREEN ALGAE
prostaglandins

FUNGI
steroidal gamones
cAMP PKA

YEASTS
steroids

SLIME MOULDS
cAMP PKA

PLATYHELMINTHS

NEMATODES

ANNELIDS

ARTHROPODS
ecdysteroids

VERTEBRATES
insulin, steroids (their receptors) prostaglandins, cAMP PKA, etc

PRIMITIVE CHORDATES

ECHINODERMS

MOLLUSCS
serotonin

COELENTERATES

SPONGES

PROTOZOA
catecholamines ACTH

AMOEBOIDS
opioids

ARCHAEA
prostaglandins

EUBACTERIA
insulin CG, cAMP

FIG 1.2 Emergence of signalling hormones. Adapted from Baulieu.[10]

important exceptions (the steroid hormones, thyroid hormone), they do this without ever needing to penetrate their target cells.

Protohormones

The first messengers (which include the hormones), and their related intracellular (second) messengers, are of great antiquity on the biological timescale. It is interesting to consider which came first: the hormones or the receptors that they control. Substances exhibiting the actions of hormones in animals first made their appearance at early stages of evolution (**Figure 1.2**). Chemical structures closely related to thyroid hormones have been discovered in algae, sponges, and many invertebrates. Steroids such as estradiol are present in microorganisms and also in ferns and conifers. Catecholamines have been found in protozoa,[7,8] and ephedrine, which is closely related, can be isolated from the stems and leaves of the Chinese herb Ma Huang (*Ephedra sinica*). Ephedrine is still in use as an oral stimulant in the treatment of hypotension (low blood pressure) and in the relief of asthma. There are claims, based on immunological detection, for the presence of peptides related to insulin and the endorphins in protozoa, fungi, and even bacteria,[9] although no messenger-like function has been discerned and it is likely that the receptors that mediate their effects in animals evolved much later.

The a- and α-type mating factors of yeast, which certainly act as messengers, are very similar in structure to gonadotropin-releasing hormone (GnRH) which controls the release of gonadotropins from the anterior pituitary in mammals.[11,12] Factors resembling mammalian atrial natriuretic factor (ANF) are present in the cytosol of the single-cell eukaryote *Paramecium multimicronucleatum*[13] and in the leaves of many species of plants, where they act as regulators of solvent and solute flow and of the rate of transpiration.[14] ACTH and β-endorphin are present in protozoa. These organisms also contain high molecular mass precursors of these peptides, reminiscent of the vertebrate pro-opiomelanocortin (POMC).[15,16] It is striking that pathways for the biosynthesis of these 'protohormones', often complex, were established early on, well before the evolution of membrane receptors.

Receptor-like proteins in non-animal cells have been much harder to identify. A recently described example is a protein expressed in the plant *Arabidopsis* that shares extensive sequence homology with the ionotropic glutamate receptor of mammalian brains.[17] A corollary is the possibility that the potent neurotoxins thought to be generated in defence against herbivores may have their origin as specific agonists, and were only later selected and adapted in some species as poisons. Molecules having a close relationship to the receptors for epidermal growth factor (EGF) and insulin apparently evolved in sponges before the Cambrian Explosion (more than 600 million years ago) and it has been proposed that they may have contributed to the rapid development of the higher metazoan phyla.[18]

Although invertebrates express some members of the nuclear receptor family (such as the receptors for thyroid hormone and vitamin D), nuclear receptors for adrenal and sex steroid hormones (cortisol, aldosterone, testosterone, estradiol, progesterone, etc.) are absent.[19] The ancestral steroid hormone receptor probably made its first appearance in a cephalochordate such as *Amphioxus*. Receptors for estradiol, progesterone, and cortisol have been cloned from lamprey. From this point in evolution onwards, the steroid hormones would have allowed for a ligand-based mechanism for the regulation of gene transcription and this could have promoted the complex processes of differentiation and development found in the higher vertebrates.[20] Thus, the *hox* genes that are critical for development and differentiation, including the brain of *Amphioxus*,[21] are regulated by oestrogens and progestins.

The lancelets (subphylum Cephalochordata, traditionally known as **amphioxus**) are fish like creatures 5–7 cm long, members of a group of primitive chordates usually found buried in sand in shallow parts of temperate or tropical seas. The body is pointed at both ends and it can swim both forwards and backwards. They split from the vertebrates more than 520 million years ago and although lacking discrete vertebrae, they are regarded as similar to the vertebrate archetype.

In general, it appears that many of the molecules that we regard as hormones arose long before the receptors that they control. An important consequence of this is that the responses to a given hormone can vary widely across different species and even within species. Numerous actions of prolactin have been identified.[22] It is the regulator of mammary growth and differentiation and of milk protein synthesis in mammals. In birds, it acts as a stimulus to crop milk production and in some species as a controlling factor for fat deposition and as a determinant of migratory behaviour.[23,24] It is a regulator of water balance in urodeles (newts and salamanders) and of salt adaptation and melanogenesis in fish.[22] Serotonin (5-hydroxytryptamine), a neurotransmitter that controls mood in humans, is reported to stimulate spawning in molluscs, probably as a consequence of its conversion to melatonin (naturally, one wonders whether it affects their mood as well).

Protoendocrinologists

Despite excellent anatomical descriptions, almost nothing was known about the functions of the various organs which constitute the endocrine system (glands) until the last decade of the 19th century. Indeed, in the standard textbook of the period (*Foster's Textbook of Physiology*, 3 volumes and more than 1200 pages), consideration of the thyroid, the pituitary, the adrenals ('suprarenal bodies'), and the thymus is confined to a brief chapter of less than 10 pages, having the title 'On some structures and processes of an obscure nature'.

The initial impetus prompting the systematic investigations which led to the discovery of the hormones can be ascribed to a series of papers that were much misunderstood. However, here we are confronted with the work of Charles Edouard Brown-Séquard, the successor to Claude Bernard at the Collège de France and also a member of leading scientific academies in England and the USA. He had held professorial appointments at both Harvard and Virginia; in London he was appointed physician at the National Hospital for the Paralysed and Epileptic (now the National Hospital for Neurology and Neurosurgery). He was an associate of Charles Darwin and Thomas Huxley. He wrote over 500 papers relating to many diverse fields such as the physiology of the nervous system; the heart, blood, muscles, and skin; the mechanism of vision; and much more. He was an outstanding experimentalist making fundamental contributions. Starting with his doctoral thesis, he described the course of motor and sensory fibres in the spinal cord, a field to which he returned many times. He was in constant demand as lecturer, teacher, and physician on both sides of the Atlantic, crossing the ocean on more than 60 occasions. Of direct relevance to us must be his demonstration that the adrenal glands are essential to life.[25–27]

In view of all this, it is curious that Brown-Séquard is now all but forgotten. On the rare occasions when he is recalled, it is generally in connection with a

In writing these paragraphs, we have relied heavily on the work of Victor Medvei[25], Michael Aminoff,[26] and Horace Davenport.[27]

SECONDE NOTE SUR LES EFFETS PRODUITS CHEZ L'HOMME PAR DES INJECTIONS SOUS-CUTANÉES D'UN LIQUIDE RETIRÉE DES TESTICULES FRAIS DE COBAYE ET DE CHIEN.

par M. BROWN-SÉQUARD. (*Communication faite le 15 juin*)

Non seulement il n'y a pas à s'étonner que l'introduction dans le sang de principes provenant de testicules de jeunes animaux soit suivie d'une augmentation de vigueur, mais encore on devait s'attendre à obtenir ce résultat. En effet, tout montre que la puissance de la moelle épiniére et aussi, mais à un moindre degré, celle du cerveau, chez l'homme adulte ou vieux, des fluctuations liées à l'activité fonctionnelle des testicules. Aux faits que j'ai mentionnés, à cette égard, dans la séance du 1er juin, je crois devoir ajouter que les particularités suivantes ont été observées un trés grand nombre de fois, pendant plusieurs années, chez deux individus âgés de quarante-cinq à cinquante ans. Sur mon conseil, chaque fois qu'ils avaient à exécuter un grand travail physique ou intellectuel, ils se mettaient dans un état de vive excitation sexuelle, en évitant cependant toute éjaculation spermatique. Les glandes testiculaires acquéraient alors temporairement une grande activité fonctionnelle, qui était bientôt suivie de l'augmentation désirée dans la puissance des centres nerveux.

series of brief reports, published in 1889, in which he described the self-administration by injection, of testicular extracts, which he considered had the effect of reinforcing his bodily functions. Some brief quotations from his paper in the *Lancet*[28] must suffice:

> I am seventy-two years old. My general strength, which has been considerable, has notably and gradually diminished during the last ten or twelve years. Before May 15th last, I was so weak that I was always compelled to sit down after about half an hour's work in the laboratory. Even when I remained seated all the time I used to come out of it quite exhausted after three or four hours of experimental labour . . .

> The day after the first subcutaneous injection, and still more after the two succeeding ones, a radical change took place in me, and I had ample reason to say and to write that I had regained at least all the strength I possessed a good many years ago . . .

> My limbs, tested with a dynamometer for a week before my trial and during the month following the first injection, showed a decided gain of strength . . .

> I have measured comparatively, before and after the first injection, the jet of urine in similar circumstances – i.e. after a meal in which I had taken food and drink of the same kind and in similar quantity. The average length of the jet during the ten days that preceded the first injection was inferior

by at least one quarter of what it came to be during the twenty following days. It is therefore quite evident that the power of the spinal cord over the bladder was considerably increased . . .

I will simply say that after the first ten days of my experiments I have had a greater improvement with regard to the expulsion of faecal matters than in any other function. In fact a radical change took place, and even on days of great constipation the power I long ago possessed had returned.

Finally, in a footnote:

It may be well to add that there are good reasons to think that subcutaneous injections of a fluid obtained by crushing ovaries just extracted from young or adult animals and mixed with a certain amount of water, would act on older women in a manner analogous to that of the solution extracted from the testicles injected into old men.

Possibly Brown-Séquard should be regarded as the father of hormone replacement therapy , but he was certainly not the inventor of organotherapy, the attempt to cure human disease by the introduction of glandular extracts from other animals. Indeed, Gaius Plinius Secundus (known as Pliny the Elder, 23–79 ce) recorded his contempt of the Greeks who used human organs therapeutically, although he did recommend the use of animal tissues, in particular that the testicles of animals should be eaten in order to cure impotence or to improve sexual function in men, and that the genitalia of a female hare should be eaten by women in order to achieve pregnancy.[25,29]

Fritz Spiegl[29] reminds us that in Chinese medicine the effect of HRT is allegedly achieved by the administration of rhinoceros horn (RHT?)

Ever mindful of the possibility of autosuggestion, Brown-Séquard pleaded that others should examine his claims and to consider them with objectivity. What a hope! But if any doubts remained, he went on record saying:[30]

Not only is there no occasion for surprise that the introduction into the blood of principles derived from the testicles of young animals should be followed by an increase in vigour: such a result should indeed have been foretold. For in point of fact all the evidence shows that in the adult or elderly male the power of the spinal cord and also, albeit to a lesser degree, that of the brain is subject to variations linked to the functional activity of the testicles. In addition to the facts mentioned in this regard at the meeting of June 1st, I believe I should add that the following particularities have been observed on many occasions during the course of several years in the case of two individuals aged between 45 and 50 . . . On my advice, each time they had to undertake a great physical or intellectual task they put themselves into a state of sexual excitement, while nevertheless avoiding ejaculation. The testicles became substantially active for a time and this was soon followed by the desired increase in the power of the nervous system.

A pentacle is a five-pointed star, of supposed supernatural significance: see http://www.witchvox.com/basics/pentacle.html

News travels fast, and news of this sort travels faster still. Within weeks this was the subject of an editorial paragraph in the *British Medical Journal* under the heading THE PENTACLE OF REJUVENESCENCE:[31]

> On two occasions this month Dr. Brown-Séquard has made communication of a most extraordinary nature to the Societé de Biologie of Paris. The statements he made – which have unfortunately attracted a good deal of attention in the public press – recall the wild imaginings of mediaeval philosophers in search of an elixir vitae . . . MM Féré and Dumontpallier, in commenting on M. Brown-Séquard's statements, observed that they would require to be rigidly tested and fully confirmed by other self-experimenters before they were likely to meet with general acceptance, and in this opinion we fully concur.

This rather measured account by the editors of the *BMJ* contrasts with the reaction on the other side of the Atlantic, where it was taken up as:

> THE NEW ELIXIR OF YOUTH. – Dr. Brown-Séquard's rejuvenating fluid is said to have been tried at Indianapolis on a decrepit old man with marvellous effect. Four hours after the injection the patient walked over a mile in twenty-five minutes. He declared that he felt more vigorous than he had done for twenty-five years. He read a newspaper without glasses, a thing he had not been able to do for thirty years. It should be added that no medical authority is given for this story.[32]

With all this happening, possibly Brown-Séquard began to harbour some doubts, as he wrote:

> In the United States especially, and often without knowing what I did or the most elementary rules regarding injections of animal materials, several physicians or rather the medicasters and charlatans have exploited the ardent desires of a great number of individuals and have made them run the greatest risks, if they have not done worse.

In England the reaction remained negative:

> Dr. Brown-Séquard's Experiments
>
> A MEMBER OF THE MEDICAL PROFESSION writes: You have not done me justice by your note on my circular relative to Dr. Brown-Séquard's experiments. Much as I disapprove of the animal torture in question, I should not have felt it my duty to print and post some 6,000 copies of my protest had that been all. I consider the idea of injecting the seminal fluid of dogs and rabbits into human beings a disgusting one, and when the treatment also involves the practice of masturbation, I think it is time for the medical profession in England to repudiate it. One may be a vivisector without also encouraging a loathsome vice, but as far as I could discover no word of protest was uttered by the leaders of the medical opinion, and I determined to take the thing

into my own hands, with the effect, as I have good reason to believe, of making it very improbable that this method of pretended rejuvenescence will be introduced into England. I have a great number of encouraging letters from eminent Englishmen, who will take good care of this, at least. Vivisection may be an open question, but self abuse is not. [31]

Dr Edward Berdoe, who was also a member of the profession, left little doubt about his feelings on the matter:

> the object of these abominable proceedings is to enable broken down old libertines to pursue with renewed vigour the excesses of their youth, to rekindle the dying embers of lust in the debilitated and aged, and to profane the bodies of men which are the temples of God, by an elixir drawn from the testicles of dogs and rabbits by a process involving the excruciating torture of innocent animals . . . We may have also a new race of beings intermediate between man and the lower animals as a remoter consequence of the boon to humanity conferred by French physiology. [33]

Like it or not, many physicians, aware of Brown-Séquard's publications, were eager to test the possibilities of applying various organ extracts in their practice. Within a year or two, organotherapy was becoming respectable. In a letter to his colleague Jacques Arséne d'Arsonval (quoted by Borell[36]) Brown-Séquard wrote that 'the thing' was 'in the air'. What made it especially so may have been George Redmayne Murray's report[37] on the treatment of myxoedema. He described a patient who was treated with regular injections of sheep thyroid extract and went on to survive for a further 28 years. [34,35]

Henry Dale's description of the discovery of adrenaline cannot be bettered: [38,39]

> Dr George Oliver, a physician of Harrogate, employed his winter leisure in experiments on his family, using apparatus of his own devising for clinical measurements. In one such experiment he was applying an instrument for measuring the thickness of the radial artery; and, having given his son, who deserves a special memorial, an injection of an extract of the suprarenal gland, prepared from material supplied by the local butcher, Oliver thought that he detected a contraction, or, according to some who have transmitted the story, an expansion of the radial artery. Which ever it was, he went up to London to tell Professor Schäfer what he thought he had observed, and found him engaged in an experiment in which the blood pressure of a dog was being recorded; found him, not unnaturally, incredulous about Oliver's story and very impatient at the interruption. But Oliver was in no hurry, and urged only that a dose of his suprarenal extract should be injected into a vein when Schäfer's own experiment was finished. And so, just to convince Oliver that it was all nonsense, Schäfer gave the injection, and then stood amazed to see the mercury mounting in the arterial manometer till the recording float was lifted almost out of the distal limb.

And respectable it remained, anyway in France, where in the 1920s Serge Voronoff was treating elderly gentlemen by implanting sections of chimpanzee or baboon testicles, the so-called 'monkey glands'.[34] R. V. Short[35] speculates that his patients may have been the first humans to become infected by HIV-1. Were it not for the fact that they remained impotent, the disease might have exploded on a world even less prepared than it was at the century's end. One might reasonably imagine that an AIDS epidemic originating at that time could have changed the course of 20th century history.

Thus the extremely active substance formed in one part of the suprarenal gland, and known as adrenaline, was discovered.

Within a few months, Oliver and Schäfer had demonstrated that the primary effect of the extract is a profound arteriolar constriction with a resulting increase in the peripheral resistance.[40,41] Their colleague Benjamin Moore reported that the activity could be transferred by dialysis through membranes of parchment paper, that it is insoluble in organic solvents but readily soluble in water, resistant to acids and to boiling, etc.[42]

Schäfer's own account of his first meeting with George Oliver relates events that may have taken place on a another planet from Dale's version:[39]

In the autumn of 1893 there called upon me in my laboratory in University College a gentleman who was personally unknown to me, but with whom I had a common bond of interest – seeing that we had both been pupils of Sharpey, whose chair at that time I had the honour to occupy. I found that my visitor was Dr George Oliver, already distinguished not only as a specialist in his particular branch of medical practice, but also for his clinical applications of physiological methods. Dr Oliver was desirous of discussing with me the results which he had been obtaining from the exhibition by mouth of extracts of certain tissues, and the effects which these had in his hands produced upon the blood vessels of man, as investigated by two instruments which he had devised – one of them, the haemodynamometer, intended to read variations in blood pressure, and the other, the arteriometer, for measuring with exactness the lumen of the radial or any other superficial artery. Dr Oliver had ascertained, or believed that he had ascertained, by the use of these instruments, that glycerin extracts of some organs produce diminution in calibre of the arteries, and increase of pulse tension, of others the reverse effect.

Hormones: a definition

These are blood-borne 'first messengers', usually secreted by one organ (or group of cells) in response to an environmental demand to signal a specific response from another. The first such messenger to be endowed with the title of hormone was secretin, later shown to be a peptide released into the blood stream from cells in the stomach lining, indicating the presence of food and alerting the pancreas. In the words of William Maddock Bayliss (co-discoverer with Ernest Henry Starling),

There are a large number of substances, acting powerfully in minute amount, which are of great importance in physiological processes. One class of these consists of the hormones which are produced in a particular organ, pass into the blood current and produce effects in distant organs. They provide, therefore, for a chemical coordination of the activities of the organism, working side by side with that through the nervous system. The internal secretions, formed by ductless glands, as well as by other tissues, belong to the class of hormones.[43]

In fact, adrenaline, the signal for 'fright, fight, and flight', is a much better candidate for the accolade of 'first hormone'.[44] Together with noradrenaline it is secreted into the bloodstream in consequence of emotional shock, physical exercise, cold, or when the blood sugar concentration falls below the point tolerated by nerve cells. Extracts having the activity of adrenaline, enhancing the force and volume of the heart output had been reported 10 years before the discovery of secretin, almost simultaneously, by George Oliver and Edward Schäfer in London,[40,41] and by Napoleon Cybulski and Szymonowicz in Krakow.[45]

What's in a name?

It has been customary in Europe to give the substance 4-[1-hydroxy-2-(methylamino)ethyl]-1,2-benzenediol, alternatively 3,4-dihydroxy-α-[(methylamino)methyl]benzyl alcohol, the name adrenaline. In the USA the same substance is called epinephrine. Why have the Europeans preferred the Latin roots while the Americans go for the Greek? Behind this linguistic conundrum lie hints of scientific skulduggery.

John Jacob Abel, America's first professor of pharmacology, initially at the University of Michigan at Ann Arbor and then at the new Johns Hopkins Medical School in Baltimore, is credited with the isolation of the 'first hormone' as a pure crystalline compound. As a part of his procedure, he treated the acidified and deproteinated extract of adrenal glands with alkaline benzoyl chloride. Then, and after further steps including hydrolysis, he obtained a crystalline material which he reported as being very active. Chemical analysis yielded an elementary formula of $C_{17}H_{15}NO_4$. What is not so clear is whether the Japanese industrial chemist Jokichi Takamine, who had an arrangement with the Parke, Davis Company (Detroit, Michigan), gained some advantage from his visit to Abel's laboratory sometime in 1899 or 1900. Certainly, Abel appears to have been quite candid with his visitor, while Takamine, accustomed to the practices of commerce, never published reports of any sort until his preparation was well protected by patents and – importantly in this case – a trademark, with the name *Adrenalin* (no terminal 'e'). What is clear is that Takamine used a simpler procedure for purifying the hormone, omitting the benzoylation step. T. B. Aldrich, also at the Parke, Davis Co. but recently of the Department of Pharmacology at Johns Hopkins and therefore a late colleague of Abel's, determined the correct elementary formula of Takamine's preparation as $C_9H_{13}NO_3$.

The Parke, Davis Co. lost little time in marketing the preparation, and continued to do so until 1975 when they replaced it with a synthetic product. It became evident that Abel's failure to provide the correct formula was due

The European Pharmacopoeia now also indicates the use of the term epinephrine. We justify our preference for adrenaline not only on its historical primacy but for reasons of logic and common usage. Who ever heard of epinephric receptors? Who ever used the expression 'that really gets my epinephrine up'? (It is said that George Bush *père* may have tried this on but the words actually used were 'that really gets my cholesterol up'. Had he stuck with common usage, his meaning might have been clearer) to incomplete hydrolysis of the benzoyl residues, and indeed, the compound to which he gave the name *epinephrin* (again, no terminal '*e*') was the monobenzoylated derivative, which nonetheless retained some biological activity. His revised formula appeared in the very first paper to be published in the first volume of the new *Journal of Biological Chemistry,* of which he was editor. In a long and contentious account,[46] mainly devoted to explaining why he had got it right and that the others had it wrong, he still managed to end up with a structure having 10 carbon atoms, $C_{10}H_{13}NO_3$. Of course, the material isolated from adrenal medulla is a mixture of adrenaline and noradrenaline and it is likely that the crystalline material was composed of both compounds, so that even an assignment of 9 carbon atoms would have been an overestimate. The correct structure and confirmation that adrenaline is a derivative of catechol was not long in coming. Independent reports from Dakin and from Stolz of complete chemical syntheses of the racemic mixture of the two optically active isomers came in 1904/5.[47,48]

The effects and the actions of adrenaline provide a pattern that has been and still remains useful in general discussions of hormone action.

Neurotransmitters

In comparison with the ready acceptance of the principle of blood-borne transmission of chemical signals between organs, the idea of chemical, as opposed to electrical transmission of impulses between nerves and nerves, and between nerves and muscles, had a long and fraught gestation. The phenomenon of vagal stimulation causing a slowing of the heart, had been described in 1845 by Weber (see Bacq[49]) and the possibility of chemical transmission of this signal was proposed as early as 1877 by Dubois-Reymond:

> Of known natural processes that might pass on excitation, only two are, in my opinion, worth talking about: either there exists at the boundary of the contractile substance a stimulatory secretion in the form of a thin layer of ammonia, lactic acid, or some other powerful stimulatory substance; or the phenomenon is electrical in nature.

It took all of 60 years for the principle of chemical transmission to achieve acceptance as the general means of communication between nerves and muscles. Otto Loewi recorded [50] that he had conceived the idea of chemical transmission between nerves as early as 1903, but that at that time:[27]

> I did not see a way to prove the correctness of this hunch, and it entirely slipped my memory until it emerged again in 1920... The night before Easter Sunday of 1920, I awoke, turned on the light, and jotted down a few notes on a tiny slip of paper. Then I fell asleep again. It occurred to me at six o'clock in the morning that during the night I had written down something

The fascinating history of the debates surrounding the issue of chemical versus electrical transmission at the synapse is excellently related by Horace Davenport.[27]

Henry Hallett Dale and Otto Loewi shared the Nobel Prize in 1936 for their discoveries relating to chemical transmission of nerve impulses.

most important, but I was unable to decipher the scrawl. The next night, at three o'clock, the idea returned. It was the design for an experiment to determine whether or not the hypothesis of a chemical transmission that I had uttered seventeen years ago was correct. I got up immediately, went to the laboratory, and performed a simple experiment on a frog heart according to the nocturnal design.

Loewi relates how, in his experiment, he induced contractions by electrical stimulation of the vagal nerve in an isolated heart. He then transferred some of the fluid from this heart into the ventricle of a second heart undergoing similar stimulation. The result was to slow it down and reduce the force of contraction. He gave the name 'vagusstoff' to the inhibitory substance, later identified as acetylcholine.

One might have thought that this demonstration of chemical, not electrical communication between hearts should have settled the issue. However, it is doubtful whether the technique that he used could have delivered the results that he described. There were problems of reproducibility which may relate both to the temperature of the laboratory and to seasonal variations in the response of the amphibian heart. In the winter months the inhibitory fibres predominate so that electrical stimulation suppresses the rate and force of contraction. In the summer the situation is reversed. Other problems arise from the transient nature of the pulse of the neurotransmitter. Eventually, with the efforts of many others, these difficulties were overcome. Even so, in order to prove the role for acetylcholine in neurotransmission it remained necessary to demonstrate its presence in the relevant presynaptic nerve endings. Also, it was essential to establish that it is actually released when the nerve is stimulated electrically.

One of the main problems confronting all ideas concerning chemical transmission was the instability of the transmitter substance acetylcholine. This had already been synthesized in the mid 19th century. René de M. Taveau, previously associated with J. J. Abel, showed that of 20 derivatives of choline, the acetyl ester is the most active in reducing heart rate and blood pressure, an effect opposed by atropine.[51] It was first isolated from natural sources in 1914 by Arthur Ewins, a member of the laboratory of Henry Dale, as a component present in an extract of ergot. At the time of his appointment to a post at the Wellcome Physiological Laboratories in 1904, Dale remarks that his employer requested that

> when I could find an opportunity for it without interfering with plans of my own, it would give him a special satisfaction if I would make an attempt to clear up the problem of ergot, the pharmacy, pharmacology and therapeutics of that drug being then in a state of obvious confusion . . . I was, frankly, not at all attracted by the prospect of making my first excursion into (pharmacology) on the ergot morass.[52]

Ergot

Although it was recognized as the 'noxious pustule in the ear of grain' over 2500 years ago, written descriptions of ergot poisoning did not appear until the Middle Ages. It is a product of the fungal parasite *Claviceps purpurea* that affects grains, particularly rye. As described, the symptoms included burning pains in the limbs followed by a long and painful gangrene, the tissue becoming dry and being consumed by the Holy Fire, blackened like charcoal. Further complications included abortion and convulsions.

The St Vitus dance and the dance of death (*danse macabre*) were peculiar to the Middle Ages, and it has been suggested[53] that the St Vitus dance may have been a manifestation of the seasonal starvation which occurred during the early summer months. This was the time of the 'hungry gap' when the grain stocks were awaiting replenishment with the August harvests. The peasants, forced to prepare meal from the most marginal sources, inevitably consumed grains laced with ergot alkaloids and other mind-bending substances gathered from the fruit and flowers of wild plants. The St Vitus dance became a public menace, and particularly in the Netherlands, Germany, and Italy during the 14th and 15th centuries, crazed mobs wandered from city to city. The scenes, and the psychedelic visions induced in the minds of the affected, are well represented in the paintings of Pieter Bruegel (the Elder) (**Figure 1.3**).

> It is in this social panorama, traversed by profound anxieties and fears, alienating frustrations, devouring and uncontrollable infirmities and dietary chaos that adulterated and stupefying grains contributed to delirious hypnotic states and crises, which could explode into episodes of collective possession or sudden furies of dancing. The forbidden zones, those most contaminated by the ambiguous, ambivalent magic of the sacred, seemed to emanate perverse influences and unleash dark energies. Psychological destitution, together with the torment of an ailing body, acted as detonator of the epileptic attacks and tumultuous and surging group fits, in which men were attracted and repelled by centres of sacrifice, like the altar.[53]

Relief, of course, was obtained at the shrine of St Anthony. Here, perhaps fortuitously, the sufferers received a diet free of contaminated grain.

Ergot was also known as a medicinal herb and was used by midwives to suppress postpartum bleeding, though it did not find its way into regular clinical practice until the 19th century. Today, the best-known products of ergot must be the hallucinogen lysergic acid diethylamide (LSD) and ergotamine which is used in the relief of the symptoms of migraine. For Henry Dale, the extracts of ergot contained an abundance of active substances to which he returned repeatedly over several years. It was an impurity in a sample sent to him in 1913 for routine quality control (see Bacq[49]), that led him back to the question of transmission at the contacts between nerves

St Vitus dance, so called because sufferers sought to obtain relief at shrines dedicated to St Vitus, is also the name given to the unrelated condition later called Sydenham's chorea. This is mainly a disease of young people, generally associated with rheumatic fever.

The impurity in the sample was probably due to contamination by *Bacillus acetylcholini*: fresh ergot does not contain acetylcholine.

FIG 1.3 *One hundred and one Netherlandish proverbs*, by Pieter Breugel the Elder (1559).
It is high summer, though the hayfield seen at the top of the painting is not being harvested. Instead, it is being laid waste by the pigs. The farmer and all the other villagers are beyond caring. In this hungry month of July, just before the staple crops come to fruition, sustenance is found in the 'bread of dreams', clearly laid out for display on the roof tiles. Containing a rich mixture of alkaloids derived from ergot infested grains, this was the cause of communal madness and wild manifestations (St Vitus dance) among the peasantry of the Middle Ages. Several hundred years later, Henry Dale and his colleagues isolated the neurotransmitter acetylcholine from that 'veritable cornucopia' that is ergot.
Courtesy of the Gemäldegalerie Staatliche Museen zu Berlin Preußischer Kulturbesitz.

and cells. When injected into the vein of an anaesthetized cat, the extract caused profound inhibition of the heartbeat. As it was obviously unsuitable for release as a drug, he obtained the whole batch for further investigation in his laboratory. The first thoughts were that the active constituent might be the stable compound muscarine, but on isolation it was found to be the profoundly labile acetylcholine.

At that time, and even later when it was identified as a chemical component in non-neural tissues, there was never a hint that acetylcholine might have physiological functions. This was not even suggested by the finding as late as 1930 that arterial injections of acetylcholine could induce contractures in denervated muscles. There were still a number of real problems to be overcome. Chief among these was the transient nature of the pulse of neurotransmitter. Also, it was not sufficient to show that acetylcholine applied from a pipette was capable of inducing a response. To prove its role in neurotransmission it still remained necessary to demonstrate its presence in the relevant presynaptic nerve endings and then to show that it is released upon electrical stimulation.

Feldberg describes his introduction of eserine and the use of an eserinized leech muscle preparation as a specific and sensitive device for measuring the acetylcholine present in the various effluents (blood, perfusate, etc.). This was the key that opened the way to the eventual conversion of that most obdurate sceptic, the electrophysiologist John Eccles. Even so, without naming any names, Zenon Bacq[49] reports that even as late as 1950, certain eminent physiologists were still refusing to incorporate the theory of chemical transmission into their teaching.

Receptors and ligands

Among the numerous proteins inserted in the plasma membranes of cells are the receptors. These possess sites, accessible to the extracellular milieu, that bind, with specificity, soluble molecules often referred to as *ligands*. The binding of just a few ligand molecules may then bring about remarkable changes within the cell as it becomes 'activated' or 'triggered'. Although our knowledge of these interactions is quite extensive, it is not so very long ago that the very notion of a receptor was merely conceptual, indicating the propensity of a cell or tissue to respond in a defined manner to the presence of a hormone or other ligand. Nowadays, the receptors are familiar to us as products of the molecular biology revolution. They can be synthesized in the laboratory in milligram quantities as recombinant proteins; they can be modified by point mutations, deletions, or insertions, and by the formation of chimeric structures. They can be expressed in the membranes of cells from which they are normally absent.

The concept of a specific binding site for a ligand certainly predates the discovery of the first hormones and can be ascribed to John Newport Langley.[54] Based on the mutual antagonism of the poisons atropine and pilocarpine (later found to be active at muscarinic cholinergic receptors), he proposed that these substances form 'compounds' in their target tissues to an extent based on the rules of mass action.

> I think it is quite clear that if either atropin or pilocarpin is present in the blood of the sub-maxillary gland, then either pilocarpin or atropin respectively is able in sufficient quantity to produce the effects it produces when present alone in certain other quantity.

> The greater the quantity of atropin the greater is this certain other quantity of pilocarpin; when a large quantity of pilocarpin overcomes

the correspondingly large quantity of atropin it restores the effect of the secretory fibres less, and causes less secretion than when a smaller dose of pilocarpin overcomes a still paralysing but smaller dose of atropin.

Until some definite conclusion as to the point of action of the poisons is arrived at it is not worth while to theorise much on their mode of action; but we may, I think, without much rashness, assume that there is some substance or substances in the nerve endings or gland cells with which both atropin and pilocarpin are capable of forming compounds. On this assumption then the atropin and the pilocarpin compounds are formed according to some law of which their relative mass and chemical affinity for the substances are factors.

Langley was careful to acknowledge the work of others, particularly Luchsinger who had already described the antagonistic actions of atropine and pilocarpine on the sweat glands of the foot of the cat in almost graphic terms:

there exists between pilocarpin and atropin a true mutual antagonism, their actions summing themselves algebraically like wave crests and hollows, like plus and minus. The final result depends simply and solely upon the relative number of molecules of the poisons present.

Twenty-eight years later, in his Croonian lecture of 1906, Langley stated:

Since neither curari nor nicotine, even in large doses, prevents direct stimulation of muscle from causing contraction, it is obvious that the muscle substance which combines with nicotine or curari is not identical with the substance which contracts. It is convenient to have a term for the specially excitable constituent, and I have called it the receptive substance. It receives the stimulus and, by transmitting it, causes contraction . . . The mutual antagonism of nicotine and curari on muscle can only satisfactorily be explained by supposing that both combine with the same radicle of the muscle, so that nicotine-muscle compounds and curari-muscle compounds are formed. Which compound is formed depends on the mass of each poison present and the relative chemical affinities for the muscle radicle.

Curari is the old spelling of curare (*d*-tubocurarine), an arrow poison used by native South Americans. It is a competitive antagonist of acetylcholine at the neuromuscular junction.

References

1. Hildebrand E. What does *Halobacterium* tell us about photoreception? *Biophys Struct Mech* 1977;3:69–77.
2. Springer MS, Goy MF, Adler J. Protein methylation in behavioural control mechanisms and in signal transduction. *Nature*. 1979;280:279–284.
3. Koman A, Harayama S, Hazelbauer GL. Relation of chemotactic response to the amount of receptor: evidence for different efficiencies of signal transduction. *J Bacteriol*. 1979;138:739–747.

4. Kenny JJ, Martinez MO, Fehniger T, Ashman RF. Lipid synthesis: an indicator of antigen-induced signal transduction in antigen-binding cells. *J Immunol*. 1979;122:1278–1284.

5. Rodbell M. The role of hormone receptors and GTP-regulatory proteins in membrane transduction. *Nature*. 1980;284:17–22.

6. Starling EH. On the chemical correlations of the functions of the body. *Lancet*. 1905;2:339–341.

7. Janakidevi K, Dewey VC, Kidder GW. Serotonin in protozoa. *Arch Biochem Biophys*. 1966;113:758–759.

8. Janakidevi K, Dewey VC, Kidder GW. The biosynthesis of catecholamines in two genera of protozoa. *J Biol Chem*. 1966;241:2576–2578.

9. Le Roith D, Shiloach J, Roth J, Lesniak MA. Evolutionary origins of vertebrate hormones: substances similar to mammalian insulins are native to unicellular eukaryotes. *Proc Natl Acad Sci USA*. 1980;77: 6184–6188.

10. Baulieu E-E. Hormones: A complex communication network. In: Baulieu E-E, Kelly PA, eds. *Hormones: From Molecules to Disease*. London: Chapman & Hall; 1990:3–169.

11. Le Roith D, Shiloach J, Berelowitz M, et al. Are messenger molecules in microbes the ancestors of the vertebrate hormones and tissue factors? *Fed Proc* 1983;42:2602–2607.

12. Hunt LT, Dayhoff MO. Structural and functional similarities among hormones and active peptides from distantly related eukaryotes. In: Gross E, Meienhofer J, eds. *Peptides: Structure and Biological Function*. Rockford.Il: Pierce Chemical Co; 1979:757–760.

13. Vesely DL, Giordano AT. Atrial natriuretic factor-like peptide and its prohormone within single cell organisms. *Peptides*. 1992;13:177–182.

14. Vesely Jr DL, Gower WR, Giordano AT. Atrial natriuretic peptides are present throughout the plant kingdom and enhance solute flow in plants. *Am J Physiol*. 1993;265:E465–E477.

15. Le Roith D, Shiloach J, Roth J, et al. Evolutionary origins of vertebrate hormones: material very similar to adrenocorticotropic hormone, β-endorphin, and dynorphin in protozoa. *Trans Assoc Am Physicians*. 1981;94:52–60.

16. Le Roith D, Liotta AS, Roth J, et al. ACTH and β-endorphin-like materials are native to unicellular organisms. *Proc Natl Acad Sci USA*. 1982;79: 2086–2090.

17. Lam H-M, Chiu J, Hsieh M-H, et al. Glutamate-receptor genes in plants. *Nature*. 1998;396:127–128.

18. Skorokhod A, Gamulin V, Gundacker D, Kavsan V, Muller IM. Origin of insulin receptor-like tyrosine kinases in marine sponges. *Biol Bull*. 1999;197:198–206.

19. Escriva H, Delaunay F, Laudet V. Ligand binding and nuclear receptor evolution. *Bioessays*. 2003;22:717–727.

20. Baker ME. Evolution of adrenal and sex steroid action in vertebrates: a ligand-based mechanism for complexity. *Bioessays*. 2003;25:396–400.

21. Manzanares M, Wada H, Itasaki N, Krumlauf R, Holland PWH. Conservation and elaboration of Hox gene regulation during evolution of the vertebrate head. *Nature*. 2003;418:854–857.

22. Nicoll CS. Prolactin and growth hormone: specialists on one hand and mutual mimics on the other. *Perspect Biol Med*. 1982;25:369–381.

23. Meier AH, Farner DS, King JR. A possible endocrine basis for migratory behaviour in the white-crowned sparrow, *Zonotrichia leucophrys gambelii*. *Animal Behavior*. 1965;13:453–465.

24. Meier AH. Daily rhythms of lipogenesis in fat and lean white-throated sparrows *Zonotrichia albicollis*. *Am J Physiol*. 1977;232:E193–E196.

25. Medvei VC. *The History of Clinical Endocrinology*. Carnforth, Lancs, UK: Parthenon Publishing; 1993.

26. Aminoff MJ. *Brown-Séquard: A visionary of science*. New York: Raven Press; 1993.

27. Davenport H. Early history of the concept of chemical transmission of the nerve impulse. *The Physiologist*. 1991;34:129–139.

28. Brown-Séquard CE. The effects produced on man by subcutaneous injections of a liquid obtained from the testicles of animals. *Lancet*. 1889;2:105–107.

29. Spiegel F. *Sick Notes: An Alphabetical Browsing Book of Derivations, Abbreviations, Mnemonics and Slang for the Amusement and Edification of Medics, Nurses, Patients and Hypochondriacs*. Carnforth, Lancs, UK: Parthenon Publishing; 1996.

30. Brown-Séquard CE. Seconde note sur les effets produits chez l'homme par les injections sous-cutanées d'un liquide retiré des testicules frais de cobaye et de chien. *C R Soc Biol (Paris)*. 1889;41:420.

31. Editorial annotation. The new elixir of youth. *Brit Med J*. 1889;1:1416.

32. Editorial annotation. The pentacle of rejuvenescence. *Brit Med J*. 1889;2:446.

33. Berdoe E. Circulated letter: a serious moral question. ms 980/67. *Archives of the Royal College of Physicians*. 1889.

34. Editorial annotation. Can old age be deferred? An interview with Dr Serge Voronoff, the famous authority on the possibilities of gland transplantation. *Scientific American*. 1925:226–227.

35. Short RV. Did Parisians catch HIV from 'monkey glands'? *Nature* 1999;398:657.

36. Borrell M. Organotherapy, British physiology and the discovery of internal secretions. *J Hist Biol*. 1976;9:236–268.

37. Murray GR. Note on the treatment of myxoedema by hypodermic injections of an extract of the thyroid gland of a sheep. *Brit Med J*. 1891;2:796–797.

38. Dale HH. Accident and opportunism in medical research. *Brit Med J*. 1948;2:451–455.

39. Schäfer EA. Present condition of our knowledge regarding the functions of the suprarenal capsules. *Brit Med J*. 1908;1:1277–1281.

40. Oliver G, Schäfer EA. On the physiological action of extract of the suprarenal capsules. *Proc Physiol Soc*. 1894;9:i–iv.

41. Oliver G, Schäfer EA. On the physiological action of extract of the suprarenal capsules. *Proc Physiol Soc*. 1895;17:ix–xiv.

42. Moore BM. On the chemical nature of a physiologically active substance occurring in the suprarenal gland. *Proc Physiol Soc*. 1895;17:xiv–xvii.

43. Bayliss WM. *Principles of General Physiology* p. 739. London: Longmans; 1924.

44. Cannon WB. *Bodily Changes in Pain, Hunger, Fear and Rage. An account of researches into the function of emotional excitement* Reprinted, 1963. New York: Harper Torch Books; 1929.

45. Bilski R, Kaulbersz J. Napoleon Cybulski, 1854–1919. *Acta Physiol Pol*. 1987;38:74–90.

46. Abel JJ, de M. Taveau R. On the decomposition products of epinephrin hydrate. *J Biol Chem*. 1905;1:1–32.

47. Stolz F. über adrenalin und alkylaminoacetobrenzcatechin. *Ber Deutch Chem Ges*. 1904;37:4149–4154.

48. Dakin HD. The synthesis of a substance allied to adrenalin. *Proc Roy Soc Lond B*. 1905;76:491–497.

49. Bacq ZM. Oxford: Pergamon; 1975.

50. Loewi O. An autobiographical sketch. *Perspect Biol Med*. 1960;4:3–25.

51. Hunt R, de R, Taveau M. On the physiological action of certain cholin derivatives and new methods for detecting cholin. *Brit Med J*. 1906;2:1788–1791.

52. Dale HH. *Adventures in Physiology*. London: Pergamon; 1953.

53. Camporesi P. Bread of dreams. *History Today*. 1989;39:14–21.

54. Langley JN. On the physiology of the salivary secretion: Part II. On the mutual antagonism of atropin and pilocarpin, having especial reference to their relations in the sub-maxillary gland of the cat. *J Physiol*. 1878;1:339–369.

CHAPTER 2

First Messengers

The natural extracellular ligands that bind and activate receptors have been called first messengers, although they may be subdivided as hormones, neurotransmitters, cytokines, lymphokines, growth factors, chemoattractants, etc. Each of these terms attempts to define a class of agents that take effect in a particular setting. However, there can be much overlap, as none of them adheres to a strict definition. This is illustrated by Table 2.1, which provides examples of extracellular first messengers that function as hormones, growth factors, and inflammatory mediators, and Table 2.2, which lists a selection of substances that can act as neurotransmitters.

Particular examples of multiple functions are:

- The coenzyme ATP and the cellular metabolite glutamate are neurotransmitters when they are secreted at synapses.
- The gut hormones gastrin, cholecystokinin, and secretin are also present in the central nervous system (CNS), where they have diverse functions as neuromodulators (influencing the release of other neurotransmitters).

TABLE 2.1 A selection of first messengers found in the circulation

Class	Messenger	Origin	Target	Major effects
Messengers derived from amino acids	adrenaline (epinephrine)	adrenal medulla	heart smooth muscle liver and muscle adipose tissue	increase in pulse rate and blood pressure contraction or dilatation glycogenolysis lipolysis
	noradrenaline (norepinephrine)	adrenal medulla	arteriolar smooth muscle	vasoconstriction
	serotonin (5-hydroxy tryptamine)	platelets	arterioles and venules	vasodilatation and increased vascular permeability
	Histamine	mast cells, basophils	arterioles and venules	vasodilatation and increased vascular permeability
Peptide hormones	insulin	pancreatic β cells	multiple tissues	glucose uptake into cells carbohydrate catabolism protein synthesis lipid synthesis (adipose tissue) Note: also a growth factor
	glucagon	pancreatic α cells	liver adipose tissue	glycogenolysis lipolysis
	gastrin	intestine	stomach	secretion of HCl and pepsin
	secretin	small intestine	pancreas	secretion of digestive enzymes
	cholecystokinin	small intestine	pancreas	secretion of digestive enzymes
	atrial natriuretic factor (ANF or ANP)	heart	kidney	Na^+ and water diuresis
	adrenocorticotropic hormone (ACTH)	anterior pituitary	adipose tissue adrenal cortex	lipolysis production of cortisol and aldosterone
	follicle stimulating hormone (FSH)	anterior pituitary	oocyte/ovarian follicles ovarian follicles	growth oestrogen synthesis
	luteinizing hormone (LH)	anterior pituitary	oocyte ovarian follicles	maturation oestrogen and progesterone synthesis
	thyroid- stimulating hormone (thryotropin or TSH)	anterior pituitary	thyroid	generation and release of thyroid hormones
	parathyroid hormone (PTH)	parathyroid	bone kidney	release of Ca^{2+} and phosphate Ca^{2+} reabsorption

Continued

TABLE 2.1 Continued

Class	Messenger	Origin	Target	Major effects
	vasopressin (antidiuretic hormone or ADH)	posterior pituitary	kidney arteriolar smooth muscle	water reabsorption from urine vasoconstriction to increase blood pressure
	Oxytocin	posterior pituitary	uterine smooth muscle milk ducts	dilation of the cervix milk ejection
Growth factors	epidermal growth factor (EGF)	multiple cell types	epidermal and other cells	growth
	Somatotropin (growth hormone or GH)	anterior pituitary	liver	production of somatomedins
	somatomedins 1 and 2 (insulin-like growth factors, IGF1 and 2)	liver	bone, muscle and other cells	growth
	Transforming growth factor β (TGFβ)	multiple locations	multiple	growth control
Eicosanoids	Prostaglandins PGA$_1$,PGA$_2$, PGE$_2$	most body cells	multiple	inflammation, vasodilatation
Hormones with intracellular receptors	Thyroid hormone (T3, tri-iodothyronine)	thyroid	multiple	metabolic regulation
	progesterone	corpus luteum placenta	uterine endometrium	preparation of endometrial layer maintenance of pregnancy

- Somatostatin, identified originally as a hypothalamic agent suppressing the secretion of growth hormone from the pituitary, also operates in the CNS as a neurotransmitter or neuromodulator. More than this, it is a paracrine agent in the pancreas and a hormone for the liver.
- The transforming growth factor TGFβ also acts as a chemoattractant and as a growth inhibitor.
- Insulin acts not only to regulate glucose metabolism, but also as a growth factor and a regulator of food intake.
- Noradrenaline (norepinephrine), depending on whether it is released at a synapse or from the adrenal medulla, can be considered as a neurotransmitter or as a hormone.
- Thrombin is a growth factor but is also involved in blood clotting as an activator of platelet function.

TABLE 2.2 A selection of compounds that act as neurotransmitters

Type	Transmitter	Structure
	acetylcholine	
Tyrosine or tryptophan-derived	adrenaline (epinephrine)	
	noradrenaline (norepinephrine)	
	dopamine	
	serotonin	
Amino acids	glutamate	
	glycine	
	GABA	
Purine based	ATP	
Neuropeptides	enkephalins	Tyr-Gly-Gly-Phe-Leu Tyr-Gly-Gly-Phe-Met
	substance P	Arg-Pro-Lys-Pro-Gln-Gln-Phe-Phe-Gly-Leu-Met
	angiotensin II	Asp-Arg-Val-Tyr-Ile-His-Pro-Phe
	somatostatin	Ala-Gly-Cys-Lys-Asn-Phe-Phe-Trp-Lys-Thr-Phe-Thr-Ser-Cys

Hormones

Hormones are commonly released in small amounts at sites remote from the organs they target. On entering the circulation, they are diluted enormously and are subject to enzymes that break them down. Many of them circulate as complexes with specific binding proteins, reducing their free concentration still further. The result of all this is that their level in the vicinity of a target cell is likely to be extremely low and accordingly, the cell receptors must possess high affinity. Another consequence is that although a target cell may react in milliseconds to hormone binding, overall response times are in the range of seconds to hours.

Examples of water-soluble hormones are included in the list of first messengers shown in Table 2.1. Adrenaline and noradrenaline (epinephrine and norepinephrine) are released from the adrenal gland and can act within seconds. The peptide hormones vary considerably in size, ranging from a few amino acids to full-scale proteins. For example, the hypothalamic factor TRH (TSH-releasing hormone) that induces the release of thyroid stimulating hormone (TSH), consists of only 3 amino acid residues, whereas FSH (follicle stimulating hormone) and TSH are heterodimeric glycoproteins each of about 200 residues.

Growth factors

The first reports that fragments of biological tissue could be maintained in a living state *in vitro* appeared nearly 100 years ago,[1] but routine culturing of dispersed cells did not become established until the 1950s. The successful maintenance of dividing mammalian cells depends on the composition of the culture medium. This is traditionally a concoction of nutrients and vitamins in a buffered salt solution. A key ingredient is an animal serum, such as fetal bovine serum. Without such a supplement, most cultured cells will not replicate their DNA and therefore will not proliferate. An important early discovery was that the best mitogenic stimulus (i.e. inducing mitosis or cell division) is provided by natural clot serum and not plasma. A 30 kDa polypeptide released by platelets was later isolated and shown to have mitogenic properties. It was named platelet-derived growth factor, PDGF (see page 303).

PDGF is a member of a class of messengers, the growth factors, comprising over 40 polypeptides, ranging in size from insulin (5.7 kDa) and EGF (6 kDa) to transferrin (78 kDa). As with the hormones, cells have specific high-affinity receptors for growth factors, though binding may elicit multiple effects. Apart from stimulating (or inhibiting) growth, they may also initiate programmed cell death (apoptosis), differentiation, and gene expression. The effects of growth factors, unlike those for hormones, may last for days. Also, unlike the hormones their effects are short range, generally influencing the growth and function of neighbouring cells.

Endocrine (Greek ἐνδον within, κρίνει to separate) denotes the 'action at a distance' of hormones secreted from the ductless glands that may pervade the whole organism, searching out specific target tissues.

Paracrine denotes the action of an extracellular messenger that takes effect only locally.

When a substance affects the same cell from which it has been released, the activity (perhaps part of a negative feedback pathway) is termed *autocrine*.

When a cell affects another cell with which it is in contact, usually through membrane-bound proteins, the mode of communication is termed *juxtacrine*.

Cytokines

In parallel with the discovery of growth factors, several extracellular signalling proteins were identified that interact with cells of the immune system. Because they activate or modulate the proliferative properties of this class of cells, they were initially termed 'immunocytokines'. As it became apparent that they also act upon cells outside the immune or inflammatory systems, the name became shortened to *cytokine*. The functions of growth factors and cytokines are so diverse that a clear distinction cannot be made. Thus cytokines include the growth factors already mentioned, along with such molecules as the interferons, tumour necrosis factor (TNF), numerous interleukins, granulocyte–monocyte colony stimulating factor (GM-CSF) and many others. Chemokines are cytokines that bring about local inflammation by recruiting inflammatory cells by chemotaxis and subsequently activating them.

Vasoactive agents

Physical damage to tissues or damage caused by infection generates an inflammatory response. This is a defence mechanism in which specialized cells (mostly leukocytes), recruited to the site, act in a concerted fashion to remove, sequester or dilute the cause of the injury. The process is complex, involving many interactions between cells and numerous extracellular messengers. Among these are cytokines that induce inflammation (pro-inflammatory mediators), or reduce it (anti-inflammatory mediators). To enable the recruitment of leukocytes and to retain a local accumulation of (diluting) fluid, there is vasodilatation and a regional increase in vascular permeability. Agents that bring this about include histamine (released by mast cells and basophils), serotonin (released by platelets), and pro-inflammatory mediators such as bradykinin.

The eicosanoids are another important family of vasoactive compounds. They are derived from the polyunsaturated fatty acid, arachidonic acid (5,8,11,14-eicosatetraenoic acid) (Figure 2.1). Because arachidonic acid and many of its derivatives possess 20 carbon atoms, they are termed *eicosanoids* (from the Greek word for 20). They include the prostaglandins, thromboxanes, and leukotrienes and they operate over short distances as potent paracrine or autocrine agents, controlling many physiological and pathological cell functions. In the context of inflammation, they cause plasma leakage, skin

FIG 2.1   arachidonic acid   prostaglandin E_2

reddening, and the sensation of pain. The drug aspirin possesses anti-inflammatory and pain-relieving properties because it inhibits a key enzyme in the pathway generating prostaglandins, such as the vasoconstrictor prostaglandin E_2 (**Figure 2.1**).

Neurotransmitters and neuropeptides

Neurotransmitters are also first messengers, but their release and detection at chemical synapses contrasts strongly with that of endocrine signals. In the presynaptic cell, neurotransmitter-containing vesicles are organized to release their contents locally into the tiny volume defined by the synaptic cleft. Individual neurons contain only very small quantities of transmitter and this arrangement ensures that an effective concentration is quickly attained within the cleft following the secretion (by exocytosis) of the contents of relatively few vesicles. The released transmitter then diffuses across the cleft and binds to receptors on the postsynaptic cell. On the macroscopic scale, diffusion is a slow process, but across the short distance that separates the pre- and postsynaptic neurons ($\sim 0.1\ \mu m$ or less), it is fast enough to allow rapid communication between nerves or between nerve and muscle at a neuromuscular junction. Table 2.2 shows the structures of some important neurotransmitters. In the central nervous system glutamate is the major excitatory transmitter, whereas GABA and glycine are inhibitory. Acetylcholine has its most prominent role at the neuromuscular junction, where it is excitatory. Whether an excitatory or inhibitory effect is evoked depends on the nature of the ion channel that is regulated by the neurotransmitter receptor (see page 54).

First messengers with intracellular receptors

There is a class of hormones that differs in several major respects from the first messengers that we have examined so far. These are the steroid and thyroid hormones (see Figure 10.1, page 275). Firstly, since many are hydrophobic they are able to cross the plasma membrane of cells and then interact with receptors that are intracellular. The hormone-bound receptors then bind to DNA and regulate the expression of particular proteins or sets of proteins. For example, the steroid cortisol can penetrate cells and bind to internal glucocorticoid receptors. These, in turn, bring about transcription of DNA sequences and thus the expression of enzymes that lead to glycogen synthesis. Such a process is slow, requiring hours to take effect. It forms a paradigm for the action of steroid and other hormones that bind to nuclear receptors (thyroid hormone, retinoic acid, vitamin D) and many other non-hormonal lipophilic messengers. This form of signalling is discussed in detail in Chapter 10.

In addition to the transcriptional actions of hydrophobic messengers, there is increasing evidence for non-transcriptional effects, which take place rapidly and involve interactions with cell surface receptors. A recently described

Almost 40 years ago there were already indications that oestrogens can cause a rapid (within 15 s) elevation of cAMP in rat uterus.[4] Indeed, as early as 1941, Hans Selye reported that steroids introduced into the peritoneum of rats induce deep anaesthesia within 15 min.[5,6] Steroid derivatives as anaesthetics in surgery have been used since the 1950s (hydroxydione sodium or Viadril, 21-hydroxypregnane-3,20-dione sodium succinate).

example of steroid interaction with a well-established cell surface receptor is that of progesterone, essential for establishing and then maintaining pregnancy in mammals. At the cell surface it interferes with the binding of the peptide hormone oxytocin to its receptor in rat uterine tissue.[2] It is possible that it is through inhibition of oxytocin binding that progesterone maintains uterine quiescence during pregnancy. The oxytocin receptor is a member of the seven transmembrane (7TM) protein superfamily of G-protein-coupled receptors (see Chapter 4) and acts to elevate the concentration of intracellular Ca^{2+}. Progesterone can also act as an inhibitor of the nicotinic acetylcholine receptor.[3] The rapid, non-genomic effects of lipophilic messengers are also discussed in Chapter 10.

Oxytocin (from the Greek meaning swift birth) is generated in the brain of both males and females. It can be added to the examples already cited which have the actions of both hormones and neuromodulators. Not only does it act as a regulator of uterine smooth muscle contraction and milk ejection but it also acts as a regulator of emotional states, enhancing trust and bonding between pairs of individuals. Sometimes called the 'cuddle hormone', it plays a role in the expression of maternal, sexual, social, stress, and feeding behaviours, as well as in learning and memory. When the natural release of oxytocin is blocked, for instance, mothers – from sheep to rats – reject their own young, while virgin female rats injected with oxytocin nuzzle the pups of other females protecting them as if they were their own. Oxytocin in females and the closely related vasopressin in males are keys to pair bonding. These actions of the hormone, which must have their origins far back in mammalian evolution, prove to be readily adaptable in today's hectic world dominated by the god Mammon.[7,8] In a money game, a double-blind trial involving the investment of serious money, volunteers playing the roles of investor and trustee were offered a sniff of oxytocin or placebo. It was found that the hormone had the effect of increasing the trust of the investors while having no effect on the trustworthiness of the trustees. The possibility of abuse of these substances with the aim of manipulating trust in business and other (e.g. love) situations must be self evident.

Common aspects

Although we have described differences in the nature of selected first messengers, our aim in this book is to accentuate their common aspects. This will become more apparent in later chapters, where it will be seen that similar intracellular mechanisms are used to promulgate an extracellular message. The ultimate effects of these first messengers are determined, not by their chemical natures, but by the anatomical context within which they operate and by the characteristics of the receptor molecules that sense their presence in different classes of cell.

Intracellular messengers

Second messengers formed as a result of the activation of receptors are discussed in subsequent chapters. They include cAMP, Ca^{2+}, diacylglycerol, and the 3-phosphorylated inositol lipids. We shall see that many, indeed most, responses to stimulation of cell surface receptors are eventually mediated by protein phosphorylation and dephosphorylation. These are the major devices by which the binding of ligand is converted into a cellular consequence. In general, these second messengers interact with a specific set of enzymes (kinases and phosphatases) either directly or indirectly, to phosphorylate or dephosphorylate target proteins at specific sites. With a few special exceptions, phosphorylations due to these second messengers occur on serine and threonine residues.

Binding of ligands to receptors

A hormone may elicit physiological effects at concentrations as low as $10^{-10}\,mol\,L^{-1}$ ($10^{-10} \times 6 \times 10^{23} = 6 \times 10^{13}$ molecules per litre). To get an idea of how many of these molecules an individual target cell might encounter, consider that a $12\,\mu m$ diameter cell occupies a volume of $10^{-12}\,L$. This volume of extracellular fluid would contain only 60 molecules of hormone. In spite of the necessary high affinity, it is important to remember that the binding interaction is never covalent and is always reversible. Thus binding of hormone (H) to receptor (R) can be written:

$$H + R \rightleftarrows HR$$

At equilibrium, the rates of the forward and backward reaction are equal (on-rate=off-rate) and by the law of mass action

$$k_{on}[H][R] = k_{off}[HR]$$

where k_{on} is the on-rate constant and k_{off} is the off-rate constant, with units of $L\,mol^{-1}\,s^{-1}$ and s^{-1}, respectively. [H] and [R] represent the concentrations of free (unbound) hormone and free (unoccupied) receptors. A high affinity interaction means that the predominant species is HR. That is, the equilibrium is to the right or the equilibrium constant K is large, where

$$K = \frac{[HR]}{[H][R]} = \frac{k_{on}}{k_{off}}$$

and the rate constant k_{on} is much greater than k_{off}. The units of K are $L\,mol^{-1}$ (i.e. reciprocal concentration). Instead of an equilibrium constant for association, the expression above can be inverted:

$$K_D = \frac{[H][R]}{[HR]} = \frac{k_{off}}{k_{on}}$$

(Equation 2.1)

The value of 60 molecules may be compared with the number of ions of electrolyte present in the same volume: $\sim 1.8 \times 10^{11}$ (for isotonic saline). Furthermore, the same volume of intracellular water will contain about 10^{10} molecules of ATP (between 5 and 10 mmol L^{-1}) and after hormonal stimulation, a cell might contain as many as 10^6 molecules of the second messenger cAMP.

K_D is the dissociation constant and has units of concentration (mol L^{-1}). It is also the concentration of hormone which, at equilibrium, causes 50% of the receptors to be occupied, i.e. [HR]=[R].

Binding heterogeneity

One might imagine that all receptors are equal in terms of the effects that they generate. This is far from being the case. Binding site heterogeneity can be caused by the presence of receptor subtypes with different affinities or by receptors of a single class undergoing regulated conformational transitions between high and low affinity states. Since such a process would be driven by intracellular mechanisms; it could be tested for by inhibiting the cells metabolically (preventing ATP synthesis/depleting intracellular ATP). An alternative approach would be to compare the hormone binding affinities of intact cells with their isolated plasma membranes. An experiment of this kind should reveal the extent to which either the affinity or the number of exposed receptor sites are regulated by cellular metabolic status, which again might be modulated by events set in train by the initial hormone-receptor interaction itself.

Measurement of binding affinity

To measure the binding parameters, the cells, or their membranes, are exposed to hormone that is radioactively labelled. After equilibration the bound and free hormone pools are separated, generally by centrifugation or filtration. The supernatant or filtrate contains the unbound material and the pellet or filtered material consists of cells or membranes bearing the bound hormone. These are briefly washed to remove residual and interstitial unbound hormone and then both samples are analysed for the presence of the label. The range of concentrations over which binding occurs and becomes saturated can be determined by varying the hormone concentration and plotting a graph of bound against free hormone concentration. For the very simplest systems, in which there is only one binding site per receptor and no interactions between receptors that might modify the affinity of binding at unoccupied sites, we can derive an equation for the binding curve, as follows.

If the total concentration of receptors in the preparation is R_{tot}, then

$$[R] = R_{tot} - [HR]$$

Substituting [R] in Equation 2.1 gives the single site equilibrium binding equation:

$$[HR] = R_{tot} \frac{[H]}{K_D + [H]}$$

Plotting [HR] against [H], we obtain a hyperbola (Figure 2.2a). As with most representations of data ranging over several orders of magnitude, it is generally more useful to plot the concentrations on a logarithmic scale. By extrapolating the graph to an infinite concentration of hormone, we may then determine the total receptor concentration in the system, $[R_{tot}]$ (Figure 2.2b). In practice, binding data are analysed by non-linear regression using computer software that takes account of multiple binding sites and other complications, yielding values of R_{tot} and K_D.

K_D and EC_{50}: receptor binding and functional consequences

Having derived R_{tot} by extrapolation or curve-fitting and knowing the number of cells, we can calculate the number of binding sites per cell. We can also ask how the estimated K_D relates to the likely hormone concentration *in vivo* and the biological responses that are evoked. More simply, is the magnitude of the tissue (or cell) response proportional to the fraction of occupied receptors? More often than not, there is no clear relationship.[9] In many cases, the maximal cellular response – whatever it is that the cell does as a consequence of stimulation – is achieved with only a tiny fraction of the receptors being occupied (Figure 2.3a). We should therefore distinguish between K_D and the EC_{50}, the concentration of hormone at which the biological response or effect is half-maximal.

There are different ways of assessing a cellular response. An example is shown in Figure 2.3b in which an early response is estimated by measuring the generation of an intracellular second messenger (cAMP). Alternatively, more distal downstream events such as secretion or contraction might be measured. Curiously, we find that the EC_{50} for each type of response tends to be different. This is because K_D is a parameter describing ligand binding at equilibrium, while EC_{50} reflects consequent events that may have multiple kinetic components determined by the rates of enzyme catalysed reactions that vary independently of receptor occupation. The concentration of a

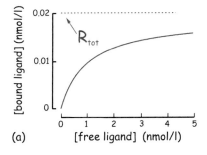

(a)

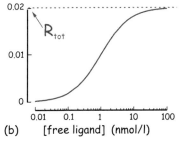

(b)

FIG 2.2 Saturation binding. Direct binding curves plotted using (a) linear and (b) semi-logarithmic scales.

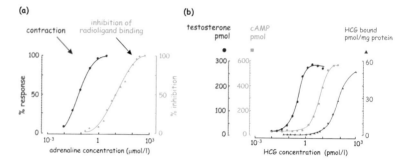

FIG 2.3 Receptor occupancy and cell response. (a) Correlation of adrenaline binding at α_1-adrenergic receptors and contraction of smooth muscle (rat vas deferens). In this experiment, receptor occupancy was determined by measuring the displacement of a specific α_1-antagonist, labelled with ^{125}I. Contraction occurred at ligand concentrations that were too low to have a measurable effect in the binding assay. Adapted from Minneman et al.[10] (b) Correlation of hormone binding (gonadotropin), cAMP generation, and testosterone production from testicular Leydig cells. Note the 10^6-fold difference in affinity of the two hormone receptors. Adapted from Mendelson et al.[11]

hormone inducing a half-maximal response can be as much as two orders of magnitude (100-fold) smaller than the concentration required to saturate 50% of the receptors, as illustrated in **Figure 2.3a**. Similarly, maximal stimulation of glucose oxidation by adipocytes is achieved by insulin when only 2–3% of its receptors are occupied.

Spare receptors

The receptors that remain unoccupied have unfortunately been termed 'spare receptors',[12] but this is not to suggest that their role is unimportant. This large apparent excess on most hormone-responsive cells helps to ensure that stimulation can occur at very low hormone concentrations. For example, imagine that an optimal cell response is only elicited when *all* of the available cell surface receptors are occupied. This condition could only be achieved at infinite hormone concentration, because the binding follows a saturation relationship as depicted in **Figure 2.2**. Conversely, if the cell response is optimal when only a small proportion of receptors are occupied, then a concentration of hormone much lower than the K_D will achieve this. The greater the receptor reserve, the lower the EC_{50} will be for a given level of occupancy. **Figure 2.4** illustrates this.

The EC_{50} can be shifted by manipulations that alter the activity of the enzymes that take part in the stimulation pathway and that are well removed from events immediately set in train by the receptors. For instance, the drug theophylline is an inhibitor of cyclic nucleotide phosphodiesterase, the enzyme that breaks down cAMP (**Figure 2.5**). Its effect is to augment the

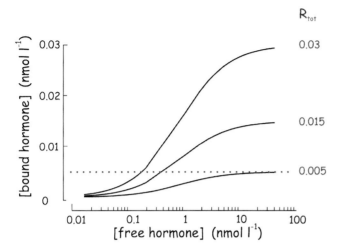

FIG 2.4 Spare receptors. Three saturation binding curves are shown for three different situations. In the upper curve the maximum extent of binding (R_{tot}) exceeds the level required for optimal stimulation (5 pmol L^{-1}, indicated by the dotted line) by 6-fold. Maximal activation is therefore achieved at the hormone concentration corresponding to the intersection of the line and the curve. If the receptor reserve is reduced (middle curve) so that R_{tot} is only three times greater than the required occupancy level, the intersection moves to the right and if the reserve approaches zero ($R_{tot} = 5$ pmol L^{-1}), the point of intersection moves towards infinity (lower curve). Under this condition it is never possible to achieve saturation of the receptors.

FIG 2.5 Action of theophylline.

elevation of cAMP induced by receptor activation. Therefore the steady state concentration of cAMP due to stimulation of receptors can be induced by much lower concentrations of hormone.

Down-regulation of receptors

Cells with receptors that are subjected to regular or persistent activation, for example by drugs that are resistant to metabolic breakdown, may become less amenable to stimulation. This down-regulation may be due to depletion of the number of exposed receptors, a reduction of their binding affinity or both. Either of these effects will increase the EC_{50} by reducing R_{tot} or increasing K_D. There is a familiar example. The instructions for use of nasal decongestants based on xylometazoline give a warning not to continue usage beyond 7 days. After this time there is a tendency for the drug to cause 'rebound congestion' and further use then exacerbates instead of relieves the condition. Xylometazoline binds and activates α_2-adrenergic receptors that suppress production of cAMP. However, the specificity though good, is not perfect and the drug has a low but nonetheless significant affinity for α_1-receptors. The result is that when the preferred targets, the α_2-receptors, have eventually been desensitized and removed from the epithelial surface, action is transduced by the α_1-receptors which now predominate. The consequence is nasal congestion, most probably mediated, not by cAMP, but through the activation of phospholipase C, Ca^{2+}, and protein kinase C.

An important mechanism of receptor down-regulation, to which we return in Chapter 4, is phosphorylation of the intracellular chains of receptor proteins by enzymes such as the protein kinases A and C and specific receptor kinases. This marks them as targets for removal from the cell surface by endocytosis.[13] It also allows the redirection of signals into different pathways.[14]

All in all, one may conclude that receptors are not static components of cells; they are in a dynamic state that is influenced by both exogenous and endogenous factors.

Discovery of the first second messenger, cAMP

Although experiments with radioactively labelled hormones and related reagents enabled the quantitative measurement of binding parameters of receptors, they told nothing of what receptors are, or what they do. One critical advance was made in 1957 by Earl W. Sutherland. With his colleagues, he showed that the activation of glycogen breakdown in liver stimulated by adrenaline or glucagon occurs in two distinct stages. The first of these is the generation of a heat-stable and dialysable factor.[15] When this was applied together with ATP to the supernatant fraction of liver homogenate, inactive glycogen phosphorylase ('dephospho liver phosphorylase', also called phosphorylase b) was phosphorylated to its active form (phosphorylase a). In a footnote they point out that

> The active factor recently has been purified to apparent homogeneity. From ultraviolet spectrum, the orcinol reaction, and total phosphate determination, the active factor appeared to contain adenine, ribose and phosphate in a ratio of 1:1:1. Neither inorganic phosphate formation nor diminution of activity resulted when the factor was incubated with various phosphatase preparations … However, the activity of the factor was rapidly lost upon incubation with extracts from dog heart, liver and brain.

With the identification of the second messenger as cyclic AMP (cAMP), it became possible to link activation of specific classes of receptors with specific biochemical responses. Of course, elevation of cAMP is not the exclusive second-messenger response to hormones. Nor is it the exclusive response to adrenaline and other closely related catecholamines. These bind, depending on the cell type or tissue, to a family of catecholamine or 'adrenergic' receptors, $\alpha_1, \alpha_2, \beta_1, \beta_2, \beta_3$, each of which has several distinct functions. These include the synthesis of cAMP (generally β), suppression of synthesis of cAMP (generally α_2) and elevation of Ca^{2+} (α_1).

Earl W. Sutherland (1915–1974), awarded the Nobel Prize in 1971 'for his discoveries concerning the mechanisms of action of hormones'.

References

1. Carrel A. On the permanent life of tissues outside of the organism. *J Exp Med*. 1912:516–528.
2. Grazzini E, Guillon G, Mouillac B, Zingg HH. Inhibition of oxytocin receptor function by direct binding of progesterone. *Nature*. 1998;392:509–512.
3. Valera S, Ballivet M, Bertrand D. Progesterone modulates a neuronal nicotinic acetylcholine receptor. *Proc Natl Acad Sci U S A*. 1992;89:9949–9953.
4. Szego CM, Davis JS. Adenosine 3′5′-monophosphate in rat uterus: Acute elevation by estrogen. *Proc Natl Acad Sci U S A*. 1967;58:1711–1718.
5. Selye H. Anesthetic effect of steroid hormones. *Proc Soc Exper Biol Med*. 1941;4:116–121.
6. Selye H. Correlations between the chemical structure and the pharmacological actions of the steroids. *Endocrinology*. 1942;30:437–453.
7. Damasio A. Brain trust. *Nature*. 2005;435:571–572.
8. Kosfeld M, Heinrichs M, Fischbacher U, Fehr E. Oxytocin increases trust in humans. *Nature*. 2005;435:673–676.
9. Stephenson RP. A modification of receptor theory. *Brit J Pharmacol*. 1956;11:379–393.
10. Minneman KP, Fox AW, Abel PW. Occupancy of α_1-adrenergic receptors and contraction of rat vas deferens. *Mol Pharmacol*. 1983;23:359–368.
11. Mendelson C, Dufau M, Catt K. Gonadotropin binding and stimulation of cyclic adenosine 3′:5′-monophosphate and testosterone production in isolated leydig cells. *J Biol Chem*. 1975;250:8818–8823.
12. Springer MS, Goy MF, Adler J. Protein methylation in behavioural control mechanisms and in signal transduction. *Nature*. 1979;280:279–284.
13. Goodman OB, Krupnick JG, Santini F, et al. β-Arrestin acts as a clathrin adaptor in endocytosis of the β_2-adrenergic receptor. *Nature*. 1996;383:447–450.
14. Daaka Y, Luttrell LM, Lefkowitz RJ. Switching of the coupling of the β_2-adrenergic receptor to different G proteins by protein kinase A. *Nature*. 1999;390:88–91.
15. Rall TW, Sutherland EW, Berthet J. The relationship of epinephrine and glucagon to liver phosphorylase: Effect of epinephrine and glucagon on the reactivation of phosphorylase in liver homogenates. *J Biol Chem*. 1957;224:463–475.

CHAPTER 3

Receptors

But when the blast of war blows in our ears, Then imitate the action of the tiger; Stiffen the sinews, summon up the blood, Disguise fair nature with hard-favour'd rage; Then lend the eye a terrible aspect …

William Shakespeare, *Henry V*, II.ii.1

Here we begin with a description of the receptors for the catecholamines (adrenaline, noradrenaline, and dopamine) and for acetylcholine. These provide the basic paradigms by which the main classes of hormone, neurotransmitter, and drug receptors have been defined.

Adrenaline (again)

Adrenaline (epinephrine) is the hormone that is secreted in anticipation of danger, preparing the body for fight or flight.[1]

> As Cannon has pointed out, this secretion (the internal secretion of the medulla of the suprarenal bodies) is poured into the blood during conditions of stress, anger, or fear, and acts as a potent reinforcement to

the energies of the body. It increases the tone of the blood vessels as well as the power of the heart's contraction, while it mobilizes the sugar bound up in the liver, so that the muscles may be supplied with the most readily available source of energy in the struggle to which these emotional states are the essential precursors or concomitants.[2]

The immediate actions are those of arousal: the sweat glands sweat, body hair stands on end, the pupils dilate to gather more light, the bronchi dilate to improve oxygen supply, the heart quickens and increases its force of contraction. Then, through dilatation of relevant vessels and constriction of others, the blood supply to the myocardium and selected skeletal muscles is increased, while the supply to the skin and visceral organs is reduced, postponing digestive and anabolic processes. Metabolic fuel stores are also mobilized (initially breakdown of glycogen in the liver and muscles, and ultimately the catabolism of fat). Thus the secretion of adrenaline is familiar in the context of emergencies. However, it important to remember that the adrenal gland continuously produces low levels of the hormone and that adrenaline is also released into the circulation during exercise.

Adrenaline belongs to the group of hormones and neurotransmitters known as catecholamines. The other members are dopamine and noradrenaline. Their structures are shown in Table 2.2, page 24. The precursor from which they are derived by a sequence of steps, is the amino acid L-tyrosine. With a few exceptions (the extracellular Ca^{2+} sensor and the receptors for the amino acid glycine are examples), receptors are stereospecific. The catecholamines are recognized only as their L-enantiomers. Adrenaline differs from noradrenaline by the presence of an *N*-methyl group on the side chain. In humans, the chromaffin cells of the adrenal medulla store and secrete both adrenaline (~80%) and noradrenaline (~20%), and noradrenaline is also secreted by sympathetic neurons.

α-and β-adrenergic receptors

The receptors that respond to adrenaline or noradrenaline are called adrenergic receptors and they are present in a wide range of tissues. They may be divided into two major classes, α and β, although as shown in Table 3.1, these elude any precise definition, either by their location or by the effects they elicit. Nor are they defined by their affinity for their natural first messengers: the differences in their sensitivity to adrenaline and noradrenaline are not large.

Adrenergic receptor agonists and antagonists

Agonists
What must characterize the different adrenergic receptors ultimately are the differences in their molecular structures. Knowledge in this area is still only partial, however, but we do have information from their interactions with

TABLE 3.1 Properties that distinguish α and β receptors

	α receptors	**β receptors**
Arteriolar smooth muscle	vasoconstriction in viscera (other than liver)	vasodilation in skeletal musculature and liver (β_2)
Bronchial smooth muscle	(not present)	bronchodilation (β_2)
Uterine smooth muscle (myometrium)	contraction (α_1)	relaxation (β_2)
Heart		inotropic effect (increased cardiac contractility) (β_1) increased heart rate (β_1)
Eye	pupillary dilation (α_1)	lowering of intraocular pressure (β_2)
Skeletal muscle		glycogenolysis (β_2)
Liver	glycogenolysis (α_1)	glycogenolysis (β_2)
Adipose tissue	inhibition of lipolysis (α_2)	lipolysis (β_1)
Pancreas	inhibition of insulin secretion (α_2)	insulin secretion (β_2)
Platelets	inhibition of aggregation (α_2)	
Order of potency	adr≥nor>>iso	iso>adr≥nor

Abbreviations: adr, adrenaline; iso, isoprenaline; nor, noradrenaline.

synthetic ligands that act as agonists or antagonists. In fact, they may be further subdivided on this basis, into four well-characterized subtypes α_1, α_2, β_1 and β_2 (Figure 3.1).

Although a particular subtype, by virtue of its dominance in a given tissue, may mediate a particular response, what really distinguishes them is their responsiveness to synthetic agonists, such as isoprenaline (isoproterenol) which causes blood vessels to constrict ($\beta>\alpha$), and phenylephrine ($\alpha>\beta$) which causes them to dilate. Adrenaline can do either, depending on the particular vessel. Clonidine (α_2) and salbutamol (β_2) are even more specific, since they can distinguish particular subtypes.

Antagonists

As indicated in **Figure 3.1**, there are also synthetic antagonists. The structures of some selected examples are shown in Figure 3.2. Propranolol inhibits the action of β-receptors. This drug finds application in the treatment of some forms of hypertension, probably because it inhibits the secretion of the peptide hormone renin from the juxtaglomerular apparatus of the kidney.

Molecular cloning has confirmed the existence of β_3 adrenergic receptors. Like β_1 and β_2 receptors, these are linked to adenylyl cyclase through the G protein G_s, but differ in not being subject to phosphorylation by protein kinase A (PKA) or G protein receptor kinase (GRK2/βARK). They possess their own characteristic profile of specific activators and inhibitors. They are expressed predominantly in adipose tissues, where

39

FIG 3.1 Adrenergic receptor agonists and antagonists.
Horizontal lines indicate approximate specificities. Note that at high concentrations the selectivities tend to be reduced. Adapted from Hanoune.[4]

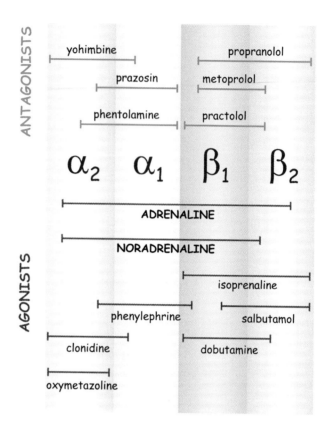

they play an important role in the regulation of lipid metabolism. For review see Strosberg.[3]

Renin is required for the formation of angiotensin from angiotensinogen. There are other β-blockers, such as practolol and metoprolol, which at low concentrations are selective for β_1-receptors. Thus, in principle, it should be possible to devise specific β_1 antagonists (to suppress cAMP and with which one could treat a heart patient without exacerbating his asthma; (activation of β_2-receptors in the lung would cause bronchoconstriction). However, this is not practicable because the selectivity between receptor subtypes is rarely perfect, especially at higher dose levels. Also, other factors such as the availability of the free form (not bound to plasma proteins), the tissue distribution, *in vivo* half-life, and toxicity have to be taken into account before drugs can be used in clinical practice. Indeed, the first β-blockers failed to reach the clinic because they were found to induce tumours in mice.

How do receptors distinguish agonists from antagonists?
Why do some compounds (such as salbutamol) cause activation whereas others (such as propranolol) are inhibitors of β-adrenergic receptor function? Certainly there are some systematic structural features that appear to determine their biological effects. In general, it is found that increasing the substitution on the

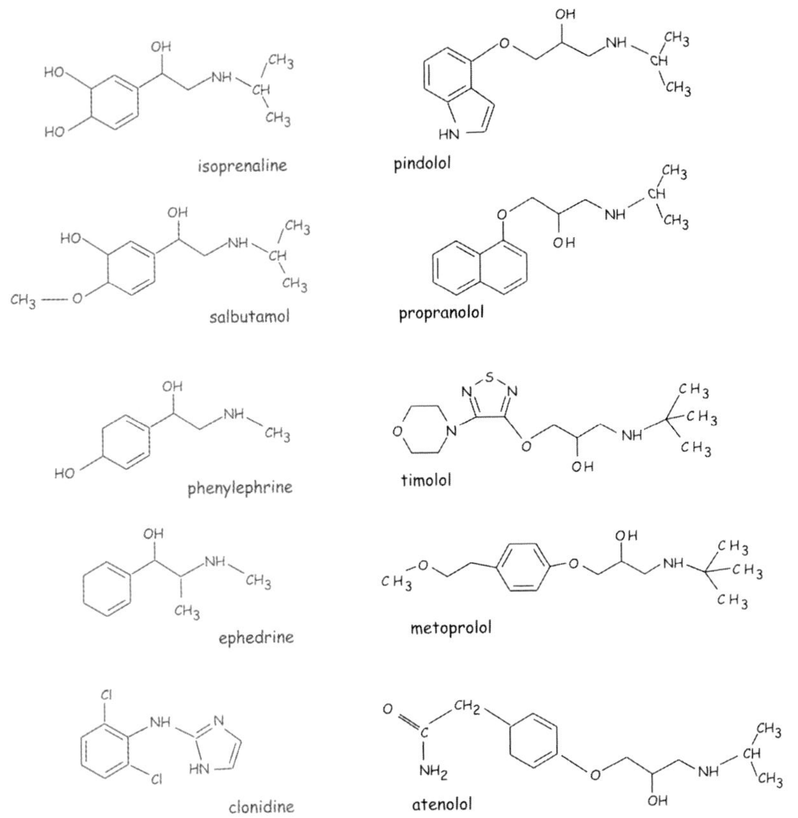

FIG 3.2 Some drugs that interact with adrenergic receptors.
As the substitution of the amine nitrogen on the agonists (in red) is increased, their selectivity for β- over α-receptors increases. Some of the α-adrenergic agonists and all the β-blocking agents (in black) lack the catechol nucleus. The antagonists are also characterized by the insertion of an ether linkage (-O(CH₂)-) in the lateral chain. The structures of the physiological catecholamines adrenaline, noradrenaline, and dopamine are shown in Table 2.1, page 22.

amine nitrogen atom of catecholamines increases the preference for β- over α-receptors. The β-blockers are all characterized by the absence of the catechol nucleus (aromatic ring with neighbouring hydroxyls) and by the presence of an $-OCH_2-$ group linking the side chain to the aromatic moiety.

An indication that the interaction of ligands with β-receptors differs for agonists and antagonists is given by the finding that reducing the temperature increases the binding affinity of agonists such as isoprenaline, but has little effect on the binding of antagonists such as practolol. A temperature-dependent increase in affinity indicates an equilibrium shift characteristic of an enthalpy-driven reaction, whereas an insensitivity to temperature suggests that a change in entropy is the main driving force. Indeed, estimates of the thermodynamic parameters reveal that the antagonists bind with an increase in entropy which is to be expected when water molecules organized around a binding site are displaced. The agonists mostly show a decrease in entropy on binding. This might mean that although the two classes of ligand occupy the same binding pockets within the receptor molecule, which incidentally would explain the competitive nature of their binding, the actual points of attachment to the lining of the pocket (i.e. to particular amino acid residues) are different. The consequences of binding are alterations such as displacement of protons and water, breakage of hydrogen bonds, disturbance of van der Waals interactions and conformational changes in the receptor itself. In the case of agonists, this results in an overall increase in order (decreased entropy).[5] Importantly, only agonists induce the conformational alterations that enable the receptors to communicate with the ensuing components of the signal transduction apparatus.

Acetylcholine receptors

At the molecular level, all the effects of the catecholamines, including dopamine, are mediated through members of the family of receptors that span the membrane seven times (7TM receptors, discussed later in this chapter) and subsequently they all activate or inhibit an enzyme, such as adenylyl cyclase or phospholipase C (Chapter 5). This is always mediated through a GTP-binding protein (GTPase) (Chapter 4). About 60% of the drugs used in clinical practice are directed at 7TM receptors. In contrast, acetylcholine interacts with two very distinct types of receptor that are quite unrelated to each other. These are the nicotinic receptors (ion channels) and muscarinic receptors (7TM receptors).

Acetylcholine

Although acetylcholine is a first messenger that interacts with receptors, it does not have the function of a hormone. It is confined to synapses between nerve endings and target cells, and it is the primary transmitter at the neuromuscular junction (between motor nerve and muscle end plate). In the autonomic nervous system, it is also the transmitter at preganglionic nerve endings and in most parasympathetic postganglionic nerves.

Parasympathetic stimulation through the vagus nerve causes the dilation of blood vessels, increases fluid secretion (e.g. from the pancreas and salivary glands), and slows heart rate. At the neuromuscular junction, acetylcholine is released from the presynaptic membrane. It diffuses across the junctional cleft and interacts with nicotinic receptors situated on the postsynaptic membrane. It is then removed from the synaptic cleft with 'flashlike suddenness' (in the words of Henry Hallett Dale). This occurs partly by diffusion, but mainly by the action of acetylcholine esterase, which converts it to choline and acetate within milliseconds. Since the affinity of nicotinic receptors for acetylcholine ($K_D \approx 10^{-7}$ mol L^{-1}) is rather moderate (at least, in comparison with some other types of receptor for their respective ligands), the rate of dissociation is sufficiently fast to allow it to detach rapidly (recall that $K_D = k_{off}/k_{on}$), following the steep decline in the local concentration due to the activity of the esterase.

Because of the extreme lability of acetylcholine , in experimental investigations it has been normal to work with stable non-hydrolysable derivatives such as carbamylcholine (carbachol). As with the natural compound, carbamycholine is an agonist at both nicotinic and muscarinic receptors.

Some of the compounds that inhibit the hydrolysis of acetylcholine are among the most toxic known (Figure 3.3). Thus, they may cause stimulation of cholinergic receptor sites throughout the central nervous system (CNS), depression at autonomic ganglia, and paralysis of skeletal muscles. This is followed by secondary depression involving irreversible receptor desensitization, discussed below. In addition to these nicotinic functions, muscarinic responses to acetylcholine also tend to persist. On the other hand, some cholinesterase inhibitors are less toxic and have found their way into clinical practice as in the treatment of myasthenia gravis and Alzheimer's disease.

The first synthesis of an irreversible inhibitor of cholinesterase was reported as early as 1854 by Clermont, predating the isolation of the alkaloid physostigmine (eserine) from Calabar beans by about 10 years (see Chapter 1). This was tetraethyl pyrophosphate. In his dedicated pursuit of science,

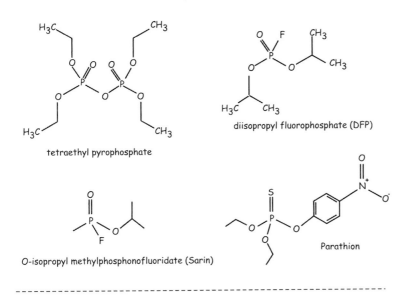

tetraethyl pyrophosphate

diisopropyl fluorophosphate (DFP)

O-isopropyl methylphosphonofluoridate (Sarin)

Parathion

physostigmine (eserine)

neostigmine

FIG 3.3 Some inhibitors of serine esterases. Organophosphate neurotoxins are shown in the upper panel. The P=O double bond is essential for toxicity. Substitution of the oxygen with sulfur, as in parathion, results in a much less toxic derivative, though the use of this class of compounds as insecticides has been banned in many countries. Also illustrated are the parasympathomimetic agents, reversible choline esterase inhibitors, physostigmine (eserine) isolated from Calabar beans and a derivative, neostigmine. Both these compounds have found wide application in clinical practice.

Clermont even went so far as to comment on its taste and one may wonder that he ever survived to tell the tale. Further organophosphorus compounds were developed with the aim of preparing nasty surprises for flies (parathion, the most widely used insecticide of this class) and subsequently even nastier surprises for soldiers (and, sadly, civilians too). These nerve gases, now regarded as weapons of mass destruction (WMDs) include sarin (isopropyl-methylphosphonofluoridate), originally developed in Nazi Germany. Four years after the attack by Saddam Hussein's forces on the city of Halabja in March 1988, an estimated 5000 people were killed and subsequent soil samples revealed traces of sarin breakdown products. The LD_{50} for humans is reckoned to be (not tested!) of the order of $10\,\mu g\ kg^{-1}$. When sarin was released on the Tokyo metro system on 20 March 1995 by zealots of the Aum cult, 12 people were killed and 5500 were injured.

All these compounds take their effects by alkylphosphorylation of active site serines, not only of acetylcholinesterase, but also of the serine esterases and serine proteinases (such as trypsin, thrombin, etc.). Clearly, the rapid removal of acetylcholine from the synapse is as essential for the process of neurotransmission as is its release. Interestingly however, some of the 'reversible' cholinesterase inhibitors, such as neostigmine, are used in the treatment of glaucoma, and myasthenia gravis, with the aim of restoring the availability of acetylcholine.

With the threat of military and terrorist abuse of these noxious agents attention has turned to the matter of antidotes. Currently atropine is administered to prevent over-occupation of the acetylcholine receptors. In addition, casualties are treated with nucleophilic agents based on 2,3-butanedione monoxime with the aim of dephosphorylating and so reactivating the poisoned acetylcholine esterase.[6,7]

Cholinergic receptor subtypes

Acetylcholine receptors fall into two classes, originally distinguished by their responses to the pharmacological agonists, nicotine and muscarine. In this section we also give brief consideration to related receptors that respond to amino acid neurotransmitters.

Nicotinic receptors

At the neuromuscular junction acetylcholine acts through nicotinic cholinergic receptors to stimulate the contraction of skeletal muscle. It also transmits signals in the ganglia of the autonomic nervous system. The receptors are cation channels which conduct both Na^+ and K^+, the principal cations present in cells and extracellular fluid. Nicotinic receptors are antagonized by tubocurarine. Neither nicotine nor tubocurarine has any

- **Nicotine**: named after Jacques Nicot, the French ambassador to Lisbon who introduced the tobacco plant *Nicotiana tabacum* into France in 1560.
- **Muscarine**: a product of the poisonous mushroom, *Amanita muscaria*.
- **Atropine**: from the Greek *a-tropos*, meaning no turning or inflexible, referring to the condition of the pupil when dilated by treatment with atropine. It is an alkaloid extracted from the berries of deadly nightshade, *Atropa belladonna*. The content of a single berry is sufficient to kill an adult human.

significant effect on muscarinic receptors. By the same token, muscarine and atropine (compounds that activate and inhibit muscarinic receptors) are without effect on the processes regulated by nicotinic receptors.

The nicotinic receptors are ligand-gated ion channels and so are termed ionotropic receptors. They are members of the Cys-loop superfamily of channels (see below), which includes the 5-HT$_3$ receptor (activated by 5-hydroxytryptamine, serotonin), the receptor for glycine and the receptors GABA$_A$ and GABA$_C$ (activated by γ-aminobutyric acid, GABA). Note however, that glycine and GABA receptors are anion-selective rather than cation-selective channels, responding to ligands that cause electrical inhibition. For all of these, the receptor and ion channel functions are intrinsic properties of the same protein.

Muscarinic receptors

Whereas acetylcholine acts through the nicotinic receptor to stimulate contraction of skeletal muscle, it decreases the rate and force of contraction of heart muscle through muscarinic receptors. These belong to the seven transmembrane-spanning (7TM) superfamily of G protein-linked receptors and they exist as five subtypes, M1, M2, M3, m4, and m5. All five subtypes are present in the CNS. M1 receptors are expressed in ganglia and a number of secretory glands. M2 receptors are present in the myocardium and in smooth muscle and M3 receptors are also present in smooth muscle and in secretory glands (Table 3.2).

Muscarine and related compounds have many actions, dependent on the tissues to which they are applied and the receptor subtypes that are expressed. The structures and the mechanisms of action of the receptors activated by muscarine are closely related to those activated by the catecholamines. Like the adrenergic receptors, the diversity of their functions first became apparent with the development of a specific antagonist. This was pirenzepine, which blocks M1 receptors and prevents gastric acid secretion while being without effect on a number of other responses elicited by muscarinic receptors.

Although the muscarinic and nicotinic receptors share a common physiological stimulus, and although the muscarinic receptors can regulate ion fluxes indirectly (for example in the heart), they are not in themselves ion channels. The difference goes much further than this. Not only are these receptors unrelated in terms of structure and evolution, but the conformation of the acetylcholine as it binds is likely to be different. While nicotine and muscarine are very different from one another (Figure 3.4) and contain ring structures that lend rigidity, the acetylcholine molecule is more flexible and is able to adopt different conformations that may be stabilized as it slots into its binding sites in the two receptors.

Tubocurarine is the main constituent of curare, a mixture of plant alkaloids obtained from *Chondrodendron tomentosum* or *Strychnos toxifera*, used as an arrow poison by some South American Indians. Claude Bernard (1856) showed that it causes paralysis by blocking neuromuscular transmission;[8] Langley (1905) suggested that it acts by combining with the receptive substance at the motor endplate (see Chapter 1).[9] Other related (synthetic) compounds such as gallamine differ mainly in their duration of action.

By convention, upper-case letters (M1, etc.) are used to assign receptors that have been pharmacologically defined, and lower-case letters (m4, etc.) are used to assign those receptors that have been revealed only by molecular cloning.

TABLE 3.2 Muscarinic receptor subtypes

Subtype	Antagonists	Tissue[a]	Transducer	Effector
M_1	atropine, pirenzipine	autonomic ganglia	G_q	phospholipase C (increases cytosol Ca^{2+})
M_2	atropine, AF-DX 384[b]	myocardium smooth muscle	G_i, G_o	activation of K^+ channels, inhibition of adenylyl cyclase
M_3	atropine	smooth muscle secretory glands	G_q	phospholipase C (increases cytosol Ca^{2+})
m_4	atropine, AF-DX 384[b]		G_i, G_o	inhibition of adenylyl cyclase
m_5	atropine		G_q	phospholipase C (increases cytosol Ca^{2+})

[a]Multiple subtypes occur in many tissues. All are present in the CNS. This column indicates tissues with high levels of expression.
[b]A benzodiazepine drug.

As with the adrenergic receptors, the response pathways operated by the muscarinic receptors for acetylcholine are all mediated through GTP-binding proteins. In the case of M1, M3, and probably the m5 subtype, the transduction is by G proteins of the G_q family, which causes activation of phospholipase C (PLC). This enzyme hydrolyses the phosphoinositide, phosphatidylinositol-4,5-bisphosphate (PI(4,5)P$_2$). One of the products is inositol-1,4,5-trisphosphate (IP$_3$) which releases Ca^{2+} ions from intracellular Ca^{2+} stores (for detailed discussion, see Chapter 7). The M2 and m4 receptors address G proteins of the G_i class with consequent inhibition of adenylyl cyclase and activation of K^+ channels (see Chapter 5).

Nicotinic receptors are ion channels

Among the receptors and over the years the nicotinic cholinergic receptor has received the most detailed attention and in many respects is the best understood. There are particular reasons for this:

- Specialized tissues exist in which nicotinic receptors are present in huge quantities.
- There are toxins that bind specifically to the receptor with high affinity, enabling its isolation.
- The channel properties of individual nicotinic receptor molecules are readily characterized by patch clamp electrophysiology.
- Three-dimensional structures of nicotinic receptors *in situ* have been obtained.

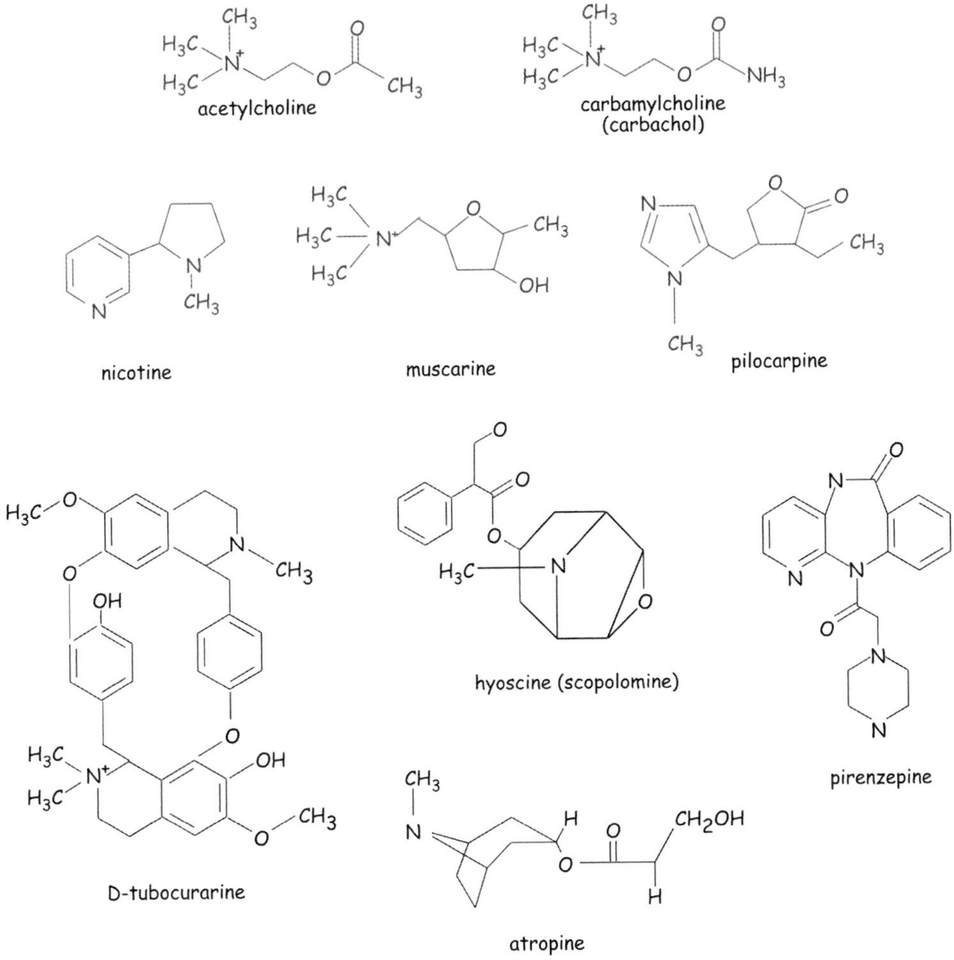

FIG 3.4 Structures of cholinergic receptor agonists (red) and antagonists (black).

Electric organs provide a source

Tissues in which nicotinic receptors are abundant include the neuromuscular junction, where there may be as many as 4×10^7 copies on a single motor endplate of skeletal muscle (Figure 3.5). When acetylcholine is secreted from the presynaptic (prejunctional) membrane into the synaptic cleft, it binds to receptors on the postsynaptic surface. This permits extracellular Na^+ to flow through the nicotinic channels, down its concentration and voltage gradient into the muscle cell, causing a local depolarization that may lead ultimately to contraction. This abundance of receptors is vastly exceeded in the electric organs of the marine ray, *Torpedo* and the freshwater electric

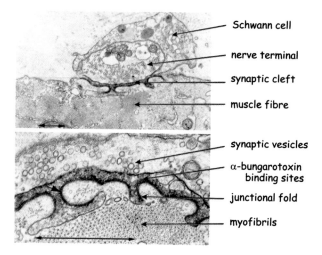

FIG 3.5 α-Bungarotoxin staining of nicotinic receptors at a neuromuscular junction: thin section electron micrographs of frog cutaneous pectoris muscle. The nerve terminal, with mitochondria and numerous synaptic vesicles, is encased by a Schwann cell and abuts a muscle fibre. The preparation was treated with α-bungarotoxin fused to horseradish peroxidase, followed by 3,3'-diaminobenzidine to reveal the acetylcholine receptors as an electron-dense reaction product. The label is mainly confined to the extracellular surface of the postjunctional membrane with some diffusion into the synaptic cleft. The scale bars are 0.5 μm From Burden et al.[10]

eel, *Electrophorus*. These electroplax organs are specialized developments of skeletal muscle and they express huge numbers of receptor molecules, about 2×10^{11} per endplate. As early as 1937, it was discovered that *Torpedo* electroplax could turn over more than its own weight of acetylcholine in an hour. The first demonstration of the biosynthesis of acetylcholine by a cell-free electroplax preparation was reported in 1944. Even today, the isolated postsynaptic membranes of the *Torpedo* electric organ provide an important source material for structural studies.

α-Bungarotoxin, a toxic peptide obtained from the venom of the snake *Bungarus multicinctus*, binds at the agonist binding site of neuromuscular-type acetylcholine receptors. There are similar snake venom α-toxins that bind to the neuronal isoforms.

α-Bungarotoxin blocks neuromuscular transmission
Because of their abundance in these specialized tissues it became possible to isolate, purify, and reconstitute nicotinic receptors, albeit in small quantities, well before the advent of cloning. The isolation was achieved by affinity chromatography using toxins that block neuromuscular transmission, in particular α-bungarotoxin. This 8 kDa peptide binds specifically and with high affinity to nicotinic receptors ($K_D = 10^{-12}$–10^{-9} mol L^{-1}), with consequent inactivation. In the laboratory, the toxin can be immobilized by attachment to a solid phase (such as Sepharose). By applying to this affinity column a solution of proteins released by detergent from muscle or electroplax membranes, it is possible to isolate receptor molecules.[11–13]

Patch clamp electrophysiology of single channels
With the introduction of the patch clamp technique by Neher and Sakmann,[14] it became possible to study individual ion channels *in situ* in a membrane. It is hard to overestimate the contribution that this advance has made to our understanding of channel mechanisms. The idea itself is simple enough.

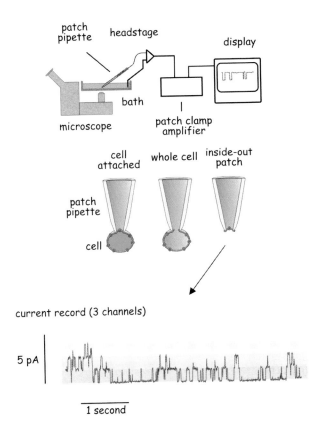

current record (3 channels)

5 pA

1 second

FIG 3.6 Using membrane patches to record single channel opening and closure.
Typically, cells are allowed to adhere to the bottom of a shallow dish. Individual cells are then attached to a patch pipette. A typical current record obtained from a patch of membrane, detached from a cell, is shown.

A patch pipette, which is a glass microelectrode with a smooth (fire-polished) tip a few micrometres in diameter, is pressed gently against the outside of a cell. With gentle suction a very tight seal (typically $>10^9\ \Omega$) forms between the pipette tip and the cell membrane. A small area of membrane can then be removed from the cell without breaking the seal (Figure 3.6). The end result is either an outside-out or an inside-out patch of membrane. These contain only a very few ion channels, but depending on the orientation, the ligand binding sites are either exposed to the exterior or to the contents of the patch-pipette. The paucity of channels within the patch and the high resistance seal then enable the recording of membrane current at a controlled voltage. The recordings reveal individual channel openings and closures.

Nicotinic receptors were among the first channels to be characterized in this way. Like other ion channels they open and close abruptly in an all-or-none fashion and they do this spontaneously. The effect of acetylcholine is to increase the probability that a particular channel is in its open state.

Erwin Neher and Bert Sakmann shared the 1991 Nobel Prize for their discoveries concerning the function of single ion channels in cells.

49

Architecture of the nicotinic receptor

Because it is difficult to grow crystals of transmembrane proteins, the three-dimensional structure of the nicotinic receptor has been studied by cryoelectron microscopy of helical tubes prepared from the receptor-rich postsynaptic membranes of *Torpedo marmorata*.[15] Although electron microscopy provides a low to medium resolution approach, applying crystallographic refinement methods to the data has yielded a structure having atomic resolution (4 Å).[16] This is shown in Figure 3.7. The receptor is pentameric and its five subunits, two of which are identical, are arranged around a central pore. The clockwise arrangement viewed from the synapse is $\alpha, \gamma, \alpha, \beta, \delta$. All five subunits must be present for the complex to function as a channel. The overall shape of the receptor is a hollow cone, approximately 80 Å in diameter at its widest point and some 160 Å long (i.e. 8×16 nm). Its position with respect to the plasma membrane is indicated in Figure 3.7. The assemblage protrudes about 80 Å into the synaptic space and within this protuberance there is a central vestibule ~20 Å in diameter. On the intracellular side of the membrane, the structure protrudes about 20 Å and it contains another, smaller cavity with side openings. The two cavities communicate through the central pore.

The sequences of the individual subunits indicate extensive homology: 35–40% amino acid identity, with many additional conservative substitutions. The four

M1–M4: an unfortunate nomenclature, easily confused with the terminology for the muscarinic receptors: M1–m5, etc.

FIG 3.7 Structure of the nicotinic receptor without ligand.
Side and top views of the pentameric structure are shown in the central figures. One of the α-subunits is redrawn on the left to show its secondary structure. The images on the right indicate the M2 helices (red) that line the channel (2bg9.[16]).

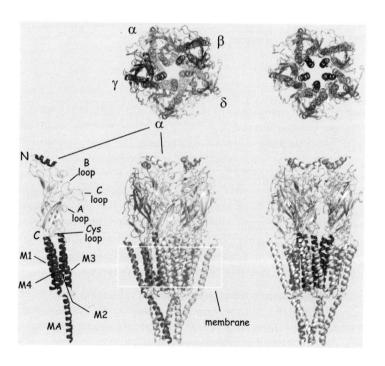

separate polypeptides evolved from a common ancestor by a process of successive gene duplication and modification that has its origin more than 1.5×10^9 years ago (thus long preceding the emergence of metazoan creatures). Before the three-dimensional structure of the nicotinic receptor was obtained, it was known from the sequences that each subunit possesses four stretches, M1, M2, M3, and M4, in which hydrophobic amino acids predominate (Figure 3.8), suggesting that these regions are membrane-spanning chains. Organized as α-helices, their lengths would be sufficient to traverse the lipid bilayer.

Subunit structure

The prediction that the M1–M4 sequences of each of the subunits are indeed transmembrane α-helices is confirmed by the structural data (Figure 3.7). The even number of traverses (4) means that the N- and C-terminals must be on the same side of the membrane. In fact they are extracellular. The exposed N-terminal portion of each subunit is glycosylated and consists of some 200 amino acids assembled around a 10-stranded β-sandwich, plus a short α-helix (Figure 3.7). For the α-subunit, loops connecting the β-strands on one side of the molecule are involved in ligand binding. These are the A, B, and C loops, the Cys-loop and the β_1–β_2 loop (not shown). The remaining residues that contribute to the binding site are provided by the adjacent subunits (δ or γ). Both binding sites must be occupied in order to activate the receptor. The inhibitor of the channel, α-bungarotoxin, binds in the close vicinity of these sites.

The loops linking the transmembrane spans, M1–M2 on the inside of the cell and M2–M3 on the outside, are relatively short, but the M3–M4 loop that

Such stretches of about 20 amino acids, having higher than average **hydrophobicity**, are present in all known membrane-spanning proteins. Furthermore, in those proteins for which structural information is available, e.g. bacteriorhodopsin, the photosynthetic reaction centre of the purple bacterium *Rhodopseudomonas*,[17] the potassium channel of *Streptomyces lividans*,[18] and a gated mechanosensitive ion channel in *Mycobacterium tuberculosis*,[19] the hydrophobic spanning segments have all been shown to be organized as α-helices.

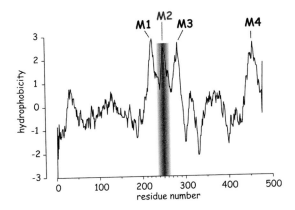
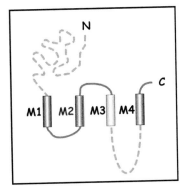

FIG 3.8 Hydropathy plot of a nicotinic receptor α-subunit.
The graph represents the hydropathy profile of the α-subunit of the nicotinic acetylcholine receptor. The diagram depicts the transmembrane chain organization. The hydrophobicity index is the average hydrophobicity of stretches of 19 amino acids centred on each position in the chain (thus each point represents the average hydrophobicity of 9+1+9 amino acids). Each amino acid is assigned a hydrophobicity value as indicated in **Figure 3.9**. Based on data for the human α_7 subunit published in the LGIC database (www.pasteur.fr/recherche/banques/LGIC/LGIC.html).[20]

forms the intracellular structure, is more extended (about 130 amino acids) and contains the intracellular MA helix. The non-helical part of this loop is unstructured in the EM data and is therefore omitted from Figure 3.7.

M2 sequences and the pore

Inspection of the sequences of the M2 regions of each of the subunits illustrates their homology (Figure 3.9). Within these short stretches there are at least eight points of identity and ten more that are conservative substitutions. Notice also that each of these four sequences contains a dozen or more strongly hydrophobic amino acids (leucine, valine, isoleucine, etc., with hydrophobicity assignments>0).

There is much evidence that residues present in the M2 helices of all five subunits are the main contributors to the pore along the central axis of symmetry of the pentameric structure. Importantly, only residues of the M2 segments bind water-soluble channel-blocking agents, such as chlorpromazine and other compounds having photoaffinity properties. The rate of binding of these compounds is 1000 times greater when the channel is in its open configuration.[22] In addition, fractionation of membrane proteins following photoaffinity labelling with lipid-soluble probes has shown that the M2 chains are fully shielded from the lipid environment of the membrane. Again, structural studies confirm that the transmembrane pore is lined by the M2 helices of each of the five subunits and

A **photoaffinity** label is a compound that becomes reactive when subjected to a pulse of UV light. This enables it to form stable derivatives with molecules in its very close vicinity

FIG 3.9 M2 sequences of the four subunits of the nicotinic receptor. (a) M2 sequence comparison (*Torpedo marmorata*) indicating identical and conserved residues, the transmembrane region and the probable location of the channel gate. The α-subunit residues in red are those that have pore-facing side chains. (b) Hydrophobicity values assigned to the 20 protein amino acids. In (a) the hydrophobicities of the M2 residues are indicated in colours that correspond to the assignments shown in (b). (These assignments were calculated from data including vapour pressures and partition coefficients by Kyte and Doolittle.[21])

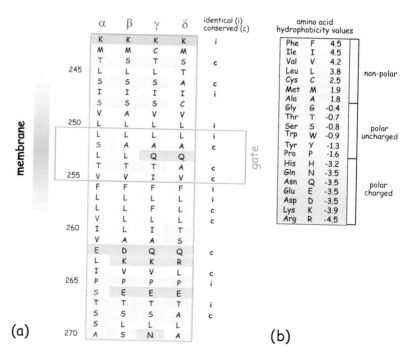

(a)

(b)

show that this assembly is surrounded by an outer ring of helices (M1, M3, M4 from each subunit), that shields it from the plasma membrane, as depicted in Figure 3.7. The rather limited number of contacts between the outer shell and the M2 chains allows the movement necessary for pore opening to occur.

By probing the open channel with chlorpromazine or other site-specific channel-blocking substances, it has been possible to map its lining. Collectively, these experiments have identified amino acids separated by three or four residues along this section of the chain, consistent with an α-helical arrangement. As the pore tapers to its smallest diameter, within the transmembrane region, the M2 segments are aligned so that the homologous residues form rings that are mostly non-polar (Figure 3.9). Polar or charged side chains mark the ends of the transmembrane region and the negatively charged groups will be selective for cations. At the narrowest point, the encircling hydrophobic side chains make contacts across the opening, keeping it effectively shut (an aperture 6–7 Å across, too small to admit a Na^+ or K^+ ion with its primary hydration shell, ~8 Å in diameter). The pore region opens out on either side of the membrane into the vestibules within the structure. The lining of the vestibules and the surfaces of the openings in the intracellular assembly, possess areas of negative charge. It is probable that these, as well as the anionic side chains exposed in the vicinity of the pore, contribute to the cationic selectivity of the channel.

Ligand binding and channel opening

Although the four different subunits (α, β, γ, δ) that constitute the nicotinic receptor are very similar, within the pentameric structure the two α-subunits are organized differently from the others. This is due to a difference in the arrangement of their inner β-sheets, resulting in an anticlockwise rotation (~10° as viewed from outside the cell) relative to the non-α-subunits. This has been termed a 'distorted' conformation.[16] Relief of this distortion is thought to lead to channel opening. Understanding the activation mechanism has been advanced by comparing the ligand-free receptor with the water-soluble acetylcholine-binding protein of the snail *Limnea stagnalis*. This is homologous to the extracellular domain of the receptor. Figure 3.10 shows part of its structure with a ligand molecule (carbamylcholine) bound between two adjacent subunits.[23] The loops that wrap around the ligand and that contribute to the binding site are shown and these correspond to the ligand-binding loops (A, B, C) indicated in Figure 3.7. (Note: unlike the nicotinic receptor, the acetylcholine-binding protein is a *homo*pentamer of five α7 subunits). A comparison of the two structures suggests that ligand binding to a receptor relieves the twisted conformation of the α-subunits and this movement is then communicated through the β-strands to the M2 helices, so perturbing the hydrophobic interactions across the gate region and allowing it to open.

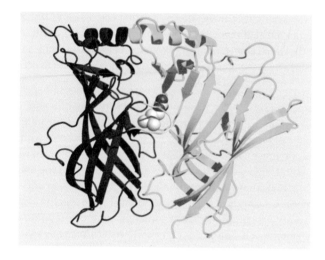

FIG 3.10 Partial structure of *Lymnea stagnalis* acetylcholine-binding protein showing bound ligand. Two of the five α-subunits are shown (blue and green) with a molecule of carbamylcholine (spheres). The B-loop (red), C-loop (yellow), and A-loop (magenta) are indicated.(1ux2.[24]).

The term **metabotropic** is used to distinguish these G-protein-linked receptors for glutamate, GABA, etc. from the **ionotropic** receptors that possess integral ion channels operated by the same ligands.

Of course things are never as simple as one might wish. The direction of the ion flux must depend on the intracellular concentration of Cl^- but in the neurons of the hypothalamic suprachiasmic nucleus (SCN) this appears to vary diurnally, being higher during the daylight hours, probably because of the operation of a cyclically modulated ion pump.[25] Depending on the time of day, the consequence

Receptor desensitization

When a receptor has been exposed to acetylcholine for prolonged periods it becomes unresponsive to further stimulation. This phenomenon, known as desensitization, is regulated by the β, γ, and δ subunits. The extended intracellular loops linking M3 and M4 of each subunit contain sites for phosphorylation by protein kinases A and C. The longer acetylcholine is bound to the α-subunits, seconds rather than milliseconds, the greater the extent of phosphorylation. This explains why the potent inhibitors of acetylcholine esterase, such as diisopropyl fluorophosphate (DFP) and sarin, are so toxic. By preventing the hydrolysis of acetylcholine, the stimulus is initially prolonged, but when it does eventually diffuse away and the channels have closed, the extensive phosphorylation ensures that they cannot then be reopened. The receptors are said to be desensitized.

Other ligand-gated ion channels

The ion channels that are activated by extracellular ligands may be divided into three superfamilies: Cys-loop receptors, glutamate-activated cationic channels, and ATP-gated channels. Members of the Cys-loop superfamily are pentameric and in addition to the nicotinic receptors, include receptors for the amino acid neurotransmitters GABA and glycine. As discussed above, these are Cl^- channels, present for example on postsynaptic membranes of neurons, and they have an inhibitory effect. When they are open, Cl^- ions enter the cells, causing a local membrane hyperpolarization. This tends to counteract the effect of an excitatory neurotransmitter by decreasing the probability that the postsynaptic cell will depolarize beyond the threshold for an action potential.

Channels regulated by glutamate are the major excitatory receptors of the CNS. Their amino acid sequences and membrane topology are quite different from the Cys-loop proteins. They are tetrameric and their subunits possess only three transmembrane segments with the consequence that the N- and C-termini are expressed on either side of the membrane. In these receptors, the equivalent M2 segment that is presumed to line the channel is rudimentary. It enters into the membrane and re-emerges on the same side (cytoplasm).[26] ATP-gated channels differ yet again from the other families, possessing three homologous subunits, each of which has two transmembrane segments.

Just as acetylcholine interacts with two distinct forms of receptor, ionotropic (nicotinic) or metabotropic (muscarinic), so does GABA and glutamate. On the one hand, there are the metabotropic GABA$_B$ and glutamate (mGlu) receptors, which communicate through G proteins to activate PLC and mobilize intracellular Ca^{2+}. On the other hand there are the ionotropic receptors GABA$_A$, GABA$_C$, and glutamate (iGlu), which are ligand-gated ion channels. Again, the ionotropic receptors and the metabotropic receptors for GABA and glutamate are antagonized by quite distinct inhibitors.

The three classes of ionotropic glutamate receptors are classified according to their susceptibility to inhibition by the synthetic glutamate analogues NMDA, AMPA, and kainate. For the NMDA receptor, opening of the non-selective cation channel requires glutamate (or aspartate) together with glycine (coagonist). If the glycine binding site is blocked (or unoccupied), glutamate alone is insufficient to effect channel opening. The AMPA receptors, present on all neurons in the central nervous system, are responsible for the fastest transmission of excitatory synaptic responses.[27] Opening of the cation channel requires the occupation of both agonist binding sites by glutamate, with relatively low affinity.

Glutamate appears to be involved in the pathogenesis of a number of severe diseases and condition such as stroke and epilepsy, due either to excessive release, reduced uptake, or altered receptor function. Antagonists at these receptors have been used as general anaesthetics (particularly in veterinary practice) and, as widely, as drugs of recreation (e.g. ketamine and phencyclidine).

The 7TM superfamily of G-protein-linked receptors

The muscarinic acetylcholine receptors and the GABA$_B$, mGlu, and catecholamine receptors are members of the same protein superfamily. Their structures are closely related to the visual pigment rhodopsin and very distantly related to the bacteriorhodopsins, which constitute ion pumps in the membranes of certain archaeans, most prominently *Halobacterium*. The feature that relates all these structures is the topological organization of the

---

is that GABA acts either as an excitatory or as an inhibitory neurotransmitter. During the day, when the concentration of Cl$^-$ in the SCN is in excess of about 15 mmol L^{-1}, GABA enhances the firing rate (8–10 Hz) and causes excitation. At night, opening of GABA$_A$ channels allows an influx of Cl$^-$ which causes membrane hyperpolarization and a reduction in the firing rate (2–4 Hz).

NMDA is *N*-methyl-D-aspartic acid, **AMPA** is α-amino-3-hydroxy-5-methyl-4-isoxazolepropionic acid, and **kainate**, derived from algae, is 2-carboxy-3-carboxymethyl-4-isopropenyl-pyrrolidine.

More than 50% of the pharmaceutical products currently on the market (or in development) are targeted at the 7TM class of receptors.[28]

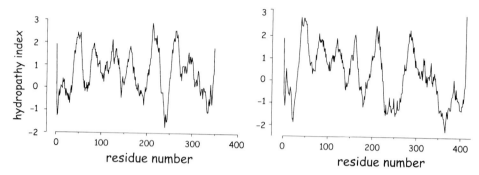

FIG 3.11 Hydropathy plots of rhodopsin and the β-adrenergic receptor.
The hydropathy plot of rhodopsin is shown on the left, the β_2-adrenergic receptor on the right. For the derivation of the hydrophobicity see legend to Figure 3.8. The sequence data were taken from the Swissprot database, accession numbers P08100 and P07550.

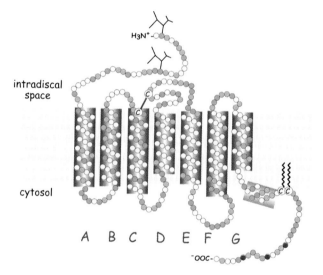

FIG 3.12 Membrane topology of bovine rhodopsin based on hydropathy data and X-ray diffraction analysis. The membrane organization of rhodopsin based on the hydropathy profile of its amino acid sequence (see Figure 3.11). The discs indicate hydrophobic residues (yellow), non-polar residues (grey), and polar residues (blue). The amino acids comprising the seven transmembrane spans (A–G) are predominantly hydrophobic, though non-polar and even charged residues are not entirely absent. The residues making contact with the ligand (11-*cis* retinal) are indicated (single letter code). The incidence of polar and charged residues is much greater in the exposed loops and terminal domains. The residues shown as red discs are targets for phosphorylation by rhodopsin kinase. From Ohguro et al.[29]

single peptide chain which in all cases traverses the membrane seven times. This is indicated by hydropathy plots, such as those shown in Figure 3.11 for rhodopsin and the β-adrenergic receptor. Topological diagrams are depicted in Figure 3.12 and Figure 3.13. The membrane-spanning segments are linked by

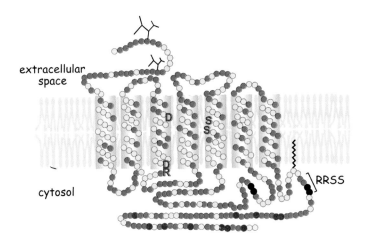

FIG 3.13 Membrane topology of the β₂-adrenergic receptor.
The membrane organization of the β₂-adrenergic receptor based on the hydropathy profile shown in Figure 3.11. The discs indicate hydrophobic residues (yellow), non-polar residues (grey) and polar residues (blue). The residues making contact with adrenaline are indicated (single-letter code). Residues shown as red discs are targets for phosphorylation by the β₂-adrenergic receptor kinase (GRK2). The black discs indicate the serine residues (in the consensus sequence -RRSS-) that are targets for phosphorylation by PKA.
From Fredericks et al.[30]

three exposed loops on either side of the membrane, with the N-terminus projected to the outside and the C-terminus in the cytosol.

Although they share obvious features of structural organization,[31] it is far from certain whether the archaean bacteriorhodopsins and the 7TM receptor proteins of eukaryotes are derived from a common ancestor. The similarity might arise from convergent rather than divergent evolution. For the receptors and the visual pigment rhodopsin, the organization of the 7TM chains was deduced mainly by analysis of the hydropathy plots of the amino acid sequences. The other essential feature that relates the 7TM receptors and the visual pigments is that nearly all of them interact inside cells, with GTP-binding proteins. Possible exceptions are members of the subgroup of EGF module-containing adhesion receptors (see page 65).

We now have the crystal structure of bovine rhodopsin, and it is clear that the organization of the transmembrane chains and the linking loops is more or less as had been inferred.[37] We may also be confident that other related 7TM receptors share the same basic structural organization. However, the idea that rhodopsin offers a predictive template for other 7TM receptors, for which structures are not yet available, may be over-optimistic.[38] Indeed, there are only a very few highly conserved residues in the whole superfamily, such as the two cysteines forming the disulfide bridge (see Figure 3.12).

Stop Press, 2008. Technical advances have recently enabled the crystallization of the β₂-adrenergic receptor, and a high-resolution structure has now been obtained.[32–34] The receptor was locked into its inactive configuration by the use of an inverse agonist. The overall conclusion is that rhodopsin and the β₂-adrenergic receptor share the main structural features with binding pockets for their ligands located deep within the transmembrane helices. However, the three extracellular loops of the receptor are clearly structured and this might give a clue of how its diffusible ligands can gain access to their binding site (for brief reviews, see Ranganathan[35] and Sprang.[36]

In comparison with other major protein families, the level of sequence homology between the members of the huge superfamily of 7TM receptors is low. However, sequence alignments do reveal some shared traits that form the basis for a classification into families.

Categories of 7TM receptor

A large proportion of the proteins acknowledged as members of the 7TM receptor family are orphans, having no known ligand and hence no known function. In addition, a similarly large proportion of the genes that code for 7TM proteins are not expressed. If we include those coding for 7TM proteins present in olfactory epithelial cells (probable receptors for odorant substances), then this family is truly enormous, approaching 2000 members, or $\sim$5% of the mammalian genome.

While the general structural plan of these proteins seems assured, there are of course important differences. The huge variety of activating ligands, encompassing ions (Ca^{2+}), small molecules, and proteins, makes this self-evident (Table 3.3). One extreme example is that of rhodopsin, which binds its ligand, 11-*cis* retinal, covalently (see Chapter 6). Unlike rhodopsin, the receptors for hormones exist in both free and bound states. The rate at which they can be activated must be limited by the availability of the ligand, its

TABLE 3.3 Categories of 7TM receptors

Receptor properties	Ligands
Ligand binds in the core region of the 7 transmembrane helices	11-*cis*-retinal (in rhodopsin) acetylcholine catecholamines biogenic amines (histamine, serotonin, etc.) nucleosides and nucleotides leukotrienes, prostaglandins, prostacyclins, thromoboxanes
Short peptide ligands bind partially in the core region and to the external loops	peptide hormones (ACTH, glucagon, growth hormone) parathyroid hormone, calcitonin
Ligands make several contacts with the N-terminal segment and the external loops	hypothalamic glycoprotein releasing factors (TRH, GnRH)
Induce an extensive reorganization of an extended N-terminal segment	metabotropic receptors for neurotransmitters (such as GABA and glutamate) Ca^{2+}-sensing receptors, for example on parathyroid cells, thyroidal C-cells (which secrete calcitonin) and on the renal juxtaglomerular apparatus
Proteinase activated receptors	receptors for thrombin and trypsin

rate of diffusion into the binding site and then its rate of attachment. What happens next is now well recognized. The activated receptors interact with G proteins and thence with intracellular effector enzymes to generate or mobilize second messengers (such as cAMP, Ca^{2+}, etc.). However, at the time that cAMP and its synthesizing enzyme adenylyl cyclase were discovered, understanding of the receptors was purely conceptual. No receptor proteins had been isolated and knowledge of membrane structure was very limited. Whether or not receptors and the enzyme activities that they control comprise a single entity had not been resolved.[39]

Receptor diversity: variation and specialization

Most signalling proteins with homologues in different species (orthologues) exhibit sequence homology, for example GTP-binding proteins, PKA, and phospholipase C in mammals, invertebrates, and yeast. The 7TM receptors, by contrast, have few conserved residues. Even within the recognized subfamilies (such as the rhodopsin-like receptors discussed below), the extent of sequence homology is exceedingly limited. Indeed, the idea that they are all derived from a single common ancestor or whether they arose as the result of convergent evolutionary processes has been repeatedly discussed. Moreover, if divergence in the face of conserved architecture is evident for the membrane-spanning regions, it is even more obvious for the various ligand-binding domains. For example, within just one family, the adhesion receptors, the exposed segments vary in length from 7 up to about 2800 residues.[41]

Although it is possible to classify these receptors on the basis of the types of ligands which they bind (see Figure 3.17), this bears little relationship to their evolutionary ancestry. Phylogenetic analysis can be used to investigate this, but it relies on single amino acid changes in conserved regions and it is unwise to base calculations on sections that are highly variable and may have undergone domain addition or replacement (see Chapter 24). In consequence, analyses of 7TM proteins have generally been based on the common, conserved transmembrane segments. Such an approach, applied to all the putative 7TM receptor proteins recognized in the year 1994, proposed six 'clans',[42,43] some of which do not exist in animals. Later, with the advent of the human genome database, a more searching analysis of the 7TM receptor-like proteins became possible. It has been concluded that there are four main human classes: A, B, C, and Frizzled (Table 3.4). The phylogenetic relationships of receptors that form these classes, based on their transmembrane segments, are illustrated in Figure 3.14. This includes a fifth class, the adhesion family, emerging from a group of mostly orphan receptors within Class B.[44] The rhodopsin class is the largest and is illustrated separately in Figure 3.15.

From these analyses, it seems certain that the members within each family do share a common evolutionary origin. It is clear, too, that the total number of

As late as 1973, Raymond Ahlquist, who made the distinction between α- and β-adrenergic receptors, still expressed his doubts about their actual existence with the words:

This would be true if I were so presumptuous as to believe that α and β receptors really did exist. There are those that think so and even propose to describe their intimate structure. To me they are an abstract concept conceived to explain observed responses of tissues produced by chemicals of various structures.[40]

TABLE 3.4 Classification of 7TM receptors

Main classes	Number	Comments	Ligands
Class A Rhodopsin family	271 + 388 olfactory	There are four main non-olfactory subgroups. Nearly all share common motifs such as DRY (D/E-R-Y/F) between transmembrane segment II and inner loop 2. Ligand binding occurs mostly in the cavity between the transmembrane segments (glycoprotein receptors are exceptions)	prostaglandins thromboxanes serotonin histamine (H2) catecholamines acetylcholine rhodopsin cone opsins melanopsin melatonin
Class B Secretin family	15	Generally bind large polypeptides that share considerable sequence identity. Often acting in a paracrine manner. N-terminal sequences (60–80 residues) contain conserved sulfhydryl bridges	glucagons GnRH PTH CRF secretin
Class C Glutamate and GABA family	22	The N-terminus forms two distinct lobes. Said to resemble a 'Venus fly trap' which snaps around the bound ligand	glutamate, GABA Ca^{2+} sweet tastes (TAS1 subgroup)
Frizzled class	11 frizzled + 25 TAS2	Includes the mammalian counterparts of the *frizzled* receptor of *Drosophila* and Wnt receptors. Also taste receptors of the TAS2 subgroup, possibly sensitive to bitter tastes	Wnt Hedgehog TAS2 stimuli
Adhesion family	33	Only recently recognized as a family distinct from the secretin group. The N-terminal 200–2800 residues are often rich in glycosylation sites, proline residues, mucin stalks, and motifs expected to take part in cell adhesion. Includes brain-specific angiogenesis inhibitor, CELSR (cadherin-EGF LAG receptor), and EMR (EGF-like module-containing mucin-like hormone receptor) present on macrophages and other cells of the immune system	mostly unidentified CD55 (regulates complement assembly) chondroitin sulfate

7TM receptors is very large. A recent study of known 7TM proteins, avoiding polymorphisms and redundancies has identified a dataset of 791 unique, full length human receptors, of which some 400 are non-olfactory.[45] Extensive information is published at http://www.iuphar-db.org/index.jsp and http://www.gpcr.org/7tm/. An olfactory receptor database is provided at http://senselab.med.yale.edu/ordb/default.asp.

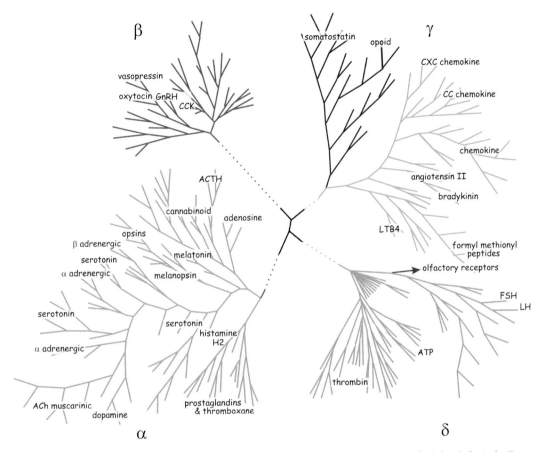

FIG 3.14 Phylogenetic relationship between the 7TM receptor-like proteins in the human genome: secretin, glutamate, Frizzled, and adhesion families. EMR, egf module-containing receptors; CELSR, cadherin-EGF LAG receptor. Adapted from Fredriksson et al.[44]

Binding of low-molecular-mass ligands

The binding sites for most low-molecular-mass ligands such as acetylcholine, those derived from amino acids (catecholamines, histamine, etc.) and the eicosanoids (see Table 2.1, page 22) are located deep within the hydrophobic cores of their receptors (see Figure 3.17a). Attachment of catecholamines is understood to involve hydrogen bonding of the two catechol OH groups to serine residues 204 and 207 on the fifth membrane spanning (E) α-helix (Figure 3.16, also Figure 3.13) and by electrostatic interaction of the amine group to an aspartate residue (113) on the C helix. The binding of acetylcholine similarly forms a cross-link between two membrane-spanning helices of muscarinic receptors. In this way, the bound hormone molecules are oriented

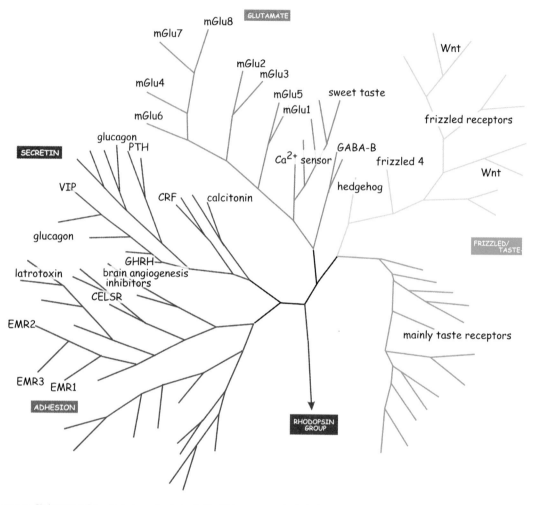

FIG 3.15 Phylogenetic relationship between the rhodopsin-like 7TM receptor proteins in the human genome. Adapted from Fredriksson et al.[44]

in the plane parallel to the membrane. They form bridges between two transmembrane spans of their receptors, perturbing their orientation relative to each other. This is the origin of the signal that is conveyed to the cell. β-Blockers, such as propranolol, also bind at this location with high affinity and so impede the access of activating ligands. However, lacking the catechol group, they fail to establish the critical link between membrane spanning segments. They fail to activate the receptor.

In contrast to the receptors for low-molecular-mass ligands, the binding sites for peptide hormones such as ACTH and glucagon are situated on the

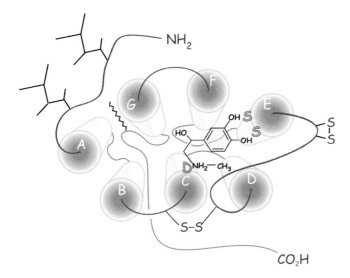

FIG 3.16 Top-down view of a hormone receptor with an adrenaline molecule bound between the membrane spanning segments.

The receptor is viewed from the extracellular surface. The N-terminal domain is glycosylated. The two catechol hydroxyls of adrenaline interact with the serine residues present in the membrane spanning α-helix E. The binding serines, S204 and S207, are separated by three residues (one turn) and so they are both projected towards the ligand, shifted by about 0.15 nm along the helix axis. The ligand amino group binds to an aspartate residue on the membrane spanning α-helix C.

Adapted from Ostrowski et al.[46]

N-terminal segment and on the exposed loops linking the transmembrane helices (Figure 3.17b). The receptors for glycoprotein hormones have been adapted by elongated N-terminal chains that extend well out into the extracellular aqueous environment (Figure 3.17c).

Calcium sensors and metabotropic receptors

The sites of attachment for neurotransmitters (such as glutamate and γ-aminobutyrate) on metabotropic receptors, and for Ca^{2+} ions on Ca^{2+}-sensing receptors, are on specialized extracellular N-terminal extensions of up to 600 residues[47] (Figure 3.17d). Binding of Ca^{2+} causes a pincer-like conformational change in the lobes of the extended extracellular domain. This exposes residues which then interact with the transmembrane core of the receptor. In this way, the extracellular domain acts as an auto-ligand. The Ca^{2+}-sensing receptor confronts a particular problem since it has to sense and then respond to very small changes in Ca^{2+} concentration against a high basal level (± 0.025 mmol L^{-1} in 2.5 mmol L^{-1}). The cells of the parathyroid gland react by secretion of parathyroid hormone whenever the concentration of circulating

FIG 3.17 Ligand binding sites. The 7TM receptor is a jack of many trades, regulating a variety of effectors and responding to diverse ligands with sizes that range from 40 Da (Ca^{2+}) to more than 100 kDa. (a) Most of the common low-molecular-mass hormones (such as adrenaline and acetylcholine) bind to sites within the hydrophobic core. (b and c) Peptide and protein ligands are accommodated on the exterior face of the receptor. (d) Although of low molecular mass, Ca^{2+} and the amino acids glutamate and GABA bind to large N-terminal extensions. This induces a conformational change in the extension, which then interacts with the receptor. (e) Proteinase-activated receptors are cleaved and the newly exposed N-terminus acts as an auto-ligand. The freed peptide may also interact separately with another receptor. Adapted from Ji et al.[41]

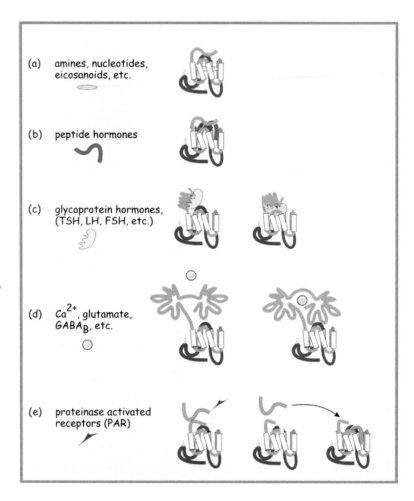

(a) amines, nucleotides, eicosanoids, etc.

(b) peptide hormones

(c) glycoprotein hormones, (TSH, LH, FSH, etc.)

(d) Ca^{2+}, glutamate, $GABA_B$, etc.

(e) proteinase activated receptors (PAR)

Ca^{2+} dips below about 2.2 mmol L^{-1}. Obviously, the affinity of the sensor has to be very low or it would be fully saturated at all times and under all conditions. On the other hand, the system must be sensitive to proportional changes in concentration that are minute compared with those sensed by conventional hormone receptors which react to changes ranging over orders of magnitude. The Ca^{2+} sensor is endowed with an extracellular domain which acts as a low affinity chelator, binding or releasing Ca^{2+} as its concentration varies within the (extracellular) physiological range.[47] It may operate as a dimer, the two components being joined by a disulfide bond between cysteine residues present in the extracellular domain.[48] A monomer–dimer equilibrium could underlie the special binding properties of this receptor.[49]

Proteinase-activated receptors (PARs)

Although blood platelets respond to many ligands (e.g. collagen exposed at sites of tissue damage, ADP released from damaged cells and, more importantly, secreted from activated platelets), the most potent activator is thrombin. Platelet-dependent arterial thrombosis triggers most heart attacks and strokes. Thrombin activates blood platelets, causing them to aggregate within seconds. It also has numerous longer-term functions related to inflammation and tissue repair which are mediated by a wide range of cell types. Thrombin is a serine proteinase enzyme related to trypsin and chymotrypsin and also to acetylcholine esterase. It has a unique specificity, cleaving peptide chains between arginine and serine, only if they are embedded in particular peptide sequences. As with the other serine esterases, the proteolytic activity of thrombin can be inhibited by the organophosphorus compounds mentioned earlier, and its action in stimulating blood platelets is also inhibited by compounds of this class. By cleaving the N-terminal exodomain of the thrombin receptor a new N-terminus is revealed which itself acts as a tethered ligand interacting with the exposed loops of the receptor (Figure 3.17). A synthetic pentapeptide, equivalent to the five N-terminal amino acids revealed after thrombin cleavage, also has agonist activity for the thrombin receptor.[50] In addition to the tethered ligand, the cleaved 41-residue peptide acts as a strong agonist for platelets.[51] Activation results in the generation of prostaglandin E2, causing bronchodilation.

Although conventional in the sense that these PARs are coupled to G proteins, there is the particular problem that their activation by proteolytic cleavage is necessarily irreversible. Of course, for the functioning of platelets there is no problem since stimulation initiates a sequence of events which terminates in their demise. For PARs in other tissues, specialized mechanisms ensure their desensitization and removal.[56] In addition to the usual processes of receptor phosphorylation and endocytosis, these include cleavage of the tethered ligand in lysosomes. Because the receptors undergo cleavage both as a consequence of stimulation and again in the process of desensitization, re-sensitization of the system necessarily occurs by *de novo* protein synthesis.

The adhesion receptor subfamily

Also called the LNB-7TM (large N-terminal family B) family of receptors, these are characterized by N-terminal extensions of 200–2800 residues that contain motifs characteristic of those present in cell-surface adhesion molecules. For example, the N-terminal extensions of the EMR group of 7TM molecules expressed predominantly in leukocytes all contain multiple EGF-like domains. In addition there is an RGD (arginine–glycine–aspartate) motif, a mucin-like

The thrombin receptor is now recognized as a member of a larger family of protease activated receptors (PAR1–4), some of which are additionally activated by trypsin.[52,53] In the epithelia of the upper intestine, PAR2 receptors confer protection against self-digestion by proteolytic enzymes through the production of prostaglandins.[54] Similarly, in the bronchial airways, the presence of proteolytic enzymes released by inflammatory cells (mast cells) appears to be signalled by PARs.[55]

PARs should not be confused with **Par proteins** (see page 385)

EMR: epidermal growth factor module-containing mucin-like receptor.

65

stalk,[57] and a cysteine box.[58] These appear to mediate cellular adhesion and for two members of this family, EMR2 and CD97, the cellular ligand has been identified as chondroitin sulfate, a sulfated glycosaminoglycan.[59] Interestingly, it is uncertain whether signalling due to the subgroup of EGF-containing LNB-7TM receptors is conveyed through GTP-binding proteins, or even if it is mediated through the 7TM domain. For other LNB-7TM molecules that are likely also to mediate cellular interactions through their large extracellular domains, a role for GTP-binding proteins in signal transmission seems likely.[60] Included in this family are also the adhesion molecules that qualify as cadherins (see Figure 13.10, page 389).

Frizzled

Frizzled, the family of receptors for the Wnt family of ligands, plays a role in the regulation of cellular differentiation signal through a number of different routes only one of which, the canonical Wnt pathway, involves heterotrimeric G proteins (G_o) (see page 426 and Figure 14.8, page 429).

Receptor–ligand interaction and receptor activation

A two-state equilibrium description of receptor activation

It is well recognized that ion channels are proteins that can exist in discrete states, typically 'open' or 'closed'. For ligand-activated channels the open-state probability is increased enormously when the ligand (e.g. acetylcholine) is bound.

A similar two-state equilibrium applies to the activation of the 7TM receptors. However, since Langley's first description of a 'receptive substance' towards the end of the 19th century,[61] it has been generally accepted that activation requires the actual binding of an agonist. Fundamental to this thinking is that the extent of the biological response is determined by the law of mass action linking the stimulus and the reactive tissue (i.e. the ligand and the empty binding site). The binding of a ligand is understood to induce a change in receptor conformation, such that only in this state does it communicate with its affiliated GTP-binding protein.

Receptors may be activated in the absence of ligand

The activity of adenylyl cyclase does not decline to zero in the absence of stimulating hormones. Of course, it is difficult to be quite sure that stimulating hormones are fully excluded even when a system has been extensively washed and then doped with inhibitors. Even so, there always remains a residual low level of basal (or constitutive) activity. Now we know that this is for real. However, even if activating ligands are not absolutely required, the coupling of a receptor to the cyclase is obligatory. Quite simply, the unoccupied receptors themselves provide the necessary stimulus. For example, the synthesis of cAMP in Sf9 insect cells, which are normally unresponsive to catecholamines, can be enhanced by transfecting them with β_2-adrenergic receptors. These cells now generate cAMP at a rate which is directly proportional to the level of receptor expression and they do this in the absence of any stimulating ligand[62] (see Figure 4.24, page 118), though of course the activity of the system is greatly increased if catecholamines are provided. (We return to the matter of reconstituting the adrenergic receptor/cyclase system in Sf9 cells in Chapter 4.)

The activation of enzyme activity in the absence of stimulating ligands can be understood if the receptors exist in two conformational states, one of which can initiate downstream events (R), and the other cannot (r). The equilibrium between these two states exists regardless of the presence or absence of a stimulating ligand (Figure 3.18). Conventional agonists are those which bind to the active conformation (R) and as a result, increase the proportion of active receptors (R+LIG.R). Conversely, there are 'inverse' agonists (lig) that bind selectively to the receptor in its inactive conformation. These increase the proportion in the form (r+lig.r) in which they are unable to transmit signals. Thus the number of active receptors is reduced and the activity of the system is suppressed below its normal basal state. In between the extremes of full (conventional) and inverse agonists are those agonists that bind to receptors in both the active and inactive conformations. These are the partial agonists. Such compounds are unable to achieve maximal stimulation of effector enzymes (e.g. of adenylyl cyclase) even when all the receptor binding sites are occupied, because they bind to both the activating and the inhibitory states of the receptor.

Sf9 insect cells and the baculovirus transfection system are used as an alternative to bacteria. Being eukaryotic and moreover animal cells, newly synthesized proteins are post-translationally modified as in other multicellular organisms.

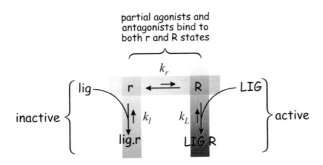

FIG 3.18 Receptor states and inverse agonists.
This diagram illustrates the equilibrium between the inactive (blue, r) and active (magenta, R) states of a receptor. Agonistic ligands (LIG) bind exclusively to the active form of the receptor and thus increase the proportion and the total amount (R+LIG.R). Conversely, inverse agonists (lig) bind exclusively to the inactive form of the receptor and thus increase the total amount of the receptor (r+lig.r) in the form that is incapable of transmitting signals.

The classification of drugs as agonists or antagonists, conventional, neutral, or inverse, is a pharmacological minefield. In practice, many of the compounds in use in the clinic and in the laboratory, and depending on the circumstances, have activities that could enable them to be classified both as agonists and antagonists. A further complicating factor is that the spectrum of activities varies from tissue to tissue. For the right atrium of the (rat) heart, in which β_2-adrenergic receptors play a greater role than in other cardiac regions, it appears that almost all the beta blockers behave as inverse agonists.

Other diseases that might be amenable to treatment with appropriate inverse agonists include inherited conditions such as retinitis pigmentosa, congenital night blindness (both due to mutations of rhodopsin), familial male precocious puberty, and familial hyperthyroidism. An inverse agonist active at the receptor for parathyroid hormone, recognized as being constitutively activate in

Inverse agonism offers the potential of developing new drugs that attenuate the effects of mutant receptors that are constitutively activated. The neutral antagonists (β-blockers such as pindolol) bind to both active and inactive conformations and are better regarded as passive antagonists. These impede the binding of both agonists and inverse agonists.

Neutral antagonists prevent the stimulation of adenylyl cyclase by catecholamines, but they also oppose the inhibitory effect of inverse agonists. Examples of inverse agonists for β-adrenergic receptors are propranolol and timolol, originally classified as β-blockers (see Figure 3.2) and used as such in the treatment of glaucoma and hypertension. However, unlike the β-blockade due to neutral antagonists, which mainly affects the heart during exercise and stress, these compounds also suppress the resting heart rate.[63,64]

Cell activation by over-expression of receptors

From this, one might imagine that if a receptor could be sufficiently over-expressed in a responsive tissue, the system would be fully activated even in the absence of a stimulating agonist. The equilibrium between the inactive and active species (r and R) would be unaltered, but the increased total would ensure a sufficient level of the active form to induce cell activation. Indeed, 200-fold over-expression of the β_2-adrenergic receptor in the hearts of transgenic mice is sufficient to maximize cardiac function in the absence of any stimulus.[65] Under these conditions, no hormone is required to increase the level of active receptors to the point at which they can stimulate the tissue.

There are indications that receptor over-expression may play a role in the aetiology of some disease states, and this offers the prospect that inverse agonists could provide more specific therapeutic approaches than the currently available neutral antagonists. Indeed, some forms of schizophrenia may be associated with an elevation of the D4 (dopamine) receptor in the frontal cortex.[66-68] Although the measured extent of the elevation is not large, about three-fold compared with controls, it is possible that this could be a reflection of much greater changes in limited focal regions, which could be of importance in determining the disease state. A number of compounds (raclopride, haloperidol) that find widespread application as antipsychotic drugs, appear to possess inverse agonist activity at dopamine receptors.[69-71] Similarly, the action of progesterone in reducing oxytocin-induced uterine excitability, essential for the maintenance of pregnancy (see Chapter 2), is also best understood as the action of an inverse agonist rather than as a conventional inhibitor.[72]

Of course, there is one receptor that is expressed at a level out of all proportion to all others. This is opsin, the major element of the visual pigment rhodopsin. It is legitimate to ask by what mechanism the photoreceptor system ensures that dark really means dark, that we see nothing when the lights are turned off. Even if the equilibrium favoured the inactive state to a degree much greater than any hormone receptor, the presence of such an enormous amount of rhodopsin would surely elicit some activating response, however minimal. Yet, dark really does mean dark. This important question is discussed in connection with the mechanism of visual transduction in Chapter 6.

patients suffering from Jansen-type metaphyseal chondrodysplasia, has recently been described.[73]

Receptor dimerization

It is increasingly apparent that many (if not most) 7TM receptors function not as monomers, but as dimers or oligomeric clusters. The first hints concerning oligomerization of β-adrenergic receptors emerged more than a quarter of a century ago.[74]

The metabotropic $GABA_B$ receptor is incapable of transmitting signals unless two subtypes of the receptor (splice variants $GABA_{B1}$ and $GABA_{B2}$) are both expressed.[75–77] They are understood to operate as a dimeric unit, regulating the activity of G-protein-activated K^+ channel (GIRK). In the relevant neural tissues, both are present so there is no problem. However, successful reconstitution of K^+ channel activity in cells that lack these receptors requires the expression of both subtypes. The $GABA_B$ receptor operates as a obligatory heterodimer (or larger oligomer). Only the B1 component binds the agonist,[78] while onward signalling (to the GTP-binding proteins G_i and G_o, see Chapter 4) is conveyed through the four intracellular segments of the B2 component.[79] (These are the three intracellular loops and the C-terminal stretch.) Indeed, coexpression of $GABA_{B1}$ and $GABA_{B2}$ is not only a requirement for effective ligand binding and signal propagation, but is also a prerequisite for maturation and transport of the receptor to the cell surface.

Homodimers

It is likely that the $GABA_{B1}$–$GABA_{B2}$ heterodimer exists as a stable entity and is unaffected by the state of receptor activation. Other receptors, such as those for glutamate (mGluR5[80]), vasopressin (V2[81]), adenosine, opiates (δ-opioid [82]) and even the well investigated β_2-adrenergic[81] and muscarinic receptors may be linked as homodimers. For some of these at least, the monomer–dimer equilibrium appears to reflect the state of receptor activation.

Stimulation of β_2-adrenergic receptors with catecholamines favours the formation of dimers[81] while, conversely, the application of an inverse agonist, timolol, favours monomer formation. When isolated, the dimers remain stable even under the denaturing conditions applied for analysis by SDS gel electrophoresis. Of the 7TM helices, the F helix (Figure 3.16) may offer the site of intermolecular attachment determining dimer formation. A synthetic peptide based on the sequence of this segment, that is therefore expected to impede the access of the second protein molecule, was found to reduce dimer formation.[81] It also inhibits the activation of adenylyl cyclase by isoprenaline.

Does it follow that dimerization provides an alternative signal for activation of the β_2-adrenergic receptor? The picture becomes more complicated when receptors are visualized on the cell surface by labelling with membrane-impermeable dyes. The evidence points three ways. For a large number of 7TM receptors dimerization appears to be constitutive, unaffected by the presence of agonists. Examples of constitutively formed dimers include the β_2-adrenergic receptor, muscarinic receptors,[83] the chemokine receptor CXCR2[84] (see page 494), and Frizzled 1[85] (see page 426). On the other hand, the extent of dimerization of the chemokine receptors CCR2, CXCR4, and CXCR5[86] increases in response to stimulation with agonists, and the dimer content of δ-opioid receptors actually diminishes. So while it is not possible to draw any general conclusions regarding a role of receptor homodimerization as a signal, there is little doubt that it is important in the processes of receptor maturation, folding, and trafficking to and from the cell surface, with dimer formation occurring early in the secretory pathway. For reviews of this field, see Hansen and Sheikh,[87] Terrillon and Bouvier.[88]

Other receptors

For receptors of other types, dimerization is well established as the initiating step in the signalling processes that they control. These include receptors with intrinsic tyrosine kinase activity (e.g. growth factor receptors such as those for EGF, Chapter 12), with intrinsic serine/threonine kinase activity (Chapter 20), receptors which interact with cytosolic tyrosine kinases (Chapter 17), and receptors which are tightly linked to tyrosine kinases (cytokine receptors, Chapter 17).

Transmitting signals into cells

The receptor and the effector: one and the same or separate entities?

When it first became apparent that hormones induce enzymatic activity at the cell membrane, the receptors themselves were no more than concepts and ideas used to explain phenomena. It soon became evident that a receptor and its catalytic function might not be two facets of a single entity. For example, adipocytes have receptors for many different types of hormone. Adrenaline, ACTH, and glucagon all act to increase the concentration of intracellular cAMP. However, the maximal rate of cAMP formation due to a particular stimulus cannot be further enhanced by stimulation of a second receptor type.[89,90] Other agents such as prostaglandins (PGE) and clonidine (an α_2-adrenergic reagent), adenosine, and insulin, all acting through their own specific receptors, suppress the induction of cAMP synthesis. For these cells anyway, the inference is that multiple receptors can communicate with a common pool of adenylyl cyclase (Figure 3.19). From this it follows that the receptors are

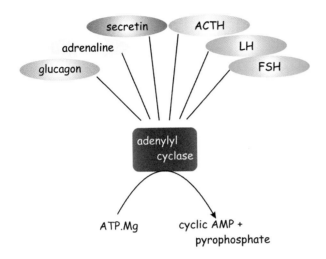

FIG 3.19 Multiple receptors coupled to a common pool of adenylyl cyclase.
A single pool of adenylyl cyclase can be accessed by many disparate hormone receptors. In general, when cyclase is maximally stimulated by one class of receptors further activation by ligands binding at other receptors is not possible. (There are however, some exceptions in which maximal activation of cyclase requires two types of receptor to act together in synergy: see Chapter 5.)
Adapted from Rodbell.[91]

likely to be discrete entities, each capable of communicating independently with the limited pool of enzyme.

To make things more complicated, there are situations in which two hormones are required together to cause stimulation. For example, in brain cortical slices the activation of adenylyl cyclase by noradrenaline (α-adrenergic stimulation which can be blocked by phentolamine, but not by propranolol) requires the simultaneous application of adenosine (blocked by theophylline).[92] Here, with full activation requiring the involvement of two quite distinct receptors, one might want to infer that these are integral with the catalytic unit itself; (although there is a much better explanation for this, which is discussed in Chapter 5; see Figure 5.4, page 139).

Mixing and matching receptors and effectors

A direct demonstration of the independent existence of receptors and their downstream effector enzymes required a procedure by which they could be obtained separately and then reconstituted into a functioning unit. In pre-cloning days, this presented a difficult problem since the receptors and the catalytic units which they control always share a common membrane location. Rather than attempt to separate them physically, the solution was to inhibit the cyclase selectively, while leaving the receptors intact, and then to use a second cellular source of cyclase, bringing the two together in order to achieve activity.

The alkylating agent *N*-ethylmaleimide was used to destroy the cyclase activity of turkey red blood cells while leaving their membrane receptors for adrenaline unimpaired. These cyclase-inactive cells were then fused to adrenal cortical cells as a source of the active enzyme (Figure 3.20). (Adrenal cortical cells lack catecholamine receptors but generate cAMP when treated with the peptide hormone ACTH.) Immediately after the fusion cyclase could only be activated by ACTH, but over a period of a few hours the sensitivity to adrenaline recovered.[93] This delay is commensurate with the rate at which the surface proteins of the two component cells diffuse laterally and merge with each other in the fused plasma membrane of the heterokaryon (the fused cell containing two nuclei). Since the cyclase to which the receptor for adrenaline is normally coupled had been irreversibly inactivated before the fusion, this could only mean that the β-adrenergic receptors derived from the turkey red blood cells were now activating the adrenal cortical cell enzyme. It follows that the receptors and the catalytic units that they activate must be situated on different protein molecules. The experiment gave no indication that there

FIG 3.20 Evidence for independent lateral diffusion of receptors and adenylyl cyclase.

By this experiment, illustrated schematically, it was shown that hormone receptors and their effector enzymes (adenylyl cyclase) are separate molecular entities capable of independent lateral diffusion in the plane of a cell membrane. For details see text.

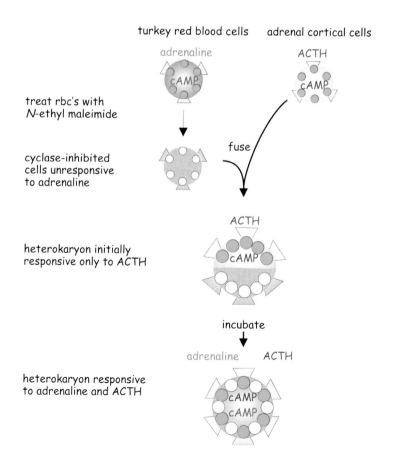

treat rbc's with *N*-ethyl maleimide

cyclase-inhibited cells unresponsive to adrenaline

heterokaryon initially responsive only to ACTH

heterokaryon responsive to adrenaline and ACTH

might be a third component intervening between the two, communicating the message from the activated receptors to the catalytic unit (see Figure 4.1, page 83). This transducing component is now known to be a GTP-binding protein.

Intracellular 7TM receptor domains and signal transmission

The onward signals for all these receptors are carried by conformational perturbations conveyed to the loops and the C-terminal chain exposed in the cytosol. Of these, the third intracellular loop (connecting the E and F chains), which is rather extended in most 7TM-receptor molecules, and also the C-terminal tail are considered to be of particular importance.

Adrenaline (yet again)

> It was ever thus. The cab-choked street, the PR men clutching their ulcers, the jewellery displayed like medals on the chest of a Soviet general, the snoozing men from Wall Street, the Sardi's supper entrance. As always in any enterprise, Americans travel hopefully, fuelled by a thirst for adrenalin not experienced by most Europeans.
>
> John Osborne, *Almost a Gentleman*
> (Faber & Faber, London, 1991)

References

1. Cannon WB. *Bodily changes in pain, hunger, fear and rage. An account of researches into the function of emotional excitement* Reprinted, 1963. New York: Harper Torch Books; 1929.
2. Starling EH. The wisdom of the body. *Brit Med J*. 1923;2:685–690.
3. Strosberg AD. Structure and function of the β_3-adrenergic receptor. *Annu Rev Pharmacol Tox*. 2000;37:421–450.
4. Hanoune J. The adrenal medulla. In: Baulieu E-E, Kelly PA, eds. *Hormones: From Molecules to Disease*. London: Chapman & Hall; 1990:309–333.
5. Weiland GA, Minneman KP, Molinoff PB. Fundamental difference between the molecular interactions of agonists and antagonists with the β-adrenergic receptor. *Nature*. 1979;281:114–117.
6. Sellin LC, McArdle JJ. Multiple effects of 2,3-butanedione monoxime. *Pharmacol Toxicol*. 1994;74:305–313.
7. van Helden HP, Busker RW, Melchers BP, Bruijnzeel PL. Pharmacological effects of oximes: how relevant are they?. *Arch Toxicol* 1996;70:779–786.
8. Bernard C. Analyse physiologique des propriétés des systèmes musculaire et nerveux au moyen du curare. *C R hebd Acad Sci*. 1856;43:829.

9. Langley JN. On the reaction of nerve cells and nerve endings to certain poisons chiefly as regards the reaction of striated muscle to nicotine and to curare. *J Physiol (Lond)*. 1905;33:374–473.

10. Burden SJ, Sargent PB, McMahan UJ. Acetylcholine receptors in regenerating muscle accumulate at original synaptic sites in the absence of the nerve. *J Cell Biol*. 1979;82:412–425.

11. Changeux JP, Kasai M, Lee CY. Use of a snake venom toxin to characterize the cholinergic receptor protein. *Proc Natl Acad Sci U S A*. 1970;67:1241–1247.

12. Meunier JC, Sealock R, Olsen R, Changeux JP. Purification and properties of the cholinergic receptor protein from *Electrophorus electricus* electric tissue. *Eur J Biochem*. 1974;45:371–394.

13. Olsen RW, Meunier JC, Changeux JP. Progress in the purification of the cholinergic receptor protein from *Electrophorus electricus* by affinity chromatography. *FEBS Lett*. 1972;28:96–100.

14. Neher E, Sakman B. Single-channel currents recorded from membrane of denervated frog muscle fibres. *Nature*. 1976;260:799–802.

15. Miyazawa A, Fujiyoshi Y, Unwin N. Structure and gating mechansim of the acetylcholine receptor pore. *Nature*. 2003;423:949–955.

16. Unwin N. Refined structure of the nicotinic acetylcholine receptor at 4 Å resolution. *J Mol Biol*. 2005;346:967–989.

17. Deisenhofer J, Epp O, Miki K, Huber R, Michel H. Structure of the protein subunits in the photosynthetic reaction centre of Rhodopseudomonas viridis at 3 Å resolution. *Nature*. 1985;318:618–621.

18. Doyle DA, Cabral JM, Pfuetzner RA, Cohen SL, Gulbis JM, Chait BT, MacKinnon R. The structure of the potassium channel: molecular basis of K conduction and selectivity. *Science*. 1998;280:69–77.

19. Chang G, Spencer RH, Lee AT, Barclay MT, Rees DC. Structure of the MscL homolog from *Mycobacterium tuberculosis*: a gated mechanosensitive ion channel. *Science*. 1998;282:2220–2226.

20. Le Novere N, Changeux JP. The ligand-gated ion channel database. *Nucleic Acids Res*. 1999;27:340–342.

21. Kyte J, Doolittle RF. A simple method for displaying the hydropathic character of a protein. *J Mol Biol*. 1982;157:105–132.

22. Heidmann T, Changeux JP. Time-resolved photolabeling by the noncompetitive blocker chlorpromazine of the acetylcholine receptor in its transiently open and closed ion channel conformations. *Proc Natl Acad Sci U S A*. 1984;81:1897–1901.

23. Brejc K, van Dijk WJ, Klaassen RV, Schuurmans M, van Der OJ, Smit AB, Sixma TK. Crystal structure of an ACh-binding protein reveals the ligand-binding domain of nicotinic receptors. *Nature*. 2001;411:269–276.

24. Celie PH, van Rossum-Fikkert SE, van Dijk WJ, Brejc K, Smit AB, Sixma TK. Nicotine and carbamylcholine binding to nicotinic acetylcholine receptors as studied in AChBP crystal structures. *Neuron*. 2004;41:907–914.

25. Wagner S, Castel M, Gainer H, Yarom Y. GABA in the mammalian suprachiasmic nucleus and its role in diurnal rhythmicity. *Nature*. 1997;387:598–603.

26. Akabas MH, Kaufmann C, Archdeacon P, Karlin A. Identification of acetylcholine receptor channel-lining residues in the entire M2 segment of the α subunit. *Neuron*. 1994;13:919–927.

27. Platt SR. The role of glutamate in central nervous system health and disease - a review. *Vet J*. 2007;173:278–286.

28. Drews J. Drug discovery: a historical perspective. *Science*. 2003;287: 1961-1964.

29. Ohguro H, Palczewsi K, Ericsson LH, Walsh KA, Johnson RS. Sequential phosphorylation of rhodopsin at multiple sites. *Biochemistry*. 1993;32:5718–5724.

30. Fredericks ZL, Pitcher JA, Lefkowitz RJ. Identification of the G-protein-coupled receptor kinase phosphorylation sites in the human β2-adrenergic receptor. *J Biol Chem*. 1999;271:13796–13803.

31. Henderson R, Unwin PN. Three-dimensional model of purple membrane obtained by electron microscopy. *Nature*. 1975;257:28–32.

32. Cherezov V, Rosenbaum DM, Hanson MA, Rasmussen SG, Thian FS, Kobilka TS, Choi HJ, Kuhn P, Weis WI, Kobilka BK, Stevens RC. High-resolution crystal structure of an engineered human β_2-adrenergic G protein-coupled receptor. *Science*. 2007;318:1258–1265.

33. Rosenbaum DM, Cherezov V, Hanson MA, Rasmussen SG, Thian FS, Kobilka TS, Choi HJ, Yao XJ, Weis WI, Stevens RC, Kobilka BK. GPCR engineering yields high-resolution structural insights into β_2-adrenergic receptor function. *Science*. 2007;318:1266–1273.

34. Rasmussen SG, Choi HJ, Rosenbaum DM, Kobilka TS, Thian FS, Edwards PC, Burghammer M, Ratnala VR, Sanishvili R, Fischetti RF, Schertler GF, Weis WI, Kobilka BK. Crystal structure of the human β_2 adrenergic G-protein-coupled receptor. *Nature*. 2007;450:383–387.

35. Ranganathan R. Signaling across the cell membrane. *Science*. 2007;318:1253–1254.

36. Sprang SR. Structural biology: a receptor unlocked. *Nature*. 2007;450: 355–356.

37. Palczewsi K, Kumasaka T, Hori T, Behnke CA, Motoshima H, Fox BA, Le Trong I, Teller DC, Okada T, Stenkamp RE, Yamamoto M, Miyano M. Crystal structure of rhodopsin: a G protein coupled receptor. *Science*. 2003;289:739–745.

38. Archer E, Maigret B, Escrieut C, Pradayrol L, Fourmy D. Rhodopsin crystal: new template yielding realistic models of G-protein-coupled receptors? *Trends Pharmacol Sci* 2003;24:36–40.

39. Robison GA, Butcher RW, Sutherland EW. Adenyl cyclase as an adrenergic receptor. *Ann N Y Acad Sci*. 1967;139:703–723.

40. Ahlquist RP. Adrenergic receptors: a personal and practical view. *Perspect Biol Med*. 1973;17:119–122.

41. Ji TH, Grossmann M, Ji I. G protein-coupled receptors. I. Diversity of receptor-ligand interactions. *J Biol Chem*. 1998;273:17299–17302.

42. Attwood TK, Findlay JB. Fingerprinting G-protein-coupled receptors. *Protein Eng*. 1994;7:195–203.

43. Kolakowski LF. GCRDb: a G-protein-coupled receptor database. *Receptors Channels*. 1994;2:1–7.

44. Fredriksson R, Lagerstrom MC, Lundin L-G, Schiöth HB. The G-protein-coupled receptors in the human genome form five main families: phylogenetic analysis, paralogon groups, and fingerprints. *Mol Pharmacol*. 2003;63:1256–1272.

45. Bjarnadottir TK, Gloriam DE, Hellstrand SH, Kristiansson H, Fredriksson R, Schioth HB. Comprehensive repertoire and phylogenetic analysis of the G protein-coupled receptors in human and mouse. *Genomics*. 2006;88: 263–273.

46. Ostrowski J, Kjelsberg MA, Caron MG, Lefkowitz RJ. Mutagenesis of the β_2-adrenergic receptor: how structure elucidates function. *Annu Rev Pharmacol Toxicol*. 1992;32:167–183.

47. Brown EM, Gamba G, Riccardi D, Lombardi M, Butters R, Kifor O, Sun A, Hediger MA, Lytton J, Hebert SC. Cloning and characterization of an extracellular Ca(2+)-sensing receptor from bovine parathyroid. *Nature*. 1993;366:575–580.

48. Ward DT, Brown EM, Harris HW. Disulfide bonds in the extracellular calcium-polyvalent cation-sensing receptor correlate with dimer formation and its response to divalent cations in vitro. *J Biol Chem*. 1998;273:14476–14483.

49. Pin J-P, Galvez T, Prézeau L. Evolution, structure, and activation mechanism of family 3/C G-protein-coupled receptors. *Pharmacology & Therapeutics*. 2003;98:325–354.

50. Scarborough RM, Naughton MA, Teng W, Hung DT, Rose J, Vu TKH, Wheaton VI, Turck CW, Coughlin SR. Tethered ligand agonist peptides. Structural requirements for thrombin receptor activation reveal mechanism of proteolytic unmasking of agonist function. *J Biol Chem*. 1994;269:32522–32527.

51. Furman MI, Liu L, Benoit SE, Becker RC, Barnard MR, Michelson AD. The cleaved peptide of the thrombin receptor is a strong platelet agonist. *Proc Natl Acad Sci USA*. 1998;95:3082–3087.

52. Dery O, Corvera CU, Steinhoff M, Bunnett NW. Proteinase-activated receptors: novel mechanisms of signaling by serine proteases. *Am J Physiol*. 1998;274C:1429–1452.

53. Kahn ML, Zheng YW, Huang W, Bigornia V, Zeng D, Moff S, Farese-RV J, Tam C, Coughlin SR. A dual thrombin receptor system for platelet activation. *Nature*. 1998;394:690–694.

54. Hara K, Yonezawa K, Sakaue H, Ando A, Kotani K, Kitamura T, Kitamura Y, Ueda H, Stephens L, Jackson TR. 1-Phosphatidylinositol 3-kinase activity is required for insulin-stimulated glucose transport but not for RAS activation in CHO cells. *Proc Natl Acad Sci*. 1994;91:7415–7419.

55. Cocks TM, Fong B, Chow JM, Anderson GP, Frauman AG, Goldie RG, Henry PJ, Carr MJ, Hamilton JR, Moffett S. A protective role for protease-activated receptors in the airways. *Nature*. 1999;398:156–160.

56. Trejo J, Hammes SR, Coughlin SR. Termination of signaling by protease-activated receptor-1 is linked to lysosomal sorting. *Proc Natl Acad Sci USA*. 1998;95:13698–13702.

57. Kristiansen K. Molecular mechanisms of ligand binding, signaling, and regulation within the superfamily of G-protein-coupled receptors: molecular modeling and mutagenesis approaches to receptor structure and function. *Pharmacol Therapeut*. 2004;103:21–80.

58. Stacey M, Lin H-H, Gordon S, McKnight AJ. LNB-TM7, a group of seven-transmembrane proteins related to family-B G-protein-coupled receptors. *Trends Biochem Sci*. 2000;25:284–288.

59. Stacey M, Chang G-W, Davies JQ, Kwakkenbos MJ, Sanderson RD, Gordon S, Lin H-H. The epidermal growth factor-like domains of the human EMR2 receptor mediate cell attachment through chondroitin sulfate glycosaminoglycans. *Blood*. 2003;102:2916–2924.

60. Lelianova VG, Davletov BA, Sterling A, Rahman MA, Grishin EV, Totty NF, Ushkaryov YA. α-latrotoxin receptor, latrophilin, is a novel member of the secretin family of G protein-coupled receptors. *J Biol Chem*. 1997;272:21504–21508.

61. Langley JN. On the physiology of the salivary secretion: Part II. On the mutual antagonism of atropin and pilocarpin, having especial reference to their relations in the sub-maxillary gland of the cat. *J Physiol*. 1878;1:339–369.

62. Chidiac P, Hebert TE, Valiquette M, Dennis M, Bouvier M. Inverse agonist activity of β-adrenergic antagonists. *Mol Pharmacol*. 1994;45:490-499.

63. Varma DR, Shen H, Deng XF, Peri KG, Chemtob S, Mulay S. Inverse agonist activities of β-adrenoceptor antagonists in rat myocardium. *Br J Pharmacol*. 1999;127:895–902.

64. Varma DR. Ligand-independent negative chronotropic responses of rat and mouse right atria to β-adrenoceptor antagonists. *Can J Physiol*. 1999;77:943–949.

65. Bond RA, Leff P, Johnson TD, Milano CA, Rockman HA, McMinn TR, Apparsundaram S, Hyek MF, Kennakin TB, Allen LF, Lefkowitz RJ. Physiological effects of inverse agonists in transgenic mice with myocardial overexpression of the β-adrenoceptor. *Nature*. 1995;374:272–276.

66. Seeman P, Guan HC, Van-Tol HH. Dopamine D4 receptors elevated in schizophrenia. *Nature*. 1993;365:441–445.

67. Marzella PL, Hill C, Keks N, Singh B, Copolov D. The binding of both [3H]nemonapride and [3H]raclopride is increased in schizophrenia. *Biol Psychiatry*. 1997;42:648–654.

68. Stefanis NC, Bresnick JN, Kerwin RW, Schofield WN, McAllister G. Elevation of D4 dopamine receptor mRNA in postmortem schizophrenic brain. *Brain Res Mol Brain Res*. 1998;53:112–119.

69. Griffon N, Pilon C, Sautel F, Schwartz JC, Sokoloff P. Antipsychotics with inverse agonist activity at the dopamine D3 receptor. *J Neural Transm*. 1996;103:1163–1175.

70. Malmberg A, Mikaels A, Mohell N. Agonist and inverse agonist activity at the dopamine D3 receptor measured by guanosine $5'$-γ-thio-triphosphate-^{35}S- binding. *J Pharmacol Exp Ther*. 1998;285:119–126.

71. Hall DA, Strange PG. Evidence that antipsychotic drugs are inverse agonists at D_2 dopamine receptors. *Br J Pharmacol*. 1997:731–736.

72. Grazzini E, Guillon G, Mouillac B, Zingg HH. Inhibition of oxytocin receptor function by direct binding of progesterone. *Nature*. 1998;392:509–512.

73. Gardella TJ, Luck MD, Jensen GS, Schipani E, Potts JT, Juppner H. Inverse agonism of amino-terminally truncated parathyroid hormone (PTH) and PTH-related peptide (PTHrP) analogs revealed with constitutively active mutant PTH/PTHrP receptors. *Endocrinology*. 1996;137:3936–3941.

74. Atlas D, Levitzki A. Tentative identification of β-adrenoceptor subunits. *Nature*. 1978;272:370–371.

75. Jones KA, Borowsky B, Tamm JA, Craig DA, Durkin MM, Dai M, Yao W-J, Johnson M, Gunwaldsen C, Huang LY, Tang C, Shen Q, Salon JA, Morse K, Laz T, Smith KE, Nagarthnam D, Noble SA, Branchek TA, Gerald C. GABA$_B$ receptors function as a heteromeric assembly of the subunits GABA$_B$R1 and GABA$_B$R2. *Nature*. 1998;396:670–674.

76. White JH, Wise A, Main MJ, Green A, Fraser NJ, Disney GH, Barnes AA, Emson P, Foord SM, Marshall FH. Heterodimerization is required for the formation of a functional GABA$_B$ receptor. *Nature*. 1998;396:679–682.

77. Kaupmann K, Malitchek B, Schuler V, Heid J, Froestl W, Beck P, Mosbacher J, Bischoff S, Kulik A, Shigemoto R, Karschin A, Bettler B. GABA$_B$-receptor subtypes assemble into functional heteromeric complexes. *Nature*. 1998;396:683–687.

78. Margeta-Mitrovic M, Jan YN, Jan LY. Ligand-induced signal transduction within heterodimeric GABA(B) receptor. *Proc Natl Acad Sci USA*. 2001;98:14643–14648.

79. Margeta-Mitrovic M, Jan YN, Jan LY. Function of GB1 and GB2 subunits in G protein coupling of GABA(B) receptors. *Proc Natl Acad Sci USA*. 2001;98:14649–14654.

80. Romano C, Yang WL, O'Malley KL. Metabotropic glutamate receptor 5 is a disulfide-linked dimer. *J Biol Chem*. 1996;271:28612–28616.

81. Hebert TE, Moffett S, Morello J-P, Loisel TP, Bichet DG, Barrett C, Bouvier M. A peptide derived from a β_2-adrenergic receptor transmembrane

domain inhibits both receptor dimerization and activation. *J Biol Chem*. 1996;271:16384–16392.

82. Cvejic S, Devi LA. Dimerization of the δ opioid receptor: implication for a role in receptor internalization. *J Biol Chem*. 1997;272:26959–26964.

83. Zeng FY, Wess J. Identification and molecular characterization of m3 muscarinic receptor dimers. *J Biol Chem*. 1999;274:19487–19497.

84. Trettel F, Di Bartolomeo S, Lauro C, Catalano M, Ciotti MT, Limatola C. Ligand-independent CXCR2 dimerization. *J Biol Chem*. 2003;278: 40980–40988.

85. Kaykas A, Yang-Snyder J, Heroux M, Shah KV, Bouvier M, Moon RT. Mutant Frizzled 4 associated with vitreoretinopathy traps wild-type Frizzled in the endoplasmic reticulum by oligomerization. *Nat Cell Biol*. 2004;6:52–58.

86. Rodriguez-Frade JM, del Real G, Serrano A, Hernanz-Falcon P, Soriano SF, Vila-Coro AJ, de Ana AM, Lucas P, Prieto I, Martinez A, Mellado M. Blocking HIV-1 infection via CCR5 and CXCR4 receptors by acting in trans on the CCR2 chemokine receptor. *EMBO J*. 2004;23:66–76.

87. Hansen JL, Sheikh SP. Functional consequences of 7TM receptor dimerization. *Eur J Pharm Sci*. 2004;23:301–317.

88. Terrillon S, Bouvier M. Roles of G-protein-coupled receptor dimerization. *EMBO Rep*. 2004;5:30–34.

89. Birnbaumer L, Rodbell M. Adenyl cyclase in fat cells. II. Hormone receptors. *J Biol Chem*. 1969;244:3477–3482.

90. Bar HP, Hechter O. Adenyl cyclase and hormone action. I. Effects of adrenocorticotropic hormone, glucagon, and epinephrine on the plasma membrane of rat fat cells. *Proc Natl Acad Sci USA*. 1969;63:350–356.

91. Rodbell M. The role of GTP-binding proteins in signal transduction: from the sublimely simple to the conceptually complex. *Curr Top Cell Regulation*. 1992;32:1–47.

92. Sattin A, Rall TW, Zanella J. Regulation of cyclic adenosine 3′,5′-monophosphate levels in guinea-pig cerebral cortex by interaction of α-adrenergic and adenosine receptor activity. *J Pharmacol Exp Ther*. 1975;192:22–32.

93. Schulster D, Orly J, Seidel G, Schramm M. Intracellular cyclic AMP production enhanced by a hormone receptor transferred from a different cell: β-adrenergic responses in cultured cells conferred by fusion with turkey erythrocytes. *J Biol Chem*. 1978;253:1201–1206.

CHAPTER 4

GTP-binding Proteins and Signal Transduction

Nucleotides as metabolic regulators

In addition to providing the alphabet of the genetic code, nucleotides play many other roles. The pyrimidine bases act as identifiers for metabolites. For instance, CTP is a reactant in phospholipid biosynthesis in which the CMP moiety is transferred to phosphatidate in the formation of CDP-diglyceride (see Figure 8.5). Similarly, it reacts with choline phosphate in the formation of CDP-choline. Uridine nucleotides are involved in the assembly of polysaccharides:

$$\text{glucose-1-P} + \text{UTP} \xrightarrow{\text{UDP-glucose pyrophosphorylase}} \text{UDP-glucose} + \text{PP}_i$$

Purine nucleotides play their main regulatory roles in association with proteins, not metabolites. ATP acts as a link between metabolic processes and cellular activities. It is present in most cells at high concentration, $5\text{–}10\,\text{mmol L}^{-1}$ in the

Estimates of cytosol ATP concentration have been somewhat lower for other (non-muscle) cell types ($1-3$ mmol L^{-1}). The determination of its concentration depends on a good estimate of cytosol volume.

cytosol of muscle and nerve. It is also subject to rapid turnover. The human body typically turns over 90% of its entire content of ATP within 90 s; thus, an amount equivalent to 75% of body weight, about 40 kg, is turned over every day. In spite of this, the cellular concentration of ATP generally remains remarkably constant, even in working muscle. In the extracellular environment ATP also acts as a neurotransmitter.

The possibility that GTP might act as a factor necessary for cellular processes first became apparent in the action of phosphoenolpyruvate carboxylase in gluconeogenesis and for succinyl CoA synthase in the operation of the tricarboxylic acid cycle. It is a necessary component, together with mRNA, activated amino acids (aminoacyl-tRNA), and ribosomes, in the initiation and elongation reactions of protein synthesis. In this sense, the initiation factors and elongation factors should be regarded as the original GTP-binding proteins.

ATP is not quite what it seems

In the early years following the discovery of cAMP, all that was required (beyond a great deal of skill and dedication) to detect the activity of adenylyl cyclase in cell membrane preparations was an activating hormone and the substrate ATP (as its Mg^{2+} salt) (Figure 4.1a). The membranes, containing an appropriate receptor and the catalytic unit, did the rest. At least, that was the general idea.

The economical manufacture of ATP, in a high state of purity and in large quantities, has been a notable achievement of the commercial suppliers. It is now possible to purchase a gram of crystalline ATP, better than 99% pure, for less than the price of a pint of beer, but this has not always been the case. In the 1960s commercially produced ATP was good enough as a substrate, but sometime around 1970 it became erratic, occasionally registering zero activity in the cyclase reaction. A similar impasse had been encountered previously in the investigation of fatty acid biosynthesis. It had been found that an essential impurity was lost as the chemical purity of the ATP was improved. This was CTP (cytidine 5′-triphosphate). In the case of adenylyl cyclase experiments it was the exclusion of contaminating traces of GTP that caused the loss of activity. Addition of GTP now allowed hormone-induced generation of cAMP to proceed.[1] Pertinently, it also had the effect of reducing the affinity of agonist (but not of antagonist) binding at receptors.[2-4] It was clear that GTP plays a central role in the cyclase reaction, not as a substrate but as a cofactor. The two established components, in Rodbell's terminology the discriminator (receptor) and the effector, were now joined by a third, the transducer that binds GTP (Figure 4.1b).

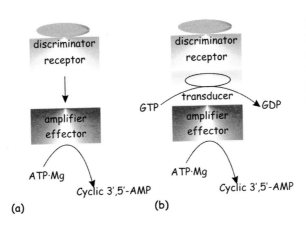

FIG 4.1 Introduction of the idea of a transducer in the relay of the receptor signal to the effector.(a) The early view. (b) A transducing GTP-binding protein relays the signal.Adapted from Rodbell.[5]

GTP-binding proteins, G proteins, or GTPases

The current terminology of GTP-binding proteins embraces G proteins and GTPases. All are capable of hydrolysing GTP, so, technically, all are GTPases. The term G protein is generally reserved for the class of GTP-binding proteins that interact with 7TM receptors. All of these are composed of three subunits, α, β, and γ, and so it is also common to refer to the *heterotrimeric G proteins*. This distinguishes them from the other main class of GTP-binding proteins involved in signalling. These are monomeric and are related to the protein products of the ras proto-oncogenes. Collectively, the GTP-binding proteins are everywhere.[6] The basic cycle of GTP binding and GTP hydrolysis, switching them between active and inactive states, is common to all of them and it is coupled to many diverse cellular functions.[7]

Historically, the G proteins linked to the 7TM receptors were discovered before the Ras-related proteins and this is the order in which we consider them here.

G proteins

Heterotrimeric structure

The structural organization of the component α, β, and γ-subunits in a heterotrimeric G protein (G_i) is illustrated in Figure 4.2. The β-subunit, which has seven regions of β structure organized like the blades of a propeller, is tightly associated with the γ-chain and together they behave as a single entity, the $\beta\gamma$-subunit. The α-subunit contains the nucleotide binding site and its contacts with the β-subunit involve the so-called switch I and II loops of the α-subunit and its N-terminal helix. These contact the β propeller so that SWII is not accessible to solvent. The whole assembly is anchored to the membrane by lipid attachments, one at the N-terminal of the α-subunit and the other at

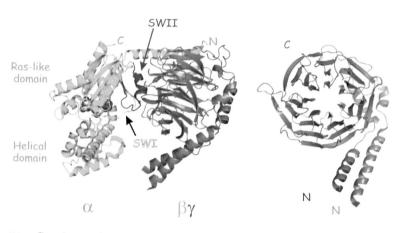

FIG 4.2 Three-dimensional structure of the α- and βγ-subunits of G_i. The α-subunit is shown on the left (green) with a molecule of GDP (spheres) in its nucleotide binding site. It is associated with a βγ-subunit. On the right, a separate view of the βγ-subunit, rotated 90° about a vertical axis and tilted, shows its β propeller structure (β, blue; γ, magenta). The α-subunit SWI and SWII loops are coloured red and yellow respectively. The surface of the heterotrimeric structure that is in contact with the membrane is at the top of the figure. The hydrophobic attachments responsible (not shown) involve the N-terminal helix of the α-subunit and the C-terminus of the γ-subunit. (1gp2[8])

Older accounts of G protein activation state that the activated heterotrimer dissociates into two functional components, α-GTP and βγ·αGTP then seeks out the effector enzyme and activates it.[10] However, with some exceptions, such as transducin (the G protein that mediates visual transduction), the evidence for the physical separation of α-subunits from βγ-subunits has been slim (though activated heterotrimers are certainly more readily dissociated in test-tube experiments). The dissociation of all G proteins when activated has been a hypothesis without a strong experimental basis that became unfortunately a paradigm. Evidence against it is discussed below (page 111). It may just be fortuitous that G protein α-and βγ- subunits *in vitro* retain the activities of activated heterotrimeric G proteins.[12]

the C-terminal of the γ-subunit. The structures of these lipid modifications are shown in **Figure 4.12**, and specific modifications are listed in Table 4.2.

The GTPase cycle: a monostable switch

The cycle of events regulated by all GTP-binding proteins starts and ends with GDP situated in the guanine nucleotide binding site of the α-subunits (Figure 4.3). Throughout the cycle, the G protein remains associated with the effector enzyme through its α- or βγ-subunits, but its association with activated receptors is transient[9,10] ('collisional coupling'[11]). Upon stimulation, and association of the receptor with the G protein/effector complex, the GDP dissociates. It is replaced by GTP. The selectivity of binding and hence the progress of the cycle is determined by the 10-fold excess of GTP over GDP in cells (GTP is present at ~100 μmol L^{-1}). Following activation, the contact between the G protein and the agonist–receptor complex is weakened, so allowing the receptor to detach and seek out further inactive G protein molecules. This provides an opportunity for signal amplification at this point.

The irreversibility of the cycle is determined by cleavage of the terminal phosphate of GTP (the GTPase reaction). Following the restoration of GDP in the nucleotide binding site of the α-subunits, the G protein and its effector enzyme return to their inactive forms. G proteins are often referred to as molecular switches. Actually, this may not be the best description, because

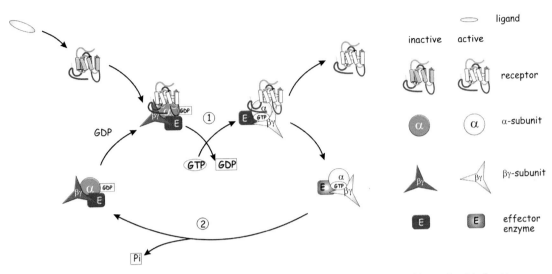

FIG 4.3 Activation and deactivation of G proteins by guanine nucleotide exchange and GTP hydrolysis. Activation and deactivation of the G protein α-subunit by (1) guanine nucleotide exchange and (2) GTP hydrolysis. The receptor binds ligand and is activated to become a catalyst, enhancing the rate of detachment of bound GDP from the α-subunit of the G protein. The vacant site is rapidly occupied by GTP and this causes activation of the α- and βγ-subunits enabling them in turn to activate specific effector enzymes. The interaction between receptor and G protein is transient, allowing the receptor to catalyse guanine nucleotide exchange on a succession of G protein molecules. The system returns to the resting state following hydrolysis of the bound GTP on each of these.

they do not necessarily switch off abruptly. A better analogy might be the light switch on a landlady's staircase, the kind which, when pressed (GDP/GTP exchange) switches on the light for just the time it takes to make the dash up to the next landing (when GTP hydrolysis occurs). In electronic parlance this would be called a *monostable switch*. The activation is thus kinetically regulated, positively by the initial rate of GDP dissociation and then negatively by the rate of GTP hydrolysis. In this way, the state of activation can be approximated by the ratio of the rate constant for dissociation of GDP to the rate constant for GTP hydrolysis (k_{diss}/k_{cat}). Both of these 'constants' are subject to modification by proteins that interact with the α-subunits (for example an activated receptor)[13]

$$\text{active/inactive} = \text{on/off}$$

$$= [G\alpha \cdot GTP]/[G\alpha \cdot GDP]$$

$$= k_{diss}/k_{cat} \qquad \text{(Equation 4.1)}$$

where k_{diss} is the rate constant for the dissociation of GDP and k_{cat} is the rate constant for the hydrolysis of GTP.

The lower maximal activation (e.g. of adenylyl cyclase) achieved by the so-called partial agonists can be ascribed to lower rates of GDP dissociation.[13] Unlike full agonists, partial agonists fail to distinguish perfectly between the active and the inactive conformations of the receptor, thus they have some of the character of inverse agonists (see page 66).

The GTPase cycle, first perceived in the processes of visual phototransduction[14] and β-adrenergic activation of adenylyl cyclase, is central to many important signal transduction mechanisms.[7] The GTP-bound form of the α-subunit is required to activate effector enzymes such as adenylyl cyclase and phospholipase C. The duration of this interaction lasts only as long as the bound GTP remains intact. As soon as it is hydrolysed to GDP communication ceases and the effectors revert to their inactive state.

Switching off activity: switching on GTPase

It was initially surprising to find that the rate of GTP hydrolysis catalysed by isolated G proteins is far too slow to account for the transient nature of some of the known G-protein-mediated responses. The extreme example must be the visual transduction process which, of necessity, must be turned off rapidly. When the lights go off, we perceive darkness promptly. If transducin (G_t, the G protein responsible for visual transduction) were able to persist in its active form (α_t-GTP), the illumination would appear to dim gradually as the hydrolysis of GTP proceeded. It is now clear that the rate of the GTPase reaction is stepped up very considerably as soon as activated α-subunits come into contact with effector enzymes (increased k_{cat}). This is certainly the case for the GTP hydrolysis catalysed by α_t[15,16] (see page 172). Likewise, the GTPase activity of the α-subunit of G_q is accelerated as a result of its association with its effector enzyme, phospholipase Cβ,[17,18] so ensuring that the α-subunit retains its activity just for as long as it takes to make productive contact. More generally, the GTPase activity of the main classes of G proteins, (with the exception of the G_s subgroup) are subject to regulation through the family of RGS proteins (Regulator of G protein Signalling, 20 members). These interact with specific α-subunits to accelerate the rate of GTP hydrolysis (k_{cat}) and, in addition, act as acute regulators of a wide variety of physiological processes.[19] In the context of down-regulation, the RGS proteins act to terminate the signals already generated at the level of G proteins. We shall see later that there are also receptor kinases that act upstream to prevent further receptor activation by ligand.

The smaller RGS proteins (<220 residues) appear to have no particular homologies apart from their conserved catalytic RGS homology domain. Others, however, are much larger (370–1400 residues), and possess multiple structural domains (DH, PH, PTB, PDZ, etc.) that might allow them to link up with other proteins involved in signalling. An example is the desensitization of a Ca^{2+} channel (N-type) which is mediated through $GABA_B$ receptors and G_o, and opposed by RGS12.[20] This provides yet another level of control over the processes regulated by G proteins. Also, there is the possibility that the RGS proteins might themselves be subject to regulation through the level of their expression.

There are other proteins having similar activity that act to enhance the GTPase activity of the small monomeric GTP-binding proteins such as Ras.

These are known as GTPase-activating proteins or GAPs and are discussed below.

Modulation of receptor affinity by G proteins

One of the earliest indications of a role for GTP in the activation of receptor-mediated processes was the reduction of the affinity for activating ligands by stable chemical analogues of GTP (GTPγS, GppNHp, etc.; Figure 4.4).[21] Because these analogues are not readily hydrolysed they ensure that the α-subunits remain persistently activated. This, coupled with the observation that application of agonistic (but not antagonistic) ligands accelerates the rate of GTP hydrolysis,[22] gave the clue that the communication between receptors and GTP-binding proteins is a two-way affair. The receptors speak to the G proteins and the G proteins speak to the receptors. Following activation, the contact between the G protein and the agonist–receptor complex is weakened, so allowing the receptor to seek out another G protein molecule. The situation for inverse agonists is the converse (Figure 4.5).[23–26] We referred earlier (page 68) to the example of the oxytocin receptor, for which progesterone acts as an inverse agonist, binding to the receptor in its inactive conformation. Here GTPγS characteristically causes the affinity for the peptide hormone to decline while the affinity for steroid hormone increases.[27]

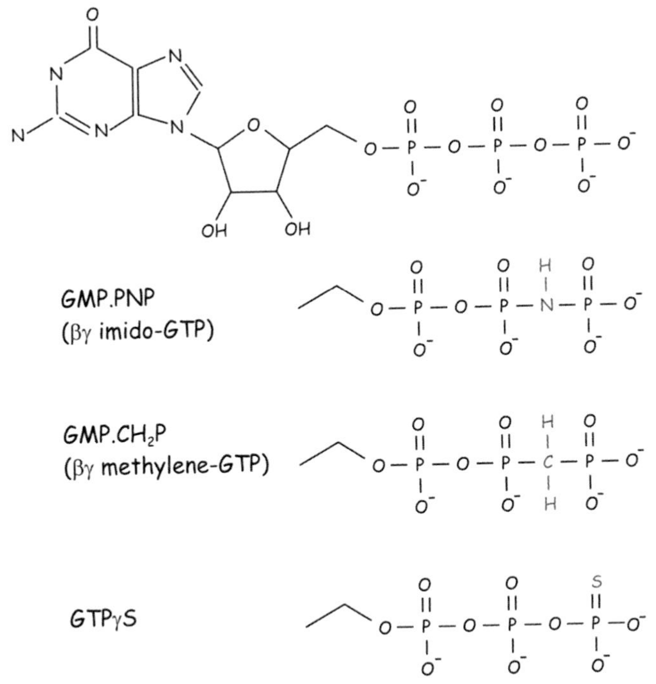

FIG 4.4 Structures of stable analogues of GTP.

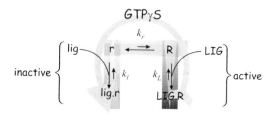

FIG 4.5 Opposing effects of guanine nucleotides on the binding of agonists and inverse agonists to receptors. The diagram illustrates the equilibrium between the inactive (blue, r) and active (magenta, R) states of a receptor. Agonistic ligands (LIG) bind exclusively to the active form of the receptor and thus increase the proportion and the total amount (R + LIG.R) of the receptor in the form that can transmit a signal. Conversely, inverse agonists (lig) bind exclusively to the inactive form of the receptor and thus increase the total amount of the receptor (r + lig.r) in the form that is incapable of transmitting signals. The stable analogue, GTP-γS, depresses the affinity of receptors for agonists while enhancing their affinity for inverse agonists. There is no effect on the affinity for the common antagonists. Compare this figure with Figure 3.18, page 67.

α-Subunits

Historically, an experimental challenge has been to manipulate receptors, G proteins, and effector enzymes independently of each other. Because the receptors and the effectors are either intrinsic or associated membrane proteins, they cannot readily be added or withdrawn in an experimental system. Even though they can be mixed and matched in cell fusion experiments, as discussed in Chapter 3, this system does not allow the manipulation of the different subunits of the heterotrimeric G proteins. Instead, it was the use of the lymphoma cell line, S49 cyc$^-$, that allowed the identification of the α-subunits as the GTP-binding component relaying signals from receptors to adenylyl cyclase.[28]

Of course, **cyc$^-$**, indicating an absence of cyclase, is a misnomer, based on the initial thought that these cells might be devoid of adenylyl cyclase.

In the wild-type S49 cells, elevation of cAMP has the effect of arresting the cell cycle in the G1 phase and then promoting cell death.[29,30] By growing the cells in the presence of isoprenaline to elevate cAMP and then rescuing the few survivors, and then repeating this procedure, a resistant line, cyc$^-$, was obtained[31] (Figure 4.6). Although endowed with β_2-adrenergic receptors and an adenylyl cyclase which could be activated by the terpenoid forskolin (a direct activator of adenylyl cyclase, see page 142), these cells are unresponsive to agonists binding to the β_2-adrenergic receptor. The signal transduction pathway may be said to have become uncoupled. Something stands in the way (or is missing), so that the communication between the receptor and the effector enzyme is blocked.

Communication between the receptor and the cyclase in the membranes of these cyc$^-$ cells can be re-established by provision of the GTP-binding component (α-subunit) of the G protein G$_s$.[32-34] By failing to express α_s,

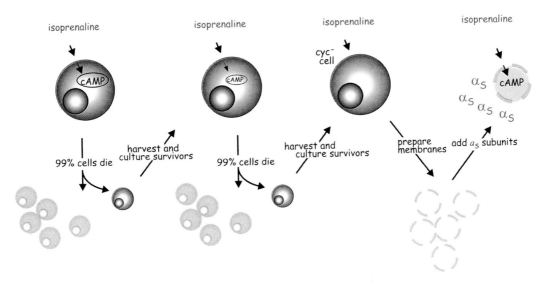

FIG 4.6 Reconstitution of hormone-stimulated cyclase activity in S49 cyc⁻ lymphoma cells by addition of α_s-subunits. The figure illustrates the principle underlying the selection of a line of cells (S49 cyc⁻) that fail to generate cAMP in response to stimulation by ligands that bind to receptors linked to G_s. Activity can be restored to the isolated membranes by addition of G protein α_s-subunits.

the cyc⁻ cells avoid the possibility of lethal elevation of cAMP. This system was widely used as a test bed to study the effects of different GTP-binding proteins: for example, those proteins having selected point mutations, or chimeric variants containing components of different G proteins such as G_s and G_i (inhibitory). This opened the way to understanding the roles and interactions of G proteins in the sequence of steps leading from receptor occupation to cyclase activation.

α-Subunits determine G protein diversity

The diversity of the heterotrimeric G proteins is principally a function of their α-subunits (Table 4.1). Unlike the 7TM receptors, of which there are thousands, there are only 16 genes specifying the α-subunits in animals; with alternative splice variants of G_s and G_o, these provide about 20 gene products in all (Figure 4.7). Most animal cells express 9 or 10 of the 16 possible α-subunit gene products. Some, such as the α_t, α_{olf}, and α_{gust} genes are expressed only in single classes of sensory cells (photoreceptors, olfactory epithelial neurons and taste receptors). Others are found in cells that share a common embryonic origin. Thus, α_o-subunits are expressed in cells derived from the neural crest and in endocrine tissues such as pancreatic β-cells and cells of the pituitary gland. α_z is found primarily in neurons, and α_{16} is expressed exclusively in cells of haematopoietic origin. Only α_s, α_{i2}, and α_{11} are universally expressed in animal cells, and most cells also express α_q and either α_{i1} or α_{i3}.

TABLE 4.1 Some functions of G protein α-subunits

	Generation of second messengers	Other processes
α_s	stimulation of adenylyl cyclase	
α_{olf}	stimulation of adenylyl cyclase	
α_{i1}	inhibition of adenylyl cyclase	
α_{i2}	inhibition of adenylyl cyclase	Mitogenic signalling[35,36] positive regulator of insulin action[37]
α_{i3}	inhibition of adenylyl cyclase	
α_o	inhibition of type I adenylyl cyclase[38]	potentiates EGF-R and Ras signalling on MAP kinase activity[39] sequesters Rap1GAP[40] activates STAT3 signalling[41] inhibition of cardiac L-type currents[42]
α_t	stimulation of cyclic GMP phosphodiesterase	
α_z	K^+ channel closure	
α_q	stimulates PLCβ[43,44] stimulation of Btk[45]	affects migratory responses of fibroblasts to thrombin platelet function disrupted in $\alpha_q^{-/-}$ (knock out) mice[46]
α_{12}	stimulation of Btk[47] RasGAP (Gap1m)[47] suppresses α_{13} on RhoGEF	
α_{13}	stimulates activity of RhoGEF[48]	regulates developmental angiogenesis,[49] affects migratory responses of fibroblasts to thrombin[49] regulates growth factor induced cell migration (not linked to 7TM receptor)[50]

When the amino acid sequences of the various α-subunits are aligned, it can be seen that about 40% of the residues are invariant. Moreover, the proteins are strongly conserved in evolution. It is likely that this high degree of conservation has been driven by the need to maintain the very specific, yet multiple physiological functions of each of these protein gene products.

The α_s subunits of rats and humans differ by only a single amino acid out of a total of 394; α_{i1} is identical in cows and humans and α_{i2}, α_{i3}, α_o, and α_z retain more than 98% identity among all mammalian species.

We can classify the family of α-subunits in four main classes; each of these contains a number of closely related isotypes (Figure 4.7). Thus, the α-subunits of mammalian α_s and α_{olf} differ in only 12% of their amino acids and α_{i1} and α_{i3} differ in only 6%. Within these groups, the functions are, in the main, closely related. All the three α_i proteins inhibit adenylyl cyclase; all α-subunits of the $G_{q/11}$ family activate PLCβ. Other α-subunits function not in the regulation of second messengers, but to modulate the activity of proteins in other signal transduction pathways. The opposing effects of α_{12} and α_{13} on the activity of RhoGEF is an example of this[48,51] (see Chapter 14). Also, α_q

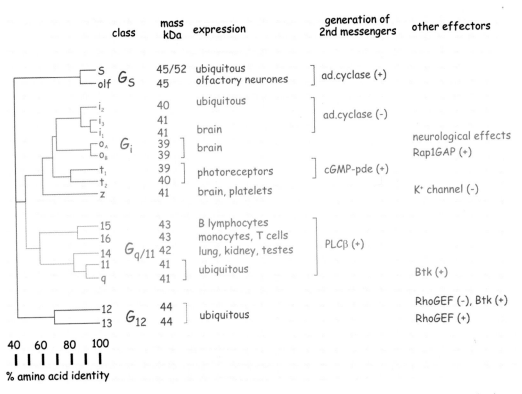

class	mass kDa	expression	generation of 2nd messengers	other effectors
s	45/52	ubiquitous	ad.cyclase (+)	
olf G_S	45	olfactory neurones		
i_2	40	ubiquitous	ad.cyclase (-)	
i_3	41			
i_1	41	brain		
o_A G_i	39	brain		neurological effects
o_B	39			Rap1GAP (+)
t_1	39	photoreceptors	cGMP-pde (+)	
t_2	40			
z	41	brain, platelets		K⁺ channel (-)
15	43	B lymphocytes		
16	43	monocytes, T cells		
14 $G_{q/11}$	42	lung, kidney, testes	PLCβ (+)	
11	41	ubiquitous		
q	41			Btk (+)
12	44	ubiquitous		RhoGEF (-), Btk (+)
13 G_{12}	44			RhoGEF (+)

40 60 80 100
| | | | | | |
% amino acid identity

FIG 4.7 Evolutionary relationship of G protein α-subunits. All proteins that bind GTP have related sequences in those segments that interact directly with the nucleotide (see Figure 4.16). The sequence identity of the α-subunits of heterotrimeric G proteins is greater than 40%. In the figure, examples of downstream effectors that generate second messengers are given. Some other effector proteins that are discussed elsewhere in the book are also listed. Note that G$_s$ is expressed in short and long forms due to alternative splicing. Btk, Bruton's tyrosine kinase,[45,47] is a non-receptor protein tyrosine kinase (see Chapter 12). It is likely that G$_q$ and G$_{12}$ also regulate other related protein tyrosine kinases. RhoGEF,[48,51] see Chapter 14. Rap1GAP interaction with G$_o$,[40] see Chapter 9. The dendrogram is adapted from Simon et al.[52]

and α_{12} are able to activate Btk (Bruton's tyrosine kinase) and possibly other related non-receptor protein tyrosine kinases (see Chapter 17).

The downstream effectors of some of the α-subunits still remain very uncertain. In particular, the functions of G$_o$ have so far evaded any sort of precise description. A problem here is that G$_o$ appears to have multiple effects. One might have thought that ablation of the gene coding for such an abundant protein, as much as 2% of membrane protein in brain, would generate a drastically affected phenotype that would offer a ready diagnosis. Yet, $\alpha_o^{-/-}$ (knockout) mice, survive some weeks. They are hyperactive, tending to run in circles for hours on end, and display hypersensitivity in standard pain tests.[42] There appears to be some involvement of G$_o$ in the regulation of voltage-sensitive (L-type) Ca^{2+} channels but this is likely to be indirect.

G$_{olf}$ regulates adenylyl cyclase in olfactory epithelial neurons.

RhoGEF is a guanine nucleotide exchange catalyst for the Ras-related GTPase Rho.

Among other functions of G_o, it regulates the activity of Rap1, a Ras-related protein[40] (page 251). It also enhances signalling from the EGF receptor, potentiating the activity of ERK (extracellular signal regulated protein kinase),[39] and it activates signalling by STAT3[41] (see Chapter 12).

Sites on α-subunits that interact with the membrane and with other proteins

The C-terminal region of α-subunits dictates the specificity of interaction with receptors. It is the site of ADP-ribosylation by pertussis toxin (see page 143). The N-terminal sequence is the site of interaction with βγ-subunits (**Figure 4.2**). When removed, this is prevented. The N-terminal glycines of the α-subunits of G_i and G_o are the sites of attachment of myristic acid ($CH_3(CH_2)_{12} CO_2H$), a modification that facilitates heterotrimer formation. Antibodies raised against synthetic peptides having the same sequences that are present in the C-terminal domain of α-subunits block the line of communication from receptors.[53,54] Point mutations in this region also determine an 'uncoupled' (*unc*) phenotype[55] in which downstream processes can be stimulated by stable analogues of GTP (see **Figure 4.4**). They also fail to respond to receptors.

It has been much harder to determine the sites at which α-subunits communicate with effectors. This is because the protein interfaces are not composed of just a few residues on a single chain. Instead, they are distributed as clusters present on loops, separated by long distances in the sequence. After an extensive analysis of chimeric proteins constructed from $α_{i2}$ and $α_s$, three distinct regions were perceived to be necessary for the interaction with adenylyl cyclase.[56] These are the switch regions shown in **Figure 4.2**. Subsequently, detailed crystallographic structures of $α_t$ and $α_{i1}$ revealed the conformational changes that occur in these switch regions, when GDP is exchanged for GTP.[57] These are illustrated for $α_i$ in Figure 4.8. The main conformational changes that take place are highlighted in colour. These loops move towards the terminal phosphate of the nucleotide. The SWII chain undergoes the largest rearrangement and the overall result is the modification of the α-subunit surface that contacts the β-subunit allowing the α- and βγ-subunits to detach.

Structural investigations of α-subunits indicate that they possess two apparently independent domains.[8,58,59] One of these has a structure closely related to that of the monomeric GTPase Ras (Figure 4.9). The Ras-like domains of the different α-subunits are similar, possessing a nucleotide binding site and binding surfaces for βγ-subunits, receptors, effectors, and RGS proteins. The other domain is a compact six-helix bundle that is absent from the monomeric GTP-binding proteins (actually an insertion in the sequence of the Ras-like domain). These helical domains are more variable across α-subunits, implying a link to function. They cover the nucleotide binding site on the Ras-like

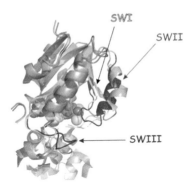

FIG 4.8 The effect of nucleotide exchange on α-subunit conformation. The two superimposed grey structures show the core region of α_i in its GDP- and GTP-bound states. The nucleotide (spheres) is the GTP analogue GTP-γS with a bound Mg^{2+} (green). GDP is not shown. Upon exchange the following movements of the switch loops occur: SWI (white to yellow), SWII (pink to red), and SWIII (light blue to dark blue). The β subunit is not shown. (1gp2[8] and 1gia[58]).

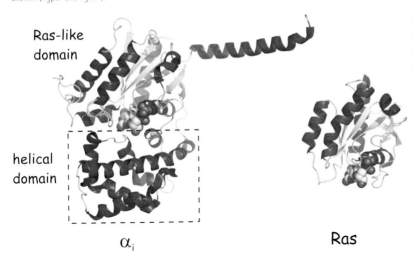

Ras-like domain

helical domain

α_i

Ras

FIG 4.9 The α-subunits of G proteins and the monomeric GTPases exhibit structural similarities. The Ras-like domain of α_i (left) resembles the GTPase Ras (right). The helical domain of α_i covers the nucleotide binding site. Each molecule shown has a bound GDP (spheres). (1gp2[8] and 4q21[63–66]).

domain and must be displaced to allow GDP dissociation. Not surprisingly, they can contribute to nucleotide binding[60] and assist the hydrolysis of GTP.[61] It also appears that the helical domain of the transducin α-subunit (α_t) acts as an allosteric modulator, enhancing the efficiency of the interaction of the activated α_t with its effector, cyclic GMP-phosphodiesterase[62] (see page 170). The helical domains of the other α-subunits may behave similarly.

Following the binding of GTP and the detachment of the βγ-subunit, a surface on the α-subunit is revealed that includes the modified SWII region. This new surface can interact with an effector molecule, but the area of contact is localized. This is illustrated for α_s and adenylyl cyclase in Figure 4.10.

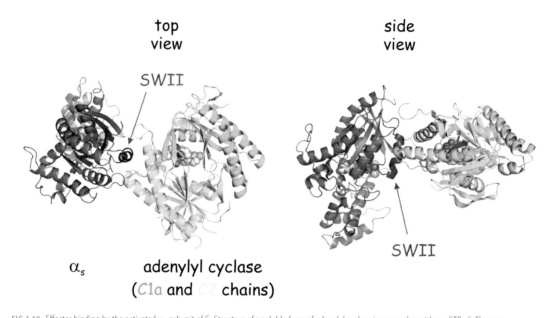

top
view

side
view

SWII

SWII

α_s

adenylyl cyclase
(C1a and C2 chains)

FIG 4.10 Effector binding by the activated α-subunit of G_s.Structure of a soluble form of adenylyl cyclase in a complex with α_s.GTPγS. The non-hydrolysable GTP analogue (spheres) maintains the G protein α-subunit (magenta) in its active conformation, allowing it to interact through its SWII chain (red) with its effector, adenylyl cyclase, provided here in fragmentary form; it consists of the C1a domain (green) of canine adenylyl cyclase 5 and the C2 domain (cyan) of rat adenylyl cyclase 2, a combination that is catalytically active *in vitro*. A molecule of forskolin (spheres) is bound to the cyclase chains. The surface of the complex that is predicted to be in contact with the plasma membrane is at the top of the left hand structure (1azs[67]).

The binding of an effector at localized regions on the surface of an α-subunit allows other molecules, such as a regulatory RGS protein, to attach to a neighbouring site at the same time. For example RGS9 (which acts here as a GTPase activating protein) attaches to the Ras-like domain of the α-subunit of G_t (rod transducin) at a location adjacent to, but separate from the site that binds its effector, the inhibitory γ-subunit of cGMP-phosphodiesterase (PDEγ). Figure 4.11 illustrates this, showing an activated α_t bound to PDEγ, shown as a fragment, and to the RH domain of RGS9.

Note that the acceleration of GTPase activity here requires the binding of both RGS9 and PDEγ, as well as other proteins. Also, RGS proteins do not appear to affect the rate of GTP hydrolysis of some α-subunits and they can interact directly with a variety of effectors and receptors.[68]

βγ-Subunits

β- and γ-subtypes

There are five β subtypes. Of these, β_1–β_4 share 50–90% identity, having molecular masses of either 35 kDa or 39 kDa. β_5 is distinct having only 53% sequence identity with the others[70] and associates with α_q. There are least 12 subtypes of γ (5–10 kDa). The sequences of γ are more diverse. Not all of the potential βγ pairs exist in nature; for example, there is $\beta_1\gamma_2$, but not $\beta_2\gamma_1$.

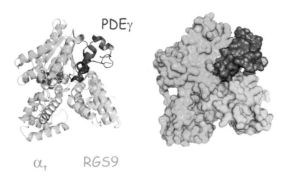

FIG 4.11 Separate binding sites for effector and RGS protein on an activated α-subunit. The structure on the left is the activated α-subunit of G_t (rod transducin, green) bound to a fragment of its effector, phosphodiesterase γ (PDEγ, magenta), and to the RGS domain of RGS9 (cyan). To aid crystallization, the α-subunit is a chimeric protein 60% identical to α_t. The remainder is identical to α_i. The activating nucleotide (spheres) is GDP.AlF$_4$ which is structurally similar to GTP (see page 142). The SWII region of the α-subunit is shown in red. The calculated molecular surfaces on the right show that the binding sites for the effector and for RGS9 are neighbouring, but separate (1fqj[69]).

In spite of great difficulty in demonstrating specificity of $\beta\gamma$ pairing in reconstitution experiments, it has become clear that the identity of the $\beta\gamma$ dimer contributes, together with the α-subunit, to the coupling of G proteins to individual receptors. How else could a cell possessing 11 different receptors, 7 different α-subunits, 5 β-subunits, 8 γ-subunits, and 6 effector enzymes ever be able to make sense out of anything? The example here is the rat ventricular myocyte, for which it has been shown that the angiotensin receptor (AT$_{1A}$), that mediates the elevation of cytosol Ca^{2+}, acts through a G protein comprising $\alpha_{13}\beta_1\gamma_2$.[71] Again, rat basophilic leukaemia cells (RBL-2H3) offer the choice of four β subtypes and five γ subtypes, yet the activation of phospholipase Cβ by M1 muscarinic receptors is mediated by a G protein comprising $\alpha_{q/11}\beta_{1/4}\gamma_4$.[72]

There are many other examples of such heterotrimer specificity.[73] In particular, α_t, the α-subunit of transducin associates exclusively with $\beta_1\gamma_1$, uniquely expressed in photoreceptors. Also in the retina, β_5 (actually, its long-splice variant, Gβ_5L) exists in association with the GAP protein RGS9 that contains a γ-like domain,[74,75] and the two proteins are expressed coordinately.[76] As a result, mice lacking the gene for RGS9 fail to express a functional Gβ_5L in their retinal rods (see Chapter 6). In other parts of the nervous system the short-splice variant of Gβ_5 is found in association with other members of the RGS family that share a common structural γ-like domain with RGS9.[77]

Mammalian γ-subunits are post-translationally modified at the C-terminus by the addition of the 20-carbon geranylgeranyl group (4 isoprene units) or the 15-carbon farnesyl group (3 isoprene units) (Table 4.2, Figure 4.12). This hydrophobic adduct ensures that they are tethered to the membrane and with them, their associated β-subunits. Although mutated γ-subunits

TABLE 4.2 Heterotrimeric G protein post-translational lipid modifications

Subunit	Modification
α_q	palmitoyl
α_s	palmitoyl
α_i	myristoyl
α_o	myristoyl
α_t	myristoyl
α_{olf}	myristoyl
α_z	myristoyl, palmitoyl
$\gamma1, \gamma8, \gamma11$	farnesyl
$\gamma2-7, \gamma9-11$	geranylgeranyl

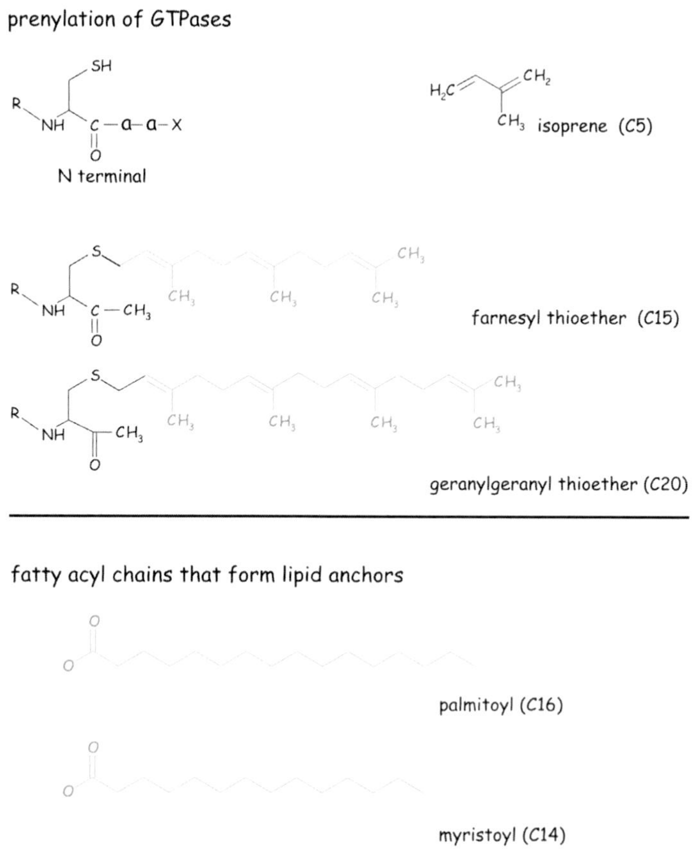

prenylation of GTPases

N terminal

isoprene (C5)

farnesyl thioether (C15)

geranylgeranyl thioether (C20)

fatty acyl chains that form lipid anchors

palmitoyl (C16)

myristoyl (C14)

FIG 4.12 Membrane tethers of $\beta\gamma$-subunits.

unable to undergo the prenylation reaction still associate with β-subunits, the resulting heterodimers are soluble and unable to interact with α-subunits such as α_s.

$\beta\gamma$-Subunits as signalling proteins

The first indication that $\beta\gamma$-subunits can transmit information independently of α-subunits and of second messengers (cAMP, Ca^{2+}, etc.) was in connection with the process by which acetylcholine reduces cardiac output through parasympathetic stimulation.[78] As related in Chapter 1, this very effect had provided the first firm evidence in favour of the idea of chemical transmission at synapses. Some 70 years later, it was found that this muscarinic response is mediated by a G protein.[79] No soluble second messenger is involved and the G protein, likely to be G_i or G_o (sensitive to pertussis toxin: see page 143), is understood to interact directly with a cardiac K^+ channel. As with the early idea of chemical transmission, the proposal that $\beta\gamma$-subunits are capable of signal transduction was regarded as heretical and generated much fractious debate.[80] It then emerged that purified $\beta\gamma$ dimers induce the opening of K^+ channels when applied to the inside surface of isolated membrane patches.[78,81] Counter-arguments came from those who insisted that the phenomenon was more likely due to the presence of contaminating α-subunits or even to the detergents used to maintain the G proteins in solution. However, similar experiments using purified α-subunits produced erratic results at best. The dust only settled when it was shown that coexpression, in *Xenopus* oocytes, of $\beta\gamma$-subunits together with the muscarinic K^+ channel results in constitutive (persistent) activity. In contrast, application of α_i-subunits containing GTPγS or coexpression of α-subunits with the K^+ channel is without effect.[82] By the time the argument had been finally resolved (at least to the satisfaction of most people), it had also been shown that $\beta\gamma$-subunits are activators of phospholipase A2[83] and some β-isoforms of phospholipase C.[84]

The cardiac muscarinic K^+ channel is a member of the family of G-protein activated inwardly rectifying K^+ channels, GIRKs. The functional channels are composed of four subunits and bind four $\beta\gamma$ subunits. In this respect, they are similar to the distantly related cyclic nucleotide-gated channels, which are also tetramers and also exhibit 1:1 binding stoichiometry with their ligands.[84]

It is now clear that at the functional level, G_s and all the other heterotrimeric G proteins behave as if they were a dimeric combination of an α-subunit and an inseparable $\beta\gamma$ pair. $\beta\gamma$-subunits have the following functions.

- They ensure the localization, effective coupling, and deactivation of the α-subunits.
- They regulate the affinity of the receptors for their activating ligands.
- They reduce the tendency of GDP to dissociate from α-subunits (thus stabilizing the inactive state).
- They are required for certain α-subunits (G_i and G_o) to undergo covalent modification by pertussis toxin (ADP-ribosylation, see page 143).
- They interact directly with some downstream effector systems.
- They regulate receptor phosphorylation by specific receptor kinases.

TABLE 4.3 G protein receptor kinase family

	Expression	Former name
GRK1	retina	rhodopsin kinase
GRK2	ubiquitous	β-adrenergic receptor kinase 1
GRK3	ubiquitous	β-adrenergic receptor kinase 2
GRK4	testis, cerebellum, kidney	
GRK5	ubiquitous	
GRK6	ubiquitous	
GRK7	retina	cone opsin kinase

The G protein receptor kinase family

The phosphorylation of residues in the C-terminal regions of 7TM receptors keeps activation under control. One of the ways in which this occurs is through the phosphorylation of activated receptors by the G protein receptor kinases (GRK 1–7, see Table 4.3). The members of this family have similar structures comprising a kinase domain, an RGS homology (RH) domain and a more variable C-terminal membrane-targeting region. For instance, the C-terminal residues of the visual GRKs (1 and 7) are hydrophobically modified (farnesylated), while the ubiquitous β-adrenergic receptor kinases (GRKs 2 and 3) possess PH domains (see Chapter 24). These bind to the phosphoinositide $PI(4,5)P_2$ and to G protein βγ-subunits, enabling the recruitment of the enzyme from the cytosol.

Receptor phosphorylation, down-regulation and pathway switching

The βγ-subunits also serve to present negative feedback signals as demonstrated first through the activation of rhodopsin kinase and later with GRK2/β-adrenergic receptor kinase. Receptors such as muscarinic and adrenergic receptors, phosphorylated at serines and threonines in the C-terminal domain (Figure 4.13), may then recruit adaptor molecules called arrestins. While this is sufficient to block the normal line of communication through G proteins (receptor desensitization), the bound arrestin can also act as a docking site for a range of signalling molecules, including MAP kinases (Chapter 12), PI 3-kinase (Chapter 18), PKB/Akt (Chapter 18) as well as the soluble protein tyrosine kinase Src. These can initiate new signalling pathways,

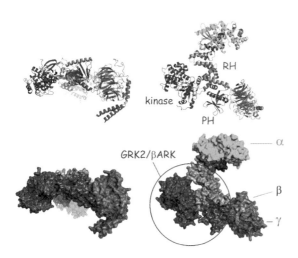

FIG 4.13 The G protein receptor kinase GRK2/βARK1 binds α and β-γ subunits at separate sites. Two views of the Gα$_{i/q}$·GRK2/Gβγ complex are shown as cartoons (top) and as molecular surfaces (bottom). The left-hand structures are side views with the membrane-binding surface at the top. The right-hand structures have been rotated 90° about a horizontal axis. GRK2 possesses a PH domain (orange), an RH domain (pink), and a kinase domain (dark pink). The PH domain binds a β-subunit (blue), the RH domain binds the α-subunit (mostly green) at sites close to its switch regions (SWI dark blue, SWII red). GDP.AlF$_4$ is shown as spheres in the upper structures (2bcj[86]).

for example the activation of ERK (see Chapter 12).[87] The arrestins can also bind the membrane coat protein clathrin and the associated machinery that allows the receptors to be removed from the plasma membrane by endocytosis.[88,89] It is thought that this mechanism is of particular importance in locations, such as sympathetic synapses, where agonist concentrations have a tendency to soar.

As the concentration of the stimulus is raised, two phosphorylating mechanisms, catalysed by protein kinase A (PKA, cAMP-activated protein kinase, see Chapter 9) and by GRK2 are called into action. Since these two kinases have different substrate specificities (consensus sequence selectivities), reacting with different residues in the C-terminal domain of the receptor, the pattern of phosphorylation is different. It depends on the intensity of stimulation: a mild stimulus inducing PKA activity, or a strong stimulus additionally activating GRK2. The consequence of phosphorylation at the PKA-reactive sites provides a classic feedback mechanism which acts to disrupt the line of communication with the effector, adenylyl cyclase, and so shuts down the production of cAMP. The phosphorylation due to PKA, which may be maximally activated when less than 10% of the receptors are occupied will also act to down-regulate other receptors having appropriate phosphorylation sites. These will also be prevented from activating cyclase and this is termed heterologous desensitization (*other* receptors) (Figure 4.14a). Phosphorylation

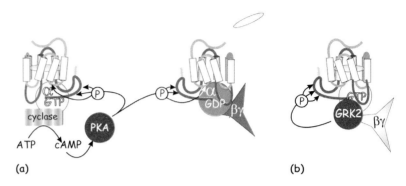

(a)   (b)

FIG 4.14 Heterologous and homologous desensitization by phosphorylation of receptors. (a) Heterologous desensitization. Under mild stimulation the generation of cAMP results in the phosphorylation of all proteins (including receptor molecules) having the PKA substrate consensus motif (-RRSS-). The phosphorylation of receptors occurs regardless of their state of occupation and so the consequence is a generalized down-regulation of all the receptors that regulate cAMP production. (b) Homologous desensitization. Under strong stimulation, the $\beta\gamma$-subunits associated with the receptor act as an anchor for a soluble receptor kinase (GRK2 in the case of β_2-adrenergic receptors) which phosphorylates only the attached receptor.

may also occur through the action of other second messenger-activated, broad spectrum kinases such as PKC.

By contrast, phosphorylation by receptor-specific kinases affects only activated receptors, for example, GRK2, which is activated by $\beta\gamma$-subunits. This form of inhibition is termed homologous desensitization (*same* receptor) (Figure 4.14b).

It must be evident that the whole sequence of control, from the first interaction of a hormone with its receptor right through to the generation of second messengers, and then back again to the receptors, is tightly regulated at all stages. Throughout, it remains flexibly sensitive to the needs of the cell.

Receptor mechanisms obviating G proteins

There are a number of apparently anomalous situations in which 7TM receptors are clearly involved, but which appear to cause activation of their target systems without involving G proteins.[90] Neither do they involve arrestins or receptor kinases as described in the previous section. Many of these have been described in neuronal cells and identified initially as changes in the conductance of ion channels. Examples of such non-G-protein-mediated processes are to be found in the activation of non-receptor protein kinases that lead to the activation of the MAP kinase pathway by ligands acting at β-adrenergic, acetylcholine (muscarinic), and glutamatergic receptors (for a fuller description of these pathways, see Chapters 12 and 17).

TABLE 4.4 Monomeric GTPase post-translational lipid modifications

GTPase	Target sequence	Modification
H-Ras	CVLS	farnesyl and palmitoyl
K-Ras	CVIM	farnesyl
N-Ras	CVVM	farnesyl and palmitoyl
Rho family (Rac and Rho)	CxxL	farnesyl, geranyl and palmitoyl
Rab	CC/CxC	geranylgeranyl
ARF	N-terminal glycine	myristoyl

Monomeric GTP-binding proteins

Ras proteins discovered as oncogene products

The Ras proteins are often referred to as proto-oncogene products. This is because they were first discovered as the transforming products[91,92] of a group of related retroviruses, including the Harvey murine (H) virus[93] and Kirsten sarcoma (K) virus.[94] The transforming genes are fusions of the viral *gag* gene (see page 324) and one of the *ras* genes derived from the rats through which the virus had been passaged.

N-Ras was discovered as a transforming gene product having sequence homology to the other Ras proteins present in a neuroblastoma cell line.[95,96] They are all single-chain polypeptides, 189 amino acids in length, bound to the plasma membranes of cells by post-translational lipid attachments at their C-termini (Table 4.4). They all bind guanine nucleotides (GTP and GDP) and they are GTPases.[97,98] Evidence for a link to human tumours came with the finding that cultured fibroblasts transfected with DNA derived from a human tumour cell line contain a mutated form of Ras.[99]

In a high proportion of human tumours one of the three endogenous cellular forms of Ras is altered by somatic mutations that inhibit the rate of GTP hydrolysis.[100] This ensures that they are in a persistently activated state. On the other hand, non-oncogenic forms, c-Ras, are present in all cells. These are regulators of cell growth and differentiation (Chapter 12).

Subfamilies of Ras

The sequences of the Ras proteins are closely related (Figure 4.15). The first 164 amino acids of human H-Ras and chicken Ras differ in only 2 positions, and the sequences of the first 80 amino acids of human N-Ras and *Drosophila* D-Ras

Gag (glycosylated antigen) is the gene encoding the internal capsid of the viral particle. Crk (C10 regulator of kinase).

The incidence of mutated Ras proteins varies among different types of tumour. 90% of human pancreatic adenocarcinomas and 50% of colon adenocarcinomas are associated with Ras mutations. They are rarely found in adenosarcoma of the breast.

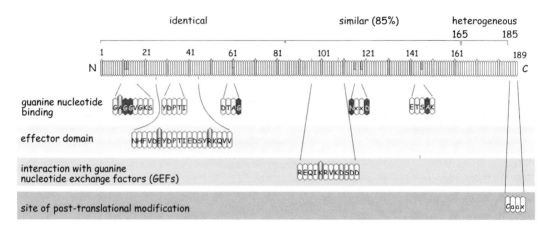

FIG 4.15 Main features of the Ras primary sequence. The mammalian Ras proteins (H, K, and N) share very close identity for the first 164 amino acids (green indicates identity, mauve indicates a conservative substitution). With the exception of the cysteine residue (186), the C-terminal segment is highly divergent (yellow). The lower sections of the figure illustrate the details of sequence motifs involved in binding to the guanine nucleotide, the effector domain, and the C-terminal Caax box that forms the substrate for post-translational modification by isoprenylation. The residues marked in red are associated with oncogenic mutations

Yeast *RAS* activates adenylyl cyclase.[101] The proteins RAS1 and RAS2, having 321 and 309 residues, are longer than their mammalian counterparts, though the first 80 residues still maintain a high (80%) degree of homology. Yeast cells lacking both RAS1 and RAS2 are non-viable, but viability can be restored if they are induced to express the homologous mammalian H-Ras.[102,103] (By convention, capital letters are used to indicate yeast genes.)

are identical. These close similarities are supported by many conservative substitutions. As with the α-subunits, it is likely that this high degree of conservation has been driven by the need for these proteins to communicate with a large number of other components, including the activators, inhibitors and effectors (see **Figure 4.19**).

The Ras proteins are archetypes of a large superfamily. All members share some sequence homology to Ras and then fall into distinct groups, called Ras, Rho, Rab, Ran, Arf, and Kir/Rem/Rad (see Table 4.6). Within each subfamily, the homologies are rather strong. Beyond the more immediate subgroups of the Ras superfamily, all these proteins share some limited sequence homologies with short 'fingerprint' sequences present in the bacterial elongation factors (involved in protein synthesis) and also in the α-subunits of the heterotrimeric G proteins (Table 4.5 and Figure 4.16). Indeed, the presence of these short motifs, appropriately distributed along the chain of a protein, can be taken as a fairly sure indication that it will be a GTPase. More than this, the presence of β-strands immediately adjacent to these highly conserved motifs is an invariant feature, whether they are Ras-related, elongation factors, or α-subunits. Not too surprisingly, these conserved motifs constitute the sites of contact with guanine nucleotides.

Structure

A stereoscopic image of Ras is shown in Figure 4.17. Also shown are the elements of secondary structure and connecting regions, G1–G5, that form

TABLE 4.5 Conserved nucleotide binding motifs in H-Ras and bovine α_i. Binding contacts in red bold; non-conserved residues in lower case

Contact	Residues	Sequence	
G1	Ras (10-17) α_i (40-47)	GaggvGKS GagesGKS	binds to α- and β-phosphates
G2	Ras (32-36) α_i (178-182)	ypdTi rvkTt	Mg^{2+} coordination, effector loop Note: The equivalent arginine (R201) in the G2 loop of α_s is the site of ADP-ribosyl attachment by cholera toxin
G3	Ras (57-60) α_i (200-203)	DtaG DvgG	binds to Mg^{2+}, glycine binds to the γ-phosphate of GTP
G4	Ras (116-119) α_i (269-272)	NKcD NKkD	confers specificity for guanine over other nucleotides
G5	Ras (144-146) α_i (324-326)	tsA tcA	buttresses guanine base recognition site

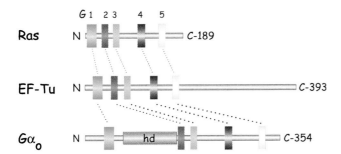

FIG 4.16 Conserved nucleotide binding motifs in Ras, Gα, and EF-Tu. Three families of GTPases have generally divergent sequences but possess short stretches (G1–G5) that are similar to each other. When folded, these segments are almost superimposable and comprise the guanine nucleotide binding pocket. The motifs G2 and G3 (blue and cyan) lie within the switch regions 1 and 2 that undergo a conformational change when GDP is exchanged for GTP. hd indicates the α-helical domain. From Bourne.[6]

the nucleotide binding pocket in colours corresponding to Table 4.5 and **Figure 4.16**. These motifs are also present in the elongation factors and the α-subunits of heterotrimeric G proteins. Residues 26–45, encompassing the G2 contact, comprise the effector region that communicates with downstream proteins.

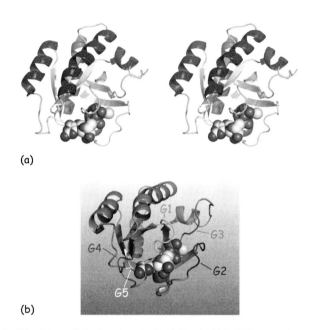

(a)

(b)

FIG 4.17 RasGDP and the motifs that form the nucleotide binding site. (a) RasGDP structure (stereo pair). (b) The conserved motifs G1–G5 that make contact with the nucleotide. They are depicted in colours that match those of Table 4.5. The bound nucleotide is GDP (spheres). The Mg^{2+} ion is coloured green. Instructions for viewing stereo-images can be found on page xxv (4q21[63–65,106]).

Within this region, residues 30–40 are conserved in all forms of Ras, from yeast to mammals, and when mutated, the products are generally inactive (as measured in cell transformation assays: see page 306). Ras proteins mutated in the effector domain (switch region, see below) retain the ability to bind and catalyse the hydrolysis of GTP. Even when combined with a second (transforming) mutation that suppresses GTP hydrolysis and therefore prolongs the lifetime of the activated GTP bound state, the effector mutants remain biologically inactive.[104] The sequence of amino acids between residues 97 and 108 are responsible for interaction with guanine nucleotide exchange proteins (GEFs).[105]

Two regions of Ras, called the switch regions, change their conformation when the GDP is exchanged for GTP. These are indicated in Figure 4.18; here, the nucleotide binding site is occupied by the non-hydrolysable GTP analogue GppNHp. Both the switch I and switch II regions are implicated in the binding to effectors (such as the serine/threonine kinase Raf-1: see page 335) and to the GTPase-activating protein RasGAP. Switch I corresponds to the G2 loop that forms a part of the binding site for Mg^{2+} and the switch II region includes G3 which makes contact with the γ-phosphate of GTP, and the following

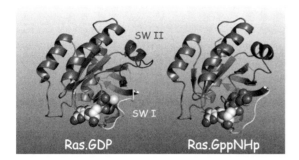

FIG 4.18 Structural differences in the switch regions of GDP- and GTP-bound Ras. Structures of the N-terminal residues of H-Ras complexed with GDP and with GppNHp (an analogue of GTP). The switch regions are indicated. The nucleotides are depicted as spheres. The green sphere is Mg^{2+}. ($4q21^{63-65,106}$ and $5p21^{66}$).

helix (α2-helix) (**Figure 4.18** and Table 4.5). The conformational transitions in these two switch regions are coupled so that binding of GTP brings about an ordered helix $\rightarrow$ coil transition at the N-terminus (closest to the nucleotide) of switch II. Subsequent changes to the α2-helix effectively reorganize the components of the effector-binding interface. All this is reversed when the bound GTP is hydrolysed to GDP.

Post-translational modifications

Beyond residue 165, the sequences become highly divergent. However, at the C-terminal extremities there is again homology in the form of Caax (Cysteine, aliphatic, aliphatic, anything: CVLS in human H-Ras, CVVM in N-Ras and CIIM in K-Ras: see Table 4.4). The Caax motif is post-translationally modified by the addition of a farnesyl extension (C15, 3 isoprene units: see **Figure 4.12**) through stable thioether linkage to the cysteine. This is followed by carboxymethylation, cleavage of the final three amino acids, and methylation of the newly exposed C-terminus (now cysteine). A second lipid modification occurs by palmitoylation of another cysteine residue within the C-terminal hypervariable region.

These modifications ensure strong attachment to the internal surface of the plasma membrane. When the post-translational prenylation reaction is prevented, either by mutation of the Caax sequence or by farnesyl transferase inhibitors, the normal functions of Ras are suppressed.[107] The non-farnesylated Ras accumulates in the cytoplasm and as a result, it scavenges the associated kinase Raf (see page 334), preventing it from coming into contact with the membrane.[108]

K-Ras-2B is different. It is not palmitoylated but instead there is a positive charge cluster which is thought to assist its association with negatively charged phospholipids on the inner surface of the membrane.

Farnesyl transferase inhibitors such as **simvastatin, lovastatin,** and **pravastatin** are used to suppress cholesterol biosynthesis in patients having a history of stroke or heart disease. These drugs inhibit the conversion of 3-hydroxy-3-methylglutaryl CoA to mevalonate (catalysed by HMG CoA reductase). This is the key step in the pathway to the formation of isoprene units which then polymerize to form C_{10} (geranyl), C_{15} (farnsesyl), C_{20} (geranylgeranyl), and C_{30} (squalene). The last of these is the precursor in the biosynthesis of cholesterol. Farnesyl transferase inhibitors are also potential anticancer agents.

GTPases everywhere!

There are hundreds of monomeric GTPases. With few exceptions they have been identified by screening cDNA libraries for Ras-like proteins and their functions have necessarily been assigned (and are still being assigned) subsequent to their discovery. This is an ongoing activity. The normal strategy is to express (or introduce) mutant forms, expected to possess altered functions which might be revealed as an altered phenotype. Thus, the sequence comprising the effector domain of Ras was established long before any effectors were identified. By comparison, the first discovered heterotrimeric G proteins (G_s, G_i, G_o, G_t (transducin), etc.) were isolated by classical purification strategies seeking proteins that bind and hydrolyse GTP and which interact with appropriate receptors and downstream effectors known to be affected by GTP.

To date, representative functions have been assigned for members of all the groups of Ras-related proteins in vertebrates and also in other eukaryotic organisms such as yeasts, flies, nematodes, and slime moulds (Table 4.6). Some of the Ras-like proteins appear to have more than one function, possibly determined by the cells in which they are expressed or by their subcellular localization. Doubtless, more will become evident. Here we concentrate on the Ras proteins. Others are considered at appropriate points in the text.

Mutations of Ras that promote cancer

As with the heterotrimeric G proteins, the proportion of Ras in the activated form is given by the ratio k_{diss}/k_{cat} (see Equation 4.1). Some of the oncogenic viral gene products (vRas) differ from the wild-type cellular forms by single point mutations which inhibit the GTPase activity (reducing k_{cat}) and this

TABLE 4.6 Functions of Ras-related proteins

Family	Regulatory function
Ras	cell growth, proliferation and differentiation
Rho	actin cytoskeleton, cell proliferation
Rab	intracellular vesicle trafficking
Ran	regulates transport between nucleus and cytoplasm, cell cycle transitions (S & M phases)
Arf	activation of PLD, vesicle formation in trafficking and endocytosis
Kir/Rem/Rad	cytoskeleton, cell shape, voltage-dependent channel regulation

prolongs the active state. For this reason, a higher proportion of the activated GTP-bound form exists in cells that have been transformed by vRas. The most common mutations promoting the transformed phenotype occur at residues 12, 13, and 61 (see Figure 4.15).

Other mutations promoting spontaneous activation of Ras are those which enhance the rate of dissociation (k_{diss}) of bound nucleotide (GDP). In the unmodified protein, this occurs at a very slow rate indeed (about 10^{-5} mol s^{-1} per mol of protein), but there are numerous mutations which increase this rate significantly. Many of these are at or in the close vicinity of the residues interacting with the nucleotide. In particular, substitution of Ala-146 increases the rate of guanine nucleotide exchange 1000-fold. As with the GTPase mutants, these mutations ensure that a higher proportion of the protein exists in the activated GTP-bound form. In consequence these too, are transforming mutations.

Functions of Ras

While the identification of receptors and their catalytic activities generally preceded the discovery of the heterotrimeric G proteins that link their operations, the proteins operating immediately upstream or downstream of Ras at first eluded definition. Even so, a place for Ras in the regulation of mitogenesis was not in doubt. Cellular expression of the transformed *ras* genes,[109,110] or direct introduction of their products into single cells by microinjection,[111,112] promotes cell detachment and loss of contact inhibition, characteristics of the transformed phenotype (see page 306). Conversely, microinjection of antibodies (to neutralize the native cellular Ras), or of a dominant inhibitory mutant form (S17N) prevents normal growth and cell division.[113–115] However, it should not be assumed that mitogenesis is the exclusive outcome of activating Ras. Other cell types undergo differentiation as the consequence of Ras activation. In PC12 cells the introduction of the Ras oncogene product provides a signal for neurite outgrowth.[116,117]

S17N: Serine replaces asparagine at position 17.

PC12 cells are adrenal medullary cell line derived from a phaeochromocytoma.

How Ras controls all this has been far from clear. It is true to say that Ras lies at the centre of a network of interacting pathways, is activated and modulated directly and indirectly by several receptors, and in turn, has its influence on a large number of downstream processes (see Figure 4.19 and Chapter 12). Indeed, it is amazing that a protein as small as Ras can interact with so many other proteins. Among functions directly linked to Ras are the activation of the protein kinase Raf and the activation of phosphatidylinositol 3-kinase (PI 3-kinase, see Chapter 18). It is the first member of the chain sequence of phosphorylating enzymes that leads to the activation of extracellular receptor kinase (ERK) and thence to the transcription of genes controlled through the serum response element, SRE (Chapter 12).

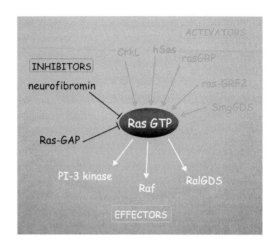

FIG 4.19 Some of the many influences and interactions of Ras.

RasGAPs

In cell-free preparations, the half-life of the GTP bound to Ras is between 1 and 5 h. With a rate constant for hydrolysis of about $4.6 \times 10^{-4}\,s^{-1}$, wild-type Ras hardly functions as a GTPase and one might expect it to be almost permanently activated. It would also be hard to understand why the differences in the rates of GTP hydrolysis by the wild-type and the oncogenic mutants should be so crucial. However, when the rate of GTP hydrolysis in cells is measured, things are very different.[118] RasGTP injected into *Xenopus* oocytes is converted to its inactive GDP form within 5 min. Cells contain a protein or proteins that accelerate the rate of GTP hydrolysis enormously. By contrast, GTP bound to GTPase-defective transforming mutants of Ras remains intact. These observations led to the discovery and isolation of the GTPase activating protein RasGAP (also known as p120[GAP]). The GTPase activating proteins (GAPs) were the first proteins found to interact directly with Ras.

RasGAP

This GAP protein interacts specifically with wild-type Ras, accelerating the GTPase activity by up to five orders of magnitude. When over-expressed,[119,120] or injected into fibroblasts at a concentration in excess of the normal level,[121] RasGAP inhibits the mitogenic action of growth factors. In this respect it acts, as one might expect, as a negative regulator of Ras. However, in T- and B-lymphocytes and in adipocytes a mitogenic signal is mediated by a decrease in the activity of RasGAP.[122–124]

In addition to its (Ras-interacting) catalytic function, RasGAP contains a number of identifiable domains[125,126] (Figure 4.20). These include a PH

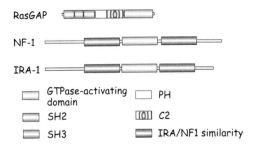

FIG 4.20 The domain organization of RasGAP and related GTPase activating proteins.

domain,[127] a C2 domain,[128] two SH2 domains, and one SH3 domain[129] (these and other domain structures are described in Chapter 24). This implies the potential for a multiplicity of interactions with other signalling proteins, scaffolds, adaptors and other molecules. The Ca^{2+}-sensitivity of RasGAP is discussed in Chapter 8.

GAPs and effectors

Does RasGAP do more than regulate GTPase activity? Is it, like the effectors of heterotrimeric G proteins, also an effector of downstream processes?[130] The fact that Ras-GAP interacts only with RasGTP and not with RasGDP is certainly consistent with the idea of an effector function. Furthermore, the interaction is mediated through its contact with the effector domain of Ras. Some (though not all) mutations in this domain prevent the interaction with GAP. Against this is the finding that RasGAP suppresses the transformation of fibroblasts induced by over-expression of normal wild-type Ras. Here, the GAP appears to play the role of a negative regulator, simply accelerating the rate of GTP hydrolysis.

An alternative possibility would be that RasGAP might be the mediator of just a subset of Ras functions, the rest of which are linked to other effector proteins. For instance, it can stimulate transcription of the c-fos promoter,[131] but here its catalytic domain plays no part since deletion mutants comprising just the SH2 and SH3 domains are equally effective. However, Ras still appears to play a part since the induction of c-fos by the GAP deletion mutant is prevented by N17Ras, a dominant inhibitory mutant. It would appear that the GAP deletion mutant cooperates with another signal emanating from activated Ras. Also, as a result of complex formation with Rho-GAP (specific for the Ras related GTPase Rho), RasGAP acts to inhibit the formation of stress fibres and focal adhesions, both of which are Rho-dependent processes[132] (see figure 16.11 page 499). The assignment of GAP functions will only be resolved when we can identify the relevant protein–protein interactions in normal cells.

Mechanism of GTPase activation

In trying to understand the mechanism of activation of the intrinsic GTPase activity of Ras by GAP proteins, two main ideas have been examined.[133] Firstly, there is the possibility that the GAP acts simply by driving the Ras protein into a conformation active for GTP hydrolysis, without itself forming a part of the active site. In this case, the action of GAP on Ras would be catalytic and hence non-stoichiometric. Now, however, it is thought that the GAP interacts with the Ras stoichiometrically, contributing cationic residues (arginines) that stabilize the transition state of the reacting GTP (Figure 4.21). This allows access of Q61 which directs an incoming water molecule for the hydrolysis of GTP. Mutations of this residue, frequently found in human tumours, prevent the enhancement of GTPase activity by GAP and lock the protein in its active conformation. The analogous interaction of Rho with Rho-GAP defines the catalytic centre of the Rho GTPase. For both Rho and Ras, the interaction with their respective GAPs is transient. They form active heterodimeric enzymes and then separate as the GTPases take up the conformations determined by the presence of GDP in the nucleotide binding pocket.

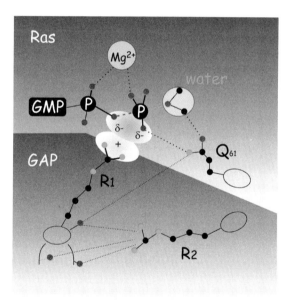

FIG 4.21 Organization of the active site of Ras and its interaction with RasGAP. The point of contact between the catalytic GTPase site of Ras and its GTPase activating protein (RasGAP). A primary arginine finger extends across the cleft between the two proteins and neutralizes the negative charge (pink) that develops in the transition state of the reaction. A second arginine residue (R2) stabilizes the primary arginine finger. The departing phosphate is depicted as a pentacoordinate intermediate. The Ras Q61 also contributes to the transition state. Adapted from Scheffzek et al.[133]

Other molecules having GAP activity are more likely to be pure negative regulators of Ras. One of these is neurofibromin, the product of the *NF1* gene, which is associated with neurofibromatosis type 1 (or von Recklinghausen's neurofibromatosis). In neurofibroma-derived cell lines deficient in this protein, 30–50% of the Ras is present in the GTP-bound form (as opposed to less than 10% in normal cells).

Neurofibromin contains a segment clearly related to the catalytic domain of RasGAP, but otherwise the *NF1* gene product possesses no identifiable domains or motifs and shares no sequence similarity with any other mammalian signalling proteins (see Figure 4.20). These findings are all consistent with the idea that neurofibromin is purely and simply a GTPase activator.

Guanine nucleotide exchange factors (GEFs)

The identification of proteins that catalyse GTP/GDP exchange on Ras, (guanine nucleotide exchange factors, or GEFs) was achieved by studying mutations in yeast and other simple eukaryotes, in which an exchange factor is deficient.

In yeast (*S. cerevisiae*), the RAS gene products regulate adenylyl cyclase,[134] mediating the response to starvation and leading to spore formation. By mutating non-RAS genes that destroy the ability to activate cyclase, it is possible to identify other proteins that contribute to this signalling pathway. Some of these mutations can be overridden (bypassed) by the presence of constitutively activated RAS and are therefore likely to lie upstream of the GTPase. The CDC25 gene product is such a protein.[135] Deletion is lethal, but can be compensated by expression of constitutively activated RAS, strongly suggesting that the CDC25 gene product acts as an upstream activator, probably a RAS-GEF.[136] In addition, the effect of deletion can be overcome by expression of the C-terminal portion of CDC25, which is understood to house its catalytic activity. The catalytic domain of a human GEF Sos1 (see page 332) can restore cyclase activity in yeast lacking CDC25.[137]

Essay: Activation of G proteins without subunit unit dissociation

The standard textbook description of the G-protein cycle assumes that the α-subunits dissociate entirely from their βγ partners and become independent entities. However, the evidence for this idea, based mainly on biochemical experiments carried out in cell-free systems, and never strong, is becoming ever slimmer.[138] This is, at best, an oversimplification. We wish to ask whether

von Recklinghausen's neurofibromatosis is one of the most common of human hereditary disorders, having an estimated incidence of 1 in 3500 individuals worldwide. Almost half of the patients have no previous family history of the disease, so with 1 in 10 000 individuals harbouring a new mutation, it follows that *NF1*, of all human genes, must be one of the most prone to mutation. It predisposes to benign and malignant tumour formation, especially in cells derived from the neural crest.

Most work in this field has been confined to the investigation of the activation (and inhibition) of adenylyl cyclase, initially in turkey red blood cell membranes and later in other cell-derived and reconstituted systems. The turkey red cell membrane cyclase is particularly amenable since it is totally dependent on the presence of a stimulating hormone and is rather slow, allowing the collection of many data points. There have been no comparable attempts to elucidate the mechanism of activation of phospholipase C.

α and $\beta\gamma$ do truly dissociate in every case. The main arguments favouring dissociation are:

- Over-expression, or provision of purified α-subunits to isolated membrane preparations, causes activation of downstream effector enzymes such as adenylyl cyclase or phospholipase C.
- Provision or over-expression of $\beta\gamma$-subunits tends to oppose the activation due to α-subunits in some experimental systems such as platelet membranes.
- Provision of fluoride ions, or stable analogues of GTP (GTPγS, etc.) to G proteins in detergent solution in the presence of high concentrations of Mg^{2+} enhances their tendency to dissociate. These manipulations are an essential component in the strategies applied in subunit purification.
- The α- and $\beta\gamma$-subunits of the retinal G protein transducin certainly dissociate from each other but also from the membrane when activated by rhodopsin.[139] However, ~8% remains attached to the membrane as the intact heterotrimer and it is possible that it is this fraction that activates the effector, cyclic GMP phosphodiesterase.
- Structural studies of GDP-bound, heterotrimeric G proteins and of complexes of activated α-units with their effectors have shown (1) that the switch regions are obscured by the $\beta\gamma$-subunits in the inactive G protein and (2) that contact with the effectors is made through the same regions.
- $\beta\gamma$-Subunits are unable to bind to α-subunits and effector molecules such as adenylyl cyclase or phospholipase Cβ simultaneously. The binding regions for both classes of molecule overlap each other.[140]

Although it is clear that G protein subunits can dissociate under conditions promoting activation and certainly do so when pressed hard enough (fluoride plus detergent plus Mg^{2+}), it is far from clear that this is what actually happens in cell membranes.

In support of the idea of non-dissociation is that the process of adenylyl cyclase activation is a first-order reaction, dependent only on the amount of activated receptor. From this it follows that the G_s is permanently coupled to adenylyl cyclase and indeed, is maintained through a 3000-fold purification of the enzyme. When this is carried out in the presence of an activating guanine nucleotide (GppNHp), a 1:1:1 stoichiometry of cyclase to α_s to β is retained.[141,142] There is no need to invoke the dissociation of the $\beta\gamma$-subunits from the complex in order for α_sGTP to cause activation, even though this might deliver a stronger stimulus. The important possibility that arises is that the presence of both subunits may be required in order for the G protein to fulfil its authentic physiological role. What would happen if the subunits are prevented from dissociating?

Pheromone-induced mating response in yeast

An excellent system in which to challenge the ideas concerning the dissociation of G protein subunits and their relationship to receptors is provided by the mating response of the yeast (*S. cerevisiae*) in which two haploid cells fuse to form a diploid cell. The haploid cells must be of opposite mating type, **a** and α (analogous to sex). Mating is initiated by the reciprocal binding of pheromones called mating factors, a factor from **a** cells binding to receptors on α cells and vice versa. The **a**- and α-type mating factors are oligopeptides (12 and 13 residues) similar to mammalian gonadotrophin releasing hormone (GnRH).[143] In response to binding, the haploid cells undergo cell cycle arrest at the late G1 phase. Cells of opposite mating type attach to each other, fuse, and eventually give rise to diploid α/**a** cells (Figure 4.22a). The receptors for the secreted pheromones are products of the genes *STE2* and *STE3*. The signals are transduced by the products of the genes *GPA*, *STE4*, and *STE18*. The homology between these products and mammalian signalling molecules is shown in Figure 4.22b.

Deletion of GPA is lethal because the free $\beta\gamma$-subunits then activate a pathway leading to growth inhibition. However, mammalian α-subunits of any class (also chimeric mammalian/yeast α-subunits), that can bind to the $\beta\gamma$-subunits, are able to restore viability. However, they do not restore the signal transduction pathway leading to the generation of diploid cells. Indeed, the presence of genes coding for all of the G protein subunits is required. Consequently, using the production of diploid cells as a read-out, it is possible to ask searching questions concerning the status and interactions of the various components of the signal transduction pathway. This has been achieved by the use of mutants in which any one of the genes has been substituted by a fusion construct, for example composed of the N- and C-terminal segments of different absent proteins (Figure 4.22b, c).

Using such an approach, the mating response of *STE4*$^{-/-}$ (β-subunit knockout) cells was restored by genes coding for chimeras composed of the N-terminus of Ste4 coupled to the C-terminus of Gpa1.[12] These were fully functional in the transduction of pheromone signals inducing growth arrest and mating. Since the chimeric construct allows no possibility of subunit dissociation, it is likely that the activated receptor acts to induce conformational changes in the heterotrimer, exposing hidden binding interfaces that allow communication with downstream effectors. The implication is that the fusion construct is fully competent to convey signals ascribed to both the α- and the $\beta\gamma$-subunits. It follows that the wild-type proteins, although capable of dissociating, may not actually do so. If they do tend towards detachment, then it is likely that they remain in very close proximity to each other throughout the cycle. Indeed, it begins to appear that most of the molecules involved in signal

113

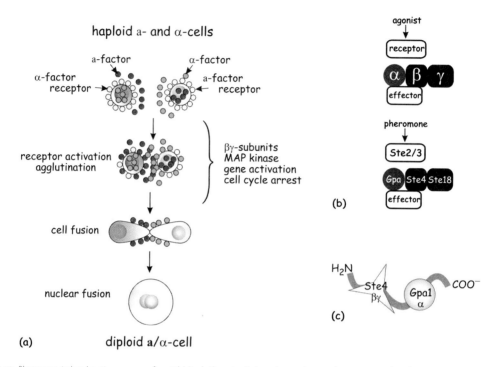

FIG 4.22 Pheromone-induced mating response of yeast. (a) Haploid yeast cells (**a** and α mating types) generate **a** and α pheromone mating factors that bind to specific receptors on cells of the opposite type. This initiates a series of events that include agglutination, cell cycle arrest and fusion to create diploid cells. In the presence of nutrients, the diploid cells may undergo multiple divisions. In starvation, they undergo meiosis forming four haploid spores. Since α and **a** pheromone receptors are G-protein-linked receptors coupled to adenylyl cyclase, the production of diploid cells can be used to test the integrity of the signal transduction pathway. (b) Representation of a construct containing Ste2 (receptor) fused to a chimera of Gpa1 (yeast, N-terminal segment) and α_s (rat, C-terminal segment). This supports efficient signal transduction. (c) Representation of a construct containing Ste4 (β-subunit) fused to Gpa1 (α-subunit). This transmits signals as efficiently as the native G protein subunits.

transduction – receptors, G proteins, effectors – are close neighbours at all times and under all conditions of activation.

Monitoring subunit interactions in living cells by FRET

A direct way of detecting the association or dissociation of fluorescent molecules is provided by the phenomenon of *fluorescence resonance energy transfer* (FRET). Excitation energy may be transferred from one fluorophore (a fluorescent molecule or moiety) to another. The requirements are (1) that the two fluorophores should be in close proximity to one another (within 10 nm), (2) that the emission spectrum of the donor fluorophore should overlap the excitation spectrum of the acceptor, and (3) that the emission dipole moment of the donor should be aligned sufficiently with the excitation dipole moment of the acceptor. When these conditions are met, excitation of the donor fluorophore results in emission from the acceptor. The transfer of excitation energy is a radiationless process and no fluorescence is emitted by the donor.

A biological application of FRET is illustrated in Figure 4.23 for cyan and yellow fluorescent proteins (CFP and YFP, mutated forms of green fluorescent protein). These have spectra that overlap, such that the fluorescence emission spectrum of CFP (solid line: 450–600 nm) falls within the excitation spectrum of YFP (dotted line: 450–550 nm). An energy transfer event occurs when a photon is absorbed by the donor (CFP) followed by the emission of a photon from the acceptor (YFP). If the donor and acceptor move apart, the energy transfer efficiency will decrease and vice versa. In this example, emission from YFP, in response to the absorption of violet light by CFP, will decline. At the same time, blue fluorescence emission from CFP will begin to appear.

To investigate the association of G protein α- and $\beta\gamma$-subunits, it is necessary to tag them with different fluorophores. This is most effectively achieved by transfecting cells so that the subunits are expressed as fusion proteins, for example with YFP and CFP. In an experiment illustrated in Figure 4.23c,[144] HEK cells expressing $G\alpha_{i1}$ fused with YFP and $G\beta_1$ fused with CFP, exhibit FRET (yellow emission from $G\alpha_i$ in response to illumination at 436 nm). If, during stimulation, the α- and $\beta\gamma$-subunits separate from one another, the YFP emission would be expected to drop and the CFP fluorescence to rise as the transfer efficiency falls. However, what we see is the converse. FRET *increases* during stimulation (yellow line) and CFP fluorescence *decreases* (blue line). This indicates that the fluorophores, and thus the subunits, move closer together (or become better aligned for FRET). The $t_{1/2}$ of the onset of the fluorescence changes (~ 1 s) and of their reversal on withdrawal of the stimulating hormone (~ 38 s) (Figure 4.23c) closely match the rates of opening and closure of GIRK channels introduced as effectors into the same cells, indicating that it is the non-dissociated subunits that are functional.

The details of the actual molecular rearrangement are not clear. While FRET increases upon stimulation when CFP is located at the N-terminus of the β-subunit, it decreases when it is sited at the C-terminus of the γ-subunit (see Figure 4.2). The simplest explanation involves a rearrangement in which the N-termini of β and γ move closer to the helical domain of α, while the C-terminal region of γ moves away. At all events, the α- and $\beta\gamma$-subunits certainly do not dissociate.

Similar FRET investigations indicate that G_{i2}, G_{i3}, and G_z also undergo inter-subunit rearrangements without dissociating when stimulated. On the other hand, G_o and the G_s-like proteins of the slime mould *Dictyostelium* tell different stories that can be ascribed to subunit dissociation (or very different forms of rearrangement).[145,146] The B/C helical domains of the α_i and α_z subunits appear to be essential for stable tethering to $\beta\gamma$-subunits, and substitution of this region with the homologous sequence from G_o allows

HEK cells (human embryonic kidney cells): an epithelial cell line that is easy to culture and transfect.

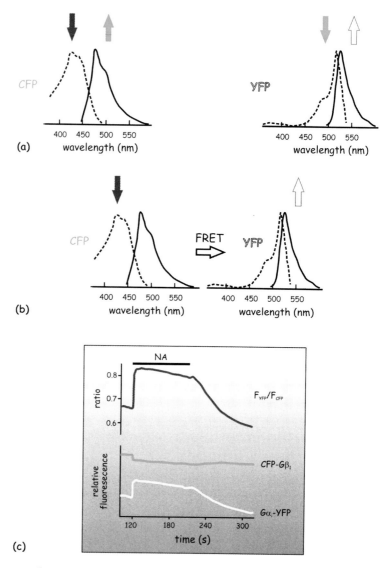

FIG 4.23 Fluorescence resonance energy transfer (FRET) used to determine subunit dissociation.
(a) Fluorescence excitation and emission spectra of cyan and yellow fluorescent protein (CFP and YFP). (b) FRET results in the emission of yellow light by YFP after the absorption of violet light by CFP. (c) Fluorescence time course of CFP-Gβ_1(blue)and YFP-Gαi$_1$ (yellow) emission from a single HEK cell during noradrenaline stimulation (NA). Excitation was provided at 436 nm. While the YFP emission registers energy transfer, the emission ratio yellow:cyan (red line) takes account of experimental artefacts (e.g. light scattering) and is a more sensitive measure of FRET.Adapted from Bunemann et al.[144]

dissociation (fluorescence ratio decrease in the FRET experiment). On the other hand, the converse situation, insertion of the B/C domain from α_{i1} into α_o, fails to induce G_i-like behaviour, so other regions of α_i must be involved in the stabilization of the intact trimeric formation.[146]

The stability of the G_i and G_z heterotrimers may provide the key to understanding the specificity of signal transmission between receptors and those effectors, such as the GIRK channels, which are activated by $\beta\gamma$-subunits. As mentioned previously (see page 94), when tested in reconstitution experiments, it has been hard to discern much selectivity, let alone specificity for particular $\beta\gamma$ pairs in the activation of effector systems. The stability of the G_i and G_z heterotrimers allows for specificity to be determined by the identity of the α-subunit, while the activation signal is carried by the $\beta\gamma$-subunit.

So the best answer to the question of whether heterotrimeric G proteins dissociate on activation is that some do, at least sometimes, and others don't.

Constructing the mammalian β-adrenergic transduction system in insect cells

Earlier we described how the use of S49 cyc⁻ lymphoma cells was instrumental in determining the role of the α_s-subunit of the signal transduction process. More versatile are the Sf9 insect cells.[147] With these cells and the baculovirus vector it is possible to manipulate the expression of all the components – mammalian receptors, α-subunits, and $\beta\gamma$-subunits – independently of each other, with regard to identity, specified mutations, and levels of expression. The native adenylyl cyclase can be used to read out the information transfer from the activated receptors.

When expressed in Sf9 cells, the affinity of the mammalian β-adrenergic receptor for its ligands is not only very low, but also insensitive to the presence of guanine nucleotides (**Figure 4.24**a–c). The response of the native (insect) cyclase to the mammalian receptor is also low, and in general the transduction of the signals from the mammalian receptor by the native insect G_s is inefficient. The system displays features of an 'uncoupled' (*unc*) phenotype.[148] All this changes when the Sf9 cells express mammalian α_s-subunits in addition to the mammalian receptor (Figure 4.24d). The affinity of ligand binding (isoprenaline) is now high, but importantly, it becomes sensitive to the presence of guanine nucleotides, just as in mammalian cells. In this aspect of receptor function, it appears that the insect $\beta\gamma$-subunits are able to cooperate with the mammalian α-subunits. However, the GTPase activity remains insensitive to the presence of stimulating ligands. It requires the additional expression of mammalian $\beta\gamma$-subunits for the system to take on the full characteristics of the mammalian signalling pathway (Figure 4.24). β_2-Adrenergic receptors expressed together with mammalian G_s heterotrimers

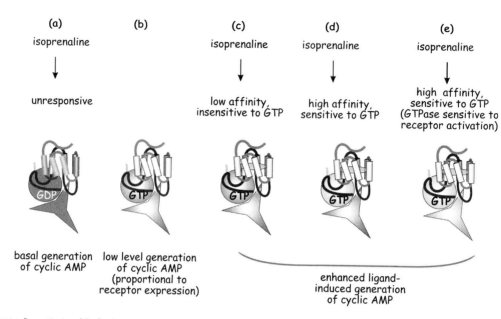

(a) isoprenaline → unresponsive

(b)

(c) isoprenaline → low affinity, insensitive to GTP

(d) isoprenaline → high affinity, sensitive to GTP

(e) isoprenaline → high affinity, sensitive to GTP (GTPase sensitive to receptor activation)

basal generation of cyclic AMP

low level generation of cyclic AMP (proportional to receptor expression)

enhanced ligand-induced generation of cyclic AMP

FIG 4.24 Reconstitution of the β-adrenergic response in insect cells. (a) Although they possess G protein (grey disc) coupled to adenylyl cyclase, Sf9 (insect) cells lack β-adrenergic receptors and are therefore unresponsive to catecholamines. (b) When transfected with the human β-adrenergic receptor (pale green), Sf9 cells generate cAMP at a low rate that is proportional to the extent of receptor expression. No activating ligand is needed for this low level activity. (c) The rate of cAMP generation is greatly enhanced when the transfected cells are stimulated with isoprenaline (yellow ellipse). However, in membrane preparations, the affinity of the receptor is not sensitive to the presence of GTP. The native G protein does not transmit a feedback signal to the mammalian receptor. Nor is the GTPase activity of the insect α-subunit sensitive to the presence of the agonist. (d) By additionally transfecting the cells with mammalian α_s (blue), the affinity of the receptor becomes sensitive to the activation state of the G protein. However, GTPase activity remains insensitive to the presence of the hormone. (e) To establish the complete pathway of forward and backward control, it is necessary to express both the mammalian α- and βγ-subunits (pink). The GTPase activity of the α-subunit is now sensitive to the presence of the activating hormone. For details of this experiment, see Lachance et al.[148]

allow high signal throughput, high ligand binding affinity, and now, ligand-enhanced GTPase activity. The conclusion from all this is that the α- and the βγ-subunits communicate with the receptor throughout the period of activation. They never really lose sight of each other, nor of the receptors.

That this might really be the case, not just in transfected insect cells, finds support in the frequent reports of the coimmunoprecipitation of receptors together with G proteins and of G proteins with their effector enzymes. Similarly, the purification of 7TM receptors on affinity supports can result in copurification of G protein subunits, frequently of several different classes. As examples, the receptors for somatostatin and for opioids (δ-opioid) copurify with α- and βγ-subunits to an extent that depends on the state of their activation.[149-151] This raises the possibility that receptors, rather than communicating in a linear fashion through one G protein to one catalytic

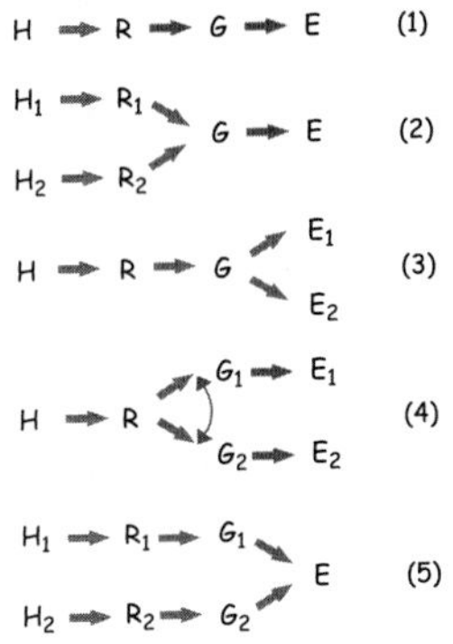

$$H \longrightarrow R \longrightarrow G \longrightarrow E \quad (1)$$

$$\begin{array}{l} H_1 \longrightarrow R_1 \searrow \\ \qquad\qquad\qquad G \longrightarrow E \quad (2) \\ H_2 \longrightarrow R_2 \nearrow \end{array}$$

$$H \longrightarrow R \longrightarrow G \begin{array}{l} \nearrow E_1 \\ \searrow E_2 \end{array} \quad (3)$$

$$H \longrightarrow R \begin{array}{l} \nearrow G_1 \longrightarrow E_1 \\ \searrow G_2 \longrightarrow E_2 \end{array} \quad (4)$$

$$\begin{array}{l} H_1 \longrightarrow R_1 \longrightarrow G_1 \searrow \\ \qquad\qquad\qquad\qquad\qquad E \quad (5) \\ H_2 \longrightarrow R_2 \longrightarrow G_2 \nearrow \end{array}$$

FIG 4.25 Pathways of information flow through receptors and G proteins. The communication of signals from receptor to effector is not necessarily a simple linear sequence of steps (1). There are other more complex modes by which receptors, transducers and effectors are linked. Some of these may seem self evident: different receptors accessing one class of G proteins (2) and a single class of G protein regulating more than one type of effector enzyme (3). The regulated switching of attention of adrenergic receptors between G_s and G_i (4) is considered in Chapters 9 and 12. The synergistic interaction of two receptors and two G proteins in the activation of some isoforms of adenylyl cyclase (5) is discussed in Chapter 5.

effector, may communicate, depending on their state of activation, with different G proteins and thence with different effector enzymes (Figure 4.25).

For yeast cells devoid of the endogenous *STE2* (pheromone receptor) and *GPA1* (α-subunit) genes, transduction of the mating response can be achieved by a fusion protein generated from the N-terminus of Ste2 linked to a chimera composed of the N-terminus of Gpa1 and the C-terminus of rat α_s[152] (**Figure 4.22b**). The presence of the Gpa1/α_s construct (i.e. not linked to the receptor), while capable of restoring viability to haploid cells lacking Gpa1, fails to restore mating competence. This is due to its inability to recognize the receptor, but there is no problem when the two are fused together as Ste2-Gpa1/α_s. It follows that the C-terminus of the α-subunit operates mainly to bring the G protein into the proximity of the receptor, allowing guanine nucleotide exchange and ensuring efficient coupling. Questions related to the communication of signals to downstream effectors are not raised in this experiment since for yeast this is a function of the $\beta\gamma$-subunits.

Contrary to standard descriptions of G protein activation, couched in terms of fleeting interactions between receptors and dissociated subunits, the evidence now shows that receptors, G proteins and their effectors remain together as operational ensembles. Some of them may dissociate when activated, but the extent of this under physiological conditions remains uncertain.

References

1. Rodbell M, Birnbaumer L, Pohl SL, Krans HM. The glucagon-sensitive adenyl cyclase system in plasma membranes of rat liver: An obligatory role of guanylnucleotides in glucagon action. *J Biol Chem.* 1971;246:1877–1882.

2. Rodbell M, Krans HM, Pohl SL, Birnbaumer L. The glucagon-sensitive adenyl cyclase system in plasma membranes of rat liver: Effects of guanylnucleotides on binding of ^{125}I-glucagon. *J Biol Chem.* 1971;246:1872–1876.

3. Maguire ME, Van Arsdale PM, Gilman AG. An agonist-specific effect of guanine nucleotides on binding to the β-adrenergic receptor. *Mol Pharmacol.* 1976;12:335–339.

4. Lefkowitz RJ, Mullikin D, Caron MG. Regulation of β-adrenergic receptors by guanyl-5′-yl imidodiphosphate and other purine nucleotides. *J Biol Chem.* 1976;251:4686–4692.

5. Rodbell M. Nobel Lecture. Signal transduction: evolution of an idea. *Biosci Rep.* 1995;15:117–133.

6. Bourne HR. GTPases everywhere! In: Dickey BF, Birnbaumer L, eds. Berlin: Springer Verlag; 1993:3–15. *GTPases in Biology*; Vol. 1.

7. Bourne HR, Sanders DA, McCormick F. The GTPase superfamily: a conserved switch for diverse cell functions. *Nature.* 1990;348:125–132.

8. Wall MA, Coleman DE, Lee E, et al. The structure of the G protein heterotrimer Gi α1β1γ2. *Cell.* 1995;83:1047–1058.

9. Levitzki A. From epinephrine to cyclic AMP. *Science.* 1988;241:800–806.

10. Gilman AG. G proteins and dual control of adenylate cyclase. *Cell.* 1984;36:577–579.

11. Arad H, Rosenbusch JP, Levitzki A. Stimulatory GTP regulatory unit Ns and the catalytic unit of adenylate cyclase are tightly associated: mechanistic consequences. *Proc Natl Acad Sci U S A.* 1984;81:6579–6583.

12. Klein S, Reuveni H, Levitzki A. Signal transduction by a nondissociable heterotrimeric yeast G protein. *Proc Natl Acad Sci U S A.* 2000;97:3219–3233.

13. Arad H, Levitzki A. The mechanism of partial agonism in the β-receptor dependent adenylate cyclase of turkey erythrocytes. *Mol Pharmacol.* 1979;16:749–756.

14. Liebman PA, Parker KR, Dratz EA. The molecular mechanism of visual excitation and its relation to the structure and composition of the rod outer segment. *Annu Rev Physiol.* 1987;49:765–791.

15. Arshavsky VY, Dumke CL, Zhu Y, et al. Regulation of transducin GTPase activity in bovine rod outer segments. *J Biol Chem.* 1994;269:19882–19887.

16. He W, Cowan CW, Wensel TG. RGS9, a GTPase accelerator for phototransduction. *Neuron.* 1998;20:95–102.

17. Biddlecombe GH, Berstein G, Ross EM, et al. Regulation of phospholipase C-β1 by G$_q$ and m1 muscarinic receptor. *J Biol Chem*. 1996;271:7999–8007.

18. Chidiac P, Ross EM. Phospholipase C-β1 directly accelerates GTP hydrolysis by Gαq and acceleration is inhibited by G$\beta\gamma$ subunits. *J Biol Chem*. 1999;274:19639–19643.

19. de Vries L, Farquhar MG. RGS proteins: more than just GAPs for heterotrimeric G proteins. *Trends Cell Biol*. 1999;9:138–144.

20. Schiff ML, Siderovski DP, Jordan JD, et al. Tyrosine-kinase-dependent recruitment of RGS12 to the N-type calcium channel. *Nature*. 2000;408:723–727.

21. Rodbell M, Krans HM, Pohl SL, Birnbaumer L. The glucagon-sensitive adenyl cyclase system in plasma membranes of rat liver: Binding of glucagon: method of assay and specificity. *J Biol Chem*. 1971;246:1861–1871.

22. Cassel D, Selinger Z. Catecholamine-stimulated GTPase activity in turkey erythrocyte membranes. *Biochim Biophys Acta*. 1976;452:538–551.

23. Burgisser E, De Lean A, Lefkowitz RJ. Reciprocal modulation of agonist and antagonist binding to muscarinic cholinergic receptor by guanine nucleotide. *Proc Natl Acad Sci U S A*. 1982;79:1732–1736.

24. Green RD. Reciprocal modulation of agonist and antagonist binding to inhibitory adenosine receptors by 5'-guanylylimidodiphosphate and monovalent cations. *J Neurosci*. 1984;4:2472–2476.

25. Westphal RS, Sanders Bush E. Reciprocal binding properties of 5-hydroxytryptamine type 2C receptor agonists and inverse agonists. *Mol Pharmacol*. 1994;46:937–942.

26. Sundaram H, Newman Tancredi A, Strange PG. Characterization of recombinant human serotonin 5HT1A receptors expressed in Chinese hamster ovary cells. [3H]spiperone discriminates between the G-protein-coupled and -uncoupled forms. *Biochem Pharmacol*. 1993;45:1003–1009.

27. Grazzini E, Guillon G, Mouillac B, Zingg HH. Inhibition of oxytocin receptor function by direct binding of progesterone. *Nature*. 1998;392:509–512.

28. Haga T, Ross EM, Anderson HJ, Gilman AG. Adenylate cyclase permanently uncoupled from hormone receptors in a novel variant of S49 mouse lymphoma cells. *Proc Natl Acad Sci U S A*. 1977;74:2016–2020.

29. Coffino P, Bourne HR, Tomkins GM. Mechanism of lymphoma cell death induced by cyclic AMP. *Am J Pathol*. 1975;81:199–204.

30. Bourne HR, Coffino P, Tomkins GM. Somatic genetic analysis of cyclic AMP action: characterization of unresponsive mutants. *J Cell Physiol*. 1975;85:611–620.

31. Bourne HR, Coffino P, Tomkins GM. Selection of a variant lymphoma cell deficient in adenylate cyclase. *Science*. 1975;187:750–752.

32. Musacchio A, Cantley LC, Harrison SC. Crystal structure of the breakpoint cluster region-homology domain from phosphoinositide 3-kinase p85 α subunit. *Proc Natl Acad Sci U S A*. 1996;93:14373–14378.

33. Ross EM, Howlett AC, Ferguson KM, Gilman AG. Reconstitution of hormone-sensitive adenylate cyclase activity with resolved components of the enzyme. *J Biol Chem*. 1978;253:6401–6412.

34. Northup JK, Sternweis PC, Smigel MD, Schleifer LS, Ross EM, Gilman AG. Purification of the regulatory component of adenylate cyclase. *Proc Natl Acad Sci U S A*. 1980;77:6516–6520.

35. Gupta SK, Gallego C, Lowndes JM, et al. Analysis of the fibroblast transformation potential of GTPase-deficient gip2 oncogenes. *Mol Cell Biol*. 1992;12:190–197.

36. Lyons J, Landis CA, Harsh G, et al. Two G protein oncogenes in human endocrine tumors. *Science*. 1990;245:655–659.

37. Malbon CM, Moxham CC. Insulin action impaired by deficiency of the G-protein subunit Giα2. *Nature*. 1996;379:840–844.

38. Sunahara RK, Dessauer CW, Gilman AG. Complexity and diversity of mammalian adenylyl cyclases. *Annu Rev Pharmacol Toxicol*. 1996;36:461–480.

39. Antonelli V, Bernasconi F, Wong YH, Vallar L. Activation of B-Raf and regulation of the mitogen-activated protein kinase pathway by the G(o)α chain. *Mol Biol Cell*. 2000;11:1129–1142.

40. Jordan JD, Carey KD, Stork PJ, Iyengar R. Modulation of rap activity by direct interaction of Gα_o with Rap1 GTPase-activating protein. *J Biol Chem*. 1999;274:21507–21510.

41. Ram PT, Horvath CM, Iyengar R. Stat3-mediated transformation of NIH-3T3 cells by the constitutively active Q205L Gαo protein. *Science*. 2000;287:142–144.

42. Jiang M, Gold MS, Boulay G, et al. Multiple neurological abnormalities in mice deficient in the G protein G$_o$. *Proc Natl Acad Sci U S A*. 1998;95:3269–3274.

43. Singer WD, Brown HA, Sternweis PC. Regulation of eukaryotic phosphatidylinositol-specific phospholipase C and phospholipase D. *Annu Rev Biochem*. 1997;66:475–509.

44. Taylor SJ, Chae HZ, Rhee SG, Exton JH. Activation of the β1 isozyme of phospholipase C by α subunits of the Gq class of G proteins. *Nature*. 1991;350:516–518.

45. Bence K, Ma W, Kozasa T, Huang XY. Direct stimulation of Bruton's tyrosine kinase by G$_q$-protein α-subunit. *Nature*. 1997;389:296–299.

46. Offermanns S, Toombs CF, Hu YH, Simon MI. Defective platelet activation in Gα(q)-deficient mice. *Nature*. 1997;389:183–186.

47. Jiang Y, Ma W, Wan Y, Kozasa T, Hattori S, Huang XY. The G protein Gα12 stimulates Bruton's tyrosine kinase and a rasGAP through a conserved PH/BM domain. *Nature*. 1998;395:808–813.

48. Hart MJ, Jiang X, Kozasa T, et al. Direct stimulation of the guanine nucleotide exchange activity of p115 RhoGEF by Gα13. *Science*. 1998;280:2112–2114.

49. Offermanns S, Mancino V, Revel JP, Simon MI. Vascular system defects and impaired cell chemokinesis as a result of Gα13 deficiency. *Science*. 1997;275:533–536.

50. Shan D, Chen L, Wang D, Tan YC, Gu JL, Huang XY. The G protein Gα(13) is required for growth factor-induced cell migration. *Dev Cell*. 2006;10:707–718.

51. Kozasa T, Jiang X, Hart MJ, et al. p115 RhoGEF, a GTPase activating protein for Gα$_{12}$ and Gα$_{13}$. *Science*. 1998;280:2111.

52. Simon MI, Strathmann MP, Gautam N. Diversity of G proteins in signal transduction. *Science*. 1991;252:802–808.

53. McKenzie FR, Kelly EC, Unson CG, Spiegel AM, Milligan G. Antibodies which recognize the C-terminus of the inhibitory guanine-nucleotide-binding protein (Gi) demonstrate that opioid peptides and foetal-calf serum stimulate the high-affinity GTPase activity of two separate pertussis-toxin substrates. *Biochem J*. 1988;249:653–659.

54. McKenzie FR, Milligan G. δ-Opioid-receptor-mediated inhibition of adenylate cyclase is transduced specifically by the guanine-nucleotide-binding protein Gi2. *Biochem J*. 1990;267:391–398.

55. Sullivan KA, Miller RT, Masters SB, Beiderman B, Heideman W, Bourne HR. Identification of receptor contact site involved in receptor-G protein coupling. *Nature*. 1987;330:758–760.

56. Berlot CH, Bourne HR. Identification of effector-activating residues of G$_s$α. *Cell*. 1992;68:911–922.

57. Iiri T, Farfel Z, Bourne HR. G-protein diseases furnish a model for the turn-on switch. *Nature*. 1998;394:35–38.

58. Coleman DE, Berghuis AM, Lee E, Linder ME, Gilman AG, Sprang SR. Structures of active conformations of Giα1 and the mechanism of GTP hydrolysis. *Science*. 1994;265:1405–1412.

59. Lambright DG, Noel JP, Hamm HE, Sigler PB. Structural determinants for activation of the α-subunit of a heterotrimeric G protein. *Nature*. 1994;369:621–628.

60. Remmers AE, Engel C, Liu M, Neubig RR. Interdomain interactions regulate GDP release from heterotrimeric G proteins. *Biochemistry*. 1999;38:13795–13800.

61. Markby DW, Onrust R, Bourne HR. Separate GTP binding and GTPase activating domains of a Gα subunit. *Science*. 1993;262:1895–1901.

62. Liu W, Northup JK. The helical domain of a G protein α subunit is a regulator of its effector. *Proc Natl Acad Sci U S A*. 1998;95:12878–12883.

63. Milburn MV, Tong L, deVos AM, et al. Molecular switch for signal transduction: structural differences between active and inactive forms of protooncogenic ras proteins. *Science*. 1990;247:939–945.

64. Prive GG, Milburn MV, Tong L, et al. X-ray crystal structures of transforming p21 ras mutants suggest a transition state stabilization mechanism for GTP hydrlysis. *Proc Natl Acad Sci U S A*. 1992;89:3649–3653.

65. de Vos AM, Tong L, Milburn MV, et al. Three dimensional structure of an oncogene protein catalytic domain of human c-H-ras p21. *Science.* 1988;239:888–893.

66. Pai EF, Krengel U, Petsko GA, Goody RS, Kabsch W, Wittinghofer A. Refined crystal structure of the triphosphate conformation of H-ras p21 at 1.35 A resolution: implications for the mechanism of GTP hydrolysis. *EMBO J.* 1990;9:2351–2359.

67. Tesmer JJ, Sunahara RK, Gilman AG, Sprang SR. Crystal structure of the catalytic domains of adenylyl cyclase in a complex with $G_s\alpha.GTP\gamma S$. *Science.* 1997;278:1907–1916.

68. Abramow-Newerly M, Roy AA, Nunn C, Chidiac P. RGS proteins have a signalling complex: interactions between RGS proteins and GPCRs, effectors, and auxiliary proteins. *Cell Signal.* 2006;18:579–591.

69. Slep KC, Kercher MA, He W, Cowan CW, Wensel TG, Sigler PB. Structural determinants for regulation of phosphodiesterase by a G protein at 2.0 Å,. *Nature.* 2001;409:1071–1077.

70. Lindorfer MA, Myung C-S, Savino Y, Yasuda H, Khazan R, Garrison JC. Differential activity of the G protein $\beta 5\gamma 2$ subunit at receptors and effectors. *J Biol Chem.* 1998;273:34429–36344.

71. Macrez-Leprêtre N, Kalkbrenner F, Morel J-L, Schultz G, Mironneau J. G protein heterotrimer $G\alpha 13\beta 1\gamma 3$ couples the angiotensin AT1A receptor to increases in cytoplasmic Ca^{2+} in rat portal vein myocytes. *J Biol Chem.* 1997;272:10095–10102.

72. Rhee SG, Bae YS. Regulation of phosphoinositide specific phospholipase C isozymes. *J Biol Chem.* 1997;272:15045–15048.

73. Rebois R, Hébert TE. Protein complexes involved in heptahelical receptor-mediated signal transduction. *Recept Chann.* 2003;9:169–194.

74. Makino ER, Handy JW, Li T, Arshavsky VY. The GTPase activating factor for transducin in rod photoreceptors is the complex between RGS9 and type 5 G protein β subunit. *Proc Natl Acad Sci U S A.* 1999;96: 1947–1952.

75. Snow BE, Krumins AM, Brothers GM, et al. A G protein γ subunit-like domain shared between RGS11 and other RGS proteins specifies binding to $G\beta_5$ subunits. *Proc Natl Acad Sci U S A.* 1998;95:13307–13312.

76. Chen CK, Burns ME, He W, Wensel TG, Baylor DA, Simon MI. Slowed recovery of rod photoresponse in mice lacking the GTPase accelerating protein RGS9-1. *Nature.* 2000;403:557–560.

77. Sondek J, Siderovski DP. $G\gamma$-like (GGL) domains: new frontiers in G-protein signaling and β-propeller scaffolding. *Biochem Pharmacol.* 2001;61:1329–1337.

78. Logothetis DE, Kurachi Y, Galper J, Neer EJ, Clapham DE. Purified subunits of GTP-binding proteins regulate muscarinic K^+ channel activity in heart. *Nature.* 1987;325:321–326.

79. Pfaffinger PJ, Martin JM, Hunter DD, Nathanson NM, Hille B. GTP-binding proteins couple cardiac muscarinic receptors to a K channel. *Nature*. 1985;317:536–538.

80. Wickman K, Clapham DE. Ion-channel regulation by G-proteins. *Physiol Rev*. 1995;75:865–885.

81. Neer EJ, Clapham DE. Roles of G protein subunits in transmembrane signalling. *Nature*. 1988;333:129–134.

82. Reuveny E, Slesinger PA, Inglese J, et al. Activation of the cloned muscarinic potassium channel by G protein $\beta\gamma$ subunits. *Nature*. 1994;370:143–146.

83. Jeselma CL, Axelrod A. Stimulation of phospholipase A_2 activity in bovine rod outer segments by the $\beta\gamma$ subunits of transducin and its inhibition by the α-subunit. *J Biol Chem*. 1987;84:3623–3627.

84. Camps M, Hou C, Sidiropoulos D, Stock JB, Jakobs KH. P. Gierschik, Stimulation of phospholipase C by guanine-nucleotide-binding protein $\beta\gamma$ subunits,. *Eur J Biochem*. 1992;206:821–831.

85. Corey S, Clapham DE. The stoichiometry of $G\beta\gamma$ binding to G-protein-regulated inwardly rectifying K^+ channels (GIRKs). *J Biol Chem*. 2001;276:11409–11413.

86. Tesmer VM, Kawano T, Shankaranarayanan A, Kozasa T, Tesmer JJ. Snapshot of activated G proteins at the membrane: the $G\alpha q$-GRK2-$G\beta\gamma$ complex. *Science*. 2005;310:1686–1690.

87. Daaka Y, Luttrell LM, Lefkowitz RJ. Switching of the coupling of the β_2-adrenergic receptor to different G proteins by protein kinase A. *Nature*. 1999;390:88–91.

88. Menard L, Ferguson SS, Barak LS, et al. Members of the G protein-coupled receptor kinase family that phosphorylate the β_2-adrenergic receptor facilitate sequestration. *Biochemistry*. 1996;35:4155–4160.

89. Goodman OB, Krupnick JG, Santini F, et al. β-Arrestin acts as a clathrin adaptor in endocytosis of the β_2-adrenergic receptor. *Nature*. 1996;383:447–450.

90. Heuss H, Gerber U. G-protein-independent signaling by G-protein-coupled receptors. *Trends Neurosci*. 2000;23:469–475.

91. Scher CD, Scolnick EM, Siegler R. Induction of erythroid leukaemia by Harvey and Kirsten sarcoma viruses. *Nature*. 1975;256:225–226.

92. Zheng B, de Vries L, Farquhar MG. Divergence of RGS proteins: Evidence for the existence of six mammalian RGS subfamilies. *Trends Biochem Sci*. 2001;24:411–414.

93. Harvey JJ. An unidentified virus which causes the rapid production of tumours in mice. *Nature*. 1964;204:1104–1105.

94. Kirsten WH, Carter RE, Pierce MI. Studies on the relationship of viral infections to leukemia in mice: The accelerating agent in AKR mice. *Cancer*. 1962;15:750–758.

95. Shimizu K, Goldfarb M, Suard Y, et al. Three human transforming genes are related to the viral ras oncogenes. *Proc Natl Acad Sci U S A*. 1983;80:2112–2116.

96. Strathmann MP, Simon MI. Gα12 and Gα13 subunits define a fourth class of G protein. *Proc Natl Acad Sci U S A*. 1991;88:5582–5586.

97. Shih TY, Papageorge AG, Stokes PE, Weeks MO, Scolnick EM. Guanine nucleotide-binding and autophosphorylating activities associated with the p21src protein of Harvey murine sarcoma virus. *Nature*. 1980;287:686–691.

98. Gibbs JB, Sigal IS, Poe M, Scolnick EM. Intrinsic GTPase activity distinguishes normal and oncogenic ras p21 molecules. *Proc Natl Acad Sci U S A*. 1984;81:5704–5708.

99. Parada LF, Tabin CJ, Shih C, Weinberg RA. Human EJ bladder carcinoma oncogene is homologue of Harvey sarcoma virus *ras* gene. *Nature*. 1982;297:474–478.

100. Valencia A, Chardin P, Wittinghofer A, Sander C. The ras protein family: evolutionary tree and role of conserved amino acids. *Biochemistry*. 1991;30:4637–4648.

101. Broek D, Toda T, Michaeli T, et al. The *S. cerevisiae* CDC25 gene product regulates the RAS/adenylate cyclase pathway. *Cell*. 1987;48:789–799.

102. DeFeo-Jones D, Tatchell K, Robinson LC, et al. Mammalian and yeast *ras* gene products: biological function in their heterologous systems. *Science*. 1985;228:179–184.

103. Kataoka T, Powers S, Cameron S, et al. Functional homology of mammalian and yeast *RAS* genes. *Cell*. 1985;40:19–26.

104. Stacey DW, Marshall MS, Gibbs JB, Feig LA. Preferential inhibition of the oncogenic form of RasH by mutations in the GAP binding/'effector' domain. *Cell*. 1991;64:625–633.

105. Segal M, Willumsen BM, Levitzki A. Residues crucial for Ras interaction with GDP-GTP exchangers. *Proc Natl Acad Sci U S A*. 1993;90:5564–5568.

106. Brunger AT, Milburn MV, Tong L, et al. Crystal structure of an active form of RAS protein, a complex of a GTP analog and the HRAS p21 catalytic domain. *Proc Natl Acad Sci U S A*. 1990;87:4849–4853.

107. Gibbs JB. Ras C-terminal processing enzymes: New drug targets?. *Cell* 1991;65:1–4.

108. Reuveni H, Geiger T, Geiger B, Levitzki A. Reversal of the Ras-induced transformed phenotype by HR12, a novel ras farnesylation inhibitor, is mediated by the Mek/Erk pathway. *J Cell Biol*. 2000;151:1179–1192.

109. Perucho M, Goldfarb M, Shimizu K, Lama C, Fogh J, Wigler M. Human-tumor-derived cell lines contain common and different transforming genes. *Cell*. 1981;27:467–476.

110. Seeburg PH, Colby WW, Capon DJ, Goeddel DV, Levinson AD. Biological properties of human c-Ha-ras1 genes mutated at codon 12. *Nature*. 1984;312:71–75.

111. Feramisco JR, Gross M, Kamata T, Rosenberg M, Sweet RW. Microinjection of the oncogene form of the human H-ras (T-24) protein results in rapid proliferation of quiescent cells. *Cell*. 1984;38:109–117.

112. Stacey DW, Kung H-F. Transformation of NIH 3T3 cells by microinjection of Ha-ras p21 protein. *Naturem*. 1984;310:508–511.

113. Mulcahy LS, Smith LR, Stacey DW. Requirement for *ras* proto-oncogene function during serum-stimulated growth of NIH 3T3 cells. *Nature*. 1985;313:241–243.

114. Kung H-F, Smith MR, Bekesi E, Manne V, Stacey DW. Reversal of transformed phenotype by monoclonal antibodies against Ha-ras p21 proteins. *Exp Cell Res*. 1986;162:363–371.

115. Stacey DW, Feig LA, Gibbs JB. Dominant inhibitory Ras mutants selectively inhibit the activity of either cellular or oncogenic Ras. *Mol Cell Biol*. 1991;11:4053–4064.

116. Bar-Sagi D, Feramisco JR. Microinjection of the ras oncogene protein into PC12 cells induces morphological differentiation. *Cell*. 1985;42:841–848.

117. Altin JG, Wetts R, Bradshaw RA. Microinjection of a p21ras antibody into PC12 cells inhibits neurite outgrowth induced by nerve growth factor and basic fibroblast growth factor. *Growth Factors*. 1991;4:145–155.

118. Trahey M, McCormick F. A cytoplasmic protein stimulates normal N-ras p21 GTPase, but does not affect oncogenic mutants. *Science*. 1987;238:542–545.

119. Gibbs JB, Marshall MS, Scolnick EM, Dixon RA, Vogel US. Modulation of guanine nucleotides bound to Ras in NIH3T3 cells by oncogenes, growth factors, and the GTPase activating protein (GAP). *J Biol Chem*. 1990;265:20437–20442.

120. Nori M, Vogel US, Gibbs JB, Weber MJ. Inhibition of v-src-induced transformation by a GTPase-activating protein. *Mol Cell Biol*. 1991;11:2812–2818.

121. al-Alawi N, Xu G, White R, Clark R, McCormick F, Feramisco JR. Differential regulation of cellular activities by GTPase-activating protein and NF1. *Mol Cell Biol*. 1993;13:2497–2503.

122. Downward J, Graves JD, Warne PH, Rayter S, Cantrell DA. Stimulation of p21ras upon T-cell activation. *Nature*. 1990;346:719–723.

123. Lazarus AH, Kawauchi K, Rapoport MJ, Delovitch TL. Antigen-induced B lymphocyte activation involves the p21ras and ras, GAP signaling pathway. *J Exp Med*. 1993;178:1765–1769.

124. DePaolo D, Reusch JE, Carel K, Bhuripanyo P, Leitner JW, Draznin B. Functional interactions of phosphatidylinositol 3-kinase with GTPase-activating protein in 3T3-L1 adipocytes. *Mol Cell Biol*. 1996;16:1450–1457.

125. Lowy DR, Willumsen BM. Function and regulation of Ras. *Annu Rev Biochem*. 1993;62:851–891.

126. Haubruck H, McCormick F. Ras p21: Effects and regulation. *Biochim Biophys Acta*. 1991;72:215–229.

127. Musacchio A, Gibson T, Rice P, Thompson J, Saraste M. The PH domain: a common piece in the structural patchwork of signalling proteins. *Trends Biochem Sci*. 1993;18:343–348.

128. Weissbach L, Settleman J, Kalady MF, et al. Identification of a human rasGAP-related protein containing calmodulin-binding motifs. *J Biol Chem*. 1994;269:20517–20521.

129. Martin GA, Yataani A, Clark R, et al. GAP domains responsible for ras p21-dependent inhibition of muscarinic atrial K+ channel currents. *Science*. 1992;255:192–194.

130. McCormick F. ras GTPase activating protein: signal transmitter and signal terminator. *Cell*. 1989;13:5–8.

131. Medema RH, de Laat WL, Martin GA, McCormick F, Boss JL. GTPase-activating protein SH2-SH3 domains induce gene expression in a Ras dependent fashion. *Mol Cell Biol*. 1992;12:3425–3430.

132. McGlade J, Brunkhorst B, Anderson D, et al. The N-terminal region of GAP regulates cytoskeletal structure and cell-adhesion. *EMBO J*. 1993;12:3073–3081.

133. Scheffzek K, Ahmadian MR, Wittinghofer A. GTPase-activating proteins: Helping hands to complement an active site. *Trends Biochem Sci*. 1999;23:257–262.

134. Toda T, Uno I, Ishikawa T, et al. In yeast, RAS proteins are controlling elements of adenylate cyclase. *Cell*. 1985;40:27–36.

135. Robinson LC, Gibbs JB, Marshall MS, Sigal IS, Tatchell K. CDC25: a component of the RAS-adenylate cyclase pathway in *Saccharomyces cerevisiae*. *Science*. 1987;235:1218–1221.

136. Daniel J, Becker JM, Enari E, Levitzki A. The activation of adenylate cyclase by guanyl nucleotides in *Saccharomyces cerevisiae* is controlled by the CDC25 start gene product. *Mol Cell Biol*. 1987;7:3857–3861.

137. Chardin P, Camonis JH, Gale NW, et al. Human Sos1: a guanine nucleotide exchange factor for Ras that binds to GRB2. *Science*. 1993;260:1338–1343.

138. Lambert NA. Dissociation of heterotrimeric G proteins in cells. *Sci Signal*. 2008;1:re5.

139. Fung BK. Characterization of transducin from bovine retinal rod outer segments. I. Separation and reconstitution of the subunits. *J Biol Chem*. 1983;258:10495–10502.

140. Panchenko MP, Saxena K, Li Y, et al. Sites important for PLCβ_2 activation of the G protein subunit map to the sides of the propeller structure. *J Biol Chem*. 1998;273:28298–28304.

141. Marbach I, Bar-Sinai A, Minich M, Levitzki A. β subunit copurifies with GppNHp-activated adenylyl cyclase. *J Biol Chem*. 1990;265:9999–10004.

142. Bar-Sinai A, Marbach I, Shorr RG, Levitzki A. The GppNHp-activated adenylyl cyclase complex from turkey erythrocyte membranes can be isolated with its $\beta\gamma$ subunits. *Eur J Biochem*. 1992;207:703–708.

143. Hunt LT, Dayhoff MO. Structural and functional similarities among hormones and active peptides from distantly related eukaryotes. In: Gross E, Meienhofer J, eds. *Peptides: Structure and Biological Function*. Rockford, Il: Pierce Chemical Co; 1979:757–760.

144. Bunemann M, Frank M, Lohse MJ. Gi protein activation in intact cells involves subunit rearrangement rather than dissociation. *Proc Natl Acad Sci U S A*. 2003;100:16077–16082.

145. Janetopoulos C, Jin T, Devreotes P. Receptor-mediated activation of heterotrimeric G-proteins in living cells. *Science*. 2001;291:2408–2411.

146. Hein P, Frank M, Hoffmann C, Lohse MJ, Bunemann M. Dynamics of receptor/G protein coupling in living cells. *EMBO J*. 2005;24:4106–4114.

147. Hartman JL, Northup JK. Functional reconstitution in situ of 5-hydroxytryptamine2c (5HT2c) receptors with αq and inverse agonism of 5HT2c receptor antagonists. *J Biol Chem*. 1996;271:22597.

148. Lachance M, Ethier N, Wolbring G, Schnetkamp PPM, Hebert TE. Stable association of G proteins with β_2AR is independent of the state of receptor activation. *Cell Signal*. 1999;11:523–533.

149. Brown PJ, Schonbrunn A. Affinity purification of a somatostatin receptor-G-protein complex demonstrates specificity in receptor-G-protein coupling. *J Biol Chem*. 1993;268:6668–6676.

150. Law SF, Reisine T. Changes in the association of G protein subunits with the cloned mouse δ opioid receptor on agonist stimulation. *J Pharmacol Exp Ther*. 1997;281:1476–1486.

151. Law SF, Reisine T. Agonist binding to rat brain somatostatin receptors alters the interaction of the receptors with guanine nucleotide-binding regulatory proteins. *Mol Pharmacol*. 1992;42:398–402.

152. Medici R, Bianchi E, Di Segni G, Tocchini-Valentini GP. Efficient signal transduction by a chimeric yeast-mammalian G protein α subunit Gpa1-Gsα covalently fused to the yeast receptor Ste2. *EMBO J*. 1997;16:7241–7249.

CHAPTER 5

Effector Enzymes Coupled to GTP Binding Proteins: Adenylyl Cyclase and Phospholipase C

Adenylyl cyclase

Cyclic AMP: the first second messenger

Looking back to the time when most wisdom in mammalian biochemistry was derived from experiments using rat liver (perfusions, slices, homogenates, etc.), one gets a sense of the happy circumstance that led Sutherland to use dog liver in his investigations of hormonal activation of glycogenolysis[1] (see page 34). During the growth and maturation of rats (6–60 days), the expression of β-adrenergic receptors in the liver declines while that of the α_1-adrenergic receptors increases.[2] Adrenergic stimulation of glycogen breakdown in the liver of adult rats is therefore mediated primarily through the α-receptors, which induce an elevation in the concentration of cytosol Ca^{2+} (see Chapter 7). Using dog, Sutherland showed in 1957 that the breakdown of glycogen, in response to adrenaline or glucagon, is consequent on the generation of a heat-stable and dialysable factor.[1] This factor proved to be cyclic AMP, the first second messenger to be identified. The basic reactions of glycogen synthesis and breakdown are summarized in Figure 5.1.

Although cyclic AMP (henceforth cAMP) is most famously the intracellular signal for the breakdown of liver glycogen (and correspondingly for the shut-down of glycogen synthesis), it is a key mediator for many other important receptor responses. We return to these questions in Chapter 9. cAMP is also present in both the Eubacteria and Archaea, the other fundamental domains of life.[3,4] Although its functions as a regulator in these organisms are very

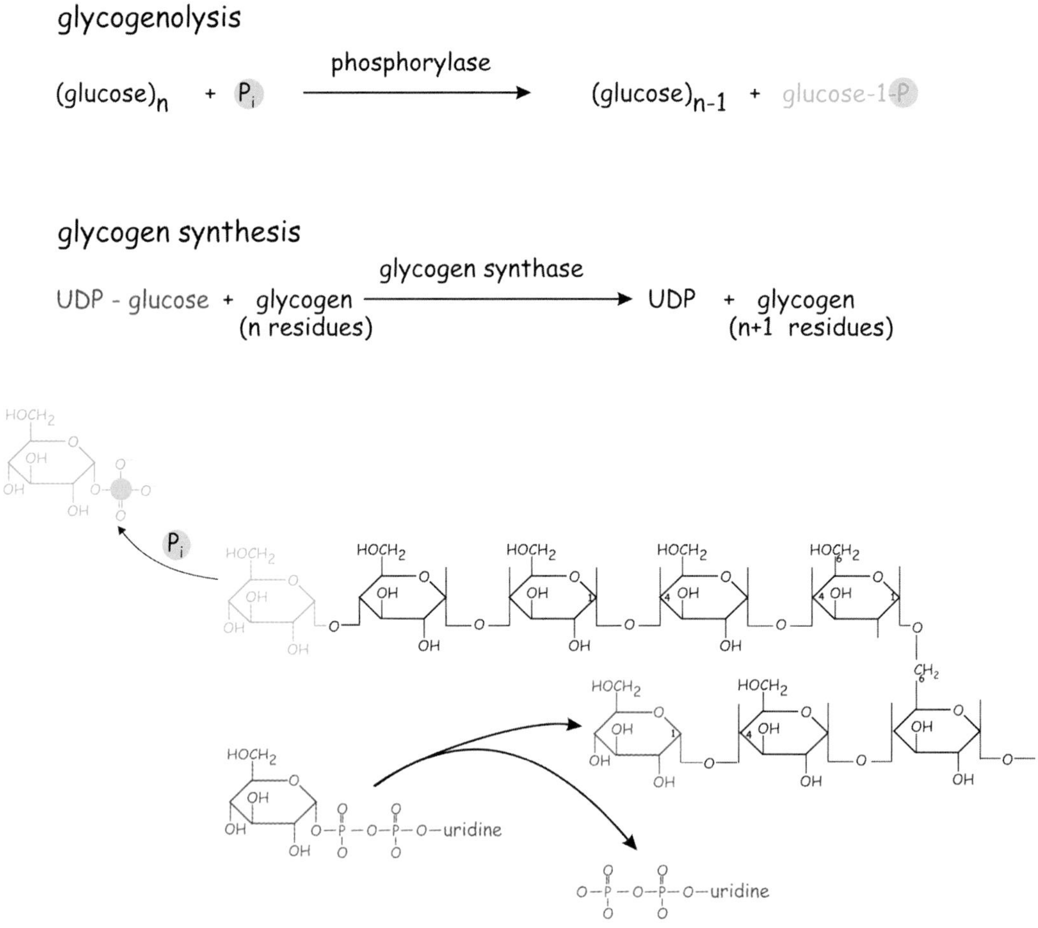

glycogenolysis

$$(\text{glucose})_n \;+\; P_i \;\xrightarrow{\;\text{phosphorylase}\;}\; (\text{glucose})_{n-1} \;+\; \text{glucose-1-}P$$

glycogen synthesis

$$\text{UDP - glucose} \;+\; \underset{(n\ \text{residues})}{\text{glycogen}} \;\xrightarrow{\;\text{glycogen synthase}\;}\; \text{UDP} \;+\; \underset{(n+1\ \text{residues})}{\text{glycogen}}$$

FIG 5.1 The basic reactions of glycogen synthesis and breakdown.
Both the incoming (glycogen synthesis) and leaving (glycogenolysis) residues are attached to the non-reducing termini of the formed glycogen.

different from those in eukaryotes, its generation has signified a response to starvation throughout evolution. cAMP is synthesized in *E. coli* starved of glucose and it induces expression of sugar-metabolizing enzymes such as β-galactosidase. In this case, it interacts with promoter elements in the vicinity of the RNA polymerase site and is directly involved in the regulation of gene transcription. In yeast (unicellular, eukaryotic), cAMP is a growth signal. The presence of nutrients signals the activation of adenylyl cyclase through the RAS proteins. In starvation conditions, the ratio of GTP- to GDP-bound RAS declines, either because of down-regulation of CDC25 (the guanine nucleotide exchange factor) or because of activation of IRA (GTPase activating protein). As a result, the cellular content of cAMP declines, growth ceases, and the cells switch to meiosis and spore formation. In the slime mould *Dictyostelium discoides* under conditions of starvation, cAMP is secreted by the individual cells as a signal for assemblage into a slug. Although it has generally been hard to discern functions for cAMP in plants,[5] there is now good evidence that it acts as a regulator of ion channels for both K^+ and Ca^{2+}. The K^+ channels in the guard cells of *Vicia faba* (broad bean) leaves appear to be regulated by a cAMP-dependent kinase (protein kinase A, PKA[6,7]) and are indirectly sensitive to the actions of cholera and pertussis toxins that modulate the activity of GTP-binding proteins (see below).

cAMP is formed from ATP

cAMP (strictly, 3′,5′-cyclic adenosine monophosphate) is synthesized from ATP which is present in all cells at high concentration (5–10 mmol L^{-1}). The formation of cAMP is catalysed by adenylyl cyclases and its removal, by conversion to 5′-AMP (adenylate), is catalysed by the action of phosphodiesterases (cAMP-PDE) (Figure 5.2).

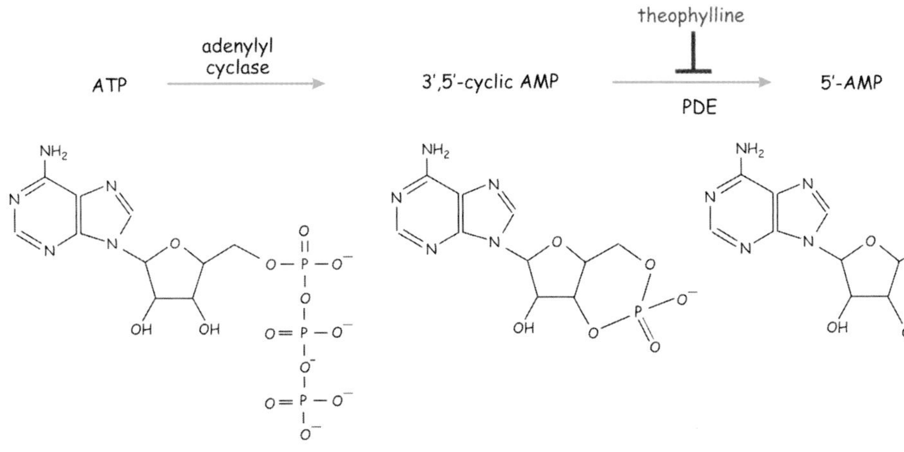

FIGURE 5.2 Cyclic AMP is generated from ATP.

The preparation of adenyl cyclase in a simplified and purified form has been hampered by the association with particulate material in the 'nuclear' fraction, by its lability and by its close association with the detergent Triton after solubilization. It is hoped that the enzyme may be solubilized by other procedures and that more stable soluble preparations can be achieved.

Earl Sutherland, 1962

It is simple now to measure the level of cAMP with the use of kits, binding proteins, and specific antibodies. Not so simple for Sutherland and his colleagues who measured the product cAMP using a two-stage assay procedure. The first involved incubation together with ATP, phosphorylase b, and fractions of liver homogenate (containing phosphorylase kinase and PKA). This was followed up by measurement of the activity of the active product phosphorylase a on its substrate glycogen.[8,9]

Adenylyl cyclase and its regulation

The enzyme

Adenylyl cyclase remained an activity without a proper molecular description for about 30 years, until 1990. At first, the possibility was entertained that more than one enzyme might be involved in the cyclization reaction. It was the importance of adenylyl cyclase in metabolic regulation and the simplicity of the reaction it catalyses, that provoked the intense investigation of its properties.

A number of proteins and other agents can activate or inhibit adenylyl cyclase. The known physiological activators and inhibitors include the subunits of the heterotrimeric G proteins. α_s and $\beta\gamma$-subunits can activate, while α_i, α_o, and $\beta\gamma$-subunits can inhibit. Also, Ca^{2+}, through various mechanisms, can activate or inhibit. The plant product forskolin (described below) is a potent activator and this poses the interesting question of whether there might exist an endogenous counterpart in animal cells.

A major problem for research in this area was the fact that adenylyl cyclase, an integral membrane protein, generally comprises no more than 0.01–0.001% of total membrane protein. (An exception is the olfactory epithelium in which it comprises more than 0.1%.) The poor solubility of membrane proteins in aqueous media and their dependence upon the phospholipid environment for the maintenance of their tertiary structure makes them extremely difficult to isolate, purify and reconstitute into a form upon which sensible measurements can be made. Nevertheless the G_s-coupled form of adenylyl cyclase, present in many tissues, and a brain form of the enzyme have been purified.

From these preparations it became possible to obtain sequence information and thence a cDNA encoding the full length of adenylyl cyclase 1. The recombinant protein catalyses the conversion of ATP to cAMP and it can be activated by α_s.GTP and by Ca^{2+}-calmodulin. Altogether, nine membrane-bound isoforms of mammalian adenylyl cyclase (AC1 to 9) have been cloned. (There is also a soluble isoform that is restricted to testis).

Structural organization of adenylyl cyclases

The structure reveals a number of distinct domains (Figure 5.3). Starting from the N-terminus, there is a predominantly hydrophobic domain, M1, organized as six membrane-spanning α-helices linked on alternate sides of the membrane by short hydrophilic loops. Next, there is an extensive (40 kDa) domain, C_1, then another transmembrane domain M2, similar to M1, and finally at the C-terminus a second extensive cytoplasmic domain, C_2 similar to C_1.[10] Beyond the general conservation of topology, the sequences of all forms of adenylyl cyclase are similar, having 40–60% identity overall. Within the C_1 and C_2 domains are the subdomains C_{1a} and C_{2a} that share 50% or even

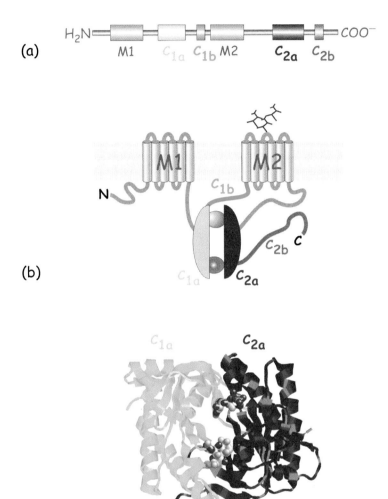

(a)

(b)

(c)

FIG 5.3 Organization of adenylyl cyclases, depicting the main structural features of the mammalian adenylyl cyclases. (a) Domain architecture of adenylyl cyclase (b) General topology of the enzyme. With an even number of membrane spanning α-helices, six each in the domains M1 and M2, the N- and C- termini are both in the cytosol. The catalytic unit comprises the two domains C_{1a} and C_{2a}. These contact each other to form binding sites for ATP, forskolin, and the α-subunits of G_s and G_i. (c) The molecular structure of the C_{1a} and C_{2a} dimeric cluster containing ATP⌒S (upper spheres) and forskolin (lower spheres).

higher similarity. Residues from both C_1 and C_2 contribute to ATP-binding and catalysis.

Organized in head-to-tail fashion and having extensive contacts with each other, C_{1a} and C_{2a} comprise the catalytic centre of the enzyme.[11] The interface between C_{1a} and C_{2a} accommodates two highly conserved nucleotide binding sites, only one of which appears to be crucial for catalysis. The other is the site at which the activating substance forskolin binds. The binding site for α_s is distant (some 3 nm) from the catalytic site and has been localized to a small region at the N-terminus of C_{1a} and to a larger region composed of a negatively charged surface and a hydrophobic groove on C_{2a}. The binding sites for the α-subunits of G_i and G_o do not compete with α_s. They too are formed by parts of both the C_{1a} and C_{2a} subdomains, but probably in a location much closer to the catalytic centre. Although nucleotides, forskolin, and α_s can bind to residues on both C_{1a} and C_{2a}, it appears that $\beta\gamma$-subunits potentiate the activity of AC2 by interacting only with the C_{2a} subdomain. This causes a conformational change that indirectly enables it to modulate the conformation of the C_{2a} and so indirectly promotes optimal alignment at the catalytic site. The inhibitory actions of $\beta\gamma$-subunits on AC1 must occur elsewhere since the sequences are not conserved in this region of the two subtypes.

The C_1 and C_2 domains, shorn of their extensive membrane attachments, are together sufficient to catalyse the conversion of ATP to cAMP.[12,13] Although the activity is very low, the association between them is enhanced and the rate of conversion is increased more than 100-fold by forskolin or the presence of α_s.[14] These two activators are synergistic, the presence of the one enhancing the affinity of the other.

Although the M1 and M2 domains clearly anchor the enzyme to the membrane, one would imagine that one or two transmembrane spans would suffice for this purpose and it is reasonable to ask why these enzymes are endowed with 12. The answer probably lies in their evolutionary ancestry. It is certainly intriguing that the general membrane topology of the mammalian cyclases, particularly the two groups of six membrane-spanning segments, bears a strong resemblance to known pore-forming structures such as the cystic fibrosis transmembrane conductance regulator (CFTR) and the P-glycoproteins. However, no ion channel-like activity has been detected for any of the mammalian adenylyl cyclases, nor do they exhibit any sequence similarity with known channels.

Sequence similarities certainly extend to the guanylyl cyclases and to some non-mammalian forms of the enzyme expressed in lower organisms, such as the slime mould *Dictyostelium*. In general, a common ancestry can be inferred. More distantly, the yeast and bacterial (*E. coli*) enzymes are membrane-attached, but are not true intrinsic membrane proteins. Sequence similarities between these and the mammalian cyclases have been hard to discern.

Regulation of adenylyl cyclase

With the emergence of the metazoans, the regulation of cyclase came under the control of heterotrimeric G proteins. In the few systems in which it has been possible to examine stoichiometry of the components leading to the synthesis of cAMP, it appears that α_s is present in enormous excess over both its controlling receptors and the cyclase. Thus, for rat ventricular myocytes, the level of effector enzyme is three times that of the receptor, while the level of $G_s\alpha$ is in great excess over either (some 200-fold).[16] Because the level of cyclase is so low, the signal amplification that occurs with the activation of the

---

The human multidrug-resistant **P-glycoprotein**, encoded by the *MDR1* gene, is a member of the family of ABC (ATP-binding cassette) transporters. Over-expression causes resistance of human cancers to chemotherapy.

CFTR and P-glycoproteins have the signature of an ATP-driven transport protein, also known as pumps. This may explain why adenylyl cyclase has maintained an ATP binding site. Strangely enough, CFTR behaves as a chloride ion channel (for reasons not understood), but P-glycoproteins behave as real pumps, even to the point of operating with a kind of piston-like mechanism with two cylinders.[15]

While G_s originally denoted a stimulatory activity and G_i denoted inhibition with respect to cyclase, these G proteins also regulate many other systems.

G protein is not maintained. This sets an upper limit to the rate of synthesis of cAMP.[17]

All isoforms of mammalian adenylyl cyclase are activated by the α-subunit of G_s and all are inhibited by the so-called P-site inhibitors (analogues of adenosine such as 2'-deoxy-3'-AMP). Beyond this, their responses to the many and various inhibitory ligands vary widely. With the exception of AC9, all are activated by forskolin which acts in synergy with α_s. They are influenced, both positively and negatively, by the α- and $\beta\gamma$-subunits of other G proteins, by Ca^{2+}, and through phosphorylation by PKA and PKC (see Table 5.1). It appears that their evolutionary development has determined their specialized sensitivities towards the various regulatory influences.

Broadly, they can be grouped in three main classes:

- AC2, AC4, and AC7, which are activated synergistically by α_s- and $\beta\gamma$-subunits
- AC5 and AC6, which are inhibited by α_i and Ca^{2+}
- AC1 and AC8, which are activated synergistically by α_s together with Ca^{2+}-calmodulin (AC3 has been included in this class, but its activation by Ca^{2+} *in vivo* is uncertain. It is inhibited by Ca^{2+}-calmodulin kinase II).

So far, there is limited information concerning the precise roles of these individual isoforms.[18] A lack of good antibodies has prevented attempts to determine their subcellular localization and only for AC1, AC3, and AC8 do we have information from knock-outs. This is a tricky field in which to venture, as illustrated by RNA_i knock-down experiments that have revealed a remarkable degree of interdependence in the expression of G-protein subunits and AC isoforms, so that elimination of one component may cause compensatory expression or increased activity of others. For example, knock-down of $G\beta_1$ and $G\beta_2$ increased the expression of $G\beta_4$. It also increased the expression of some AC isoforms and reduced that of $G\alpha_i$. The overall result was an unexpected increase rather than a decrease in cyclase activity.[19] It is noteworthy that no human disease has so far been linked to mutations in the adenylyl cyclases.

Regulation by GTP binding proteins

Since the G_i proteins are generally expressed at a level greatly exceeding that of G_s, stimulation of receptors linked to G_i can generate considerable quantities of activated $\beta\gamma$-subunits. It has been proposed that this excess of $\beta\gamma$-subunits should be able to sequester free α_s-subunits, opposing their ability to activate cyclase:

$$\beta\gamma + \alpha_s \rightarrow \alpha_s\beta\gamma$$

TABLE 5.1 Regulatory signals

Isoform	Distribution	βγ	α_i	α_0	α_z	PKA	PKC	Forskolin	Ca^{2+}–CaM*	Ca^{2+} (CaM indep)
AC2	brain, lung	↑	n/e	n/e			↑	↑	n/e	
AC4	widely distributed:	↑					↑	↑		
AC7	widely distributed	↑					↑	↑	n/e	
AC5	heart>brain	n/e	↓	n/e	↓	↓	↑	↑	n/e	→
AC6	heart, brain>other tissues	n/e	↓	n/e	↓	↓	↓	↑	n/e	→
AC8	neural tissue	↓					n/e	↑	↑	
AC1	neural tissue	↓	↓	↓				↑	↑	
AC3	predominantly olfactory	n/e						↑	↑	

CaM, calmodulin; n/e, no effect.

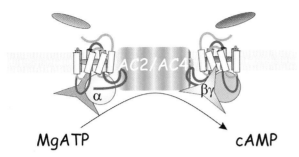

MgATP   cAMP

FIG 5.4 Adenylyl cyclase as a coincidence detector, integrating the signals from two G proteins.

For this to operate as an inhibitory mechanism, it would be necessary for the subunits to exist as independent entities, but as we have argued, it is unlikely that α_s ever separates from $\beta\gamma$ (see page 111). However, AC1 and AC8 are inhibited by $\beta\gamma$-subunits, but here the effect is primarily directed through interaction with the enzyme, not by sequestration of α-subunits. On the other hand, $\beta\gamma$-subunits synergize with α_s to activate AC2 and AC4 cyclases. As one might expect, there is little if any effect of the $\beta\gamma$-subunits in the absence of α_s. The level of $\beta\gamma$-subunits required is much greater than could possibly be derived from G_s alone, but it is fully commensurate with the amounts that could be generated through the activation of G_i or G_o. This implies that two sets of hormone receptors must be activated simultaneously in order to ensure maximal stimulation (Figure 5.4). It is possible that the dimeric structure of many receptors (including the β-adrenergic receptor; see page 69) enables them to operate as a coincidence detectors, inducing a powerful response when the two signals are generated, but only a weak or insignificant response to unsynchronized signals. This is the likely explanation for the costimulatory effect of noradrenaline and adenosine on the cyclase activity in brain cortical slices,[20] mentioned in Chapter 3 (see page 70). The presence of particular subtypes of adenylyl cyclase could then determine distinct patterns of response due to the signals from different classes of receptors. Figure 5.5 presents several possible schemes that explain how the converging pathways might interact to cause activation or inhibition. This could be of particular importance in synapses, allowing them to coordinate their responses to incoming stimuli. One of the signals would be transmitted by α_s, the other would be coupled to the more prolific G_i or G_o that would generate the required amounts of $\beta\gamma$-subunits.

So far, with the exception of $\beta_1\gamma_1$ which is located exclusively in the retina, it has been hard to establish any specific functional correlates of the diverse $\beta\gamma$-pairs. Combinations of β_1 or β_2 with any of the four γ-isoforms γ_2, γ_3, γ_5, or γ_7, appear to inhibit AC1 and to activate AC2 with equal effect. They also activate the β-isoforms of phospholipase C (see below). This lack of specificity is surprising and could be an artefact, arising from the high concentrations of $\beta\gamma$-subunits which must be provided in reconstitution experiments.

Alternatively, the lack of specificity could be real. Though never demonstrated, $\beta\gamma$-subunits might be subject to a transient covalent modification following interaction with activated receptors (ADP-ribosylation: see below). This could allow $\beta\gamma$ dimers to impose their effects on downstream effectors at much lower concentrations.[21] Modified $\beta\gamma$-subunits could then, in principle, remain active beyond the time that the α-subunits have reverted to their GDP state. It might also allow for specificity among the different $\beta\gamma$ pairs, depending on their association with particular α-subunits and particular classes of receptors.

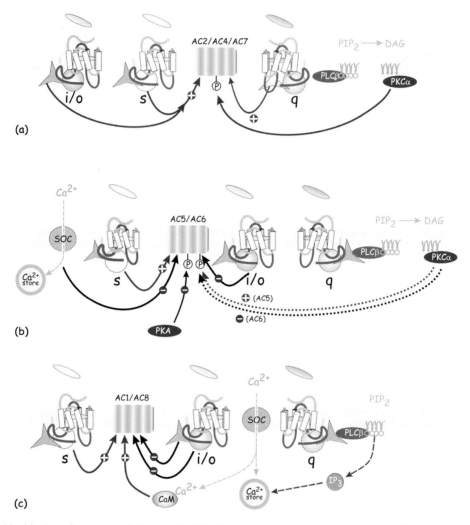

FIG 5.5 Adenylyl cyclase isoforms: patterns of activation and inhibition. The adenylyl cyclases are subject to regulation by multiple signals. These may cooperate or oppose each other. (a) In some instances the activation of AC2, 4, and 7 by α_s together with $\beta\gamma$-subunits occurs with a high level of synergy. Activation is only achieved when two distinct classes of receptor are activated simultaneously. On the other hand, the phosphorylation of these same enzymes by PKC is likely to be of longer duration, so that the enzyme is set up in readiness for stimulation by α_s. (b) AC5 and AC6 are subject to indirect feedback inhibition by phosphorylation of the enzyme by PKA. They are also inhibited by Ca^{2+} and by receptors linked through the G_i proteins. (c) AC1 and AC8 are activated by Ca^{2+}-calmodulin and inhibited by $\beta\gamma$-subunits. (SOC indicates a store-operated Ca^{2+} channel.)

Clearly, there are several routes to the activation of adenylyl cyclase and the elevation of cAMP, but equally important are those influences which inhibit its activity. There are many receptors (for example, the α_2-adrenergic receptors) that are linked to the group of G proteins that have α_i-subunits. These inhibit AC5 and AC6.

Regulation by Ca^{2+}

Ca^{2+} is also an important intracellular second messenger and an important regulator of adenylyl cyclases.[22] The signalling mechanisms that cause an elevation of intracellular Ca^{2+} are considered in detail in Chapters 7 and 8. Briefly, cytosol $[Ca^{2+}]$ increases when the ion enters cells through plasma membrane channels, such as voltage-activated Ca^{2+} channels. Alternatively, it may be released into the cytosol from internal stores when phospholipase C (PLC) is activated by a receptor mechanism. For example, receptors that are coupled to G_q regulate PLCβ. In association with the protein calmodulin (see Chapter 8), Ca^{2+} increases the activity of AC1 and AC8. A binding site for calmodulin is present on the C1b sub-domain of AC1 and there is also a candidate site on AC8. In contrast, AC5 and AC6 are both inhibited by Ca^{2+} in a calmodulin-independent manner. The predominantly olfactory AC3 is also Ca^{2+}-sensitive, but the outcome is unclear. It is activated by Ca^{2+} in the presence of calmodulin *in vitro*, but it lacks an identifiable calmodulin binding site and it is inhibited by Ca^{2+} in human embryonic kidney cells (HEK293).

In non-excitable cells, the increase in concentration achieved following activation of Ca^{2+}-mobilizing receptors is not high enough to have more than a limited effect on the Ca^{2+}-sensitive adenylyl cyclases. Rather, it is Ca^{2+} entering the cell through plasma membrane channels to replenish depleted intracellular stores that is responsible; (store-operated Ca^{2+} entry or SOCE is described in Chapter 7). In excitable cells, Ca^{2+} entry through voltage-activated channels is also effective. It thus appears that location of the cyclases close to the high levels of Ca^{2+} entering through open channels is important.[22]

In general, the distribution over the plasma membrane of channels, receptors, mediators, and effectors is not uniform. Some occupy or may be recruited to dynamic plasma membrane microdomains rich in cholesterol and glycosphingolipids, known as lipid rafts (see page 522). The Ca^{2+}-regulated ACs (AC1, 3, 5, 6, 8) are concentrated in raft regions, where they coreside with a set of regulators and effectors, such as G-protein-coupled receptors, G proteins, putative SOCE components, and recruited phosphodiesterases. In contrast, the Ca^{2+}-insensitive AC2, AC4, and AC7 are excluded from rafts. The compartmentalization of the Ca^{2+}-sensitive ACs with relevant signalling proteins and the phosphodiesterases (PDEs) that break down cAMP provides a machinery that is able to create localized pockets of briefly elevated cAMP.

The ability of the Ca^{2+}-dependent ACs to be activated by both G_s and Ca^{2+} provides another mechanism of coincidence detection.

Regulation by phosphorylation

Although one might expect that phosphorylation by PKA (activated by cAMP) might cause feedback inhibition, this has only been detected for AC5 and AC6. On the other hand, phosphorylation by PKC enhances the activity of AC2,

AC4, AC7, and AC5 (AC6 is inhibited). This is a consequence of the activation of receptors linked through G proteins of the $G_{q/11}$ family to the activation of phospholipase , generating diacylglycerol, the activator of PKC. The precise timing of signals transmitted through α_s is likely to be less critical because a phosphorylated state of the enzyme is expected to persist for longer than other more transient influences, such as those activated by α- or $\beta\gamma$-subunits, or elevated Ca^{2+}. Rather than requiring a coincidence of signals, phosphorylation may set the enzyme up in readiness for transient stimulation by α_s.

Aluminium fluoride

Adenylyl cyclase can be activated experimentally by manipulations that bypass the receptors. A number of substances cause activation through interaction with G_s, others with the cyclase itself. An example is the fluoride ion.[23,24] Here is yet another instance of a development in which the emergence of highly purified reagents led first to the loss of biological activity and subsequently to the identification of its real nature. With better chemicals, well-dialysed tissue extracts, the use of laboratory plastic instead of glassware, and the availability of high-quality water, it became harder to stimulate cyclase with fluoride. The real activator, it turns out, is not the fluoride ion but a complex of aluminium and fluoride ($[AlF_4{}^-]$).[25] Aluminium is present as a contaminant of solutions that have been in contact with glass and of commercially produced nucleotides. It has been thought that the $[AlF_4{}^-]$ complex, resembling a phosphate group in its shape and charge distribution, becomes trapped adjacent to GDP in the nucleotide binding pocket of α-subunits in the site that would be occupied by the γ-phosphate of GTP.

At higher concentrations, rather than activating, fluoride acts to suppress the formation of cAMP. This finding gave the first inkling of the G proteins that signal the inhibition of cyclase activity.[27,28]

Forskolin

Another activator that has already been mentioned is forskolin (Figure 5.6), a diterpene isolated from the roots of *Coleus forskohlii*. Unlike aluminium fluoride, this interacts directly with the cyclase enzyme (see Figure 5.3). Physiologically, forskolin has a positive inotropic action: it increases heart rate and lowers blood pressure. It also exhibits non-specific spasmolytic activity on gastrointestinal smooth muscle. At large doses it has a depressant action on the central nervous system. Some of these phenomena could arise as a consequence of the direct action of forskolin on adenylyl cyclase, bypassing all upstream influences including receptors and GTP-binding proteins. Alternatively, they could be due to interactions with other intrinsic membrane proteins of related structure, including glucose transporters, voltage-gated K^+ channels, and the P-glycoproteins that confer multidrug resistance. The existence of natural products that react potently and specifically as regulators

---

More recent examination of the crystal structure of $AlF_4{}^-$ has revealed that in association with GDP, the complex is a square-planar entity.[26] Rather than behaving strictly as an appendage, forming a GTP look-alike, it resembles the transition state in the GTPase reaction. It is widely accepted that the interaction with the $[AlF_4{}^-]$ is characteristic only of the heterotrimeric G proteins. However, it has become apparent that monomeric GTPases such as Ras also interact with $AlF_4{}^-$ when they are associated with GTPase-activating proteins such as p120GAP and neurofibromin.[29,30]

A member of the mint family, *Coleus forskohlii* grows wild on the mountain slopes of Nepal, India, and Thailand. The tuberous roots have long been used as a spice and by practitioners of Ayurvedic medicine in the treatment of many diverse conditions related to heart disease, spasmodic pain, painful micturition, and convulsions.

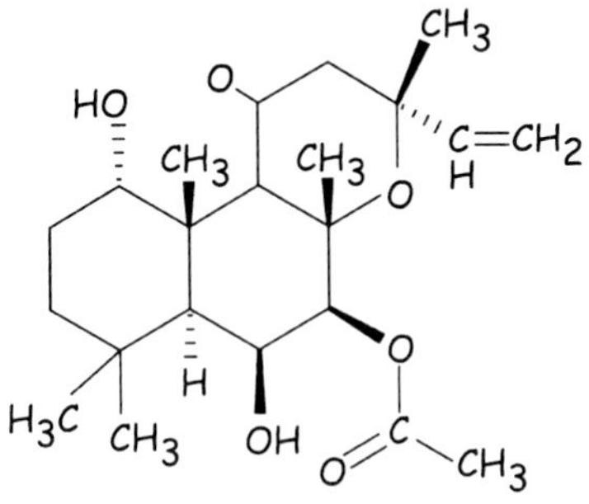

FIG 5.6 Forskolin.

of enzyme activity, begs questions relating to their endogenous physiological counterparts. So far however, no endogenous agent having the activity of forskolin has been identified.[31]

Cholera and pertussis toxins and ADP ribosylation

The accumulation of cAMP is enormously promoted in tissues that have been treated with cholera or pertussis toxins (respectively the exotoxins of *Vibrio cholerae* and *Bordetella pertussis*). Neither of these toxins has its effect directly on the cyclase enzyme. Instead, cholera toxin acts to prolong the activated state of the α_s. This situation arises because it inhibits the GTPase activity so that the bound GTP remains intact. The system becomes persistently activated. It is widely (but possibly mistakenly) accepted that the watery diarrhoea of cholera, the consequence of increased ion flux through the chloride channels of the intestinal epithelium, results from the elevated concentration of cAMP due to the maintained activation of adenylyl cyclase.

Pertussis toxin also has the effect of elevating the level of cAMP, but its mechanism of action is quite different. Intoxication by pertussis toxin plays a role in the aetiology of whooping cough.[33] The affected cells include the β-cells of pancreatic islets (hence its former name, islet activating protein), lymphocytes, leukocytes, cardiac cells, and others that cause paroxysms and neurological disturbance. The effects on these cells are irreversible, so that the restoration of function depends on their replacement. Unlike cholera toxin, which acts to inhibit GTPase activity, pertussis opposes the signal transmission by inhibitory receptors communicating with the G_i proteins.

Both cholera and toxin are enzymes, modifying their target proteins (substrates) by covalent attachment of the ADP-ribosyl moiety of NAD$^+$ (Figure 5.7).[34,35] However, in keeping with their distinct mechanisms of action, their site specificities are quite different. Cholera toxin transfers ADP-ribose to α_s, forming a covalent attachment with an arginine residue (R201) situated in the close vicinity of the γ-phosphate of the bound GTP (see Table 4.5, page 103). This modification prevents the GTPase activity and so causes the α-subunit to remain

The mechanism of **cholera-induced diarrhoea** is certainly more complicated than this. Indeed, although the level of cAMP is undoubtedly elevated, it may not even involve cAMP at all. Introduction of dibutyryl cAMP, a cell-permeant analogue which mimics the effect of the natural compound, is without effect on the rate of water flux through isolated pieces of intestinal tissue. Similarly, theophylline, that inhibits phosphodiesterase and so allows the accumulation of cAMP, is also without effect. Alternative candidates are the prostaglandins,[32] products of the cyclooxygenase pathway of arachidonate metabolism. Arachidonate is released from membrane phospholipids by the action of phospholipase A2.

Similar to **pertussis** and **cholera toxin**, **diphtheria toxin**, from *Corynebacterium diphtheriae*, is also an ADP-ribosylating enzyme. It acts on the elongation factor EF2. Unlike these other toxins it gains entry to cells by a mechanism involving endocytosis. One molecule of the toxin is sufficient to block the entire protein synthesizing machinery of a cell.

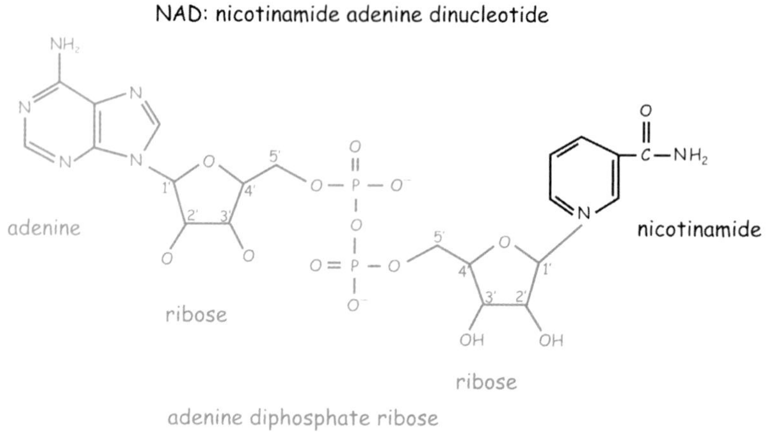

NAD: nicotinamide adenine dinucleotide

adenine

ribose

nicotinamide

ribose

adenine diphosphate ribose

FIG 5.7 ADP-ribose.

persistently activated (reduction of k_{cat}: see Equation 4.1 on page 85). Pertussis toxin transfers ADP-ribose to α_i at a cysteine situated just four residues from its C-terminus. ADP-ribosylation by pertussis toxin has the effect of interrupting the communication between the receptor and G_i, freezing it in its GDP-bound state. In this way it prevents the negative signal from inhibitory receptors. Either way, the consequence is persistent activation of adenylyl cyclase and elevated levels of cAMP. With the exception of α_z, all members of the G_i/G_o group of α-subunits (**Figure 5.7**) serve as substrates for pertussis toxin. In most cells and tissues, the α-subunit of G_s is the unique substrate for cholera toxin.

ADP-ribosylating enzymes such as cholera and pertussis toxins are complex structures. They bind to gangliosides that are ubiquitously expressed on the external surface of animal cells. Pertussis toxin is composed of five non-covalently linked subunits which are organized around a single catalytic subunit S1 (**Figure 5.8**). Two subunits (S2 and S3) bind with high affinity (K_D ~0.6 µmol L^{-1}) to the monosialoganglioside Gd1a. As a result, hydrophobic domains are exposed on the single S1 subunit, which is thus enabled to penetrate the cell. Here, the reducing environment in the cytosol causes its separation into two components, A1 and A2. The A1 subunit catalyses the ADP-ribosylation reaction:

$$NAD^+ + protein \rightarrow ADP\text{-ribosyl-protein} + nicotinamide + H^+$$

Like pertussis toxin, cholera toxin is composed of five non-covalently-linked subunits which are arranged in a ring-like pentameric formation surrounding a single catalytic-subunit. The B-subunits bind with high affinity (K_D ~1 nmol L^{-1}) to the monosialoganglioside GM1, ubiquitously expressed on the surface of all animal cells. (Cholera toxin also induces ADP-ribosylation of GTPases in plant cells.[36,37]) On gaining entry, the A1 subunit catalyses ADP-ribosylation.

Note that in addition to the ADP-ribosylating toxins, both *Vibrio cholerae* and *Bordetella pertussis* synthesize and release many other cell-penetrating toxins. Interestingly, one of the major virulence factors of *B. pertussis* is a secreted form of adenylate cyclase which becomes activated in contact with cellular calmodulin.

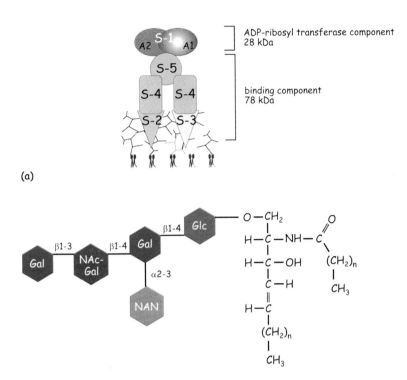

(a)

(b)

FIG 5.8 Pertussis toxin structure.
(a) The subunit organization of
pertussis toxin. The S2 and S3 subunits
bind to the exposed sugar residues
of glycolipids (gangliosides) in the
plasma membrane. The S1 subunit
penetrates the cell and, in the reducing
environment, dissociates. The A1
component catalyses the cleavage
of NAD$^+$ and effects the transfer of
the ADP-ribosyl group to the cysteine
residue at the C-terminus of α_i.
(b) Structure of ganglioside GM1. Gal,
galactose; NAc, N-acetyl-; Glc, glucose;
NAN, N-acetylneuraminic acid.

When this reaction was first carried out in the presence of radioactively labelled NAD$^+$, the label was found to be transferred to a single membrane-bound protein, which, when isolated, proved to be the nucleotide-binding component (α-subunit) of G$_s$.[34,38–40]

As reagents, cholera and pertussis toxins have been instrumental in assigning particular receptors to particular G proteins in signalling pathways. The substrate arginine (R201) which is modified by cholera toxin is a crucial component of the GTPase catalytic site of α_s. Mutants generated by substituting R201, hydrolyse GTP at a greatly reduced rate (reduced k_{cat}) and are in consequence persistently activated. Acquired (somatic) mutations at this site sometimes occur in non-metastatic endocrine tumours (adenomas). In spite of being non-invasive and restricted to a single clone of cells, adenomas have many dire consequences.

The word **somatic** refers to any cell of a multicellular organism that *does not* contribute to the production of gametes.

As an example, the physiological stimulus to the secretion of growth hormone from pituitary somatotrophs is provided by pulses of groth hormone relasing hormone (GHRH) released from the hypothalamus, especially during periods of deep sleep. This provides both the signal to secrete growth hormone and the trophic stimulus by which these cells maintain their normal number and size. GHRH binds to 7TM receptors on the somatotrophs and the signals are transduced through G_s. In some patients presenting with the symptoms of giantism or acromegaly, it is found that R201 of the α_s in the somatotrophs has been replaced by cysteine or histidine. The consequent and persistent activation of adenylyl cyclase and the elevated level of cAMP causes hypersecretion of growth hormone and gross enlargement of the gland.[41] Another somatic mutation of α_s in thyroid cells causes excess secretion of thyroid hormone. Mutations in the equivalent position (R179: see Table 4.5, page 103), suppressing the GTPase activity of G_{i2} are associated with adrenocortical and ovarian tumours.[42]

ADP-ribosylation of $\beta\gamma$-subunits

A normal physiological (i.e. not toxin-induced) cycle of ADP-ribosylation and de-ribosylation may control the activity of $\beta\gamma$-subunits and constitute another arm of control by 7TM receptors. When cells are incubated with radiolabelled adenine (as a metabolic precursor of NAD^+), the most heavily labelled protein is the β-subunit of heterotrimeric G proteins. The extent of labelling under basal conditions, about 0.2% of the total pool of β-subunits in CHO cells, depends on the ratio of activities of the transferase (present in the plasma membrane) and the hydrolase (present in the cytosol). The point of attachment of the ADP-ribosyl group is an arginine present in the N-terminal segment of the β-subunit.[43] This portion of the β-subunit provides a contact surface that abuts α-subunits and also its effectors including adenylyl cyclase, phospholipase Cβ2 and the muscarinic K^+ channel.

As pointed out earlier, the Ca^{2+}-calmodulin sensitive AC1 is normally subject to inhibition by $\beta\gamma$-subunits. However, ADP-ribosylated $\beta\gamma$-subunits are without effect.[43] Likewise, activation of phosphatidylinositol-3-kinase by $\beta\gamma$-subunits (see page 547) fails when they have been modified by ADP-ribosylation. If $\beta\gamma$-subunits were to become ADP-ribosylated as part of a normal process regulated by agonist intervention at receptors, it would have the effect of maintaining the effector enzyme in an activated state. A number of ligands (e.g. cholecystokinin, UTP, and thrombin) that are active at receptors that activate phospholipase C through G_q, increase the extent of β-subunit ADP ribosylation.[44] Similarly, activation of the serotonin receptor that couples through a G_i- or G_o-like system to induce activation of the MAP kinase pathway (Chapter 12), also causes ADP-ribosylation of β-subunits.

ADP-ribosylation and deribosylation: a general mechanism of cell control

In addition to G-protein $\beta\gamma$-subunits, the cycle of ADP-ribosylation and deribosylation, catalysed by endogenous intracellular enzymes, also regulates the activities of other systems. These include mitochondrial glutamate dehydrogenase and the chaperone BiP, resident in the endoplasmic reticulum.

A separate class of cell surface ADP-ribosylases targets the integrins, defensins, and the purinergic receptor P_2X_7.[45]

Phospholipase C

First hints of a signalling role for inositol phospholipids

The first hints of a role for the inositol-containing phospholipids in cell regulation emerged in 1953,[46] but 20 years elapsed before a viable proposal concerning their role was forthcoming. Michell[47] then pointed to the striking correlation between the activation of receptors that cause an increase in the metabolism of phosphatidylinositol (PI) and activation of processes such as secretion and the contraction of smooth muscle that are dependent on Ca^{2+}. Although a phospholipase C (PLC)-catalysed reaction was inferred, the precise identities of the substrate and the resulting second messenger(s) remained elusive. Then it was found that the rate of depletion of the polyphosphoinositides, in particular phosphatidylinositol-4,5-bisphosphate ($PI(4,5)P_2$) (Figure 5.9) and not phosphatidylinositol itself, correlates with the onset of cellular activation.[48] The water-soluble product IP_3, resulting from the hydrolysis of $PI(4,5)P_2$, was found to release Ca^{2+} from intracellular storage sites when introduced into permeabilized pancreatic cells.[49] Within a very short time it was shown that this mechanism is ubiquitous and that the immediately accessible Ca^{2+} stores are present in the endoplasmic reticulum which is endowed with receptors for IP_3 (see Chapter 7). The hydrophobic product of $PI(4,5)P_2$ breakdown is diacylglycerol (DAG) and this is retained in the membrane bilayer. It causes the activation of protein kinase C (see Figure 7.10, page 199 and Chapter 9).

The phospholipase family

Phospholipids possessing a glycerol backbone contain four ester bonds, all of which are potentially susceptible to enzyme-catalysed hydrolysis by specific phospholipases (Figure 5.10). Hydrolysis at three of these positions gives rise to products that are either second messengers or substrates for enzymes that yield further signalling molecules.

PLA_2 and PLD

Phospholipase A_2 (PLA_2) hydrolyses mostly phosphatidylcholine yielding a lysophospholipid and releasing the fatty acid bound to the 2-position, generally arachidonate. Arachidonate is the substrate for the formation of prostaglandins and leukotrienes (potent signalling molecules with autocrine and paracrine effects, involved in pathological processes such as inflammation). Phospholipase D (PLD) provides another, less transient source of DAG by cleaving phosphatidylcholine (the major membrane phospholipid) to form phosphatidate and water-soluble choline. Removal of the phosphate by phosphatidate phosphohydrolase produces DAG.

147

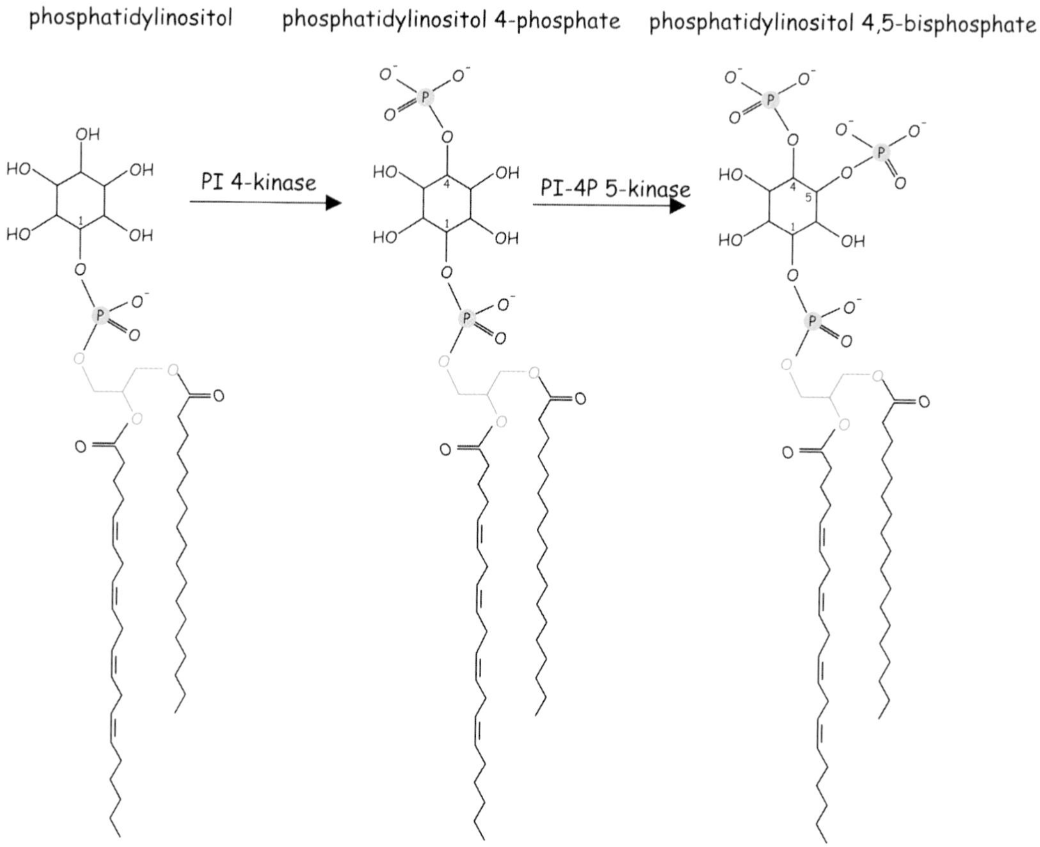

phosphatidylinositol   phosphatidylinositol 4-phosphate   phosphatidylinositol 4,5-bisphosphate

FIG 5.9 Phosphatidylinositol (PI) and its derivatives PI(4)P and PI(4,5)P$_2$

Phospholipase C

The process by which IP$_3$ is formed in response to agonist stimulation requires the activation of a plasma membrane-associated enzyme of the phospholipase C (PLC) family. Its substrate, PI(4,5)P$_2$, is derived from phosphatidylinositol through the action of lipid kinases (see Figure 5.9). In addition to the PLCs, PI(4,5)P$_2$ is also the substrate of the phosphoinositide 3-kinases that cause further phosphorylation, generating the lipid second messenger PI(3,4,5)P$_3$ (see Figure 18.1, page 546).

The isoenzymes of PLC

Eleven mammalian isoforms of PLC have been purified and cloned. These fall into four main classes: β, γ, δ, and the recently discovered ε.[50,51] Examination of the amino acid sequence reveals that all of the PLCs possess multiple structural domains. The δ isoform is the simplest, containing in addition to

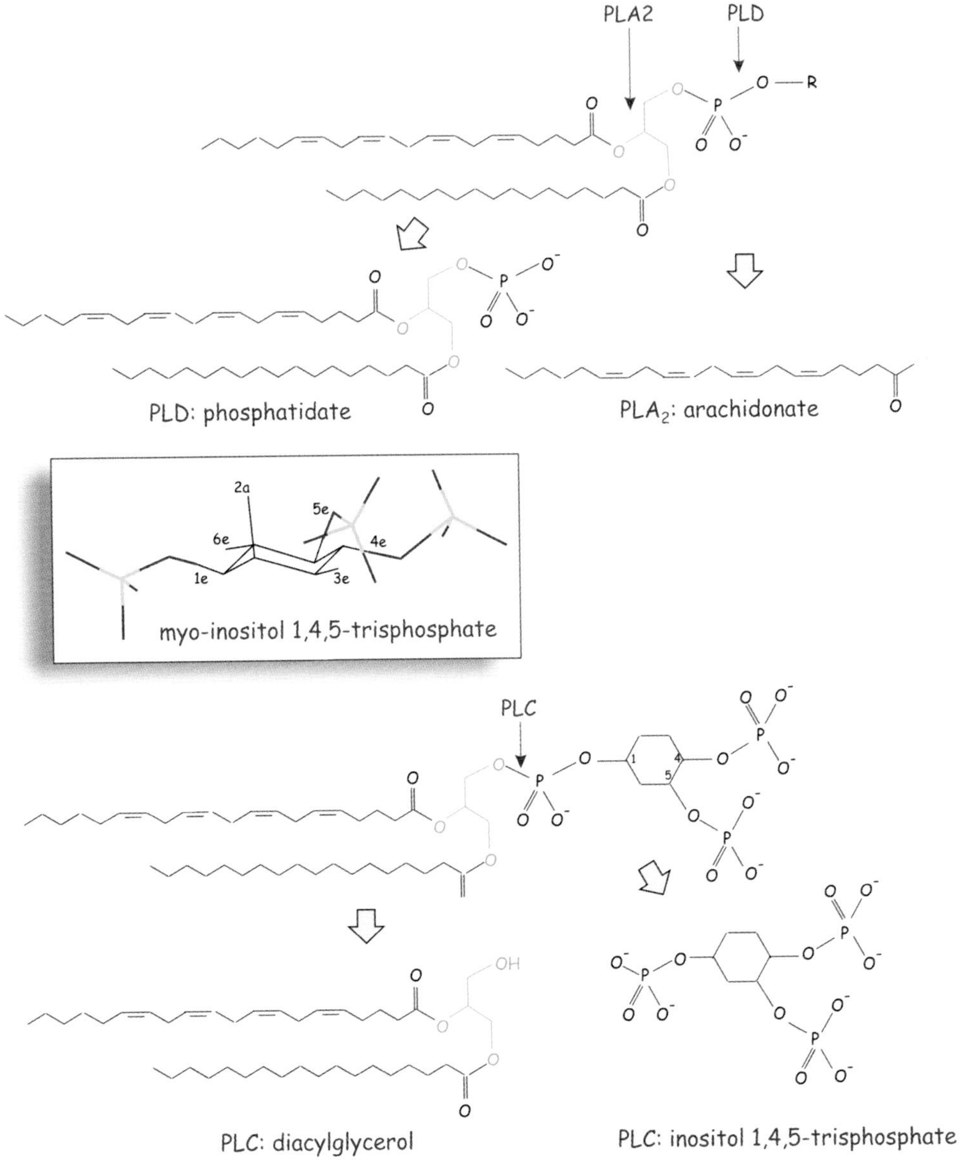

PLD: phosphatidate

PLA₂: arachidonate

myo-inositol 1,4,5-trisphosphate

PLC: diacylglycerol

PLC: inositol 1,4,5-trisphosphate

FIG 5.10 Phospholipases cleave specific phospholipid ester bonds. Top: arachidonate (together with a lysophospholipid) is the product of PLA2. Phosphatidate (and choline) are the products of PLD (the main substrate is phosphatidylcholine). The biological form of inositol (inset), based on cyclohexane, is *myo* -inositol, in which the hydroxy substitutions at positions 1, 3, 4, 5, and 6 are all equatorial (e), while the 2-position is substituted axially (a). Bottom: the structures of phosphatidylinositol 4,5-bisphosphate (specifically, 1-stearoyl, 2-arachidonyl PI(4,5)P₂) and its breakdown products produced by PLC: inositol trisphosphate and diacylglycerol. The glycerol nucleus is blue.

The status of a possible α isoform is uncertain since the cDNA originally associated with this enzyme was later found to code for a protein disulfide isomerase. Isoenzymes also exist within each class, as PLCβ$_{1-4}$, PLCγ$_{1-2}$, PLCδ$_{1-4}$.

the catalytic domain only PH and C2 domains (see Chapter 24) and a stretch containing EF-hand motifs (Chapters 8 and 24). All forms of PLC possess a C2 domain in the vicinity of the C-terminus. The catalytic centre, split into two separated subdomains, X and Y, possesses high sequence similarity. In the case of PLC-γ, the sequence intervening between the X and Y domains extends to more than 500 residues and encompasses two SH2 domains and a single SH3 domain (Figure 5.11, see also Table 24.1, page 769). In spite of their considerable separation in the primary sequence, in the folded structure the X and Y segments come together forming two halves of a single structural unit which forms the catalytic centre. The ε isoform, lacking a PH domain, contains two unique domains that allow it to bind and activate Ras.

PLCδ: a prototype

The δ isoforms of PLC are likely to be ancestral to the other animal PLC isoenzymes. They are of lower molecular weight and possess no domains that are not present in one or more members of the other classes. Some organisms such as the yeasts, *Dictyostelium*, and flowering plants contain only the δ forms. It is has been speculated that PLCδ emerged initially as a Ca^{2+}-dependent generator of DAG, the activator of PKC.[52] In this case, the water-soluble by-product, IP$_3$, would have had no role to play in signalling. Later in evolution, as it too became a second messenger regulating Ca^{2+}, more complex forms of PLC (β and γ) emerged that are activated by receptors.

The relevance of the domains that are common to all the known isoforms is unclear. Although EF-hand motifs may bind Ca^{2+} (Chapter 8), the N-terminal EF-hand-containing structure does not appear to be the site through which regulation by Ca^{2+} is exerted. However, it is certainly of importance, possibly in a structural context, since the activity of the enzyme is abolished when about half of this domain is deleted. X-ray diffraction of PLCδ cocrystallized with IP$_3$, which mimics the head group of PI(4,5)P$_2$, has revealed that the regulating Ca^{2+} ions are bound at the active site of the enzyme and also to the IP$_3$ molecule.[53] It is likely that the PH domain tethers the enzyme to the membrane by high affinity attachment to PI(4,5)P$_2$. The C2 domain may then reinforce this attachment in a Ca^{2+}-dependent manner. In this way it may optimize the orientation so that the active site can access its substrate in the membrane (Figure 5.12).

While the β and γ isoforms of PLC are certainly more central to the regulation of cellular activity by receptors, a basis for the understanding of their regulation is provided by knowledge of the simpler δ isoform. It may also offer clues about the operation of other signal transduction enzymes (such as PKC (Chapter 9), cytosolic PLA2, and PI 3-kinase (Chapter 18) that use similar structural modules for reversible association with membranes.

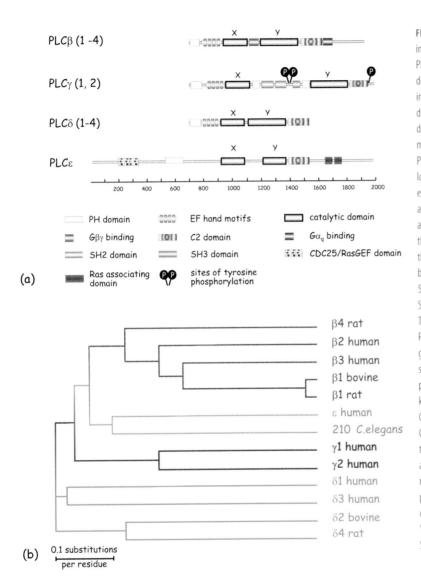

FIG 5.11 Organization of domains in the amino acid sequences of the PLC enzymes. (a) The main structural domains are indicated. In all of the inositide-specific PLCs, the catalytic domain is divided into two parts, designated X and Y. PLCδ has a minimalist design, having only the PH, C2, and EF-hand domains. The long (~500 amino acids) C-terminal extension on PLCβ enables its association with membranes and allows for regulation by α-subunits of the G$_{q/11}$ family of G proteins. In PLCγ the X and Y components are separated by an extended stretch of more than 500 amino acids which includes two SH2 and one SH3 domain. These mediate the interaction of the PLCγ enzymes with phosphorylated growth factor receptors and other signalling molecules. The sites of phosphorylation by receptor tyrosine kinases are indicated. Phospholipase Cε contains a RasGEF domain (see Chapter 4) at the N-terminus and two Ras associating (RA) domains at the C-terminus. (b) Evolutionary relationships of PI-specific phospholipases C based on sequence comparisons of the conserved X and Y catalytic domains. Adapted from Singer et al.[51]

Regulation of PLC

There are two ways in which PLC may be activated to hydrolyse PI(4,5)P$_2$. One of these involves G proteins and the other a protein tyrosine kinase.

G-protein-coupled activation of PLCβ

The β isoforms of PLC are activated by G-protein-mediated mechanisms involving members of the G$_{q/11}$, G$_i$, or G$_o$ families. For proteins of the G$_{q/11}$ family it is the α-subunit that conveys the activation signal, interacting at

151

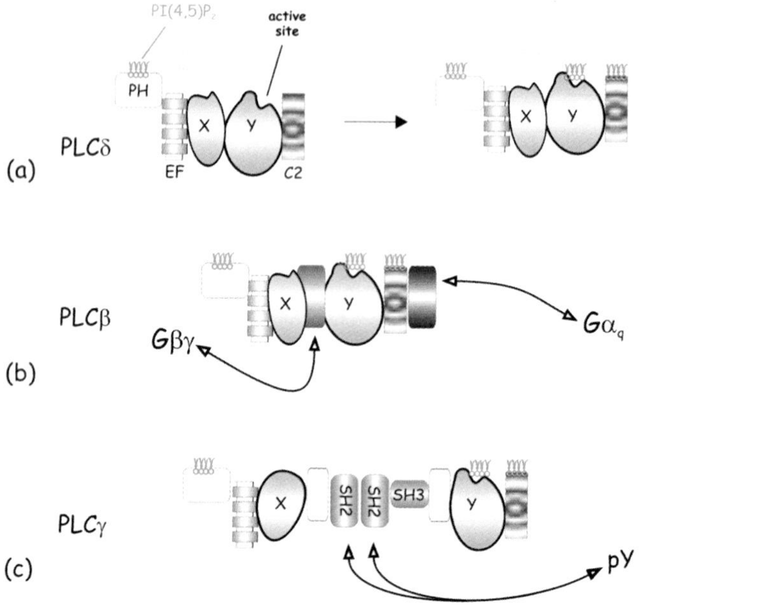

FIG 5.12 Membrane-association of PLC. (a) Initial attachment of PLCδ is through its PH domain. Subsequently, the C2 domain brings the catalytic site into contact with its substrate. (b) PLCβ interacts additionally with Gβγ or Gα$_q$ to initiate catalysis. (c) PLCγ also interacts with phosphotyrosines through its SH2 domains. The domains are colour-coded as in Figure 5.11. Adapted from Katan and Williams.[54]

Ligands that activate PLCβ through the α-subunits of G proteins of the G$_{q/11}$ family include α$_1$-adrenergic ligands, acetylcholine (M1 and M3), histamine, vasopressin, bradykinin, thromboxane A$_2$, angiotensin II, bombesin, TSH, CXC chemokines, and endothelin.

Ligands that activate PLCβ2 through βγ subunits derived from G$_i$ and G$_o$ include β$_1$- and β$_2$-adrenergic agonists, vasopressin (V2 receptor), acetylcholine (M2), formylmethionyl peptides, interleukin 8, and complement factor C5a.[57]

a site in the extended C-terminal region of these enzymes. In the case of receptors which communicate through G$_i$ and G$_o$ it is the βγ heterodimer that activates PLCβ2 (and to a lesser extent PLCβ1 and β3), interacting at a site between the X and Y catalytic subdomains.[55] Deletion of the C-terminal section of PLCβ results in enzymes that fail to respond to α$_q$ but still respond to βγ subunits.[56] In cells, this may be prevented by pertussis toxin which catalyses ADP-ribosylation of the α-subunits of G$_i$ and G$_o$, but not the members of the G$_{q/11}$ family. In addition, it is important to remember that the G proteins, their subunits, and the PLC effector enzymes are all membrane-associated. Also that the substrate, PI(4,5)P$_2$, and one of the products of reaction, DAG, are hydrophobic and are retained as integral components of the membrane phospholipid bilayer. In contrast, the other product, IP$_3$, is water soluble and is released into the cytosol where it acts as a ligand to induce release of Ca^{2+} from intracellular stores (see Chapter 7).

Regulation of PLCε

PLCε can be activated by α-subunits of G$_{12}$. Although relatively poorly characterized, G$_{12}$ and G$_{13}$ appear to be involved in the regulation of cytoskeletal changes and may themselves be activated by thrombin, lysophosphatidic acid, and some other mitogenic agents. A persistently activated mutant of G$_{12}$ (GTPase-deficient Q229L) promotes cell growth and transformation characteristic of cancer cells (see Chapter 11). This phospholipase, which possesses a RasGEF domain and other

Ras-associating (RA) domains, is intermediate between receptor activation and the MAP-kinase pathway leading to mitogenesis.[51] The PLCε isoform is a multifunctional entity that senses and mediates cross-talk between the pathways set in train by both the heterotrimeric and the Ras-like GTPases.[58]

PLCγ is activated through interaction with a protein tyrosine kinase

The γ isoforms of PLC possess an additional region, absent from the β and δ forms, inserted between the X and Y sequences of the catalytic domain (see Figure 5.11). This region contains one SH3 domain and two SH2 domains. PLCγ is present in cells that have receptors with intrinsic tyrosine kinase activity (Chapter 12) and which bind growth factors (e.g. EGF, PDGF, FGF, and IGF) and also in cells with receptors of the immunoglobulin superfamily, which recruit non-receptor protein tyrosine kinases (T and B lymphocytes and mast cells; Chapter 17). The SH2 domains of PLCγ enable it to bind to motifs on the intracellular chains of receptors that have become phosphorylated on tyrosine residues following the binding of a stimulatory ligand (Chapter 12). This brings PLCγ into a signalling complex, where it becomes phosphorylated itself by the tyrosine kinase activity of the receptor (or accessory, non-receptor tyrosine kinase). This causes its activation.

References

1. Rall TW, Sutherland EW, Berthet J. The relationship of epinephrine and glucagon to liver phosphorylase: effect of epinephrine and glucagon on the reactivation of phosphorylase in liver homogenates. *J Biol Chem.* 1957;224:463–475.
2. Nakamura T, Tomomura A, Kato S, Noda C, Ichihara A. Reciprocal expressions of α1- and β-adrenergic receptors, but constant expression of glucagon receptor by rat hepatocytes during development and primary culture. *J Biochem Tokyo.* 1984;96:127–136.
3. Botsford JL, Harman JG. Cyclic AMP in prokaryotes. *Microbiol Rev.* 1992;56:100–122.
4. Leichtling BH, Rickenberg HV, Seely RJ, Fahrney DE, Pace NR. The occurrence of cyclic AMP in archaebacteria. *Biochem Biophys Res Commun.* 1986;136:1078–1082.
5. Bolwell GP. Cyclic AMP, the reluctant messenger in plants. *Trends Biochem Sci.* 1995;20:492–495.
6. Li WW, Luan S, Schreiber SL, Assmann SM. CyclicAMP stimulates K^+ channel activity in mesophyll cells of *Vicia faba* K. *Plant Physiol.* 1994;105:957–961.
7. Wu WH, Assmann SM. A membrane delimited pathway of G-protein regulation of the guard cell inward K^+ channel. *Proc Natl Acad Sci U S A.* 1994;91:6310–6314.

8. Sutherland EW, Rall TW, Menon T. Adenyl cyclase: distribution, preparation and properties. *J Biol Chem.* 1962;237:1220–1227.

9. Sutherland EW. Studies on the mechanism of hormone action. *Science.* 1972;177:401–408.

10. Sunahara RK, Dessauer CW, Gilman AG. Complexity and diversity of mammalian adenylyl cyclases. *Annu Rev Pharmacol Toxicol.* 1996;36:461–480.

11. Tang W-J, Hurley JH. Catalytic mechanism and regulation of mammalian adenylyl cyclase. *Mol Pharmacol.* 1998;54:231–240.

12. Dessauer CW, Gilman AG. Purification and characterization of a soluble form of mammalian adenylyl cyclase. *J Biol Chem.* 1996;271:16967–16974.

13. Yan SZ, Hahn D, Huang ZH, Tang WJ. Two cytoplasmic domains of mammalian adenylyl cyclase form a Gs α- and forskolin-activated enzyme in vitro. *J Biol Chem.* 1996;271:10941–10945.

14. Whisnant RE, Gilman AG, Dessauer CW. Interaction of the two cytosolic domains of mammalian adenylyl cyclase. *Proc Natl Acad Sci U S A.* 1996;93:6621–6625.

15. van Veen HW, Margolles A, Muller M, Higgins CF, Konings WN. The homodimeric ATP-binding cassette transporter LmrA mediates multidrug transport by an alternating two-site (two-cylinder engine) mechanism. *EMBO J.* 2000;19:2503–2514.

16. Post SR, Hilal-Dandan R, Urasawa K, Brunton LL, Insel PA. Quantification of signalling components and amplification in the β-adrenergic-receptor-adenylate cyclase pathway in isolated rat ventricular myocytes. *Biochem J.* 1995;311:75–80.

17. MacEwan DJ, Kim GD, Milligan G. Agonist regulation of adenylate cyclase activity in neuroblastoma×glioma hybrid NG109-15 cells transfected to co-express adenylate cyclase type II and the β2-adrenoceptor. Evidence that adenylate cyclase is the limiting component for receptor-mediated stimulation of adenylate cyclase activity. *Biochem J.* 1996;318:1033–1039.

18. Hanoune J, Defer N. Regulation and role of adenylyl cyclase isoforms. *Annu Rev Pharmacol Toxicol.* 2001;41:145–174.

19. Krumins AM, Gilman AG. Targeted knockdown of G protein subunits selectively prevents receptor-mediated modulation of effectors and reveals complex changes in non-targeted signaling proteins. *J Biol Chem.* 2006;281:10250–10262.

20. Sattin A, Rall TW, Zanella J. Regulation of cyclic adenosine 3′,5′-monophosphate levels in guinea-pig cerebral cortex by interaction of α-adrenergic and adenosine receptor activity. *J Pharmacol Exp Ther.* 1975;192:22–32.

21. Sternweis PC. The active role of $\beta\gamma$ in signal transduction. *Curr Opinion Cell Biol.* 1994;6:198–203.

22. Willoughby D, Cooper DM. Organization and Ca^{2+} regulation of adenylyl cyclases in cAMP microdomains. *Physiol Rev.* 2007;87:965–1010.

23. Rodbell M. Metabolism of isolated fat cells. V. Preparation of 'ghosts' and their properties; adenyl cyclase and other enzymes. *J Biol Chem.* 1967;242:5744–5750.

24. Birnbaumer L, Pohl SL, Rodbell M. Adenyl cyclase in fat cells. 1. Properties and the effects of adrenocorticotropin and fluoride. *J Biol Chem.* 1969;244:3468–3476.

25. Sternweis PC, Gilman AG. Aluminum: a requirement for activation of the regulatory component of adenylate cyclase by fluoride. *Proc Natl Acad Sci U S A.* 1982;79:4888–4891.

26. Mittal R, Ahmadian MR, Goody RS, Wittinghofer A. Formation of a transition-state analog of the Ras GTPase reaction by Ras-GDP, tetrafluoroaluminate, and GTPase-activating proteins. *Science.* 1996;273:115–117.

27. Harwood JP, Low H, Rodbell M. Stimulatory and inhibitory effects of guanyl nucleotides on fat cell adenylate cyclase. *J Biol Chem.* 1973;248:6239–6245.

28. Harwood JP, Rodbell M. Inhibition by fluoride ion of hormonal activation of fat cell adenylate cyclase. *J Biol Chem.* 1973;248: 4901–4904.

29. Daemen FJ. Vertebrate rod outer segment membranes. *Biochim Biophys Acta.* 1973;300:255–288.

30. Kuhn H, Bennett N, Michel-Villaz M, Chabre M. Interactions between photoexcited rhoddopsin and GTP-binding protein: kinetic and stoichiometric analyses from light scattering changes. *Proc Natl Acad Sci USA.* 1981;78:6877.

31. Laurenza A, Sutkowski EM, Seamon KB. Forskolin: a specific stimulator of adenylyl cyclase or a diterpene with multiple sites of action? *Trends Pharmacol Sci* 1989;10:442–447.

32. Peterson JW, Ochoa LG. Role of prostaglandins and cAMP in the secretory effects of cholera toxin. *Science.* 1989;245:857–859.

33. Pittman M. The concept of pertussis as a toxin-mediated disease. *Pediatr Infect Dis.* 1984;3:467–486.

34. Gill DM, Meren R. ADP-ribosylation of membrane proteins catalyzed by cholera toxin: basis of the activation of adenylate cyclase. *Proc Natl Acad Sci U S A.* 1978;75:3050–3054.

35. Cassel D, Pfeiffer T. Mechanism of cholera toxin action: covalent modification of the guanyl nucleotide-binding protein of the adenylate cyclase system. *Proc Natl Acad Sci U S A.* 1978;75:2669–2673.

36. Seo HS, Kim HY, Jeong JY, Lee SY, Cho MJ, Bahk JD. Molecular cloning and characterization of RGA1 encoding a G protein α subunit from rice (*Oryza sativa* L. IR-36). *Plant Mol Biol.* 1995;27:1119–1131.

37. Wu WH, Assmann SM. A membrane-delimited pathway of G-protein regulation of the guard-cell inward K$^+$ channel. *Proc Natl Acad Sci U S A.* 1994;91:6310–6314.

38. Katada T, Ui M. Direct modification of the membrane adenylate cyclase system by islet activating protein due to ADP-ribosylation of a membrane protein. *Proc Natl Acad Sci U S A*. 1982;79:3129–3133.
39. Pfeuffer T. GTP-binding proteins in membranes and the control of adenylate cyclase activity. *J Biol Chem*. 1977;252:7224–7234.
40. Pfeuffer T, Helmreich EJ. Activation of pigeon erythrocyte membrane adenylate cyclase by guanylnucleotide analogues and separation of a nucleotide binding protein. *J Biol Chem*. 1975;250:867–876.
41. Landis CA, Masters SB, Spada A, Pace AM, Bourne HR, Vallar L. GTPase inhibiting mutations activate the α chain of Gs and stimulate adenylate cyclase in human pituitary tumours. *Nature*. 1989;340:690–696.
42. Lyons J, Landis CA, Harsh G, et al. Two G protein oncogenes in human endocrine tumors. *Science*. 1990;245:655–659.
43. Lupi R, Corda D, Di Girolamo M. Endogenous ADP-ribosylation of the G-Protein β subunit prevents the inhibition of type 1 adenylyl cyclase. *J Biol Chem*. 2000;275:9418–9424.
44. Lupi R, Dani N, Dietrich A, et al. Endogenous mono-ADP-ribosylation of the free G$\beta\gamma$ prevents stimulation of phosphoinositide 3-kinase-γ and phospholipase C-β2 and is activated by G-protein-coupled receptors. *Biochem J*. 2002;367:825–832.
45. Di GM, Dani N, Stilla A, Corda D. Physiological relevance of the endogenous mono(ADP-ribosyl)ation of cellular proteins. *FEBS J*. 2005;272:4565–4575.
46. Hokin MR, Hokin LE. Enzyme secretion and the incorporation of ^{32}P into phospholipides of pancreas slices. *J Biol Chem*. 1953;203:967–977.
47. Michell RH. Inositol phospholipids in cell surface receptor function. *Biochim Biophys Acta*. 1975;415:81–147.
48. Creba JA, Downes CP, Hawkins PT, Brewster G, Michell RH, Kirk CJ. Rapid breakdown of phosphatidylinositol 4-phosphate and phosphatidylinositol 4,5-bisphosphate in rat hepatocytes stimulated by vasopressin and other Ca^{2+}-mobilizing hormones. *Biochem J*. 1983;212:733–747.
49. Streb H, Irvine RF, Berridge MJ, Schultz I. Release of Ca^{2+} from a non-mitochondrial intracellular store in pancreatic acinar cells by inositol-1,4,5-trisphosphate. *Nature*. 1983;306:67–69.
50. Song C, Hu C-D, Masago M, et al. Regulation of a novel human phospholipase C, PLCε, through membrane targeting by ras. *J Biol Chem*. 2001;276:2752–2757.
51. Lopez I, Mak EC, Ding J, Hamm HE, Lomasney JW. A novel bifunctional phospholipase C that is regulated by Gα12 and stimulates the Ras/mitogen-activated protein kinase pathway. *J Biol Chem*. 2001;276:2758–2765.
52. Irvine R. Phospholipid signaling. Taking stock of PI-PLC. *Nature*. 1996;380:581–583.

53. Essen L-O, Perisic O, Cheung R, Katan M, Williams RL. Crystal structure of a mammalian phosphoinositide-specific phospholipase Cδ. *Nature*. 1996;380:595–602.

54. Katan M, Williams RL. Phosphoinositide-specific phospholipase C: structural basis for catalysis and regulatory interactions. *Semin Cell Dev Biol*. 1997;8:287–296.

55. Kuang Y, Wu Y, Smrcka A, Jiang H, Wu D. Identification of a phospholipase C β2 region that interacts with G$\beta\gamma$. *Proc Natl Acad Sci U S A*. 1996;93:2964–2968.

56. Lee SB, Shin SH, Hepler JR, Gilman AG, Rhee SG. Activation of phospholipase C β_2 mutants by G protein αq and $\beta\gamma$ subunits. *J Biol Chem*. 1993;268:25952–25957.

57. Rhee SG. Regulation of phosphoinositide-specific phospholipase C. *Annu Rev Physiol*. 2001;70:281–312.

58. Wing MR, Bourdon DM, Harden TK. PLC-ε: a shared effector protein in Ras-, Rho-, and G$\alpha\beta\gamma$-mediated signaling. *Mol Interv*. 2003;3:273–280.

CHAPTER 6

The Regulation of Visual Transduction and Olfaction

Phototransduction

> Do not the Rays of Light in falling upon the bottom of the Eye excite Vibrations in the Tunica retina? Which Vibrations, being propagated along the solid Fibres of the optick Nerves into the Brain, cause the sense of seeing.
>
> Sir Isaac Newton, *Opticks* (1704)

The visual system provides an exceptional opportunity to investigate the transduction mechanism of a 7TM receptor at the level of a single molecule. The molecule is the photoreceptor pigment rhodopsin. This consists of the protein opsin to which is bound the photosensitive compound 11-*cis*-retinal. The stimulus is light and the second messengers are cyclic GMP (cGMP) and Ca^{2+}.[1]

Light is defined by its wave properties – frequency (ν) and wavelength (λ) where the velocity $c = \nu \times \lambda$. The intensity of a beam of light is the rate at which energy is delivered per unit area. It is measured in $W\,m^{-2}$. Light also possesses particle properties. The quanta are called **photons** and the energy (in J) of a single photon is given by $E = h \times \nu$, where h is Planck's constant (6.626×10^{-34} J s).

Sensitivity of photoreceptors

Significant advances in the understanding of molecular mechanisms in biology have, from time to time, emerged from opportunities that allow us to observe unitary events. For example, appreciation of ion channels has benefited from the observation of the opening and closing of individual channels. In a similar way, our understanding of the first steps of the visual transduction mechanism has followed from the investigation of the interaction of light with single photoreceptor molecules. To detect this, it is necessary to illuminate photoreceptors with very low light intensities so as to establish the minimal conditions for excitation. The first attempts were made in the 19th century, well before it had been realized that light possesses both wave and particle properties. Only then could estimates of the minimal intensity in terms of quanta be made.

With little knowledge of the basic physiology of the human eye, Hecht, Shlaer, and Pirenne in 1942 set about measuring its quantum sensitivity.[2] Using dark-adapted human subjects responding verbally to precisely calibrated single flashes (1 ms duration at 510 nm, close to the optimal wavelength for vision in dim light), they determined that the eye's detection limit corresponds to energies, incident at the cornea, in the range $2-6 \times 10^{-17}$ J, equivalent to 54–148 photons. To calculate the energy actually absorbed by the retinal receptors, they applied corrections to take account of losses due to reflection (4% at the cornea) and absorption by the ocular medium (50%). Also, at least 80% of the light falling on the retina passes through unabsorbed. After all this, they estimated that number of photons absorbed by the visual pigment was in the range 5–14. Since this is very small in comparison with the number of retinal photoreceptors in the field illuminated, it was concluded that they are of such sensitivity that the coincidence of single photons impinging simultaneously on five cells is sufficient to strike consciousness in a human being.[2]

> The fact that for the absolute visual threshold, the number of quanta is small makes one realize the limitation set on vision by the quantum structure of light. Obviously the amount of energy required to stimulate any eye must be large enough to supply at least one quantum to the photosensitive material. No eye need be so sensitive as this. But it is a tribute to the excellence of natural selection that our own eye comes so remarkably close to the lowest level.[2]

Much later, estimation of the quantum sensitivity of individual photoreceptors was performed electrophysiologically using suction electrodes applied to single rod cells obtained from the toad.[3] This approach avoids the subjectivity of human psychophysical experimentation and many of the assumptions made concerning the proportion of the light signal that actually reaches the photoreceptors. An outward membrane current was recorded during illumination and when dim flashes of light were applied, the current fluctuated in a quantal manner. Similar quantal responses have been observed in photocurrent records from primate rod cells; examples are shown in Figure 6.1. Each unitary event is considered to be the result of an interaction between a single photon and a single pigment molecule.

Although the human eye is able to sense fluxes of just a few photons per second, it can also detect subtle intensity differences under conditions of very bright illumination. This gives it a remarkable dynamic range. Additionally, the human eye can detect transient events and recover rapidly. For example, when exposed to a train of dim flashes, we can distinguish individual events up to a frequency of 24 Hz. The flickering image is an enduring (and endearing) feature of old movies, in which the frame rate is slower than the flicker-fusion frequency of the audience. The faster frame rate of modern cinematography (and television) is just sufficient to allow successive images to fuse and give the impression of continuous motion. The flicker-fusion frequency of other species, such as some insects, can exceed 100 Hz.

> And when a Coal of Fire moved nimbly in the circumference of a Circle, makes the whole circumference appear like a Circle of Fire; is it not because the Motions excited in the bottom of the Eye by the Rays of Light are of a lasting nature, and continue till the Coal of Fire in going round returns to its former place? And considering the lastingness of the Motions excited in the bottom of the eye by Light, are they not of a vibrating nature?
>
> Sir Isaac Newton, *Opticks*.

Rhodopsin, initially known as visual purple, was first isolated and described in 1878 by Willy Kühne, professor of physiology at the University of Heidelberg. He realized that its purple colour is due to a chemical group (chromophore) distinct from those present in egg yolk and β-carotene. He showed that it fades from a deep red to pink when illuminated by light of appropriate wavelengths. In addition to rhodopsin, Kühne also coined the words trypsin, enzyme, and myosin.[5]

Photoreceptor mechanisms

We know more about signalling through vertebrate photoreceptors than through any of the other 7TM receptors interacting with heterotrimeric GTP-binding proteins. However, a warning is in order. It should not be imagined that what we learn here is a general reflection of G-protein-coupled receptor mechanisms. Almost everything that happens in vertebrate phototransduction is the very converse of the normal sequence of events by which receptors signal the generation of second messengers and then their interactions with effectors. This is a world turned upside down. 11-*cis*-retinal is the prosthetic group in rhodopsin that absorbs visible light (see **Figure 6.3**). It is not only the chromophore, but also the ligand and this is already bound *covalently* to rhodopsin before receipt of the light signal. After photoisomerization it detaches and diffuses into a neighbouring cell. The mechanism of transduction is coupled to membrane *hyper*polarization and a *reduction* rather than *de*polarization, and an elevation of the concentration

FIG 6.1 Detecting light responses in single rod cells.
(a) Responses of a single retinal rod cell to flashes of light. These evoke transient photocurrents in the membrane of a single rod cell outer segment. The amplitude of the transient outward currents increases with intensity up to a saturating level of 34 pA (wavelength 500 nm, flash duration 11 ms, photon density 1.7–503 photons μm^{-2}, species monkey, *Macaca fascicularis*). (b) Quantal responses of single retinal rod cells to dim flashes. The membrane current responses of a single rod cell to a train of dim flashes are variable in amplitude. This variability together with the presence of background noise gives the appearance of a sequence of successes and failures. Amplitude histograms (not shown) reveal a quantal response to light. The unitary events have an average amplitude of 0.7 pA and they correspond to the detection of a single photon (0.6 photons μm^{-2}, other conditions similar to panel a). (c) Response of a single rod cell to steady illumination. Records of photocurrents correspond to the superposition of random photon responses. The photon flux density is indicated below each trace (photons $\mu m^{-2} s^{-1}$).
Adapted from Baylor et al.[4]

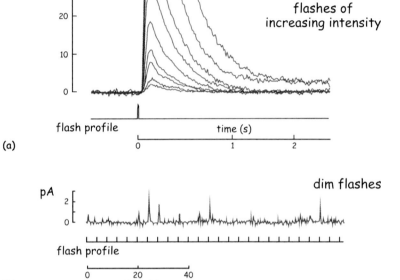

(a)

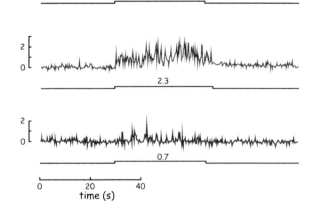

(b)

(c)

of cytosol Ca^{2+}. The mechanism of down-regulation and preparation of the system for a subsequent bout of illumination continues in this vein, generally the very reverse of what we have learned about hormone receptor systems. By contrast, in the invertebrate photoreceptor system we appear to be on rather more familiar ground. Here, the excitation of rhodopsin is coupled mostly through G_q to the activation of PLC and elevation of cytosol Ca^{2+}.

Photoreceptor cells

The photoreceptor cells of the retina (illustrated in Figure 6.2) are of two types. There are the cones which collectively provide colour discrimination (*photopic vision*), and the rods which are responsible for sensing low levels of light (*scotopic vision*). The human retina contains ~120 million rods and ~6 million cones. It is common experience that the appearance of colour is lost when objects are viewed in dim light and this is because the image is then detected only by the rods. The mechanism of phototransduction in cones is very similar to that of rods, differing only in terms of colour specificity. There are three types of cone, each containing one of three different pigments. Each of these, like rhodopsin, consists of 11-*cis*-retinal embedded in an opsin molecule. The three different cone opsins tune the absorption spectrum of the attached retinal to sense either red, green, or blue light. Rods, on the other hand, are tuned to match to the spectral distribution of dim natural light, with an optimum at ~500 nm. Subsequent discussion will concentrate on rod cells.

Rod cells

Photoreceptor cells are highly differentiated epithelial cells in which the light-sensing region is segregated from the main cell body as the outer segment

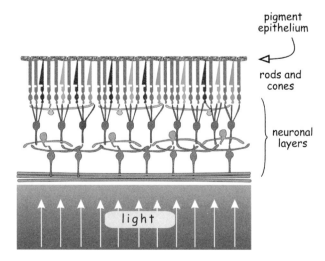

pigment
epithelium

rods and
cones

neuronal
layers

light

FIG 6.2 Organization of mammalian retina.

(Figure 6.3). It is a salient (if not another perverse) feature of the vertebrate visual system that the light enters these cells through the end of the cell body distal to the photosensitive outer segment. Not only this, before it enters the photoreceptors, it first traverses a network of blood vessels and then several layers of neuronal cells. The outer segments of the photoreceptors contain arrays of 1000–2000 intracellular discs, flattened vesicular structures about 16 nm thick, the membranes of which contain the photopigment, rhodopsin. Each disc is formed by the invagination of the plasma membrane to produce a structure that eventually becomes detached, so that the intradiscal space is topologically equivalent to the extracellular space. By weight, 50% of the disc membrane is protein, mostly rhodopsin. A single human rod cell contains on average 10^8 molecules of rhodopsin.[6] This protein is organized in the same way as the conventional 7TM receptors that are found in plasma membranes. The glycosylated N-terminus projects into the intradiscal space and the C-terminal domain, containing several potential phosphorylation sites (serine/threonine), is exposed to the cytosol.

Transducin, the G protein that transduces the light signal,[7] is also very abundant, though clearly outnumbered by rhodopsin. It comprises ~20% of the total (50% of the soluble) protein and unlike other G-proteins, it is

FIG 6.3 Rod photoreceptor cell and the rhodopsin molecule.

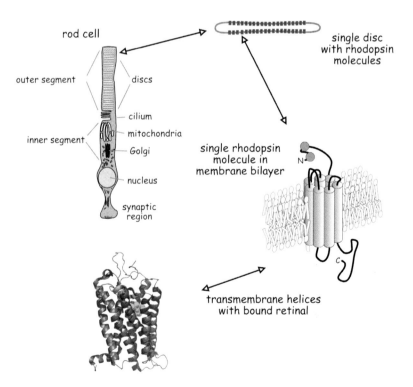

soluble. Photoreceptor cells provide the only instance in which the number of receptors greatly exceeds the number of G protein molecules to which they couple. The need for such vast amounts of rhodopsin is probably determined by the fact that photons travel in straight lines. Unlike normal soluble ligands, they are unable to diffuse in the extracellular medium until they are captured by their favoured receptor.

The chromophore, 11-*cis* retinal, is coupled through its aldehyde group to the α-amino group of a lysine in the centre of the G transmembrane helix, as a protonated Schiff base (Figure 6.4). Although covalent, this linkage is broken subsequent to the photoisomerization to all-*trans*-retinal, which is released for reprocessing.

The extremities of the rod outer segments protrude into the layer of cells that form the pigment epithelium. This layer serves a number of purposes. Most significantly in connection with phototransduction, its cells contain the metabolic enzymes that regenerate the active 11-*cis* isomer from the inactive all-*trans*-retinal following its detachment from the visual pigment in the neighbouring rods. As the name suggests, the cells of this layer contain their own pigment, which is melanin. This absorbs light that has penetrated past the arrays of discs in the rod cells and so prevents it from being scattered back into the photoreceptor layer. A failure to synthesize melanin (associated with the inherited conditions described as albinism) causes problems associated with bright lights and glare.

Beyond these difficulties, in **albinism** the fovea (the central region of the retina responsible for visual acuity) may also fail to develop properly, so that the eye cannot process sharp images well.

The principal enzymes of the transducing cascade are the G protein transducin (G_t) and the effector enzyme cyclic GMP phosphodiesterase. Rhodopsin is an integral membrane protein, but the phosphodiesterase is soluble. Transducin is attached to the membrane but dissociates and is

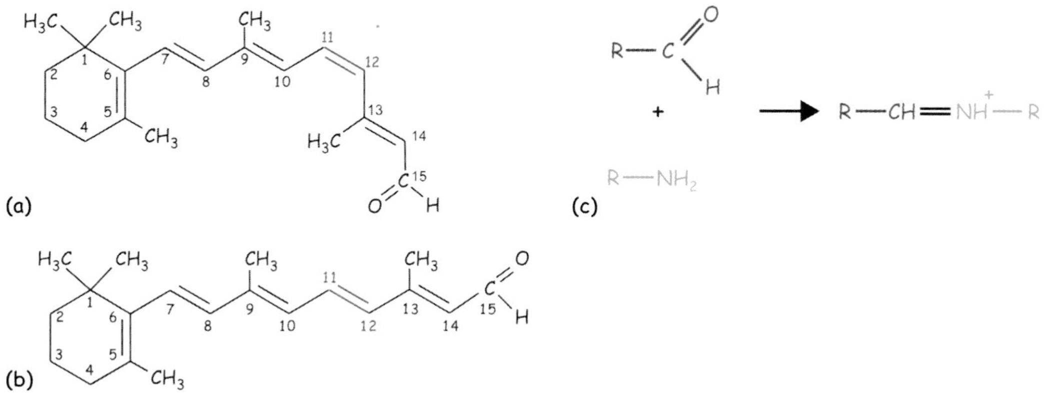

FIG 6.4 Structures of retinal.
(a) 11-*cis*-retinal. (b) all-*trans*-retinal. (c) Formation of a protonated Schiff base from an aldehyde and an amine.

released into the cytosol as the rhodopsin becomes activated following illumination.[8,9]

The cell body contains the nucleus, mitochondria, and the protein synthetic apparatus. This is very active, enabling the photosensory cells to keep pace with the loss of the outer segment discs which have a life of just a few weeks.[10] Over this period, the discs migrate progressively to the distal end of the outer segments where they are shed and phagocytosed by the pigment epithelial cells.[11] The inner ends of the photosensory cells form synapses with intermediate neurons. The neurotransmitter is glutamate.

Dark current and signal amplification

In the absence of illumination, the membrane potential of the photoreceptor cells is about $-30\,mV$. This is much less negative than the potential of excitable nerve and muscle cells and is due to a continuous photoreceptor dark current caused by an influx of Na^+ and Ca^{2+} through non-specific cation channels. The ions are extruded from the cells by pumps (Na^+, K^+-ATPase) and exchangers (Na^+/Ca^{2+}) which are located in the plasma membrane of the inner segment of the photoreceptor cells. Thus there is a constant flow of cations into and out of the cell (Figure 6.5).

Cyclic GMP: a second messenger in reverse

In the dark, the cation channels are held in their open configuration by the presence of cGMP. The effect of light is to initiate a series of molecular

FIG 6.5 Principal cation movements across a rod cell plasma membrane in the dark.

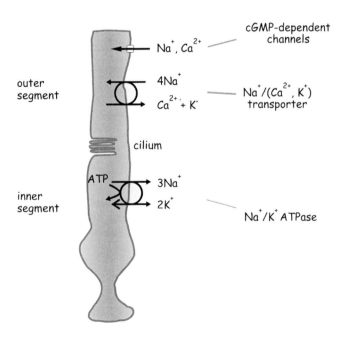

rearrangements in rhodopsin that generate activated rhodopsin (Rh*). This catalyses the exchange of GTP for GDP on transducin and provokes the dissociation of the complex, releasing the components α_t.GTP and $\beta\gamma$-subunits. The α-subunit then activates cGMP phosphodiesterase and cGMP is hydrolysed to 5'-GMP.

In the dark-adapted eye, cGMP is present in the cytosol of rod outer segments at a high concentration (40–80 μM, about 300-fold higher than in the nerve cells of brain). There are effectively 10–20 molecules of cGMP for each molecule of rhodopsin. It binds directly to the cation channels to keep them in the open state. The sequence of events following illumination ensures enormous amplification of the signal such that a single photon, causing a single photoisomerization, can lead during the next second to the hydrolysis of more than 10^5 molecules of cGMP, leading to the closure of $\sim$500 cation channels, so blocking the influx of as many as 10^7 Na$^+$ ions.

The cGMP-regulated channels are ligand-gated, but they differ from the ion channels operated by neurotransmitters in that the ligand is applied from the inside of the cell and it is normally in place in the unstimulated eye in the dark. Three molecules of cGMP are required to maintain the channel in its open state, with the result that the activation curve for channel closure is very steep indeed. The consequence of this is that once it is committed to opening or closing, as a result of a sufficient change in cGMP concentration, there is very little tendency for indeterminacy. Like an electrical switch, the channel tends to be either open or shut. The 500 or so cation channels in a rod cell that close following the transduction of a single photon represent 3% of the total number that are open in the dark. The resultant hyperpolarization (membrane potential more negative) is $\sim$1 mV and lasts $\sim$1 second. This is sufficient to depress the rate of neurotransmitter release at the synapse which impinges on the nerve cells that transmit the onward signals.

Adaptation: calcium acts as a negative regulator

Given that the incidence of a singe photon can lead to the closure of 3% of the channels, it might seem that steady illumination of even modest intensity would cause closure of the whole lot. Indeed, if the response of the system was directly proportional to the *number* of photoreceptor molecules activated, then full saturation would be achieved at very low light levels. Yet, the human eye is capable of sensing small differences in the intensity of light against high background levels. Clearly, there must be an adaptive mechanism that reduces the amplification, so that there are always some open channels remaining even when the background is very bright. Overall the eye is responsive over an intensity range of about 11 orders of magnitude,

For example, this response allows us to perceive light–dark contrasts at levels of illumination that range from that of a dark overcast night sky ($\sim 10^{-6}$ cd m^{-2}) to the brilliance of sunlight reflected from snow fields ($\sim 10^{7}$ cd m^{-2}). The candela (cd) is an SI unit of measure of the intensity of visible light, called the luminous intensity. 1 cd is the luminous intensity, in a given direction, of a source that emits monochromatic radiation of frequency 540×10^{12} Hz (green light) and that has a radiant intensity in that direction of 1/683 W per steradian.

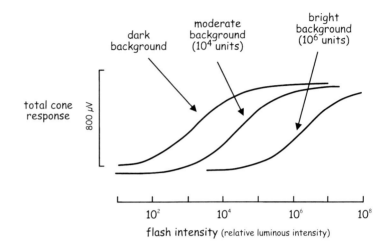

FIG 6.6 The shift in the flash response curve of mammalian cone photoreceptors as the background light level increases. Cone responses (instantaneous illumination plus background) measured with a bipolar electrode. Modified from Valeton and Van Norren.[13]

a range that matches the extremes of the intensity of illumination at the Earth's surface during the normal day–night cycle.[12] Much of this range can be accounted for by the ability of individual photoreceptors to adapt by decreasing their sensitivity. Thus rod cells can adapt over 2 orders of magnitude and cones can adapt even more (over 7–9 orders; see Figure 6.6).

An important component of the adaptive mechanism by which the sensitivity is adjusted against ambient light levels is contributed by the cytosol concentration of free Ca^{2+}. As the cation channels close in response to the hydrolysis of cGMP, the concentration of Ca^{2+} declines from its dark level of ~ 300 nmol L^{-1} to 10 nmol L^{-1} under strong illumination. As always, the effects of Ca^{2+}, whether stimulatory (e.g. causing the contraction of muscle) or inhibitory (as in the present case), are mediated through specific Ca^{2+}-binding proteins. Here, the synthesis of cGMP by guanylyl cyclase is inversely and cooperatively regulated by Ca^{2+}. This is mediated through its interaction with guanylyl cyclase activating protein (GCAP), a member of the EF-hand group of Ca^{2+}-binding proteins (see page 233) permanently bound to its effector, guanylyl cyclase. Thus, in the dark, when the level of Ca^{2+} in the cytosol is high, the rate of cGMP synthesis is low (Figure 6.7). This means that cGMP hydrolysed under low light conditions is not so readily replenished and the system is kept at its most sensitive. In strong light, the conversion of GTP to cGMP is accelerated due to the reduction in the concentration of free Ca^{2+}. The free GCAP binds to the cyclase and stimulates the resynthesis of cGMP, which in turn opposes the closure of the membrane channels. The effect is to counteract the long-term effects of the light-activated hydrolysis of cGMP.

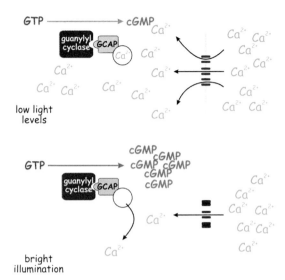

FIG 6.7 Photo-adaptation: a fall in intracellular Ca^{2+} promotes cGMP synthesis, opposing the effect of cyclic GMP phosphodiesterase.

Calcium also acts as a negative signal at a number of other points in the chain of events. It controls the influx of cations and restricts the extent of membrane hyperpolarization. Although the reaction rate of the phosphodiesterase reaction appears to be unaffected, the concentration of Ca^{2+} acts to regulate the lifetime of the activated enzyme. In the dark, through its interaction with another EF-hand Ca^{2+}-binding protein, recoverin (see page 233), Ca^{2+} prolongs the signal through inhibition of rhodopsin kinase. The decline in cytosol Ca^{2+} under intense illumination is permissive of rhodopsin phosphorylation and this necessarily curtails the lifetime of the activated phosphodiesterase (see **Figure 6.11**). Other sites of action of Ca^{2+} may include transducin (low Ca^{2+} might accelerate the GTPase reaction and so curtail the activation signal at this level) and the cGMP-regulated ion channels. As cytosol free Ca^{2+} declines during illumination, the affinity of the channels for cGMP increases. This must oppose the effects of the fall in the concentration of cGMP, favouring the open state.

To summarize, these highly integrated forms of automatic gain control arise in response to the generation of two signals as the membrane cation channels close in response to light. One of these is purely electrical and is conveyed as a pulse of hyperpolarization to the synaptic body, depressing the rate of release of the transmitter, glutamate. The other signal is chemical and results from the reduction in the concentration of cytosol Ca^{2+}. In essence, the light signal to these cells acts as a negative stimulus. In the dark, they are partially depolarized and the rate of transmitter release at the synaptic body is maximal. In the light, they become hyperpolarized and the rate of transmitter release is depressed.

In the long term, as in the diurnal cycle, it appears that adaptation to intense illumination is also mediated by a redistribution of transducin between the outer and the inner segments of the retinal rod cells. As much as 90% of the transducin present in the outer segments in dark-adapted conditions can translocate into other compartments of the cell on a time scale of tens of minutes.[14] This is accompanied by a corresponding 10-fold reduction in amplification of the signal, allowing the eye to maintain its sensitivity to changes in light against the brighter daytime background.

Photo-excitation of rhodopsin

The initial light-induced isomerization of 11-*cis*-retinal to all-*trans*-retinal, which occurs within picoseconds, does not immediately trigger rhodopsin to catalyse the exchange of guanine nucleotides on transducin. The first rapid event is followed by a series of dark reactions. These have been characterized spectrally as the stepwise shift of the peak wavelength of light absorption of the photoreceptor protein to shorter wavelengths as the ligand and the opsin form a series of intermediates (Figure 6.8).

The early forms – bathorhodospin, lumirhodopsin, and metarhodopsin-I – are unstable. It is metarhodopsin-II, also called photo-excited rhodopsin (Rh*), formed within milliseconds of light absorption, which initiates the transduction cascade by inducing guanine nucleotide exchange on transducin.[15–17] However, in contrast with 'conventional' G-protein

FIG 6.8 Steps following isomerization of retinal.

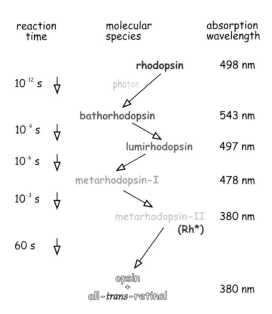

transduction mechanisms, the ensuing step takes place without amplification. The effector, cGMP phosphodiesterase, is a heterotetramer having two independent catalytic subunits, α and β, and two inhibitory γ-subunits. These impede the access of the substrate, cGMP, to the catalytic sites.[18] The nucleotide exchange catalysed by Rh* occurs on two transducin molecules that are bound to the inhibitory γ-subunits of the phosphodiesterase (Figure 6.9). When GDP is replaced by GTP, the α_t.GTP relieves the inhibition. As pointed out in Chapter 4, this is a cooperative process that involves both the conserved Ras-like (rd) domain of the α_t that interacts with the γ-subunits of the phosphodiesterase, and the unique helical domain (hd: see Figure 4.16, page 103) that interacts with its catalytic units. The α_t remains associated with the phosphodiesterase, so preventing it from interacting with further effectors and by its proximity ensuring rapid deactivation of the complex following GTP hydrolysis.[19,20]

Switching off the mechanism

We have described how a transient light signal rapidly initiates the phototransduction pathway and we have seen how the transduction of a single photon results in a hyperpolarization that lasts for about a second. However, this does not accord with our perception of light flashes that can be much briefer than this; the flicker-fusion frequency of 24 Hz corresponds to an image every 40 ms. This implies that there is a mechanism that shuts down visual transduction as soon as the stimulus is removed. Thus it is important

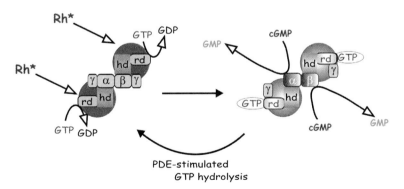

FIG 6.9 Activation of cyclic GMP phosphodiesterase by the α-subunit of transducin is a cooperative process. In the resting state α_t.GDP (shown as a disc) interacts through its helical domain (hd) with the catalytic α- and β-subunits of the phosphodiesterase (grey). This reduces their affinity for their inhibitory γ-subunits. Interaction of the Ras-like domain (rd) of α_t with the γ-subunits then acts as the switch, enabling access of the substrate cGMP to the active sites on the α- and β- subunits. The return to the resting state is determined by the hydrolysis of GTP on the α-subunit of transducin. This is assisted by the immediate proximity of the phosphodiesterase, acting as a component of the GTPase activating protein complex (GAP).

for photoreceptor cells to be able to terminate signals emanating from the activated pigment and also to be able to terminate any signals that are in progress further down the cascade. These terminations therefore involve inhibition of the activities of the photopigment (the receptor), transducin (the transducer), and cGMP phosphodiesterase (the catalytic unit or effector). cGMP is replenished through activation of guanylyl cyclase.

The conversion of metarhodopsin-I to metarhodopsin-II involves the loss of the proton from the Schiff base attaching the ligand to the protein. After metarhodopsin-II has activated transducin and over the next minute or so, the linkage is hydrolysed to yield all-*trans*-retinal and the colourless apoprotein opsin. The pigment is said to be *bleached*. Later on, rhodopsin will be reformed by the binding of 11-*cis*-retinal regenerated from all-*trans*-retinal in the adjacent cells of the pigment epithelium.[21,22]

Retinal, an inverse agonist?

A potential problem here is that opsin, although insensitive to visible light, is capable in the presence of all-*trans*-retinal of catalysing guanine nucleotide exchange on transducin, albeit to a small extent.[23–25] Furthermore, as noted in Chapter 3, the very high levels of rhodopsin in photoreceptor cells must pose the risk that spontaneous activation might give rise to some form of spurious 'dark vision' (see page 68). Of course, the catalytic potency of opsin is very small, only 10^{-6} that of photo-excited rhodopsin.[25] However, with 10^8 molecules of rhodopsin per cell, even a very low degree of spontaneous activation must be sufficient to cause the sensation of light even in total darkness. It is now thought that this is prevented by the ligand, 11-*cis*-retinal, acting as an inverse agonist that stabilizes the inactive conformation of the photoreceptor molecule[23] (Figure 6.10). Indeed, in the cell-free situation, the

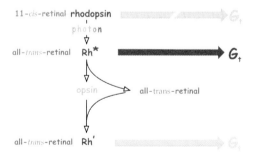

FIG 6.10 Four-state model of rhodopsin.
Of the four states of the rhodopsin molecule that are shown, both Rh* and Rh' can activate transducin. Rh' is a non-covalent complex of opsin and all-*trans*-retinal and it is a weak activator of the G protein. Rhodopsin itself does not activate G_t because of the presence of 11-*cis*-retinal, acting as an inverse agonist.
After Surya et al.[23]

apoprotein opsin can catalyse guanine nucleotide exchange on transducin and this is inhibited by 11-*cis*-retinal. Conversely, all-*trans*-retinal behaves as a somewhat inefficient activator[24] in a manner reflecting the process of activation of other receptors by their specific ligands.

Activation of transducin by the apoprotein is avoided when it is phosphorylated. Rhodopsin kinase is activated by the transducin $\beta\gamma$-subunits in a manner similar to the action of the receptor kinases (see page 98). As a result, the phosphorylated C-terminal domain of rhodopsin acts as a binding site for arrestin, the most abundant protein in the cytosol of the outer segments. This effectively blocks any further interaction with transducin and prevents any signals emanating from the light-insensitive opsin (Figure 6.11).

The rate of decline of the photoresponse is primarily determined by the deactivation of transducin which constitutes the rate limiting step in the sequence of reactions leading from activated rhodopsin to the cGMP phosphodiesterase.[26] The persistence of the active form of the α-subunits must depend entirely on the rate of hydrolysis of the bound GTP. The hydrolysis rate by isolated α_t subunits is far too slow to account for the physiological rate of recovery, but this is accelerated about 100-fold when it is reconstituted *in vitro* together with the phosphodiesterase, approaching the rate that can be measured in disrupted retinal rod outer segments.[27] In this sense, the phosphodiesterase appears to act not only as the downstream effector of transducin, but also as a GTPase-activating protein

However, this is not the full story. RGS9, a member of the RGS family of GAP proteins, is present uniquely in rod outer segments.[28] For knock-out mice that lack RGS9, recovery is not only slow, but is insensitive to the presence of the phosphodiesterase.[29] From this it appears that GAP activity also resides in the RGS–Gβ_5 complex, which lies in wait until the α_t has established communication with its effector phosphodiesterase, before it pounces. In single rods from RGS9 knock-out mice, the time constant of recovery from a light flash, normally ~0.2 s, is increased to ~9 s.[29] This system has the advantage that it allows α_t to remain active until it scores an effective hit on its

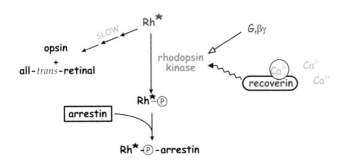

FIG 6.11 Phosphorylation of activated rhodopsin and the binding of arrestin shuts off the signal.

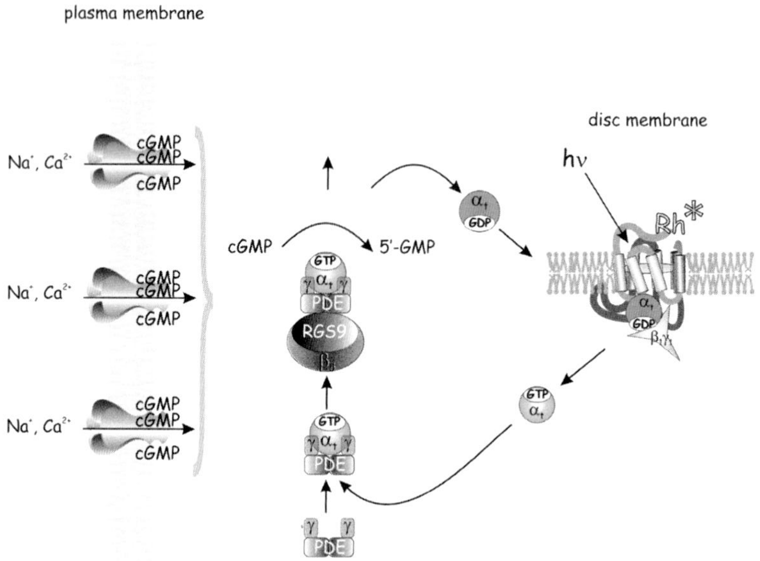

FIG 6.12 Activation and inactivation of transducin.
Photoactivated rhodopsin Rh* catalyses guanine nucleotide exchange on the α-subunit of transducin, which is then released into the cytosol. Here it activates phosphodiesterase, causing hydrolysis of cGMP. The consequent reduction in the concentration of cGMP allows the membrane ion channels to close, curbing the inflow of Na^+ and Ca^{2+}. The resulting membrane hyperpolarization acts as the signal to restrict the secretion of transmitter from the synaptic terminal. Inactivation of transducin occurs through hydrolysis of the bound GTP. The basal catalytic rate of the GTPase is enhanced 100 times through interaction with the phosphodiesterase, together with the heterodimeric complex of RGS9 and $G\beta_5$.

target enzyme, the phosphodiesterase, which also initiates GTPase activation. However, under no circumstances must the activated transducin be allowed to linger. In mammalian rods, the secondary interaction of the transducin–PDE complex with the RGS9–$G\beta_5$ complex ensures that inactivation occurs within 200 ms. These mechanisms are outlined in Figure 6.12. A full and detailed description of the inactivation process may be found in Arshavsky et al.[30]

Note on phototransduction in invertebrates

We have described the basic elements of the signal transduction process as it occurs in the rod cells of vertebrate eyes. The situation in invertebrates is very different, as summarized in Figure 6.13.

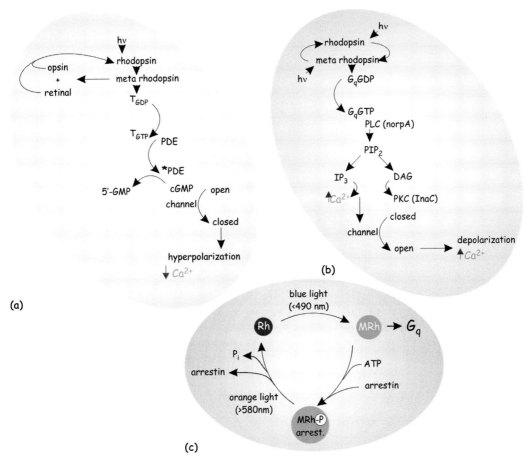

FIG 6.13 Comparing the main features of retinal phototransduction in vertebrates and flies.
The main elements of visual phototransduction in vertebrates (a) and in *Drosophila* (b, c) are illustrated. For further detail, see text. The diagram in (c) illustrates the rhodopsin cycle in *Drosophila* in which the chromophore remains attached to the receptor throughout the process of activation and subsequent recovery. As with vertebrates, activation is initiated by a photon that causes the conversion of 11-*cis*-retinal to the all-*trans* form. In the invertebrate however, recovery is effected by a second, longer-wavelength pulse of light following phosphorylation of the rhodopsin.

In some ways, phototransduction in flies and other spineless creatures appears to follow more familiar pathways. Here, the light-activated rhodopsin is coupled through the G protein G_q, not transducin. This regulates phospholipase Cβ, producing IP_3 and DAG, and results in the elevation of intracellular Ca^{2+} and the activation of protein kinase C. As with vertebrates, these events initiate both electrical and chemical signals, but here the consequence is the opening, not the closure of plasma membrane ion channels. This causes a very transient depolarization (instead of hyperpolarization) and further elevation of the concentration of cytosol

175

Ca^{2+}. As in vertebrates, Ca^{2+} plays several roles ensuring the deactivation of the photosignal. In addition, PLC is both the target and a regulator (GTPase activator, see page 86) of G_q.[31] PKC is a negative regulator of the Ca^{2+} channels, Trp and TrpL (Trp-like, see page 210).

As in vertebrates, the chromophore retinal is an integral component of rhodopsin, but with the difference that it remains attached to the opsin throughout the cycle of activation and recovery. Following illumination by blue light (below 490 nm), the retinal undergoes rapid isomerization to the all-*trans* form and the resulting metarhodopsin catalyses guanine nucleotide exchange on G_q. The retinal remains firmly attached and the activated metarhodopsin is so stable that it would have a half-life of more than 5 h.[32,33] This would not be much help to flies but for the fact that phosphorylation and then the binding of arrestin sensitizes the system to a second photon, this one of longer wavelength (580 nm, orange). This triggers the reinstatement of 11-*cis*-retinal.

> Is not Vision perform'd chiefly by the Vibrations of this medium, excited in the bottom of the Eye by the Rays of Light, and propagated through the solid, pellucid and uniform Capillamenta of the optick Nerves into the place of Sensation? And is not Hearing perform'd by the Vibrations either of this or some other Medium, excited in the auditory Nerves by the Tremors of the Air, and propagated through the solid, pellucid and uniform Capillamenta of those Nerves into the place of Sensation? And so of the Other Senses.
>
> Sir Isaac Newton, *Opticks*

Olfaction

It is said that St Jerome remarked of St Hilarion that he had the gift of knowing what sins and vices anyone was inclined to by smelling either the person or their garments. By the same faculty, he could discern good feelings and virtuous propensities.[34] This may, or may not be true: we have no way of knowing. What is true is that dogs can distinguish between the smells of T-shirts worn by non-identical but not by identical twins.

The human olfactory system is capable of sensing and distinguishing several hundred different molecules (odorants) and a seemingly limitless number of different smells (effectively, combinations of odorants). All of these compounds are volatile, none has a molecular weight much greater than 300 and all have the ability to partition between lipid and aqueous phases. The human olfactory epithelium is formed from ~6 million olfactory receptor neurons (ORNs; 50 million in the rat), together with the supporting cells of Bowman's glands that secrete mucus into the upper reaches of the nasal cavity (Figure 6.14). The dendrites of the receptor neurons terminate as whip-like ciliary extensions, 30–200 μm long and ~20 on each cell, bearing the odorant receptors which project into the mucus layer lining the upper reaches of the nose. These dendritic terminations represent the single point at which cells of the central nervous system are exposed to the environment outside the body.

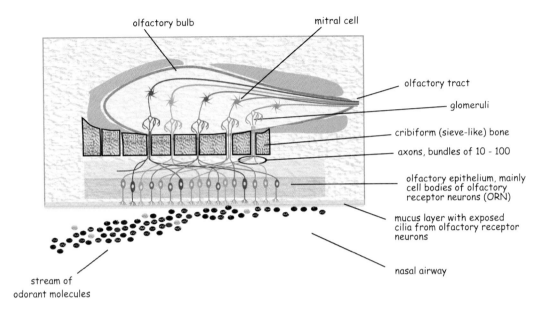

FIG 6.14 Olfactory apparatus, schematic view.

Olfactory receptor cells

The ORNs are bipolar cells. Their single axons project into the olfactory bulbs (extensions of the forebrain, one located above each nasal cavity), where they terminate in glomeruli. These are spherical bundles 50–100 µm in diameter, composed of neuronal and glial processes (known as neuropil). Regardless of their location on the surface of the nasal epithelium, the axons of ORNs bearing the same receptor are gathered together and cross the cribriform plate, terminating in a single glomerulus within the bulb. Here they form synapses with the dendrites of a much smaller number of mitral cells. These are the most numerous cells in the bulb (~50 000) and they interact laterally with each other and with other cells. It is in this layer that the inputs of individual ORNs are integrated. The axons of the mitral cells form the lateral olfactory tract that conveys signals to the primary olfactory cortex in the brain (Figure 6.14). The convergence is such that a single mitral cell may receive input from ~1000 ORNs. This means that although the amount of transmitted information is reduced, a very high sensitivity is ensured.

Each day large numbers of ORNs must be replaced and each one of these must be correctly wired up to its own specific target glomerulus. It is evident that the receptors, in addition to providing the means of chemoperception, themselves provide the signals that guide the axons through the maze to their individual target glomeruli. Thus, in mice engineered to express a

Mitral: pertains to the form of a mitre, a bishop's hat.

177

FIG 6.15 Specificity of axonal targeting to olfactory glomeruli. Gene targeting technology was used to generate a strain of mice in which expression of a particular odorant receptor gene is coupled to that of an axonal marker, which is revealed as a blue stain.
From Mombaerts.[35]

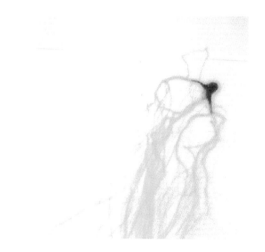

defined receptor coupled to an axonal marker (coloured blue in the image shown in Figure 6.15), it was found that the axons project to specific pairs of glomeruli, situated symmetrically in each of the olfactory bulbs.[35,36]

Olfactory receptors

The odorant receptors are 7TM structures, collectively forming by far the largest subgroup of the rhodopsin-like class of receptors[37,38] (see Figure 3.14, page 61). In mammals, the odorant receptors are coded by ~1000 genes and comprise by far the largest single family in the genome (though in humans a considerable proportion, about 60%, are pseudogenes, and are not transcribed). Within this family the sequence identity ranges from less than 40% to over 90% (near identity). Certain features, such as the elongated second extracellular loop (~35 residues), appear to be characteristic of this particular class of receptor (see Figure 6.16). Indeed, within the class, the sequence of this stretch is highly conserved, a feature not shared by the pair of transmembrane helices to which it is linked. Lacking the possibility of carrying out the ligand-binding experiments that have played such an important role in classical pharmacology, clues such as this have been exploited to indicate the zones responsible for odorant binding in these receptors. Indeed, it has been speculated that hypervariable regions in transmembrane spans D, E, and F may form the sites of odorant attachment. There is no reason to think that the olfactory receptors possess the high specificity towards their ligands characteristic of most other receptors. In general, it appears that they can bind multiple odorants of related structure and that most odorants are capable of activating several different receptors to varying degrees. Thus, different odorants activate different sets of glomeruli.[39]

The failure to express pseudogenes may arise from frameshifts, nonsense mutations, or deletions. This may be the basis of our inferior sense of smell in comparison with other animals (possibly compensated to some degree by visual and auditory senses). The considerable diversity in olfactory perception among humans may be due to individuals having different non-expressible pseudogenes.[35]

Linda Buck and Richard Axel were awarded the Nobel Prize in 2004 for their discoveries of 'odorant receptors and the organization of the olfactory system'.

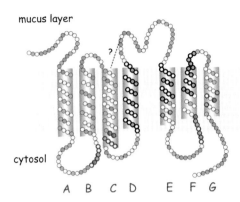

mucus layer

cytosol

A B   C D   E   F G

FIG 6.16 Membrane topology of an olfactory receptor (M71). The most variable residues are outlined in black, the most conserved are outlined in red.

Transduction of olfactory signals

The sequence of events following activation of these receptors by odorants both resembles and differs from the signalling sequence that operates in vertebrate phototransduction. Of course, the most obvious difference is that these receptors are activated conventionally, by the binding of the incoming ligands, while rhodopsin is activated when 11-*cis*-retinal, effectively a resident inverse agonist, is displaced. The activated odorant receptors interact with G_{olf}, a G protein closely related to G_s that activates the Ca^{2+}-calmodulin-sensitive adenylyl cyclase 3 (see Table 5.1, page 138). The cAMP formed interacts with a cyclic nucleotide-gated (CNG) channel, allowing it to conduct Na^+ and Ca^{2+} ions into the cell. Unlike the cGMP-regulated channel in the retina, which is relatively unresponsive to cAMP, the olfactory channel is equally responsive to both nucleotides. Also, unlike the ion channel of photoreceptor cells, the olfactory channel in the retina opens in response to cyclic nucleotides, causing the cells to depolarize. Cloning of CNGs has shown that they are closely related and probably share common ancestry with the superfamily of voltage-regulated ion channels. As the membrane potential drops below 20 mV, the cell generates an action potential which is conveyed along the axon into the specific glomerulus in the olfactory bulb. From here, the signal is conveyed by the mitral cells to the higher centres.

As with the mechanism of phototransduction, the olfactory process allows enormous amplification of the initial signal. A collision of an odorant molecule with its receptor can activate many molecules of G_{olf}. Each of these will activate a single molecule of adenylyl cyclase that can generate large amounts of cAMP. In isolated ciliary membranes, this becomes apparent within 25 ms of applying an odorant and peaks within 500 ms,[40] although recorded odorant-induced membrane currents occur after a latency of several hundreds of milliseconds. Some odours, particularly those of a fruity or floral nature, are effective when applied as solutes at concentrations as low as

Although olfactory and retinal CNGs share more than 80% sequence identity in their cyclic nucleotide binding sites, their channel properties are very different. The olfactory channels hardly discriminate between cAMP and cGMP whereas the retinal channels have a much higher affinity for cGMP. The olfactory channel has a larger pore size, a higher single channel conductance and a lower degree of selectivity among monovalent cations.[38]

The olfactory receptor neurons contain a highly specialized isoform of phosphodiesterase having high affinity for Ca^{2+} ($K_M = 1.4\,\mu mol$ L^{-1}). More than this, it has a higher affinity for cAMP than any other brain isoform.[42] Thus, the binding of odorants to receptors not only initiates the cAMP signal pathway, but also signals its termination.

$10\,nmol\,L^{-1}$. Others, particularly those with a putrid smell, have no effect in this isolated ciliary membrane preparation and it is thought that these may utilize an additional pathway involving $\beta\gamma$-subunits derived from G_o and a Ca^{2+} signal derived from the production of IP_3.[41]

Similar to the channels in the retinal rod outer segment membranes, three molecules of cyclic nucleotide are required to open a CNG and this allows hundreds of thousands of ions to enter the cell. A single odorant molecule is capable of inducing a measurable electrical signal, although the synchronous stimulation of several receptors is probably needed to generate a sensation of smell.

cAMP is certainly the main second messenger that mediates the early stages of odorant perception (Figure 6.17). To illustrate this, adenylyl cyclase 3 knock-out mice cannot smell[43] (said to be anosmic). They are insensitive to odorants that induce signals, mediated by phospholipase C and protein kinase C, that activate adenylyl cyclase 3 (see Table 5.1). However, the signals induced by cAMP are certainly affected by Ca^{2+} and by cGMP. The effects of Ca^{2+} are complex, as it plays a part in the enhancement, modulation, and termination of the odorant-induced signals. As with the retinal system, Ca^{2+}-calmodulin rapidly acts to decrease the affinity of

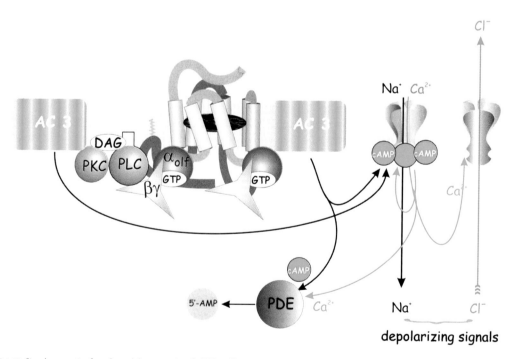

FIG 6.17 Signals emanating from G_{olf} and the generation of cAMP in olfactory receptor neurons.

the channel for the activating cyclic nucleotide.[44] In this way it acts as a regulator of adaptation, allowing the system to detect odorants over a very wide dose range. Ca^{2+} also acts to terminate signals through the activation of cAMP phosphodiesterase. On the other hand, Ca^{2+} can enhance the signal generated by cAMP by opening a Cl^- channel present in the apical membranes of the olfactory neurons.

Rather unusually, these cells maintain a rather high concentration of Cl^- ions so that when the channels are opened, and the Cl^- ions flow out, the cells tend to depolarize. This mechanism is thought to act as a fail-safe, in case the availability of Na^+ in the nasal mucus, a very uncontrolled environment, is insufficient to generate a sufficient inward current of positive charges through the CNG to cause membrane depolarization.

cGMP is also elevated in olfactory neurons following receptor activation, though this occurs on a slower time scale than the elevation of cAMP and then over a more sustained time course. It is thought to act as a determinant of adaptation and desensitization, suppressing the gain of the cAMP pathway and helping to prevent saturation of the olfactory response to high and repeated stimuli.[45]

In most tissues intracellular Cl^- is kept low and so negatively charged ions tend to flow inwards through Cl^- channels to maintain a depolarized state. Cl^- channels, such as those gated by GABA or glycine, are regarded as inhibitory receptors.

References

1. Lagnado L, Baylor DA. Signal flow in visual transduction. *Neuron*. 1992;8:995–1002.
2. Hecht S, Shlaer S, Pirenne MP. Energy, quanta and vision. *J Gen Physiol*. 1942;25:819–840.
3. Baylor DA, Lamb TD, Yau KW. Responses of retinal rods to single photons. *J Physiol Lond*. 1979;288:613–634.
4. Baylor DA, Nunn BJ, Schnapf JL. The photocurrent, noise and spectral sensitivity of rods of the monkey *Macaca fascicularis*. *J Physiol Lond*. 1984;357:575–607.
5. Kühne W. *On the Photochemistry of the Retina and on Visual Purple*. London: Macmillan; 1878.
6. Daemen FJ. Vertebrate rod outer segment membranes. *Biochim Biophys Acta*. 1973;300:255–288.
7. Wheeler GL, Matuto Y, Bitensky MW. Light-activated GTPase in vertebrate photoreceptors. *Nature*. 1977;269:822–824.
8. Kuhn H. Light and GTP-regulated interaction of GTPase and other proteins with bovine photoreceptor membranes. *Nature*. 1980;283:589.
9. Heck M, Hofmann KP. Maximal rate and nucleotide dependence of rhodopsin-catalyzed transducin activation: initial analysis based on a double displacement mechanism. *J Biol Chem*. 2001;276:10009.
10. Young RW. The renewal of photoreceptor cell outer segments. *J Cell Biol*. 1967;33:61–72.

11. Young RW. Shedding of discs from rod outer segments in the rhesus monkey. *J Ultrastruct Res*. 1971;34:190–203.

12. Rodieck RW. *The First Steps in Seeing*. Sunderland, Massachussetts: Sinauer Associates; 1998.

13. Valeton JM, Van Norren D. Light adaptation of primate cones: an analysis based on extracellular data. *Vision Res*. 1983;23:1539–1547.

14. Sokolov M, Lyubarsky AL, Strissel KJ, et al. Massive light-driven translocation of transducin between the two major compartments of rod cells: A novel mechanism of light adaptation. *Neuron*. 2002;34:95–106.

15. Liebman PA, Pugh EN. The control of phosphodiesterase in rod disk membranes: kinetics, possible mechanisms and significance for vision. *Vision Res*. 1979;19:375–380.

16. Pappone MC, Hurley JB, Bourne HR, Stryer L. Functional homology between signal-coupling proteins. Cholera toxin inactivates the GTPase activity of transducin. *J Biol Chem*. 1982;257:10540–10543.

17. Fung BK, Hurley JB, Stryer L. Flow of information in the light-triggered cyclic nucleotide cascade of vision. *Proc Natl Acad Sci U S A*. 1981;78:152–156.

18. Granovsky AE, Natochin M, Artemyev NO. The γ subunit of rod cGMP-phosphodiesterase blocks the enzyme catalytic site. *J Biol Chem*. 1997;272:11686–11689.

19. Liu W, Northup JK. The helical domain of a G protein α subunit is a regulator of its effector. *Proc Natl Acad Sci U S A*. 1998;95:12878–12883.

20. Liu W, Clark WA, Sharma P, Northup JK. Mechanism of allosteric regulation of the rod cGMP phosphodiesterase activity by the helical domain of transducin α subunit. *J Biol Chem*. 1998;273:34284–34292.

21. Flannery JG, O'Day W, Pfeffer BA, Horwitz J, Bok D. Uptake, processing and release of retinoids by cultured human retinal pigment epithelium. *Exp Eye Res*. 1990;51:717–728.

22. Bridges CD, Alvarez RA. The visual cycle operates via an isomerase acting on all-trans retinal in the pigment epithelium. *Science*. 1987;236:1678–1680.

23. Surya A, Stadel JM, Knox BE. Evidence for multiple, biochemically distinguishable states in the G protein-coupled receptor, rhodopsin. *Trends Pharmacol Sci*. 1998;19:243–247.

24. Surya A, Foster KW, Knox BE. Transducin activation by the bovine opsin apoprotein. *J Biol Chem*. 1995;270:5024–5031.

25. Melia TJ, Cowan CW, Angleson JK, Wensel TG. A comparison of the efficiency of G protein activation by ligand-free and light-activated forms of rhodopsin. *Biophys J*. 1997;73:3182–3191.

26. Sagoo MS, Lagnado L. G-protein deactivation is rate-limiting for shut-off of the phototransduction cascade. *Nature*. 1997;389:392–395.

27. Arshavsky VY, Dumke CL, Zhu Y, et al. Regulation of transducin GTPase activity in bovine rod outer segments. *J Biol Chem*. 1994;269:19882–19887.

28. He W, Cowan CW, Wensel TG. RGS9, a GTPase accelerator for phototransduction. *Neuron*. 1998;20:95–102.

29. Chen CK, Burns ME, He W, Wensel TG, Baylor DA, Simon MI. Slowed recovery of rod photoresponse in mice lacking the GTPase accelerating protein RGS9-1. *Nature*. 2000;403:557–560.

30. Arshavsky VY, Lamb TD, Pugh EN. G proteins and phototransduction. *Annu Rev Physiol*. 2002;64:153–187.

31. Cook B, Bar-Yaacov M, Cohen Ben-Ami H, et al. Phospholipase C and termination of G-protein-mediated signalling in vivo. *Nature Cell Biol*. 2000;2:296–301.

32. Scott K, Zuker C. Lights out: deactivation of the phototransduction cascade. *Trends Biochem Sci*. 1997;22:350–354.

33. Feiler R, Bjornson R, Kirschfeld K, et al. Ectopic expression of ultraviolet-rhodopsins in the blue photoreceptor cells of *Drosophila*: visual physiology and photochemistry of transgenic animals. *J Neurosci*. 1992;12:3868.

34. Room A. *Brewer's Dictionary of Phrase and Fable*. London: Cassell; 1999.

35. Mombaerts P. Seven-transmembrane proteins as odorant and chemosensory receptors. *Science*. 1999;286:707–711.

36. Feinstein P, Mombaerts P. A contextual model for axonal sorting into glomeruli in the mouse olfactory system. *Cell*. 2004;117:817–831.

37. Buck L, Axel R. A novel multigene family may encode odorant receptors: a molecular basis for odor recognition. *Cell*. 1991;65:175–187.

38. Ronnett GV, Moon C. G proteins and olfactory signal transduction. *Annu Rev Physiol*. 2002;64:189–222.

39. Firestein S. A code in the nose. *Sci STKE*. 2004;227:pe15.

40. Breer H, Boekhoff I, Tareilus E. Rapid kinetics of second messenger formation in olfactory transduction. *Nature*. 1990;345:65–68.

41. Schandar M, Laugwitz KL, Boekhoff I, et al. Odorants selectively activate distinct G protein subtypes in olfactory cilia. *J Biol Chem*. 1998;273:16669–16677.

42. Yan C, Zhao AZ, Bentley JK, Louchney K, Ferguson K, Beavo J. Molecular cloning and characterization of a calmodulin-dependent phosphodiesterase enriched in olfactory sensory neurons. *Proc Natl Acad Sci U S A*. 2003;92:9677–9681.

43. Wong ST, Trinh K, Hacker B, et al. Disruption of the type III adenylyl cyclase gene leads to peripheral and behavioral anosmia in transgenic mice. *Neuron*. 2000;27:487–497.

44. Chen TY, Yau KW. Direct modulation by Ca^{2+}–calmodulin of cyclic nucleotide-activated channel of rat olfactory receptor neurons. *Nature*. 1994;368:545–548.

45. Leinders-Zufall T, Shepherd GM, Zufall F. Modulation by cyclic GMP of the odour sensitivity of vertebrate olfactory receptor cells. *Proc Roy Soc Lond B*. 1996;263:803–811.

Intracellular Calcium

Calcium is an abundant element. It constitutes 3% of the Earth's crust and it is prominent throughout the biosphere. It is present in fresh and sea water at levels that range from micromolar to millimolar. Ca^{2+} ions and Ca^{2+}-binding proteins are essential for many biochemical processes in prokaryotic and eukaryotic cells. In this chapter and in Chapter 8, we shall see that Ca^{2+} has special roles in the signalling mechanisms that regulate the activities of eukaryotic cells and that this depends upon the sensitivity of some proteins to the concentration of free Ca^{2+} in their surroundings.

A new second messenger is discovered

A number of the more enduring truths of science have been discovered when a failure of vigilance is coupled with uncommon perspicacity. Sydney Ringer's discovery of a requirement for calcium in a biological process must surely be placed in this category.[1,2]

Henry Dale's account:

> Ringer was a physician to University College Hospital, and, in such time as he could spare from his practice, one of the pioneers of pharmacological

research in this country. In his early experiments he had found that a solution containing only pure sodium chloride, common salt, in the proportion in which it is present in the serum of frog's blood would keep the beat of the heart in action for only a short time, after which it weakened and soon stopped. And then suddenly the picture changed: apparently the same pure salt solution would now maintain the heart in vigorous activity for many hours. Ringer was puzzled, and thought for a time that the difference must be due to a change in the season of the year – until he discovered what had really happened. Being busy with other duties, he had trusted the preparation of the solutions to his laboratory boy, one Fielder; and as Fielder himself, who I knew as an ageing man, explained to me, he didn't see the point of spending all that time distilling water for Dr Ringer, who wouldn't notice any difference if the salt solution was made up with water straight out of the tap. But, as we have seen, Ringer did notice the difference; and when he discovered what had happened he did not merely become angry and insist on having distilled water for his saline solution, he took full advantage of the opportunity which accident had thus offered him and soon discovered that water from the tap, supplied then to North London by the New River Water Company, contained just the right small proportion of calcium ions to make a physiologically balanced solution with his pure sodium chloride . . .'

Hints of a wider role for calcium soon followed with Locke's demonstration[3] that removal of calcium could block the transmission of impulses at the neuromuscular junction in a frog sartorius muscle preparation. The realization that this requirement for calcium is due to its role in controlling the secretion of a chemical messenger (neurotransmitter) had to wait nearly 50 years.[4,5]

The direct introduction of Ca^{2+} ions into muscle fibres to cause contraction was first reported by Kamada and Kinoshita in 1943.[6] At this time Japan and the USA were at war, and as a result of the breakdown of communication, the credit for this finding has generally been ascribed to the Americans Heilbrunn and Wiercincki who reported their results 4 years later.[7] Other cations such as Na^+, K^+, and Mg^{2+} were found to be without effect. It had been Heilbrunn's contention, at that time at least, that the effect of calcium was on the 'general colloid properties of the protoplasm' and that the effects of ions on isolated proteins would lack biological relevance.[8] However, Otto Loewi (see Chapter 1) who attended their presentation at the New York Academy of Sciences was heard to growl 'Kalzium ist alles' (calcium is everything). So it remained, until the biochemical understanding of signalling mechanisms became more developed, taking in cyclic AMP (Chapter 5), GTPases (Chapter 4), tyrosine kinases (Chapter 11), PI 3-kinases (Chapter 13), and much more. Even with his foresight, however, had Loewi lived to the end of the century, he would surely have been astounded by the prominence of calcium in contemporary biology.

In previous chapters we have seen how the idea of second messengers followed from the work of Sutherland and Rall,[9] showing that the generation of cAMP represents the essential link between membrane events and the metabolic process in the signalling of glycogenolysis. Reflecting this history, cAMP was said to be the first second messenger and Ca^{2+} the second. In fact, elevation of cytosol Ca^{2+} is more ubiquitous than elevation of cAMP and it regulates a very diverse range of activities, including secretion, muscle contraction, fertilization, gene transcription, and cell proliferation.

Calcium and evolution

The evolution of the multicompartment structure of eukaryotic cells from their prokaryotic forerunners is thought to have occurred in stages, involving successive invaginations and internalizations of the cell membrane. Some modern bacteria provide evidence for this (Figure 7.1). The internalization of elements of the plasma membrane had the effect of segregating specific functions on to specialized intracellular structures, such as the nuclear envelope, the smooth and rough endoplasmic reticulum, the Golgi membranes, lysosomes, etc. A probable early outcome of this process was the formation of a reticular compartment responsible for protein synthesis – a 'protoendoplasmic' reticulum. The concentration of free Ca^{2+} within prokaryotic cells is kept very low by powerful ion translocation mechanisms that expel Ca^{2+} ions. The early endoplasmic reticulum (and other organelles)

FIG 7.1 Thiovulum, a sulfur-oxidizing bacterium possessing endoplasmic membrane structures.
This exceptionally large bacterium (diameter ~6 μm) possesses an intracellular membrane system, possibly including a rudimentary rough endoplasmic reticulum. The black dots inside the cell are probably ribosomes. The dark blobs at the top left are other bacteria.
By courtesy of Tom Fenchel, Marine Biological Laboratory Helsingør, Denmark.

inherited these mechanisms allowing it to draw the ion into its lumen from the cytosol, to provide an intracellular Ca^{2+} store.

The formation of specialized membranes and organelles imposed a need for eukaryotic cells to elaborate the means of communication between their various internal compartments and the external world. Furthermore, the later evolution of complex metazoan organisms having differentiated cells with individual specialized functions necessitated increasingly complex signalling systems. These would enable the coupling of extracellular signals to intracellular signalling pathways and thence to specific downstream effectors. Calcium is well suited for the role of an intracellular messenger, for two main reasons. One is that its immediate proximity provides the opportunity for it to enter the cytosol directly, through membrane channels, either from the extracellular environment or from the internal reticular compartment; the other is provided by calcium's distinctive coordination chemistry.

Distinguishing Ca^{2+} and Mg^{2+}

The term **ligand** was first coined by chemists to describe an electron-donating group that forms coordination complexes with metal ions.

While nature seems to have taken advantage of the presence of Ca^{2+}, its suitability for signalling depends upon its ability to form stable complexes with particular biological molecules. The propensity of metal ions such as Ca^{2+} and Mg^{2+} to form coordination complexes with anionic or polar ligands can be described by equilibria of the form

$$M + L \xrightleftharpoons{K} ML$$

The **stability constant** K is the reciprocal of the dissociation constant K_D, which represents the concentration of free cation M at which the ligand is half saturated.

in which ML represents the complex of the metal ion M with the ligand L. The stability (equilibrium) constant K is a measure of the stability of the complex. All ions in aqueous solution are surrounded by a hydration shell, so in order for coordination to occur, water molecules must be displaced from both the metal ion and the ligand. The factors that determine the stability therefore include not just the electrostatic attraction between cation and ligand, but also attractive and repulsive interactions between adjacent coordinating ligands. Importantly, water is itself a ligand that stabilizes the ionized state. Not all of the water molecules present in the hydration shell need necessarily be removed to form a complex. The stability of the complex will therefore depend upon the energy of association of M and L, set against the energies of association of M and L separately with water.

Ca^{2+} is remarkably different from Mg^{2+} in these respects.[10] This is principally because Ca^{2+} is larger and has less difficulty accommodating ligand atoms around its surface. The number of atoms that bind to it in a complex (the coordination number) is variable, in the range 6–12, and the geometry of the arrangement around the coordination sphere is rather flexible. Conversely, Mg^{2+} forms complexes in which the ligand atoms are nearly always arranged in an octahedral formation (a coordination number of 6). For small anions that can be accommodated in such a structure, the strength of association with Mg^{2+} is greater than with Ca^{2+}. For bulky and irregularly shaped anions, for which the octahedral requirement cannot be satisfied, the preference is reversed.

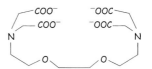

FIG 7.2 EGTA (ethylene-bis(2-aminoethylether)-*N,N,N',N'*-tetraacetic acid).

Chelating agents are synthetic compounds that consist of an array or cage of several ligand atoms. Such multidentate ('many-toothed') assemblies, particularly those that form an irregular structure, can be very selective for Ca^{2+} over Mg^{2+}, because of the ability of Ca^{2+} to tolerate an irregular geometry. One of the most specific Ca^{2+} chelators is EGTA (Figure 7.2). This molecule possesses four carboxyl groups and two ether oxygens and exhibits a 10^5-fold selectivity for Ca^{2+} over Mg^{2+}. (Note that Ca^{2+} overwhelmingly prefers oxygen atom donors.) EGTA and similar compounds are used in the laboratory as Ca^{2+} buffers to control the concentration of free Ca^{2+} in experiments.

> The word **chelate** is derived from the Greek *chela*, meaning the claw of a crab or lobster.

Free, bound, and trapped Ca^{2+}

In the biological world, Ca^{2+} may be considered to exist in three main forms: free, bound, and trapped. In vertebrates, calcified tissues such as bones and teeth account for the major proportion of body calcium. Within these tissues calcium is trapped in mineral form (with phosphate) as hydroxyapatite. Apart from its structural function, bone provides a reservoir of slowly exchangeable calcium that can be mobilized, when needed, to maintain a steady extracellular concentration (Table 7.1). The total amount of calcium in extracellular fluid, in comparison with the mineralized tissues, is very small and of this, only about half can be regarded as free. The remainder is bound mostly to proteins in the extracellular milieu. Within cells, the concentration of free Ca^{2+} inside the cytosolic compartment is still lower, by about four orders of magnitude. A considerable proportion of intracellular calcium is also bound

TABLE 7.1 Approximate levels of free Ca^{2+} and Mg^{2+}

	$[Ca^{2+}]$		$[Mg^{2+}]$
Plasma, extracellular fluid	1–2 mM	pCa ~3	1 mM
Intracellular cytosolic (resting cells)	50–100 nM	pCa ~7	0.5–1 mM
Intracellular lumenal (ER)	30–300 μM	pCa 5–4	

to proteins and these exhibit a wide range of affinities. More importantly, many are able to bind Ca^{2+} ions in the presence of a huge excess of Mg^{2+}. This is because the arrangement of the amino acid side chains that bear the coordinating atoms tends to create sites of irregular geometry. There is also structural flexibility. This has important consequences for the 'on–off' kinetics of binding.

In the extracellular environment, free Ca^{2+} and Mg^{2+} are both present at millimolar concentrations (Table 7.1). Low affinity Ca^{2+}-binding proteins provide a short-term local buffer, while in vertebrates at least, long-term (homeostatic) control of free extracellular Ca^{2+} is maintained by a hormonal mechanism that utilizes the mineralized tissue reservoirs. Failure to maintain a stable level of extracellular Ca^{2+} can have serious consequences. Excess circulating Ca^{2+} (hypercalcaemia) reduces neuromuscular transmission and causes myocardial dysfunction and lethargy. Hypocalcaemia affects the excitability of membranes and if left untreated leads to tetany, seizures, and death.

Cytosol Ca^{2+} is kept low

Ca^{2+}-binding proteins within prokaryotic and eukaryotic cells fulfil a wide range of functions (see Chapter 8). Some, such as the membrane pumps and ion exchanger, are employed to keep cytosol $[Ca^{2+}]$ low. Others act as Ca^{2+} buffers. As already mentioned, a low Ca^{2+} level in the cytosol of eukaryotic cells is maintained by ATP-dependent transporters in the plasma membrane that eject Ca^{2+} from the cell, while others transport the ion into the lumen of the endoplasmic reticulum (ER) or in muscle cells the sarcoplasmic reticulum (a specialized type of ER). The pumps are transmembrane proteins that move Ca^{2+} ions across membranes against their electrochemical potential gradients. They are called Ca^{2+}-ATPases and they provide primary active transport. Those in the membrane of the ER and SR are called sarcoplasmic/ endoplasmic reticulum Ca^{2+} ATPases (SERCAs). Extrusion of Ca^{2+} from the cytosol by plasma membrane pumps is supported by secondary active transport through Na^+–Ca^{2+} exchangers: transmembrane proteins that use the inward Na^+-electrochemical gradient to transport Ca^{2+} ions out of the cell.

Electrochemical potential gradient: the combined effect of the concentration and voltage gradients.

Changes in cytosol Ca^{2+} concentration are also resisted by resident, high-affinity Ca^{2+}-binding proteins, some of which may be confined to particular locations within the cytosolic compartment. All of them may act as buffers that tend to restrict local changes of Ca^{2+}, although their capacity is not high and the speed at which they take effect will depend on the binding kinetics. Within the ER, low-affinity Ca^{2+}-binding proteins ($K_D \sim 2\,mmol\,L^{-1}$) such as calreticulin, can retain up to 20 mol of Ca^{2+} per mol of protein, greatly increasing the capacity of this intracellular store. Estimates of ER $[Ca^{2+}]$ vary between $100\,\mu mol\,L^{-1}$ and $800\,\mu mol\,L^{-1}$.

Over the long term, the coordination of these mechanisms underpins cellular Ca^{2+} homeostasis. Passive inward leaks are balanced by the action of the Ca^{2+}-ATPases which provide a low capacity extrusion pathway. Their rather high affinity ($K_D \sim 0.2\,\mu mol\,L^{-1}$) enables them to maintain low cytosolic Ca^{2+} in resting cells. During cell activation, cytosol Ca^{2+} levels can rise considerably and the pumps and exchangers then operate to restore Ca^{2+} to its resting level. The plasma membrane Na^+-Ca^{2+} exchangers, present in most tissues and of particular importance in heart, although of low affinity ($K_D = 0.5-1\,\mu mol\,L^{-1}$), provide a high-capacity extrusion mechanism for this purpose.

Extracellular calcium and activation

The notion that Ca^{2+} might have a role in cell activation was based mainly on the experience that some tissue responses are suppressed when Ca^{2+} is withdrawn from the surrounding medium, much as Locke and Ringer had demonstrated over 100 years ago. However, many other tissues, in which we now know that an elevation of cytosol Ca^{2+} occurs, are seemingly unaffected because the cells can call upon their own intracellular stores. Confirmation of a direct role for Ca^{2+} in cell activation therefore required more direct means of controlling or sensing changes in its level *in situ*.

Using Ca^{2+} ionophores to impose a rise in Ca^{2+}

The Ca^{2+} ionophores are lipid-soluble, membrane-permeant ion carriers that are specific for Ca^{2+}. They were originally isolated from cultures of *Streptomyces* and include the antibiotics A23187 (524 Da) and ionomycin (709 Da). They may be used to convey Ca^{2+} ions into cells, and during the 1970s they were applied to numerous preparations, particularly in the field of secretion. Among the observed effects were responses similar to those that follow biological stimulation, but without the involvement of receptor-associated mechanisms. This revealed a widespread involvement of Ca^{2+} in activation mechanisms, establishing it as a legitimate second messenger alongside cAMP. However, ionophores are blunt weapons and although they certainly enable the introduction of Ca^{2+}, seemingly to induce biological responses, it proved hard to exert control over the concentrations achieved.

Sensing changes in intracellular Ca^{2+} concentration

To detect and follow the progress of changes in $[Ca^{2+}]$ in living cells has required the introduction of a Ca^{2+}-sensing agent. This presented a formidable challenge. It was first attempted in 1928, by injecting the dye alizarin sulfonate into an amoeba. Apparent evidence of a local increase in Ca^{2+} was obtained when a visible precipitate formed at the site of pseudopod formation.[11] Unfortunately, it later emerged that alizarin is not specific for Ca^{2+}. More unfortunately, the development of practical Ca^{2+} indicators suitable for use with mammalian cells took a further 50 years. In the meantime, intracellular calcium levels were mostly investigated by measuring transmembrane fluxes of the radioisotope ^{45}Ca. It is possible to measure changes in *total* cell Ca^{2+} by this means, but impossible to determine how concentrations have changed within particular intracellular compartments, such as the cytosol. To make matters worse, the bulk of cell calcium is retained within organelles and there is very little in the cytosolic pool.

Ca^{2+}-sensitive photoproteins

Ca^{2+}-sensitive photoproteins, extracted in their active form from marine invertebrates such as the cnidarians *Aequorea*[15] and *Obelia*,[16] undergo an internal molecular rearrangement activated by Ca^{2+}. The conformational change oxidizes a luminophoric prosthetic group, coelenterazine, emitting a pulse of blue light.

Among the more colourful fauna of the oceans are luminescent jellyfish from which the Ca^{2+}-activated, light-emitting protein aequorin can be extracted.[12] Aequorin was the first effective indicator for intracellular Ca^{2+}. Its molecular mass is 22 kDa and it cannot cross membranes, so before the era of transfection techniques its use was limited to larger cell types into which it could be microinjected. Once in the cytosol, an increase in free Ca^{2+} concentration causes the protein to release a pulse of light. Analysis of the time course of the emission can be used to follow changes in Ca^{2+} concentration (up to $100\,\mu mol\,L^{-1}$). In important early work, aequorin was used with preparations such as squid giant axon[13] and permeabilized skeletal muscle.[14] These studies provided the first direct observations of so-called 'Ca^{2+} transients', the biphasic rise and fall of cytosol Ca^{2+} following a stimulus.

Fluorescent Ca^{2+} indicators

The measurement of free intracellular Ca^{2+} became a practical reality with the design by Roger Tsien of the first of a series of synthetic Ca^{2+} indicators. The fluorescence of these compounds is sensitive to Ca^{2+} concentration in the range $50\,nmol\,L^{-1}$ to $1\,\mu mol\,L^{-1}$.[17,18] The first of these to be developed was called Quin2. Its structure (**Figure 7.3**) is based on that of the Ca^{2+} chelator EGTA (**Figure 7.2**) and like its forebear, it has high affinity for Ca^{2+} and low affinity for Mg^{2+}. Because it is anionic and cannot cross membranes, a trick was needed to deliver it to the cytosolic compartment without resorting to microinjection.

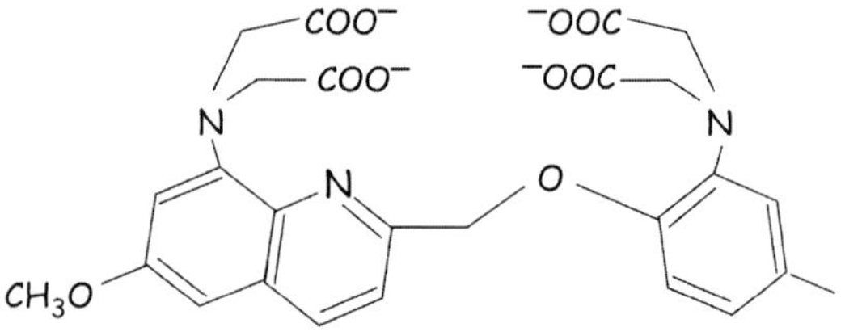

FIG 7.3 Quin2.

This problem, which affects all Ca^{2+} indicators of this type, was overcome when they were provided as uncharged acetoxymethyl ester derivatives.[19] In this form they can diffuse freely across the plasma membrane. Upon entry, endogenous esterase activity releases the acidic form of the indicator, which then accumulates in the cytosol of the intact cells, its fluorescence reporting on changes in the level of cytosolic free Ca^{2+}.

This remarkable technical advance enabled the sensing and monitoring of cell activation in real time. A further landmark came with the development of another Ca^{2+}-indicator dye, Fura2.[20] While Quin2 can provide uncalibrated estimates of changes in cytosol calcium, Fura2, loaded into cells in the same way, can be used to estimate the actual concentration of Ca^{2+} in the cytosolic compartment. This dye exhibits progressive wavelength shifts in its fluorescence excitation spectrum as Ca^{2+} is bound (Figure 7.4). The ratio of the fluorescence emission at two selected excitation wavelengths can then be calibrated and used quantitatively to measure local $[Ca^{2+}]$.

Following these developments, many other Ca^{2+} indicator dyes that operate on the same principles have been devised. These offer improved spectral properties and different Ca^{2+} affinities. The affinity determines the range of sensitivity, and dyes have been developed with K_D values extending from $100\,nmol\,L^{-1}$ up through the micromolar range.

Ca^{2+} mobilization and Ca^{2+} entry

The major significance of all this has been that it is possible to monitor intracellular Ca^{2+} in almost every kind of cell, large or small, simply by recording fluorescence. Initially, measurements were performed on suspensions of isolated

Spectral properties and sensitivity. Cell autofluorescence (uv/blue region) is a problem. Modern indicators, such as Calcium Green and Calcium Crimson, emit visible light, avoiding this background.

Visualizing and monitoring Ca^{2+} changes in single cells has benefited from the introduction of dyes that undergo up to a 100-fold increase in fluorescence emission upon binding Ca^{2+} (Fluo3 , Fluo4). At resting Ca^{2+} levels these dyes are almost non-fluorescent.

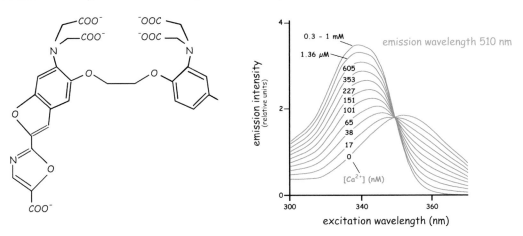

FIG 7.4 Molecular structure and fluorescence excitation of Fura2.

The graph shows how the excitation spectrum of Fura2 varies with Ca^{2+} concentration. For example, if excitation is provided at 340 nm (in the near UV region), the intensity of emission (green fluorescence, measured here at 510 nm) will increase with Ca^{2+} concentration as indicated.

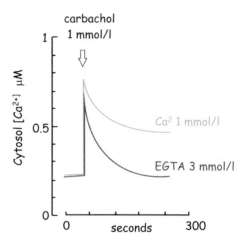

FIG 7.5 Changes in the concentration of cytosol Ca^{2+}: effect of extracellular Ca^{2+}.
The traces show elevations of Ca^{2+} in parotid acinar cells measured using Fura2. The stimulus (carbachol, 1 mmol L^{-1}) was applied at the time indicated by the arrow to cells in the presence of extracellular Ca^{2+} (1 mmol L^{-1}, blue line) or in very low Ca^{2+} ($<$1 nmol L^{-1}, red line). (The low Ca^{2+} medium had no added Ca^{2+} salts and contained EGTA to remove residual Ca^{2+}.)
Adapted from Merrit et al.[21]

cells undergoing activation in a fluorimeter cuvette. In this situation, the recorded emission represents the sum of contributions from each cell in the light beam. Very soon, studies on a wide variety of non-excitable cells revealed characteristic response patterns as illustrated in Figure 7.5. Application of a stimulatory ligand typically produces a biphasic change, in which the concentration rises from the resting level (40–100 nmol L^{-1}) towards the micromolar range and then falls back to a level somewhat higher than the resting concentration. Since the transient phase of the signal is not affected when Ca^{2+} is removed from the external solution, its source must be Ca^{2+} ions mobilized from internal stores. Its duration varies from seconds to minutes, depending on the type of cell. The residual component of the signal, on the other hand, is eliminated by the effective removal of extracellular Ca^{2+} prior to stimulation. It must therefore require entry of Ca^{2+} into the cell across the plasma membrane.

Monitoring cytosol Ca^{2+} in individual cells

The problem with measurements (of any type) on cell suspensions is that the signal represents the summation of individual, unsynchronized contributions. When Ca^{2+} indicator dyes are used with single cells observed under a fluorescence microscope, the time courses are very different from that shown in Figure 7.5. Following activation, the signals may appear as trains of sharp spikes or as a series of smooth oscillations (Figure 7.6). The

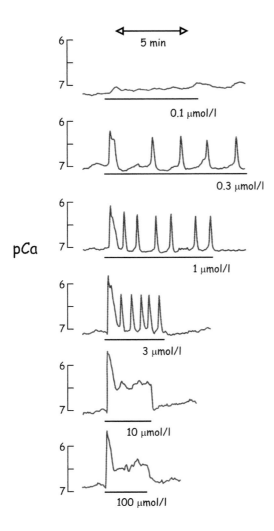

pCa

6 — 7   0.1 μmol/l

6 — 7   0.3 μmol/l

6 — 7   1 μmol/l

6 — 7   3 μmol/l

6 — 7   10 μmol/l

6 — 7   100 μmol/l

5 min

FIG 7.6 Time course of Ca^{2+} concentration changes induced by histamine in a single vascular endothelial cell.

The effect of different concentrations of histamine on intracellular $[Ca^{2+}]$ in a single human endothelial cell in the presence of 1 mmol L^{-1} Ca^{2+}. The bars indicate the application of histamine. The Ca^{2+} indicator was Fura2.

Adapted from Jacob et al.[22]

periodicity and amplitude tend to vary with the strength of the stimulus.[22] Low concentrations or brief exposure to an activating ligand may elicit short episodes of oscillations. These then die away and the concentration of Ca^{2+} reverts to the basal level. Stronger stimulation may evoke longer trains of waves that can merge to give a more persistent elevation. Possible explanations of these phenomena are discussed later in this chapter.

Detecting and imaging subcellular Ca^{2+} changes

Although Ca^{2+} ions are distributed uniformly in aqueous solutions, the spatial distribution of free Ca^{2+} within the cytosolic compartment is often far from uniform. During activation the pattern can fluctuate rapidly. For example,

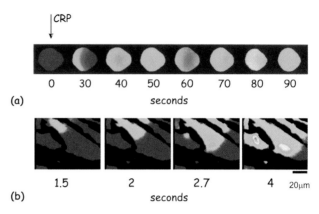

CRP

| 0 | 30 | 40 | 50 | 60 | 70 | 80 | 90 |

seconds

(a)

1.5   2   2.7   4   20μm

seconds

(b)

FIG 7.7 Ca^{2+} waves in a megakaryocyte and in astrocytes in response to stimulation by soluble ligands. Time courses of Ca^{2+} levels in cells preloaded with a Ca^{2+}-sensitive fluorescent dye. a) Activation of a single megakaryocyte in response to a collagen peptide (CRP). The increase in Ca^{2+} peaks at 50 s. This is followed by a sustained increase lasting up to 5 min (not shown). The extracellular Ca^{2+} concentration was 200 μmol L^{-1}. The cell is ~40 μm in diameter. From Mountford et al.[23] (b) Ca^{2+} waves in single astrocytes stimulated with ATP which activates purinergic receptors. From Michael Duchen.[24]

Confocal microscopy.
By eliminating the out-of-focus contributions to the image from regions of the specimen above and below the focal plane, the confocal microscope provides an image of a thin slice. The resolution and quality are higher than those of a standard fluorescence microscope. Moreover, a set of such two-dimensional images, obtained by moving the plane of focus successively through the specimen, can provide a three-dimensional image of cell fluorescence.

the levels in the close vicinity of open Ca^{2+} channels can be much higher than those in more distal regions. Under the fluorescence microscope, the mapping of local Ca^{2+} levels is restricted by the limit of spatial resolution of the objective lens, (theoretically ~0.2 μm, but usually of the order of 0.5 μm). This is usually sufficient to distinguish between nuclear and cytosolic pools and to observe the propagation of Ca^{2+} increases in larger cell types. These are called Ca^{2+} waves and examples are shown in Figure 7.7. To interpret these data in terms of intracellular Ca^{2+} levels, images are frequently presented as intensity maps, using a pseudocolour display in which 'warmer' colours indicate higher Ca^{2+} levels.

Better-quality images are provided by laser confocal microscopy which enables the imaging of a section through the specimen. Confocal images of living cells loaded with Ca^{2+}-sensitive fluorophores can provide a much more detailed picture of the spatial complexity of Ca^{2+} fluctuations. In many microscopes confocal images may be obtained at video rates, enabling calcium imaging at both high temporal and spatial resolution. An example of a transient Ca^{2+} signal evoked in a polarized secretory cell by a hormone is shown in Figure 7.8.

Targeted fluorophores and genetically encoded indicators
The resolution of subcellular Ca^{2+} events reported by soluble indicators is limited by the optical system and by microdomains of Ca^{2+} that may exist in different regions of the cytosol or within particular organelles. These are usually not well revealed. However, it is possible to target Ca^{2+}-sensors to specific

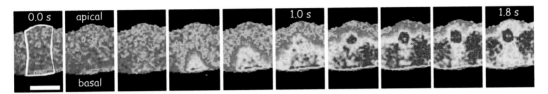

FIG 7.8 Exposure of exocrine secretory cells to 5-hydroxytryptamine (5HT) evokes a transient increase in cytosolic Ca^{2+} that propagates across the cell. Confocal images of blowfly salivary cells exposed to a low dose (3 nmol L^{-1}) of 5HT, applied to the basal surface of the cells (in a Ca^{2+}-free medium). The central cell is outlined in white and the warmer colours indicate higher Ca^{2+} concentrations. The transients last about 10 s and progress from the basal membrane towards the apical surface. The signal also propagates to surrounding cells.[25]

subcellular compartments. For example, the Ca^{2+}-sensitive bioluminescent protein aequorin, described above, has been expressed in transfected mammalian cells. When the N-terminal targeting sequence of cytochrome *c* oxidase is incorporated in the transferred DNA, the photoprotein is expressed in mitochondria, enabling the detection of lumenal Ca^{2+} changes.[26]

A drawback is that the light yield from aequorin is rather low, making it difficult to use with single cells. However, the fluorescent proteins, such as GFP, can also be transfected into mammalian cells and although their emission is not Ca^{2+}-dependent, it can become so when they are expressed as fusion constructs with the small protein calmodulin and a short peptide from one of its binding targets, myosin light chain kinase (see Chapter 8). When Ca^{2+} rises above its resting level, it binds to the calmodulin moiety which then undergoes a pronounced conformational change as it interacts with the peptide sequence (Figure 8.2, page 226). This movement has the effect of altering the fluorescence efficiency of the coexpressed fluorescent protein, rendering it Ca^{2+}-dependent. Alternatively, the calmodulin/peptide construct can be used to alter the fluorescence resonance energy transfer (FRET, see page 114) between two coexpressed fluorophores (such as EYFP and CFP). Examples include the pericam probes that utilize a single (permuted) fluorophore sequence[27] and the 'cameleon' probes that are FRET-based.[28] In both cases the transfected cells synthesize the Ca^{2+} probes *in situ* and these may be expressed in designated cytosolic regions or compartments by including appropriate targeting sequences.

A greater challenge is to monitor Ca^{2+} levels in the close vicinity of the Ca^{2+}-binding effector proteins that direct downstream signals. Here, the sensor must be localized at or close to a target protein, without affecting its function. One way to achieve this is to engineer a pair of dicysteine motifs into the target.[29] Application to the transfected cells of a bisarsenical derivative of Calcium Green (Calcium Green FIASH) then results in the indicator binding covalently to the four cysteine side chains. This probe can report on highly localized Ca^{2+} changes immediately related to downstream signalling and it has the added

GFP and the fluorescent proteins.

Green fluorescent protein (GFP) is a 20 kDa protein found in the same jellyfish (*Aequoria victoria*) that is the source of aequorin. Mammalian cells may be transfected to express GFP alone or as a fusion partner to another protein, rendering it fluorescent and allowing its location to be tracked in living cells.

Engineered forms of GFP with improved fluorescence and emitting other colours, have extended the utility of this powerful tool, e.g. Cyan (CFP), yellow (YFP), enhanced GFP (EGFP), etc.

advantage of a rapid response. It has been used to observe the transient increases in Ca^{2+} close to active L type Ca^{2+} channels in single cells (Figure 7.9).[30]

Mechanisms that elevate cytosol Ca^{2+} concentration

Ca^{2+}-sensitive enzymes first appeared in eukaryotic cells at around the same time as the ion channels that allow Ca^{2+} to flow into the cytosolic compartment on demand. This provided the basis for a signalling system, as these enzymes responded to the rise in intracellular Ca^{2+} by activating a variety of downstream pathways. Although Ca^{2+}-dependent processes take place in both bacteria and eukaryotes, it is only in eukaryotic cells that extensive use of Ca^{2+} as an intracellular signal has evolved.

Ca^{2+}-binding effectors are discussed in Chapter 8; in the following sections the mechanisms that elevate cytosol Ca^{2+} will be described. These involve its

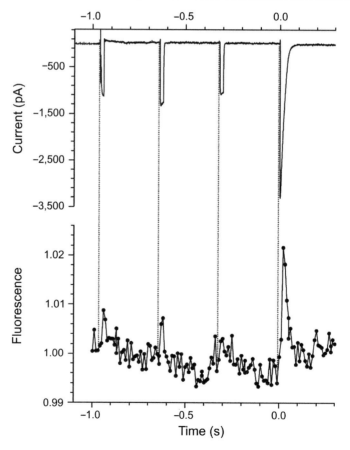

FIG 7.9 Ca^{2+} transients in the immediate vicinity of plasma membrane Ca^{2+} channels. Rapidly changing Ca^{2+} levels (lower trace) sensed by Calcium Green FlAsH. The probe was bound covalently to the intracellular chains of L-type Ca^{2+} channels expressed in HEK293 cells. Opening and closing of the voltage-sensitive channels was effected under voltage clamp by a series of 20 ms depolarizing pulses. The evoked currents are shown in upper trace. The fluorescence signal is corrected for photobleaching. Modified from Tour et al.[30]

release from Ca^{2+}-storing organelles or its entry into cells from the exterior. In non-excitable cells, release of Ca^{2+} from internal stores commonly follows a ligand–receptor interaction at the cell surface that leads to the activation of phospholipase C (described in Chapter 5). It hydrolyses the phospholipid, phosphatidylinositol 4,5-bisphosphate $PI(4,5)P_2$) to produce two second messengers, one of which is inositol trisphosphate (IP_3) which triggers release of Ca^{2+} into the cytosol from the intracellular stores in the endoplasmic reticulum (Figure 7.10). When the stores become depleted, they are replenished from the extracellular Ca^{2+} pool by a mechanism that involves the opening of yet other ion channels in the plasma membrane.

In electrically excitable cells, Ca^{2+} elevation is initiated by the opening of voltage-dependent Ca^{2+}-channels in the plasma membrane following a depolarizing stimulus. It is generally more rapid and it achieves high concentrations of Ca^{2+} in the immediate vicinity of the channel. Although localized, this increase may itself cause release of further quantities of the ion from intracellular stores, in a regenerative manner, so that the signal spreads away from the original source.

It is important to note that IP_3 is not the sole Ca^{2+}-mobilizing second messenger, nor are the stores in the endoplasmic or sarcoplasmic reticulum the only intracellular sources of Ca^{2+}. Other messenger molecules that can release Ca^{2+} into the cytosol include the two nucleotides cyclic ADP-ribose and nicotinic acid adenine dinucleotide phosphate, and the lipid sphingosine-1-phosphate. These are discussed below.

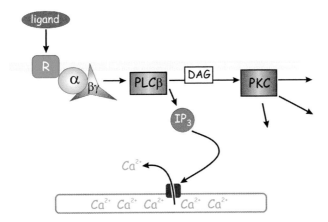

FIG 7.10 Activation of phospholipase C and the mobilization of Ca^{2+} from intracellular stores. Calcium mobilizing receptors are coupled through G_q or G_i to PLCβ (for details, see Chapter 5) which hydrolyses $PI(4,5)P_2$ releasing two products. DAG is retained in the membrane and activates protein kinase C. The water-soluble IP_3 binds to ligand-gated channels present in the membranes of intracellular Ca^{2+} stores in the endoplasmic reticulum, releasing Ca^{2+} into the cytosol.

Ca²⁺ release from intracellular stores, IP₃ and ryanodine receptors

Ca^{2+} release into the cytosol from intracellular stores is brought about by ligand-activated ion channels in the membrane of the ER. There are two types of structurally related channel, the inositol trisphosphate receptors (IP$_3$Rs) and the ryanodine receptors (RyRs). IP$_3$Rs are widely expressed, whereas RyRs, which were first identified in the sarcoplasmic reticulum (SR) of skeletal and cardiac muscle, are most evident in excitable cells.

Ryanodine receptors

RyRs are large assemblies of four 560 kDa subunits arranged in a homotetrameric structure. Some 20% of this is submerged in the membrane and contains the Ca^{2+} release channel. The remainder extends into the cytoplasm. Of the three isoforms, RyR1s are predominant in skeletal muscle SR. Here the SR and the plasma membranes are in close apposition so that the receptors interact directly with the voltage-sensing subunits (dihydropyridine receptors) of the plasma membrane Ca^{2+} channel complexes (see Figure 8.6, page 238). In contrast, RyR2s expressed in cardiac muscle and some nerve cells, do not interact with plasma membrane Ca^{2+} channels in such a direct manner. A third isoform, RyR3, is present in brain, muscle and some non-excitable cells.

A three-dimensional reconstruction of a RyR1 obtained by cryoelectron microscopy is shown in Figure 7.11. (RyR2 and RyR3 have similar structures). The transmembrane assembly contains the Ca^{2+} release channel, while the major portion of the molecule extends into the cytoplasm where it receives and coordinates signals that control the opening and closing of the channel. The assembly possesses binding sites for a number of physiological modulators. It also has a number of phosphorylation sites. Ligands include Ca^{2+}, ATP, and calmodulin (both apocalmodulin and Ca^{2+}-calmodulin, see Chapter 8). Also, there is a 12 kDa protein, FKBP12, that copurifies with the receptor and appears to be an integral part of its structure. FKBP12 binds the immunosuppressive drug FK506 (tacrolimus) and is termed an immunophilin. It interacts with the Ca^{2+}-dependent protein phosphatase calcineurin (see page 683). Although the sites at which many of these ligands bind are known, it is unclear how their actions regulate channel activity. The most important regulator of the channel is Ca^{2+} itself and this is discussed below.

Inositol trisphosphate receptors

Like the RyRs, IP$_3$Rs are also large assemblies of four major subunits, although the molecular mass of each is smaller (~310 kDa). There are three similar isoforms of the basic subunit, but unlike the RyRs they may assemble as either homo- or heterotetramers. Type 1 IP$_3$Rs are present at high levels in

Ryanodine is a plant alkaloid that affects the release of Ca^{2+} from the sarcoplasmic reticulum in muscle. It binds to ryanodine receptors with high affinity. .At low concentrations the channel open probability is increased, at high concentrations (micromolar) it decreases.

Sarcoplasmic reticulum (SR) denotes the smooth endoplasmic reticulum of muscle cells.

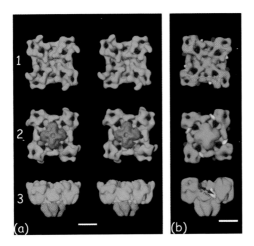

FIG 7.11 Three-dimensional structure of the type 1 ryanodine receptor obtained by cryoelectron microscopy. (a) Three-dimensional reconstruction of RyR1 from skeletal muscle. Stereoscopic views of (1) the surface that faces the T-tubule; (2) that facing the lumen of the SR; (3) a side view. The transmembrane part is pink and the cytoplasmic region is green. (Instructions for viewing stereo-images can be found on page xxv). (b) A depiction of the receptor (cyan) indicating the locations of calmodulin (yellow) and FKBP12 (magenta). The orange shading indicates regions likely to interact with dihydropyridine receptors. The scale bar indicates 10 nm. Adapted from Samsó et al.[31]

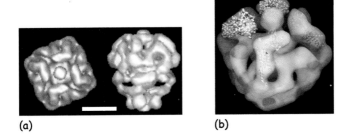

FIG 7.12 IP$_3$ receptor calcium release channel.
(a) Electron density map of IP$_3$R1 as viewed from the cytosol (left) and from the side (right). Four L-shaped density regions are apparent in the cytosol view. The yellow band approximates the predicted transbilayer region. The scale bar indicates 10 nm. (b) Three-dimensional reconstruction of the receptor showing the predicted positions of the four IP$_3$-binding cores, shown as either space-filled or ribbon structures, within the L-shaped electron densities.
Adapted from Sato et al.[32]

cerebellar tissue, but apart from the CNS, nearly all mammalian cells express more than one isotype. **Figure 7.12** shows a three-dimensional reconstruction of the whole assembly in its Ca^{2+}-free form, obtained by cryoelectron microscopy. The structure contains many cavities and most of it protrudes into the cytoplasm.[32] Both the N- and C-termini of each subunit lie within the cytosol and it is the 2000 amino acid N-terminal chain that accounts for most of the cytoplasmic structure (some 85% of the protein). This region possesses binding sites for IP_3, Ca^{2+}, and other ligands. The IP_3 binding region lies at the N-terminus and consists of a 'core' and a 'suppressor' domain. Both are necessary for channel function and it has been proposed that the core domain, bearing the IP_3 binding site, lies in the L-shaped structures on the surface distal to the membrane (see Figure 7.12).

The cytoplasmic component also binds more than a dozen other ligands, implying a capacity to integrate multiple inputs, though cellular evidence for many of these is so far limited.[33] There are also several phosphorylation sites, some of which may have regulatory roles. A number of protein ligands have been shown to affect activity. These include the calmodulin-related Ca^{2+}-binding proteins CaBP1/2 (expressed in brain and retina) and the widely expressed CIB1 (Ca^{2+} and integrin-binding protein 1, calmyrin). Their modes of action, whether as activators or inhibitors of receptor function, are not well understood. Other protein ligands include the adaptor protein RACK1 (see page 581) that increases sensitivity to IP_3 and IRBIT (IP_3R-binding protein released with inositol 1,4,5-trisphosphate), which when phosphorylated, masks the binding sites for IP_3. Finally, as with the RyRs, the immunophilin FKBP12 (see above) binds to the IP_3 receptor N-terminal region. FKBP12 interacts with the Ca^{2+}-dependent phosphatase calcineurin and so may affect the phosphorylation state of the receptor.

Ca^{2+}-induced Ca^{2+}-release

With so many ligand binding sites on the cytoplasmic chains of ryanodine and IP_3 receptors, it is clear that the regulation of Ca^{2+} release by both channels is complex. To appreciate how channel opening is controlled, it is important to grasp that in both cases, the main determinant of activation is the concentration of free Ca^{2+} in the cytosol. Since activated channels remain open for only brief periods, Ca^{2+} release is determined by the rate at which they open, which is itself dependent on $[Ca^{2+}]$. Under resting conditions (50–100 nmol L^{-1}), the rate of opening is very low and any Ca^{2+} entering the cytosol from the stores is quickly removed, but as the concentration rises, more channels open and so more Ca^{2+} is released into the cytosol. The increased local $[Ca^{2+}]$ then causes neighbouring channels to open progressively, giving rise to the phenomenon of Ca^{2+}-induced Ca^{2+}-release, or CICR. It is a form of positive feedback and it underlies the regenerative nature of many Ca^{2+} signals.

IP_3 and CICR

The key modulator of IP_3Rs is, predictably, IP_3 itself. Its main effect is as an allosteric modulator of channel Ca^{2+} sensitivity. The IP_3R cytoplasmic domain has both high- and low-affinity Ca^{2+}-binding sites and it is the binding of Ca^{2+}

to the high-affinity sites that increases the rate of channel opening, causing CICR. To prevent a runaway situation which might empty the stores, negative feedback is introduced when Ca^{2+} binds to the low-affinity sites that inhibit the channel. IP_3 brings about Ca^{2+} release, not by increasing the Ca^{2+}-sensitivity of channel activation, but by decreasing the Ca^{2+}-sensitivity of channel inhibition. This causes the dependence of channel open probability on local $[Ca^{2+}]$ to follow a bell-shaped curve that broadens as $[IP_3]$ increases. This is illustrated in **Figure 7.13** in which experimental data, fitted to a biphasic Hill equation, provide an empirical demonstration of this dependence.

There are however, many other factors that act to complicate the picture.[33] For example, the Ca^{2+}-dependent inhibition of release may be both independent as well as dependent on IP_3 and there is evidence that the receptor has several distinct types of Ca^{2+} binding site. Other ligands that also influence channel sensitivity to $[Ca^{2+}]$ include free ATP (ATP^{4-}, i.e. ATP not bound to Mg^{2+}).

IP_3R Isoforms

The existence of multiple splice variants of the different IP_3R isoforms suggests that a wide diversity of channels may be expressed. Attempts to detect physiological differences between channels formed from the three isoforms have produced somewhat contradictory data. This has been in part due to the relative inaccessibility of the channels (in the ER), their sensitivity to different environments and their tendency to form *hetero*tetramers. In fact, electrophysiological studies of IP_3R channels expressed in native membranes (albeit nuclear, rather than ER) indicate that all three isoforms have very similar gating and ion permeation characteristics.[35]

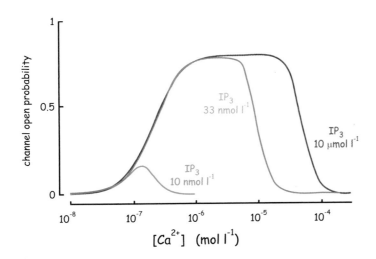

FIG 7.13 IP_3 and Ca^{2+}-dependence of type 1 IP_3 receptor channel open probability.
This graph shows the open probability of IP_3 receptors in the outer membrane of nuclei of *Xenopus laevis* oocytes at different levels of IP_3. Adapted from Mak et al.[34]

Elevation of Ca^{2+} by cyclic ADP-ribose and NAADP

While IP$_3$ is the most widespread and certainly the best understood Ca^{2+}-mobilizing second messenger, there are other small molecules that can also cause the release of Ca^{2+} from intracellular stores. Two of these are adenine nucleotides, cyclic ADP-ribose (cADPR), and nicotinic acid adenine dinucleotide phosphate (NAADP). Both were first identified as Ca^{2+}-mobilizing agents in sea urchin eggs.[36] Their structures are shown in Figure 7.14. In contrast to IP$_3$, it has been difficult to establish their credentials as messenger molecules in mammalian cells and many questions remain about their synthesis and their intracellular targets. They are thought to be produced from the coenzymes NAD$^+$ and NADP$^+$: cADPR by cyclization of NAD$^+$ and NAADP by exchange of the nicotinamide moiety of NADP$^+$ for nicotinic acid, so-called base exchange.

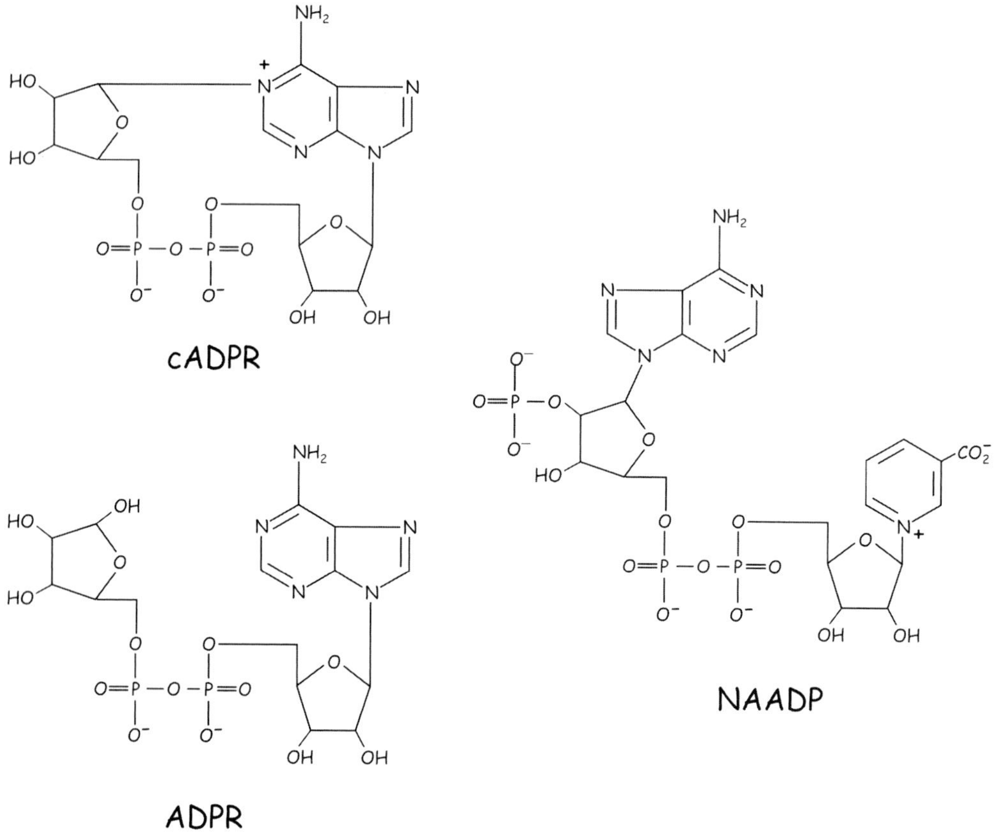

FIG 7.14 Structures of cyclic ADP-ribose, NAADP, and ADP-ribose.

Cyclic ADP-ribose

Evidence for Ca^{2+} mobilization by cADPR exists for a range of cell types that includes oocytes, neuronal cells, muscle cells (skeletal, smooth and cardiac), certain secretory cells, and lymphocytes.[37] When cADPR is introduced into cells at low concentrations (10–50 nmol L^{-1}), it initiates CICR from a ryanodine-sensitive Ca^{2+} pool. Its mode of action is rather like that of IP_3 interacting with IP_3Rs in that its elevation promotes CICR. In effect, it sets the Ca^{2+}-sensitivity of type 2 and 3 RyR channels (but not type 1). However, it appears that an accessory protein may be required, so that it may not act alone. In sea urchin eggs this is calmodulin.[38]

Activation of ADP-ribosyl cyclase to form cADPR occurs in response to a cell surface ligand–receptor interaction, but how these events are coupled is unclear and there is uncertainty about the location of the enzyme. The only well-characterized form of ADP-ribosyl cyclase is CD38, an ectoenzyme present on the extracellular surface of the plasma membrane. The clearest evidence for the formation of cADPR and for its action as a second messenger comes from pancreatic β cells (endocrine cells). In response to a glucose stimulus, the level of intracellular cADPR rises and this is accompanied by elevation of cytosol Ca^{2+} followed by insulin secretion.[39] Also, stimulation of T lymphocytes through their T cell receptor complexes (see page 515) activates ADP-ribosyl cyclase to produce a sustained elevation of the level of cADPR and a long-lasting Ca^{2+} signal (a series of oscillations and waves).[40] Sustained Ca^{2+} elevations appear to be necessary for lymphocytes to become committed to a full proliferative response.

Nicotinic acid adenine dinucleotide phosphate

NAADP (Figure 7.14) is also a potent Ca^{2+}-releasing second messenger. Initially it was found to bring about release from intracellular stores in sea urchin egg homogenates, when provided at concentrations in the range 10–50 nmol L^{-1}.[41] More recently, it has been shown that its level rises in certain mammalian cells following receptor activation. For instance, in acinar cells from pancreas (exocrine secretory cells), there is a rapid and transient rise in the level of NAADP following activation by the peptide hormone cholecystokinin.[42] (Note that cADPR also rises in response to this ligand, but with a slower time course.) Further direct evidence for a second messenger function for NAADP has been obtained from studies of T lymphocytes.[43] When these cells are activated through the T cell receptor/CD3 complex, there is a ~7-fold increase in the level of NAADP lasting some 20 s, followed by a slower component that lasts many minutes. The transient increase is accompanied by an increase in cytosolic $[Ca^{2+}]$ that peaks after a few minutes.

The mechanism by which NAADP is synthesized is poorly understood. It has been assumed that the enzyme that forms NAADP is the same ADP-ribosyl cyclase (the ectoenzyme CD38) that was implicated in cADPR synthesis, albeit by carrying out base exchange on $NADP^+$. However, a very different NAADP synthase activity has recently been identified on the surface of sea urchin sperm.[44] This is also an ectoenzyme, but it differs from CD38 in that it performs only base exchange and is unable to cyclize NAD^+ to form cADPR. Its substrate specificity is quite unlike that of any member of the ADP-ribosyl cyclase family. Whether there is a mammalian equivalent remains to be seen. A challenge to these ideas, revealed by experiments with CD38 knock-out mice, is that the production of NAADP in myometrial cells is independent not only of CD38, but also of the base exchange reaction. Thus it appears that there is yet another enzyme pathway that can generate NAADP.[45]

While IP_3R and cADPR regulate Ca^{2+} release through IP_3Rs and RyRs, the target of NAADP is less clear. One of its physiological effects is to induce CICR by causing an initial release of Ca^{2+} that is then amplified and propagated by neighbouring RyRs or IP_3Rs. In some cells NAADP interacts with RyRs, either directly or indirectly. In others it releases Ca^{2+} by a mechanism that is insensitive to inhibitors of IP_3Rs and RyRs (heparin and ryanodine respectively). This release is typically localized and seems to originate from acidic organelles rather than the ER or SR. The identity of the putative NAADP receptor is unknown.

A general picture emerges in which cells may use different Ca^{2+}-mobilizing messengers to release Ca^{2+} from different reserves, to produce and orchestrate a variety of responses. If an ectoenzyme is involved in messenger synthesis, then both the substrate and the product must be transported across the plasma membrane at some stage. This raises the possibility of a paracrine function for NAADP. Alternatively, its synthesis might take place within an acidic organelle.

Elevation of Ca^{2+} by sphingosine-1-phosphate

Many lipids have key roles in regulatory mechanisms, either as substrates of receptor-activated effectors, as tethering points for signalling proteins or as membrane-resident second messengers. They include for instance $PI(4,5)P_2$ (see also Chapter 5), $PI(3,4,5)P_3$ (Chapter 18), DAG (Chapter 5), arachidonic acid (Chapter 5), and lysolipids. Classification of this group as 'biologically active' has meant that changes in their levels or locations affect cell function. To this list must be added the sphingolipids – the precursors and metabolites of the membrane lipid sphingomyelin. Because these compounds through their metabolism are readily interconvertible, and because they are mostly membrane restricted, they give rise to a highly complex network of regulatory pathways that feature prominently in a huge range of cellular processes, many of which are pathophysiological.[46]

The structures of the most important sphingolipids are shown in Figure 7.15. Ceramide lies at centre of the network and through variations in saturation,

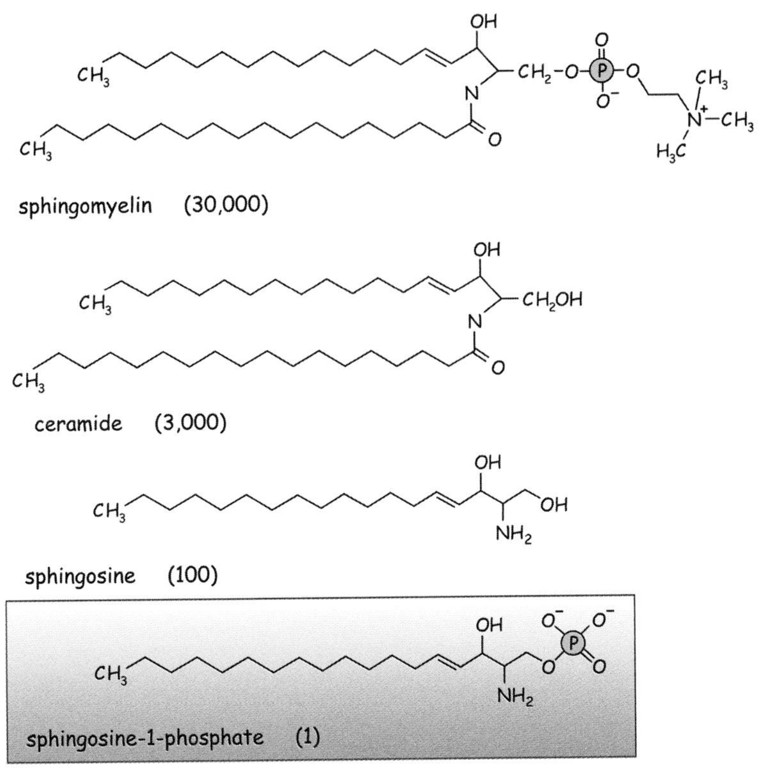

sphingomyelin   (30,000)

ceramide   (3,000)

sphingosine   (100)

sphingosine-1-phosphate   (1)

FIG 7.15 Sphingomyelin, ceramide, sphingosine, and sphingosine -1-phosphate. Relative cellular levels of each lipid are shown in parentheses.[46]

chain length, and substitution, exists in some 50 distinct molecular forms. It participates in numerous pathways, most notably in cellular stress responses. Sphingosine, the single-chain lipid formed from ceramide, is less hydrophobic and so can move between different cell membranes, where it affects multiple protein targets (e.g. PKC). Phosphorylation of sphingosine produces the amphipathic sphingosine-1-phosphate (S1P), which is even more soluble. It exists at low (nanomolar) levels in resting cells and it can cross the plasma membrane through ABCC1 transporters. Outside cells it binds to high-affinity G-protein-coupled receptors (S1P1 to S1P5) and has roles in cell growth, survival and migration, and it is important in inflammatory processes. Inside cells it acts as a Ca^{2+} mobilizing messenger.

S1P is formed through the recruitment to membranes of the cytosolic sphingosine kinases (SK1 and SK2) which occurs in response to activation of wide range of cell surface receptors[47] including immunoglobulin receptors (FcεRI in mast cells, FcγRI in macrophages) and receptors for growth factors (PDGF, VEGF) and cytokines (TNF, IL1). For example, the immune receptors signal through recruited tyrosine kinases (see Chapter 17) with consequential activation of ERK, PKC and phospholipase D. Sphingosine kinase is then recruited to the plasma membrane by the phosphatidate formed and by phosphatidylserine.

ABC transporters are ATP-binding cassette transporter proteins that use ATP hydrolysis to carry impermeant molecules across membranes. Example are the multidrug resistance transporters (MDRs) which eliminate toxic substances from cells.

The actions of S1P released from cells are well established, but because its receptors, which may be present on the same cell, can couple through PLC, it is difficult to distinguish its action as a Ca^{2+}-mobilizing second messenger from that of an extracellular ligand. None the less, there is considerable evidence for its intracellular action and its ability to mobilize Ca^{2+} independently of S1P receptors has been demonstrated.[48]

The intracellular levels of S1P are tightly regulated. They can never rise too high, in part because its surface active properties would disrupt membranes. Because its concentration is kept low, it must act on Ca^{2+} stores through high-affinity receptors, though these remain to be identified.

Ca^{2+} influx through plasma membrane channels

Voltage-operated channels

In the context of activation there are a number of different pathways by which Ca^{2+} ions can enter cells from the exterior. Nerve and muscle cells possess a variety of cation channels that are sensitive to changes in membrane potential. Some of these conduct Ca^{2+} and have roles in the initiation of action potentials, the control of excitability and the activation of exocytosis. Such voltage-operated calcium channels (VOCCs) open in response to membrane depolarization and allow Ca^{2+} entry. They are essentially transducers that convert an electrical signal into a second messenger (Ca^{2+}) signal. VOCCs have been categorized according to their electrophysiological and pharmacological properties into several classes (L, N, P/Q, R, and T types). They are all tightly regulated. L-type channels are primarily modulated by phosphorylation and N, P/Q, and R-type channels by G proteins.

L-type Ca^{2+} channels are expressed by many vertebrate excitable cells and are the major VOCCs of skeletal muscle. They are a target for 1,4-dihydropyridine drugs (such as nifedipine), which act as channel blockers. The part of the channel complex that binds the drug is termed the dihydropyridine receptor (DHPR), (although this term is often used to denote the whole channel). In vertebrate skeletal muscle, plasma membrane DHPRs and RyRs on the SR are in close apposition. Possible points of contact are indicated in Figure 7.11. This occurs at the triad regions, where the terminal cisternae of the SR almost contact the T-tubules (see Figure 8.6, page 238). Interaction between these proteins initiates the CICR that underlies excitation–contraction coupling (see page 237).

VOCCs also play important roles at synapses. In the presynaptic neuron, they are situated close to the sites at which neurotransmitter-containing vesicles await the stimulus to undergo exocytosis. Membrane depolarization causes the entry of Ca^{2+} ions that interact locally with (relatively) low affinity Ca^{2+}-binding effectors (see below and page 236). VOCCs are also present in certain endocrine cells, such as pituitary cells and pancreatic β cells, where elevated intracellular Ca^{2+} is also a trigger for secretion. Common characteristics of Ca^{2+} entry through VOCCs are the speed, brevity, and intensity of the observed Ca^{2+} transients. The rather low affinity of the effectors is matched to the high concentrations of Ca^{2+} achieved and the limited spatial volume in which it is confined. Rapid dissipation may then follow, ensuring that the subsequent effects are local and transient.

Receptor-operated channels

In some neuronal cells, low-specificity ion channels in the plasma membrane may open as a consequence of activation and allow Ca^{2+} to enter. Glutamate receptors sensitive to NMDA are important postsynaptic ion channels that mediate excitatory transmission in the CNS. In order for these channels to open, two conditions must be satisfied: they must bind glutamate and the membrane in which they reside must already be depolarized in order to remove a blocking Mg^{2+} ion.[49] This may be achieved by the simultaneous effects of two different neurotransmitters, one of which depolarizes the cell in preparation for the action of glutamate itself. Alternatively, a sustained release of glutamate from the presynaptic cell could provide both signals, by first activating AMPA receptors to depolarize the postsynaptic cell. The ability of NMDA receptors to act as coincidence detectors, coupled with other processes, gives them a role in the long term potentiation of synaptic signalling.[50,51]

NMDA: *N*-methyl-D-aspartic acid, a glutamate analogue that activates a subclass of glutamate receptors.

TRPM2 channels

TRPM2 is a plasma membrane cation channel that is Ca^{2+} permeable. It is a member of the TRP superfamily, which is discussed below. It is activated from within cells by the adenine nucleotide, ADP-ribose (ADPR, not to be confused with cADPR, see Figure 7.14). The role of ADPR as a signalling molecule has yet to be fully established. Its principal function seems to be as a messenger that causes Ca^{2+} influx in cells exposed to oxidative stress. It is not clear how its synthesis is initiated and it may take place at different locations, involving nuclear, mitochondrial or cytosolic enzymes, as well as the ectoenzyme CD38 acting on NAD^+.[37] The ADPR-sensitivity of TRPM2 is Ca^{2+}-dependent and probably involves calmodulin.[52] TRPM2 does not interact with STIM1 (see below).

Replenishing depleted stores

The rapid transmission of signals between nerves and the fast responses of skeletal muscles, all of which depend on signalling through VOCCs, are essential for survival. For other types of tissue, there is often less urgency. In hormone-secreting cells, the changes in Ca^{2+} concentration may take the form of a series of oscillations that continue over a period of minutes (see

Figure 7.6). In cells stimulated by growth factors or cytokines, there may be a need for a protracted period (hours) of Ca^{2+} elevation in order to ensure full commitment to a proliferative response.[53] Such demands present a problem since repetitive or sustained elevation of cytosol Ca^{2+} must lead to depletion of the Ca^{2+} stores. This is prevented by a mechanism that allows extracellular Ca^{2+} to enter the cell through plasma membrane cation channels in response to store depletion. This is called *store-operated Ca^{2+} entry* (SOCE). Originally proposed in 1986 and termed capacitative Ca^{2+} entry,[54,55] it took some 20 years before the sensor of ER/SR Ca^{2+} depletion was identified. Although the proteins likely to form the channel are now known, the details of the mechanism are still not clear.

Store-operated Ca^{2+} channels

The first hint of a protein with potential SOCE activity was obtained from the photoreceptor cells of *Drosophila*. A spontaneous mutation caused the light response to decay to zero on continuous exposure to bright illumination. Patch clamp studies of isolated receptor cells revealed a reduced Ca^{2+} current in the mutants,[56, 57] and it was later concluded that lack of Ca^{2+} in the stores could be responsible. In these cells, regions of the plasma membrane, close to elements of the ER, possess a Ca^{2+} channel encoded by the *trp* gene (transient receptor potential).[58] This finding stimulated a search for a vertebrate homologue.

The *Drosophila trp* gene product is a member of a superfamily of TRP proteins that are widely expressed across species. These transmembrane proteins are predicted to assemble in tetramers to form Ca^{2+} channels. There are six families and they have diverse functions. TRPV channels, which include the vanilloid receptors, and some TRPM channels are involved in sensory functions, such as sensitivity to temperature, osmolarity, odorants, and mechanical stress. The so-called classical or canonical TRP channels, TRPC1–7, are activated following receptor-induced $PI(4,5)P_2$ hydrolysis and seem most likely to be involved in SOCE. (Note that TRPC2 is a pseudogene in humans.) Mammalian cells expressing cloned TRPCs have shown enhanced SOCE and, conversely, ablation of TRP genes or application of anti-TRPC antibodies reduces SOCE activity. Despite some conflicting data, there is evidence that TRPCs can contribute towards SOCE in a variety of cells, especially TRPC1.[59] *Drosophila* Trp has ~40% amino acid identity to TRPC1. However, there is no firm evidence of a homomeric TRPC1 channel activated by store depletion. Alternatively SOCE channels might be heteromers of different TRP subunits, which might account for the variation of channel properties that is observed.

There is, however, a more powerful contender for the role of store-operated channel that is not a TRP protein. First detected electrophysiologically in rodent mast cells, in which an inward current is activated within about 10 s of store depletion, the channel is of low conductance and specific for Ca^{2+}. It was termed I_{CRAC} (calcium release-activated calcium current).[60] Ca^{2+} entry through CRAC channels is a requirement for the activation of mast cells through the IgE receptor FcεRI, and for the expression of NFAT-controlled genes in lymphocytes. The channels are also present in other cells.

A gene encoding a transmembrane protein essential for SOCE was identified by genome-wide RNAi screening in *Drosophila* S2 cells (particularly suitable for RNAi studies). The assay used a fluorescent Ca^{2+}-sensing dye to detect knockdown of SOCE and involved over 145 000 fluorescence measurements. The gene, called *Orai*, is highly conserved and has three mammalian homologues.[61,62] A single mutation of a human homologue, Orai1, accompanies a rare form of severe combined immune deficiency (SCID) in which I_{CRAC} is absent from lymphocytes.[63] Expression of wild type Orai1 in T cells from the SCID patients was found to restore CRAC channel activity.

Orai1 is a 33 kDa protein with four transmembrane stretches. It exists only on the plasma membrane and the tetramer has all the credentials of a CRAC channel (low conductance, Ca^{2+}-selective, inwardly rectifying). For instance, mutations of acidic residues in the putative pore region alter its ion selectivity and other channel characteristics, showing that it is able to act as a Ca^{2+} selectivity filter. Over-expression of Orai1 together with its activator STIM1 (see below) produces very large I_{CRAC} responses.[64] In consequence, Orai1 is considered to be an essential component of SOCE channels. Whether it forms homotetrameric channels or heteromeric channels with its homologues Orai2 and Orai3 is uncertain, but this could account for some of the observed variation in Ca^{2+}-selectivity of CRAC responses.

The apparent and surprising lack of a requirement for any TRP protein raises many questions and has been the cause of some disagreement. It has been suggested that TRPC1 might provide a parallel or additional form of SOCE, or that it might function as a channel component together with Orai1 subunits, or it might be a channel regulator.[59, 65] Furthermore, activation of Orai1 requires the clustering of its activator STIM1 (see below), but STIM1 clustering also proves to be necessary for agonist activation of all of the TRPC channels, except for TRPC7.[66] (Note that TRPC7 is also known as TRPM2 and it is activated by ADP-ribose.)

The sensor that activates SOCE

The identification of Orai1 was made possible by RNAi screening. At about the same time, large-scale RNAi screening also led to the identification of a protein that senses the Ca^{2+} level within the ER, and that activates the SOCE channels. This was achieved by screening *Drosophila* S2 cells with siRNA from a panel of 170 genes known to encode proteins with potential roles in SOCE. These included the TRPs and other proteins with channel properties.[67] In a similar, parallel study, human cells (HeLa) were screened against 2304 proteins with signalling motifs.[68] These investigations identified Stim

Mast cells are associated with the immune system. They secrete inflammatory mediators in response to an antigen challenge. The antigen cross-links immunoglobulin E molecules (IgE) bound to FcεRI receptors. If the antigen is also an allergen the response can present a particular problem for allergic individuals.

Orai1 is also known as CRACM. Orai is named after the keepers of the gates of heaven (in Greek mythology).

(stromal interaction molecule) as the only Ca^{2+} sensor necessary for SOCE in *Drosophila,* and STIM1 as a key component of SOCE in mammalian cells. (The related molecule STIM2 was also identified, but despite early indications, it is now thought to function as a regulator of the resting level of cytosolic Ca^{2+}.[69])

STIM1 is widely expressed in vertebrates. It has a molar mass of 77 kDa and possesses a single transmembrane domain. It is predominantly present (75%) in the membrane of the ER, with its N-terminus within the lumen and C-terminus in the cytosol. The N-terminal chain contains a domain sterile α motif (SAM) and a single Ca^{2+}-sensing EF-hand motif (note: unpaired EF-hands are unusual, see Chapter 8). The cytosolic chain has a coiled coil region and a number of phosphorylation sites. The protein has also been detected at the plasma membrane with its N-terminus outside the cell.

The evidence for STIM1's function as a Ca^{2+} sensor and activator of SOCE is strong. Knockdown of STIM1 abolishes SOCE and mutation of its EF-hand from high to low Ca^{2+} affinity, causes activation of the Orai1 channel, even when the stores are replete. STIM1 is not only necessary for SOCE, it is sufficient if Orai1 is present in the plasma membrane. Also, as already mentioned, when STIM1 and Orai1 are co-over-expressed, a huge increase in SOCE activity ensues.

A number of mechanisms by which STIM1 might activate the plasma membrane channels have been proposed. These include (1) direct interaction, across the cytosol, of the C-terminal chains of ER-resident STIM1 with the intracellular chains of the channel, (2) movement of STIM1 from the ER membrane to the plasma membrane, where it can interact with the channels, and (3) the release from the ER of a soluble messenger that activates the channels.

Immunofluorescence studies show that the distribution of STIM1 changes upon store depletion. When the store is replete it is dispersed across the ER membrane. When depleted, it forms clusters (visible as punctuate fluorescence) that are close to (or possibly even on) the plasma membrane. It has been proposed that the clustering enables the coiled coil regions of the cytoplasmic tail of STIM1 to interact directly with Orai1 channels at the regions where the ER juxtaposes the plasma membrane (**Figure 7.16**). This results in the opening of the channels which then admit Ca^{2+} into a confined region of the cytosol, allowing it to be taken up into the ER through nearby SERCA pumps. This should happen before the ion can diffuse into the bulk cytosol where it would be buffered. This hypothesis accords with evidence that SOCE can occur without any change in the overall cytosol Ca^{2+} concentration.[70]

A second proposal suggests that upon store depletion, STIM1 is transferred from the ER and inserted into the plasma membrane, presumably by vesicular trafficking. Once there, it activates the channels. A third idea concerns a rather mysterious substance called calcium influx factor (CIF), first discovered in 1993.[71] CIF is a readily diffusible small molecule that is produced in cells when store

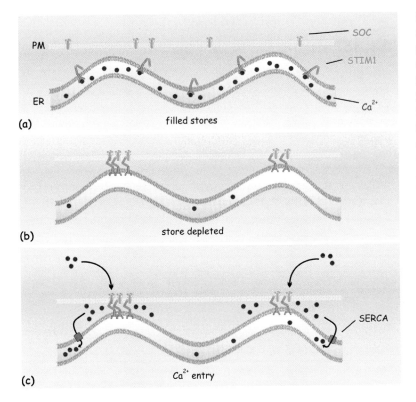

FIG 7.16 Model showing how STIM1 may sense store depletion and activate plasma membrane Ca^{2+} channels. (a) When Ca^{2+} stores in the ER are replete, the lumenal EF-hands of STIM1 bind Ca^{2+}. The distribution of STIM1 (blue) across the ER membrane is diffuse. (b) When the store is depleted, the STIM1 molecules remain in the ER but extend hydrophobic chains into the plasma membrane. They cluster at the regions where the ER is close to the cell surface and they interact with the SOC channels. (c) Ca^{2+} enters through the SOC channels and is taken up into the ER by sarcoplasmic/endoplasmic reticulum ATPase pumps (SERCA). Modified from Wu et al.[70]

Ca^{2+} is depleted. Its production ensues within 20–30 s of Ca^{2+} depletion and it is detected in the ER fraction.[72] Since its discovery, evidence has accumulated for CIF production in a wide variety of cell types.[73] It is remarkable that in the 15 years following its discovery, its molecular identity remains unknown. CIF can only be obtained as a purified extract from cells with depleted stores and its activity is assayed by injecting it into *Xenopus laevis* oocytes and then measuring the rise in intracellular Ca^{2+}. It has a molecular mass of ~600 Da and seems most likely to be a phosphorylated sugar nucleotide (but not cADPR or NAADP, nor is it S1P). The effect of CIF is to activate I_{CRAC} and it is thought that this is mediated by a Ca^{2+}-independent phospholipase A_2 at the plasma membrane (iPLA2β). It has recently been shown that CIF production is closely linked to the expression of STIM1. Knockdown of STIM1 prevents the formation of CIF in vascular smooth muscle cells as well as inhibiting SOCE. Conversely, cells over-expressing STIM1 produce CIF extracts three times more potent than controls. If STIM1 is directly responsible for CIF production, the clustering of STIM1 upon store depletion would allow CIF to be delivered directly into the vicinity of the SOCE channels.

Clearly there have been major advances in identifying the molecules likely to be responsible for SOCE, but many questions remain.

Ca^{2+} microdomains and global cellular signals

Digital imaging techniques and confocal microscopy have added considerable detail to our knowledge of Ca^{2+} signals. Within cells, the initial increments in Ca^{2+} concentration tend to be confined to regions called *microdomains*. These are mostly close to locations where effector molecules are gathered and where action takes place, for example in the vicinity of the triad structures in skeletal muscle or in neurons near to neurotransmitter release sites or confined within individual dendritic spines. These localized increases in Ca^{2+} may remain restricted and transient or they may be amplified by CICR and then spread further. They may eventually merge to generate a global increase in Ca^{2+} concentration which permeates the whole cell, especially under conditions of strong stimulation, and may last for minutes or longer.

When these changes are visualized using fluorescent Ca^{2+} indicators, a variety of optical phenomena may be observed. These range from 'Ca^{2+} sparks', the punctate flashes seen in cardiac muscle cells (10 μm or more across and lasting tens of milliseconds), to broad waves that spread within seconds right across the cytoplasm. These have been observed in a wide range of cells that include pancreatic acinar cells, smooth muscle cells, megakaryocytes, and astrocytes (see **Figure 7.7**). Such Ca^{2+} waves can also pass from cell to cell through gap junctions. In both electrically excitable and non-excitable cells, cytoplasm that is permeated by elements of the ER or SR bearing IP$_3$Rs and RyRs may form an 'excitable medium'[74] in terms of its ability to release Ca^{2+}. That is, once initiated, the process of Ca^{2+}-induced Ca^{2+}-release can produce an expanding, regenerative response that progresses at a rate faster than that achievable by diffusion of Ca^{2+} ions.

Ca^{2+} signals in electrically excitable cells

Skeletal muscle

Ca^{2+} release in skeletal muscle commences when dihydropyridine receptors (DHPR, see L-type channels above) in the plasma membrane sense a depolarization. The DHPRs are directly coupled to RyRs on the SR membrane in functional units of four DHPRs and one RyR1. Depolarization leads to allosteric activation of the RyRs and generates a localized release of Ca^{2+} that spreads as neighbouring RyRs that are not coupled to DHPRs are activated by CICR. The DHPR/RyR complexes and surrounding RyRs are arranged so that Ca^{2+} is released into a microdomain close to the contractile machinery. This is not only in the interests of efficiency, but also to avoid uncontrolled activation through further CICR. The Ca^{2+}-sensitivity of RyRs, like that of IP$_3$Rs, is bell-shaped (see Figure 7.13), so that as the concentration of Ca^{2+} approaches micromolar levels the rate of channel opening declines. This prevents the stores from emptying completely and it also sets a limit to the magnitude of the Ca^{2+} signal. These factors shape the Ca^{2+} transients that are seen in skeletal muscle. They are less intense and less punctate than the spark events seen other cells such as cardiac muscle cells, and may last for hundreds of microseconds. They have been termed 'embers'. Contraction of skeletal and smooth muscle is also discussed in Chapter 8.

Cardiac muscle

Ca^{2+} signalling in cardiac muscle is different. In both atrial and ventricular cells the type 2 RyRs are not directly coupled to DHPR/L-type channels as they are in skeletal muscle. Instead they rely on CICR for activation. In ventricular cells, the T-tubule system that extends into the cell from the plasma membrane comes within a few nanometres of the SR at some 10 000 junctional zones. At each zone there are ~10 L-type channels on the T-tubule and ~100 RyR2s on the juxtaposed SR. A Ca^{2+} spark occurs when one or more L-type channels open to activate 10–15 RyR2s [75]. The high Ca^{2+}-conductance and the longer open times of RyR2 channels, synchronized in all the zones by the plasma membrane depolarization, creates a rapid increase in Ca^{2+} that is localized over the myofibrils.

In atrial cells there is no T-tubular system and the junctional zones are only present at the cell surface. However, there are many non-junctional RyR2s elsewhere on the SR. Depolarization causes sparks near the surface, but this excitation is prevented from propagating through the cell by a wall of mitochondria situated just under the plasma membrane. The mitochondria begin to take up Ca^{2+} when it rises to high levels, and together with the SERCA pumps, they have the effect of preventing the propagation of CICR beyond the surface. Thus, under normal conditions contractions are relatively weak. In the event of β-adrenergic stimulation, more sparks are generated at the periphery and the Ca^{2+} level becomes high enough to breach the wall. Then CICR can generate a global Ca^{2+} signal that causes a strong contraction. [76]

Nerve cells

The morphology of nerve cells is complex and varied, and many pre- and postsynaptic interactions depend upon Ca^{2+} signals. Postsynaptic signalling in the dendrites and dendritic spines that receive excitatory inputs involves VOCC or glutamate or AMPA receptors (see page 54). Brief openings of these channels produce Ca^{2+} microdomains that are restricted to the immediate environment of the channel. Subsequent amplification can involve either RYRs or IP_3Rs or both. [77] Signals tend to be spatially restricted however, and this enables individual spines to process inputs independently.

At presynaptic nerve endings, Ca^{2+} is the signal for exocytosis. VOCCs open transiently in response to depolarization. A microdomain of Ca^{2+} is created that may be amplified by CICR, again through the opening of RyR and IP_3R channels. Activation of neurotransmitter secretion is discussed in Chapter 8.

Calcium signals in non-excitable cells

The Ca^{2+} release process in electrically non-excitable cells is generally slower than its counterpart in nerve and striated muscle. In part, this is because there is no equivalent of the depolarization signal to synchronize Ca^{2+} entry and its release from internal stores. It mostly relies on IP_3Rs, but RyRs also take part in some cells. Elementary Ca^{2+} release events are less conspicuous in non-excitable cells. Imaging of Ca^{2+} transients reveals phenomena that range from tiny 'blips' (possibly release from single IP_3Rs), through more extensive 'puffs' (IP_3R clusters), up to full-scale regenerative calcium waves, as neighbouring channels are successively recruited. These waves can propagate across the entire cytoplasm in a few seconds (see Figure 7.7). Whether a signal is localized or global depends on the relative locations of the receptors that bind the activating ligand and the Ca^{2+}-sensitive effector proteins that transduce the signal.

References

1. Ringer S. Regarding the action of the hydrate of soda, hydrate of ammonia, and hydrate of potash on the ventricle of the frog's heart. *J Physiol (Lond)*. 1883;3:195–202.
2. Ringer S. A further contribution regarding the influence of the different constituents of the blood on the contraction of the heart. *J Physiol (Lond)*. 1884;4:29–42.
3. Locke FS. Notiz über den Einfluss, physiologischer Kochsalzläsung auf die Errebarkeit von Muskel und Nerv. *Zentralblat Physiol*. 1894;8:166–167.
4. Mann PJG, Tennenbaum M, Quastel JH. Acetylcholine metabolism in the central nervous system: the effects of potassium and other cations on acetylcholine liberation. *Biochem J*. 1939;33:822–835.
5. Harvey AM, MacIntosh FC. Calcium and synaptic transmission in a sympathetic ganglion. *J Physiol (Lond)*. 1940;97:408–416.
6. Kamada T, Kinoshita H. Disturbances initiated from naked surface of muscle protoplasm. *Japan J Physiol*. 1943;10:469–493.
7. Heilbrunn LV, Wiercinski FJ. The action of various cations on muscle protoplasm. *J Cell Comp Physiol*. 1947;29:15–32.
8. Heilbrunn LV. The action of calcium on muscle protoplasm. *Physiol Zool*. 1940;13:88–94.
9. Sutherland EW. Studies on the mechanism of hormone action. *Science*. 1972;177:401–408.
10. Levine BA, Williams RJP. The chemistry of calcium ion and its biological relevance. In: Anghilieri LJ, Tuffet-Anghilieri A-M, eds. Florida: CRC Press; 1982:3–26. *The Role of Calcium in Biological Systems*; Vol 1.
11. Pollack H. Micrurgical studies in cell physiology: calcium ions in living protoplasm. *J Gen Physiol*. 1928;11:539–545.

12. Shimomura O, Johnson FH. Properties of the bioluminescent protein aequorin. *Biochemistry*. 1969;8:3991–3997.

13. Llinas R, Blinks JR, Nicholson C. Calcium transient in presynaptic terminal of squid giant synapse: detection with aequorin. *Science*. 1972;176:1127–1129.

14. Ashley CC. Aequorin-monitored calcium transients in single *Maia* muscle fibres. *J Physiol Lond*. 1969;203:32P–33P.

15. Shimomura O, Johnson FH, Saiga Y. Extraction, purification and properties of aequorin, a bioluminescent protein from the luminous hydromedsan *Aequorea*. *J Cell Comp Physiol*. 1962;59:223–239.

16. Campbell AK. Extraction, partial purification and properties of the Ca^{2+}-activated luminescent protein obelin from the hydroid *Obelia geniculata*. *Biochem J*. 1974;143:411–481.

17. Tsien RY. New calcium indicators and buffers with high selectivity against magnesium and protons: design, synthesis, and properties of prototype structures. *Biochemistry*. 1980;19:2396–2404.

18. Tsien RY, Pozzan T, Rink TJ. Calcium homeostasis in intact lymphocytes: cytoplasmic free calcium monitored with a new, intracellularly trapped fluorescent indicator. *J Cell Biol*. 1982;94:325–334.

19. Tsien RY. A non-disruptive technique for loading calcium buffers and indicators into cells. *Nature*. 1981;290:527–528.

20. Grynkiewicz G, Poenie M, Tsien RY. A new generation of Ca^{2+} indicators with greatly improved fluorescence properties. *J Biol Chem*. 1985;260:3440–3450.

21. Merritt JE, Rink TJ. Rapid increases in cytosolic free calcium in response to muscarinic stimulation of rat parotid acinar cells. *J Biol Chem*. 1987;262:4958–4960.

22. Jacob R, Merritt JE, Hallam TJ, Rink TJ. Repetitive spikes in cytoplasmic calcium evoked by histamine in human endothelial cells. *Nature*. 1988;335:40–45.

23. Mountford JC, Melford SK, Bunce CM, Gibbins J, Watson SP. Collagen or collagen-related peptide cause $[Ca^{2+}]_i$ elevation and increased tyrosine phosphorylation in human megakaryocytes. *Thromb Haemost*. 1999;82:1153–1159.

24. Boitier E, Rea R, Duchen MR. Mitochondria exert a negative feedback on the propagation of intracellular Ca^{2+} waves in rat cortical astrocytes. *J Cell Biol*. 1999;145:795–808.

25. Zimmermann B. Subcellular organization of agonist-evoked Ca^{2+} waves in the blowfly salivary gland. *Cell Calcium*. 2000;27:297–307.

26. Rizzuto R, Simpson AW, Brini M, Pozzan T. Rapid changes of mitochondrial Ca^{2+} revealed by specifically targeted recombinant aequorin. *Nature*. 1992;358:325–327.

27. Nagai T, Sawano A, Park ES, Miyawaki A. Circularly permuted green fluorescent proteins engineered to sense Ca^{2+}. *Proc Natl Acad Sci USA*. 2001;98:3197–3202.

28. Miyawaki A, Llopis J, Heim R, et al. Fluorescent indicators for Ca^{2+} based on green fluorescent proteins and calmodulin. *Nature*. 1997;388:882–887.

29. Griffin BA, Adams SR, Tsien RY. Specific covalent labeling of recombinant protein molecules inside live cells. *Science*. 1998;281:269–272.

30. Tour O, Adams SR, Kerr RA, et al. Calcium green FlAsH as a genetically targeted small-molecule calcium indicator. *Nat Chem Biol*. 2007;3:423–431.

31. Samsó M, Wagenknecht T. Contributions of electron microscopy and single particle techniques to the determination of the ryanodine receptor three-dimensional structure. *J Struct Biol*. 1998;121:172–180.

32. Sato C, Hamada K, Ogura T, et al. Inositol 1,4,5-trisphosphate receptor contains multiple cavities and L-shaped ligand-binding domains. *J Mol Biol*. 2004;336:155–164.

33. Foskett JK, White C, Cheung KH, Mak DO. Inositol trisphosphate receptor Ca^{2+} release channels. *Physiol Rev*. 2007;87:593–658.

34. Mak DO, McBride S, Foskett JK. Inositol 1,4,5-tris-phosphate activation of inositol tris-phosphate receptor Ca^{2+} channel by ligand tuning of Ca^{2+} inhibition. *Proc Natl Acad Sci USA*. 1998;95:15821–15825.

35. Mak DO, McBride S, Raghuram V, Yue Y, Joseph SK, Foskett JK, . Single-channel properties in endoplasmic reticulum membrane of recombinant type 3 inositol trisphosphate receptor. *J. Gen Physiol*. 2000;115:241–256.

36. Lee HC, Walseth TF, Bratt GT, Hayes RN, Clapper DL. Clapper, structural determination of a cyclic metabolite of NAD^+ with intracellular Ca^{2+}-mobilizing activity. *J Biol Chem*. 1989;264:1608–1615.

37. Guse AH. Regulation of calcium signaling by the second messenger cyclic adenosine diphosphoribose (cADPR). *Curr Mol Med*. 2004;4:239–248.

38. Lee HC, Aarhus R, Graeff RM. Sensitization of calcium-induced calcium release by cyclic ADP-ribose and calmodulin. *J Biol Chem*. 1995;270:9060–9066.

39. Kato I, Yamamoto Y, Fujimura M, Noguchi N, Takasawa S, Okamoto H. CD38 disruption impairs glucose-induced increases in cyclic ADP-ribose, $[Ca^{2+}]_i$ and insulin secretion. *J Biol Chem*. 1999;274:1869–1872.

40. Guse AH, da-Silva CP, Berg I, et al. Regulation of calcium signalling in T lymphocytes by the second messenger cyclic ADP-ribose. *Nature*. 1999;398:70–73.

41. Lee HC, Aarhus R. A derivative of NADP mobilizes calcium stores insensitive to inositol trisphosphate and cyclic ADP-ribose. *J Biol Chem*. 1995;270:2152–2157.

42. Yamasaki M, Thomas JM, Churchill GC, et al. Role of NAADP and cADPR in the induction and maintenance of agonist-evoked Ca^{2+} spiking in mouse pancreatic acinar cells. *Curr Biol*. 2005;15:874–878.

43. Gasser A, Bruhn S, Guse AH. Second messenger function of nicotinic acid adenine dinucleotide phosphate revealed by an improved enzymatic cycling assay. *J Biol Chem*. 2006;281:16906–16913.

44. Vasudevan SR, Galione A, Churchill GC. Sperm express a Ca^{2+}-regulated NAADP synthase. *Biochem J*. 2008;411:63–70.

45. Soares S, Thompson M, White T, et al. NAADP as a second messenger: neither CD38 nor base-exchange reaction are necessary for in vivo generation of NAADP in myometrial cells. *Am J Physiol Cell Physiol*. 2007;292:C227–C239.

46. Hannun YA, Obeid LM. Principles of bioactive lipid signalling: lessons from sphingolipids. *Nat Rev Mol Cell Biol*. 2008;9:139–150.

47. Chalfant CE, Spiegel S. Sphingosine 1-phosphate and ceramide 1-phosphate: expanding roles in cell signaling. *J Cell Sci*. 2005;118:4605–4612.

48. Meyer zu HD, Liliom K, Schaefer M, et al. Photolysis of intracellular caged sphingosine-1-phosphate causes Ca^{2+} mobilization independently of G-protein-coupled receptors. *FEBS Lett*. 2003;554:443–449.

49. Nowak L, Bregestovski P, Ascher P, Herbet A, Prochiantz A. Magnesium gates glutamate-activated channels in mouse central neurones. *Nature*. 1984;307:462–465.

50. Malenka RC, Kauer JA, Perkel DJ, et al. An essential role for postsynaptic calmodulin and protein kinase activity in long-term potentiation. *Nature*. 1989;340:554–557.

51. Bliss TVP, Collingridge GL. A synaptic model of memory: long-term potentiation in the hippocampus. *Nature*. 1993;361:31–39.

52. Starkus J, Beck A, Fleig A, Penner R. Regulation of TRPM2 by extra- and intracellular calcium. *J Gen Physiol*. 2007;130:427–440.

53. Wacholtz MC, Lipsky PE. Anti-CD3-stimulated Ca^{2+} signal in individual human peripheral T cells. Activation correlates with a sustained increase in intracellular Ca^{2+}. *J Immunol*. 1993;150:5338–5349.

54. Putney JWJ. Capacitative calcium entry revisited. *Cell Calcium*. 1990;11:611–624.

55. Putney JWJ. A model for receptor-regulated calcium entry. *Cell Calcium*. 1986;7:1–12.

56. Hardie RC, Minke B. The trp gene is essential for a light-activated Ca^{2+} channel in *Drosophila* photoreceptors. *Neuron*. 1992;8:643–651.

57. Hardie RC, Minke B. Calcium-dependent inactivation of light-sensitive channels in *Drosophila* photoreceptors. *J Gen Physiol*. 1994;103:409–427.

58. Pollock JA, Assaf A, Peretz A, et al. TRP, a protein essential for inositide-mediated Ca^{2+} influx is localized adjacent to the calcium stores in *Drosophila* photoreceptors. *J Neurosci*. 1995;15:3747–3760.

59. Ambudkar IS, Ong HL, Liu X, Bandyopadhyay B, Cheng KT. TRPC1: the link between functionally distinct store-operated calcium channels. *Cell Calcium*. 2007;42:213–223.

60. Hoth M, Penner R. Calcium-release-activated calcium current in rat mast cells. *J Physiol*. 1993;465:359–386.

61. Zhang SL, Yeromin AV, Zhang XH, et al. Genome-wide RNAi screen of Ca^{2+} influx identifies genes that regulate Ca^{2+} release-activated Ca^{2+} channel activity. *Proc Natl Acad Sci USA*. 2006;103:9357–9362.

62. Vig M, Peinelt C, Beck A, et al. CRACM1 is a plasma membrane protein essential for store-operated Ca^{2+} entry. *Science*. 2006;312:1220–1223.

63. Feske S, Gwack Y, Prakriya M, et al. A mutation in orai1 causes immune deficiency by abrogating CRAC channel function. *Nature*. 2006;441:179–185.

64. Peinelt C, Vig M, Koomoa DL, et al. Amplification of CRAC current by STIM1 and CRACM1 (Orai1). *Nat Cell Biol*. 2006;8:771–773.

65. Liao Y, Erxleben C, Yildirim E, Abramowitz J, Armstrong DL, Birnbaumer L. Orai proteins interact with TRPC channels and confer responsiveness to store depletion. *Proc Natl Acad Sci USA*. 2007;104:4682–4687.

66. Yuan JP, Zeng W, Huang GN, Worley PF, Muallem S. STIM1 heteromultimerizes TRPC channels to determine their function as store-operated channels. *Nat Cell Biol*. 2007;9:636–645.

67. Roos J, DiGregorio PJ, Yeromin AV, et al. STIM1, an essential and conserved component of store-operated Ca^{2+} channel function. *J Cell Biol*. 2005;169:435–445.

68. Liou J, Kim ML, Heo WD, et al. STIM is a Ca^{2+} sensor essential for Ca^{2+}-store-depletion-triggered Ca^{2+} influx. *Curr Biol*. 2005;15:1235–1241.

69. Brandman O, Liou J, Park WS, Meyer T. STIM2 is a feedback regulator that stabilizes basal cytosolic and endoplasmic reticulum Ca^{2+} levels. *Cell*. 2007;131:1327–1339.

70. Wu MM, Luik RM, Lewis RS. Some assembly required: constructing the elementary units of store-operated Ca^{2+} entry. *Cell Calcium*. 2007;42:163–172.

71. Randriamampita C, Tsien RY. Emptying of intracellular Ca^{2+} stores releases a novel small messenger that stimulates Ca^{2+} influx. *Nature*. 1993;364:809–814.

72. Csutora P, Peter K, Kilic H, et al. Novel role for STIM1 as a trigger for calcium influx factor production. *J Biol Chem*. 2008;283:14524–14531.

73. Bolotina VM, Csutora P. CIF and other mysteries of the store-operated Ca^{2+}-entry pathway. *Trends Biochem Sci*. 2005;30:378–387.

74. Lechleiter JD, Clapham DE. Molecular mechanisms of intracellular calcium excitability in X. laevis oocytes. *Cell*. 1992;69:283–294.

75. Wang SQ, Song LS, Lakatta EG, Cheng H. Ca^{2+} signalling between single L-type Ca^{2+} channels and ryanodine receptors in heart cells. *Nature*. 2001;410:592–596.

76. Berridge MJ, Lipp P, Bootman MD. The versatility and universality of calcium signalling. *Nat Rev Mol Cell Biol*. 2000;1:11–21.

77. Berridge MJ. Calcium microdomains: organization and function. *Cell Calcium*. 2006;40:405–412.

Calcium Effectors

Calcium-binding by proteins

In the previous chapter we have seen how transient elevations of the concentration of free Ca^{2+} may occur in cells in response to receptor activation or membrane depolarization. We also saw how these signals may be confined at subcellular locations or may propagate and how they may be temporally encoded as spikes or oscillations. Now we ask how changes in $[Ca^{2+}]$ within the cytosol are sensed and converted into downstream signals. Not surprisingly, Ca^{2+}-binding proteins are involved. However, molecules that bind Ca^{2+} are not necessarily mediators of signalling. For instance, cytosol proteins with very high Ca^{2+} affinity will already be saturated and will remain unaffected by a rise in its concentration above the resting level (40–100 nmol L^{-1}). Lower affinity proteins with dissociation constants >0.1 μmol L^{-1} will certainly bind more Ca^{2+}, but again it does not necessarily follow that they will take part in signalling. Instead, they may be important as buffers that stabilize Ca^{2+} levels or help to shape Ca^{2+} transients. On the other hand, there is a diversity of signalling proteins that bind Ca^{2+} at regulatory sites and that are activated by increases in its level. These are the Ca^{2+} effectors. A selection is listed in Table 8.1.

TABLE 8.1 Examples of effector proteins of vertebrate origin that are activated by Ca^{2+}

Protein	Type	Ca^{2+}-binding domains	Function	Main location
Calreticulin			Ca^{2+} buffering	ER and SR
Calsequestrin			Ca^{2+} buffering	SR in muscle
Parvalbumin		EF hand	cytosolic Ca^{2+} buffering	muscle/nerve
Ca^{2+}-ATPase (PMCA)	CaM dependent		pumps Ca^{2+} out of cell	plasma membrane
Ca^{2+}-ATPase (SERCA)			pumps Ca^{2+} into stores	ER membrane
Calmodulin (CaM)		EF hand	multipurpose Ca^{2+} sensor	
Troponin C		EF hand	Ca^{2+}-sensor mediating contraction	striated muscle
Calmodulin kinases I, II, IV	CaM is a regulatory subunit		multipurpose signalling	
Myosin light chain kinase	Ca^{2+}/CaM dependent		phosphorylates myosin II	smooth muscle
Adenylyl cyclases 1, 8	Ca^{2+}/CaM dependent		makes cyclic AMP	
Adenylyl cyclases 5, 6			makes cyclic AMP (Ca^{2+} inhibits)	

Protein	Property	Domain	Function	Location
Cyclic nucleotide phosphodiesterase (1A–C)	Ca^{2+}/CaM dependent		breaks down cyclic AMP	
Phosphorylase b kinase	CaM is a regulatory subunit		phosphorylates glycogen phosphorylase	skeletal muscle
Recoverin	Ca^{2+}- myristoyl switch	EF hand	Ca^{2+}- sensing mediator	photoreceptor cells
Calpain		EF hand	protease	
α-Actinin		EF hand	cytoskeleton	
Gelsolin			actin severing and capping	
Synaptotagmin	Putative Ca^{2+} sensor	C2	signalling	secretory cells
Calcineurin (protein phosphatase 2B)	Ca^{2+}/CaM dependent	EF hand	signalling, e.g. transcription	
Protein kinase C (α, β1, β2, γ)		C2	signalling	
Phospholipase C (all isoforms)		EF hand, C2	signalling	
Diacylglycerol kinase		EF hand	makes phosphatidate	
Nitric oxide synthase	CaM is a regulatory subunit		production of NO for signalling	

CaM, calmodulin; PMCA, plasma membrane Ca^{2+}-ATPase; SERCA, sarco/endoplasmic reticulum Ca-ATPase.

Polypeptide modules that bind Ca^{2+}

Common Ca^{2+}-binding sites on proteins include the EF-hand motif and the C2 domain. Their structures are described in Chapter 24. EF-hand motifs occur in pairs and a single motif binds a single Ca^{2+} ion with K_D in the range 10^{-7}–10^{-5} mol L^{-1}. However, some EF-hands have lost or never even acquired the ability to bind Ca^{2+}, for example the EF-hands of phospholipase C (see Chapter 5). The Ca^{2+} dissociation constants of C2 domains also vary widely (10^{-6}–10^{-3} mol L^{-1}) and again, some do not bind Ca^{2+} at all.

Many Ca^{2+} effectors are without either C2 domains or EF-hands, but acquire their Ca^{2+} dependence through the Ca^{2+} sensor protein calmodulin, which in some cases forms an integral subunit (Table 8.1). Other Ca^{2+}-binding proteins lack C2 domains or EF-hands and function independently of calmodulin. They may be found among the wide range of channels and ATPases that conduct Ca^{2+} ions across membranes. Further examples are the buffering proteins calreticulin and calsequestrin, present in the ER and SR, the extracellular adhesion molecule cadherin, and the actin-modifying protein gelsolin.

Decoding Ca^{2+} signals

When Ca^{2+} signals are transient, an effector must be able detect the change and initiate a response before the concentration subsides. When Ca^{2+} rises transiently, sensing the change does not just depend on the stability constant of the binding, but also the rates of the 'on' and 'off' reactions. The forward reaction requires the successive displacement, one at a time, of water molecules from the solvation shell of the cation. In general, if a multidentate coordinating ligand is to associate with or dissociate from a cation rapidly, then its framework needs to be flexible. The evolution of such coordination sites in proteins has led to the emergence of Ca^{2+}-activated regulatory enzymes that can bind and respond very rapidly to changes in Ca^{2+} concentration.

Ca^{2+} signals are often repetitive, taking the form of trains of spikes or pulses with periods of the order of minutes (see Figure 7.6, page 195) and it is likely that there are specific effectors that can respond to this form of temporal encoding. While a sustained rise in $[Ca^{2+}]$ might be damaging because it could activate downstream effectors indiscriminately, an oscillatory signal would favour those with low effective off-rates, allowing them to retain their bound Ca^{2+} during a downswing. It would also favour effector pathways that remain active for just long enough to ensure throughput during dips in $[Ca^{2+}]$. Evidence is accumulating in support of these ideas. For instance, in T lymphocytes a cytoplasmic Ca^{2+} signal that oscillates can be more effective at activating transcription than one that is steady. Moreover, high-frequency oscillations can activate three different transcription factors, while at low frequencies only NF-κB is activated[1,2] (see page 521).

Calmodulin and troponin C

Calmodulin

Calmodulin and its isoform troponin C are the most prominent Ca^{2+}-sensing proteins in animal cells. Calmodulin itself is highly conserved and present at significant levels in all eukaryotic cells. In mammalian brain it comprises about 1% of the total protein. Calmodulin, through its interaction with Ca^{2+}, can cause the activation of more than 100 different enzymes, that are not necessarily themselves Ca^{2+}-sensitive (see Table 8.1). The most important effectors of Ca^{2+}-calmodulin are shown in **Figure 8.1**.

Calmodulin first came to light in association with cyclic nucleotide phosphodiesterase.[3,4] It is an acidic protein of modest size (17 kDa), consisting of a single, predominantly helical polypeptide chain with four Ca^{2+}-binding EF-hands, two at each end (**Figure 8.2b**). The affinities of the individual Ca^{2+} binding sites are in the range 10^{-5}–10^{-6} mol L^{-1} and adjacent sites bind Ca^{2+} with positive cooperativity, so that attachment of the first Ca^{2+} ion enhances the affinity of its neighbour. This has the effect of making the protein sensitive to small changes in the concentration of Ca^{2+} within the signalling range. Ca^{2+}-calmodulin itself has no intrinsic catalytic activity. Its actions depend on its close association with a target enzyme, in some instances, as in phosphorylase kinase and calcineurin, acting as a permanent component of a multisubunit complex.

In the absence of bound Ca^{2+}, the helices of calmodulin pack so that their hydrophobic side chains are not exposed. In this form it is unable to interact with its targets (**Figure 8.2a**). Binding of Ca^{2+} to the four sites induces a large conformational change causing the terminal helices to expose hydrophobic

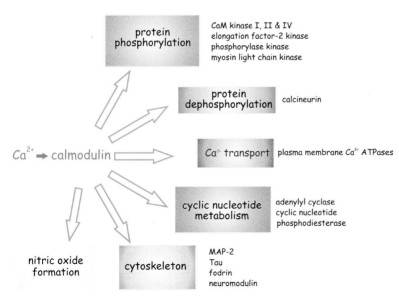

FIG 8.1 Multiple signal transduction pathways activated by calmodulin. Calmodulin bound to Ca^{2+} interacts with and activates many enzymes, opening up a wide range of cellular responses. Abbreviations: MAP-2, microtubule-associated protein-2; Tau, tubulin assembly unit.

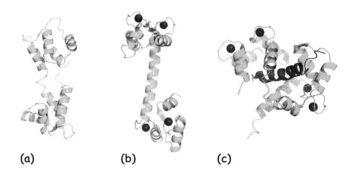

(a) (b) (c)

FIG 8.2 Three-dimensional structure of apocalmodulin, Ca^{2+}-bound calmodulin, and calmodulin bound to a target peptide.
(a) Ca^{2+}-free or apocalmodulin) (b) Calmodulin with four bound Ca^{2+} ions (blue spheres). The terminal helices now expose hydrophobic residues (not indicated). (c) Calmodulin bound to a peptide corresponding to the calmodulin-binding domain of smooth muscle myosin light chain kinase (1cfd,[5] cll1,[6] and 2bbm[7]).

surfaces and also a long central α-helical segment (Figure 8.2b).[8] Ca^{2+}-bound calmodulin binds to its targets with high affinity ($K_D \sim 10^{-9}\,mol\,L^{-1}$). To form the bound state, the central residues of the link region unwind from their α-helical arrangement to form a hinge that allows the molecule to bend and wrap itself around the target. The N- and C-terminal regions approach each other and by their hydrophobic surfaces bind to it, rather like two hands holding a rope. This encourages the target sequence to adopt an α-helical arrangement so that it occupies the centre of a hydrophobic tunnel (**Figure 8.2c**). The consequence of this interaction is a conformational change in the target, a state that persists only as long as the Ca^{2+} concentration remains high. When it falls, Ca^{2+} dissociates and calmodulin is quickly released, inactivating the target. However, at least one important target protein is an exception to this rule. This is CaM-kinase II which can retain its active state after it has been activated by calmodulin (see below).

Troponin C

Troponin C, which regulates the interaction of actin and myosin in striated muscle, is effectively an isoform of calmodulin. It also possesses two pairs of Ca^{2+}-binding EF-hands located at opposite ends of a peptide chain. The affinities of these sites for Ca^{2+} lie between 10^{-5} and $10^{-7}\,mol\,L^{-1}$. Its function is described below (page 237).

Kinases regulated by calmodulin

Among the enzymes controlled by calmodulin are a number of serine/threonine kinases. These include phosphorylase kinase, myosin light chain kinase (MLCK), and members of the family of Ca^{2+}-calmodulin-dependent kinases (CaM-kinases). Each of these interacts with calmodulin to convert a Ca^{2+} signal into a phosphorylation signal.

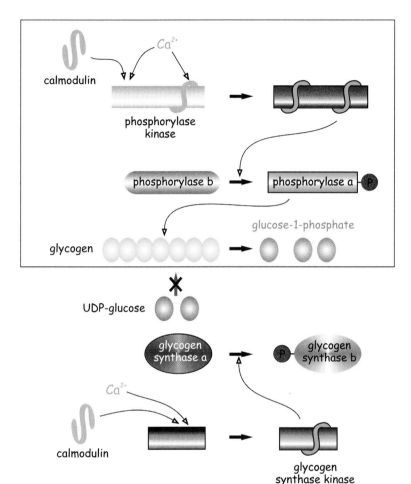

FIG 8.3 Calmodulin and Ca^{2+} stimulate the breakdown of glycogen and inhibit its synthesis in skeletal muscle.

Phosphorylase kinase and glycogen synthase kinase

Phosphorylase kinase was the first protein kinase to be discovered.[9] It is controlled by protein kinase A under β-adrenergic stimulation and also through the effect of Ca^{2+} on calmodulin (Figure 8.3). Even in the absence of Ca^{2+}, calmodulin forms a permanent, integral subunit of phosphorylase kinase.[10] When $[Ca^{2+}]$ rises above the resting level, the kinase is activated both through its endogenous calmodulin and through the binding of an exogenous Ca^{2+}-calmodulin. The enzyme catalyses glycogenolysis. To reinforce this, glycogen synthesis is simultaneously inhibited through the action of Ca^{2+}-calmodulin on glycogen synthase kinase, which phosphorylates and inactivates glycogen synthase (Figure 8.3). These themes will be developed further in Chapters 9 and 21 (page 679), where we shall see that phosphorylase kinase is also activated by a phosphorylation mediated by the second messenger cyclic AMP (cAMP).

Myosin light chain kinase

MLCK functions in smooth muscle, where it phosphorylates the light chain of myosin II to activate contraction. This is described in more detail below (see page 239).

Ca^{2+}-calmodulin activated protein kinases

On the basis of *in vitro* phosphorylation assays the CaM kinases have been termed multifunctional, suggesting that each member of the family phosphorylates a broad range of physiological targets. The range of substrates observed *in vivo*, however, is not so wide. All of the CaM kinases possess an autoinhibitory domain that folds over the active site. The calmodulin binding site overlaps this domain and when Ca^{2+}-calmodulin binds, the domain is released and the inhibition is lifted. Like many other kinases, the CaM kinases are also regulated by phosphorylation.[11] The different isoforms (CaMKI–IV) are as follows.

CaMKI

This is ubiquitously expressed in mammalian cells where it is predominantly cytoplasmic. Activation occurs through release of the autoinhibitory domain by Ca^{2+}-calmodulin and also requires phosphorylation on a threonine in the activation loop by calmodulin kinase kinase. This upstream enzyme is itself activated by Ca^{2+}-calmodulin. CaMKI has a number of targets that include the cystic fibrosis transmembrane regulator (CFTR) Cl$^-$ channel. Activation loop phosphorylation may not be essential for some substrates.

CaMKII

A broad-spectrum serine/threonine kinase, CaMKII is the most prominent member of the family. It is widely expressed, with particularly high levels in brain. Unlike other kinases of this family, it has an additional association domain that lies C-terminal to its regulatory domain. This enables it to assemble as a multimeric cylindrical structure consisting of 8–12 copies of the enzyme as subunits. CaMKII is encoded by four genes, α, β, γ, and δ; the α and β subtypes are expressed only in brain. There are also a number of splice variants. As the multimeric structures can assemble as either homomers or heteromers, there is potential for considerable diversity.

Activation of CaMKII occurs when one of the subunits in a multimeric structure binds Ca^{2+}-calmodulin. Its autoinhibitory domain is released from its active site and this allows it to phosphorylate the autoinhibitory domain of the neighbouring subunit at T286. (Note that unlike the other CaM kinases, this is not an activation loop phosphorylation.) Phosphorylation of the neighbouring subunit not only releases its autoinhibitory domain, it also increases its affinity for Ca^{2+}-calmodulin by a factor of more than 1000. This effectively 'traps' a second Ca^{2+}-calmodulin and the enzyme can then remain active or 'autonomous' for tens of minutes after the Ca^{2+} stimulus has been removed. It remains autonomous until the bound calmodulin is eventually released and the phosphate at T286 is cleaved by a phosphatase. Moreover, in the face of a train of spikes or oscillations, its level of activation, presumably related to the number of subunits activated, is matched to the frequency and duration of the Ca^{2+} spikes. This ability to decode Ca^{2+} signals and to become autonomous is thought to underlie the role of CaMKII in learning and memory.

CaMKIII

The protein formerly known as CaMKIII specifically phosphorylates eukaryotic elongation factor 2 (eEF2), a GTPase necessary for the elongation step in protein translation. This kinase has no sequence similarity with the other members of the CaM kinase family and it is now termed eEF2 kinase.

CaMKIV

This is expressed in a limited range of tissues, principally neuronal cells, male germ cells, and lymphocytes, where it regulates gene expression by phosphorylating transcription factors. It is found mostly in the nucleus. It is activated on binding of Ca^{2+}-calmodulin to the autoinhibitory domain, followed by autophosphorylation in the activation loop at by CaM kinase kinase. An important target is the transcription factor CREB which is activated by CaMKIV, but inhibited by CaMKII. CaMKIV can become partly Ca^{2+}-calmodulin-independent when autophosphorylated.

Other Ca^{2+}-calmodulin dependent enzymes

Plasma membrane Ca^{2+} ATPase (PMCA)

The plasma membrane Ca^{2+} ATPases that are responsible for Ca^{2+} homeostasis in the cell were introduced in Chapter 7 (page 190). There are four isoforms. PMCA1 and 4 are ubiquitous, while PMCA2 and PMCA3 are predominant in neuronal cells.[12] Together with the SERCAs and the Na^+-Ca^{2+} exchangers, they maintain cytosol $[Ca^{2+}]$ in the range 40–$100\,nmol\,L^{-1}$. Under these conditions, the Ca^{2+}-affinity of the PMCAs is low ($K_D > 10\,\mu mol\,L^{-1}$) and therefore they are barely active. When Ca^{2+} is elevated, Ca^{2+}-calmodulin binds to the C-terminal tail of the ATPase, increasing its affinity for Ca^{2+} by a factor of 10, and as a consequence its activity. However, acidic phospholipids, such as phosphatidylserine and $PI(4,5)P_2$, can also increase Ca^{2+}-affinity to an even greater extent, implying that the most active pumps might be localized to membrane domains rich in these lipids. Other potential modulators also interact with PMCAs, such as calcineurin (see below).

Ras guanine nucleotide exchange factor

The principal signalling pathway through the GTPase Ras, described in Chapter 4, involves the targeting of the guanine nucleotide exchange factor Sos to the plasma membrane, where it catalyses the exchange of GDP for GTP on Ras. RasGTP activates the ERK/MAPkinase cascade (see page 337) until the bound GTP has been hydrolysed to GDP through the action of a GTPase-activating protein or GAP. While Sos is not in itself Ca^{2+}-dependent, other exchange factors, such as Ras-GRF1 and Ras-GRF2, are. These bring about activation of Ras in response to an elevation of cytosol Ca^{2+} and there is evidence that they require calmodulin in order to respond to a Ca^{2+} signal, although additional inputs are necessary.[13] Other Ca^{2+}-sensitive Ras exchange factors (RasGEFs) and RasGAPs that are not calmodulin-dependent are discussed below.

Calcineurin

Calcineurin is a calmodulin-dependent serine/threonine phosphatase. It exists as a heterodimer in mammalian cells (Figure 8.4) and it has a number of specific targets, most notably the transcription factor NFAT (nuclear factor of activated T cells), which functions in immune, muscle, and neuronal cells (see page 521). The catalytic subunit, calcineurin A, is also familiar as protein phosphatase 2B. The regulatory subunit, calcineurin B, possesses four EF-hands and binds Ca^{2+}, but Ca^{2+}-calmodulin is also required to release an autoinhibitory chain from the active site on calcineurin A.[14]

The immunosuppressive drugs ciclosporin A and tacrolimus (originally known as FK506) are used to prevent organ rejection after transplant surgery. They bind to cytosolic proteins termed immunophilins (cyclophilin A and

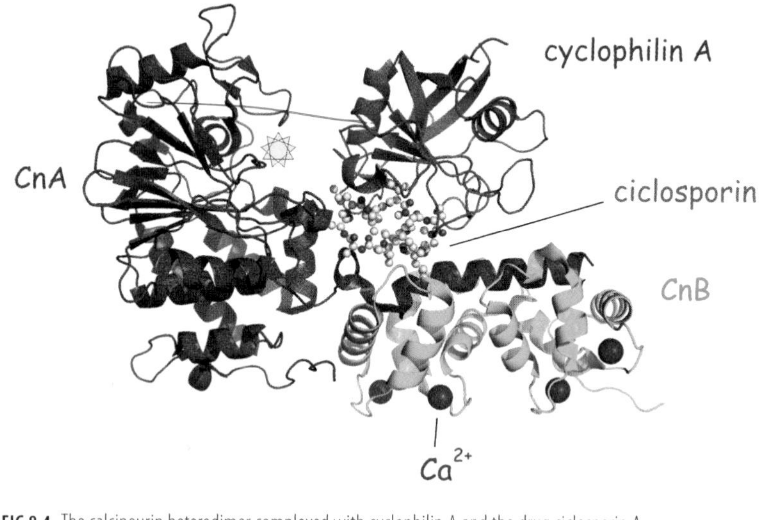

> "The nomenclature police insist that the names cyclosporin and cyclosporine should be replaced by ciclosporin. Whether the spelling of its binding protein ciclophilin remains moot."

FIG 8.4 The calcineurin heterodimer complexed with cyclophilin A and the drug ciclosporin A. The regulatory subunit, calcineurin B (CnB, green), resembles calmodulin, having EF-hand binding sites for four Ca^{2+} ions (blue spheres). The catalytic site on the calcineurin A subunit (CnA, red) is indicated by a star. The autoinhibitory helix that normally covers this site and the sequence that binds Ca^{2+}-calmodulin are not shown. It is not necessary for all four Ca^{2+} binding sites on CnB to be occupied for activity. (1mf8.[16])

FKBP12 respectively). The drug–protein complex binds to calcineurin as shown in Figure 8.4 and prevents substrate from gaining access to the active site. In this way they prevent the activation of T lymphocytes through the dephosphorylation of the transcription factor NFAT. It should be noted that immunophilins are not restricted to cells of the immune system. They also influence cell growth and repair in nerves[15] and they are abundant in other cells as components of ryanodine and IP_3 receptor complexes (see page 200).

Another target of calcineurin is the inhibitory-1(I-1) subunit of protein phosphatase 1 (PP1). I-1 acts as an inhibitory pseudosubstrate and when it is dephosphorylated by calcineurin, it is released from the phosphatase and the inhibition is lifted. As PP1 has multiple targets, this expands the range of phosphoproteins that may be dephosphorylated when calcineurin is activated (see page 679).

The effective Ca^{2+} affinity of calcineurin is high, allowing it to be activated by a mild increase in concentration. It then remains active until the level subsides. By contrast, CaMKII (see above) has low Ca^{2+} affinity and responds most effectively to one or more sharp transients, after which it can remain autonomous for tens of minutes. Thus CaMKII responds best to the intense Ca^{2+} transients that occur close to the mouths of open channels, but not to a mild more generalized elevation of Ca^{2+} in the bulk cytosol. Conversely, calcineurin is most effectively activated by a sustained Ca^{2+} increase in the cytosol, where its target transcription factors await dephosphorylation. Thus the two effectors are sensitive to Ca^{2+} signals that have different temporal and spatial characteristics. For example, in neurons undergoing mild stimulation, there will only be local activation of Ca^{2+}-effectors, such as the CaM kinases, because Ca^{2+} ions entering through channels are restricted to microdomains by local buffers and pumps. Alternatively, strong stimulation, generating a global and longer-lasting Ca^{2+} increase, may activate transcription that leads to the modulation synaptic strength.

Ca^{2+}-calmodulin-sensitive adenylyl cyclases and phosphodiesterase

Ca^{2+}- and cAMP-mediated signalling pathways are linked through the Ca^{2+}-dependent isoforms of adenylyl cyclase (AC) described in Chapter 5. AC1 and AC8 are activated by Ca^{2+}-calmodulin, while AC5 and AC6 are inhibited by Ca^{2+}, independently of calmodulin. The Ca^{2+}-sensitive ACs occupy regions of the plasma membrane rich in cholesterol and glycosphingolipids (lipid rafts, see page 522). This keeps them close to the channels that admit Ca^{2+} and to other signalling proteins. They also encounter molecules of cyclic nucleotide phosphodiesterase 1 (PDE1), a Ca^{2+}- and calmodulin-dependent terminator of cAMP signalling. Consequently, when Ca^{2+} enters neurons through voltage-dependent plasma membrane channels (e.g. L-type channels), it can stimulate AC1 or AC8 to produce cAMP, which is then broken down by PDE1. Thus, a localized Ca^{2+} signal can generate a brief and localized cAMP signal. AC1 and AC8 are also sensitive to oscillatory Ca^{2+} signals and are thought to have roles in learning and memory.

Nitric oxide synthase

It was initially surprising to find that nitric oxide (NO), familiar as a noxious gas, has an important role as an *inter*cellular messenger. NO was first perceived as endothelium-derived relaxing factor (EDRF) in the vascular system.[17] It is formed by the oxidation of l-arginine by the haem protein nitric oxide synthase (NOS). There are three members of the NOS family: the membrane-bound endothelial enzyme called eNOS (or NOS III), a soluble enzyme first characterized in brain called nNOS (or NOS I), and in macrophages a cytokine-inducible form called iNOS (or NOS II).[18] nNOS is most apparent in nerve and skeletal muscle, though not restricted to these cell types. eNOS is present in a wide variety of cells. Both eNOS and nNOS are Ca^{2+}-calmodulin-dependent enzymes and produce NO in response to a rise in cytosol Ca^{2+}. By contrast, the transcriptionally regulated iNOS binds calmodulin at resting Ca^{2+} levels and is constitutively active. It has high output and, for example in macrophages, it can remain activated for many hours, maximizing the cytotoxic damage that these cells can inflict.

NO is by no means a classical messenger molecule. It diffuses rapidly in solution and because it is short-lived *in vivo*, it has mostly short-range (paracrine) effects.[19] It can readily cross cell membranes to reach neighbouring cells and these can be activated without the need for plasma membrane receptors. A wide range of cells are affected by NO. For example, as mentioned above, it is a potent smooth muscle relaxant, not only in the vasculature, but also in the bronchioles, gut, and genitourinary tract. In intestinal tissue, nNOS in varicosities of myenteric neurons is activated by an influx of Ca^{2+}. The NO formed diffuses into neighbouring smooth muscle cells, where at nanomolar concentrations it activates a soluble guanylyl cyclase (also a haem protein) that forms cyclic GMP (cGMP).[20,21] This then activates cGMP-dependent kinase I (G kinase) which brings about relaxation by interacting with the mechanisms that regulate the cytosol Ca^{2+}, essentially keeping it low.[22] In the cardiovascular system, eNOS in endothelial cells responds to Ca^{2+} in a similar way, to relax vascular smooth muscle and reduce blood pressure.[23,24] (Note: as well as stimulating vasodilatory responses, NO has a homeostatic role in the regulation of vascular tone).

In the central nervous system, nNOS is tethered close to NMDA-type glutamate receptors (see page 209) so that it can respond to the transient but intense increases in Ca^{2+} concentration that occur in the vicinity of the open channels.[26] The NO that is formed has the potential of acting as either an anterograde or retrograde messenger.[27] Because of its diffusibility, it may also influence other cells in the vicinity such as glial cells. All this may have implications for synaptic plasticity (the modulation of synaptic potency).[28]

Trinitroglycerol, as a precursor of NO, has been in use as a palliative for angina (also as an explosive) since the 1870s.[25] It is now commonly applied sublingually for acute treatment, or as skin patches. The NO is released at the site of application and diffuses to the heart (and other organs) within minutes.

An important downstream consequence of NOS activation is the effect of NO on cellular Ca^{2+} homeostasis.[29] Many of the effects of NO are mediated by cGMP and the consequent activation of G kinase. This can phosphorylate and inactivate PLC and IP_3 receptors,[30] and modulate SOCE.[31] It also inhibits Ca^{2+} release from intracellular stores in a variety of cells,[32] (but not endothelial cells or liver). Finally, in sea urchin eggs, the generation of cyclic ADP-ribose (see page 204), an activator of ryanodine receptors in these cells, is cGMP-dependent.[33] How these actions are coordinated in the control of Ca^{2+} concentration is not yet clear.

Calcium-dependent enzymes that are not regulated by calmodulin

While calmodulin is a widely distributed mediator of enzymes that are not in themselves Ca^{2+}-binding proteins, it is not the only Ca^{2+}-sensing intermediate. Furthermore, there are many enzymes that respond to Ca^{2+} changes directly. Such Ca^{2+}-sensitive proteins usually have one or more high affinity binding sites and these may be in the form of EF-hand motifs, C2 domains, or other structures. We now consider some specific examples.

Neuronal calcium sensors

The complex morphology of nerve cells and the confinement to microdomains of pre- and postsynaptic Ca^{2+} signals requires localized Ca^{2+} sensors and effectors. While calmodulin is a multipurpose sensor, a set of more specific Ca^{2+} sensor proteins exists. These are the neuronal Ca^{2+} sensors or NCS proteins, and although not restricted to neurons, they are most diverse in mammalian neuronal tissue (14 genes). They fulfil a wide range of functions, some of which are quite specific. They all possess four EF-hand motifs, but their sequence similarity with calmodulin is very limited (<20% identity) and unlike calmodulin, they remain compact structures when they have bound Ca^{2+}.[34]

The majority of mammalian NCS proteins bear an N-terminal myristoyl group (others are palmitoylated) and some of them, most notably the retinal protein recoverin, conceal this group when Ca^{2+} is not bound. Such proteins are cytosolic at resting Ca^{2+} levels, but adhere to membranes when Ca^{2+} is elevated. This effect is reversible and the conformational change is termed a Ca^{2+}/myristoyl switch. Those that do not conceal the acyl group are permanently associated with the membrane. When the Ca^{2+}/myristoyl switch is operated by Ca^{2+} in retinal rod cells, recoverin is recruited to the plasma membrane exposing a binding site for rhodopsin kinase on its surface. This prevents the kinase from phosphorylating rhodopsin and so extends the lifetime the photoexcited state. Under bright illumination, at low Ca^{2+} levels, the excited rhodopsin is rapidly deactivated, allowing adaptation (see Figure 6.11, page 173). Also among the human NCS proteins are three guanylyl cyclase-activating proteins (GCAP1–3) that are also expressed only in retina. These are permanently associated with their guanylyl cyclase effectors and in the light at low Ca^{2+} levels, they stimulate cGMP formation. In the dark, at high Ca^{2+} levels, they have the opposite effect and inhibit cGMP formation (Figure 6.7, page 169).

Calpain

Calpain is a member of a widely distributed family of cytosolic, Ca^{2+}-activated cysteine proteases, possessing EF-hand sites.[35] It cleaves a wide range of intracellular proteins, modifying their functions, often destructively. It operates in cell death pathways and is involved in neurodegeneration and apoptosis. It also degrades cytoskeletal and other proteins in the vicinity of the plasma membrane and is thought to be involved in processes where remodelling of the cytoskeleton takes place.

Although selective, the effects of this neutral protease are mostly irreversible, which is perhaps one reason why cells do not permit Ca^{2+} levels to remain high for a prolonged period. To make additionally sure, calpain is also held in check by calpastatin.[35] This not only inhibits its activity, but also prevents it from binding to membranes.

Synaptotagmin

Exocytosis is the fusion of the membrane of a secretory vesicle with the plasma membrane, allowing the vesicle contents to be released to the exterior without affecting the integrity of the cell.

In neurons and in many types of endocrine cell, the release of a neurotransmitter or hormone by exocytosis is activated by an increase in intracellular Ca^{2+}. This event is mediated by a Ca^{2+}-sensitive protein (or proteins). Among the numerous proteins that are present on the surface of secretory vesicles or secretory granules is synaptotagmin.[36,37] This is a highly conserved transmembrane protein having two C2 domains in its cytosolic chain, and it is thought to be the Ca^{2+} sensor for exocytosis. Indeed, it has been shown to bind Ca^{2+} ions at physiological concentrations. This is discussed below (page 236).

DAG kinase

The hydrophobic second messenger diacylglycerol (DAG) formed by the phospholipases PLC and (indirectly) by PLD (see Chapter 5) is the activator of protein kinase C (PKC, see Chapter 9). It is short-lived because it is rapidly phosphorylated to form phosphatidate by diacylglycerol kinase. This reaction is part of the cycle that regenerates phosphatidylinositol (Figure 8.5) and since it removes DAG, it acts to terminate the activation of PKC. Of the eight known isoforms of DAG kinase, α, β, and γ possess EF-hand motifs and are Ca^{2+}-dependent.[38]

Ras GEFs and GAPs

Regulation of the GTPase Ras involves effectors that sense Ca^{2+} directly and those that are activated by Ca^{2+}-calmodulin. The calmodulin-dependent

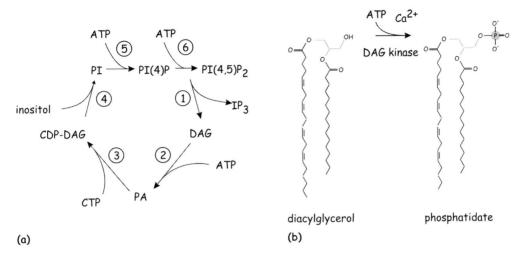

FIG 8.5 Phosphoinositide formation and metabolism.

(a) The inositol lipid cycle. Numbers indicate enzymes as follows: (1) phospholipase C, (2) diacylglycerol kinase, (3) CDP-diacylglycerol synthase, (4) phosphatidylinositol synthase, (5) phosphatidylinositol 4-kinase, (6) phosphatidylinositol-4 phosphate 5-kinase. (b) The formation of phosphatidate from DAG.

RasGRF proteins that are guanine nucleotide exchange factors (GEFs) are mentioned above. In addition to these, there is a family of five calmodulin-independent GEFs that act on both the Ras and Rap GTPases. These possess EF-hands, but although they show Ca^{2+} sensitivity their main stimulus appears to be diacylglycerol. For this reason they are called calcium and DAG-regulated GEFs or CalDAG-GEFs.[13] In order for an exchange factor to engage Ras, it must locate to the plasma membrane and the cytosolic CalDAG-GEFs have DAG-binding C1 domains that enable them to do this. (RasGRP2, a member of this family, is an exception, and uses myristoylation and palmitoylation to locate at the membrane surface).

While GTPase signalling is initiated by GEFs, it is turned off by the GTPase activating proteins (GAPs) that activate the hydrolysis of GTP to GDP. The most prominent Ca^{2+}-sensitive GAP for Ras is p120RasGAP (see page 108). This has a C2 domain which could enable the protein to bind to the plasma membrane in a Ca^{2+}-dependent way. However, it has recently been shown that the enzyme attaches to the membrane by associating with the Ca^{2+}-sensing protein annexin VI.[39]

Two members of the GAP1 family are Ca^{2+}-dependent. These are CAPRI (Ca^{2+}-promoted Ras inactivator) and RASAL (Ras GTPase-activating-like), both of which have tandem C2 domains at the N-terminus, followed by a RasGAP domain, a PH domain, and a Btk domain. The C2 domains enable the GAPs to translocate from the cytosol to the membrane when Ca^{2+} is elevated. Binding at the membrane switches on the GAP activity, so terminating Ras activation

The **annexins** are a family of soluble intracellular proteins that bind to phospholipids in a Ca^{2+}-dependent fashion. They interact with intracellular membranes and some associate with the cytoskeleton.[40] They have mostly α-helical structures and bind Ca^{2+} at loops on one surface. This enables them to bind to negatively charged membrane surfaces.

of the ERK/MAPK pathway. An interesting property of RASAL is that its membrane attachment follows the progress of the Ca^{2+} signal. A train of Ca^{2+} spikes causes it to oscillate on and off the membrane in step with the spikes. By contrast, CAPRI binds rapidly but transiently and does not appear to sense Ca^{2+} oscillations.[41]

Our knowledge of the molecules and mechanisms by which Ca^{2+} regulates Ras signalling remains incomplete, but there are hints of a machinery that is adapted to the pulsatile nature of the Ca^{2+} signals evoked by receptor activation. Repeated surges of Ca^{2+} will activate the Ras GEFs and the GAPs in a periodic fashion. The resultant effect on Ras activation must depend on the relative kinetics of the two processes, but it will most likely be pulsatile if the GAPs act more rapidly than the GEFs. If the opposite is the case, the activation of Ras will build during a train of Ca^{2+} spikes, perhaps reaching a threshold for activation of downstream pathways. Different Ras effectors may have different requirements in this respect, so that the system may decode complex Ca^{2+} signals.

Cytoskeletal proteins

Cytoskeletal proteins are responsible for the maintenance of cell shape and for motile functions. Almost every form of cellular activation is either preceded, accompanied, or followed by a rearrangement of at least part of the cytoskeleton. Such a change can be an essential or even defining component of the cellular response. In non-muscle cells, the cytoskeletal arrangement of microfilaments (F-actin) is controlled by a large array of proteins, some of which bind actin.[42] A number of these, e.g. α-actinin and gelsolin, are sensitive to the concentration of cytosol Ca^{2+}. These proteins have various effects on the cytoskeleton. α-Actinin is a cross-linker, while gelsolin is a Ca^{2+}-regulated actin-severing (and capping) protein. However, the cross-linking action of α-actinin (which has two EF-hands) is inhibited by Ca^{2+} and its consequent removal from F-actin allows access for gelsolin.

Paradigms of calcium signalling

Triggering neurotransmitter secretion

For many species the speedy transmission of neural signals is crucial for survival. The exchange of information between nerve cells (and between nerve and muscle) involves the release, by exocytosis, of neurotransmitter substances at synapses and neuromuscular junctions. Ca^{2+} plays a critical role in mediating this process. Voltage-sensitive Ca^{2+} channels in the plasma membrane of the presynaptic cell admit Ca^{2+} directly into the region where secretory vesicles containing neurotransmitter are located, primed to undergo exocytosis. In fast exocytosis from neurons, secretion is triggered within a millisecond of channel opening. Within the presynaptic cell, the sites of exocytosis lie on average only 50 nm from the Ca^{2+} channels and the concentration of Ca^{2+} in this microdomain may rise well into the micromolar range. Cytosolic Ca^{2+} buffers and efficient extrusion mechanisms ensure that this Ca^{2+} signal is local and abrupt. The resting level is resumed within a few tens of milliseconds.

The mechanism by which the brief elevation of Ca^{2+} leads to the fusion of the membrane of the synaptic vesicle with the plasma membrane, involves one or more Ca^{2+}-sensing proteins with fast binding kinetics. For fast exocytosis, the best candidate for this task is synaptotagmin I, a transmembrane protein of synaptic vesicles. Its single cytosolic chain consists of two consecutive C2 domains, C2A and C2B, linked by a short sequence. The C2A domain binds to acidic phospholipids in a Ca^{2+}-dependent manner[43] and this helps the protein to become tethered to the plasma membrane (see page 780). Synaptotagmin also associates with components of the SNARE complex of proteins in Ca^{2+}-dependent secretory cells.[44] This consists of the cytosolic chains of vesicle and plasma membrane proteins that, together with cytosolic proteins, form a tightly bound complex which draws the vesicle and plasma membranes together. It is only when Ca^{2+} binds to synaptotagmin that the two membranes merge to form a fusion pore that leads to exocytosis.

Initiation of contraction in skeletal muscle

The contraction of all types of muscle depends on an increase in intracellular Ca^{2+}. In skeletal muscle acetylcholine released at the neuromuscular junction binds to nicotinic receptors on the muscle end-plate. These are non-specific cation channels and the binding causes them to open, producing a depolarization that propagates across the plasma membrane (sarcolemma) of the muscle cell and throughout the transverse tubular invaginations (T-tubules), which penetrate the cell in the vicinity of the contractile machinery. The voltage-sensing subunits of the sarcolemmal voltage-operated Ca^{2+} channels are termed dihydropyridine receptors (DHPRs: dihydropyridine is a drug). These are distributed along the T-tubules and are directly coupled to ryanodine receptors (RyR1) on the membrane of sections of the sarcoplasmic reticulum that lie very close to the tubular membrane. The assemblage of T-tubules with the SR membranes forms the junctional triads. The ryanodine receptors are visible in electron micrographs of triads as 'junctional feet' (**Figure 8.6**).

The direct interaction between the DHPRs and the RyRs results in substantial Ca^{2+} release from the SR (page 214) and a steep rise in concentration of Ca^{2+} in the close vicinity of the contractile apparatus. This focusing of the Ca^{2+} signal ensures a rapid response and has the further advantage that the resting state can be rapidly resumed, since only a small quantity of Ca^{2+} ions have to be removed.

Within each skeletal muscle cell, the contractile machinery of actin and myosin filaments is controlled by the proteins tropomyosin and troponin. The attachment of myosin heads to actin, which causes contraction, is prevented

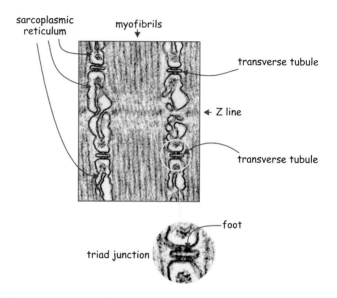

FIG 8.6 Triad junctions and junctional feet.
Electron micrographs of the swimbladder muscle from the toadfish (*Opsanus tau*). The section is cut perpendicular to the T-tubule axis and the 'junctional feet', which are ryanodine receptors, are indicated in the magnified image. Micrograph courtesy of Clara Franzini-Armstrong.[45]

by the presence of threads of tropomyosin organized on the surface of the actin filaments. Troponin, a complex of three subunits – I, T, and C – is distributed at intervals along the actin filaments and it acts to mediate the effects of Ca^{2+}. Troponin C closely resembles calmodulin, possessing four EF-hand Ca^{2+} binding sites. The C-terminal sites are of high affinity ($K_D \sim 10^{-7}$ mol L^{-1}) and the two sites in the N-terminal or 'regulatory' lobe are of low affinity ($K_D \sim 10^{-5}$ mol L^{-1}). The C-terminal sites can bind Ca^{2+} or Mg^{2+} and are occupied at resting levels of Ca^{2+}. As the concentration rises, it binds to the low-affinity N-terminal sites causing a conformational change. This is passed on to the rest of the troponin complex, resulting in the lifting of the inhibition of the myosin ATPase by tropomyosin and allowing contraction to occur. As the concentration of Ca^{2+} returns to the resting level, inhibition by tropomyosin is re-established and contraction ceases.[46]

The speed with which these events occur can be very fast. For example, the muscles that surround the swim bladder of the toadfish can twitch at frequencies in excess of 100 Hz, enabling it to emit a characteristic 'boatwhistle' sound. Humming birds flap their wings ~80 times per second, but this can rise to 200 in courtship flight. A failure to clear Ca^{2+} between successive stimuli (depolarizations) would cause the fusion of consecutive twitches and ultimately a permanent contraction (tetany). The achievement of such high-frequency contractions without fusion requires both the fast generation and fast removal of Ca^{2+}. In order to respond to such rapid fluctuations, the on- *and* off-rates of Ca^{2+} binding to troponin C must also be high.

Smooth muscle contraction

Compared with skeletal and cardiac muscle, smooth muscle contraction is a slow process. The enormous variability that exists between different smooth muscle cells makes it difficult to draw general conclusions concerning their signalling mechanisms. In smooth muscle there is no troponin. Contraction is achieved through the phosphorylation of the P chain of myosin by MLCK and this is activated by Ca^{2+}-calmodulin. (Note: calmodulin-dependent kinase also phosphorylates MLCK in skeletal muscle but here it does not cause contraction. Instead, it appears to affect the force of contraction.)

Some smooth muscle cells in arteries and veins display a mechanism similar to that in the heart, where entry of calcium through VOCCs is further amplified by Ca^{2+}-induced Ca^{2+} release from internal stores.[47–49] In contrast to cardiac cells, however, these smooth muscle cells operate a low gain mechanism so that the entry of external calcium contributes a larger proportion of the global calcium signal.

In addition to depolarization-dependent mechanisms, many smooth muscle cells are also activated by receptors that activate PLC, forming IP_3 and releasing Ca^{2+} from internal stores.[50] Calcium signals are generated by both IP_3Rs and RyRs and there are indications that these two release channels might cooperate with each other in some types of smooth muscle.

References

1. Dolmetsch RE, Xu K, Lewis RS. Calcium oscillations increase the efficiency and specificity of gene expression. *Nature*. 1998;392:933–936.
2. Li W-H, Llopis J, Whitney M, Zlokarnik G, Tsien RY. Cell-permeant caged InsP$_3$ ester shows that Ca^{2+} spike frequency can optimize gene expression. *Nature*. 1998;392:936–941.
3. Cheung WY. Cyclic 3',5'-nucleotide phosphodiesterase, Demonstration of an activator. *Biochem Biophys Res Commun*. 1970;38:533–538.
4. Kakiuchi S, Yamazaki R. Calcium dependent phosphodiesterase activity and its activating factor (PAF) from brain studies on cyclic 3',5'-nucleotide phosphodiesterase. *Biochem Biophys Res Commun*. 1970;41:1104–1110.
5. Kuboniwa H, Tjandra N, Grzesiek S, Ren H, Klee CB, Bax A. Solution structure of calcium-free calmodulin. *Nat Struct Biol*. 1995;2:768–776.
6. Chattopadhyaya R, Meador WE, Means AR, Quiocho FA. Calmodulin structure refined at 1.7 Angstrom resolution. *J Mol Biol*. 1992;228:1177–1192.

7. Ikura M, Clore GM, Gronenborn AM, Zhu G, Klee CB, Bax A. Solution structure of a calmodulin-target peptide complex by multidimensional NMR. *Science*. 1992;256:632–638.

8. Babu YS, Sack JS, Greenhough TJ, Bugg CE, Means AR, Cook WJ. Three-dimensional structure of calmodulin. *Nature*. 1985;315:37–40.

9. Cori GT, Cori CF. The enzymatic conversion of phosphorylase a to b. *J Biol Chem*. 1945;158:321–332.

10. Cohen P, Burchell A, Foulkes JG, Cohen PT. Identification of the Ca^{2+}-dependent modulator protein as the fourth subunit of rabbit skeletal muscle phosphorylase kinase. *FEBS Lett*. 1978;15:287–293.

11. Hook SS, Means AR. Ca^{2+}/CaM-dependent kinases: from activation to function. *Annu Rev Pharmacol Toxicol*. 2001;41:471–505.

12. Di LF, Domi T, Fedrizzi L, Lim D, Carafoli E. The plasma membrane Ca^{2+} ATPase of animal cells: Structure, function and regulation. *Arch Biochem Biophys*. 2008.

13. Cullen PJ, Lockyer PJ. Integration of calcium and Ras signalling. *Nat Rev Mol Cell Biol*. 2002;3:339–348.

14. Stemmer PM, Klee CB. Dual calcium ion regulation of calcineurin by calmodulin and calcineurin B. *Biochemistry*. 1994;33:6859–6866.

15. Snyder SH, Sabatini DM, Laiho M, et al. Neural actions of immunophilin ligands. *tips*. 1998;19:21–26.

16. Jin L, Harrison SC. Crystal structure of human calcineurin complexed with cyclosporin A and human cyclophilin. *Proc Natl Acad Sci U S A*. 2002;99:13522–13526.

17. Palmer RM, Ferrige AG, Moncada S. Nitric oxide release accounts for the biological activity of endothelium-derived relaxing factor. *Nature*. 1987;327:524–526.

18. Andrew PJ, Mayer B. Enzymatic function of nitric oxide synthases. *Cardiovasc Res*. 1999;43:521–531.

19. Balligand JL, Cannon PJ. Nitric oxide synthases and cardiac muscle, Autocrine and paracrine influences. *Arterioscler Thromb Vasc Biol*. 1997;17:1846–1858.

20. Rodgers KR. Heme-based sensors in biological systems. *Curr Opin Chem Biol*. 1999;3:158–167.

21. Foster DC, Wedel BJ, Robinson SW, Garbers DL. Mechanisms of regulation and functions of guanylyl cyclases. *Rev Physiol Biochem Pharmacol*. 1999;135:1–39.

22. Lucas KA, Pitari GM, Kazerounian S, Ruiz-Stewart I, Park J, Schulz S, Chepenik KP, Waldman SA. Guanylyl cyclases and signaling by cyclic GMP. *Pharmacol Rev*. 2000;52:375–414.

23. Huang PL, Huang Z, Mashimo H, Bloch KD, Moskowitz MA, Bevan JA, Fishman MC. Hypertension in mice lacking the gene for endothelial nitric oxide synthase. *Nature*. 1995;377:239–242.

24. Gyurko R, Kuhlencordt P, Fishman MC, Huang PL. Modulation of mouse cardiac function in vivo by eNOS and ANP. *Am J Physiol*. 2000;278: H971–H981.

25. Marsh N, Marsh A. A short history of nitroglycerine and nitric oxide in pharmacology and physiology. *Clin Exp Pharmacol Physiol*. 2000;27:313–319.

26. Duchen MR. Mitochondria and calcium: from cell signalling to cell death. *J Physiol*. 2000;529:57–68.

27. Park JH, Straub VA, O'Shea M. Anterograde signaling by nitric oxide: characterization and in vitro reconstitution of an identified nitrergic synapse. *J Neurosci*. 1998;18:5463–5476.

28. Gudi T, Hong GK, Vaandrager AB, Lohmann SM, Pilz RB. Nitric oxide and cGMP regulate gene expression in neuronal and glial cells by activating type II cGMP-dependent protein kinase. *FASEB J*. 1999;13:2143–2152.

29. Martinez-Serrano A, Borner C, Pereira R, Villalba M, Satrustegui J. Modulation of presynaptic calcium homeostasis by nitric oxide. *Cell Calcium*. 1996;20:293–302.

30. Rooney TA, Joseph SK, Queen C, Thomas AP. Cyclic GMP induces oscillatory calcium signals in rat hepatocytes. *J Biol Chem*. 1996;271:19817–19825.

31. Xu X, Star RA, Tortorici G, Muallem S. Depletion of intracellular Ca^{2+} stores activates nitric-oxide synthase to generate cGMP and regulate Ca^{2+} influx. *J Biol Chem*. 1994;269:12645–12653.

32. Clementi E. Role of nitric oxide and its intracellular signalling pathways in the control of Ca^{2+} homeostasis. *Biochem Pharm*. 1998;55:713–718.

33. Galione A, White A, Willmott N, Turner M, Potter BV, Watson SP. cGMP mobilizes intracellular Ca^{2+} in sea urchin eggs by stimulating cyclic ADP-ribose synthesis. *Nature*. 1993;365:456–459.

34. Burgoyne RD. Neuronal calcium sensor proteins: generating diversity in neuronal Ca^{2+} signaling. *Nat Rev Neurosci*. 2007;8:182–193.

35. Kawasaki H, Kawashima S. Regulation of the calpain-calpastatin system by membranes. *Mol Membr Biol*. 1996;13:217–224.

36. Geppert M, Goda Y, Hammer RE, Li C, Rosahl TW, Stevens CF, Sudhof TC. Synaptotagmin I: a major Ca^{2+} sensor for transmitter release at a central synapse. *Cell*. 1994;79:717–727.

37. Li C, Davletov BA, Sudhof TC. Distinct Ca^{2+} and Sr^{2+} binding properties of synaptotagmins. Definition of candidate Ca^{2+} sensors for the fast and slow components of neurotransmitter release. *J Biol Chem*. 1995;270:24898–24902.

38. Sakane F, Kanoh H. Molecules in focus: diacylglycerol kinase. *Int J Biochem Cell Biol*. 1997;29:1139–1143.

39. Grewal T, Evans R, Rentero C, et al. Annexin A6 stimulates the membrane recruitment of p120GAP to modulate Ras and Raf-1 activity. *Oncogene*. 2005;24:5809–5820.

40. Gerke V, Creutz CE, Moss SE. Annexins: linking Ca^{2+} signalling to membrane dynamics. *Nat Rev Mol Cell Biol*. 2005;6:449–461.

41. Walker SA, Kupzig S, Bouyoucef D, et al. Identification of a Ras GTPase-activating protein regulated by receptor-mediated Ca^{2+} oscillations. *EMBO J*. 2004;23:1749–1760.

42. Holt MR, Koffer A. Cell motility: proline-rich proteins promote protrusions. *Trends Biochem Sci*. 2001;11:38–46.

43. Brose N, Petrenko AG, Sudhof TC, Jahn R. Synaptotagmin: a calcium sensor on the synaptic vesicle surface. *Science*. 1992;256:1021–1025.

44. Brunger AT. Structural insights into the molecular mechanism of Ca^{2+}-dependent exocytosis. *Curr Opin Neurobiol*. 2000;10:293–302.

45. Franzini-Armstrong C, Nunzi G. Junctional feet and particles in the triads of a fast-twitch muscle fibre. *J Muscle Res Cell Motil*. 1983;4:233–252.

46. Vassylyev DG, Takeda S, Wakatsuki S, Maeda K, Maeda Y. Crystal structure of troponin C in complex with troponin I fragment at 2.3-Å resolution. *Proc Natl Acad Sci USA*. 1998;95:4847–4852.

47. Ganitkevich VY, Isenberg G. Contribution of Ca^{2+}-induced Ca^{2+} release to the $[Ca^{2+}]_i$ transients in myocytes from guinea-pig urinary bladder. *J Physiol (Lond)*. 1992;458:119–137.

48. Ya. Ganitkevich V, Isenberg, G. Efficacy of peak Ca^{2+} currents (I_{Ca}) as trigger of sarcoplasmic reticulum Ca^{2+} release in myocytes from the guinea-pig coronary artery. *J Physiol (Lond)*. 1995;484:287–306.

49. Gregoire G, Loirand G, Pacaud P. Ca^{2+} and Sr^{2+} entry induced Ca^{2+} release from the intracellular Ca^{2+} store in smooth muscle cells of rat portal vein. *J. Physiol (Lond)*. 1993;472:483–500.

50. Somlyo AP, Somlyo AV. Signal transduction and regulation in smooth muscle. *Nature*. 1994;372:231–236.

CHAPTER 9

Phosphorylation and Dephosphorylation: Protein Kinases A and C

Protein phosphorylation as a switch in cellular functioning

The first indications that protein phosphorylation might be an important regulator of enzyme activities emerged at a time when the regulatory roles of non-covalent interactions of various substances (substrates, cofactors, end products, etc.) were already well established. In particular, the allosteric influence of end products, interacting with regulatory subunits of multisubunit enzymes provides feedback control over metabolic pathways comprising many steps. The regulation of the activation state of phosphorylase by 5'-AMP (positive feedback) and glucose-6-phosphate (negative feedback) is a pertinent example. Although neither of these metabolites takes part in the reaction catalysed by phosphorylase, the presence of glucose-6-phosphate signals metabolic sufficiency, while 5′-AMP signals hunger. Both can be described as allosteric effectors, interacting with the enzyme at sites remote from its catalytic centre. Then it emerged that the addition of a single phosphate group to phosphorylase b also switches the enzyme between inactive and active states[1] (**Figure 9.1**). Thus, in addition to allosteric influences, covalent modification by phosphorylation (or dephosphorylation) also affects enzyme activity. Although these two regulatory processes operate through similar conformational changes, phosphorylation is primarily a response to extracellular influences, expressed through hormone receptors, while the allosteric effectors allow the system to respond to intracellular conditions.

The importance of phosphorylation is underlined by the fact that the genome of the budding yeast *Saccharomyces pombe* specifies more than 120 different protein kinases.[3] And this is only yeast! The human genome specifies 518. One in three cytoplasmic proteins contains covalently bound phosphate. Phosphorylation modifies proteins by the addition of negatively charged groups to serines, threonines, and less commonly, tyrosines. These neutral hydroxyamino acid residues are often exposed on surfaces and in the interfaces between the subunits of regulatory proteins. Phosphorylation of a protein can alter its properties substantially. It can result in the recognition, binding, activation, deactivation, phosphorylation, or dephosphorylation of its substrate. Phosphorylation of an enzyme can switch it on or off.

The mammalian kinases fall into seven main groups or classes:

AGC: includes PKA, PKG, and PKC families (this chapter)

CAMK: includes Ca^{2+}/calmodulin-dependent protein kinase (Chapter 8)

CK: casein kinase 1 (Chapter 14)

CMGC: containing CDK, MAPK, GSK3, CLK families (Chapters 12, 14, and 18)

STE: homologues of yeast Sterile 7, Sterile 11, Sterile 20 kinases (Chapter 12)

TK: tyrosine kinase (Chapter 12)

TKL: tyrosine kinase-like.

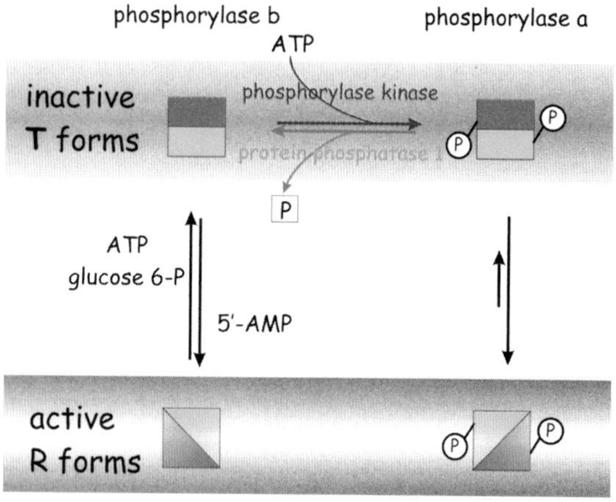

FIG 9.1 Regulation of phosphorylase through phosphorylation and by the action of allosteric effectors. Catalytically inactive (T) and active (R) conformations of the enzyme. When phosphorylase is phosphorylated (phosphorylase a), its 'relaxed' R form is favoured. Phosphorylase b adopts a 'tense' T conformation except under conditions of starvation, signalled by the presence of 5′-AMP, when it adopts its active R form. The activity of phosphorylase is mainly determined by its phosphorylation status. The terms tense and relaxed refer to the alternative quaternary structures of allosteric enzymes.[2] This terminology was originally applied in the analysis of oxygen binding to haemoglobin.

Things can go badly wrong when the balance of phosphorylation and dephosphorylation is disturbed. Examples are the explosive diarrhoea and hepatotoxicity caused by the inhibition of protein phosphatases by toxins present in infected shellfish (okadaic acid) and in cyanobacteria (microcystins). The reverse is even more grim: for example, the consequences of uncontrolled stripping of phosphate groups which occurs in cells affected by the virulence factor of *Yersinia pestis* (the agent of bubonic plague, still a major killer). This factor is a protein tyrosine phosphatase.[4] The prospects are just as bad when tyrosine kinases become constitutively activated (oncogenic mutations, Chapters 11 and 12). On a happier note, however, a number of important therapeutic procedures (tumour suppression, immunosuppression, and treatment of inflammation) are based on drugs that either promote or suppress phosphorylation reactions. For example, ciclosporin A, used as an immunosuppressive agent, interacts with calcineurin, a subunit of which is protein phosphatase 2B (see page 230).

Cyclic AMP and the amplification of signals

Most but not all of the signals conveyed by cAMP are mediated through phosphorylation reactions catalysed by protein kinase A (PKA). For example, following stimulation of β-adrenergic receptors on liver cells by adrenaline, at concentrations around 10^{-10} mol L^{-1}, the intracellular concentration of cAMP increases to about 1 μmol L^{-1}. As a rule of thumb, such 'typical' receptors are present at a density of $\sim 10^5$ per cell and, in general, only a small proportion of these need to be occupied or activated to maximize the response (see page 32). This small signal can activate a great deal more of the enzyme glycogen phosphorylase (100 μmol L^{-1}) and this can liberate truly enormous quantities of glucose into the bloodstream.

Intermediate amplification of the signal provided by cAMP is provided through two rounds of protein phosphorylation. The first of these is catalysed by the cAMP-dependent protein kinase (PKA). This phosphorylates and activates phosphorylase kinase, which in turn phosphorylates and activates glycogen phosphorylase (**Figure 9.2**). During exercise or under conditions of starvation, the elevated level of 5'-AMP acts as an allosteric signal for the activation of phosphorylase b, regardless of its state of phosphorylation. Thus, phosphorylation of the enzyme and its activated state are not synonymous. However, it is fair to say that phosphorylase b is generally inactive and that phosphorylase a is generally active. On the other hand, if there is no need to break down glycogen, then why do so? Under resting conditions, the catalytic activity of phosphorylase b is suppressed by the presence of glucose-6-phosphate and ATP (which competes at the binding site for 5'-AMP) (see Figure 9.1). In effect, an elevated level of 5'-

Okadaic acid, named from the marine sponge *Halichondria okadai*, from which it was fist isolated, is actually the product of a symbiotic dinoflagellate. It accumulates in bivalves and causes diarrheoretic shellfish poisoning. Its toxicity, expressed as LD_{50}, in p388 and Ll1210 cells is 17 and 1.7 nmol L^{-1} respectively. It is a potent inhibitor of serine/threonine phosphatases, having a K_i of 30 pmol L^{-1} against protein phosphatase 2A.

The microcystins are a group of cyclic heptapeptides. The peptide ring is made up of two protein amino acids and five non-protein amino acids. Microcystins accumulate in the liver and affect the organization of the cytoskeleton, causing the hepatocytes to collapse inwards. Although not intrinsically carcinogenic, the microcystins appear to hasten the progress of cancer. It has been suggested that the high incidence of liver cancer in Qidong near Shanghai in China may be due to the presence of cyanobacterial toxins in drinking-water.[5]

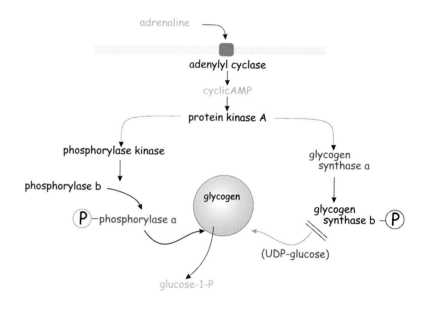

FIG 9.2 Control of glycogen breakdown. The effect of a receptor activating PKA and the phosphorylation of phosphorylase (active) and glycogen synthase (inactive)

AMP indicates metabolic insufficiency. Phosphorylase a is insensitive to allosteric regulation by 5′-AMP, ATP, or glucose-6-phosphate.

The net result of all this is the transduction of a very weak first signal (low receptor occupancy by hormone) into a large kinetic signal (rapid glycogenolysis). However, although activated target enzymes can process their substrates rapidly (in our example the release of glucose-1-phosphate from liver glycogen), the penalty for all this stage-by-stage complexity, control, and amplification is a delay in the onset of the response. The really fast systems, such as the signalling for neurotransmitter release or the contraction of skeletal muscle, are not regulated in this way. In skeletal muscle, phosphorylase kinase is activated by elevation of cytosol Ca^{2+} which is also the signal that initiates contraction. The effect of phosphorylation by PKA is not only to maximize the rate of conversion of phosphorylase b to phosphorylase a, but also to increase its affinity for Ca^{2+}. This ensures that muscle phosphorylase kinase is operative even under the ambient conditions of low Ca^{2+}. In terms of the fright, fight, and flight response, adrenaline readies the muscle for maximal activity even before it receives its signal to contract (acetylcholine). The cAMP signal also acts to inhibit dephosphorylation of phosphorylase and hence its deactivation. We return to this topic in Chapter 21.

In general, the concentration of ATP is unaffected even after prolonged periods of acute exercise or starvation, but the level of **5′-AMP** is elevated as the cell attempts to maintain its level of ATP by transphosphorylation of ADP in the reaction catalysed by myokinase: $2ADP = ATP + AMP$

Protein kinase A

Protein kinase A is a broad-spectrum kinase. The catalytic subunit is directed at serine or threonine residues embedded in the sequence RRxS/Tx (x is

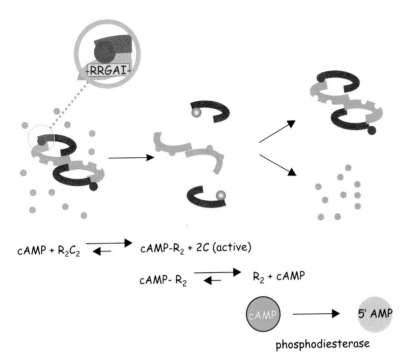

$$cAMP + R_2C_2 \rightleftharpoons cAMP\text{-}R_2 + 2C \text{ (active)}$$

$$cAMP\text{-} R_2 \rightleftharpoons R_2 + cAMP$$

$$cAMP \longrightarrow 5' \ AMP$$

phosphodiesterase

FIG 9.3 Activation of protein kinase A by cAMP.

The holoenzyme is composed of two regulatory components (green) that each bind two molecules of cAMP and two catalytic units (red). The components are linked between the catalytic sites and a 'pseudo-substrate' sequence on the regulatory units. This has the sequence -RRGAI- which closely resembles the consensus sequence for phosphorylation (-RRGSI-). On binding four molecules of cAMP the catalytic units are released and their catalytic sites exposed for interaction with real as opposed to 'pseudo' substrates. The affinity of the regulatory subunits declines allowing the bound cAMP to dissociate. It is converted to 5'-AMP by a phosphodiesterase.

variable) and this motif is widespread. What distinguishes the different forms of PKA are the tissue-specific regulatory subunits with which they are associated.

In the absence of cAMP, PKA is inactive and exists as a stable tetramer ($K_D \sim$ 0.2 nmol L^{-1}), R_2C_2, composed of two regulatory subunits (R) and two catalytic subunits (C) (**Figure 9.3**). The tetramer is stabilized through an interaction between the catalytic sites and a pseudosubstrate sequence present in the regulatory subunits, which closely resembles the substrate phosphorylation consensus sequence.[6, 7] For type I regulatory subunits, isolated from skeletal muscle, this is a true pseudosubstrate motif, in which the serine is replaced by glycine or alanine, RRxG/Ax, neither of which can be phosphorylated. In type II isoforms (from cardiac muscle) the serine is retained and the regulator is itself a substrate and undergoes phosphorylation.[8]

Autoinhibition by such motifs is a common feature of the serine/threonine protein kinases (see, for example, page 557). Here, the regulatory subunits, tightly and stably attached to each other as R_2, both possess two similar binding sites for cAMP (35% identity). The binding of cAMP is cooperative, so that occupation of the first site by cAMP enhances the affinity of the vacant site. With both binding sites occupied, the stability of the R_2C_2 complex declines by a factor of $10^4 - 10^5$, so liberating the catalytic units that can then phosphorylate their target substrates.[9]

The concentration of PKA in mammalian cells is so high, in the range $0.2 - 2\ \mu\text{mol L}^{-1}$,[10] that one might expect that little or no cAMP would ever be free in the cytosol and accessible to the phosphodiesterases which catalyse its hydrolysis. However, following the dissociation of the inactive R_2C_2 complexes, the cooperative nature of the binding of cAMP to the regulatory subunits is lost, and with it the high-affinity state.[9] As a result, the cAMP dissociates and becomes subject to conversion to 5'-AMP by phosphodiesterase (Figure 9.3). Furthermore, the activity of this enzyme may be enhanced through phosphorylation by PKA, so ensuring abrupt termination of the signal.[11, 12]

The specific actions of protein kinase A and other phosphorylating enzymes are controlled by targeting of the regulatory subunits. Specific anchoring proteins are present in particular locations such as the Golgi membranes, centrioles (microtubule organizing centres), cytoskeleton, etc.[6, 9, 13] With four isoforms of the regulatory subunits that can form heterodimers (as in $R\alpha R\beta$), and three isoforms of the catalytic subunit, it is possible that up to 24 different forms of the holoenzyme exist, although not all isoforms are expressed in all tissues.

Protein kinase A and the regulation of transcription

Activation of the CREB transcription factor

The action of PKA in the control of glycogen metabolism is an example of short-term regulation, as between one meal and the next. PKA also acts to regulate events in the longer term by switching on the transcription of specific genes. For this purpose, the signals have to be conveyed into the nucleus. The first identified example of such long-term control by PKA was the expression of the hypothalamic peptide somatostatin. Other well-established examples are the stimulation, by catecholamines, of the synthesis of mRNA coding for the β_2-adrenergic receptor and by glycoprotein hormones that generate receptors for TSH and LH. The phosphorylation substrate is the transcription factor cAMP response element binding protein (CREB, see **Figure 9.4**). Following phosphorylation by PKA, the CREB dimer interacts with DNA at the cAMP response element, CRE. This is an 8 base pair palindromic sequence (TGACGTCA) generally located within 100 nucleotides of the TATA box (see page 287).

However, transcriptional partners or coactivators are necessary for subsequent activation of transcription. Such factors are CREB binding protein, CBP, and p300, both large proteins that only interact with the phosphorylated form of CREB and direct the transcription factors to the transcriptional machinery situated at the TATA box.[14]

Somatostatin, a tetradecapeptide, is best known as an inhibitor of growth hormone secretion, but it appears to inhibit the release of other of pituitary hormones and also insulin from pancreatic β-cells. In view of this it has been suggested that its name should be changed to *panhibin*.

Palindrome: a string or phrase that reads the same backwards and forwards. In the present case it refers to ACTGCAGT with its complementary sequence TGACGTCA. For more examples, see Table 19.1, page 579.

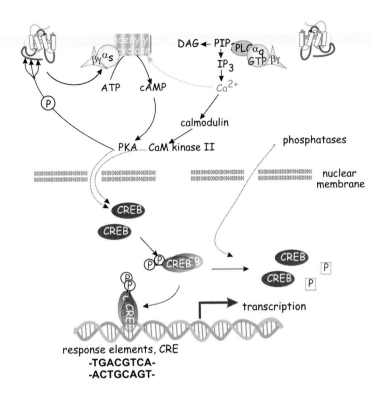

response elements, CRE
-TGACGTCA-
-ACTGCAGT-

FIG 9.4 Pathways leading to activation of CREB-mediated gene expression.
The figure illustrates the basic cytosolic and nuclear events leading to the activation of the cAMP response elements (CRE). Both PKA and (Ca^{2+} activated) CaM kinase can diffuse into the nucleus and phosphorylate CREB at the same serine residue. The phosphorylation promotes CREB dimer formation and in this form the active transcription factor binds specifically at the CRE resulting in gene transcription. Dephosphorylation of CREB leads to attenuation of the response.

Not surprisingly, the activation of CREB can also occur as a consequence of those hormonal stimuli that cause an elevation in the cytosol concentration of Ca^{2+} acting through AC1 (Ca^{2+}-sensitive adenylyl cyclase 1).[15] Less expected was the finding that in some cells, Ca^{2+} can activate CREB directly as a result of phosphorylation by CaM kinases at the same residue (S133).[16, 17] These effects of Ca^{2+} may have important consequences in the medium and long term, in cells which also react abruptly to transient elevations in cytosol Ca^{2+}. Clearly, for cells that must be able to secrete proteins repeatedly and 'on demand', such as pancreatic endocrine (islet) and exocrine (acinar) cells, it is essential to restore the complement of secretory proteins after the event. The Ca^{2+} that provides the stimulus to secretion may thus also provide the stimulus to specific protein synthesis. In pancreatic α cells, Ca^{2+} entering through voltage-sensitive channels acts both as the stimulus to the secretion of glucagon and also, through the action of CaM kinase II, as a stimulus to transcription of the glucagon gene.[18] The CREB is also understood to regulate many of the long-term effects of stimulus-induced plasticity at synapses and to underlie the process of long-term potentiation.[17, 19]

In addition to phosphorylation, the effect of CREB on cellular responses may be also be regulated through the expression of CREB itself. In testicular Sertoli cells, follicle stimulating hormone (FSH) acting through cAMP not only induces phosphorylation to activate the CREB, but also results in the accumulation of CREB-specific mRNA.[20] As with other hormones of pituitary origin, FSH is released in a pulsatile manner under the command of the hypothalamus. This observation therefore suggests that the resulting pulses of cAMP could prime the cells to become more sensitive to subsequent stimulation by the hormone.

Attenuation of the cAMP response elements by dephosphorylation

Over the 15–20 min following hormonal activation, the catalytic units of PKA diffuse into the nucleus to cause phosphorylation of CREB[21] and the initiation of transcription. Then, over 4–6 h (the attenuation phase), transcription of the target genes gradually declines. This is probably due to dephosphorylation of CREB since it can be prolonged by application of phosphatase inhibitors (such as okadaic acid[22]). Furthermore, fibroblasts that over-express phosphatase inhibitor-1 (PP1) manifest an enhanced transcriptional response to CRE.[23] In hepatic cells, manipulation of the activity of the levels of PP2A causes dephosphorylation of CREB.[24] It appears that the magnitude of CRE-induced responses is regulated, at least in part, through dephosphorylation, but that different phosphatases are involved in different cells.

As mentioned above, specificity of PP1 is determined by its association with regulatory subunits, which target it to particular cellular locations. In the nucleus PP1 is attached to the chromatin by its association with NIPP, the endogenous nuclear inhibitor of PP1.[25] Phosphorylation of NIPP by PKA drastically reduces its affinity for PP1 and so cAMP may act to down-regulate its own signals through the activation of the phosphatase (see Chapter 21).

As will become apparent in later chapters, the binding of a dimerized transcription factor complex to a palindromic enhancer element on DNA is a recurring feature in the regulation of specific gene expression.

Protein kinase A and the activation of ERK

Most of the actions of the β_2-adrenergic receptor are transduced through G_s and mediated by phosphorylation catalysed by PKA. However, following phosphorylation of the receptor by PKA (heterologous desensitization: see Figure 4.14, page 100), its specificity alters so that it shifts its attention from G_s to G_i. The mechanism that uncouples the receptor from its normal transducer also enables it to couple with G_i. This opens up a whole new range of possibilities, not least because of the much greater quantities of G_i proteins expressed in most cells, and hence the much greater availability of

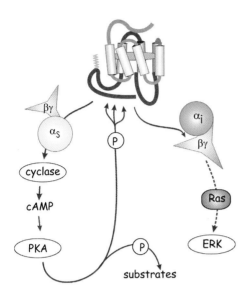

FIG 9.5 All change trains. Down-regulated receptors are not necessarily down-regulated, merely looking the other way. Phosphorylation of the β_2-adrenergic receptor by PKA diverts its attention from G_s to G_i. The resulting activation of $\beta\gamma$-subunits initiates a pathway of reactions leading to the activation of ERK (extracellular signal regulated kinase).
Adapted from Daaka et al.[26]

$\beta\gamma$-subunits. An important end result of this switching is the involvement of Ras leading to activation of ERK (extracellular regulated kinase) and the generation of nuclear signals (**Figure 9.5**). Further details will be given in Chapter 12.

This line of communication is interrupted in cells that have been treated with pertussis toxin, a characteristic of the involvement of G_i proteins, not G_s. It can also be prevented by over-expression of a peptide containing a PH domain, derived from the C-terminus of the receptor kinase GRK2 (GRK2/βARK: see page 98) which binds to $\beta\gamma$-subunits. All this effectively places cAMP and PKA upstream of the receptor, conditioning it to communicate with G_i and then Ras, leading to the activation of ERK (Chapter 12).

Actions of cAMP not mediated by PKA

Regulation of ion channels by cyclic nucleotides

Important among the actions of cAMP that are not mediated through PKA is the regulation of a gated cation channel, a component of the signalling mechanism in olfactory cells[27–30] (see page 179).

Epac, a guanine nucleotide exchange factor directly activated by cAMP

Rap1 is a monomeric GTPase involved in a number of cellular processes that include the activation of platelets, cell proliferation, differentiation, and

The **cAMP gated K^+ channel** is composed of four similar protein chains that are assembled together so that the transmembrane domains (generally spanning the membrane five or six times) combine to form a central pore. In the voltage-gated Na^+ and Ca^{2+} channels the subunits are linked as a single protein chain.

morphogenesis.[31] Although initially identified as a K-Ras-related protein,[32] Rap1 does not transform but rather antagonizes the action of Ras[33] (**Figure 9.6**). Unlike Ras, Rap1 is confined to organelle membranes. This may explain how, despite the two proteins sharing some downstream effectors, their effects can be antagonistic. Depending on the cell type, several ligands are able to activate Rap1, through interaction with their receptors and following the generation of second messengers such as Ca^{2+}, diacylglycerol (DAG), and cAMP. It is important to note, however, that this works equally well in cells that express a mutant form of PKA that has a low affinity for cAMP.

The activation of Rap1 is mediated through the direct action of cAMP with a guanine nucleotide exchange factor (Epac, exchange protein directly activated by cAMP).[34] This was identified by screening sequence databases for those proteins that have homology with GEFs for Ras and Rap1 and also binding sites for cAMP. The nucleotide binding site is similar to those present in the regulatory subunits of PKA and the cyclic nucleotide-regulated ion channels. Using purified Epac it was shown that cAMP is required to initiate the release of GDP from Rap1. The interaction of a mutant form of Epac, lacking the cyclic nucleotide binding site, is constitutively active. It can release GDP from Rap1 in the absence of cAMP.

As with to other monomeric GTPases, the return to the GDP-bound state is promoted by interaction with a GTPase activating protein. The specific Rap1GAP is itself regulated by the heterotrimer G_o in its resting state.[35] Thus, α_o.GDP can sequester Rap1GAP and in this way regulate the various activities of Rap1. We return to Rap1 and integrin activation in Chapter 16 (page 498).

FIG 9.6 Activation of Rap1 by Epac, a guanine nucleotide exchange factor activated by cAMP. Rap1 is a monomeric GTPase that is confined to organelle membranes and that counteracts the oncogenic effects of Ras. Its activation is an example of a process mediated by cAMP without the participation of PKA.

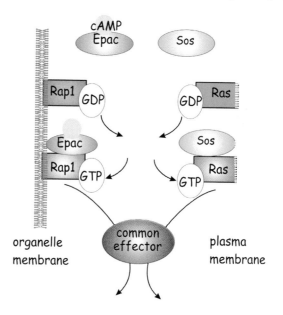

Protein kinase C

Discovery of a phosphorylating activity independent cAMP

The enzymes that we call protein kinase C (PKC) became known during the late 1970s as a cyclic nucleotide-independent protein kinase present in bovine cerebellum.[36] As first described, this activity appeared to be the product of limited proteolysis by a calcium-dependent protease,[37] but shortly afterwards it was found to be activated by Ca^{2+} (50 μmol L^{-1}) and phospholipids, with no need for proteolysis.[38] More significantly, the activity of this protein kinase could also be stimulated by the addition of a small quantity of diacylglycerol (DAG, product of the phospholipase-C reaction, together with phospholipids (generally phosphatidylserine) and Ca^{2+}, now at concentrations not far above the physiological range (2–6 μmol L^{-1}).[39]

The new enzyme became a focus of interest when it was found that it constitutes the predominant intracellular target for the active principles of croton oil, long recognized as skin irritants with tumour-promoting capacities.[40] Such phorbol esters[41] activate PKC, substituting for, and competing with, the physiological activator DAG.[42,43] Phorbol ester (**Figure 9.7**) has since been applied to every cell and system imaginable. It is apparent that PKC is involved in an enormous number of cellular processes. These range from tumour formation, host defence, embryological development, pain perception, neurite outgrowth, and the development of long-term memory. We consider the roles of protein kinase C in tumour formation and cell polarity in Chapter 19.

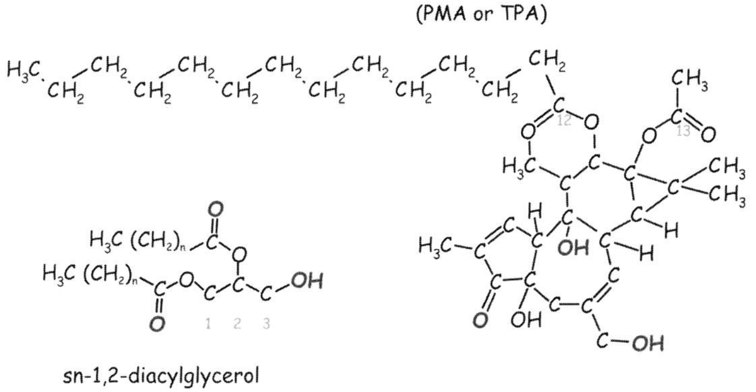

FIG 9.7 Structure of phorbol myristate acetate (PMA/TPA) and diacylglycerol.
Note the similar orientation of the three critical oxygens (red) which are essential for activation of PKC.

Croton oil is a poisonous, viscous liquid obtained from the seeds of a small Asiatic tree, *Croton tiglium*, a member of the spurge family (Euphorbiaceae). The tree is native to India and the Indonesian archipelago. Croton oil is pale yellow to brown and is transparent, with an acrid persistent taste and disagreeable odour. It is a violent irritant. The use of croton oil as a drug was introduced to the West from China by the Dutch in the 16th century. The 19th century physician (and pioneer in the field of signal transduction) Sidney Ringer (see page 185) describes its use as a purgative and its topical application in the treatment of ringworm.[44] It is now considered too dangerous for medicinal use.

The **phorbol ester** most often used in the laboratory is phorbol 12-myristate 13-acetate or 12-*O*-tetradecanoyl phorbol-13-acetate. The alternative abbreviations PMA and TPA appear widely in the literature. In general, we have preferred the term PMA but TPA cannot be avoided as it gives rise to the TRE, the TPA responsive element (see page 578).

The protein kinase C family

Molecular cloning has revealed a family of 9 genes giving rise to 11 distinct enzymes.[45,46] Not all of these are activated by phorbol esters.[47] They can be subdivided into three sub-families (see Table 9.1 and **Figure 9.8**). In addition, there are more distantly related enzymes such as PKC-related kinase (PRK, alternatively PKN),[48] not considered here. The various isoforms are also referred to as isozymes, but this term is better reserved for an enzyme such as lactate dehydrogenase, since the isoforms of PKC have separate and characteristic substrates.

TABLE 9.1 PKC isoforms

Subfamily	Isoforms	Requirements for activation
Conventional (cPKC)	α, β1, β2, γ	DAG, PS, Ca^{2+}
Novel (nPKC)	δ, ε, η, θ	DAG, PS
Atypical (aPKC)	ζ, ι (human)/λ (mouse) ξ, μ (PKD)	PS

DAG, diacylglycerol; PS, phosphatidylserine.

FIG 9.8 The family of protein kinases C. Mammalian PKC comprises a family of nine distinct members subdivided, on the basis of their modes of activation, into three subfamilies: conventional (c), novel (n), and atypical (a). They are identified as PKC by the conserved domains C1, C2, C3, and C4. The protein is made up of regulatory and catalytic domains, connected by a hinge region. cPKC β1 and β2 are splice variants of the same gene. An alternative transcript exists for PKCζ which only encodes the catalytic domain and is referred to as PKMζ (not shown). We return to the matter of communication of atypical PKCs with Par-6, p62, and Dsh on page 585.

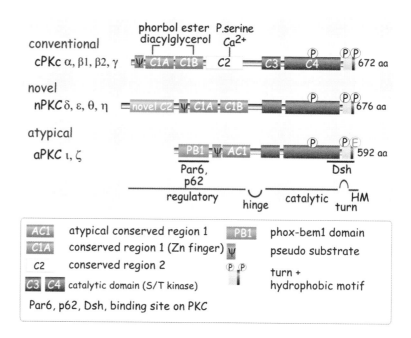

The three subfamilies of PKC are recognized on the basis of sequence similarities and their modes of activation. The conventional PKCs are all activated by phospholipid (in particular phosphatidylserine), DAG, and Ca^{2+}. The novel PKC isoforms require phospholipid and DAG, but not Ca^{2+}. This is explained by the presence of an alternative C2 domain, which does not bind Ca^{2+}. The atypical PKC isoforms respond neither to DAG nor Ca^{2+} and seem not to require phospholipids. They carry an atypical C1 domain and lack the C2 domain. They also have an additional PB1 protein–protein interaction domain. All the PKCs share a common serine/threonine kinase catalytic domain, comprising two highly conserved subdomains, C3 (N-lobe) and C4 (C-lobe). Furthermore, they are all characterized by the presence of a pseudosubstrate sequence which suppresses kinase activity in the absence of a stimulus. PKCs α, δ, and ε are widely expressed, whereas expression of the others is largely cell-type specific.

Phosphorylation of substrate by PKC occurs only at serine and threonine residues in the close vicinity of arginine residues situated in the consensus sequence, Rxx**S/T**xRx.[49] This is present in many proteins (for a detailed discussion, see Nishikawa et al.[50]), and as a result the PKCs can be regarded as a family of broad-specificity kinases, with differences in substrate recognition between the subfamilies. Thus, all forms of PKC (with the exception of PKCζ) are capable of phosphorylating MARCKS and GAP-43, while ribonucleoprotein A1 (hnR A1) is only efficiently phosphorylated by PKCζ. The different expression of the various PKC isoforms in the tissues may contribute to particular tissue- or cell-specific responses to hormones, growth factors, cytokines, or neurotransmitters.[51]

Genes coding for *pkc2* (conventional), *pkc1* and *tpa-1* (novel), and *pkc3* (atypical) are present in *C. elegans*.[52] Four isotypes, dPKC1, dPKC13, InaC, and a putative dPKCδ, are present in *Drosophila*.[53] Genetic screening of these organisms has provided insights into the functioning of PKC, in particular with respect to the role of assembling proteins in the formation of large signalling complexes (see below).

Structural domains and activation of protein kinase C

The C1–C4 regions

The deduced amino acid sequences reveal four reasonably conserved domains,[54] C1–C4, having similarities to those present in other signalling proteins.[45,55] (**Figure 9.9** and Chapter 24). Proceeding from the N-terminus, C1 and C2 constitute the regulatory domains and then C3 and C4 together constitute the catalytic domain characteristic of all kinases (see Chapter 24).

C1 domains are present in a wide range of other proteins, some of which bind phorbol esters (so-called typical) while others do not (atypical). Examples of phorbol ester binding proteins that lack kinase activity include the chimaerins (a family of Rac-GAPs), CalDAG-GEF (a Ras/Rap1 guanine exchange factor (see 'Ras GEFs and GAPs,' page 234 and Figure 9, page 498)), diglyceride kinases, protein kinase D, and Unc-13/Munc-13 (involved in exocytosis) (for review, see Colon-Gonzalez and Kazanietz[54]).

C1 domain as a protein–protein interaction domain.

Recent data suggest non-equivalent roles for the C1A and C1B domains in targeting to intracellular compartments such as the Golgi apparatus, mitochondria, and the nucleus. The C1B domain of PKC α interacts with fascin, a cell matrix protein, and this may have an inhibitory effect on cell migration. The C1B domain of PKC γ binds 14-3-3τ, important in the regulation of solute transport across gap junctions. The C1A domain of PKC βII binds to pericentrin, a scaffold protein of the centrosome. Loss

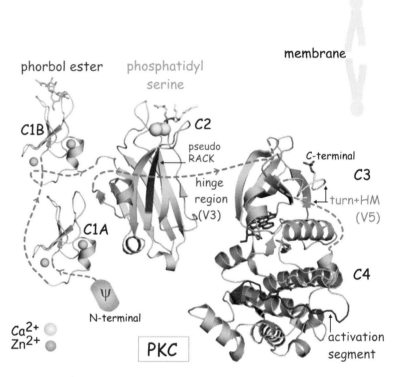

FIG 9.9 Structure of PKC.

The conserved domains C1–C4 are functional modules. C1 binds DAG or phorbol ester, C2 is involved in the attachment to phospholipid, which is reinforced by the binding of Ca^{2+}, and C3 + C4 constitute the kinase domain which is linked to C2 by a hinge region. Regulation of activity and interaction with the upstream kinase PDK1 occurs through the turn- and hydrophobic-(HM)-motif (V5). The structures shown are a compilation of conventional and novel PKC isoforms: C1 (A + B) domain of PKCδ, a C2 domain of PKC-α, and C3/C4 domains of PKC θ (1ptr,[58] 1dsy,[59] 1xjd[60]).

C1 domain

The C1 domain contains a zinc finger motif (see page 781) that forms the binding site for DAG and phorbol esters in the context of phospholipids. This domain is present in all isoforms (singly or doubly). The stimulus-mediated generation of DAG effectively plugs a hydrophilic site in the C1 domain, making the surface more hydrophobic, so allowing C1 to become buried in the membrane (Figure 9.9). In addition to the PKCs, other proteins containing typical C1 domains are also regulated by DAG, or phorbol esters.[56] The C1 domains of the novel PKCs also bind DAG, whereas the atypical PKCs ι, ζ have C1 domain s (termed AC1 but not related to adenylyl cyclase 1) that bind to neither diacylglycerol nor phorbol ester.

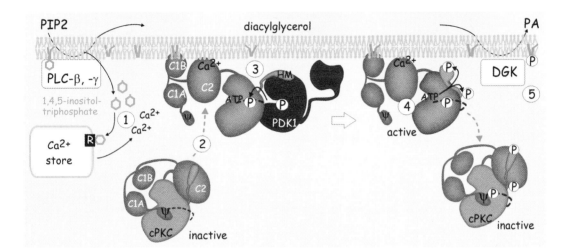

FIG 9.10 Activation of conventional protein kinase C occurs in four steps.
Newly synthesized cPKC is processed to render it catalytically competent. This involves three phosphorylations all catalysed by PDK1, most likely at the membrane following generation of DAG and the liberation of Ca^{2+} (1). Membrane binding of PKC (2) allows unfolding. The hydrophobic motif (HM) binds to the N-terminal lobe of PDK1 which causes its activation (3). In return, PDK1 phosphorylates the activation segment of PKC (Table 9.2). This is followed by two autophosphorylations, one on the turn, the other on the hydrophobic motif, which now binds the PKC N-terminal loop (4). At this point, the enzyme is active and said to be fully competent. With depletion of DAG (conversion to phosphatidate) and Ca^{2+}, PKC detaches from the membrane and folds (5). This is caused by the pseudosubstrate (ψ) binding in the catalytic cleft and the pseudo-RACK motif with the HM motif. The enzyme remains competent for some time and re-addition of DAG and Ca^{2+} suffice to bring it back to life.

C2 domain

The C2 domain binds to negatively-charged phospholipid head groups, such as that of phosphatidylserine. The C2 domain of novel PKC ('novel C2') lacks key residues involved in Ca^{2+} binding, and as a consequence it binds neither Ca^{2+} nor phospholipids (although PS is till needed as a cofactor). Surprisingly, in the case of PKCδ and PKCθ, the C2 domain binds phosphotyrosine residues instead,[57] so providing an opportunity to link tyrosine receptor kinase pathways with those of PKC. C2 domains are also present in a number of other signalling molecules (see Chapter 24).

of this interaction prevents mitotic spindle formation, and since this structure is necessary for the segregation of chromosomes, it prevents mitosis.[54]

C3 and C4 domains

The C3 and C4 domains together constitute a serine/threonine kinase domain with a characteristic ATP-binding N-lobe and a C-lobe containing an activation segment that must be phosphorylated on a threonine to enable catalysis.

The regulatory (C1 and C2) and the catalytic (C3 and C4) domains are linked by a hinge region. When the enzyme is membrane-bound, the hinge is vulnerable to proteolysis. If this occurs, the fragment (protein kinase m) containing the kinase domain becomes detached from the membrane and constitutively active.

TABLE 9.2 Phosphorylation sites that render PKC isoforms catalytically competent. The first phosphorylation occurs in the activation segment (T-loop) of the catalytic domain, near the APE sequence. This involves the action of PDK1. The phosphorylations of the turn- and hydrophobic motif are autophosphorylations. Note the substitution of the hydrophobic motif serines by a glutamate (E) in PKC ζ and PKC ι.

Isoform	Activation segment		Turn motif		Hydrophobic motif	
hPKCα	GVTTRTFCGTPDYIAPE	T497	RGQPVLTPPDQLVI	T638	QSDFEFGSYVNPQ	S657
hPKCβ1	GVTTKTFCGTPDYIAPE	T500	RQPVELTPTDKLFI	T642	QNEFAGFSYTNPE	S661
hPKCβ2	GVTTKTPCGTPDYIAPE	T500	RHPPVLTPPDQEVI	T640	QSEFEGFSFVNSE	S659
hPKCγ	GTTTRTFCGTPDYIAPE	T514	RAAPAVTPPDRLVL	T655	QADFQGFTYVNPD	T674
hPKCδ	ESRASTFCGTPDYIAPE	T507	NEKARLTYSDKNLI	S645	QSAFAGFSFVNPK	S664
hPKCε	GVTTTTFCGTPDYIAPE	T566	REEPVLTLVDEAIV	T710	QEEFKGFSYFGED	S729
hPKCζ	GDTTSTFCGTPNYIAPE	T410	SEPVQLTPDDEDAI	T560	QSEFEGFEYINPL	E579
hPKCη(L)	GVTTATFCGTPDYIAPE	T512	KEEPVLTPIDEGHL	T655	QDEFRNFSYVSPE	S674
hPKCθ	DAKTNTFCGTPDYIAPE	T538	NEKPRLSPADRALI	S676	QNMFRNFSFMNPG	S695
hPKCι(λ)	GDTTSTFCGTPNYIAPE	T403	NEPVQLTPDDDDIV	T555	QSEFEGFEYINPL	E582

Adapted from Newton.[52]

The turn and C-terminal hydrophobic motif

The turn and C-terminal motifs are involved in the binding and the activation of the upstream PDK1 (Figure 9.10). In addition, they control the conformation of the small N-lobe of the PKC and therefore the positioning of the ATP-binding site. Both motifs must be phosphorylated in order to render the kinase fully competent; (exceptions are the atypical PKCι and PKCζ, which have a glutamate in the hydrophobic motif and only require phosphorylation in the turn motif).

Pseudosubstrate (ψ) and pseudoRACK

PKC is kept in an inactive, folded state by (at least) two intramolecular interactions. One involves the segment immediately N-terminal to the C1 domain (residues 19–36 in PKC α) that constitutes the autoinhibitory pseudosubstrate. The sequence resembles the consensus phosphorylation sites present in target proteins that are phosphorylated by PKC. However, the pseudosubstrate possesses an alanine at the position occupied by serine in the substrates and is consequently not amenable to phosphorylation.[61] In the absence of a stimulus, the catalytic domain binds to the pseudosubstrate, causing the enzyme to fold about the hinge linking C2 and C3, and kinase activity is suppressed. The second interaction is a contact between the RACK-binding site (see below), located in the C-terminal region, and a pseudo-RACK motif, located in the C2 domain (Figure 9.10).

Activation of protein kinase C

PKC is synthesized as a non-phosphorylated precursor probably associated with the cytoskeleton. At this stage the catalytic site is accessible to ATP but the enzyme is catalytically inactive.[62] Although a number of phosphorylations control its activation state, so far only one 'upstream' kinase, PDK1, has been identified. It is thought that membrane association, through binding to DAG (and Ca^{2+}), both increases the chance of meeting PDK1 and then renders the catalytic domain accessible to its action.[63] In this respect, the role of DAG seems equivalent, though not identical, to that of $PI(3)P_2$ in the activation of PKB (see page 550 and Figure 18.6, page 553).

Binding of PKC to the membrane brings about a conformational change in the catalytic domain that separates the catalytic and pseudosubstrate domains. The hydrophobic motif (HM) next binds to the hydrophobic groove in the N-terminal lobe of PDK1, triggering activation segment phosphorylation. Although the conformation remains suboptimal for catalysis, this phosphorylation permits autophosphorylation of the turn and hydrophobic motifs (Figure 9.10) so rendering full competence. The phosphorylation activity of PKC is now directed towards its substrates.[52]

Things are different for the atypical PKCs ι and ζ because they do not bind DAG and are insensitive to Ca^{2+}. The activation process may be initiated by other lipid messengers such as $PI(3, 4, 5)P_3$, or the process may be initiated through interaction with proteins such as the Rho-family GTPase Cdc42 in complex with the polarity protein Par-6[64] (see Figure 9.8 and Figure 19.5, page 586).

Multiple sources of diacylglycerol and other lipids activate protein kinase C

In view of the requirement for diacylglycerol, the regulation of PKC is clearly linked to the activity of PLC (see Chapter 5).[65] The hydrophobicity and small size of DAG enables it to diffuse laterally within membranes very rapidly. Virtually all ligands, growth factors, hormones, or neurotransmitters promote the production of DAG (and IP_3) in one way or another, and this means that PKC is implicated in a large number of cellular responses. Sustained activation of conventional PKC requires the simultaneous presence of elevated levels of DAG and high-frequency Ca^{2+}-spikes.[66]

Several other lipid second messengers and mediators either potentiate the effect of DAG and Ca^{2+} or activate the PKCs directly (Figure 9.11). In particular, the 3-phosphorylated inositol lipids ($PI(3,4)P_2$ and $PI(3,4,5)P_3$, products of the phosphoinositide 3-kinases (see Chapter 18), activate both the novel (δ, ε, θ, and η) and the atypical (ι, ζ) PKCs.[67] Unsaturated fatty-acids (particularly arachidonic acid and lysophosphatidic acid) and lysophosphatidylcholine

In test-tube experiments, highly purified PKC can be activated *in vitro* by the addition of phospholipids, phorbol ester (or diacylglycerol), and Ca^{2+} (no requirement for PDK1). Moreover, normal phosphorylation occurs at all three processing sites. Based on the reversibility of this activation following depletion of Ca^{2+} and diacylglycerol, it is suggested that in the cellular environment PKC returns to a closed conformation binding to its pseudosubstrate. It is not known how rapidly and under what conditions phosphatases remove these critical phosphates, although dephosphorylation has been proposed to accompany the phorbol ester-mediated down-regulation of PKC activity that follows long exposures times.

FIG 9.11 Multiple lipid sources, and a protein, to activate protein kinase C. Distinct lipid sources, phospholipases, and kinases activate the different isoforms of PKC. Uniquely, the atypical PKCι can also be activated by the GTPase Cdc42 in the presence of Par-6.

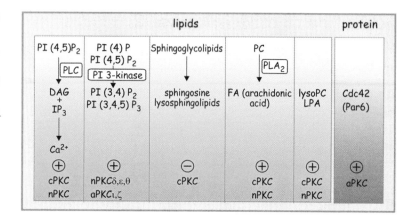

also enhance the activity of PKC, potentiating the effects of DAG at resting Ca^{2+} concentrations.[68] Conversely, the breakdown products of sphingolipids (sphingosine and lysosphingolipids) inhibit the conventional PKCs, most likely by masking their interaction with phosphatidyl serine.[69]

Differential localization of PKC isoforms

The existence of numerous isoforms, but the lack of individual substrate specificities, raises the question of whether the various PKCs might have redundant roles. Immunocytochemical analysis, however, reveals that particular isoforms are present in different cells and subcellular compartments and bind to unique protein complexes. They may have specialized functions. The first indication was provided by a study of inductive signalling in *Xenopus laevis* embryos. Here, dorsal ectoderm is more competent to develop as neural tissue than the ventral ectoderm. The difference persists when an artificial stimulus (phorbol ester) is applied. PKC α is preferentially expressed in dorsal ectoderm, whereas PKC β is uniformly distributed.[70] Over-expression of PKC α in ventral ectoderm annuls the difference, the ventral ectoderm becoming fully competent to form neural tissue.[71]

Over-expressed isoforms of PKC tend to be diffusely distributed throughout the cytoplasm of fibroblasts, with just a fraction of PKC δ and PKC θ attached to the membrane and concentrated in the Golgi apparatus. However, within minutes of adding phorbol ester an extensive redistribution occurs. The α and ε isoforms concentrate at the cell margin. α also appears in the endoplasmic reticulum, $β_2$ associates with actin-rich microfilaments of the cytoskeleton γ in Golgi, and ε attaches to the nuclear membrane.[72] This differential redistribution effectively dictates access to isoform-specific substrates and ultimately confers functional selectivity and so counteracts the basic lack of substrate selectivity among the various isotypes. The mechanisms that direct the localization are not fully understood.

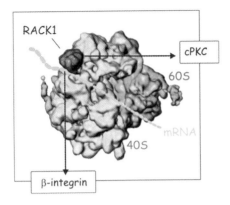

FIG 9.12 RACK1 attaches conventional PKC to the ribosome.
RACK1 binds to the ribosomal 40s subunit and acts to coordinate events arising from conventional PKCs that may have an effect on the translation rate. It also binds β-integrins enabling it to localize ribosomes to subcellular compartments.
Adapted from Nilsson et al.[76]

Different types of PKC-binding proteins

RACKs

Subcellular localization is determined by an array of PKC-binding proteins. Their nomenclature is muddled and they have been called anchors, scaffolds, and receptors. Initially, distinction was made between those that interact with active or inactive PKC and those that are or are not substrates. Proteins that interact exclusively with activated PKC are the RACKs ('receptors' for activated C kinase).

RACK1 is characterized by seven WD repeats, with a β-propeller architecture similar to that of the β-subunits of G-proteins (see Figure 4.2, page 84).[73] Of these, WD2, 3, and 6 interact with the C-terminal region of PKC (predominantly α, β, and ε). By preventing the interaction with the pseudo-RACK motif in the C2 domain, the RACKs could act to maintain the open active conformation of PKC.[74] They are thought to anchor activated PKC to particular membrane domains in the vicinity of appropriate substrate proteins. RACK1 links PKC β with the intracellular domain of the receptors for interleukin-5.[75] It is also a component of the 40S ribosomal subunit, localized near the head region where mRNA exits the ribosomal particle. RACK1-mediated recruitment of active PKC stimulates translation through phosphorylation of the initiation factor eIF-6 and possibly of mRNA-associated proteins.[76] Since RACK1 also interacts with membrane-bound receptors, including integrin β-subunits, it promotes the docking of ribosomes at sites where local translation is required, such as focal adhesions (**Figure 9.12**) and the tips of extending neural growth-cones. Finally, RACK1 links PKC with the JNK protein kinase and in this way enforces expression of cyclinD, giving rise to uncontrolled proliferation of melanocytes (see page 581). RACKII (also known as β-COP), interacts with PKCε, linking it to the Golgi membrane.[77]

Substrates that interact with inactive states of PKC

Examples of substrates that interact with PKC in its inactive states are vinculin, talin, MARCKS, α- and γ-adducin, and the annexins. These all play roles in

TABLE 9.3 PKC binding proteins

Binding protein/substrate	Cellular location	Known functions
STICKs		
Vinculin/talin	focal contacts	adhesion to extracellular matrix
Annexins I and II	–	–
MARCKS	vesicles, plasma membrane	vesicle trafficking, secretion, cell spreading
Desmoyokin, AHNAK	desmosomes/nucleus	cell-cell binding
α-Adducin	cortical cytoskeleton	cell polarity, actin capping
γ-Adducin	cortical cytoskeleton	interaction of actin with spectrin
STICK72 and gravin	plasma membrane	–
STICK34	cytoskeleton, caveolae	–
GAP43 (B-50)	membrane growth cone	neurite outgrowth, neuro transmitter release
P47phox	neutrophil cytoplasm	activation NADPH oxidase
AKAP79	synaptic densities	neurotransmitter release
PAR-3	CNS	role in cell polarity
PICKs		
InaD (*Drosophila*)	photoreceptor	connects PKC with PLC
ASIP	tight junction epithelial cells	–
PICK1	perinuclear	–
RACKs		
RACK1 (binds PKC β)	interleukin-5 receptor	–
β-COP (binds PKC ε)	Golgi	protein traffic

Adapted from Jaken and Parker.[79]

linking the actin cytoskeleton with the plasma membrane (Table 9.3). This implies that PKC is a regulator of membrane-cytoskeletal interactions. None of these is selective between different PKC isoforms. In the case of MARCKS and γ-adducin, phosphorylation disrupts both the interaction with the PKC and with phospholipids, so releasing the protein into the cytosol.[78]

Proteins that interact with atypical PKCs
Atypical PKC isoforms contain a PB1 domain that interacts with other PB1-containing proteins such as the multifunctional cytoplasmic protein p62. This

recruits atypical PKCs into TNF-α and interleukin-1 (IL-1) receptor complexes and contributes to the activation of the NF-κB pathway. p62 also links PKCζ with the potassium channel subunit Kvβ2 and to the growth factor receptor bound Grb14. PKCζ phosphorylates RelA (a component of NF-κB complex) in the nuclear localization signal motif, so promoting its transcriptional activity.[80] Par-6, to which we return later (page 585), links atypical PKC to the Rho family of GTPases, Cdc42, and Rac1 involved in the regulation of cell polarity.[81]

The primary function of all these anchoring proteins is to position individual PKCs into the precise locations in which they can respond to specific receptor-mediated activating signals and address their substrates. Given the promiscuity of PKC in test-tube kinase assays, the anchoring proteins may act to prevent inappropriate phosphorylation events in the cellular environment. A similar theme emerges with the serine/threonine protein phosphatases (see Chapter 21).

Holding back the PKC response

Diacylglycerol kinases

Phosphorylation of DAG to form phosphatidic acid provides a means to deactivate PKC. The family of diacylglycerol kinases (DAG kinases, 9 members) is subdivided on the basis of either substrate specificity or the presence of specific domains. They all have C1 or extended C1 domains (15 extra amino acids) and they all have catalytic and accessory domains (see also PI 3-kinase, in Chapter 18). Despite the presence of a C1 domain, only DAG kinases β and γ interact with phorbol ester. The type I DAG kinases (comprising α, β, and γ) are characterized by two Ca^{2+}-binding EF-hands. Type II kinases (comprising δ and η) are characterized by PH- and SAM domains (sterile α motif). The SAM domain allows binding to the endoplasmic reticulum and plays a role in vesicle trafficking. Type III (only ε) is without recognizable additional domains. Type IV kinase (comprising ζ and ι) have a MARCKS-homology sequence, which acts as a nuclear localization signal. Type V (only θ) possesses three C1 domains.

Although conclusive evidence is still lacking, the DAG kinases are believed to be quiescent until their activity is required. This is achieved by recruitment to the membrane by binding cofactors and phosphorylation, by for instance, activated PKC.[82]

Phosphatidic acid

Although DAG kinase removes DAG and thus terminates membrane localization of PKC, it does not necessarily stop membrane signalling, because the product phosphatidic acid (phosphatidate) itself has numerous effects. It stimulates DNA synthesis, it helps recruitment of Raf to the membrane, and it modifies the activity of RasGAP and of PKCζ (which does not respond to DAG).[83]

Down-regulation by phorbol esters

Prolonged activation by phorbol esters results in a loss of PKC activity. This is the consequence of both dephosphorylation of critical residues and degradation of the protein. Prior treatment of cells with phorbol ester, leading to depletion of a some PKC isoforms, has been extensively applied in studies that test for a role for PKC in physiological processes.

A matter of life or death: PKC signalling complexes in the evasion of the fly-swat

Anyone who has tried unsuccessfully to swat a fly must realize that the speed of their visual response is faster by far than ours. More than this, the sensitivity of their photosensors and the capacity to adapt far outstrips that of any vertebrate. The phototransduction cascade in *Drosophila* (see Figure 6.13, page 175) reveals how scaffold proteins operate in signal transduction through PKC (**Figure 9.13**). Photoactivation of rhodopsin causes the opening

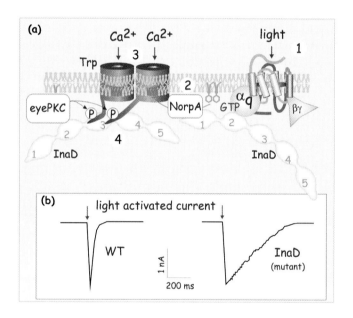

FIG 9.13 InaD scaffold protein and signalling complexes.

(a) The formation of a complex that operates in *Drosophila* phototransduction. Two copies of InaD gather a number of signalling molecules to form a signalling circuit. Light falls on rhodopsin (1) which activates $G\alpha_q$, which acts on NorpA, a PLC (2). This causes the generation of DAG, which opens a Trp Ca^{2+} channel (3), causing membrane depolarization , (see page 210). This rapid event is followed by an almost instantaneous desensitization of Trp due to phosphorylation by eye-PKC (InaC). PDZ domains are numbered 1–5. (b) A loss of InaD causes a retardation of the return to the resting state and hence a loss of visual resolution.
Adapted from Xu et al.[84]

of the light-sensitive cation influx channels Trp and TrpL. This rapid response results in a very transient depolarization of the photoreceptor cells. Most of the proteins that operate in the phototransduction cascade associate as a single supramolecular signalling complex. The key component in this complex is InaD, which has multiple PDZ domains. InaD together with InaC, a protein kinase C, emerged in genetic screens for mutations in *Drosophila* visual transduction.

InaD, present in the microvillar compartment of the photoreceptor (the rhabdomere: see Figure 12.9, page 329) is closely associated with several components including a β-type PLC (gene product of NorpA), the cation influx channels (Trp and TrpL), InaC, the eye-specific *Drosophila* variant of PKC (eye-PKC), and the photosensor rhodopsin (product of the ninaE gene).[84] They are attached through the PDZ domains of InaD. PDZ-1 attaches PLC (NorpA), PDZ-2 attaches eye-PKC and α_q, and PDZ-3 attaches to Trp (Figure 9.13). At least two InaDs are required to gather all the signal transduction components involved, linking the rhodopsin to the ion channel. This supramolecular complex ensures both the rapid opening of the Ca^{2+} channel and then its almost instantaneous desensitization due to phosphorylation of Trp by InaC.[85] PP1/PP2A-like phosphatases, preferentially expressed in photoreceptor cells, act as the reset button, and so allowing a new cycle of excitation to follow.[86]

Disruption of any one of these interacting components alters the photo-response. In InaD mutants, the various components of the signalling complex, such as the Ca^{2+}-channel, type-β PLC, and eye-PKC, are no longer exclusively located in the rhabdomeres, though rhodopsin and the Ca^{2+}-channel (intrinsic membrane proteins) of course remain. Photoreceptor cells in flies that have mutations at either InaC or InaD desensitize sluggishly, remaining depolarized. Such flies have problems distinguishing light of different intensities and flash duration.[87]

The macromolecular assembly avoids random collisions of individual components of the signalling pathway and ensures high precision.

Phorbol ester and inflammation

The phorbol ester PMA is a skin irritant and inflammatory agent. It induces prompt remodelling of the vasculature, resulting in oedema,[88] and the response is enhanced by the induced secretion of histamine from mast cells and blood-borne basophils (**Figure 9.14**).[89,90] It also activates adhesion molecules present on the surface of leukocytes, causing them to bind and then to migrate through the vascular endothelial layer into the tissues underlying the skin.[91] In the tissues, PMA potentiates antigen-mediated stimulation of T lymphocytes, enhancing the production of the cytokine interleukin-2 (IL-2), essential for the induction of clonal expansion (see Chapter 17).[92] This is mediated through activation of PKCθ and the

The symptoms of inflammation were summarized for all time by the 1st century Rom encyclopaedist Celcius as *tumor, calor, rubor,* and *dolor* (swelling, warmth, redness, a pain). Inflammatio recognized as a ty non-specific imm response, repre basic means b body reacts t irritation or i

FIG 9.14 Induction of an inflammatory response by phorbol ester; effects of PMA. The processes indicated result in tumor, calor, rubor, and dolor, four well-recognized characteristics of the inflammatory response.

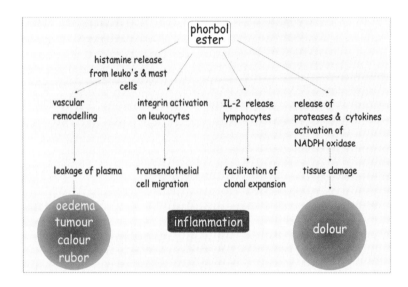

AP-1 complex.[93,94] PMA also causes degranulation of neutrophils, releasing proinflammatory cytokines and matrix proteases. It also activates the respiratory burst,[95] vital for first line host defence and also implicated in the tissue destruction associated with chronic inflammatory disease.

References

1. Krebs EG, Fischer EH. The phosphorylase b to a converting enzyme of rabbit skeletal muscle. *Biochim Biophys Acta*. 1956;20:150–157.
2. Monod J, Wyman J, Changeux JP. On the nature of allosteric transitions: a plausible model. *J Mol Biol*. 1965;12:88–118.
3. Hunter T, Plowman GD. The protein kinases of budding yeast: six score and more. *Trends Biochem Sci*. 1997;22:18–22.
4. Guan KL, Dixon JE. Protein tyrosine phosphatase activity of an essential virulence determinant in *Yersinia*. *Science*. 1990;249:553–556.
5. Ohta T, Sueoka E, Iida N, et al. Nodularin, a potent inhibitor of protein phosphatases 1 and 2A, is a new environmental carcinogen in male F344 rat liver. *Cancer Res*. 1994;54:6402–6406.
   'ancis SH, Corbin JD. Structure and function of cyclic nucleotide-endent protein kinases. *Annu Rev Physiol*. 1994;56:237–272.
      'ing TR. Protein kinases and phosphatases: regulation by 'bitory domains. *Biotechnol Appl Biochem*. 1993;18:185–200.
         l, Rosenfeld R, Rosen OM. Phosphorylation of a cyclic
            ':5'-monophosphate-dependent protein kinase from bovine
               J Biol Chem. 1974;249:5000–5003.

9. Doskeland SO, Maronde E, Gjertsen BT. The genetic subtypes of cAMP-dependent protein kinase – functionally different or redundant?. *Biochim Biophys Acta* 1993;1178:249–258.

10. Shabb JB, Corbin JD. Cyclic nucleotide-binding domains in proteins having diverse functions. *J Biol Chem*. 1992;267:5723–5726.

11. Gettys TW, Blackmore PF, Redmon JB, Beebe SJ, Corbin JD. Short-term feedback regulation of cAMP by accelerated degradation in rat tissues. *J Biol Chem*. 1987;262:333–339.

12. Gettys TW, Vine AJ, Simonds MF, Corbin JD. Activation of the particulate low K_m phosphodiesterase of adipocytes by addition of cAMP-dependent protein kinase. *J Biol Chem*. 1988;263:10359–10363.

13. Hubbard MJ, Cohen P. On target with a new mechanism for the regulation of protein phosphorylation. *Trends Biochem Sci*. 1993;18:172–177.

14. De Cesare D, Fimiaa GM, Sassone-Corsi P. Signalling routes to CREM and CREB: plasticity in transcriptional activation. *Trends Biochem Sci*. 1999;24:281–285.

15. Impey S, Wayman G, Wu Z, Storm DR. Type I adenylyl cyclase functions as a coincidence detector for control of cyclic AMP response element-mediated transcription: synergistic regulation of transcription by Ca^{2+} and isoproterenol. *Mol Cell Biol*. 1994;14:8272–8281.

16. Hanson PI, Schulman H. Neuronal Ca^{2+}/calmodulin-dependent protein kinases. *Annu Rev Biochem*. 1992;61:559–601.

17. Hu SC, Chrivia J, Ghosh A. Regulation of CBP-mediated transcription by neuronal calcium signaling. *Neuron*. 1999;22:799–808.

18. Schwaninger M, Lux G, Blume R, Oetjen E, Hidaaka H, Knepel W. Membrane depolarization and calcium influx induce glucagon gene transcription in pancreatic islet cells through the cyclic AMP-responsive element. *J Biol Chem*. 1993;268:5168–5177.

19. Bito H, Deisseroth K, Tsien RW. CREB phosphorylation and dephosphorylation: a Ca^{2+}- and stimulus duration-dependent switch for hippocampal gene expression. *Cell*. 1996;87:1203–1214.

20. Walker WH, Fucci L, Habener JF. Expression of the gene encoding transcription factor cyclic adenosine 3′,5′-monophosphate (cAMP) response element-binding protein (CREB): regulation by follicle-stimulating hormone-induced cAMP signaling in primary rat Sertoli cells. *Endocrinology*. 1995;136:3534–3545.

21. Harootunian AT, Adams SR, Wen W, Meinkoth JL, Taylor SS, Tsien RY. Movement of the free catalytic subunit of cAMP-dependent protein kinase into and out of the nucleus can be explained by diffusion. *Mol Biol Cell*. 1993;4:993–1002.

22. Hagiwara M, Alberts A, Brindle P, et al. Transcriptional attenuation following cAMP induction requires PP-1-mediated dephosphorylation of CREB. *Cell*. 1992;70:105–113.

23. Alberts AS, Montminy M, Shenolikar S, Feramisco JR. Expression of a peptide inhibitor of protein phosphatase 1 increases phosphorylation and activity of CREB in NIH 3T3 fibroblasts. *Mol Cell Biol*. 1994;14:4398–4407.

24. Wadzinski BE, Wheat WH, Jaspers S, et al. Nuclear protein phosphatase 2A dephosphorylates protein kinase A-phosphorylated CREB and regulates CREB transcriptional stimulation. *Mol Cell Biol*. 1993;13:2822–2834.

25. Van Eynde A, Wera S, Beullens M, et al. Molecular cloning of NIPP-1, a nuclear inhibitor of protein phosphatase-1, reveals homology with polypeptides involved in RNA processing. *J Biol Chem*. 1995;270:28068–28074.

26. Daaka Y, Luttrell LM, Lefkowitz RJ. Switching of the coupling of the β_2-adrenergic receptor to different G proteins by protein kinase A. *Nature*. 1999;390:88–91.

27. Nakamura T, Gold GH. A cyclic nucleotide-gated conductance in olfactory receptor cilia. *Nature*. 1987;325:442–444.

28. Liman ER, Buck LB. A second subunit of the olfactory cyclic nucleotide-gated channel confers high sensitivity to cAMP. *Neuron*. 1994;13:611–621.

29. Bradley J, Li J, Davidson N, Lester HA, Zinn K. Heteromeric olfactory cyclic nucleotide-gated channels: a subunit that confers increased sensitivity to cAMP. *Proc Natl Acad Sci USA*. 1994;91:8890–8894.

30. Chen TY, Yau KW. Direct modulation by Ca^{2+}–calmodulin of cyclic nucleotide-activated channel of rat olfactory receptor neurons. *Nature*. 1994;368:545–548.

31. Zwartkruis FJ, Bos JL. Ras and Rap1: two highly related small GTPases with distinct function. *Exp Cell Res*. 2000;1253:157–165.

32. Pizon V, Chardin P, Lerosey I, Olofsson B, Tavitian A. Human cDNAs rap1 and rap2 homologous to the *Drosophila* gene Dras3 encode proteins closely related to ras in the 'effector' region. *Oncogene*. 1998;3:201–204.

33. Kitayama H, Sugimoto Y, Matsuzaki T, Ikawa Y, Noda M. A ras-related gene with transformation suppressor activity. *Cell*. 1989;56:77–84.

34. de Rooij J, Zwartkruis FJT, Verheijen MHG, Cool RH, Wittinghofer A, Bos JL. Epac is a Rap1 guanine-nucleotide-exchange factor directly activated by cyclicAMP. *Nature*. 1998;396:474–477.

35. Jordan JD, Carey KD, Stork PJ, Iyengar R. Modulation of rap activity by direct interaction of $G\alpha_o$ with Rap1 GTPase-activating protein. *J Biol Chem*. 1999;274:21507–21510.

˥akai Y, Kishimoto A, Inoue M, Nishizuka Y. Studies on a cyclic nucleotide-'ependent protein kinase and its proenzyme in mammalian tissues, ˤfication and characterization of an active enzyme from bovine ˥lum. *J Biol Chem*. 1977;252:7603–7609.

Ҝishimoto A, Takai Y, Nishizuka Y. Studies on a cyclic nucleotide-˥t protein kinase and its proenzyme in mammalian issues, II. d its activation by calcium-dependent protease from rat ˥. 1977;252:7610–7616.

38. Takai Y, Kishimoto A, Iwasa Y, Kawahara Y, Mori T, Nishizuka Y. Calcium-dependent activation of a multifunctional protein kinase by membrane phospholipids. *J Biol Chem*. 1979;254:3692–3695.

39. Kishimoto A, Takai Y, Mori T, Kikkawa U, Nishizuka Y. Activation of calcium and phospholipid-dependent protein kinase by diacylglycerol, its possible relation to phosphatidylinositol turnover. *J Biol Chem*. 1980;255:2273–2276.

40. Van Duuren BL, Orris L, Arroyo E. Tumour-enhancing activity of the active principles of Croton tiglium L. *Nature*. 1963;200:1115–1116.

41. Crombie L, Games ML, Pointer DJ. Chemistry and structure of phorbol, the diterpene parent of the co-carcinogens of croton oil. *J Chem Soc (Perkin1.)* 1968;11:1362.

42. Castagna M, Takai Y, Kaibuchi K, Sano K, Kikkawa U, Nishizuka Y. Direct activation of calcium-activated, phospholipid-dependent protein kinase by tumor-promoting phorbol esters. *J Biol Chem*. 1982;257:7847–7851.

43. Sharkey NA, Leach KL, Blumberg PM. Competitive inhibition by diacylglycerol of specific phorbol ester binding. *Proc Natl Acad Sci USA*. 1984;81:607–610.

44. Ringer S. *A Handbook of Therapeutics*. London: H.K. Lewis; 1888.

45. Parker PJ, Coussens L, Totty N, et al. The complete primary structure of protein kinase C – the major phorbol ester receptor. *Science*. 1986;233:853–859.

46. Coussens L, Parker PJ, Rhee L, et al. Multiple, distinct forms of bovine and human protein kinase C suggest diversity in cellular signaling pathways. *Science*. 1986;233:859–866.

47. Nishizuka Y. Protein kinase C and lipid signaling for sustained cellular responses. *FASEB J*. 1995;7:484–496.

48. Dekker LV, Palmer RH, Parker PJ. The protein kinase C and protein kinase C related gene families. *Curr Opin Struct Biol*. 1995;5:396–402.

49. Klemp BE, Pearson RB. Protein kinase recognition sequence motifs. *Trends Biochem Sci*. 1990;15:342–346.

50. Nishikawa K, Toker A, Johannes FJ, Songyang Z, Cantley LC. Determination of the specific substrate sequence motifs of protein kinase C isozymes. *J Biol Chem*. 1997;272:952–960.

51. Ohno S, Kawasaki H, Imajoh S, et al. Tissue-specific expression of three distinct types of rabbit protein kinase C. *Nature*. 1987;325:161–166.

52. Newton AC. Regulation of the ABC kinases by phosphorylation: protein kinase C as a paradigm. *Biochem J*. 2003;370:361–371.

53. Rosenthal A, Rhee L, Yadegari R, Paro R, Ullrich A, Goeddel DV. Structure and nucleotide sequence of a *Drosophila melanogaster* protein kinase C gene. *EMBO J*. 1987;6:433–441.

54. Colon-Gonzalez F, Kazanietz MG. C1 domains exposed: from diacylglycerol binding to protein-protein interactions. *Biochim Biophys Acta*. 2006;1761:827–837.

55. Newton AC. Protein kinase C. Seeing two domains. *Curr Biol*. 1995;5: 973–976.

56. Hurley JH, Newton AC, Parker PJ, Blumberg PM, Nishizuka Y. Taxonomy and function of C1 protein kinase C homology domains. *Protein Sci*. 1997;2:477–480.

57. Benes CH, Wu N, Elia AE, Dharia T, Cantley LC, Soltoff SP. The C2 domain of PKCδ is a phosphotyrosine binding domain. *Cell*. 2005;121:271–280.

58. Zhang G, Kazanietz MG, Blumberg PM, Hurley JH. Crystal structure of the cys2 activator-binding domain of protein kinase C δ in complex with phorbol ester. *Cell*. 1995;81:917–924.

59. Verdaguer N, Corbalan-Garcia S, Ochoa WF, Fita I, Gomez-Fernandez JC. Ca^{2+} bridges the C2 membrane-binding domain of protein kinase Cα directly to phosphatidylserine. *EMBO J*. 1999;18:6329–6338.

60. Carrell HL, Hoier H, Glusker JP. Modes of binding substrates and their analogues to the enzyme D-xylose isomerase. *Acta Crystallogr D Biol Crystallogr*. 1994;50:113–123.

61. House C, Kemp BE. Protein kinase C contains a pseudosubstrate prototope in its regulatory domain. *Science*. 1987;238:1726–1728.

62. Dutil EM, Newton AC. Dual role of pseudosubstrate in the coordinated regulation of protein kinase C by phosphorylation and diacylglycerol. *J Biol Chem*. 2000;275:10670–10697.

63. Collins BJ, Deak M, Murray-Tait V, Storey KG, Alessi DR. In vivo role of the phosphate groove of PDK1 defined by knockin mutation. *J Cell Sci*. 2005;118:5023–5034.

64. Joberty G, Petersen C, Gao L, Macara IG. The cell-polarity protein Par6 links Par3 and atypical protein kinase C to Cdc42. *Nat Cell Biol*. 2000;2:531–539.

65. Berridge MJ, Irvine R. Inositol trisphosphate, a novel second messenger in cellular signal transduction. *Nature*. 2000;312:315–321.

66. Oancea E, Meyer T. Protein kinase C as a molecular machine for decoding calcium and diacylglycerol signals. *Cell*. 1998;95:307–318.

67. Toker A, Meyer M, Reddy KK, et al. Activation of protein kinase C family members by the novel polyphosphoinositides PtdIns-3,4-P2 and PtdIns-3,4,5-P3. *J Biol Chem*. 1994;269:32358–32367.

68. Sando JJ, Chertihin OI. Activation of protein kinase C by lysophosphatidic acid: dependence on composition of phospholipid vesicles. *Biochem J*. 1996;317:583–588.

69. Hannun YA, Bell RM. Functions of sphingolipids and sphingolipid breakdown products in cellular regulation. *Science*. 1989;243:500–507.

70. Otte AP, Kramer IM, Durston AJ. Protein kinase C and regulation of the local competence of *Xenopus* ectoderm. *Science*. 1991;251:570–573.

71. Otte AP, Moon RT. Protein kinase C isozymes have distinct roles in neural induction and competence in *Xenopus*. *Cell*. 1992;68:1021–1029.

72. Goodnight JA, Mischak H, Kolch W, Mushinski JF. Immunocytochemical localization of eight protein kinase C isozymes overexpressed in NIH 3T3 fibroblasts. Isoform-specific association with microfilaments, Golgi, endoplasmic reticulum, and nuclear and cell membranes. *J Biol Chem.* 1995;270:9991–10001.

73. McCahill A, Warwicker J, Bolger GB, Houslay MD, Yarwood SJ. The RACK1 scaffold protein: a dynamic cog in cell response mechanisms. *Mol Pharmacol.* 2002;62:1261–1273.

74. Kheifets V, Mochly-Rosen D. Insight into intra- and inter-molecular interactions of PKC: design of specific modulators of kinase function. *Pharmacol Res.* 2007;55:467–476.

75. Geijsen N, Spaargaren M, Raaijmakers JA, Lammers JW, Koenderman L, Coffer PJ. Association of RACK1 and PKCβ with the common β-chain of the IL-5/IL-3/GM-CSF receptor. *Oncogene.* 1999;18:5126–5130.

76. Nilsson J, Sengupta J, Frank J, Nissen P. Regulation of eukaryotic translation by the RACK1 protein: a platform for signalling molecules on the ribosome. *EMBO Rep.* 2004;5:1137–1141.

77. Csukai M, Chen CH, De Matteis MA, Mochly-Rosen D. The coatomer protein β-COP, a selective binding protein (RACK) for protein kinase Cε. *J Biol Chem.* 1997;272:29200–29206.

78. Dong L, Chapline C, Mousseau B, et al. 35H, a sequence isolated as a protein kinase C binding protein, is a novel member of the adducin family. *J Biol Chem.* 1995;270:25534–25540.

79. Jaken S, Parker PJ. Protein kinase C binding partners. *Bioessays.* 2000;22:245–254.

80. Moscat J, az-Meco MT, Wooten MW. Signal integration and diversification through the p62 scaffold protein. *Trends Biochem Sci.* 2007;32: 95–100.

81. Lamark T, Perander M, Outzen H, et al. Interaction codes within the family of mammalian Phox and Bem1p domain-containing proteins. *J Biol Chem.* 2003;278:34568–34581.

82. van Baal J, de Widt J, Divecha N, van Blitterswijk WJ. Translocation of diacylglycerol kinase θ from cytosol to plasma membrane in response to activation of G protein-coupled receptors and protein kinase C. *J Biol Chem.* 2005;280:9870–9878.

83. Topham MK. Signaling roles of diacylglycerol kinases. *J Cell Biochem.* 2006;97:474–484.

84. Xu XZ, Choudhury A, Montell C. Coordination of an array of signaling proteins through homo-and heteromeric interactions between PDZ domains and target proteins. *J Cell Biol.* 1998;142:545–555.

85. Popescu DC, Ham AJ, Shieh BH. Scaffolding protein INAD regulates deactivation of vision by promoting phosphorylation of transient receptor potential by eye protein kinase C in *Drosophila*. *J Neurosci.* 2006;26:8570–8577.

86. Liu M, Parker LL, Wadzinski BE, Shieh BH. Reversible phosphorylation of the signal transduction complex in *Drosophila* photoreceptors. *J Biol Chem*. 2000;275:12194–12199.

87. Smith DP, Ranganathan R, Hardy RW, Marx J, Tsuchida T, Zuker CS. Photoreceptor deactivation and retinal degeneration mediated by a photoreceptor-specific protein kinase C. *Science*. 1991;254:1478–1484.

88. Janoff A, Klassen A, Troll W. Local vascular changes induced by the cocarcinogen, phorbol myristate acetate. *Cancer Res*. 2000;30:2568–2571.

89. Katakami Y, Kaibuchi K, Sawamura M, Takai Y, Nishizuka Y. Synergistic action of protein kinase C and calcium for histamine release from rat peritoneal mast cells. *Biochem Biophys Res Commun*. 1984;121:573–578.

90. Schleimer RP, Gillespie E, Lichtenstein LM. Release of histamine from human leukocytes stimulated with the tumor-promoting phorbol diesters, I. Characterization of the response. *J Immunol*. 1981;126:570–574.

91. Kavanaugh AF, Lightfoot E, Lipsky PE, Oppenheimer-Marks N. Role of CD11/CD18 in adhesion and transendothelial migration of T cells, Analysis utilizing CD18-deficient T cell clones. *J Immunol*. 1991;146:4149–4156.

92. Rosenstreich DL, Mizel SB. Signal requirements for T lymphocyte activation, I. Replacement of macrophage function with phorbol myristic acetate. *J Immunol*. 1979;123:1749–1754.

93. Werlen G, Jacinto E, Xia Y, Karin M. Calcineurin preferentially synergizes with PKC-θ to activate JNK and IL-2 promoter in T lymphocytes. *EMBO J*. 1998;17:3101–3111.

94. Sun Z, Arendt CW, Ellmeier W, et al. PKCθ is required for TCR-induced NF-κB activation in mature but not immature T lymphocytes. *Nature*. 2000;23:402–407.

95. Repine JE, White JG, Clawson CC, Holmes BM. The influence of phorbol myristate acetate on oxygen consumption by polymorphonuclear leukocytes. *J Lab Clin Med*. 1974;83:911–920.

CHAPTER 10

Nuclear Receptors

First steps in the isolation of steroid hormones

Origins

The effects of the self-injection of testicular extracts described by Brown-Séquard, related in Chapter 1, may or may not have been due to androgenic steroids. Or they may have been a matter of self-delusion. However, his contemporary, the Scottish surgeon George Thomas Beatson, undoubtedly observed the regression of inoperable breast tumours following ovariectomy, the female equivalent of castration.[1] Of course, this manoeuvre – the removal of the source of the sex-related steroids – has a history that probably goes back to the origins of our long and close association with domestic animals.

In the management of the imperial court of medieval China, eunuchs played many important roles. Also, the castration of domestic animals was widely practised both for medicinal purposes and for the improvement of the meat. As the importance of the testes was realized, extracts were prepared and applied in various forms of organotherapy. Placental extracts get a first

The roots of scientific
development in China,
including aspects that we
now call endocrinology,
have been documented
by Joseph Needham (a
Cambridge biochemist
who turned his attention
to the science and
civilization of China).[2]

mention in a Chinese pharmacopoeia of 725 AD, though their widespread
use had to wait another seven centuries until promulgated by Chu Chen-
Heng. He encouraged their application for the treatment of many conditions.
More importantly, the Chinese had, over this period of time, developed a
remarkable set of extraction and fractionation techniques, with the aim of
obtaining active materials from urine. Pharmacists were engaged 'almost
on a manufacturing scale', using up to 200 gallons (almost 1000 L) at a time,
to prepare active products by precipitation with calcium sulfate (which
sedimented steroid conjugates along with protein), evaporation to dryness,
sublimation, and crystallization. According to Chu Chen-Heng, 'all people
suffering from impotence, sexual debility, excess ang of the burning feverish
type which no medicine can benefit, will improve. The natural precipitate of
urine can also drive out the undue fire element affecting the liver and the
bladder'. Joseph Needham had no doubt that the Chinese had, between the
11th and 17th centuries, achieved preparations of androgens and oestrogens
that were 'probably quite effective in the quasi-empirical therapy of the time.'[3]

Beginning again

The root of these
chemical names is the
word oestrous (estrus,
in American spelling),
which derives from the
Greek *oistros* meaning a
gadfly. Fritz Spiegl points
out that to be stung by
a gadfly was thought to
send people, especially
women, into a frenzy.[3]

The first steroid hormone to be isolated, purified, and characterized was
the oestrous-inducing substance oestrin (estrin), $C_{18}H_{22}O_2$, obtained in
gram quantities from the urine of pregnant women.[4, 5] Other oestrogenic
substances such as oestriol[6] soon followed. Stilboestrol, the first synthetic
oestrogen, was originally isolated from plants. It lacks the steroid nucleus,
characteristic of many hormones; its structure is based on that of stilbene.
It was soon followed by 'dynestrol', described in 1938.[7] Progesterone, first
described as a 'special hormone present in the corpus luteum responsible
for preparing the uterus for the reception of embryos'[8] required heroic
quantities of material. Adolph Butenandt, who had already isolated the male
sex hormone androsterone, used the corpora lutea from 50 000 pigs for the
isolation of a few milligrams of 'progestin'.

Adolf Friedrich Johann
Butenandt (1903–95)
was awarded the Nobel
Prize in Chemistry in
1939 for his 'work on sex
hormones.' The Nazis
forced him to decline the
award, and he was only
able to accept it in 1949.

The discovery of intracellular hormone receptors

The steroid and thyroid hormones, and other lipophilic messengers (Figure
10.1) are carried by specific binding proteins in the circulation. It has been
thought that they penetrate cell membranes and enter cells passively,[9] but
this is not always the case. For example, there is evidence of a saturable
transport mechanism for oestrogens and for thyroid hormone, suggesting
that they require the assistance of a protein for this step.[10] In the cytosol, most
of them bind to intracellular receptor molecules that then migrate to the
nucleus, attach to particular regulatory DNA sequences (response elements),

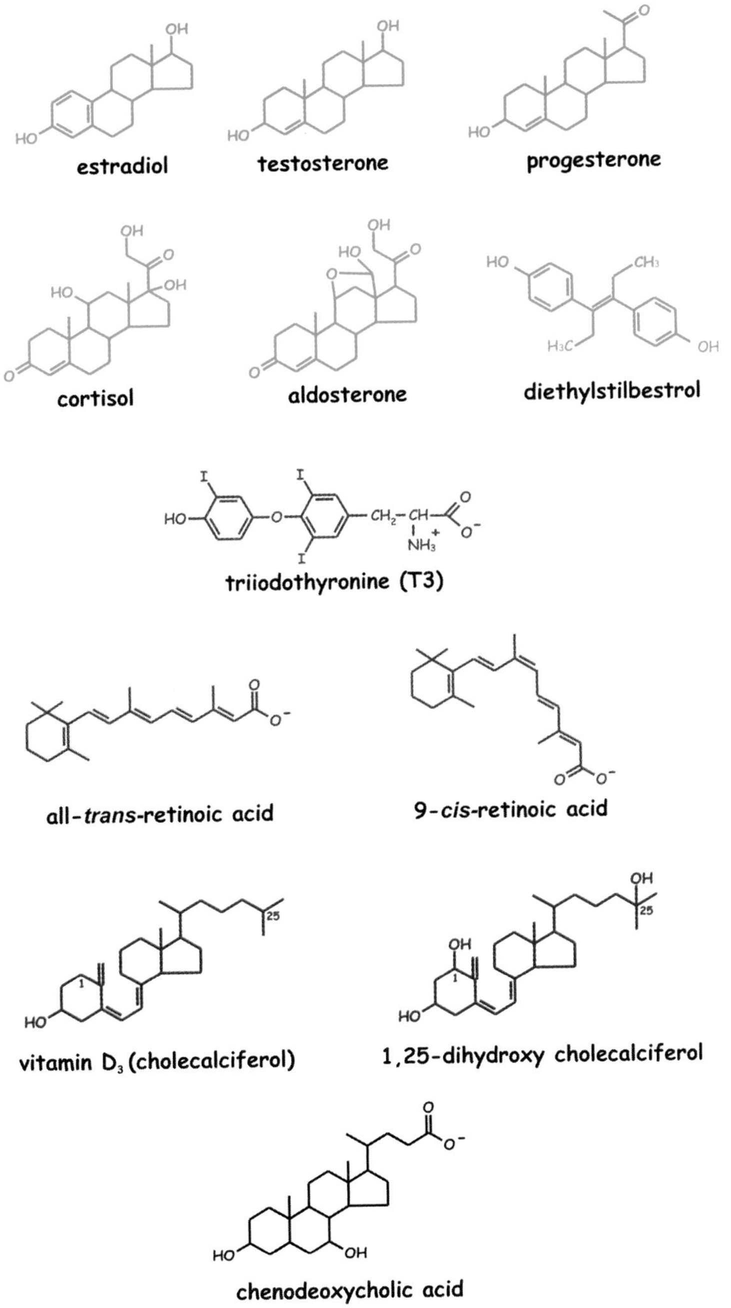

FIG 10.1 Structures of some nuclear receptor ligands.
Steroid hormones and diethylstilbestrol are shown in green, thyroid hormone is red, retinoic acids are purple, and vitamin D is blue. Chenodeoxycholic acid (in black) is a bile acid.

and activate or inhibit transcription (transactivation or transrepression). Before developing this basic theme, it should be noted that the regulation of transcription is not necessarily the unique function of the intracellular receptors. Some exert effects on cytosolic signalling pathways quite independently of transcription (see pages 290 and 292).

Evidence for intracellular receptors

Although the earlier findings of preferential binding of oestrogens in target tissues (e.g. estradiol in the uterus and vagina) supported a receptor hypothesis, the first actual demonstration of steroid binding to a receptor protein, indeed, the first demonstration of the binding of any hormone to any receptor in a cell-free system, was reported in 1966 by Toft and Gorski.[11] They injected female rats with [^{3}H]-estradiol and then carried out subcellular fractionation of the homogenized uteri. This revealed that ~50% of the radioactivity was present in the nuclear fraction, while 30% was in the soluble cytosol fraction. After centrifuging the cytosolic material through a sucrose density gradient, they recovered the radioactivity in a component that sedimented at ~9.5S, equivalent to a protein of ~200 kDa. Binding of the hormone could be antagonized by diethylstilbestrol, but not by the non-oestrogenic steroids testosterone and corticosterone. Similar high-affinity binding proteins (K_D ~0.1–10 nmol L^{-1}) are found in soluble fractions prepared from homogenates of various unstimulated hormone-responsive tissues. Following binding of hormone, these receptor proteins are found in the nuclear fraction. The key experimental observation was the accumulation of [^{3}H]-corticosterone in the nuclear fraction of hepatoma cells, at the expense of a loss of radioactivity from the cytosol.[12]

Studies of the subcellular location of steroid receptors relying on cell fractionation and immunostaining have, however, been controversial, with some conflicting data. To overcome this, it has been essential to visualize receptors in living cells and to follow their translocation by real-time imaging. This became possible with the development of green fluorescent protein (GFP) tagging (see page 196). Cells may be transfected to express functional receptors that are GFP fusion proteins. Trafficking of the tagged receptors within the cell may then be followed using high-resolution fluorescence microscopy (usually laser confocal microscopy: see page 196). This approach has provided the clearest demonstration that cytoplasmic receptors, typified by glucocorticoid receptors (GR), can translocate to the nucleus within an hour of exposure to ligand (Figure 10.2).

The pattern is broadly similar for other steroid hormone receptors, such as mineralocorticoid receptors (MR), androgen receptors (AR), and progesterone receptors (PR). However, the partition of steroid receptors into either the cytosol or the nucleus is never absolute. Their distribution between

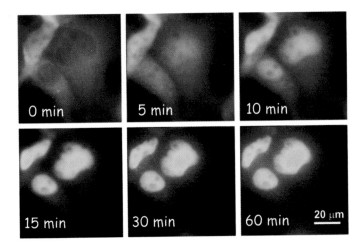

FIG 10.2 The progression of nuclear translocation of GFP-tagged glucocorticoid receptors expressed in COS-1 cells following treatment with dexamethasone, a synthetic glucocorticoid.
Image courtesy of M. Kawata.[13]

the nuclear and cytoplasmic compartments is variable, depending upon cell type, receptor subtype and possibly the stage of the cell cycle. Some, such as the oestrogen receptor (ER), are principally nuclear (but not exclusively so, see page 292). These different patterns may reflect differences in the balance between nuclear import and export, in a dynamic shuttling of receptors between the two compartments. What is clear is that, in the presence of ligand, the balance is weighted in the nuclear direction.

A superfamily of nuclear receptors

The receptors for steroid hormones are members of a very much larger class of ligand-activated transcription factors. It includes the receptors for thyroid hormones, for retinoids (i.e. all-*trans*-retinoic acid or vitamin A, and its isomer 9–*cis*-retinoic acid) and for vitamin D (cholecalciferol). Some ligand structures are shown in Figure 10.1. The family also includes receptors for numerous other lipophilic messengers, some of which exist naturally in the body, like fatty acids, bile acids, oxysterols, and eicosanoids, and others that are of external origin, such as certain pharmaceuticals, carcinogens, and environmental pollutants (known collectively as *xenobiotics*).

The family is called the nuclear receptor superfamily. Its members have a wide range of biological functions including the regulation of growth and embryonic development, the maintenance of phenotype, and the regulation of metabolic processes, such as cholesterol and bile acid metabolism. Disorders of these systems can lead to infertility, obesity, diabetes, and cancer. The receptors act principally by controlling transcription.

Attempts have been made to classify the receptors according to their ligands or their targets, but as knowledge of the family has increased, simple

TABLE 10.1 Types of intracellular receptor

Receptor			Examples of ligands
Steroid hormone receptors	ER	oestrogen receptor	estradiol
	GR	glucocorticoid receptor	cortisol
	MR	mineralocorticoid receptor	aldosterone
	AR	androgen receptor	testosterone
	PR	progesterone receptor	progesterone
Thyroid hormone receptors	TR	thyroid hormone receptor	triiodothyronine (T3)
Retinoid receptors	RAR	retinoic acid receptor	all-*trans*-retinoic acid
	RXR	retinoic acid X receptor	9-*cis*-retinoic acid
Vitamin D receptor	VDR	vitamin D receptor	1,25-hydroxy-cholecalciferol
Lipid sensors	LXR	liver X receptor	oxysterols (e.g. hydroxycholesterols)
	FXR	farnesoid X receptor	bile acids (e.g. chenodeoxycholate)
PPAR	PPAR	peroxisome proliferator activated receptor	fatty acids, eicosanoids (LTB$_4$, PGJ$_2$)

LTB$_4$, leukotriene B$_4$; PGJ$_2$, prostaglandin J$_2$.

Maybe not quite as absent from fungi as was formerly thought. A combination of sequence profile searching with structural predictions at a genomic scale revealed heterodimeric transcription factors in yeast that contain ligand-binding domains resembling those of animal receptors.[14] Although they share little sequence homology, the resemblance of the ligand-binding structural folds in yeast Oaf1 and Pip2 and the vertebrate PPAR-RXR heterodimers is evident. The ligand-binding domains can bind fatty acids such as oleate, which provoke heterodimer formation (though these may not be the preferred ligands).

classifications have become difficult to sustain. One reason is the promiscuity and rather low affinity of some receptors for their apparent biological ligands. Table 10.1 lists some of the best-characterized receptors and their known binding partners, but it is important to remember that a particular receptor may also bind other ligands. For example, MRs bind not only aldosterone but also cortisol. Another difficulty is that many of them are orphan receptors apparently lacking any physiological ligand.

In evolution, the nuclear receptors first appeared in metazoans (animals and sponges) but are absent from plants, algae, fungi, and protists (but see text box). A retinoic acid receptor, remarkably homologous with the vertebrate retinoid X receptor (RXR), is present in Cnidarians (animals with mostly radial symmetry, such as jellyfish, hydras, and corals).[15] In the worm *Caenorhabditis elegans* there are 270 nuclear receptors, exceeding by far the 48 in humans.[16] However, as with other invertebrates, this species lacks receptors for adrenal steroids (aldosterone and cortisol) and sex steroids (estradiol, progesterone, testosterone). In vertebrates, these hormones regulate reproduction, differentiation, development, and homeostasis. It is likely that receptors for steroid hormones first arose in a cephalochordate such as *Amphioxus*[17] (see page 3).

Orphan receptors and evolution

The isolation of cDNAs coding the mammalian oestrogen and glucocorticoid receptors was first reported in the 1980s.[18–20] About half of the mammalian proteins identified as nuclear receptors are currently designated as 'orphans', having no recognized ligands. In invertebrates, the proportion of orphans is higher still. Indeed, the receptors that have known ligands are only expressed in those subfamilies that evolved comparatively recently,[21] suggesting that their ancestral forms were insensitive to ligands.

The most comprehensive classification of nuclear receptors within the superfamily has been obtained by building phylogenetic trees from sequence information. There are six subfamilies.[22] In simplified form, this is illustrated in Figure 10.3, where a consensus tree of principally human nuclear receptors is set out. Such analysis shows that receptors for similar ligands are not necessarily confined to particular branches, but are distributed throughout the tree (e.g. RAR and RXR). This lack of correspondence suggests that the acquisition of the capacity to bind particular ligands may have arisen independently and at different times during evolution. Receptors without assigned ligands, such as the COUP-TF family, the testis receptors TR2 and TR4, and the oestrogen-related receptors ERR1 and ERR2, are orphans.

Orphan receptors that do not bind ligands may still be able to influence transcription. For example, NURR1 lacks a binding pocket because of the presence of several bulky hydrophobic residues,[23] but even so it is a potent transcriptional activator. Furthermore, even if a ligand is identified in the future, it may not play a role in regulation of receptor activity. Other receptors have ligands permanently bound within the cavity and so may be unable to influence activity. In this respect, HNF4 has a constitutively bound fatty acid and the overall molecular conformation corresponds to a transcriptionally active receptor.[24]

The orphan receptors may represent proteins that never had the ability to be controlled by a ligand, or that once possessed and then subsequently lost it. A possible mechanism for acquisition would involve primitive receptors existing in an equilibrium between active (able to bind DNA and thus capable of regulating gene function) and inactive conformations. This would be similar to the two-state equilibrium exhibited by cell surface receptors (see page 66). The emergence of the ability to bind ligands that shift the equilibrium towards the DNA-binding conformation would then confer a new tier of control over the expression of specific proteins.

Some orphan receptors (e.g. HNF4) stimulate gene transcription, whereas others (e.g. REV-ERBα) repress it. Although orphan receptors appear not to be regulated by ligand, they can still be subject to regulation by phosphorylation/dephosphorylation, either of the receptors themselves or of their accessory proteins (see page 290). For example, the ability of HNF4 to bind DNA is prevented by its phosphorylation by PKA.[27] Other ways of controlling orphan receptors include restriction of nuclear uptake and, within the nucleus, competitive interactions with other transcription factors for coactivator or corepressor proteins (see page 287).

NURR1 (nuclear receptor related) regulates development in the central nervous system and is essential for the development of dopaminergic neurons in brain.[25] It is linked to familial Parkinson's disease[26] and schizophrenia.

HNF4 (hepatocyte nuclear factor 4) is expressed in liver, kidney, gut, and endocrine pancreas. It regulates apolipoprotein expression, glucose metabolism, and insulin secretion. Mutation of the gene is responsible for a form of diabetes.

REV-ERBα is a component of the circadian clock.

Of course, the assignment of a particular receptor to orphan status must be done with caution. Sometimes activating ligands have become apparent later. Examples of such 'adopted orphans' are the liver X receptor (LXR) and the farnesoid receptor (FXR). These are now known to have important roles in cholesterol and bile acid metabolism and they are discussed, together with the xenobiotic receptors CAR and PXR, on page 289.

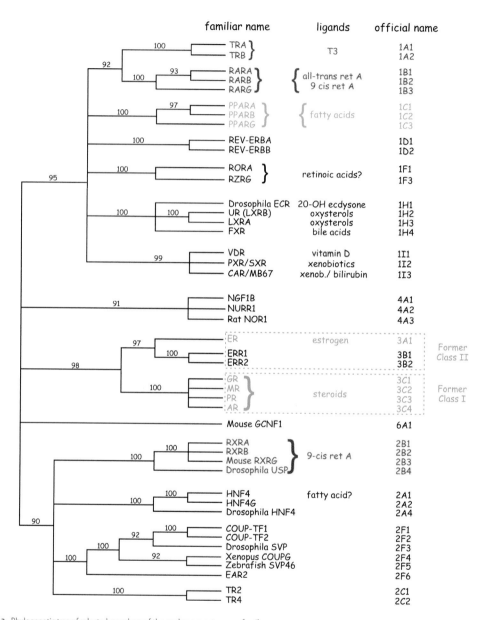

FIG 10.3 Phylogenetic tree of selected members of the nuclear receptor superfamily.

Colours correspond to those in Figure 10.1. Only known ligands are indicated. Each of the six subfamilies is indicated by the first number of its official name (subfamily 5 is not shown). Following convention, Greek characters are shown as capital letters. For example TRA is thyroid hormone receptor α and its official name is NR1A1. This simplified consensus tree was obtained from an unrooted, neighbour-joining phylogenetic distance tree, by collapsing branches with bootstrap values below 90%. The branches shown are unscaled, but bootstrap values are indicated.

Adapted from Laudet.[22]

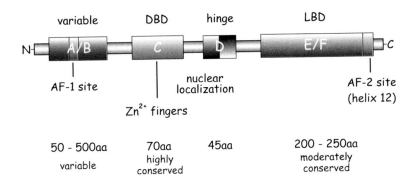

FIG 10.4 Domains of nuclear receptors.
Abbreviations: DBD, DNA-binding domain; LBD, ligand-binding domain; AF-1, activation function 1 (ligand-independent); AF-2, activation function 2 (ligand-dependent).

Nomenclature of nuclear receptors

Over the years, the study and classification of nuclear receptors has suffered from the problem of multiple names for the same gene. A systematic nomenclature has therefore been established,[28] and official names are indicated in Figure 10.3. Full details are available from the bioinformatic database, Nurebase (http://www.ens-lyon.fr/LBMC/laudet/nurebase/nurebase.html). More molecular specific information is provided by the nuclearRDB database.

Receptor structure and ligand binding

The general structure of the nuclear receptors is highly conserved, consisting of a set of common functional domains (see Figure 10.4). There is a highly variable N-terminal domain (A/B), a highly conserved DNA-binding domain (DBD) containing twinned zinc finger motifs, a hinge region (D) and a C-terminal segment (E/F) that contains the ligand-binding domain (LBD). There are two activation function sites, AF-1 and AF-2. AF-1 is in the N-terminal domain and is ligand-independent, while AF-2 is located in the LBD and is ligand-dependent. There are also sites that determine nuclear localization and dimerization.

Ligand-binding domains are molecular switches

The most common structure of the LBD is based on a compact assembly of 11 helices that form a three-layer sandwich (Figure 10.5). The middle layer does not extend all the way across, leaving a non-polar cavity at one end formed by the two outer layers. This is the binding pocket, and its size varies among the different receptors. For steroid and thyroid hormone receptors, VDR, RAR, and RXR, the cavity is matched to ligand size. These receptors bind their ligands with high affinity (K_D in the nanomolar range). Then there is the group of receptors that bind with low affinity (micromolar range). These include LXR

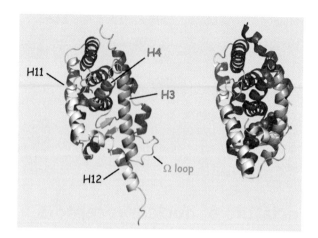

FIG 10.5 Structure of the retinoic acid X receptor (RXR) without and with ligand.
Apo-RXR is shown on the left and the ligand-bound form on the right. Upon binding 9-*cis* retinoic acid (purple spheres), a conformational change occurs, involving mainly helix H12, which is amphipathic, closing the entrance to the pocket. Rearrangements of H3 and the Ω-loop also take place (1lbd[29] and 1fby[30]).

and FXR, which have been termed metabolic sensors rather than receptors. They cope with a wider variety of ligands. The xenobiotic receptor PXR has one of the largest and most flexible cavities, with a volume of $\sim$1100 Å^3, nearly twice that of the very high affinity TRβ receptor ($\sim$600 Å^3). Conversely, the orphan NURR1 has virtually no cavity.

The effect of ligand binding is a conformational change. In general, the molecule adopts a more compact structure with a rearrangement of the chains forming the cavity. This is particularly evident when RXR binds its ligand (Figure 10.5). The amphipathic C-terminal helix H12 (AF-2) swings across to close the pocket, trapping the ligand between the outer layers of the sandwich and a short two-stranded β-sheet at the other end. For other receptors this movement is less distinct. In their unoccupied, apo forms, the region around the binding pocket (the lower end of the molecule in Figure 10.5) tends to be flexible (as a result, apo-receptors are difficult to crystallize). Binding of ligand then stabilizes the structure, which forms a more compact arrangement in which H12 becomes less mobile. The surface formed by H3 and H4 is altered so that receptor molecules bound at their DNA sites can interact with coregulator proteins (see below).

Activation of cytosol-resident receptors

Steroid receptors such as GR, which in the absence of a stimulus exist principally in the cytosol, cannot bind ligands until their ligand binding domains have been assembled in a receptive conformation. This is illustrated in Figure 10.6. Competence to bind hormone is achieved by the attachment of a series of heat shock proteins, including hsp70, hsp40, and most importantly hsp90, followed by a number of other accessory molecules. These include the regulatory cochaperone p23 and an immunophilin (e.g. FKBP52) which links the complex to the microtubule motor protein dynein. However,

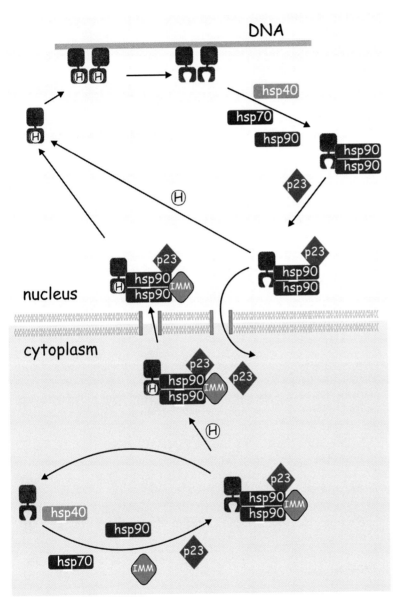

FIG 10.6 A chaperone machine enables cytosolic steroid receptors to bind hormone and enter the nucleus. The sequential recruitment of molecular chaperones, including hsp40, hsp70, and hsp90, with cochaperones, including p23 and an immunophilin (IMM), enables the binding of steroid hormone to the receptor. The complex may then enter the nucleus and release hormone-bound receptor which binds as a dimer to specific DNA sites. (The transcription complex that forms at this stage is not shown.) Finally, receptors may either be recycled within the nuclear compartment, with the help of the chaperones, or they may be exported or degraded. Based on the glucocorticoid receptor.[34, 35]

Heat shock proteins.
Here is another example of a chance error that led to an important discovery. Before the recombinant revolution, gene expression was studied by microscopic examination of the expansion (puffing) of bands in the giant polytene chromosomes of organisms like *Drosophila*. The expansion is due to the decondensation of chromatin.

Ferruccio Ritossa relates how, as a young graduate student at the University of Pavia in the early 1960s,[31] he investigated the type of nucleic acid in the chromosomal puffs of *Drosophila* salivary glands. A colleague had inadvertently increased the temperature of his incubator and he noticed another puffing pattern. New RNA was synthesized within minutes. The importance of the expressed heat shock proteins as molecular chaperones came later.[32]

Heat shock proteins also have diverse roles in unstressed cells, affecting folding, packaging, secretion, and degradation of other proteins. Failure of these systems results in a number of important human diseases.[33]

competence does not persist and a cyclic disassembly and re-assembly of the complex ensues until ligand binding occurs. Hormone-bound receptors are transported along microtubules to nuclear pores, where the entire complex is shuttled into the nuclear space. Inside the nucleus the complex dissociates, allowing the dimerized receptor to bind to its response element (in the promoter region of genes). They then engage a different set of proteins to become part of a large transcription complex. Subsequently, used or 'experienced' receptors lose their hormone and again associate with the chaperones as they are released from their DNA binding sites. They may then either be degraded, exported back into the cytosol or reused within the nucleus, where they may bind hormone once again.

This sequence of events has been demonstrated most clearly for GRs, which in the absence of ligand are principally cytosolic, but there is evidence that a similar mechanism operates for the other steroid receptors. Even ER, predominantly resident in the nucleus, shows a dependence on hsp90. Other receptors that are exclusively nuclear appear to be associated with DNA in the absence of ligand, but bind to specific sites in its presence.

DNA binding

Nuclear receptors bind to DNA as dimers at specific loci within the regulatory region of target genes. These sites, known as response elements, consist of two half-sites, each of six base pairs. For steroid receptors the consensus sequences of the sites are palindromic, inverted repeats separated by a 3 bp spacer (termed IR3). GR, PR, MR, and AR bind as homodimers to a response element having the consensus sequence 5'-AGAACA-3' (Figure 10.7). ER also binds to an IR3 inverted repeat, but the consensus sequence is different, 5'-AGGTCA-3'. Most non-steroid receptors bind to half-sites that are *direct* repeats of the AGGTCA hexad (termed DRn, where n is the number of spacers). They may bind either as homodimers (TR, VDR) or commonly with higher affinity, as heterodimers with RXR, as in the case of TR, VDR, RAR, LXR, FXR, PXR, CAR, and PPAR. The binding of heterodimers provides a combinatorial mechanism in which the two different receptors can coregulate transcription at a single response element.

Since the sites at which the different heterodimers bind have the same consensus sequence, how is specificity achieved? A measure of discrimination is provided by differences in binding site geometry determined by the number of spacers (DR1–DR5) between the half-sites (see Figure 10.7). Each additional base-pair displaces the half-sites by 3.4 Å and introduces a relative rotation of 36°. Note that RXR is always situated upstream (5') of its binding partner, except in the case of the RAR-RXR heterodimer that binds at DR1 sites. When the linker length is 2 or 5, the dimer assembles in the order RXR-RAR.

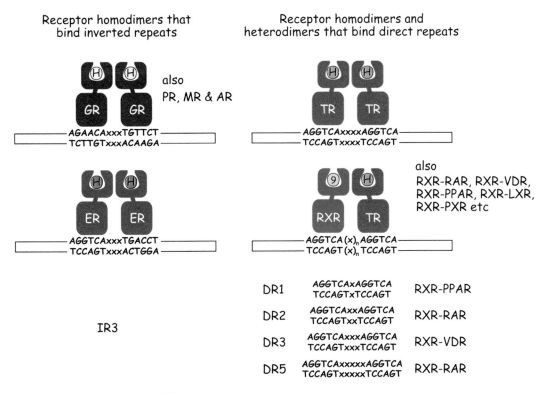

Receptor homodimers that bind inverted repeats

Receptor homodimers and heterodimers that bind direct repeats

also PR, MR & AR

GR GR

AGAACAxxxTGTTCT
TCTTGTxxxACAAGA

TR TR

AGGTCAxxxxAGGTCA
TCCAGTxxxxTCCAGT

also RXR-RAR, RXR-VDR, RXR-PPAR, RXR-LXR, RXR-PXR etc

ER ER

AGGTCAxxxTGACCT
TCCAGTxxxACTGGA

IR3

RXR TR

AGGTCA (x)ₙ AGGTCA
TCCAGT (x)ₙ TCCAGT

DR1	AGGTCAxAGGTCA TCCAGTxTCCAGT	RXR-PPAR
DR2	AGGTCAxxAGGTCA TCCAGTxxTCCAGT	RXR-RAR
DR3	AGGTCAxxxAGGTCA TCCAGTxxxTCCAGT	RXR-VDR
DR5	AGGTCAxxxxxAGGTCA TCCAGTxxxxxTCCAGT	RXR-RAR

FIG 10.7 Dimerization and binding to DNA half-sites.
Steroid hormones bind to inverted repeats as homodimers. Other receptors bind to direct repeats as homodimers or as heterodimers with RXR. Binding site geometry is affected by the number of base pairs in the linker sequence. (H=hormone, 9=9-*cis* retinoic acid).

For many receptors, further specificity comes from interactions of other residues in the DBD with flanking regions of DNA. For example, the C-terminal extension of the DBD may bind to sites upstream of the hexad sequences. Finally, it should also be stressed that the actual sequences of the response element half-sites that are recognized by nuclear receptors commonly differ from the consensus sequences at one or more positions.

Recognizing response elements

Nuclear receptor DBDs contain two C4-type zinc fingers in which four cysteine residues are linked to a tetrahedrally chelated Zn^{2+} ion (see page 781). These structures nucleate the protein fold. A schematic representation of the glucocorticoid DBD is shown in Figure 10.8a and the three-dimensional structure of the homodimer bound to DNA half-sites with the idealized AGAACA sequence is shown in Figure 10.8b. An N-terminal recognition helix inserts into the major groove of the target DNA, where it contacts specific

FIG 10.8 The zinc finger DNA-binding domain of the glucocorticoid receptor. (a) Amino acid sequence of a single DBD. (b) Structure of receptor DBD dimers binding to DNA. P-boxes are coloured red and D-boxes are yellow.[36,37] (1r4).[38]

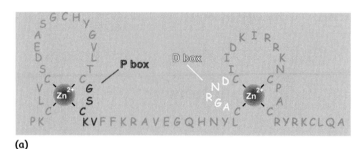

(a)

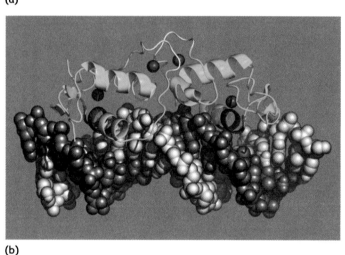

(b)

bases by means of a motif called the P-box. For the non-oestrogenic steroids, such as GR, the P-box sequence is GSCKV. For ER it is EGCKA and for receptors that form heterodimers with RXR, it is EGCKG. A mutation of the GR P-box, G→E, produces a receptor that can bind either GR or ER response elements. Complete swapping of the GR P-box for that of ER confers an ability to bind and activate at oestrogen response elements, eliminating glucocorticoid responsiveness.

For receptors binding to inverted repeats, dimerization occurs upon binding, through interactions between a D-box motif on each monomer, as illustrated in Figure 10.8. For receptor dimers that bind direct repeats, the monomers are oriented head to tail and there is no D-box interface. Instead, a more diffuse set of interactions between the DBDs ensures heterodimer stability. Dimerization may also be reinforced by interactions between sites on the two LBDs. Such interactions enable some receptors to bind as dimers to DNA without a bound ligand. In these circumstances they may act to repress transcription.

In summary, nuclear receptors may be divided into two main classes: those that bind as homodimers to inverted repeat DNA half-sites, such as the steroid hormones, and those that bind as heterodimers to direct repeat sites.

Activation and repression of transcription

The mechanism of action of transcription factors is complex. In general, the basal transcription machinery of a gene is assembled around a region of DNA upstream of the coding sequence, called the *core promoter*. Its sequence is just sufficient to specify unregulated (basal) transcription. Core promoters that possess a TATA element bind TBP (TATA-binding protein), which nucleates the binding of the other subunits (termed *general transcription factors*) to form a pre-initiation complex that includes RNA polymerase II. Activation of this complex and the initiation of transcription by RNA polymerase occur when transcriptional regulators bind at response elements further up- or downstream. The interaction of these sites with the pre-initiation complex takes advantage of the flexibility of DNA and involves yet more accessory proteins that mediate interactions between the regulatory and general transcription factors. The regulation of transcription by nuclear receptors follows this pattern, relying on key accessory proteins termed coregulators.

Coactivators

Ligand-bound nuclear receptors attached to DNA at a response element activate transcription by the recruitment of coactivators. These associate with receptors mostly via the amphipathic AF-2 regions of the LBD (see Figure 10.4). Coactivators possess a nuclear receptor box containing a number of LxxLL motifs, which bind to a hydrophobic surface on the receptor. This surface is exposed only when ligand is bound due to the rearrangement of the AF-2 region (helix 12), illustrated in Figure 10.5 for the LBD of RXR.

The effect of coactivator binding is to increase the rate of transcription. This requires interactions with components of the transcriptional machinery (the general transcription factors) and, at the same time, the opening up of the chromatin structure to provide access. Rearrangement of chromatin is brought about by ATP-dependent chromatin remodelling complexes and also by acetylation of histones. This leads to the unwinding of DNA from its compact structure on nucleosomes; (this theme recurs: see Figure 14.3, page 423 and Figure 22.5, page 710). Some coactivators possess histone acetyltransferase (HAT) activity, others form complexes that activate chromatin remodelling and yet others interact indirectly with the general transcription factors. Figure 10.9a shows a simplified activation scheme for a steroid hormone receptor (GR). The ligand-bound receptor recruits coactivators principally through its AF-2 domain (for example SRC-2, CBP/p300, and p/CAF). A complex is formed that possesses HAT activity and that interacts with

The **TATA element** is a highly conserved DNA sequence present in the promoter region of most genes. It binds transcription factors and histones, the binding of the one acting to block the binding of the other. TATA boxes, usually located 25 bp upstream to the transcription site, are involved in the initiation of transcription by RNA polymerase.

Many genes lack a TATA box, its place taken by an initiator element or downstream core promoter. Such TATA-less promoters nonetheless always involve the TATA-binding protein (TBP) which binds without sequence specificity.

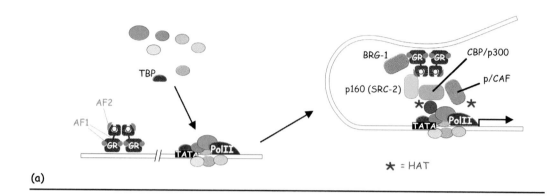

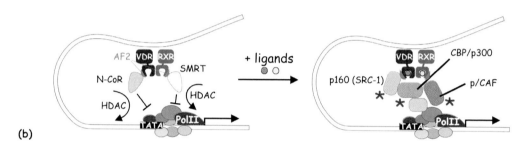

FIG 10.9 A simplified scheme of transactivation by glucocorticoid and vitamin D receptors.

the general transcription factors. A chromatin remodelling factor (BRG-1) is recruited to the AF-1 domain.

Corepressors

Nuclear receptors that bind as heterodimers to direct repeat half-sites are commonly nucleus-resident and may be located on DNA even in the absence of ligand (e.g. thyroid hormone receptors, but not steroid hormone receptors). In this state the AF-2 sites attach corepressors, the converse of coactivators, which act to repress the level of basal transcription. For example, the corepressors NCoR and SMRT recruit histone deacetylases that stabilize chromatin, keeping the DNA strands tightly wound on nucleosomes. They also interact with the general transcription factors in an inhibitory fashion (Figure 10.9b). The binding of ligand to the LBD induces the conformational change, in particular the rearrangement of H12 in AF-2 as discussed above, which releases the corepressor and allows it to be replaced by a coactivator, such as SRC-1. This leads to the recruitment of other coactivators with HAT activity, such as pCAF and p300/CBP. Also, TRAP/DRIP coactivators form a complex that interacts directly with the general transcription factors. Because of the opposing effects of corepressors and coactivators, the effect of ligand binding is to increase transcription from a fraction of the basal level (corepressor bound), to many times the basal level (coactivator bound), generating a response with a large dynamic range.

Transrepression

In the absence of ligand, nucleus-resident receptors bind corepressors which lead to the repression of transcription at a particular response element. But some receptors can work in the reverse sense, so that in the absence of ligand, transcription proceeds constitutively, but in its presence it is inhibited. This is transrepression and it is less well understood than transactivation. Here, we take the thyroid hormone receptor (TR) as an example.

Transactivation by TR involves the binding of RXR-TR heterodimers to thyroid hormone response elements and the activation of transcription in the presence of the hormone T3 as described above. In the absence of ligand the receptor represses transcription by recruitment of corepressors. However, there are also negative response elements to which the receptor can bind. Here, the receptor activates transcription in the absence of T3, but represses it when T3 is bound, silencing target genes that encode thyroid stimulating hormone (TSH) or thyrotropin-releasing hormone (TRH). The mechanism involves recruitment of HDAC activity and there is a requirement for corepressors such as SMRT. Curiously however, in the absence of ligand, the corepressor acts as a coactivator and this suggests that coregulator function may finally depend on the type of response element involved (positive or negative).

Regulatory networks

Because a single receptor, such as TR, can bind different types of response element, different nuclear receptors may vie for the same or overlapping target sequences. Moreover, they may have common ligands. All this can lead to crosstalk between different receptor mechanisms. Other ways in which receptors interfere with each other include the sequestration of coregulators or recruitment of corepressors at nearby sites. Thus, some orphan receptors can be viewed as constitutive repressors or controllers of basal transcription through corepressor recruitment and the consequent HDAC activity and interaction with the general transcription factors. Conversely, other orphans may increase responsiveness of particular nuclear receptors.

Such crosstalk forms the basis of regulatory networks that underpin important physiological functions. These transcriptional control systems are often complex. An example is provided by the nuclear receptors that regulate the breakdown of cholesterol in the liver to form bile acids. The first and rate-limiting enzyme in the catabolism of cholesterol is a member of the cytochrome P_{450} family CYP7A1 (cholesterol 7α-hydroxylase). A negative

Cytochrome P_{450} proteins. These haem proteins are oxidoreductases that catalyse the oxidation of many endogenous and exogenous hydrophobic compounds, such as fatty acids, sterols, prostaglandins, leukotrienes, and a wide range of xenobiotics. Some 40 different P_{450} protein families have been identified on the basis of their sequences. Humans express 57 different proteins. P_{450} proteins are classified with the prefix CYP.

Note: the CYPs are not related to the cyclophilins that have been assigned a similar prefix, Cyp. Nor is SHP related to SHP-1 and SHP-2, Src homology domain phosphatases (see page 644).

feedback mechanism causes the repression of the *CYP7A1* gene in response to bile acid synthesis, but this is indirect. It involves three different nuclear receptors FXR (page 281), SHP (NR0B2, an atypical receptor that lacks a DBD), and LHR-1 (NR5A2, responsible for the basal transcription of *CYP7A1*). Firstly, FXR is activated by bile acids to induce expression of SHP, which in turn, represses LHR-1 by forming a heterodimeric complex. In the opposite sense, LXR acts to increase throughput in the catabolic pathway by inducing the expression of reverse cholesterol transporters in the tissues. In mouse and rat it increases throughput through the expression of CYP7A1.

Bile acids are hepatotoxic and a further layer of control is provided to avoid excessive accumulation in the liver (cholestasis). This involves CAR and PXR. These adopted orphans were initially identified as detectors of a range of lipophilic xenobiotics (including many drugs). They induce expression of CYP3A4, which brings about hydroxylation and leads to detoxification. It later emerged that they have physiological activators that include precursors and products of cholesterol metabolism. Thus these receptors also act to repress transcription of CYP7A1 in response to bile acids, but at a 10–fold lower affinity than FXR and by a different mechanism. They also activate expression of transporters which promote export of bile acids from the liver and of CYP3A enzymes that increase bile acid metabolism and lead to excretion of the products. Even this is a very simplified picture of the network regulating bile acid metabolism. There are many other components and interactions. In general, although the circuitry of such networks may be complicated, they are often accessible to pharmacological intervention and, less fortunately, to environmental influence.

Interaction with other signalling pathways

Nuclear receptor signalling does not exist in isolation from other forms of cellular signal transduction. Not only is there cross-talk between different ligand-activated nuclear receptor pathways, there are also interactions with other signalling mechanisms. The effects can range from modulation of transcriptional activation to its activation in the absence of nuclear receptor ligand.

Phosphorylation

Although nuclear receptors exist as phosphoproteins in the absence of their ligands, their function may be modified by hyperphosphorylation.[37] The target sites are concentrated on the N-terminal A/B domain, but they may also be present on each of the other major domains. The number of sites and their locations vary with receptor type; for example, PR has 13 possible phosphorylation sites on its N-terminal domain, others have just one or two and VDR, exceptionally, is not phosphorylated in this region. Target residues in the A/B domain are mostly serines in the vicinity of prolines.

Nuclear receptors are phosphorylated by a wide range of kinases. The cyclin-dependent kinases, which play a role in regulation of the cell division cycle, are responsible for the ligand-independent, basal phosphorylation and also hyperphosphorylation in response to nuclear receptor ligand binding. This can cause the level of phosphorylation to vary throughout the cell cycle. For example, in S phase the basal level of GR phosphorylation is low and it is high in G2/M, while its hyperphosphorylation is the reverse.

Nuclear receptors are also targets of the downstream kinases of the major signalling pathways, namely PKC and PKA (activated by G-protein-coupled receptors: see Chapter 9), ERK and PKB/Akt (activated by growth factor receptors via Ras and PI 3-kinase respectively: Chapters 12 and 18), JNK/SAPK and p38-MAP kinase (activated by stress and cytokine receptors: Chapter 16).

Phosphorylation may up- or down-regulate transcription

Phosphorylation at a given site on a given receptor may facilitate or inhibit transcription. Given the number of potential sites and the variety of kinases and receptors, the picture is complicated. In general, ligand-independent phosphorylation by cyclin-dependent kinases on a receptor AF-1 region promotes transcription. Phosphorylation in the N-terminal region of receptors such as PR, ER, AR, and PPARα by members of the MAP kinase family (ERK, JNK/SAPK, p38-MAP kinase) also enhances transcription by helping the recruitment of coactivators, HAT activity, and chromatin remodelling proteins. Furthermore, ER and AR may be phosphorylated in their N-terminal domains by PKB/Akt, again enhancing transcription. Phosphorylation of the AF-2 region of ERα by the tyrosine kinase Src also up-regulates transcriptional activity. Coregulators may be also be phosphorylated by MAP kinases (and other kinases). This has the effect of promoting the binding of coactivators to nuclear receptors in the presence of ligand and inhibiting the binding of corepressors. The general trend of all this is positive regulation, ensuring the formation of an efficient transcription initiation complex.

Transcriptional activity can be down-regulated by phosphorylation of residues in the DBD by PKC or PKA, by preventing recognition of response element half-sites or by modifying receptor dimerization surfaces. MAP kinase phosphorylation of the LBD of RXRα inhibits transcription by RAR/RXR and VDR/RXR heterodimers through conformational changes that disrupt coregulator interactions.

Ligand-independent activation

Growth factors such as EGF (see page 318) or IGF (see page 554) can activate transcription by ER apparently in the absence of oestrogens. This appears to

CYP3A4, the most prominent P_{450} protein in human liver, is thought to be responsible for the metabolism of some 60% of all clinically used drugs (MIM 124010).

depend upon phosphorylation mediated by MAP kinase, affecting contact with coactivator molecules. Ligand-independent activation occurs in some cancerous conditions. In cancer of the prostate, for example, ERα and AR may be activated in the absence of an androgenic stimulus in cells exposed to the cytokine IL-6. Again, MAP kinase activation is implicated.

Non-transcriptional actions of nuclear receptors and their ligands

To add yet another layer of complexity, some responses activated by nuclear receptor ligands take place so rapidly (in seconds to minutes) that they cannot be accounted for by a transcriptional mechanism. Furthermore, membrane-impermeant ligands (e.g. steroid hormones fused to a polypeptide), are also effective, indicating the presence of some kind of receptor at the plasma membrane. Commonly, the consequence of steroid hormone binding to plasma membrane-associated receptors is to activate cytosolic signalling pathways through the production of second messengers. Such signalling effects are described as 'non-genomic' or 'non-transcriptional'.

The presence of oestrogen binding sites at the plasma membrane was first reported in 1977[39] and there are now many examples of steroid actions initiated at the cell surface.[40] The intracellular pathways that are activated vary, and our understanding of the mechanisms involved is limited. Hormone binding to plasma membrane steroid receptors has been shown to activate PLCβ, adenylyl cyclase,[41] PI 3-kinase, and MAP kinases in a variety of systems.[42, 43] For example in osteoblasts, 1,25–dihydroxycholecalciferol (calcitriol), estradiol, or progesterone causes an elevation of intracellular Ca^{2+} within 5 s by activating PLCβ in a G-protein-dependent fashion.[44, 45] In a cancer cell line, androgens can activate Src within 1 min leading to MAP kinase activation. In connection with the vascular protective affects of oestrogens, ERα has been shown to couple to the regulatory subunit of PI 3-kinase. Estradiol can activate this lipid kinase to produce 3-phosphoinositides which in turn recruit PKB, leading to the activation of nitric oxide synthase (eNOS) in vascular endothelial cells.[46, 47]

Whether the same receptors mediate both the rapid and the nuclear responses is not always clear. Prompt effects of steroids may not involve interaction with nuclear receptors at all, the ligands binding to membrane-located effector proteins or receptors. For example, progesterone modulates the effect of the peptide hormone oxytocin, by interacting directly with the oxytocin receptor (see page 27). Also, the modulation of ion channels in the nervous system by oestrogens, without involvement of ERs, is well documented.[48] In general, it seems likely that the non-transcriptional actions of steroid hormones produce effects that facilitate or might even be a prerequisite for transcriptional signalling by the same ligand.

References

1. Beatson GT. On the treatment of inoperable cases of carcinoma of the mammary: suggestions for a new method of treatment with illustrative cases. *Lancet*. 1896;2:104–107.

2. Needham J, Lu G. Proto-endocrinology in Medieval China. In: *Clerks and Craftsmen in China and the West*: Cambridge University Press; 1970:294–315.

3. Spiegel F. *Sick Notes: An Alphabetical Browsing Book of Derivations, Abbreviations, Mnemonics and Slang for the Amusement and Edification of Medics, Nurses, Patients and Hypochondriacs*. Carnforth, Lancs, UK: Parthenon Publishing; 1996.

4. Doisy EA, Veler CD, Thayer S. The preparation of the crystalline ovarian hormone from the urine of pregnant women. *J Biol Chem*. 1930;86:499–509.

5. Butenandt AF. Über Progynon, ein krystallisiertes weibliches Sexualhormon. *Naturwissenschaften*. 1929;17:879.

6. Marrain GF. The chemistry of oestrin: Preparation from urine and separation from an unidentified solid alcohol. *Biochem J*. 1929;23:1090–1098.

7. Dodds EC, Goldberg L, Lawson W, Robinson R. Estrogenic activity of certain synthetic compounds. *Nature*. 1938;141:247–248.

8. Corner GW, Allen WM. Physiology of the corpus luteum: Production of a special reaction: progestational proliferation by extracts of the corpus luteum. *Am J Physiol*. 1929;88:326–339.

9. Kilvik K, Furu K, Haug E, Gautvik KM. The mechanism of 17β-estradiol uptake into prolactin-producing rat pituitary cells (GH3 cells) in culture. *Endocrinology*. 1985;117:967–975.

10. Milgrom E, Atger M, Baulieu E-E. Studies on estrogen entry into uterine cells and on estradiol-receptor complex attachment to the nucleus – Is the entry of estrogen into uterine cells a protein-mediated process? *Biochim Biophys Acta* 1973;320:267–283.

11. Toft DO, Gorski J. A receptor molecule for estrogens: isolation from the rat uterus and preliminary characterization. *Proc Natl Acad Sci USA*. 1966;55:1574–1581.

12. Rouseau GG, Baxter JD, Higgins SJ, Tomkins GM. Steroid induced nuclear binding of glucocorticoid receptors in intact hepatoma cells. *J Mol Biol*. 1973;79:539–554.

13. Nishi M, Takenaka N, Morita N, Ito T, Ozawa H, Kawata M. Real-time imaging of glucocorticoid receptor dynamics in living neurons and glial cells in comparison with non-neural cells. *Eur J Neurosci*. 1999;11:1927–1936.

14. Phelps C, Gburcik V, Suslova E, et al. Fungi and animals may share a common ancestor to nuclear receptors. *Proc Natl Acad Sci USA*. 2006;103:7077–7081.

15. Kostrouch Z, Kostrouchova M, Love W, Jannini E, Piatigorsky J, Rall JE. Retinoic acid X receptor in the diploblast, Tripedalia cystophora. *Proc Natl Acad Sci USA*. 1998;95:13442–13447.

16. Robinson-Rechavi M, Laudet V. How many nuclear hormone receptors in the human genome? *Trends Genet* 2001;17:554–556.

17. Baker ME. Evolution of adrenal and sex steroid action in vertebrates: a ligand-based mechanism for complexity. *Bioessays*. 2003;25:396–400.

18. Yamamoto KR. Steroid receptor regulated transcription of specific genes and gene networks. *Annu Rev Genet*. 1985;19:209–252.

19. Hollenberg SM, Weinberger C, Ong ES, et al. Primary structure and expression of a functional human glucocorticoid receptor cDNA. *Nature*. 1985;318:635–641.

20. Green S, Walter P, Kumar V, et al. Human oestrogen receptor cDNA: sequence, expression and homology to v-erb-A. *Nature*. 1986;320:134–139.

21. Escriva H, Delaunay F, Laudet V. Ligand binding and nuclear receptor evolution. *Bioessays*. 2003;22:717–727.

22. Laudet V. Evolution of the nuclear receptor superfamily: early diversification from an ancestral orphan receptor. *J Mol Endocrinol*. 1997;19:207–226.

23. Wang Z, Benoit G, Liu J, et al. Structure and function of Nurr1 identifies a class of ligand-independent nuclear receptors. *Nature*. 2003;423: 555–556.

24. Wisely GB, Miller AB, Davis RG, Thornquest AD, Johnson R, Spitzer T, Sefler A, Shearer B, Moore JT, Miller AB, Willson TM, Williams SP et al. Hepatocyte nuclear factor 4 is a transcription factor that constitutively binds fatty acids. *Structure (Camb)*. 2002;10:1225–1234.

25. Saucedo-Cardenas O, Quintana-Hau JD, Le W-D, et al. Nurr1 is essential for the induction of the dopaminergic phenotype and the survival of ventral mesencephalic late dopaminergic precursor neurons. *Proc Natl Acad Sci USA*. 1998;95:4013–4018.

26. Le W-D, Xu P, Jankovic J, Appel SH, Smith RG, Vassilatis DK. Mutations in NR4A2 associated with familial Parkinson disease. *Nature Genet*. 2002;33:85–89.

27. Viollet B, Kahn A, Raymondjean M. Protein kinase A-dependent phosphorylation modulates DNA-binding activity of hepatocyte nuclear factor 4. *Mol Cell Biol*. 2005;17:4208–4219.

28. Nuclear Receptors Committee. A unified nomenclature system for the nuclear receptor superfamily. *Cell*. 1999;97:161–163.

29. Bourguet W, Ruff M, Chambon P, Gronemeyer H, Moras D. Crystal structure of the ligand-binding domain of the human nuclear receptor RXR-α. *Nature*. 1995;375:377–382.

30. Egea PF, Mitschler A, Rochel N, Ruff M, Chambon P, Moras D. Crystal structure of the human RXR α ligand-binding domain bound to its natural ligand: 9-cis retinoic acid. *EMBO J*. 2000;19:2592–2601.

31. Ritossa F. Discovery of the heat shock response. *Cell Stress & Chaperones*. 1996;1:97–98.

32. Lindquist S. The heat shock response. *Annu Rev Biochem*. 2004;55:1151–1191.

33. Thomas PJ, Qu BH, Pedersen PL. Defective protein folding as a basis of human disease. *Trends Biochem Sci*. 1995;20:456–459.

34. Pratt WB, Galigniana MD, Morishima Y, Murphy PJ. Role of molecular chaperones in steroid receptor action. *Essays Biochem*. 2004;40:41–58.

35. Pratt WB, Galigniana MD, Harrell JM, DeFranco DB. Role of hsp90 and the hsp90-binding immunophilins in signalling protein movement. *Cell Signal*. 2005;16:857–872.

36. Freedman LP, Luisi BF, Korszun ZR, Basavappa R, Sigler PB, Yamamoto KR. The function and structure of the metal coordination sites within the glucocorticoid receptor DNA binding domain. *Nature*. 1988;334:543–546.

37. Rochette-Egly C. Nuclear receptors: integration of multiple signalling pathways through phosphorylation. *Cell Signal*. 2003;15:355–366.

38. Luisi BF, Xu WX, Otwinowski Z, Freedman LP, Yamamoto KR, Sigler PB. Crystallographic analysis of the interaction of the glucocorticoid receptor with DNA. *Nature*. 1991;352:497–505.

39. Pietras RJ, Szego CM. Specific binding sites for oestrogen at the outer surfaces of isolated endometrial cells. *Nature*. 1977;265:69–72.

40. Simoncini T, Genazzani AR. Non-genomic actions of sex steroid hormones. *Eur J Endocrinol*. 2003;148:281–292.

41. Aronica SM, Kraus WL, Katzenellenbogen BS. Estrogen action via the cAMP signaling pathway: stimulation of adenylate cyclase and cAMP-regulated gene transcription. *Proc Natl Acad Sci USA*. 1994;91:8517–8521.

42. Kousteni S, Bellido T, Plotkin LI, et al. Nongenotropic, sex-nonspecific signaling through the estrogen or androgen receptors: dissociation from transcriptional activity. *Cell*. 2001;104:719–730.

43. Singer CA, Figueroa-Masot XA, Batchelor RH, Dorsa DM. The mitogen-activated protein kinase pathway mediates estrogen neuroprotection after glutamate toxicity in primary cortical neurons. *J Neurosci*. 2004;19:2455–2463.

44. Le Mellay V, Lasmoles F, Lieberherr M. $G\alpha_{q/11}$ and $G\beta\gamma$ proteins and membrane signaling of calcitriol and estradiol. *J Cell Biochem*. 1999;75:138–146.

45. Le Mellay V, Lieberherr M. Membrane signaling and progesterone in female and male osteoblasts. II. Direct involvement of $G\alpha_{q/11}$ coupled to PLC-β1 and PLC-β3. *J Cell Biochem*. 2000;79:173–181.

46. Haynes MP, Sinha D, Russell KS, et al. Membrane estrogen receptor engagement activates endothelial nitric oxide synthase via the PI3-kinase-Akt pathway in human endothelial cells. *Circulation Res*. 2000;87:956–960.

47. Haynes MP, Li L, Sinha D, et al. Src kinase mediates phosphatidylinositol 3-kinase/Akt-dependent rapid endothelial nitric-oxide synthase activation by estrogen. *J Biol Chem*. 2003;278:2118–2123.

48. Fu XD, Simoncini T. Extra-nuclear signaling of estrogen receptors. *IUBMB Life*. 2008;60:502–510.

CHAPTER 11

Growth Factors: Setting the Framework

The trails that led to the discovery of the growth factors (and related messengers) are very different from those that revealed the first hormones and neurotransmitters. For a start, people knew, more or less, what to look for and where to go looking, and in general, the tale is somewhat less romantic and less fraught with angst and vehement disagreements. However, what began with a simple search for factors that would sustain living cells in laboratory conditions has expanded into a plethora of subject areas which are subjects in their own right. These include inflammation, wound healing, immune surveillance, development, and carcinogenesis. To confront this bewildering prospect we first set out some details of how it evolved initially so that the reader is aware of how the major questions developed and have been confronted. A difficulty arises from the convergence of different disciplines, each bringing with it the baggage of its favoured nomenclature. We return to this matter at the end of this chapter.

For the ultimate in aggressive rivalry and claims to primacy in hormone discovery, coupled to barely restrained personality conflict, see Nicholas Wade's account of the discovery by Roger Guillemin and Andrew Schally of thyrotropin releasing factor and other hypothalamic peptides.[1–3]

Viruses and tumours

The first report of a tumour linked to a virus appeared in 1908, when Ellermann and Bang obtained a filterable agent from a chicken leukaemia and were able to make six passages of it, from fowl to fowl, producing the same disease each time.[4] Their report was generally disregarded. Leukaemia was not considered to be a tumour, though as should have been evident from the work of Aldred Warthin (Figure 11.1), first reported in 1904,[5, 6] a link between leukaemias and tumours, and hence cell growth and proliferation, had already been established:

> These conditions are comparable to malignant tumors. The formation of metastases, the infiltrative and destructive growth, the failure of innoculations and transplantations etc., all favor the view that they are neoplasms, and present the same problems as do the malignant tumors.

More than 20 years were to pass before the leukaemias were eventually recognized as being tumours or neoplastic diseases.

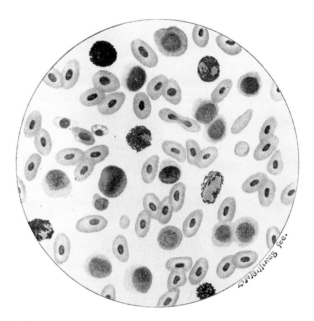

FIG 11.1 Stained blood smear from a leukaemic fowl.
'In December 1905, there came into my hands a Buff Cochin Bantam hen showing signs of illness in the way of indisposition to move about and a general weakness of a progressive character. No symptoms of ordinary fowl diseases were present . . . Examination of blood smears showed, however, a great increase of white cells of the large lymphocyte type . . . A diagnosis of leukemia was therefore made . . . A great variety of staining methods were used, including the most recent methods for the staining of spirochetes and protozoan parasites . . . No evidence of the existence of any infective agent could be obtained.[6] (Note that avian red blood cells (yellow stain) are nucleated.)

In 1910, Francis Peyton Rous described a chicken tumour, identified as a sarcoma, that could be propagated by transplanting its cells, these then multiplying in their new hosts giving rise to tumours of the same sort.[7] The cells yielded a virus, now known as Rous sarcoma virus (RSV), from which, in 1980, the first protein tyrosine kinase, v-Src, was isolated.[8,9] Writing at the end of his career in 1967, Rous[10] gives an apt description of how things were done:

> Those were primitive times in the raising of chickens. They were sold in a New York market not far from the institute, and many individual breeders brought their stock there. Every week F. S. Jones, a gifted veterinarian attached to my laboratory, went to it, not only to buy living chickens with lumps which might be tumors, but any that seemed sickly and had a pale comb, as perhaps having leukemia. Thus within less than four years we got more than 60 spontaneous tumors of various sorts . . .

The discovery of NGF . . . and EGF

> Among the fractions that I assayed in vitro the following day, there was one containing snake venom. Having not been told which of the fractions had been specially treated, I was completely stunned by the stupendous halo radiating from the ganglia. I called Stan in without telling him what I had seen. He looked through the microscope's eyepieces, lifted his head, cleaned his glasses which had fogged up, and looked again. 'Rita,' he murmured, 'I'm afraid we've just used up all the good luck we're entitled to. From now on, we can only count on ourselves . . . .' Events were to prove him wrong Rita Levi-Mpntaicini.

Nerve growth factor (NGF) may perhaps be regarded as the first identified growth factor, but there were many early clues hinting at their existence. The embryologist Hans Spemann (Nobel Prize winner in 1935, see page 618) had described the eponymous 'organizer' that directs the creation of the anteroposterior axis in the gastrula stage in the development of the amphibian embryo. This work had famously indicated that soluble factors made by the embryonic cells must be instrumental in regulating cell proliferation and differentiation.[13]

Rita Levi-Montalcini's affair with growth factors had its origin at that time. Dismissed from her university post by the Nazi racial edicts, she determined to continue on alone. She had been alerted to the work of Spemann by an article written by one of his protégés, Viktor Hamburger. He described the concept of the inductive reaction of certain tissues on others during early development. In particular, he cited the effect of ablating the embryo limb buds of chicks upon the reduction in the volume of the motor column and the spinal ganglia responsible for the innervation of the limbs. The idea was that the failure of

Francis Peyton Rous (1879–1970) shared the 1966 Nobel Prize for Physiology or Medicine for his discovery of tumour-inducing viruses.

Sarcoma is a form of cancer that arises in the supportive tissues such as bone, cartilage, fat or muscle. A **carcinoma** is defined as a malignant new growth that arises from epithelium, found in skin or, more commonly, the lining of body organs, for example: breast, prostate, lung, stomach, or bowel. Carcinomas tend to infiltrate into adjacent tissue and spread (metastasize) to distant organs, e.g. to bone, liver, lung, or brain.

v indicates a viral origin, commonly a transforming mutant. **c** indicates the cellular wild type. **Src**, sarcoma.

The quotations in this section are taken from: *In Praise of Imperfection, My life and work* by Rita Levi-Montalcini.[11] A facsimile collection of her major papers, with annotations by the author, is presented in Levi-Montalcini.[12]

Stanley Cohen and Rita Levi-Montalcini were awarded the Nobel Prize in 1986 'for their discoveries of growth factors'.

the cells to differentiate and to develop is due to the absence of an inductive factor normally released by the innervated tissues. She aimed to understand how the excision of non-innervated tissue could affect differentiation and subsequent development.

Confined to working in her bedroom she made use of just the most basic materials and instruments needed for histological investigation. In her examination of this problem, Levi-Montalcini found that nerve cell differentiation proceeds quite normally in the embryos with excised limbs, but that a degenerative process (what we would now call apoptosis) commences as soon as the cells emerge from the cord and ganglia appear at the stump of the amputated limb.[14,15] It appeared to her that the failure to develop was best explained by the absence of a trophic factor.

In 1946, she sailed for the USA in the company of Renato Dulbecco (see below), a friend from her student days. She was destined for St. Louis, he for Bloomington. Intending to pay a brief visit of a few weeks to the laboratory of Viktor Hamburger, she remained there for 30 years. The work that led eventually to the discovery of NGF had its origins in the observation by a late student of Hamburger's, Elmer Bueker. He had had reported that fragments of an actively growing mouse tumour, grafted on to chick embryos, caused a great ramification of nerve fibres into the mass of tumour cells.[16] Seeking advice and encouragement, he suggested that the tumour generated conditions favourable for the differentiation of the nerve cells which was reflected in the increased volume of the ganglia. Repeating the experiment, Levi-Montalcini describes the extraordinary spectacle of seeing bundles of nerve fibres passing between the tumour cells like rivulets of water flowing steadily over a bed of stones. In no case did they make any connection with the cells, as is the rule when fibres innervate normal embryonic or adult tissue. Later she describes how the sympathetic fibres invaded the embryonal viscera, even entering into lumen of the venous, but not the arterial blood vessels so that the smaller veins were quite obstructed.

> The penetration of the nerve fibers into the veins, furthermore, suggested to me that this still unknown humoral substance might be exerting a neurotropic effect, or what is known as a chemotactic directing force, one that causes nerve fibers to grow in a particular direction . . . . Among these, I guessed, was undoubtedly also the humoral growth factor that the cells produced. This hypothesis would explain this most atypical finding of sympathetic fibers gaining access inside the veins . . . Now, with the hindsight of the nearly forty years gone by since those moments of keenest excitement – it appears that the new field of research that was opening up before my eyes was, in reality, much vaster than I could possibly have imagined.

It was clearly necessary to develop an *in vitro* assay. These were early days in the field of cell culture and it took the best part of 6 months to develop a practical method that could be used as the basis for measuring the biological activity of fractions in protein purification. Only then was the point reached when a biochemical approach could usefully be applied in the pursuit of NGF, the name by which the factor became known shortly after.

> 'Rita,' Stan said one day, 'you and I are good, but together, we are wonderful'.

After a year's intense work, they had narrowed down the factor as a nucleoprotein, though Stanley Cohen suspected that the nucleic acid component was likely to be a contaminant. On the advice of Arthur Kornberg, he applied an extract of snake venom as a source of nuclease activity with the aim of removing nucleic acids, present as an impurity in their material. To their great surprise this yielded a preparation that enhanced neuronal growth still further. It emerged that the snake venom alone was active.[17] Making the connection between venom and saliva, they tested mouse salivary glands and found that this too is an excellent source of NGF activity. (If they had been able to purchase purified nuclease enzyme, the course of the discovery would surely have been prolonged.)

Later, Cohen discovered a new phenomenon that 'was destined to become a magic wand that opened a whole new horizon to biological studies.' A contaminating factor was present that caused precocious growth in epidermis (Figure 11.2). It was also found that the factor has a powerful proliferative effect on connective tissues and it became clear that there is a link between the mechanisms that control normal proliferation and neoplastic growth.

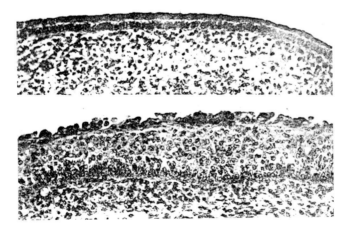

FIG 11.2 An early demonstration of the effects of EGF. Thickening of the epidermal layer in skin explants from chick embryos after 3 days in culture.[18]

Since, under culture conditions, the stimulus to proliferate could not involve systemic or hormonal influences, Cohen called the new protein *epidermal growth factor* (EGF).[18,19]. Later, it was shown that mouse EGF enhances DNA synthesis in cultured human fibroblasts. It was also found that EGF is similar to uragastrone, a peptide that had been isolated from human urine and recognized by pharmacologists because of its ability to inhibit gastric acid secretion.[20] Out of 53 residues in the amino acid sequence, 37 share a common location and the two polypeptides have similar effects on both gastric acid secretion and the growth of epidermal cells.

All this now paved the way for a more molecular approach using isolated cells. It was found that rat kidney cells, transformed with the Kirsten sarcoma virus (see Chapter 4) fail to bind EGF. This down-regulation, which is due to internalization of the receptor, is caused by the elevated expression and release of EGF by the cells themselves (an autocrine mechanism of feedback regulation). The possibility that internalization might be a necessary step initially found many adherents, but it became apparent that the transduction mechanism emanates from events at the plasma membrane. In particular, ligand binding directly induces phosphorylation (on tyrosine residues) of a membrane protein, later shown to be the EGF receptor (EGFR) itself.[21] This was an important breakthrough because tyrosine kinase activity had already been associated with virus-induced sarcomas (the v-src gene product). Thus, a firm link was established between neoplasia and the physiological regulation of cellular growth.

Further evidence for the role of growth factors in tumour generation came with the revelation that the avian erythroblastosis virus oncogene, *v-erb-B,* codes a product having similarities with the EGF receptor. Indeed, it became apparent that the transformation derives from inappropriate acquisition from the (cellular) *c-erb-B* gene, of a truncated receptor, lacking the binding site for EGF and which is constitutively activated.[22]

The development of this field has generated much excitement but also frustration. It has been exciting, because it has yielded a good understanding of cell transformation and growth factors, and frustrating because it has become clear that cancer cells do not readily lend themselves as specific targets for drugs. The main impetus behind these studies was that non-mammalian genes might be the cause of disease. The hope was to discover targets which might be exploited to kill tumour cells selectively, for example the products of the viral genes. We now realize that these are initially hijacked from the mammalian genome itself, then inaccurately transcribed by sloppy DNA or RNA polymerases in the virus which then offers them back to the host upon infection. Apart from this, the large proportion of human tumours is of non-viral origin, arising as a consequence of ageing, tumour-promoting substances, radiation, etc.

Platelet-derived growth factor (PDGF)

In 1912 Alexis Carrel[23] reported a number of experiments having the purpose 'to determine the conditions under which the active life of a tissue outside that of the organism could be prolonged indefinitely'. When he tried to maintain tissues for a few days in a simple buffered salt solution, the cells lost their growth capacity and then their viability. It was supposed that the senility and death of the cultures was due to the accumulation of catabolic substances and exhaustion of essential nutrients. Continuous and more rapid growth was achieved by supplementing the solution with diluted plasma, and then, from time to time, submerging the tissue in serum for a few hours. Interestingly, the notion that the cells might require specific factors present in the serum never appears to have crossed Carrel's mind. In conclusion, he wrote that

> fragments of connective tissue have been kept in vitro in a condition of active life for more than two months. As a few cultures are now eighty-five days old and are growing very actively, it is probable that, if no accident occurs, the life of these cultures will continue for a long time (Figure 11.3).

About 40 years later, Temin and Dulbecco, working independently, set out to define the precise requirements for cell culture with respect to amino acids, vitamins, salts (together, Carrel's 'nutrients') and importantly, growth factors. Their ambition to culture cells arose from their interest in the role of viruses in cell transformation and tumour formation. Thus an important landmark was the finding that the requirement for serum is drastically reduced in cells infected with tumour viruses. They proposed that transformation might occur as a result of the enhanced capacity of tumour cells to respond to the proliferation signals present in the serum.

An important turning point was the discovery that serum (the soluble component of *clotted* blood) supports growth and proliferation. Plasma merely allows survival, and over a period of about 2 days cells become quiescent (arrested at the G0 stage of the cell cycle). Serum contains the products of activated platelets, suggesting that they might have a necessary role in the provision of growth factors. In 1974, Russell Ross showed that factors extracted from platelets can induce quiescent smooth muscle cells to synthesize DNA,[24,26] and in the same year, Kohler and Lipton obtained a similar result with mouse 3T3 fibroblasts. The factor derived from blood platelets (platelet-derived growth factor, PDGF) could propel the quiescent cells into the cell cycle S-phase (DNA synthesis).

With the purification of PDGF from human platelets[27] the question of how it acts could be faced. PDGF exists as a disulfide-linked dimer, so that binding automatically causes the cross-linking of two receptors. This constitutes the signal for activation.[28] Subsequently, it was found that the oncogene of the simian sarcoma virus (v-*sis*) is homologous to the gene coding for PDGF.[29–31]

Alexis Carrel was awarded the Nobel Prize (in 1912) in recognition of his work on vascular suture and the transplantation of blood vessels and organs'. He tarnished his reputation by his association with the eugenics movement, calling for the establishment of institutions equipped with 'appropriate gases' in order to eliminate the insane. He gave enthusiastic support to the Vichy government in France during World War II and after the liberation he was charged for collaboration with the Nazi occupiers. He died before his case came to trial. Even his claim to have kept heart cells in culture may have been unwittingly fraudulent. Assistants may have introduced fresh cells from time to time in order to achieve the expected result . . . and to keep Alexis happy.[25]

Howard Temin and **Renato Dulbecco** shared the Nobel prize in 1975 for their discoveries concerning the interaction between tumour viruses and the genetic material of the cell.

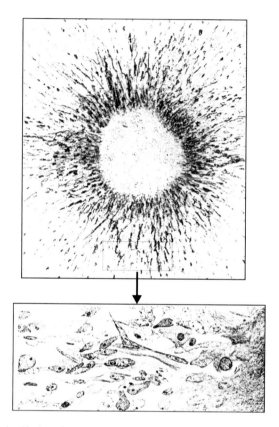

FIG 11.3 'A fifty day old culture of connective tissue. Active tissue is encircling a piece of old plasma.' From Carrel.[23]

Here was another clear link between a growth factor and a tumour virus. This time, however, the signal to uncontrolled cell proliferation is due to excessive production of growth factor, rather than expression of a constitutively activated receptor. Furthermore, v-*sis* causes cell transformation in primates.[32]

The mechanisms by which an oncogene might cause a tumour were becoming clearer. Here is another example of a mammalian gene, surreptitiously borrowed by a virus, then mutated or mutilated. On return to the host by infection, it causes cell transformation and tumour formation.

Transforming growth factors (TGFα and TGFβ)

The transforming growth factors (TGFs) were originally isolated from the conditioned medium of a virally infected mammalian fibroblast cell line, 3T3. These are proteins that can bring about transformation of phenotype.

The discovery of the TGFs followed some years after the first reports and descriptions of EGF, and it derives particularly from the work undertaken in the laboratory of George Todaro.[33]

TGFα, although quite distinct from EGF with respect to amino acid sequence, binds to the EGF receptor and signals cells in a similar fashion (see Chapter 12). A related factor isolated from these tumour cells, TGFβ, which does not compete with the binding of TGFα or EGF, can nevertheless induce transformation when provided with either of these two factors. Importantly, TGFβ is a normal cellular product and the finding of high quantities in blood platelets and its release during blood coagulation established a clear link with PDGF.[34]

In screening the transforming effect of TGFβ in numerous tumours, there was an unexpected finding. Depending on the cells and the conditions, TGFβ can either promote or suppress cell growth and transformation. It cooperates with TGFα and EGF to cause cell transformation. On the other hand, it inhibits colony formation in cells derived from human tumours. It appears that its effects are a function of the total set of growth factors and their receptors that are operational at a given time.[35] In addition, TGFβ plays a number of key roles in the process of tissue remodelling and wound healing.[36] It induces the production of fibronectin and collagen and thus regulates the deposition of the cell matrix, itself a key determinant of cell growth (see page 397).

Problems with nomenclature

As must be evident, nomenclature in this area is arbitrary, to say the least. Some growth factors were named after the cells from which they were first isolated, others from the cells which they stimulated, yet others from the principal action that they appeared to perform.[37] In immunology we hear of interleukins and colony stimulating factors. These direct the maturation and proliferation of white blood cells. In virology, we have the interferons that 'interfere' with viral infection, and in cancer research we have tumour necrosis factor (and its relatives) that can influence the growth of solid tumours. In each discipline it seemed that these factors functioned mainly in the category in which they first came to light. Of course, we now know that some growth factors have actions that are totally unrelated to growth. For instance, PDGF, released from platelets at sites of tissue damage,[38] not only supports the growth of fibroblasts, smooth muscle cells and glial cells, but also acts to regulate the distribution and migration of vascular smooth muscle cells and fibroblasts in wound healing.[39]

To add further complexity to an already complex situation, the conditions in which the cells are studied can determine the cellular response, for instance the presence of other factors, other cells and attachment to substrates. A good example of this is TGFβ. As its name implies, this emerged as a factor enhancing cell transformation, but we now recognize that it can inhibit cell

Cytokines: 'soluble (glyco)proteins, non-immunoglobulin in nature, released by living cells of the host, acting non-enzymatically in picomolar to nanomolar concentrations to regulate host cell function.'

305

proliferation[40] and that it is a very potent chemotactic factor for neutrophils[41] and fibroblasts.[42] It has been proposed that a common name for these factors should be **cytokines**.[43] While offering no clues to their various actions, this definition represents a move towards coordinating our understanding of their roles as first messengers. The unity of the cytokines is a concept as important as that of the hormones, defined by Bayliss and Starling nearly 90 years ago (see Chapter 1). However, while the pharmacology of the hormones has been so extensively (some might say exhaustively) investigated, for the protein (growth) factors, this area remains relatively unexplored. Attention here is firmly concentrated on intracellular events.

Essay: Cancer and transformation

Definitions

Cancer (kæ·nsəɪ), *sb.* ME. [L. *cancer (cancrum)* crab, also gangrene. OE. *cancer, cancor,* helped by Norman Fr. *cancre,* gave ME. CANKER. The L. form was re-introduced later for techn. use.] **1.** A crab. (Now *Zool.*) 1562. **b.** *Med.* An eight-tailed bandage 1753. **2.** *Astron.* **a.** The Zodiacal constellation lying between Gemini and Leo. **b.** The fourth of the twelve signs of the Zodiac (♋), beginning at the summer solstitial point, which the sun enters on the 21st of June ME. **3.** *Pathol.* A malignant growth or tumour, that tends to spread and to reproduce itself; it corrodes the part concerned, and generally ends in death. See also CANKER. 1601. Also *fig.* †**4.** A plant: perh. *cancer-wort* ‒1609.
2. *Tropic of C.*: the northern Tropic, forming a tangent to the ecliptic at the first point of C. **3.** C. is decidedly a hereditary disease ROBERTS. *fig.* Sloth is a C., eating up..Time KEN. Comb. (in sense 3) **c.-root,** *Conopholis (Orobanche) americana* and *Epiphegus virginiana;* **-wort,** *Linaria spuria* and *L. Elatine;* also the genus *Veronica.*

Canker (kæ·ŋkəɪ), *sb.* OE. [a. ONF. *cancre* :—L. *cancrum* crab, also gangrene. See CHANCRE.] **1.** An eating, spreading sore or ulcer; a gangrene. Used as = CANCER till *c* 1700. Now *spec.* A gangrenous affection of the mouth, with fetid sloughing ulcers; *canker of the mouth,* or *water c.* **b.** *Farriery.* A disease of a horse's foot, with a fetid discharge from the frog. **2.** Rust. Now *dial.* 1533. **3.** A disease of plants, *esp.* fruit trees, attended by decay of the bark and tissues 1555. **4.** A canker-worm ME. **5.** The dog-rose (*Rosa canina*). Now *local.* 1582. **6.** *fig.* Anything that frets, corrodes, corrupts, or consumes slowly and secretly 1564.
1. No cankar fretteth flesh so sore 1559. **4.** Cankers in the muske rose buds SHAKS. **5.** 1 *Hen. IV,* I. iii. 176. **6.** Enuie which is the c. of Honour BACON.
Comb.: **c.-berry,** the fruit of the dog-rose; also the plant *Solanum bahamense;* **-bloom,** the blossom of the dog-rose; **-blossom,** a canker (sense 4); also *fig.;* **-rash,** a form of scarlet fever in which the throat is ulcerated; **-rose,** (*a.*) the Dog-rose; (*b.*) the wild poppy (*Papaver Rhæas*).

FIG 11.4 Definitions of cancer, from the *Shorter Oxford English Dictionary* (3rd ed. 1944, corrected 1977).

The essence of cancer

The essence of cancer is that cells run out of control. They no longer respond correctly to the demands and influences of the environment in which they exist. Cancer in a particular tissue manifests itself by an increased cell mass which we describe as a tumour, meaning a swelling. The acquisition of cell mass is a consequence of new growth (neoplasm). In some cancers the primary tumour gives rise to the dissemination of cells that invade other tissues to form secondary tumours (metastasis). In addition, cancer cells tend to be poorly differentiated. As a consequence, they can lose their capacity to carry out the normal functions of the tissue from which they are derived, or they may take on an entirely inappropriate function (ectopic tumour). This means they may either fail to produce important factors or produce wrong factors in overwhelming excess. Under the microscope, cancerous cells are typically characterized by an enlarged nucleus with a very large nucleolus, less cytoplasm, and generally altered morphology. Cancer cells are often more rounded than their normal counterparts and there is generally less contact between them and their neighbours.

Most tumours develop in stages, going from benign to malignant. This is the result of a sequence of mutations that gradually enhance the sensitivity of the cells to growth factor signals, reduce the requirement of cell–cell and cell–matrix contact and render cells insensitive to the signals that determine programmed cell death. This change in phenotype is also called transformation, and so cancerous cells are often called transformed cells. The classification *benign tumour* means that there is an increase in the number of cells (hyperplasia, there may be a lump), but that the normal functions and morphology are retained. Importantly, there is no infiltration of other tissues, particularly the lymph nodes draining the areas in the vicinity of the tumour. By contrast, *m alignancy* refers to the loss or perversion of normal physiological function, altered morphology, and infiltration of other tissues, including metastasis.

Alterations dictating malignancy

Malignancy is a consequence of interplay between the transformed cell and its environment. In order for cells to proliferate in excess and to disseminate, they acquire new functions and they also induce the collaboration of the surrounding tissue.[44] For instance, transformed cells release angiogenic factors that induce the formation of new vasculature.[45] Surrounding tissue also provides activators of metalloproteinases that cause the matrix degradation necessary for tissue invasion.[46]

Fully transformed cells possess one or more of the following characteristics:

- They may not require an exogenous growth signal. In culture conditions, this means that the provision of serum is no longer required. It also means that the cells no longer have to adhere to an extracellular matrix; they can grow in soft agar. They also grow well in athymic (nude) mice.

- They may be insensitive to growth-inhibitory signals. For cells in culture, this means that proliferation of cultured cells continues beyond the point of formation of a tight monolayer. It also means that cells have become insensitive to the cell cycle checkpoints in G1, G2, and mitosis.
- They can evade programmed cell death (apoptosis). This means that these cells have an inbuilt rescue mechanism.
- They have limitless replicative potential. Normal cells can undergo ~50 divisions. This is because progressive erosion of the telomeres (several thousand repeats of 6 bp sequence elements) results in exposure of chromosomal ends. When the telomeres are exhausted, chromosome fusion ensues causing nuclear chaos, cell senescence and cell death. By contrast, tumour cells can continue to divide without limit.
- They can induce angiogenesis. During their development, tumours undergo a so-called angiogenic switch enabling them to induce the formation of new vasculature. This provides them with nutrients and oxygen and allows them to disseminate. Of course, dissemination is only possible when they have also acquired the capacity to survive when detached from the extracellular matrix (see Chapter 14).
- They may invade other tissues (metastasis). This is the ultimate consequence of the preceding alterations. The cells have acquired the capacity to detach without dying, to destroy extracellular matrix in order to migrate through tissue, to be carried by the lymph or blood stream, to attach and then colonize sites where they would not normally survive due to the absence of specific adhesion and growth factors.

Genetic alterations at the basis of malignancy

Cell transformation in cancer is the consequence of mutations of several different types, some inherited, some due to environmental factors, others due to ageing. Loss-of-function or gain-of-function mutations have been found in genes (some 380 in total)[53] that code for the following:

- Molecules involved in the signalling of growth factors. These are either growth factors themselves (c-Sis, PDGF), their receptors (ErbB, the EGF receptor), or their downstream signalling components (Ras) (see Chapter 12).
- Molecules involved in the control of the cell cycle (so-called tumour suppressors). Most important are the mutations in the retinoblastoma protein (Rb) , the p53 protein or the failure to express $p15^{INK4A}$ and $p15^{INK4B}$, inhibitors of the cyclin-dependent protein kinases.[47]
- Adhesion molecules. These are mainly implicated in cadherin-mediated cell-cell adhesion (cadherin, β-catenin, APC) and integrins (see Chapter 14).
- Molecules involved in the rescue from apoptosis. Important here are Bcl2 or Bcl-Xl, intracellular inhibitors of apoptosis, and the survival factors IGF-1 or IL-3.[48]

- Molecules involved in the signal transduction pathway that regulates cell survival (PTEN) (see Chapter 18).
- Molecules implicated in telomere maintenance. Transformed cells up-regulate the expression of the catalytic subunit of the telomerase reverse transcriptase (hTERT).[49]
- Angiogenic factors or their inhibitors. For instance, VEGF, FGF, or thrombospondin-1.[45]

Cells will normally die rather than allow somatic (non-inherited) alterations to their genetic code to persist. The natural rate of somatic mutation (due, for instance, to natural environmental carcinogens or bombardment by cosmic rays) is low, and it is unlikely that a tumour would ever develop over the decades of a human life. Yet, in western Europe and North America, one in three people will develop a cancerous growth. Certain mutations create a form of genomic instability that allows an enhanced and more generalized mutation frequency. The loss of function of the tumour suppressors p53 and Rb or other molecules involved in cell cycle control may be crucial here, the cells becoming too eager to replicate, failing to halt at the appropriate check points and failing to repair their DNA.[50] From here on things go from bad to worse.

Unlike human cells, rodent cells express telomerase activity and, as a result, there is no erosion of the telomeres during ageing. Because of this, the loss of telomeres does not represent a barrier to transformation in rodents and only two oncogenic mutations are necessary to cause transformation.

Constructing cancer in a dish

Cells possess mechanisms which protect the organism against the deleterious effects of their possible transformation. The responses that form this line of defence are set out in Table 11.1.

To understand the mechanism of transformation, it is necessary to establish its minimal requirements. How may this be achieved? Occasionally it has been possible to select rare, spontaneously arising immortalized cells. Alternatively, cells can be transformed by applying chemical and physical agents or a virus.

TABLE 11.1 Defences against transformation

Contingency	Consequence
Loss of contact with the extracellular matrix	apoptosis
Damage to DNA	cell cycle arrest (G1)
Defects in replication	cell cycle arrest (G1)
Failure of DNA repair	apoptosis
Excessive proliferation signals	senescence or apoptosis
Loss of telomeres (after ~50 cell divisions)	senescence and apoptosis (after ~50 cell divisions)

These can be said to represent shotgun approaches from which little can be learned. However, in certain human cells the imposition of just three mutations is sufficient to generate a transformed phenotype.[51]

- High telomerase activity to maintain telomere size and allow unlimited replication. Achieved by inserting the gene (hTERT) that codes for the catalytic subunit of the enzyme.
- An antitumour suppressor activity in the form of a viral oncoprotein (SV40 large-transforming antigen, Large T). This disables the function of p53 and Rb, preventing arrest of the cell cycle in G1, and enables the cells to evade apoptosis.
- A proliferative signal provided by a constitutively active form of Ras (V12G).

The sequence here is important.[52] Enhanced telomerase activity is rather easily tolerated by the cells, whereas an excessive growth factor signal is not. The construction of a transformed human cell line was achieved by inducing telomerase activity first, then large-transforming antigen, and lastly Ras (see Figure 11.5). With just these three oncogenes, cells continue to proliferate and grow well in soft agar or in immunodeficient nude mice. In human cancer six oncogenic mutations seem to be required for full transformation.

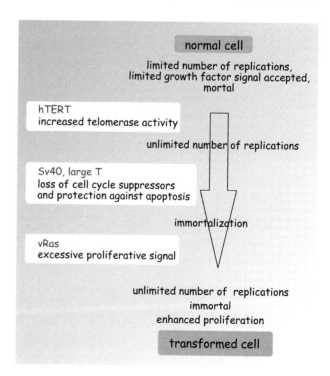

FIG 11.5 Construction of a transformed cell line.

References

1. Wade N. Guillemin and Schally: the years in the wilderness. *Science*. 1978;200:279–282.
2. Wade N. Guillemin and Schally: the three-lap race to Stockholm. *Science*. 1978;200:411–415.
3. Wade N. Guillemin and Schally: a race spurred by rivalry. *Science*. 1978;200:510–513.
4. Ellermann V, Bang O. Experimentelle leukämie bei hünern. *Zentr Bakteriol Parasitenk*. 1908;4:595–609.
5. Warthin AS. *Trans Assoc Am Physicians*. 1904;19:421.
6. Warthin AS. Leukemia of the common fowl. *J Infect Diseases*. 1907;4:369–380.
7. Rous P. A transmissable avian neoplasm. (Sarcoma of the common fowl). *J Exp Med*. 1910;12:696–705.
8. Erikson RL, Purchio AF, Erikson E, Collett MS, Brugge JS. Molecular events in cells transformed by Rous Sarcoma virus. *J Cell Biol*. 1980;87:319–325.
9. Hunter T, Sefton BM. Transforming gene product of Rous sarcoma virus phosphorylates tyrosine. *Proc Natl Acad Sci U S A*. 1980;77:1311–1315.
10. Rous P. Comment. *Proc Natl Acad Sci U S A*. 1967;58:843–845.
11. Levi Montalcini R. *In Praise of Imperfection, My Life and Work* (translation by L Attardi of Elogio dell'imperfezione). New York: Basic Books; 1988.
12. Levi-Montalcini R. *The Saga of the Nerve Growth Factor: Preliminary studies, Discovery, Further Development*. Singapore: World Scientific Publishing; 1997.
13. Hamburger V. Hans Spemann and the organizer concept. *Experientia*. 1969;25:1121–1125.
14. Hamburger V, Levi-Montalcin R. Proliferation, differentiation and degeneration in the spinal ganglia of the chick embryo under normal and experimental conditions. *J Exp Zool*. 1949;111:457–502.
15. Levi-Montalcini R. The origin and development of the visceral system in the spinal cord of the chick embryo. *J Morphology*. 1950;86:253–283.
16. Bueker ED. Implantation of tumors in the hind limb field of the embryonic chick and the developmental response of the lumbosacral nervous system. *Anat Record*. 1948;102:369–389.
17. Cohen S, Levi-Montalcini R. A nerve-growth stimulating factor isolated from snake venom. *Proc Natl Acad Sci U S A*. 1958;42:571–574.
18. Cohen S. The stimulation of epidermal proliferation by a specific protein (EGF). *Dev Biol*. 1964;12:394–407.
19. Cohen S. Nobel lecture. Epidermal growth factor. *Biosci Rep*. 1986;6:1017–1028.
20. Gregory H. Isolation and structure of urogastrone and its relationship to epidermal growth factor. *Nature*. 1975;257:325–327.
21. Ushiro H, Cohen S. Identification of phosphotyrosine as a product of epidermal growth factor-activated protein kinase in A-431 cell membranes. *J Biol Chem*. 1980;255:8363–8365.

22. Downward J, Yarden Y, Mayes E, *et al.* Close similarity of epidermal growth factor receptor and v-erb-B oncogene protein sequences. *Nature.* 1984;307:521–527.

23. Carrel A. On the permanent life of tissues outside of the organism. *J Exp Med.* 1912:516–528.

24. Ross R, Vogel A. The platelet-derived growth factor. *Cell.* 1978;14:203–210.

25. Hamilton D. The Monkey Gland Affair. London: Chatto & Windus; 1986.

26. Ross R, Glomset J, Kariya B, Harker L. A platelet-dependent serum factor that stimulates the proliferation of arterial smooth muscle cells in vitro. *Proc Natl Acad Sci U S A.* 1974;71:1207–1210.

27. Antoniades HN, Scher CD, Stiles CD. Purification of human platelet-derived growth factor. *Proc Natl Acad Sci U S A.* 1979;76:1809–1813.

28. Cooper JA, Bowen PD, Raines E, Ross R, Hunter T. Similar effects of platelet-derived growth factor and epidermal growth factor on the phosphorylation of tyrosine in cellular proteins. *Cell.* 1982;31:263–273.

29. Doolittle RF, Hunkapiller MW, Hood LE, *et al.* Simian sarcoma virus onc gene, v-sis, is derived from the gene (or genes) encoding a platelet-derived growth factor. *Science.* 1983;221:275–277.

30. Waterfield MD, Scrace GT, Whittle N, *et al.* Platelet-derived growth factor is structurally related to the putative transforming protein p28sis of simian sarcoma virus. *Nature.* 1983;304:35–39.

31. Robbins KC, Antoniades HN, Devare SG, Hunkapiller MW, Aaronson SA. Structural and immunological similarities between simian sarcoma virus gene product(s) and human platelet-derived growth factor. *Nature.* 1983;305:605–608.

32. Theilen GH, Gould D, Fowler M, Dungworth DL. C-type virus in tumor tissue of a woolly monkey (*Lagothrix* spp.) with fibrosarcoma. *J Natl Cancer Inst.* 1971;47:881–889.

33. De Larco JE, Todaro GJ. Growth factors from murine sarcoma virus-transformed cells. *Proc Natl Acad Sci U S A.* 1978;75:4001–4005.

34. Childs CB, Proper JA, Tucker RF, Moses HL. Serum contains a platelet-derived transforming growth factor. *Proc Natl Acad Sci U S A.* 1982;79:5312–5316.

35. Roberts AB, Anzano MA, Wakefield LM, Roche NS, Stern DF, Sporn MB. Type β transforming growth factor: a bifunctional regulator of cellular growth. *Proc Natl Acad Sci U S A.* 1985;82:119–123.

36. Sporn MB, Roberts AB, Wakefield LM, de-Crombrugghe B. Some recent advances in the chemistry and biology of transforming growth factor-β. *J Cell Biol.* 1987;105:1039–1045.

37. Roberts AB, Lamb LC, Newton DL, Sporn MB, De Larco JE, Todaro GJ. Transforming growth factors: isolation of polypeptides from virally and chemically transformed cells by acid/ethanol extraction. *Proc Natl Acad Sci U S A.* 1980;77:3494–3498.

38. Kaplan DR, Chao FC, Stiles CD, Antoniades HN, Scher CD. Platelet α granules contain a growth factor for fibroblasts. *Blood*. 1979;53:1043–1052.

39. Grotendorst GR, Chang T, Seppa HE, Kleinman HK, Martin GR. Platelet-derived growth factor is a chemoattractant for vascular smooth muscle cells. *J Cell Physiol*. 1982;113:261–266.

40. Cone JL, Brown DR, DeLarco JE. An improved method of purification of transforming growth factor, type β from platelets. *Anal Biochem*. 1988;168:71–74.

41. Haines KA, Kolasinski SL, Cronstein BN, Reibman J, Gold LI, Weissmann G. Chemoattraction of neutrophils by substance P and transforming growth factor-β1 is inadequately explained by current models of lipid remodeling. *J Immunol*. 1993;151:1491–1499.

42. Postlethwaite AE, Keski Oja J, Moses HL, Kang AH. Stimulation of the chemotactic migration of human fibroblasts by transforming growth factor β. *J Exp Med*. 1987;165:251–256.

43. Nathan C, Sporn M. Cytokines in context. *J Cell Biol*. 1991;113:981–986.

44. Hanahan D, Weinberg RA. The hallmarks of cancer. *Cell*. 2000;100:57–70.

45. Hanahan D, Folkman J. Patterns and emerging mechanisms of the angiogenic switch during tumorigenesis. *Cell*. 1996;86:353–364.

46. Johnsen M, Lund LR, Romer J, Almholt K, Dano K. Cancer invasion and tissue remodeling: common themes in proteolytic matrix degradation. *Curr Opin Cell Biol*. 1998;10:667–671.

47. Serrano M. The tumor suppressor protein p16INK4a. *Exp Cell Res*. 1997;237:7–13.

48. Evan G, Littlewood T. A matter of life and cell death. *Science*. 1998;281:1317–1322.

49. Prescott JC, Blackburn EH. Telomerase: Dr Jekyll or Mr Hyde? *Curr Opin Genet Dev* 1999;9:368–373.

50. Murphy KL, Rosen JM. Mutant p53 and genomic instability in a transgenic mouse model of breast cancer. *Oncogene*. 2000;19:1045–1051.

51. Iiri T, Farfel Z, Bourne HR. G-protein diseases furnish a model for the turn-on switch. *Nature*. 1998;394:35–38.

52. Weitzman JB, Yaniv M. Rebuilding the road to cancer. *Nature*. 1999;400:401–402.

53. Futreal PA, Coin L, Marshall M, Down T, Hubbard T, Wooster R, Rahman N, Stratton MR. A census of human cancer genes. *Nature Rev Cancer*. 2004;4:177–183.

CHAPTER 12

Signalling Pathways Operated by Receptor Protein Tyrosine Kinases

Introduction

Of the 90 genes that code for protein tyrosine kinases in the human genome, 58 are receptors (rPTK) and these are classified into 20 subfamilies. The remaining 32 are of the non-receptor type (nrPTK). Mouse homologues have been identified for nearly all of the human tyrosine kinases.[1] Although absent from yeasts and protozoans,[2] molecules related to the receptors for EGF and for insulin, that have integral catalytic domains, have been identified in marine sponges. It has been suggested that the insulin-receptor-like molecules evolved before the Cambrian Explosion (the name given to the seemingly rapid appearance of most of the major groups of complex animals, ~530 million years ago) and contributed to the rapid emergence of the higher metazoan phyla.[3] With respect to the transduction of signals from cell surface receptors, there are two main families of protein tyrosine kinases (PTKs). Here we consider the rPTKs that exist as integral

The many abbreviations used in this chapter are collected together at the end of the chapter.

domains of transmembrane receptors. In Chapter 17 we discuss the nrPTKs which are present in the cytosol or are attached to the plasma membrane and can be recruited to receptors.

Spotting phosphotyrosine

Here is another turning point in science that owed as much to chance as to the application of a well-prepared mind. This involved Tony Hunter and the discovery of tyrosine phosphorylation of proteins associated with malignant transformation. He was interested in identifying the transforming antigens of the tumour-causing polyoma virus, of which the main component is the so-called middle T-antigen. A report that the src-gene product (v-Src) was associated with protein kinase activity[4,5] begged the question whether other tumour virus gene products might also possess phosphorylating activities, and that this might underlie cell transformation. It became apparent that infection with the polyoma virus induces extensive phosphorylation of cellular protein, but that the transforming protein, middle T-antigen itself, also becomes phosphorylated. After proteolytic digestion of ^{32}P-labelled protein into its component amino acids, the labelled residues were separated by electrophoresis. Unexpectedly, all of the label was confined to a new spot, now known to be due to phosphotyrosine (Figure 12.1).[6]

Since the phosphate–tyrosine bond is comparatively resistant to alkali, the detection of tyrosine phosphorylation was simplified by treating ^{32}P-labelled cellular extracts with 1 M NaOH. More recently antibodies have been developed that specifically recognize individual phosphotyrosine epitopes, and this of course makes detection a very clean affair. The creation of antibodies of high specificity is now a business in its own right.

In a sense this discovery was accidental. It was common practice when carrying out electrophoretic procedures to re-use the buffers on subsequent

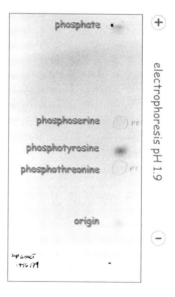

FIG 12.1 Separation of phosphotyrosine from phosphoserine and phosphothreonine by paper electrophoresis. Courtesy Tony Hunter.

occasions. Eventually the pH must alter, the anodic buffer becoming more acidic, the cathodic more alkaline. Had the pH 1.9 electrophoresis buffer been freshly prepared, the separation of phosphotyrosine would not have occurred. In the event, with reuse, it had become more acidic (pH 1.7). The phosphotyrosine migrated more slowly and was separated from the phosphothreonine.

v-Src and other protein tyrosine kinases

With electrophoresis now carried out intentionally at pH 1.7, various other labelled protein digests were tested and it was found that v-Src can phosphorylate tyrosine residues on a range of quite unrelated proteins. This identified v-Src as a protein tyrosine kinase and phosphorylation on tyrosine residues to be an authentic physiological process. Confirmation came with the finding that while labelling in non-transformed cells occurs almost exclusively upon serines and threonines, in cells transformed by Rous sarcoma virus (see page 299), phosphorylated threonine, serine, and tyrosine residues are present in almost equal amounts.

It was also found that the transforming protein of the Abelson murine leukaemia virus (v-Abl) becomes labelled on a tyrosine residue when incubated *in vitro* with ^{32}P-ATP.[7] The target in this case is the tyrosine kinase itself, an example of autophosphorylation. Importantly, a second look at the phosphorylation status of the EGF receptor (see page 299) showed that the labelling due to stimulation occurs on tyrosine, not threonine residues as previously reported.[8] The link between tyrosine phosphorylation and cell proliferation/transformation had been established.

Processes mediated through tyrosine phosphorylation

Tyrosine phosphorylation is not limited to the actions of the transforming viruses or growth factors. It regulates many other important signalling processes including:

- cell–cell and cell–matrix interactions through integrin receptors and focal adhesion sites[9,10] (Chapter 13)
- stimulation of the respiratory burst in phagocytic cells, such as neutrophils and macrophages[11]
- activation of B lymphocytes by antigen binding to the B cell receptor[12]
- activation of T lymphocytes by antigen-presenting cells through the T cell receptor complex[13] (Chapter 17)
- interleukin-2-mediated proliferation of lymphocytes[14,15]
- activation of mast cells and basophils through the high-affinity receptor for IgE[16]
- many developmental processes.

Here, we focus on the signal transduction pathway initiated by the binding of EGF, and other ligands of the EGF family, to receptors on fibroblasts or epithelial cells. A number of the principles also apply for other tyrosine kinase-containing receptors.

Tyrosine kinase-containing receptors

The ErbB receptor family and their ligands

ErbB, so named after the viral oncogene product, Erb-B, of the erythroblastoma virus to which they are related. ErbB1 represents the classic EGF receptor (EGFR). The ErbB family is an evolutionary elaboration of the ancestral gene, Let-23, present in *Caenorhabditis elegans* and its equivalent DER (also known as torpedo) in *Drosophila melanogaster*.

The first RTK discovered was the EGFR.[17] It was also the first receptor that provided evidence for a relationship between activating mutations (oncogenes) and cancer[18] (see also Chapter 11). The EGF receptor is a member of the ErbB family of proteins (ErbBs 1–4). ErbB2 lacks the capacity to interact with a ligand because its extracellular region exists in a fixed, unfolded conformation (reminiscent of the ligand-bound form of the EGFR).[19] ErbB3 lacks kinase activity.[20] These so-called non-autonomous receptors do, however, contribute to growth factor signalling because they can form heterodimeric complexes in which ErbB2 is the preferred dimerization partner. For instance, association of ErbB2 with the EGFR (ErbB1), increases both the intensity and duration of the EGF signal, by increasing the ratio of active kinase to EGF and by inhibiting receptor uptake in clathrin-coated endocytic vesicles.[21] ErbB2 can be said to act as an amplifier of the ErbB signalling network.[22]

Three groups of ligands have been identified that contain one or more copies of a conserved EGF domain (30–40 residues). All of these bind to two or more of the receptors (see Figure 12.2). Many of the ligands, including EGF, are expressed as large transmembrane proteins that can signal in a juxtacrine (cell–cell interaction) or paracrine mode, following proteolytic cleavage of their extracellular domain. In the case of human EGF, only 55 residues constitute the growth factor out of more than 1200 present in its membrane tethered precursor form.

The various ligands have redundant functions, well exemplified by deletion of TGFα, EGF, and amphiregulin. Single knockouts have little effect on developmental processes; indeed, its takes a triple deletion to manifest serious effects in mice.[25,26] By contrast, inactivation of the EGFR results in mid-gestational death.[27]

Cross-linking of receptors causes activation

Receptors containing tyrosine kinase come in several different forms, but they are unified by the presence of a single membrane-spanning domain and an intracellular kinase catalytic domain. The extracellular chains vary considerably, as indicated in Figure 12.3.

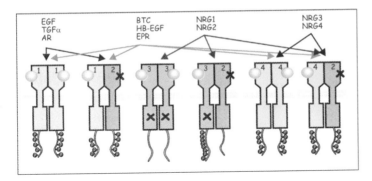

FIG 12.2 ErbB receptor family, their ligands, and some of their adaptors and effectors. Each receptor dimer recruits a different set of adaptors and effectors. A non-exhaustive list is given in Table 12.1. Abbreviations are listed at the end of the chapter. Adapted from Hynes and Lane[23] and Jones et al.[24]

TABLE 12.1 Adaptors and effectors of the ErbB family of receptors

ErbB1 (EFGR)	ErbB2	ErbB3	ErbB4
Shc	Shc	Shc	Shc
Syk	Syk	Syk	Syk
RasGAP	RasGAP	RasGAP	
Grb2	Grb2/7	Grb7	
Abl	Abl		Abl
DOK	DOK		
IRS	IRS		
Vav	Vav		
MAPK8	MAPK8		
PLCγ	PLCγ		
STAT1/3	STAT1/3		
PI 3-kinase		PI 3-kinase	
SHP2			
APBB			
		Crk	
		Nck	
		JAK	

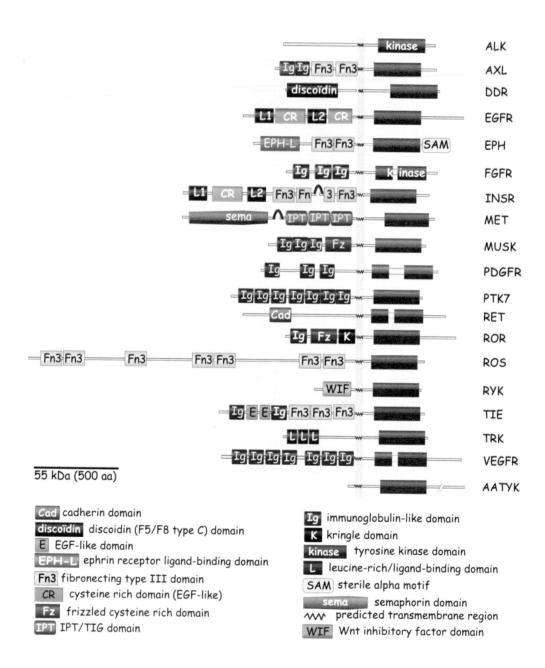

FIG 12.3 Classification of receptors containing tyrosine protein kinase. All these receptors possess a single membrane-spanning segment and all of them incorporate a tyrosine kinase catalytic domain, in some cases interrupted by an insert. The extracellular domains vary as indicated. The domain architectures are obtained from Pfam. Abbreviations are listed at the end of the chapter. Adapted from Robinson et al.[1]

A general feature of these receptors is that ligand binding causes dimerization. In addition to activation by the natural peptide ligands, some (but not all) of the functions of the EGF receptor can be elicited by cross-linking with antibodies.[28,29] Crosslinking is achieved in a number of ways. Platelet derived growth factor (PDGF) is itself a disulfide-linked dimeric ligand which cross-links its receptor upon binding. EGF is monomeric. Its binding causes the receptor to unfold, exposing a dimerization loop that allows the occupied monomers to recognize each other[30,31] (Figure 12.4). Since truncated EGF receptors that lack the extracellular ligand-binding domain are constitutively active, it follows that the unoccupied extracellular domain acts to prevent kinase activation. The insulin receptor is a dimeric molecule, already cross-linked by default, yet it still requires the attachment of a ligand in order to become activated. This permits a lateral shift and approach of the two transmembrane domains which brings the catalytic domains within reach of each other.[32,33]

To complicate matters further, dimerization may also involve direct contact between the helical transmembrane structures of the receptors. Particular mutations within the transmembrane region cause constitutive dimerization and this, in the case of the ErbB2 receptor (also known as the HER-2/neu oncogene), can induce cell transformation.[34,133]

The activation signals are, of course, more complicated than this. For activation of all the receptor functions, not only must the receptors be brought together as dimers, but they must also undergo appropriate conformational changes in relation to each other, in a way that increases the affinity for ATP within the catalytic domain and allows access of substrate.[35–37]

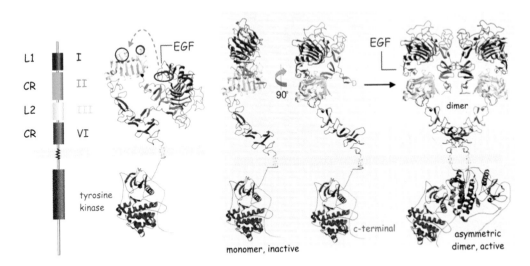

FIG 12.4 Activation of the EGF receptor. The EGF receptor is composed of four extracellular domains (I–IV) of which I and III (also called L1 and L2) are leucine-rich repeats that function in ligand binding. II and IV (also called CR1 and CR2) are furin-like, cysteine-rich domains. Binding of EGF causes the receptor to dimerize. The receptor unfolds, a dimerization arm of domain II binds to a docking site at the base of domain II of the partner, allowing close approach and activation of the two kinase domains. The first substrates are tyrosine residues in the C-terminal of the receptor itself (transphosphorylation). The asymmetrically dimerized, phosphorylated molecule constitutes the catalytically active receptor. (2gs6,[39] 1nql,[40] 1ivo[41]). Adapted from Zhang et al.[39] and Ferguson et al.[40]

Ligand binding allows both kinase domains to encounter target sequences on the partner, so enabling intermolecular cross-phosphorylation (transphosphorylation) of tyrosine residues (Figure 12.4). The phosphorylated dimer then constitutes the active receptor.

Two types of phosphorylation sites can be distinguished. One concerns those tyrosine residues that, when phosphorylated, render the tyrosine kinase catalytically competent. The other concerns tyrosines, often in the region of the C-terminus that act as docking sites for other proteins. These phosphotyrosines bind adaptors and effectors bearing SH2 or PTB domains (see below and Chapter 24) to assemble receptor signalling complexes[24,38] (Figure 12.5). In addition, dimerized and phosphorylated receptors have the potential to phosphorylate their targets.

Growth factor receptor dimers can further aggregate into oligomers of several hundreds or even thousands of units. This phenomenon, already

Adaptors are composed of protein interaction domains (SH2, SH3, PTB, etc.) and are without catalytic activity. They act to bring components of signalling pathways into close proximity. Examples are Grb2 and Drk: see page 332. By definition, effectors are catalysts, kinases, phospholipases, etc.

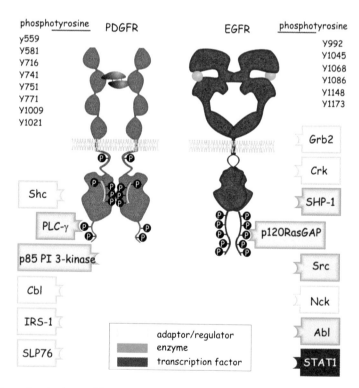

FIG 12.5 Phosphorylation and formation of receptor signalling complexes. Activated EGF and PDGF receptors associate with numerous effectors, including effector enzymes (PLC-γ, GAP etc.), adaptor proteins (p85 subunit of PI 3 kinase, Grb2, etc.) and transcription factors (STATs) to form receptor signalling complexes. See also Table 12.1.

recorded at the time when receptor dimerization first came to light,[42] has now been visualized by the use of fluorescent protein tagging.[43] Different types of receptors can be recruited, so that PDGF receptors have been found associated with EGF receptors[44] and these aggregates give rise to multiprotein signalling platforms that may contain numerous effectors.

Receptor dimerization, the concomitant conformational changes, and the prolonged juxtaposition are prerequisites for a successful transphosphorylation reaction. In the case of the insulin or PDGF receptor, the first phosphorylations occur in the activation segment (see page 782) and from here on the kinase is activated, able to phosphorylate other substrates. Surprisingly, the EGFR (ErbB1) has only a single tyrosine phosphorylation site in the activation segment, but its phosphorylation is without effect on kinase activity. It appears that catalytic competence requires the formation of an asymmetric dimer, with one kinase domain activating the other. Cross-linking increases the probability of a productive interaction between the kinase domains.[39,45]

Assembly of receptor signalling complexes

The formation of signalling complexes and the subsequent signal transduction events have been studied in the following ways.

Measurement of enzyme activity, second messenger production and tyrosine-phosphorylation

The activated receptors for EGF and PDGF stimulate PLCγ. This results in the generation of DAG and IP3 and leads within seconds to the activation of protein kinase C and a rise in the concentration of cytosol free Ca^{2+}.[46–48] All of these can be measured. Furthermore, PLCγ itself becomes phosphorylated on tyrosine residues (see page 153) indicating that it interacts directly with the catalytic domain of the receptor. Activation of PDGF and insulin receptors also causes activation of phosphatidylinositol 3-kinase (PI 3-kinase).[49] This phosphorylates $PI(4,5)P_2$ forming $PI(3,4,5)P_3$ (see Chapter 18). In addition, a number of serine/threonine kinases are also activated. These include ribosomal S6-kinase (implicated in protein synthesis), C-Raf kinase (see below), and the mitogen-activated protein kinases (MAP kinase, see below). Most importantly, Ras becomes activated.[50–52]

Coimmunoprecipitation of proteins with activated receptors

To investigate the specific interactions of activated receptors, cells pre-labelled with ^{35}S-methionine are stimulated and then solubilized with detergent. The receptors, together with any associated proteins, are precipitated using an anti-receptor antibody. The associated proteins are detected by gel electrophoresis and autoradiography. Identification is achieved by microsequencing, immunoblotting, and other techniques. This was how the associations of protein tyrosine kinase receptors with Ras GAP, PLC-γ, and PI 3-kinase were originally demonstrated.[53,54] Using a similar approach, but with cell lysates and purified receptors, it was shown that the p85 subunit of PI 3-kinase binds to the PDGF receptor.[55]

Detecting protein association in a cell-free system and cloning of receptor targets

Proteins expressed by a lambda-phage library in a bacterial host are screened for binding to the cytoplasmic domain of a receptor, labelled with ^{32}P-phosphate. The relevant bacterial clones are identified using autoradiography and the DNA sequence of the phage insert is determined. This was how the association of the EGF receptor with the adaptor Grb2 and the p85-subunit of PI 3-kinase were discovered.[56]

Detecting protein association in microarrays

Large-scale protein microarrays have shown that phosphopeptide binding to SH2 or PTB domains occurs even when the peptide sequence does not match the consensus sequence for the binding site. Furthermore, it has become apparent that increasing the phosphopeptide concentration (the equivalent of increasing the expression levels of cell surface receptors) increases the variety of SH2/PTB domains that are recognized. This suggests that the level of receptor expression regulates not only the intensity of the signal, but also the range of effector proteins that are activated.[24]

Detecting protein–protein interaction in a yeast two-hybrid assay

This method has shown that insulin receptor substrate-1 (IRS-1) binds through its PTB domain to a phosphorylated tyrosine residue present in the juxtamembrane region of the receptor.[57] The two-hybrid assay has also been instrumental in showing that H-Ras interacts directly with a conserved 81-residue segment at the N-terminus of the serine/threonine kinase B-Raf.[58]

Protein domains that bind phosphotyrosines and the assembly of signalling complexes

The assembly of receptor signalling complexes depends on protein–protein interactions that involve phosphorylated growth factor receptors. The first clues to the mechanism came from studies of p47$^{gag-crk}$, a transforming protein devoid of any catalytic activity.[60] It relies on the presence of a motif similar to a conserved region of Src (and many other tyrosine kinases), designated Src homology-2 (SH2) that interacts with tyrosine-phosphorylated proteins.[60,61] Indirect evidence for a role of SH2 domains in interactions with phosphorylated growth factor receptors emerged with the observation that all proteins that operate downstream of these receptors, or that are involved in cell transformation, possess one or two copies of this domain. Significantly,

Yeast two-hybrid assay. Transcription of the ectopic bacterial *lacZ* gene in engineered yeast utilizes the transcription factor GAL4 which codes for β-galactosidase. In the assay, this factor is substituted by two proteins: the DNA-binding segment of GAL4, fused to the protein of interest, and the transcription machinery-binding segment of GAL4 fused to fragments of proteins derived from a cDNA library. Hundreds of yeast clones expressing different cDNAs are thus created. Those that exhibit β-galactosidase activity (easily detected) are selected and analysed for the inserted cDNA.[59]

PLCγ, the only isoform directly activated by the EGF receptor, possesses an SH2 domain (see page 299).[38] Further evidence was provided by the finding that SH2 domains bind to phosphotyrosine ligands, including those present in activated EGF receptors.[62] The basic structure and properties of SH2 domains are described in Chapter 24 (see page 768).

Another domain, structurally unrelated, that binds to phosphotyrosines, also present in many signalling molecules, is the PTB (phosphotyrosine-binding) domain (see page 771). This was first identified in the adaptor protein Shc, which also contains an SH2 domain.[63] These protein interaction domains recognize phosphotyrosines within specific oligopeptide motifs. The optimal sequences of these short stretches are similar for the same type of domain, but those recognized by SH2 and PTB domains differ considerably.[64,65]

Many proteins having SH2 or PTB domains (or both) associate with rPTKs to initiate the assembly of signalling complexes (Figure 12.6 and Figure 12.7). Some of these proteins themselves then become phosphorylated as a result of this association. For PLCγ, this is a necessary activation step, but it is not clear whether this is the case for other effectors.

The human genome has 109 proteins predicted to have SH2 domains and 44 with PTB domains.[24] Some of these occur as tandem pairs (Figure 12.6). Most of the proteins that have these domains are either adaptors with no catalytic activity, or effectors such as phospholipases and protein/lipid kinases; others are transcription factors.

The proteins that associate with EGF receptors exhibit preferences for the sequence of amino acid residues in the close vicinity of the phosphorylated tyrosine. In consequence, for a given motif some proteins bind more tightly than others. This enables receptors to signal through panels of SH2-containing proteins. For example, a screen of SH2 and PTB domains (66 phosphopeptides) derived from the ErbB family, revealed that some could bind a wide range of effectors (such as PI 3-kinase, phospholipase Cγ, the tyrosine kinase Syk, the adaptors Shc1 and Crk, and RasGAP), while others are more specific.[24] Significantly, for the EGFR and ErbB2 (but not ErbB3), the interaction becomes less selective as the concentration (in the assay) of the phosphopeptide ligand increases. Thus, experimental over-expression of these rPTKs broadens the possible range of downstream signalling pathways. High expression of the EGFR or ErbB2 (but not ErbB3) certainly occurs in breast cancer (see Chapter 23).

Branching of the signalling pathway

A number of signal transduction pathways branch out from the signalling complex (Figure 12.8). Two such branches are described in the following paragraphs and others are considered later (STATs in Chapter 17, PI 3-kinase in Chapter 18).

Gag: glycosylated antigen, the gene encoding the internal capsid of the viral particle.

Crk: C10 regulator of kinase. This is an SH2 and SH3 domain-containing adaptor protein and is implicated in pathogenesis of chronic myelogenous leukemia.

p47$^{gag-crk}$ is a good example of how viruses disrupt genes resulting in chimeric proteins. In this case, viral Gag sequences are fused to the cellular CrkL. The mammalian cells express the gene product once it is inserted into their genome.

The affinity (K_D) of SH2 or PTB domains for tyrosine-phosphorylated peptides is not particularly strong, of the order of $2\,\mu mol\,L^{-1}$. In comparison, the affinity of growth factors for their receptors is typically $K_D \sim 0.1\,\mu mol\,L^{-1}$. As a consequence, the interactions have a half-life of less than 10 s, so that within the duration of receptor responses (minutes), numerous adaptor and effector complexes form, detach and reform.[24] This dynamic interplay allows tyrosine phosphatases to intervene, removing the phosphoryl groups from the rPTKs (see Chapter 21).

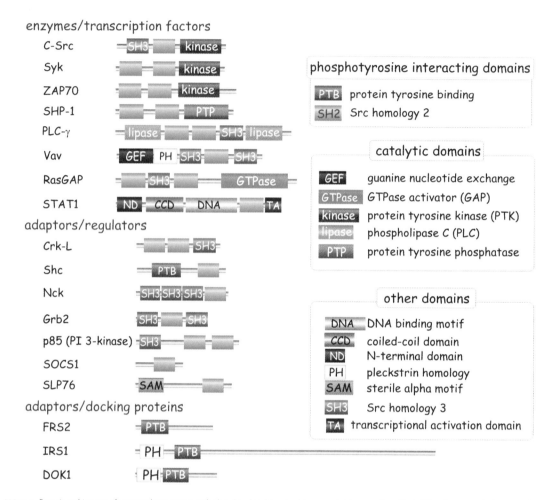

FIG 12.6 Domain architecture of proteins that associate with phosphorylated tyrosine kinase-containing receptors. Most of the effectors and adaptors and all docking proteins are also substrates of the receptor kinases. Phosphorylated docking proteins offer novel phosphotyrosine docking sites for yet other adaptors and effectors. Many of the proteins presented in this figure are discussed elsewhere in this book. Abbreviations are listed at the end of the chapter.

The PLCγ–PKC signal transduction pathway

Among the activities set in train by activation of the EGF and PDGF receptors is the generation of DAG and IP$_3$ by PLCγ. DAG remains in the membrane and acts as a stimulus for PKC. The consequence is the transformation of a phosphotyrosine signal, through activation of PLCγ, into a phosphoserine/phosphothreonine signal.

One of the first substrates of PKC is the EGF receptor itself. It becomes phosphorylated on a serine situated close to the transmembrane domain,

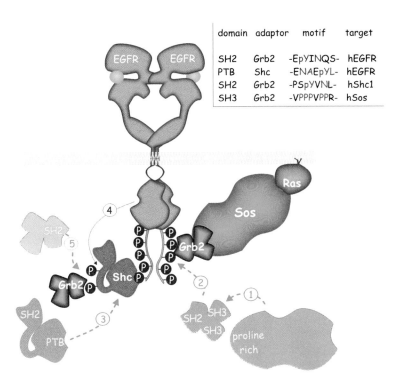

domain	adaptor	motif	target
SH2	Grb2	-EpYINQS-	hEGFR
PTB	Shc	-ENAEpYL-	hEGFR
SH2	Grb2	-PSpYVNL-	hShc1
SH3	Grb2	-VPPPVPPR-	hSos

FIG 12.7 Formation of receptor signalling complexes. SH2 and PTB domains bind to specific motifs containing a phosphotyrosine. (1) The SH3 domain of Grb2 attaches to a proline rich sequence in Sos, a guanine nucleotide exchange factor. (2) The SH2 domain of Grb2 binds the tyrosine phosphorylated EGFR, leading to recruitment of the adaptor/effector complex to the membrane, there to find its target, Ras. (3) An alternative route for recruitment of Grb2 occurs through the intervention of Shc. The PTB domain of Shc binds the tyrosine phosphorylated EGF. (4) This leads to its phosphorylation, (5) creating a novel binding site for Grb2. The discovery of Grb2 as an adaptor and Sos as a guanine nucleotide exchange factor for Ras is described below on page 332. Key residues within the binding motifs are shown in red in the table.

causing its inactivation. Although numerous proteins have proved to be substrates of the PKC enzymes, we still lack full understanding of how these kinases determine the changes in gene transcription that occur after stimulation. As discussed in Chapter 19 (see page 578), one consequence of PKC activation is actually *de*phosphorylation of c-Jun, a component of the inducible transcription factor AP-1.

The Ras signalling pathway

From the tyrosine kinase to Ras

For 40 years it has been known that infection of rats with murine leukaemia viruses can provoke the formation of a sarcoma.[66] A major advance was

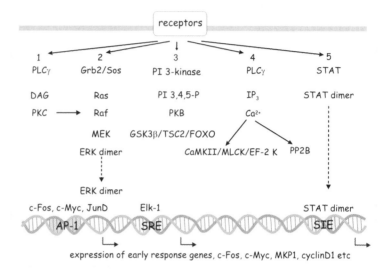

FIG 12.8 Branching of the signal transduction pathways. Following activation of a receptor PTK, several signal transduction pathways can be activated. Five are indicated.

the discovery that the Harvey murine sarcoma virus encodes a persistently activated form of H-ras in which valine is substituted for glycine at position 12 (see page 106). Expression of this mutant in quiescent rodent fibroblasts results in altered cell morphology, stimulation of DNA synthesis and cell proliferation.[67] When over-expressed, (normal) H-Ras also induces oncogenic transformation,[68] as does micro-injection of the mutant protein.[69] Conversely, injection of neutralizing antibodies to inhibit normal Ras function reverses cell transformation.[70] Furthermore, stimulation of quiescent cells with serum or with growth factors promotes the binding of GTP to Ras.[71] These findings showed that Ras is an important component in the signalling pathways that regulate cell proliferation, but how it fits into the known pathways emanating from growth factor receptors remained unclear for a considerable time. The first clues came from genetic analysis of signal transduction pathways that operate in the invertebrates *Drosophila melanogaster* and *Caenorhabditis elegans* (in plain English, flies and worms).

Photoreceptor development in the fruit fly

The compound eyes of insects are arrays of small hexagonal units called *ommatidia*. The fruit fly has approximately 800 of these 'small eyes', arranged as shown in Figure 12.9. Each is composed of eight photoreceptor cells (R1–R8) and 12 accessory cells. On the basis of their morphology, order of development, axon projection pattern, and spectral sensitivity, the photoreceptor cells can be classified into three functional classes. R8 is the first to appear, followed

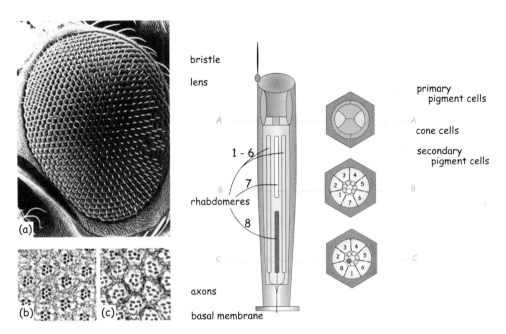

FIG 12.9 The *sevenless* mutation in fly eyes. The events leading to the development of cell R7 in *Drosophila* eye have provided a key to understanding the pathways initiated by rPTKs. Genes acting downstream of the *sevenless* receptor were revealed by screening for mutations that affect the development of cell R7. (a) A scanning electron microscope image showing the geometrical arrangement of ommatidia, units of eight photoreceptor cells. (b), (c) Thin sections, representative of cuts through section B–B in the drawing. Note that in (b), taken from a wild-type fly, seven cells are evident, while in (c), taken from a fly having the *sevenless* mutation, there are only six. The drawing illustrates the basic anatomy of a single ommatidial unit in longitudinal section. Sections cut at A, B, and C are shown in transverse section on the right. Since two of the cells, R7 and R8, do not extend the full length of the ommatidial unit, the transverse sections B–B and C–C only reveal seven, not all eight cells. For further information, see http://flybase.bio.indiana. edu/ Picture adapted from Dickson and Hafen.[73]

by R1–R6 and then R7. The photosensitive pigment resides in a stack of membranes, the rhabdomere. The larger rhabdomeres of cells R1–R6 are arranged as a trapezoid surrounding the rhabdomeres of cells R7 and R8. The R8 rhabdomere is located below R7 (Figure 12.9). The development of R7 requires the products of two genes, *sevenless* (*sev*) and *bride-of-sevenless* (*boss*). The phenotypes generated by loss-of-function mutations in either of these genes are identical, R7 failing to initiate neuronal development. These mutations are readily detected in a behavioural test. Given a choice between a green and a UV light, normal (WT) flies will move rapidly towards the UV source.[72] Failure to develop cell R7, the last of the photoreceptor cells to be added to the ommatidial cluster, correlates with the lack of this fast phototactic response and the flies move towards the green light.[72]

While the *sev* product is required only in the R7 precursor, *boss* function must be expressed in the developing R8. Cloning revealed the Boss product as a

329

100 kDa glycoprotein with seven transmembrane spans that is related to the metabotropic receptors (see page 63).[74] Although ultimately expressed on all of the photoreceptor cells, at the time that R7 is being specified it is only present on the oldest, R8.[75] The product of the *sev* gene is an rPTK (member of the insulin receptor family).[76] Evidence for direct interaction between these two gene products came with the demonstration that cultured cells expressing the *boss* product tend to form aggregates with cells expressing *Sev*.[75]

It is now understood that the binding of Boss (the ligand) to Sev (the rPTK) leads to the activation of kinase activity and that this ultimately determines the fate of R7 as a neuronal cell. Since a reduction in the gene dosage of the fly Ras1 impairs signalling by Sev, and persistent activation of Ras1 obviates the need for the *boss* and *sev* gene products, it follows that the activation of Ras1 is an early consequence of Sev activity.[77]

Further genetic screens of flies expressing constitutively activated Sev led to the identification *Drk* and *Sos* as intermediate components of this pathway (see column 1 of Figure 12.10). The Sos protein shows substantial homology with the yeast CDC25 gene product, a guanine nucleotide exchange factor for RAS.[78] Whereas reduction in the gene dosages of *Drk* and *Sos* impairs the

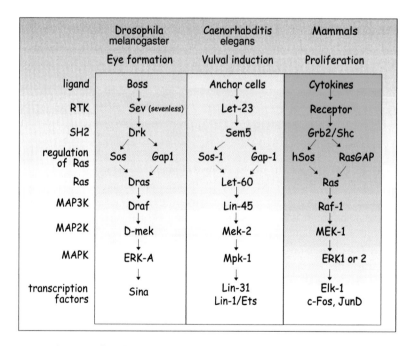

	Drosophila melanogaster	Caenorhabditis elegans	Mammals
	Eye formation	Vulval induction	Proliferation
ligand	Boss	Anchor cells	Cytokines
RTK	Sev (sevenless)	Let-23	Receptor
SH2	Drk	Sem5	Grb2/Shc
regulation of Ras	Sos　　Gap1	Sos-1　　Gap-1	hSos　　RasGAP
Ras	Dras	Let-60	Ras
MAP3K	Draf	Lin-45	Raf-1
MAP2K	D-mek	Mek-2	MEK-1
MAPK	ERK-A	Mpk-1	ERK1 or 2
transcription factors	Sina	Lin-31 Lin-1/Ets	Elk-1 c-Fos, JunD

FIG 12.10 Comparison of signalling pathways activated by a receptor protein tyrosine kinase in species from different phyla. The striking homologies that exist between the proteins operating downstream of rPTKs in distant phyla has enabled the elucidation of the EGF receptor pathway.

signal from constitutively activated Sev, there is no effect on signalling from constitutively activated Ras. Thus, in the pathway of activation, this places the functions of the *Drk* and *Sos* products into positions intermediate between Sev and Ras. The *Drk* gene codes for a small protein having no catalytic activity, but consists exclusively of Src homology domains, two SH3 flanking a single SH2 domain. An adaptor, it binds to the tyrosine-phosphorylated receptor and links it to the proline-rich domains of Sos.[79]

Vulval cell development in worms

In the nematode *C. elegans*, a similar pathway of activation involving autophosphorylation of an rPTK leads to activation of the GTPase Let-60, a homologue of Ras (column 2 of Figure 12.10). This determines the development of vulval cells (Figure 12.11). Again, these proteins were first identified by genetic analysis of mutants, namely lethal mutations (let), mutations affecting vulval development (sem, sex muscle mutants) and alterations in cell lineage (lin, lineage mutants).[80] They constitute the components of a signal transduction pathway based on Lin-3 (a product of the anchor cell), Let-23 (a tyrosine kinase receptor of the p6 p cell), and Sem-5 that associates with a (Sos-like) guanine nucleotide exchange protein. This brings about nucleotide exchange on Let-60 (Figure 12.10). Lin-3 and Let-23 are, respectively, members of the neuregulin (ligand) and ErbB (receptor) family.

More than 20 000 individual species of roundworm in the phylum *Nematoda* (named from the Greek for 'thread-like') have so far been described, though it has been suggested that there might be more than 500 000. They inhabit all terrains: sea and fresh water, the polar regions, the tropics, and deep oceanic trenches. A large proportion are parasitic, including pathogenic forms that affect plants and animals.

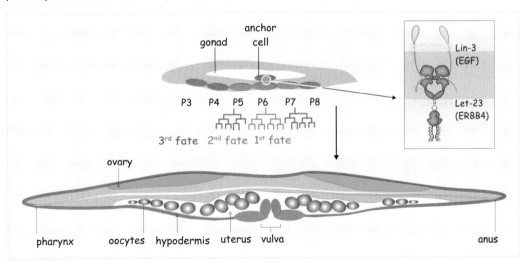

FIG 12.11 Vulval development in *C. elegans*. Because it is a relative simple structure, formed from just a few cells, the vulva is well suited for the genetic analysis of cell differentiation during embryological development. It is the product of just three cell lineages, the descendents of cells p5.p, p6.p, and p7.p. Development is initiated by a signal from the anchor cell that lies adjacent to p6.p. The ligand, lin-3 (homologue of EGF), produced by the anchor cell, binds its receptor Let-23 (homologous to the EGF-R) on the surface of cell p6.p. Cell p6.p in turn releases signals to its neighbours, p5.p and p7.p, and initiates a sequence of events involving the MAP kinase pathway, which determines the fate of these cells as components of vulval tissue. For more information, consult http://www.wormbook.org/

Neuregulin 1 (NRG1) and its receptors ErbB are implicated in the pathophysiology of schizophrenia. Among NRG1 receptors, ErbB4 plays crucial roles in neural development and in the modulation of NMDA receptor signalling. In the prefrontal cortex of schizophrenic individuals, there is a marked increase in NRG1-induced activation of ErbB4, though expression appears unaltered. NRG1 stimulation suppresses NMDA receptor activation and this is more pronounced in schizophrenic subjects.[81]

The hermaphroditic *C. elegans* is a free-living roundworm, ~1 mm in length, that inhabits temperate soils. It is used extensively as a model organism. The entire genome has been sequenced, and the developmental fate of every one of the 959 somatic cells is known. When crowded or starved of food, *C. elegans* enter a larval stage called the dauer state (see page 608). Dauer larvae are resistant to stress and do not undergo ageing.

In both worms and flies, the Ras protein acts as a switch that determines cell fate. In *C. elegans*, the activation of Ras determines the formation of vulval as opposed to hypodermal (skin) cells. In *Drosophila* photoreceptors, the activation of Ras determines the development of R7 as a neuronal as opposed to a cone cell. In both cases, Ras proteins operate downstream of rPTKs that are activated by cell–cell interactions.

Regulation of Ras in vertebrates

The elucidation of the Ras pathway in vertebrates was based on the identification of proteins having sequence homologies with those present in *Drosophila* and *C. elegans* (column 3 of Figure 12.10).[82–84] Expression or microinjection of these proteins (and appropriate reagents such as peptides, antibodies, etc.) were used to restore or modulate the activity of this pathway in cells derived from mammals, flies, or worms, bearing loss-of-function mutations. A vertebrate protein Grb2, lacking catalytic activities but having SH2 and SH3 domains, was found to be capable of restoring function in Sem-5 deficient mutants. In addition, Grb2 was found to associate with a protein that is recognized by an antibody raised against the *Drosophila* protein, Sos. In this way the sequence of events became apparent. Grb2 is an adaptor protein, linking the phosphorylated tyrosine kinase receptor to the guanine nucleotide exchanger in vertebrates (Figure 12.12). The mammalian Sos homologue, hSos, is likewise a guanine nucleotide exchange factor which interacts with Ras.[85] Grb2 is composed exclusively of Src homology domains, one SH2 flanked by two SH3 domains, similar to the *Drosophila* adaptor protein Drk. Because of the nature of the interaction of SH3 with proline-rich sequences (see Chapter 24), it is likely that Grb2 and Sos remain associated even under non-stimulating conditions. The main effect of receptor activation is to ensure the recruitment of the Grb2/Sos complex to the plasma membrane.[86]

hSos comes in two flavours, 1 and 2. They both interact through their proline-rich C-termini with the SH3 domains of Grb2. The binding affinity of hSos2 for Grb2 is significantly higher than that of hSos1.[87]

The adaptor protein Shc binds through its PTB domain to phosphorylated receptors and, on phosphorylation, becomes an indirect docking site for Grb2/Sos (see Figure 12.7). A third component, Gab1, may also be involved upstream in the relay of receptor signalling. This large docking protein contains multiple tyrosines that are directly phosphorylated by activated EGF and insulin receptors. High-level expression of Gab1 enhances cell growth and facilitates cell transformation.[88]

Warning: we stress again that investigations using high-expression vectors can reveal signal transduction pathways that do not necessarily operate under physiological conditions.

It is possible that yet other adaptor/docking proteins will come to light, but the principle of action remains the same. Their roles may vary between cell

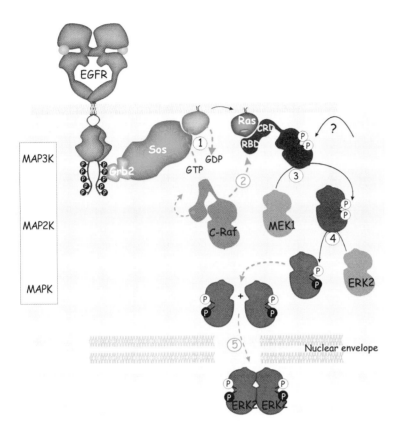

FIG 12.12 Regulation of the Ras-MAP kinase pathway by receptor protein tyrosine kinases. The adaptor Grb2, in association with Sos, attaches to the tyrosine-phosphorylated receptor through its SH2 domains. This brings the Grb2/Sos complex into the vicinity of the membrane to catalyse guanine nucleotide exchange on Ras. The activated Ras associates with the serine/threonine kinase C-Raf. Its presence at the membrane results in activation and then phosphorylation of the dual-specificity kinase MEK1, which in turn phosphorylates ERK2 on both tyrosine and threonine. Dimerization allows ERK2 to interact with proteins that guide it to the nucleus. Protein–protein interaction domains (RBD and CRD: see text) are in indicated in yellow letters.

types or may depend on the stimulus and its timing. It may also depend on the number of receptors, the concentration of the ligand and the cytosolic concentration of the different adaptors.

As already pointed out (page 108), the Ras-GTPase activating protein RasGAP also contains two SH2 domains and it too binds to phosphotyrosines on activated receptors. It is also a component of the signalling complex that assembles on activated PDGF receptors. It remains unclear what, if any, role this association of GAP with the receptor has in relaying signals. For instance, cells that express a mutant of the PDGF receptor that is unable to bind RasGAP, manifest normal activation of Ras.[89]

From Ras to MAP kinase and the activation of transcription

The events following the activation of mammalian Ras lead to the activation of a series of kinases culminating in ERK (extracellular signal regulated kinase). ERK was originally recovered as a serine/threonine phosphorylating activity present in the cytosol of EGF-treated cells and given the name mitogen-activated protein kinase, MAP kinase, or MAPK.[90] (Of the five isoforms, ERKs 1, 2, and 5 contribute to the mitogen signal transduction pathway described here (Table 12.2 and see page 335). Once activated, ERK can enter the nucleus to activate early response genes. All this operates quite independent of second messengers and, in some cell types, cyclic AMP actually opposes it (see below).[91] The earlier steps in the pathway involve the phosphorylation of each target kinase on its activation segment (Figure 12.12).

MEK is the substrate of anthrax lethal factor, LF, one of several toxins produced by *Bacillus anthracis.* It resembles a metalloprotease. By cleaving the N-terminal region of MEKs 1 and 2, LF inhibits their kinase activity and prevents signal transmission through the ERK pathway.[92,93] Whether this is the cause of lethality in anthrax infection is not clear.

The immediate activator of ERK is MEK (MAP kinase-ERK kinase, Figure 12.10). The MEKs comprise seven members, of which only MEKs 1, 2, and 5 are involved in the activation of ERK. MEKs are notable in that they act as 'dual-specificity protein kinases', phosphorylating ERKs 1 and 2 on both a threonine and a tyrosine residue. These are present in the target sequence L**TEY**VATRWYR*APE* (Table 12.2), which constitutes the activation segment. To date, the ERKs appears to be the unique substrate for phosphorylation by MEKs 1 and 2, indicating a particularly high level of specificity.[94]

Upstream of MEK, the first kinase in the cascade is C-Raf (also described as MAP-kinase-kinase-kinase or MAP3K, Figure 12.10). Like Ras, C-Raf was initially identified as an oncogene product (v-raf).[95] It phosphorylates MEK at two serine residues in the activation segment, giving rise to a catalytically competent kinase. The subsequent finding that activated Ras recruits C-Raf to the membrane and in consequence brings about kinase activation, links ERK with the Ras pathway.[96–98] Both the Ras-binding domain (RBD) and the cysteine-rich domain (CRD), present in the N-terminal region of C-Raf, are instrumental for this function. Full activation also requires association with activated Ras.[99,100] A mutant form of C-Raf, endowed with a C-terminal -Caax box that acts as a site for prenylation (page 105), and which is permanently associated with the plasma membrane, can instigate the downstream events independently of Ras (at least in part).[101]

Activation of C-Raf is complicated and still not clearly understood. It is associated with several proteins, among which are the serine/threonine phosphatase PP2A, the heat shock protein Hsp90, the scaffold protein 14-3-3, and the cochaperone Hsp50/Cdc37.[102] An essential feature is that the

TABLE 12.2 Mammalian MAP kinases (human unless otherwise indicated)

MAP kinase	Other names	P site motif in activation segment	% Seq. ident. to ERK2	UniProtKB/SwissProt code
ERK1	MK03, p44 MAPK	TEYVATRWYRAPE	88	P27361
ERK2	MK01, p42 MAPK	TEYVATRWYRAPE	100	P28482
ERK3α	MK06-Rat	SEGLVTKWYRSPR	43	P27704
ERK3β	MK06, p97 MAPK	SEGLVTKWYRSPR	42	Q16659
ERK4	MK04, p63 MAPK	SEGLVTKWYRSPR	42	P31152
ERK5	MK07, BMK1	TEYVATRWYRAPE	–	Q13164
ERK7	MK15-Rat	TEYVATRWYRAPE	–	Q9Z2A6
JNK1	MK08, SAPKγ	TEYVVTRYYRAPE	40	P45983
JNK2	MK09, SAPKα	TEYVVTRYYRAPE	41	P45984
JNK3	MK10, SAPKβ	TEYVVTRYYRAPE	40	P53779
P38α	MK14, CSBP, p38	TGYVATRWYRAPE	50	P16539
P38β	MK11, p38-2	TGYVATRWYRAPE	47	Q15759
P38γ	MK12, ERK6, SAPK3	TGYVVTRWYRAPE	44	P53778
P38δ	MK13, SAPK4	TGYVVTRWYRAPE	42	O15264

Adapted from Chen et al.[94]

N-terminal region, comprising the CRD and RBD domains, hinders the activity of the catalytic domain (CR3). Removal of this restraint requires a number of modifications, of which two will be mentioned. C-Raf is maintained in its inactive state by a locked conformation imposed by 14-3-3, tightly bound at two phosphoserine residues. The association of C-Raf with RasGTP and the plasma membrane requires dephosphorylation of a serine residue by PP2A[103] and this enables it to escape the inhibitory constraint of 14-3-3. It also renders it susceptible to phosphorylation at other residues, especially in the negatively charged region adjacent to the kinase domain (see Figure 12.13) and in the activation segment. The kinases that mediate these phosphorylations have yet to be identified. B-Raf lacks the N-region phosphorylation sites and only requires phosphorylation at the activation segment for full catalytic activity. Essential to all this is that 14-3-3 remains attached.[104]

Raf genes

The *raf* oncogenes came to light as acquired oncogenes present in the murine retrovirus 3611-MSV[105] and the avian retrovirus Mill-Hill 2.[106] Homologues

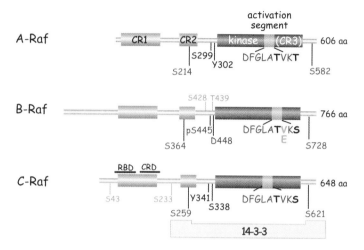

FIG 12.13 Domain architecture of the Raf proteins. The Raf isoforms, A-, B- and C-Raf, share three conserved regions, CR1, CR2, and CR3 (kinase domain, red, with two phosphorylation sites in the activation segment). CR1 contains the Ras binding (RBD) and the cysteine-rich (CRD) domains, both necessary for membrane recruitment. CR2 contains one of the 14-3-3 phosphoserine binding sites, the other is in the C-terminal region. Note that B-Raf differs with respect to the acidic regulatory region between CR2 and CR3, lacking a tyrosine which is replaced by aspartate (D448) and the S445 being constitutively phosphorylated. The V559E mutation in the activation segment of B-Raf, which renders the kinase highly active, is coloured green. Adapted from Wellbrock et al.[107]

of these proteins are present in *Drosophila* (*D-raf*) and *C. elegans* (*lin-45*). The mammalian genome contains three *raf* genes, of which C-*Raf* (also known as *raf-1*) is the most abundantly expressed and is present in all tissues. A- and B-*Raf* are also widely expressed but, apart from neural tissues, at lower levels. The Raf proteins share a common architecture (Figure 12.13) comprising three main regions. The CR1 region contains the Ras binding domain (RBD) and a cysteine-rich domain (CRD), both needed for membrane localization and activation. CR2 has a 14-3-3 protein binding site, and CR3 is the kinase domain.[107] Despite the fact that the different Raf proteins show much sequence similarity and all activate the same MEKs, mouse knock-outs indicate that they are not functionally redundant.[102]

The *raf* oncogene

Mutations in B-*raf* (but not in A- or C-*raf*), have been detected in melanoma, thyroid, colorectal, and ovarian cancers (MIM 164757). There are more than 45 mutations, of which the most common render the kinase constitutively active (gain-of-function, see page 307). Of these a valine to glutamate substitution (V599E) in the activation segment is prevalent[108] (Figure 12.13). As a consequence, the activation segment folds away from the P-loop to facilitate access of substrate without further need for phosphorylation. In a MEK1 phosphorylation assay, the V599E mutation enhances activity 700-fold, and this gives rise to a 2–5-fold increase in the activity of ERK1.[109]

The reason why only B-Raf, but not A- or C-Raf, is linked to tumour formation may be due to a difference in the N-terminal region. In B-Raf this carries an acidic sequence and phosphorylation is not needed in order to confer a negative charge. Fewer changes are needed to cause deregulation.

Beyond ERK

Docking sites and a MAP kinase phosphorylation motif

ERKs 1 and 2 phosphorylate substrates within the sequence Ser/Thr-Pro. Many proteins carry such a sequence, yet not all are ERK substrates, so there must be other determinants of specificity. Many kinases and phosphatases are regulated by interactions with their substrates at non-catalytic regions, docking motifs that help to avoid indiscriminate phosphorylations or dephosphorylations.

The idea that MAP kinases form stable links with their substrates first arose with studies of JNK2, which was isolated from cell lysates in tight association with its substrate c-Jun. The delta domain (D domain), characterized by a stretch of positively charged residues surrounded by a zone of hydrophobicity in c-Jun, is required for its stable association with JNK2.[110,111] D domains are also involved in the interaction of ERKs1 and 2 with the kinases MEK1, RSK1, MSK, MNK1, the dual specificity phosphatases MKP-3, MKP-3, MKP-7, and the transcription factor Elk-1.[112] Some proteins, such as JunD, possess both DEF and D domains.[113] In others, a DEF motif is involved in the interaction of ERKs1 and 2 with transcription factors and also with the scaffold protein KSR1. The DEF motif seems to be reserved for substrates of the ERK family (not JNK or p38).

Finally, MAP kinases themselves possess conserved domains through which they communicate with their regulators or substrates (Figure 12.14). The common docking domain (CD), which is negatively charged, interacts with proteins possessing the positively charged D domain. Another is the ERK-docking or DE motif. Exchange of only two residues within the DE motif of ERK2 alters its binding specificity to that of p38, thereby switching the attention of ERK2 to MK3.[114] Such domain swaps cause a redirection of signals so that stress signals transmute into growth factor signals and vice versa.[115]

Activation of protein kinases by ERKs 1 and 2

Besides initiating transcription, ERKs1 and 2 also activate the MAP kinase-activated protein kinases, MK (see Table 12.3). There are five subfamilies (Figure 12.15), all characterized by the presence of a CaMK-like kinase domain

ERKs 1 and 2 are widely coexpressed. They appear to phosphorylate similar substrates at the same steps in the pathway, and they are similarly controlled. In accordance with others, we will henceforth refer to **ERKs 1 and 2**. However, it should be noted that these kinases are not necessarily interchangeable.

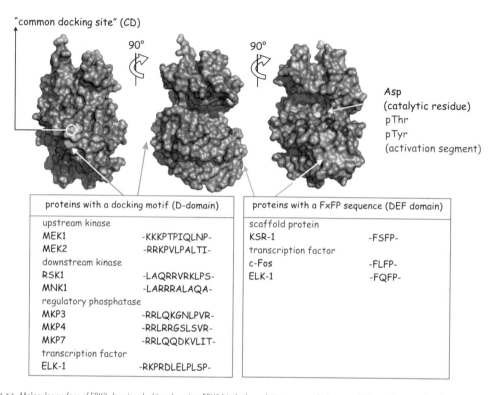

"common docking site" (CD)

90°   90°

Asp
(catalytic residue)
pThr
pTyr
(activation segment)

proteins with a docking motif (D-domain)		proteins with a FxFP sequence (DEF domain)	
upstream kinase		scaffold protein	
MEK1	-KKKPTPIQLNP-	KSR-1	-FSFP-
MEK2	-RRKPVLPALTI-	transcription factor	
downstream kinase		c-Fos	-FLFP-
RSK1	-LAQRRVRKLPS-	ELK-1	-FQFP-
MNK1	-LARRRALAQA-		
regulatory phosphatase			
MKP3	-RRLQKGNLPVR-		
MKP4	-RRLRRGSLSVR-		
MKP7	-RRLQQDKVLIT-		
transcription factor			
ELK-1	-RKPRDLELPLSP-		

FIG 12.14   Molecular surface of ERK2 showing docking domains. ERK2 binds through its common docking site (CD, green) to proteins that possess a D domain. Proteins that possess a DEF domain bind at the other site on ERK (cyan), close to its activation segment (red spots). (2erk[116]) Adapted from Dimitri et al.[117]

RSK and S6K. RSK1 was first identified in extracts of *Xenopus laevis* oocytes as a kinase that phosphorylates the ribosomal protein S6 (component of the 40S subunit).[120] Later it was shown that S6 is actually a poor substrate for RSK, but a good substrate for the S6K family of protein kinases (S6Ks 1 and 2). The S6Ks are phosphorylated and activated by mTOR and PDK1 but not by the ERKs (see Chapter 18).

and all having similar activation segment sequences that are substrates of either ERKs1 and 2 or p38 kinases, or both. Two subfamilies, RSK1–4 and MSK1 and 2 carry a second catalytic domain in the N-terminal region which resembles that of PKC. Surprisingly, the activation segment of this domain is not a substrate of ERKs1 and 2 (nor of p38) so it is unclear how these kinases are activated. It is possible that phosphorylation by ERKs 1 and 2 prepares the MK target kinase for a second, activating phosphorylation by the membrane-associated PDK1 (see page 550).

RSK

The RSKs are present in the cytosol but move partially to the nucleus when activated by phosphorylation. They have specific substrates in both locations. In the cytosol, these are initiation factors of protein translation, components of the apoptosis machinery, the oestrogen receptor, and the guanine nucleotide exchange factor Sos. In the nucleus the RSKs phosphorylate ATF4, c-Fos, and SRF and generally enforce the ERK signal.[119]

TABLE 12.3 Substrates of the MAP kinases and MAP kinase-activated kinases (MK)

"(a) Substrates of MAP kinase"

	ERK 1 & 2	p38α, β, γ, δ	JNK1, 2, 3
Membrane	CD120a, Syk, calnexin		
Transcriptional regulation (nucleus)	SRC-1, Pax6, NF-AT, Elk-1, MEF2, c-Fos, c-Myc, STAT3, JunD, Ets-2	ATF1/2, MEF2a, Sap-1, Elk-1, NF-κB, Ets-1, p53	c-jun, JunD, ATF-2, NF-Acc1, HSF-1, STAT3, p53
Actin cytoskeleton and focal adhesion	Neurofilament, paxillin, vinexin	Tau	
Signalling	MEK, Lin-1, RSK, MSK, MNK, TSC2	PLA2, MNSK, MSK, MK2	Shc

(b) Substrates of MAP kinase-activated kinases (MK)

	RSK1–4	MSK1 & 2	MNK 1 &2	MK2 & 3	MK5
Transcription	TIF-1, ER81, ERα, Mi, CBP, CREB, c-Fos, c-Jun, Nur77, SRF, ATF-4	Histone H3, HMG14, CREB, ATF1, p65, ER81, STAT3	ERβ	CREB, SRF, ER81,E47	
RNA stability, translation		4E-BP1	eIF-4E	hnRNP, TTP, PABP1, HuR, eEFkinase	
Cell cycle	P27kip, Myt1, Bub1				
Cell survival	Bad, LKB1, GSK3β, IkBα, C/EBPβ	Bad			
Signalling	Sos, TSC2	PKB			
Miscellaneous				glycogen synthase,14-3-3ζ, 5-LO	

Information from Roux and Blenis.[118]

MSK

MSKs is present mainly in the nucleus, where the substrates are CREB (see page 248), histone H3, HMGN1, and ATF1. It acts to modify DNA, making it more accessible and again, facilitates the response initiated by ERKs 1 and 2.[118] Some of the substrates are listed in Table 12.3. For further details, see Roux and Blenis[118] and Hauge and Frodin[119].

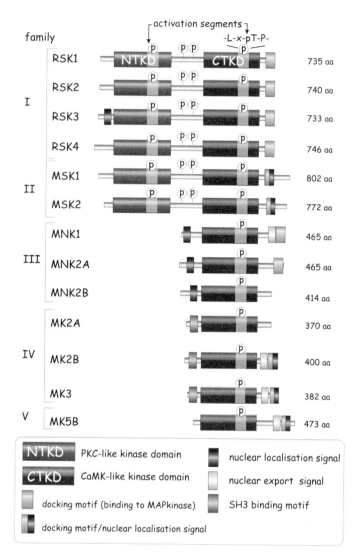

FIG 12.15 Domain architecture of the MAP kinase-activated kinases, MK. These comprise the ribosomal S6 kinases (RSK), the mitogen and stress-activated kinases (MSK), the MAP kinase-interacting kinases (MNK), MAP kinase-activated protein kinases 2 and 3 (MK-2 and -3), and MAP kinase-activated protein kinase 5B (MK5B). Phosphorylations at the indicated sites are necessary for activation. ERK and p38 only phosphorylate the C-terminal activation segment. Members of families I and II possess a pair of non-identical kinase domains. Of these, the N-terminal domains phosphorylate substrate and are classified functionally as AGC kinases. Strangely enough, this domain is not phosphorylated by ERK. PDK1 is a potential candidate and phosphorylation of the C-terminal kinase by ERK may have a permissive role.

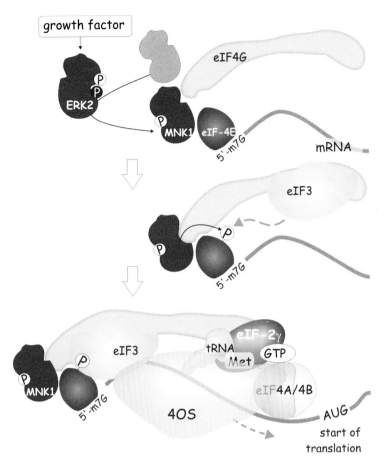

FIG 12.16 eIF-4E and the translation initiation complex. The initiation complex is formed around the eukaryotic initiation factor eIF-4E which is attached to the 5′-m7G cap of the mRNA. The initiation complex comprises tRNA^{Met} (first amino acid), numerous other eukaryotic initiation factors, and the small ribosomal particle 40S. Phosphorylation of eIF-4E, by MNK1, facilitates complex formation and therefore accelerates protein translation.

MNK

The third subfamily of kinases activated by ERKs 1 and 2 is the MAP kinase-integrating kinases, MNKs 1 and 2.[121] MNK1 exerts an indirect stimulatory effect on protein synthesis through ERKs 1 and 2 and directly through phosphorylation of initiation factor-4E (eIF-4E). This is attached to the 5′-m7G cap of mRNA and acts as an anchor for the assembly of the initiation complex. It plays an important role in the binding of the small (40S) ribosomal subunit to the mRNA and in the subsequent search for the AUG start codon (Figure 12.16). MNK1 is associated with eIF-4G, a component of the initiation complex, and

Fos, from feline osteosarcoma virus.

Myc, the cellular counterpart of the transforming gene of the avian leukosis retrovirus MC29.

Jun, from avian sarcoma virus-17: we are informed that ju-nana is 17 in Japanese.

Note: abbreviations for genes are presented in lower case: *fos*, *myc*, etc. The same abbreviations with the first letter capitalized indicates their respective gene products (proteins).

Ets and Elk. Ets (short for E-26 specific), comprises a family of transcription factors. Ets was discovered as a transforming factor, v-ets, from acute avian leukaemia retrovirus E26. This virus causes erythroblastosis mouse and myeloblastosis in chicken.[131] Thereafter, numerous mammalian homologues, c-ets, came to light.[132] One of them is ELK-1, Ets-like factor-1, which was first identified as part of a complex of three components together with p67SRF and DNA. It was therefore initially referred to as ternary complex factor or p62^{TCF} (see page 424).

through this attachment it encounters eIF-4E. How phosphorylation of eIF-4E facilitates initiation of protein synthesis is not clear.[122,123]

All in all, the stimulatory role of the MNKs on the initiation of protein synthesis is somewhat modest. Far more significant is phosphorylation through the PI 3-kinase/PKB branch of the rPTK signal transduction pathway, discussed in Chapter 18.

Interestingly, MNK1 is the target of some of the viruses that hijack the protein synthetic apparatus. The p100 gene product of adenovirus binds eIF-4G and displaces MNK1 so that it is no longer able to phosphorylate and activate eIF-4E. As a result, the cellular mRNA remains untranslated, but the translation of viral mRNA is unaffected. The cell becomes a machine for the manufacture of viral proteins.[124]

Activation of early response genes

The early response genes become activated within 20 min of receptor stimulation. Their activation is transient and can occur under conditions in which protein synthesis is inhibited. Activation of the EGF receptor causes rapid induction of c-*fos*, one of the first cytokine-inducible transcription factors to be discovered.[125] It occupies a central position in the regulation of gene expression. Other early response genes include c-*myc* and c-*jun*.

As a result of the double phosphorylation by MEK, ERKs 1 and 2 undergo dimerization and a proportion translocates into the nucleus. How this occurs is not clear. Not only does ERK itself not possess nuclear localization signal (NLS), but it is far from clear whether the proteins that bind ERK cause its nuclear localization. Both passive (diffusion) and active (Ran-GTP-mediated) transport have been demonstrated.[126] It is possible that the ERKs actually localize to the nucleus by default, but are continually returned to the cytosol by their interaction with MEK, present in both compartments and the bearer of an unmistakable nuclear export signal.[127]

The promoter region of the c-*fos* gene contains a serum response element (SRE), a DNA domain that binds the transcription factors p67^{SRF} (serum response factor) and Elk-1 (p62^{TCF}). Phosphorylation of Elk-1 by ERKs 1 and 2 increases the formation of a complex of both transcription factors with the DNA to promote transcription of the c-*fos* gene[128] (Figure 12.17). Sustained activation of ERKs 1 and 2 (and of the downstream kinase RSK1) also results in multiple phosphorylation of the newly synthesized c-*fos* that acts to increase its life time.[129,130]

The transcription factor c-Myc is yet another nuclear substrate of ERKs 1 and 2. It constitutes one of the master switches regulating protein synthesis and hence cell proliferation. Together with its partner Max, it controls the activity of DNA-dependent RNA polymerases types I, II, and III. c-Myc is phosphorylated

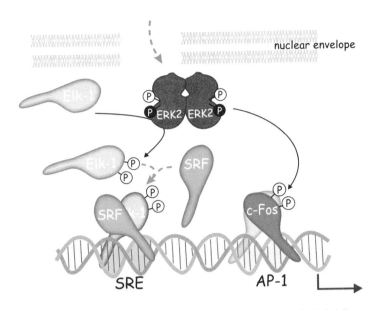

FIG 12.17 Activation of transcription by ERK2. Within the nucleus, ERK2 phosphorylates Elk-1 which then associates with SRF to form an active transcription factor complex. This binds to DNA at the serum response element (SRE). ERK2 also phosphorylates and stabilizes c-Fos, which, in complex with c-Jun, binds to DNA at the AP-1 sequence. Both SRE and AP-1 induce strong expression of c-Fos as well as genes involved in the onset of cell proliferation (e.g. cyclin D).

by ERKs 1 and 2 within the transactivation domain and this stimulates its transcriptional activity.[134,135] By regulating c-Myc activity, the Ras-MAP kinase pathway controls the quantity of ribosomes and thus protein synthesis.

Regulation of the cell cycle

Induction of cell proliferation requires that all the relevant genes are made accessible to transcription factors and all the enzymes needed for DNA replication expressed.[137] A key starting point is the expression of cyclin D, which, together with the kinase Cdk4, drives progression through the early (G1) phase of the cell cycle. Cyclin D1/Cdk4 renders DNA accessible to gene transcription by phosphorylating and inactivating retinoblastoma proteins (Rbs). Transcription of cyclin D1 is facilitated by ERKs 1 and 2 through induction of the transcription factors Fra1, Fra2, c-Jun, and Jun-B, all components of the AP-1 complex[138] that, when phosphorylated,[139] bind to the promoter of the cyclin D1 gene and drive its expression.[140] The ERKs are also instrumental in the formation of stable cyclinD1/Cdk4 complexes.[141] Finally, ERKs 1 and 2 also phosphorylate and inactivate the transcriptional corepressor Tob1, that normally silences expression of cyclin D1.[142]

RNA polymerase I, present in the nucleolus, transcribes the 45S pre-ribosomal RNA molecule (later contributing to the small and large ribosomal particles).

RNA polymerase II transcribes the ribosomal proteins (>30 of them).

RNA polymerase III transcribes the 5S ribosomal particle (which associates with the big ribosomal particle).[136]

The Rb protein was first identified as the product of a gene that is deleted in patients with retinoblastoma, a tumour originating in the retina.[143] The Rb protein is a member of the family of nuclear pocket-proteins that bind and inactivate the E2F transcription

343

factor and prevent cells from entering S phase. Inactivation of the Rb gene occurs in numerous other tumours. In the absence of Rb, cells enter into S phase more readily and they do not require the normal array of extracellular growth signals conveyed upon the expression and activation of cyclinD/cdk4 in order to proceed.

Exactly how ERKs 1 and 2 contribute to the expression of the enzymes needed for DNA replication is not known. In the previous section we pointed out that translocation of ERKs 1 and 2 to the nucleus, signals the activation of immediate early genes, many of which are themselves transcription factors. These induce expression of a second wave of genes and eventually to the expression of DNA replication genes. Examples are those encoding helicases, topo-isomerases, DNA polymerases, and ligases. Beyond their roles in transcription, ERKs 1 and 2 also phosphorylate and activate carbamoyl phosphate synthetase II (CPS II), that catalyses the rate-limiting step in the *de novo* synthesis of pyrimidine nucleotides (dCTP and dTTP).[144]

In the case of the EGF receptor, the activation of ERK and its translocation into the nucleus is an absolute requirement for cell proliferation. Forced retention in the cytoplasm suffices to block growth-factor-induced DNA replication.[145] Weak signals that fail to induce cell proliferation also cause ERKs 1 and 2 to move into the nucleus, but it then recycles to the cytosol within 15 min. A strong mitogenic signal can ensure nuclear retention for up to 6 h.[146] Importantly, the signal must also be reinforced by signals from cell adhesion complexes.[147] We return to the question of cooperation between growth factor receptors and cellular adhesion in Chapter 13 (see page 404).

Fine tuning the Ras-MAP kinase pathway: scaffold proteins

We have explained that a sequential series of protein modifications determines the propagation of cell-surface signals into the nucleus. A second tier of regulation is presented by the interaction of scaffold proteins with components of this signalling cascade. Scaffold proteins bring these signalling components together or hold them apart.

MAP kinase scaffold proteins discovered in yeast

The names of yeast genes usually have three letters plus numbers and often reflect the phenotype of a mutant. Thus *STE* denotes sterile. By convention genes are uppercase/italic (*STE5*), mutant genes are lower case/italic (*ste5*) and gene products are simply capitalized, e.g. Ste5.

The physiological relevance of scaffold proteins to MAP kinase activation was first demonstrated in *Saccharomyces cerevisiae* (baker's yeast). Here, in the response to mating pheromones (ligands that interact with 7TM receptors, see page 113), Ste5 acts as a scaffold that binds and links Ste11 (a MAP3K), Ste7 (a MAP2K), and Fus3 (a MAPK) to form a MAP kinase signalling cassette.[148,149] In addition, Ste5 also links the cassette to Ste4 and Ste18 (G protein β-and γ- subunits, respectively) and to Ste20 (equivalent to p21-activated kinase or PAK) (see Figure 12.18a).

Actually, Ste5 is more than a scaffold protein. It imposes conformational changes on components of the MAP kinase pathway which lead to kinase activation. For instance, it amplifies the signal by modifying the conformation of Ste11 following its phosphorylation by Ste20. It also

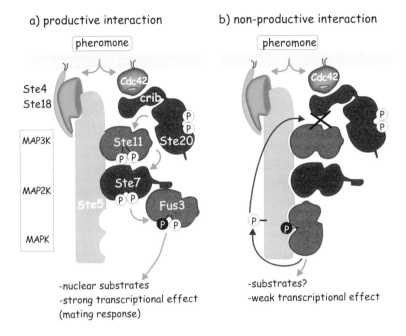

a) productive interaction

b) non-productive interaction

-nuclear substrates
-strong transcriptional effect
(mating response)

-substrates?
-weak transcriptional effect

FIG 12.18 Yeast scaffold proteins and their role in pheromone signalling. (a) Ste5 assembles the various MAP kinases (Ste11, Ste7, and Fus3) and brings them to the membrane by linking to Ste4/Ste18 (β-γ-subunit). Here it encounters the activation complex comprising Cdc42 (GTPase) and Ste20 (serine/threonine protein kinase). Fus3 attaches to Ste7 via a D domain and common docking domain interaction.[149,150] (b) The outcome is different when Fus3 binds directly to Ste5. Here, Fus3 undergoes partial activation due to autophosphorylation and then phosphorylates Ste5. This inhibits the output of the signalling complex.[151]

induces autophosphorylation in the activation segment of Fus3. Surprisingly, this activation of Fus3 inhibits the output of the signalling complex and suppresses pheromone signalling (see Figure 12.18b).

KSR, a mammalian scaffold protein that regulates MAP kinase signalling

There are no animal homologues of Ste5 but there are several other proteins that provide scaffold functions for the ERK, JNK, and p38 MAPK cascades.[152] An example is KSR, discovered in genetic screens performed in *Drosophila* and *C. elegans*.[153–155] KSR is evolutionarily conserved (two human homologues, KSR1 and 2), its site of action situated between Ras and Raf. Although KSR proteins carry the serine/threonine kinase signature, they lack the conserved lysine that normally binds ATP and no kinase activity has been detected. The KSR proteins are thought to exert their action on the Ras-MAP kinase pathway

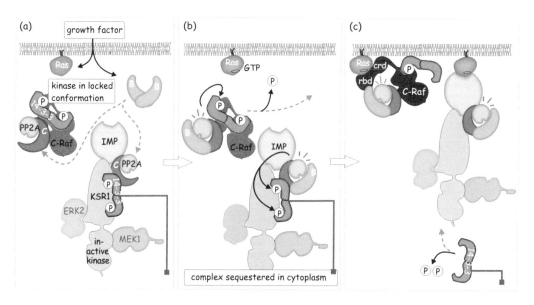

FIG 12.19 Central role of PP2A in the formation of a productive Ras-MAP kinase signalling cassette. The serine/threonine phosphatase PP2A has two roles at the onset of the Ras–ERK pathway. Both effect the relief of inhibitory constraints imposed by the phosphoserine-binding scaffold protein 14-3-3. (a) C-Raf and KSR1 are restrained by 14-3-3. Growth factors activate Ras and they complete the assembly of the active PP2A phosphatase-complex (association of the B-subunit) (b) PP2A dephosphorylates C-Raf to allow detachment of 14-3-3. This brings C-Raf into association with Ras leading to its activation (shown in panel (c)). Also, PP2A dephosphorylates two residues in KSR1, allowing detachment of 14-3-3. (c) The liberated KSRS1/MAP-kinase signalling cassette now translocates to the membrane and attaches to RasGTP through IMP.

as scaffold proteins (Figure 12.19). Although not strictly required, their presence greatly enhances signalling through this pathway.[156]

C-TAK1: Cdc25C-associated kinase-1, not to be confused with TAK1, the acronym for TGFβ-activated kinase-1.

KSR1 is permanently fixed to MEK in a complex that also contains the serine/threonine kinase C-TAK1 and two subunits of the phosphatase PP2A (PP2A-A (catalytic) and PP2A-C (regulatory). In quiescent cells the complex can be recovered from the cytosolic fraction, linked to another scaffold protein, 14-3-3 (Figure 12.19). In response to activation by growth factors, the complex containing KSR1 translocates to the cell membrane and promotes the formation of a signalling cassette.[157,158]

This translocation and subsequent assembly of the Ras–ERK signalling cassette involves detachment of 14-3-3 and then destruction of the inhibitor IMP. The first event is controlled by PP2A (Figure 12.19), the second by interaction of IMP with RasGTP (Figure 12.20). In the absence of growth factor signals, KSR1 is kept in a phosphorylated state by the kinase C-TAK1[159] and the complex retained in the cytosol through its binding to 14-3-3 (see Figure 12.19). On activation of growth factor receptors, PP2A dephosphorylates KSR1 allowing it to dissociate from 14-3-3, so leaving room for other binding partners at the membrane.

IMP is the abbreviation for 'impedes mitogenic signal propagation'. It is not inosine monophosphate!

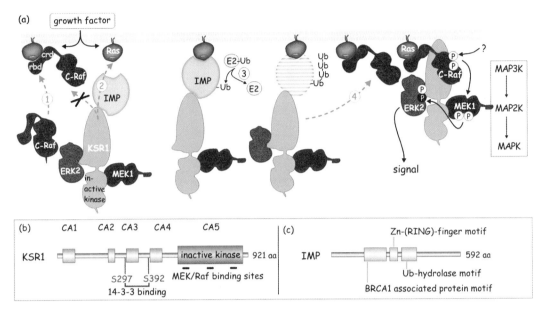

FIG 12.20 Dual effector interaction of RasGTP that leads to effective signalling to ERK. (a) In order to activate the ERK pathway, Ras has not only to recruit C-Raf (1), but also to remove IMP, the inhibitor which prevents formation of the Raf-MEK-ERK signalling cassette. RasGTP binds IMP (2). This initiates a series of ubiquitylations that mark the protein for destruction by the proteasome (3). MEK1 and ERK2, linked to the scaffold protein KSR1, now join C-Raf, enabling the signal to pass from one kinase to another (4). (b) Conserved domains in KSR1. CA2, proline-rich; CA3, cysteine-rich, resembles the C1 domain (PMA/DAG binding site) of PKC, CA4, serine/threonine-rich; CA5, a kinase domain that resembles that of Raf, but lacks an essential lysine and is therefore inactive (adapted from Matheny[160] and Kolch[152]). (c) The domain architecture of IMP reveals a RING finger, involved in ubiquitylation (see page 469).

Activation of PP2A occurs through association of its regulatory B subunit with the KSR1 complex (which already contains the A and C subunits). PP2A acts to render Raf catalytically competent by dephosphorylating a serine in the CR2 region and thereby lifting the inhibitory constraint of 14-3-3.[103]

The interaction of KSR1 and its associated MAP kinases with Raf is hindered by the presence of IMP.[160] IMP contains a RING-H2 motif, characteristic of proteins that facilitate ubiquitylation and it functions as an E3-ligase (see page 469). High expression of IMP inhibits Raf-induced activation of endogenous MEK and ERK and conversely, its depletion increases the amplitude of the Ras–ERK response. This is explained by a dual effector interaction of RasGTP, the one due to binding and activation of Raf-1 and the other due to de-repression of Raf-MEK, causing destruction of IMP due to autoubiquitylation. Its removal in the proteasome then allows the association of the KSR1/MEK/ERK complex to the membrane thereby opening up the pathway from Raf to ERK (see Figure 12.20). The role of IMP is to raise the threshold for ERK signalling.

The importance of KSR1 in MAP kinase signalling is further emphasized by the finding that mice lacking this scaffold protein are much less sensitive to oncogenic signalling through Ras[161].

347

Other proteins that regulate MAP kinase pathways

RKIP is an inhibitor of the Ras–ERK pathway.[162] It binds to C-Raf (and weakly to MEK1, but not to B-Raf) and it prevents MEK activation. When phosphorylated by PKC, it is displaced from C-Raf, relieving this inhibition and so reinstating the Ras–ERK pathway and contributing to gene transcription through the AP-1 complex (see page 580).[163]

The 7TM receptor-linked β-arrestins (see page 98) also act as scaffolds for the MAP kinase pathway. Unlike KSR1, the β-arrestins appear to retain active signalling cassettes (comprising C-Raf, MEK, and ERKs 1 and 2 or JNK3) in the cytosol and thereby prevent phosphorylation of nuclear substrates.[164]

MP-1 localizes the ERK1 signalling cassette to late endosomes through an interaction with the endosomal p14 adaptor protein.[165,166] Signalling from the receptor continues in the endosomal compartment (see below).[167,168]

Why are the signalling pathways so complicated?

And why are there so many apparently redundant components? In fact, not all pathways are long and complicated. Some, like the pathway involving Notch (Chapter 22) or the STAT proteins (see below) are relatively straightforward, entailing only the modification of a signalling protein at the membrane followed by its direct translocation into the nucleus. More generally, however, mechanisms tend to be complex and while there is no simple explanation, the reasons for complexity must stem from the need for cells to be able to sense multiple inputs, to commit to sets of appropriate responses and to conduct them in controlled and precise ways. In general:

- Growth factor signals must initiate and synchronize both transcriptional and metabolic events. This requires multiple effectors, both nuclear and cytosolic.
- Signals need to propagate among and between different cellular compartments. (For example, receptors may be removed from the cell surface by endocytosis, processed in vesicular compartments and recycled back to the plasma membrane.)
- A single signalling protein can only make contact with a limited number of effectors.
- Signals must be amplified in order to exceed downstream threshold conditions and ensure effective responses.
- Signals must be robust and their propagation should not rely on a single critical component.
- A timely and measured response requires feedback circuits in order to maintain balanced and dynamic relationships between inputs and outputs.

- Signals from different inputs must converge at sites of integration that determine appropriate output signals.

Beyond all this, one must recall that signal transduction pathways were never designed, but are the result of numerous rather messy trial and error processes. In the end, these proved, and continue to prove, to be of benefit to the organisms of which they form a part.

Termination of the ERK response

The ability to attenuate and to terminate signals initiated by growth factors is crucial, and there are numerous ways by which this is achieved. We limit the discussion to events affecting the receptors and the adaptors and effectors that bind to them.

First, there are the phosphatases, including PTP1B, SHP1/2, and DEP1, that strip phosphotyrosine residues. These, as well as others that inactivate ERK, are further discussed in Chapter 21. Secondly, phosphorylation of the guanine nucleotide exchange factor Sos1 by ERKs 1 and 2 reduces its affinity for Grb2, so suppressing the Ras signalling pathway.[169,170] Downstream of ERK, RSK can also phosphorylate Sos1 (but not Sos2, which lacks several serine/threonine phosphorylation sites).[119]

A third mechanism of negative feedback is the removal of cell surface receptors by endocytosis. In the case of EGF, endocytosis following low-level stimulation actually induces a burst of signalling before suppressing it, due to a contribution from the internalized receptor.[168] Signalling only ceases when the receptors are degraded by acid hydrolases in late endosomes. Phosphorylated receptors are preferentially endocytosed and directed by vesicular transport to the late endosomal compartment through the actions of Grb2 and Cbl (an E3-ubiquitin ligase) (see Figure 12.21). Both bind to receptors through their (atypical) SH2 domains and both contribute to the clustering of receptors and their uptake into clathrin-coated vesicles.[171,172] Cbl may also act as a handle, to prevent recycling of the activated EGFR to the plasma membrane.

A family of MAP kinase-related proteins

Once ERKs 1 and 2 were cloned, it became apparent that they are members of a substantial family, the MAP kinases. Based on sequence analysis and the composition of their activation segments, they may be classified into five groups, each operating in different signal transduction pathways (see Figure 12.22 and Table 12.2).

- **ERK1 and ERK2** are the 'prototypic' or 'classical' MAP kinases, operating mainly in mitogen activated signal transduction pathways.
- **ERK3 and ERK4** are distinguished from other ERKs by the absence of a tyrosine in the activation segment. Thus they possess only a single (serine)

Cbl and monoubiquitylation. Ubiquitylation is controlled by ubiquitin activating (E1), conjugating (E2), and ligase (E3) enzymes. It results in the attachment of the small ubiquitin protein (76 residues) to a substrate (for more detail see page 467). Ubiquitin, and ubiquitin-like peptides, are now recognized as general signalling devices with have several roles.[173] Attachment of four or more ubiquitins, polyubiquitylation, is a recognition signal for destruction by the proteasome; monoubiquitylation is a signal for endocytosis and a signal in histone regulation. In the endosome, ubiquitin serves as a molecular signature on trafficking cargoes. Cbl, an E3-ligase, catalyses ubiquitylation of the EGFR, so that it is recognized by proteins bearing ubiquitin-binding domains. For instance, the adaptor protein Eps15 has two ubiquitin-interacting motifs that interact with protein sorting machinery.

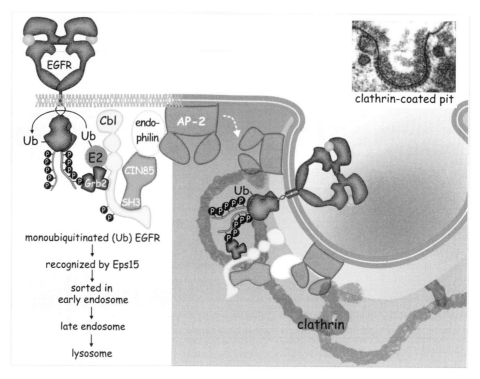

FIG 12.21 Removal of the EGFR from the cell surface and sorting into the lysosome pathway.

Activated EGF receptors are recognized by Cbl which binds either directly through a phosphotyrosine-binding motif or by interaction with the SH3 domain of Grb2. Ubiquitylation by Cbl acts as a sorting signal, directing the receptor into the lysosomal pathway for degradation. The receptor–Cbl complex is recognized by CIN85 and endophilin, which couple the receptor to a group of proteins that includes the endocytic adaptor AP-2. Clathrin monomers are then recruited and the active EGFRs accumulate in clathrin-coated membrane pits which pinch off from the plasma membrane as endocytic vesicles. Within the intracellular network of vesicular transport pathways, the receptors progress through the early and late endosomes to the lysosome and are destroyed.

Cycloheximide, a product of the bacterium *Streptomyces griseus,* is an inhibitor of protein biosynthesis in eukaryotic organisms. It interferes with the activity of peptidyl transferase activity, thus blocking translational elongation.

phosphorylation site in the motif SEG. Little is known about their upstream regulators and substrates. ERK3 has two gene variants, α and β.[94]

- **ERK5** (originally called Big MAP kinase or BMK1) is activated by mitogens, but its C-terminal domain gives rise to a protein twice the size of ERKs 1 and 2. It has transcriptional activity, either on its own, binding direct to DNA, or through its association with other transcription factors (for instance the AP-1 complex). It operates in a similar pathway as ERKs 1 and 2, and has many substrates in common. It has a role in cardiovascular development and neural differentiation.[181]

- **JNK and SAPK** phosphorylate c-Jun[174,175] and are activated in response to 'stress'.[176] They emerged in experiments in which rat livers were challenged by injection of cycloheximide.[177] Activating factors include growth factor

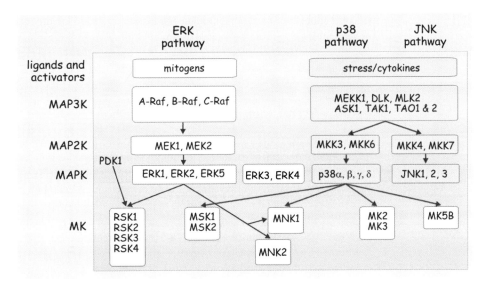

FIG 12.22 Parallel MAP kinase pathways.

The MAP kinases are classified in three groups based on the identity of the intermediate residue in their dual phosphorylation motifs (TEY, TGY, or TPY). This classification also defines the ERK, JNK, and p38 signal transduction pathways, each having unique upstream activators. The ERK pathway acts in response to mitogens, whereas the p38 and JNK pathways respond to stress and inflammatory cytokines.

deprivation, UV irradiation, or treatment with inflammatory cytokines (IL-1β, TNF). The JNKs include JNK1, 2, and 3.[178]

- **p38 kinase** came to light in genetic deletion studies in yeast (*S. cerevisiae*) as a kinase involved in the generation of glycerol in response to osmotic stress.[179] HOG1 induces expression of glycerol-3-phosphate dehydrogenase, related to glycerol synthesis.[180] In mammalian cells, it is also activated in response to stress stimuli. p38/HOG is now generally referred to as p38 kinase, of which there are four variants: p38α, β, δ, and γ.[94]

Each of the pathways shown in Figure 12.22 involves a kinase cascade, comprising a MAP3K and MAP2K and culminating in the phosphorylation and activation of the particular MAP kinase. Each contains a dual phosphorylation site, TEY, TPY, or TGY, (with the exception of ERK3 and ERK4, see above). These parallel pathways of activation may operate individually or in combination to initiate specific patterns of gene expression. Although it is generally accepted that the ERK pathway responds to mitogen stimulation, and the JNK and p38 pathways to stress and inflammation, cross-talk undoubtedly occurs. We elaborate further on the roles of JNK and p38 and their role in the regulation of the innate immune response (inflammation) in Chapter 16 (see page 493).

There is still an awful lot to learn about Ras. As our knowledge of Ras and the pathways that it regulates have advanced, so its involvements have become more complex and entangled. Ras is not merely an activator of cell growth. Indeed, in some cells it causes growth inhibition and differentiation. In others it blocks differentiation. It has become apparent that it is a regulator of multiple cell functions that depend on the cell, the state of the cell, the activation state of other GTPases and possibly the identity of the Ras isotype (N-, K-, or H-). What's more, its relationship to cell transformation is not simply a matter of cause and effect. As examples:

- Oncogenic Ras only induces transformation if the recipient cell has already endured a number of mutations in tumour suppressor genes such as p53 or Rb. Otherwise its introduction results in apoptosis (see page 306). Most of the experimental work on Ras has been carried out on rodent fibroblast cell lines that are far more susceptible to transformation than the epithelial cells in which most (human) Ras-related tumours occur. More than this, the phenotypes of the transformed cells are very different. Ras-transformed (human breast) epithelial cells are characterized by disruption of the adherens junctions and the appearance of stress fibres and focal adhesions (see Figure 13.15, page 399) but transformed fibroblasts are associated with a loss of stress fibres. While constitutively activated mutants of Ras, or its effector Raf, can induce the transformed phenotype in mouse fibroblasts, only Ras can induce transformation of rat intestinal epithelial cells.
- Due to the availability of antibodies, mutants, etc., most investigations have concentrated on H-Ras. However, although the human Ras genes N, H, and K are very similar, and in many experimental situations appear to function in the same way, there is no reason to believe that their actions are identical. The conservation of three *ras* genes in vertebrate evolution begs the question of whether or not the gene products have specific functions. Although the data are so far very sketchy, the embryonic lethality of K-*ras* (but not of N- or H-*ras*) supports the idea of non-redundancy of function.

In this chapter we have indicated the role of Ras as an activator of ERK, but this is not the only pathway implicated in its regulation of cell proliferation. In Chapter 18 we consider Ras as one of a number of activators of PI 3-kinase and the consequent activation of protein kinase B. There are several additional (and also potential) effector pathways through which the effects of activated Ras can operate.

For a more comprehensive discussion of these questions, see Shields et al.[182]

MAP kinases in other organisms
Pathways regulated by MAP kinases are present in all eukaryotic organisms.[183] In yeast (*S. cerevisiae*), processes regulated by MAP kinases include mating, sporulation, maintenance of cell wall integrity, invasive growth, pseudohyphal

growth, and osmoregulation. MAP kinase is a regulator of embryonic development and the immune response in *Drosophila* . It acts as a regulator in slime moulds, plants, and fungi.

The Ca^{2+}–calmodulin pathway

The elevation of cytosol Ca^{2+} following activation of PLC_γ results in widespread protein phosphorylation by serine/threonine kinases. These include the broad spectrum Ca^{2+}-calmodulin-dependent protein kinase II (CaM-kinase II) (see page 228), myosin light chain kinase (MLCK), phosphorylase kinase, and EF-2 kinase. All of these are activated by Ca^{2+}–calmodulin.

The level of phosphorylation of a particular substrate at any time must be determined by the rates of both phosphorylation and dephosphorylation. Ca^{2+}–calmodulin can affect this balance through activation of the protein phosphatase calcineurin. This leads to the activation of transcription factors that play essential roles in the activation of T lymphocytes (Chapter 17). Clearly, Ca^{2+} is an extremely versatile second messenger modulating numerous intracellular signals.

Activation of PI 3-kinase

Association of the p85 adaptor subunit of PI 3-kinase with the tyrosine-phosphorylated receptor positions the attached p110 catalytic subunit at the membrane where it phosphorylates inositol phospholipids at the 3 position of the inositol ring. The p110 subunit can also be activated directly by Ras. This results in activation of PKB, of prime importance in cell survival, proliferation, motility, and glucose metabolism. This pathway is considered in Chapter 18.

Direct phosphorylation of STAT transcription factors

The simplest way by which plasma membrane receptors alter gene expression is through the direct phosphorylation of transcription factors. The action of the interferons is an example. The STATs are targets of interferon receptors and they also mediate EGF and PDGF signalling.[184,185] STAT1a (p91) and STAT1b (p84) possess SH2 domains that enable them to associate with the phosphotyrosines of activated receptors (Figure 12.23). They themselves are then phosphorylated on tyrosine residues, causing them to form dimers that translocate to the nucleus where they promote transcription of early response genes such as *c-fos* (see page 525). The STAT dimer formed after phosphorylation by the PDGF receptor was originally described as Sis-inducible factor (SIF), because it was observed in cells exposed to the viral oncogene product v-Sis. This is closely related to PDGF and activates the same signal transduction pathway.[186,187]

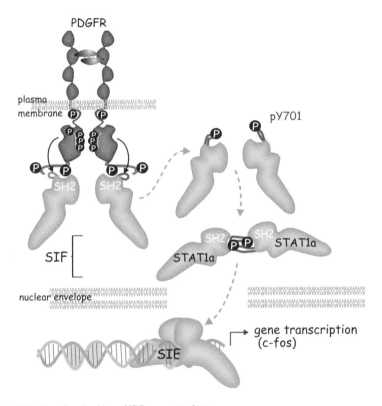

FIG 12.23 Direct phosphorylation of STAT transcription factors.
STAT1a and STAT1b bind to the tyrosine-phosphorylated receptor and themselves become phosphorylated. They form a dimer (Sis-inducible factor, SIF) which translocates to the nucleus, where it binds to a Sis-inducible element (SIE) within the fos promoter. The domain organization of STAT1 is shown in Figure 12.6.

A switch in receptor signalling: activation of ERK by 7TM receptors

Pathway switching mediated by receptor phosphorylation

G-protein-linked receptors are themselves substrates not only for PKA and PKC, but also for receptor-specific kinases, which preferentially target occupied (and therefore activated) receptors (see Figure 4.14, page 100). These receptor-specific kinases (GRKs, such as the β-adrenergic receptor kinase GRK2) are only called into action under conditions of robust stimulation. On the other hand, phosphorylation by PKA and PKC (triggered by an increase in second messenger production) affects occupied and

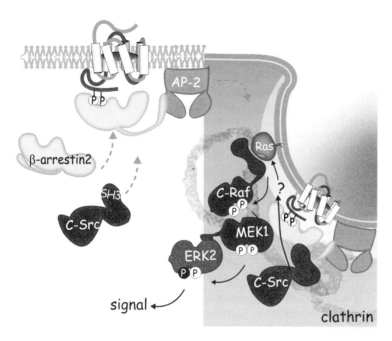

FIG 12.24 Phosphorylation of the β-adrenergic receptor by GRK2 activates the Ras–ERK pathway. Phosphorylation of the β-adrenergic receptor by GRK2/βARK1 (see Chapter 4) allows recruitment of the adaptor β-arrestin2 and terminates communication with the G protein. It also activates the Ras–ERK pathway by recruitment of Src and MEK1/ERK2. β-Arrestin2 also directs the receptor to clathrin-coated pits, there to be removed from the cell surface by endocytosis and degraded in the lysosomal pathway.[164]

unoccupied receptors alike. The effect of phosphorylation by these kinases is to switch the attention of the receptors to alternative G-proteins. As described in Chapter 4, β-adrenergic receptors phosphorylated by PKA now communicate with the G_i proteins instead of G_s, so opening the door to the Ras–ERK pathway[188,189] (see Figure 9.5, page 251). The sites phosphorylated by GRK2, though also present in the C-terminal region, are distinct from those targeted by PKA.

β-Receptors phosphorylated by GRK2 bring β-arrestin (see page 98) to a segment of the third intracellular loop that would normally interact with G-proteins, so blocking the transmission of signals through G_s and initiating receptor removal (Figure 12.24). Importantly, the bound β-arrestin offers an alternative means to activate the Ras–ERK pathway through interaction with the SH3 domain of the Src tyrosine kinase.[190] This also initiates a signalling pathway that activates ERK. The steps following the recruitment of Src are not yet clear. It is possible that it initiates a series of events resulting in the phosphorylation of the adaptor Shc-1, which in turn, binds to the Grb2/Sos complex. Finally, β-arrestins also act as scaffolds for other MAP kinases, binding JNK3, Ask1, and MKK4.[164]

Other 7TM receptors may employ different pathways to reach Ras. For instance, the lysophosphatidic acid receptor also activates ERK, but here dominant negative Src is without effect. The signal appears to involve PI 3-kinase, an unidentified tyrosine kinase, a docking protein, and finally Grb2/Sos[191].

Pathway switching by transactivation

Another way of switching, that also allows signals emanating from G-protein-linked receptors to activate the Ras–ERK pathway, is by receptor transactivation.[192,193] As an example, activation of muscarinic acetylcholine or thrombin receptors releases HB-EGF (an EGF-like factor) by cleavage from its inactive membrane-bound precursor by the metalloproteinase ADAM17. HB-EGF then acts in an autocrine/paracrine manner to stimulate the EGF receptor, resulting in the activation of ERK[193,194] (Figure 12.25). In effect, the first ligand induces the release of a second, quite unrelated ligand, which in turn sets in train its own distinct signalling pathways. Receptor transactivation enormously expands the repertoire of signalling systems that a cell can apply

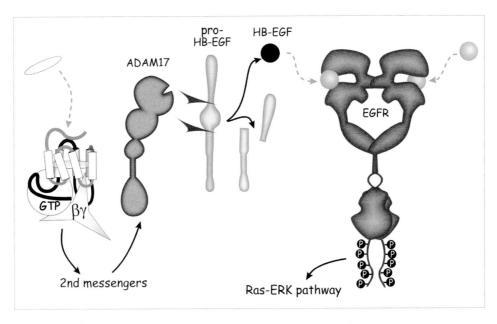

FIG 12.25 Transactivation of receptors.
Activation of the 7TM receptor results in activation of ADAM17, a transmembrane endoproteinase with release of HB-EGF, cleaved from its the membrane-bound precursor. HB-EGF then binds to the EGF receptor and induces yet another set of intracellular signalling pathways that include the Ras–ERK kinase pathway.

in mounting a response. In this example, the mechanism of transactivation depends, in the first place, upon intracellular signals that activate the transmembrane metalloproteinase which has its catalytic site situated in the extracellular domain.

Pathway switching, transactivation, and metastatic progression of colorectal cancer

Both pathway switching and transactivation make contributions in the malignant behaviour of already transformed epithelial cells of the intestinal tract. The inflammatory conditions caused by tumour cells raise the local production of prostaglandins by leukocytes, in particular PGE2 which acts through G-protein-coupled EP receptors. EP1 receptors activate G_q and raise cytosol Ca^{2+}, EP3 receptors inhibit cAMP production through G_i, and EP2 and EP4 receptors cause transactivation of EGF receptors as described in the previous paragraph.[195,196] In addition, EP2 and EP4 bind to β-arrestin to activate Src. This then, both through transactivation and pathway switching, contributes to cell migration *in vitro* and metastasis *in vivo*.[197] Perhaps aspirin, that old favourite among remedies, might offer some protection against colorectal tumour invasion through its action as an inhibitor of prostaglandin production[198].

List of Abbreviations

Abbreviation	Full name/description	SwissProt entry	Other names/OMIM
AATYK	apoptosis associated tyrosine kinase	Q6ZMQ8	LMTK1 (lemur tyrosine kinase)
ADAM17	disintegrin and metalloprotease domain 17 (convertase)	P78536	TNF-α converting enzyme
ALK	anaplastic lymphoma kinase (not to be confused with activin like kinase)	Q9UM73	CD246
AP-1	activator protein-1 (complex of two transcription factors)		
AP-2	adaptor protein-2 (complex of four proteins involved in the selection of cargo for endocytosis)		
AR	amphiregulin	P15514	colorectal cell-derived growth factor
ATF4	activating transcription factor-4 (cAMP dependent)	P18848	CREB2
AXL	anexelekto (uncontrolled, transforming gene in chronic myelogenous leukaemia)	P30530	UFO receptor tyrosine kinase

Continued

357

Abbreviation	Full name/description	SwissProt entry	Other names/OMIM
β-ARK1	see GRK2	P25098	GRK2
β-Arrestin2	arresting G protein-coupled receptor signaling	P32121	
Boss	*Drosophila* bride of sevenless (ligand for Sev)	P22815	
BTC	β-cellulin (mitogen with EGF-like domain)	P35070	betacellulin
Cbl-B	casitas B-lineage lymphoma oncogene b (E3-ubiquitin ligase)	Q13191	RING finger protein 56
cdk4	cyclin-dependent protein kinase-4	P11802	
c-Fos	cellular homologue feline osteosarcoma oncogene	P01100	
c-Jun	homologue of sarcoma virus-17 oncogene, junana = 17	P05412	
c-Myc	cellular homologue myelocytomatosis MC29 virus oncogene	P01106	
Crk	CT10 virus regulator of protein kinase, p47gag-crk viral oncogene (chimeric protein)		
CrkL	Crk-like protein,	P46109	
C-Raf	rat fibrosarcoma	P04049	
c-Src	sarcoma, cellular homologue of Rous sarcoma virus oncogene	P12931	
C-TAK1	Cdc25-associated protein kinase-1	P27448	microtubule affinity-regulating kinase 3
cyclinD1	from cell division cycle	P24385	BCL-1 oncogene
DDR	epithelial discoidin domain containing receptor	Q08345	CD167a
DEP1	density enhanced protein phosphatase-1	Q12913	RPTη, CD148
DER	*Drosophila* epidermal growth factor receptor	P04412	torpedo, gurken receptor
EGF	epidermal growth factor	P01133	urogastrone

Continued

Abbreviation	Full name/description	SwissProt entry	Other names/OMIM
EGFR	epidermal growth factor receptor	P00533	ERBB1 (avian erythroblastosis)
eIF-4E	eukaryote initiation factor 4E (translation)	P06730	mRNA cap-binding protein
EPH	erythropoietin-producing hepatoma	P21709	ephrin type-A receptor
EPR	epiregulin	O14944	EREG
ERBB1	erythroblastosis-B type 1	P00533	EGFR
ERK1	extracellular signal regulated kinase-1	P27361	MAP2 kinase, MAPK3
FGFR1	fibroblast growth factor receptor type 1	P11362	Fms-like tyrosine kinase (Flt2), CD331
FRS2	fibroblast growth factor receptor substrate-2	Q8WU20	adaptor protein SNT
Gab1	Grb2 associated binding-1	Q13480	growth factor receptor bound 2-associated protein
Grb2	growth factor receptor binding protein-2	P62993	
GRK2	G-protein coupled receptor kinase 2	P25098	β-adrenergic receptor kinase-1 (BARK)
HB-EGF	heparin-binding EGF-like growth factor	Q99075	diphtheria toxin receptor
IMP1	impedes mitogenic signal propagation	Q7Z569	BRAC1-associated protein (BRAP)
INSR	insulin receptor	P06213	CD220
IRS-1	insulin receptor substrate-1	P35568	
JNK2	c-Jun N-terminal kinase-2	P45984	MAPK9
KSR1	kinase suppressor of Ras	Q8IVT5	
Let-23	*C. elegans* lethal-mutation 23 (development)	P24348	
Let-60	*C. elegans* lethal-mutation 60 (development)	P22981	lineage-34
lin-3	*C. elegans* abnormal cell lineage	Q03345	lethal-mutation 94
MAPK3	mitogen activated protein kinase-3	P27361	ERK1, MAP2 kinase
MEK1	MAPK ERK (activator) kinase-1	Q02750	MAP2K1

Continued

359

Abbreviation	Full name/description	SwissProt entry	Other names/OMIM
MET	methyl-nitroso-nitroguanidine-induced oncogene	P08581	hepatocyte growth factor receptor (HGF), scatter factor receptor
MK	MAPkinase-activated kinase family of proteins		
MK2	MAPkinase activated protein kinase-2	P49137	MAPKAP2
MNK1	MAPkinase interacting kinase-1	Q9BUB5	
MP-1	MEK binding partner	Q9UHA4	MAPK scaffold protein 1 (MAPKSP1)
MSK1	mitogen and stress-activated kinase	O75582	ribosomal protein S6 kinase A5
MUSK	muscle receptor tyrosine kinase	O15146	
NRG1	neuroregulin 1	Q02297	neu differentiation factor, heregulin
p38α	protein kinase of 38 kDa type α	Q16539	MAPK14, SAPK2A
PDGF-A	platelet derived growth factor type A	P04085	PDGF-1
PDGFR-A	platelet-derived growth factor receptor type 1	P16234	CD140a
PDK1	3-phosphoinositide dependent kinase-1	O15530	
PI 3-kinase α	phosphatidylinositol 3-kinase, catalytic p110 subunit alpha	P42336	PIK3CA
PI 3-kinase	phosphatidylinositol 3-kinase, regulatory p85 subunit	Q92569	PIK3R3
PKBα	protein kinase B-α	P31749	Akt-1, RAC-PKα
PLC-γ	phospholipase C-γ	P19174	PLC1
PP2A-α	serine/threonine protein phosphatase 2A catalytic subunit α	P67775	PPP2CA
PTK	protein tyrosine kinase	Q13308	colon carcinoma-4 (CCK4)
PTP1B	protein tyrosine phosphatase-1B	P18031	PTPN1
Ras (K-Ras)	Kirsten rat sarcoma	P01116	

Continued

Abbreviation	Full name/description	SwissProt entry	Other names/OMIM
RasGAP	Ras GTPase activating protein	P20936	RASA1
RET	rearranged during transfection (multiple endocrine neoplasia)	P07949	
ROR	tyrosine kinase-receptor related	Q01973	not to be confused with retinoid-related orphan receptor
ROS1	Rous sarcome virus oncogene homologue-1	P08922	
RSK1	ribosomal S6 kinase	Q15418	p90S6K1, p90RSK1
RYK	receptor tyrosine kinase	P34925	
Sem5	*C. elegans* sex muscle abnormal protein-5 (vulva development)	P29355	
Sev	*Drosophila* sevenless (lack of R7)	P13368	
SH2	Src homology domain		
Shc1	Src homology collagen-like-1	P29353	
SHP1	SH2-containing protein phosphatases-1	P29350	PTP-1C, PTPN6
Sos1	son of sevenless	Q07889	
SRF	serum response factor	P11831	
STAT1	signal transducer and activator of transcription-1	P42224	ISGF-3 components p91 or p84 (isoforms)
TGF α	transforming growth factor-α	P01135	EGF-like TGF
TIE61	tyrosine kinase with Ig and EGF homology domains-1	P35590	
TOB1	transducer of ERRB2 (suppressor of transcription)	P50616	
TRK	tropomyosin receptor tyrosine kinase	P04629	neurotrophic tyrosine kinase-1 (NTRK1)
VEGFR	vascular endothelial growth factor receptor type 1	P17948	Fms-like tyrosine kinase-1 (Flt1)

References

1. Robinson DR, Wu YM, Lin SF. The protein tyrosine kinase family of the human genome. *Oncogene*. 2000;19:5548–5557.

2. Arkinstall S, Payton M, Maundrell K. Activation of phospholipase Cγ in *Schizosaccharomyces pombe* by coexpression of receptor or nonreceptor tyrosine kinases. *Mol Cell Biol*. 1995;15:1431–1438.

3. Skorokhod A, Gamulin V, Gundacker D, Kavsan V, Muller IM. Origin of insulin receptor-like tyrosine kinases in marine sponges. *Biol Bull*. 1999;197:198–206.

4. Collett MS, Erikson RL. Protein kinase activity associated with the avian sarcoma virus src gene product. *Proc Natl Acad Sci U S A*. 1978;75: 2021–2024.

5. Bazley LA, Gullick WJ. The epidermal growth factor receptor family. *Endocr Relat Cancer*. 2005;12:S17–S27.

6. Eckhart W, Hutchinson MA, Hunter T. An activity phosphorylating tyrosine in polyoma T antigen immunoprecipitates. *Cell*. 1979;18:925–933.

7. Witte ON, Dasgupta A, Baltimore D. Abelson murine leukaemia virus protein is phosphorylated in vitro to form phosphotyrosine. *Nature*. 1980;283:826–831.

8. Ushiro H, Cohen S. Identification of phosphotyrosine as a product of epidermal growth factor-activated protein kinase in A-431 cell membranes. *J Biol Chem*. 1980;255:8363–8365.

9. Giancotti FG. Integrin signaling: specificity and control of cell survival and cell cycle progression. *Curr Opin Cell Biol*. 1997;9:691–700.

10. Miettinen PJ, Berger JE, Meneses J, Phung Y, Pedersen RA, Werb Z, Derynck R. Epithelial immaturity and multiorgan failure in mice lacking epidermal growth factor receptor. *Nature*. 1995;376:337–341.

11. Naccache PH, Gilbert C, Caon AC, et al. Selective inhibition of human neutrophil functional responsiveness by erbstatin, an inhibitor of tyrosine protein kinase. *Blood*. 1990;76:2098–2104.

12. Burg DL, Furlong MT, Harrison ML, Geahlen RL. Interactions of Lyn with the antigen receptor during B cell activation. *J Biol Chem*. 1994;269:28136–28142.

13. Cantrell DA. T cell antigen receptor signal transduction pathways. *Cancer Surveys*. 1996;27:165–175.

14. Williamson P, Merida I, Greene WC, Gaulton G. The membrane proximal segment of the IL-2 receptor β-chain acidic region is essential for IL2-dependent protein tyrosine kinase activation. *Leukemia*. 1994;8(suppl 1):S186–S189.

15. Kirken RA, Rui H, Evans GA, Farrar WL. Characterization of an interleukin-2 (IL-2)-induced tyrosine phosphorylated 116-kDa protein associated with the IL-2 receptor β-subunit. *J Biol Chem*. 1993;268:22765–22770.

16. Li W, Deanin GG, Margolis B, Schlessinger J, Oliver JM. FcεR1-mediated tyrosine phosphorylation of multiple proteins, including phospholipase

Cγ1 and the receptor $\beta\gamma_2$ complex, in RBL-2H3 rat basophilic leukemia cells. *Mol Cell Biol*. 1992;12:3176–3182.

17. Carpenter G, King L, Cohen S. Epidermal growth factor stimulates phosphorylation in membrane preparations in vitro. *Nature*. 1979;276:409–410.

18. Yamamoto-Honda R, Tobe K, Kaburagi Y, et al. Upstream mechanisms of glycogen synthase activation by insulin and insulin-like growth factor-I. Glycogen synthase activation is antagonized by wortmannin or LY294002 but not by rapamycin or by inhibiting p21ras. *J Biol Chem*. 1995;270:2724–2729.

19. Cho HS, Mason K, Ramyar KX, et al. Structure of the extracellular region of HER2 alone and in complex with the Herceptin Fab. *Nature*. 2003;421:756–760.

20. Guy PM, Platko JV, Cantley LC, Cerione RA, Carraway 3rdKL. Insect cell-expressed p180erbB3 possesses an impaired tyrosine kinase activity. *Proc Natl Acad Sci U S A*. 1994;91:8132–8136.

21. Hendriks BS, Orr G, Wells A, Wiley HS, Lauffenburger DA. Parsing ERK activation reveals quantitatively equivalent contributions from epidermal growth factor receptor and HER2 in human mammary epithelial cells. *J Biol Chem*. 2005;280:6157–6169.

22. Citri A, Yarden Y. EGF-ERBB signalling: towards the systems level. *Nat Rev Mol Cell Biol*. 2006;7:505–516.

23. Hynes NE, Lane HA. ERBB receptors and cancer: the complexity of targeted inhibitors. *Nat Rev Cancer*. 2005;5:341–354.

24. Jones RB, Gordus A, Krall JA, MacBeath G. A quantitative protein interaction network for the ErbB receptors using protein microarrays. *Nature*. 2006;439:168–174.

25. Luetteke NC, Qiu TH, Peiffer RL, Oliver P, Smithies O, Lee DC. TGFα deficiency results in hair follicle and eye abnormalities in targeted and waved-1 mice. *Cell*. 1993;73:263–278.

26. Luetteke NC, Qiu TH, Fenton SE, et al. Targeted inactivation of the EGF and amphiregulin genes reveals distinct roles for EGF receptor ligands in mouse mammary gland development. *Development*. 1999;126:2739–2750.

27. Threadgill DW, Dlugosz AA, Hansen LA, et al. Targeted disruption of mouse EGF receptor: effect of genetic background on mutant phenotype. *Science*. 1995;269:230–234.

28. Defize LH, Moolenaar WH, van der Saag PT, de Laat SW. Dissociation of cellular responses to epidermal growth factor using anti-receptor monoclonal antibodies. *EMBO J*. 1986;5:1187–1192.

29. Spaargaren M, Defize LH, Boonstra J, de Laat SW. Antibody-induced dimerization activates the epidermal growth factor receptor tyrosine kinase. *J Biol Chem*. 1991;266:1733–1739.

30. Yamanaka M, Ishitani R, Nureki O, et al. Crystal structure of the complex of human epidermal growth factor and receptor extracellular domains. *Cell*. 2002;110:775–787.

31. Garrett TP, McKern NM, Lou M, et al. Crystal structure of a truncated epidermal growth factor receptor extracellular domain bound to transforming growth factor α.. *Cell* 2002;110:763–773.

32. Ottensmeyer FP, Beniac DR, Luo RZ, Yip CC. Mechanism of transmembrane signaling: insulin binding and the insulin receptor. *Biochemistry*. 2000;39:12103–12112.

33. Yip CC, Ottensmeyer P. Three-dimensional structural interactions of insulin and its receptor. *J Biol Chem*. 2003;278:27232–27329.

34. Brandt-Rauf PW, Rackovsky S, Pincus MR. Correlation of the structure of the transmembrane domain of the neu oncogene-encoded p185 protein with its function. *Proc Natl Acad Sci U S A*. 1990;87:8660–8664.

35. Syed RS, Reid SW, Li CW, et al. Efficiency of signalling through cytokine receptors depends critically on receptor orientation. *Nature*. 1998;395:511–516.

36. Ortega E, Schweitzer SR, Pecht I. Possible orientational constraints determine secretory signals induced by aggregation of IgE receptors on mast cells. *EMBO J*. 1988;7:4101–4109.

37. Jorissen RN, Walker F, Pouliot N, Garrett TP, Ward CW, Burgess AW. Epidermal growth factor receptor: mechanisms of activation and signalling. *Exp Cell Res*. 2003;284:31–53.

38. Anderson D, Koch CA, Grey L, Ellis C, Moran MF, Pawson T. Binding of SH2 domains of phospholipase $C_{\gamma 1}$, GAP, and Src to activated growth factor receptors. *Science*. 1990;250:979–982.

39. Zhang X, Gureasko J, Shen K, Cole PA, Kuriyan J. An allosteric mechanism for activation of the kinase domain of epidermal growth factor receptor. *Cell*. 2006;125:1137–1149.

40. Ferguson KM, Berger MB, Mendrola JM, Cho HS, Leahy DJ, Lemmon MA. EGF activates its receptor by removing interactions that autoinhibit ectodomain dimerization. *Mol Cell*. 2003;11:507–517.

41. Ogiso H, Ishitani R, Nureki O, et al. Crystal structure of the complex of human epidermal growth factor and receptor extracellular domains. *Cell* 2002;110:775–787.

42. Yarden Y, Schlessinger J. Epidermal growth factor induces rapid, reversible aggregation of the purified epidermal growth factor receptor. *Biochemistry*. 1987;26:1443–1451.

43. Carter RE, Sorkin A. Endocytosis of functional epidermal growth factor receptor-green fluorescent protein chimera. *J Biol Chem*. 2006;273:35000–35007.

44. Saito Y, Haendeler J, Hojo Y, Yamamoto K, Berk BC. Receptor heterodimerization: essential mechanism for platelet-derived growth factor-induced epidermal growth factor receptor transactivation. *Mol Cell Biol*. 2001;21:6387–6394.

45. Schlessinger J. Ligand-induced, receptor-mediated dimerization and activation of EGF receptor. *Cell*. 2002;110:669–672.

46. Pandiella A, Beguinot L, Velu TJ, Meldolesi J. Transmembrane signalling at epidermal growth factor receptors overexpressed in NIH 3T3 cells. Phosphoinositide hydrolysis, cytosolic Ca^{2+} increase and alkalinization correlate with epidermal-growth-factor-induced cell proliferation,. *Biochem J*. 1988;254:223–228.

47. Gilligan A, Prentki M, Knowles BB. EGF receptor down-regulation attenuates ligand-induced second messenger formation. *Exp Cell Res*. 1990;187:134–142.

48. Gonzalez FA, Gross DJ, Heppel LA, Webb WW. Studies on the increase in cytosolic free calcium induced by epidermal growth factor, serum, and nucleotides in individual A431 cells. *J Cell Physiol*. 1988;135:269–276.

49. Higaki M, Sakaue H, Ogawa W, Kasuga M, Shimokado K. Phosphatidylinositol 3-kinase-independent signal transduction pathway for platelet-derived growth factor-induced chemotaxis. *J Biol Chem*. 1996;271:29342–29346.

50. Liu XQ, Pawson T. The epidermal growth factor receptor phosphorylates GTPase- activating protein (GAP) at Tyr-460, adjacent to the GAP SH2 domains. *Mol Cell Biol*. 1991;11:2511–2516.

51. Medema RH, de Vries Smits AM, van der Zon GC, Maassen JA, Bos JL. Ras activation by insulin and epidermal growth factor through enhanced exchange of guanine nucleotides on p21ras. *Mol Cell Biol*. 1993;13:155–162.

52. Muroya K, Hattori S, Nakamura S. Nerve growth factor induces rapid accumulation of the GTP-bound form of p21ras in rat pheochromocytoma PC12 cells. *Oncogene*. 1992;7:277–281.

53. Meisenhelder J, Suh PG, Rhee SG, Hunter T. Phospholipase Cγ is a substrate for the PDGF and EGF receptor protein-tyrosine kinases in vivo and in vitro. *Cell*. 1989;57:1109–1122.

54. Koch CA, Anderson D, Moran MF, Ellis C, Pawson T. SH2 and SH3 domains: elements that control interactions of cytoplasmic signaling proteins. *Science*. 1991;252:668–674.

55. Escobedo JA, Navankasattusas S, Kavanaugh WM, Milfay D, Fried VA, Williams LT. cDNA cloning of a novel 85 kd protein that has SH2 domains and regulates binding of PI3-kinase to the PDGF β-receptor. *Cell*. 1991;65:75–82.

56. Skolnik EY, Margolis B, Mohammadi M, Lowenstein E, Fischer R, Drepps A, Ullrich A, Schlessinger J. Cloning of PI3 kinase-associated p85 utilizing a novel method for expression/cloning of target proteins for receptor tyrosine kinases. *Cell*. 1991;65:83–90.

57. O'Neill TJ, Craparo A, Gustafson TA. Characterization of an interaction between insulin receptor substrate 1 and the insulin receptor by using the two-hybrid system. *Mol Cell Biol*. 1994;14:6433–6442.

58. Vojtek AB, Hollenberg SM, Cooper JA. Mammalian Ras interacts directly with the serine/threonine kinase Raf. *Cell*. 1993;74:205–214.

365

59. Oliver SG. From gene to screen with yeast. *Curr Opin Genet Dev*. 1997;7:405–409.

60. Matsuda M, Mayer BJ, Fukui Y, Hanafusa H. Binding of transforming protein, P47gag-crk, to a broad range of phosphotyrosine-containing proteins. *Science*. 1990;248:1537–1539.

61. Sadowski I, Stone JC, Pawson T. A noncatalytic domain conserved among cytoplasmic protein-tyrosine kinases modifies the kinase function and transforming activity of Fujinami sarcoma virus P130gag-fps. *Mol Cell Biol*. 2006;6:4396–4408.

62. Moran MF, Koch CA, Anderson D, Ellis C, England L, Martin GS, Pawson T. Src homology region 2 domains direct protein–protein interactions in signal transduction. *Proc Natl Acad Sci U S A*. 1990;87:8622–8626.

63. van der Geer P, Pawson T. The PTB domain: a new protein module implicated in signal transduction. *Trends Biochem Sci*. 1995;20:277–280.

64. Kavanaugh WM, Williams LT. An alternative to SH2 domains for binding tyrosine-phosphorylated proteins. *Science*. 1994;266:1862–1865.

65. Blaikie P, Immanuel D, Wu J, Li N, Yajnik V, Margolis B. A region in Shc distinct from the SH2 domain can bind tyrosine-phosphorylated growth factor receptors. *J Biol Chem*. 1994;269:32031–32034.

66. Simons PJ, Dourmashkin RR, Turano A, Phillips DE, Chesterman FC. Morphological transformation of mouse embryo cells in vitro by murine sarcoma virus (Harvey). *Nature*. 1967;214:897–898.

67. Sweet RW, Yokoyama S, Kamata T, Feramisco JR, Rosenberg M, Gross M. The product of ras is a GTPase and the T24 oncogenic mutant is deficient in this activity. *Nature*. 1984;311:273–275.

68. Chang EH, Furth ME, Scolnick EM, Lowy DR. Tumorigenic transformation of mammalian cells induced by a normal human gene homologous to the oncogene of Harvey murine sarcoma virus. *Nature*. 1982;297:479–483.

69. Feramisco JR, Gross M, Kamata T, Rosenberg M, Sweet RW. Microinjection of the oncogene form of the human H-ras (T-24) protein results in rapid proliferation of quiescent cells. *Cell* 1984;38:109–117.

70. Feramisco JR, Clark R, Wong G, Arnheim N, Milley R, McCormick F. Transient reversion of ras oncogene-induced cell transformation by antibodies specific for amino acid 12 of ras protein. *Nature*. 1985;314:639–642.

71. Satoh T, Endo M, Nakafuku M, Nakamura S, Kaziro Y. Platelet-derived growth factor stimulates formation of active p21ras, GTP complex in Swiss mouse 3T3 cells. *Proc Natl Acad Sci U S A*. 1990;87:5993–5997.

72. Harris WA, Stark WS, Walker JA. Genetic dissection of the compound eye of *Drosophila melanogaster*. *J Physiol*. 1976;256:415–439.

73. Dickson B, Hafen E. Genetic dissection of eye development in *Drosophila*. In: Bate M, Martinez-Arias A, eds. *Development of Drosophila melanogaster*. New York: Cold Spring Harbor Press; 2000:1327–1362.

74. Hart AC, Kramer H, Van Vactor DL, Paidhungat M, Zipursky SL. Induction of cell fate in the *Drosophila* retina: the bride of sevenless protein is predicted to contain a large extracellular domain and seven transmembrane segments. *Genes Dev*. 1990;4:1835–1847.

75. Kramer H, Cagan RL, Zipursky SL. Interaction of bride of sevenless membrane-bound ligand and the sevenless tyrosine-kinase receptor. *Nature*. 1991;352:207–212.

76. Hafen E, Basler K, Edstroem JE, Rubin GM. Sevenless, a cell-specific homeotic gene of *Drosophila*, encodes a putative transmembrane receptor with a tyrosine kinase domain. *Science*. 1987;236:55–63.

77. Fortini ME, Simon MA, Rubin GM. Signalling by the sevenless protein tyrosine kinase is mimicked by Ras1 activation. *Nature*. 1992;355:559–561.

78. Jones S, Vignais M-L, Broach JR. The CDC25 protein of *S.cerevisiae* promotes exchange of guanine nucleotides bound to ras. *Mol Cell Biol*. 1991;11:2641–2646.

79. Lowenstein EJ, Daly RJ, Batzer AG, et al. The SH2 and SH3 domain-containing protein GRB2 links receptor tyrosine kinases to ras signaling. *Cell*. 1992;70:431–442.

80. Kornfeld K. Vulval development in *Caenorhabditis elegans*. *Trends Genet*. 1997;13:55–61.

81. Hahn CG, Wang HY, Cho DS, et al. Altered neuregulin 1-erbB4 signaling contributes to NMDA receptor hypofunction in schizophrenia. *Nat Med*. 2006;12:824–828.

82. Gale NW, Kaplan S, Lowenstein EJ, Schlessinger J, Bar-Sagi D. Grb2 mediates the EGF-dependent activation of guanine nucleotide exchange on Ras. *Nature*. 1993;363:88–92.

83. Stern MJ, Marengere LE, Daly RJ, et al. The human GRB2 and *Drosophila* Drk genes can functionally replace the *Caenorhabditis elegans* cell signaling gene sem-5. *Mol Biol Cell*. 1993;4:1175–1188.

84. Li N, Batzer A, Daly R, et al. Guanine-nucleotide-releasing factor hSos1 binds to Grb2 and links receptor tyrosine kinases to Ras signalling. *Nature*. 1993;363:85–88.

85. Aronheim A, Engelberg D, Li N, al-Alawi N, Schlessinger J, Karin M. Membrane targeting of the nucleotide exchange factor Sos is sufficient for activating the Ras signaling pathway. *Cell*. 1994;78:949–961.

86. Rozakis-Adcock M, Fernley R, Wade J, Pawson T, Bowtell D. The SH2 and SH3 domains of mammalian Grb2 couple the EGF receptor to the Ras activator mSos1. *Nature*. 1993;363:83–85.

87. Yang SS, Van Aelst L, Bar-Sagi D. Differential interactions of human Sos1 and Sos2 with Grb2. *J Biol Chem*. 1995;270:18212–18215.

88. Holgado-Madruga M, Emlet DR, Moscatello DK, Godwin AK, Wong AJ. A Grb2-associated docking, protein in EGF- and insulin-receptor signalling. *Nature*. 1996;379:560–564.

89. Burgering BM, Freed E, van der Voorn L, McCormick F, Bos JL. Platelet-derived growth factor-induced p21ras-mediated signaling is independent of platelet-derived growth factor receptor interaction with GTPase-activating protein or phosphatidylinositol-3-kinase. *Cell Growth Differ*. 1994;5:341–347.

90. Ray LB, Sturgill TW. Rapid stimulation by insulin of a serine/threonine kinase in 3T3- L1 adipocytes that phosphorylates microtubule-associated protein 2 in vitro. *Proc Natl Acad Sci U S A*. 1987;84:1502–1506.

91. Dumaz N, Light Y, Marais R. Cyclic AMP blocks cell growth through Raf-1-dependent and Raf-1-independent mechanisms. *Mol Cell Biol*. 2002;22:3717–3728.

92. Duesbery NS, Webb CP, Leppla SH, et al. Proteolytic inactivation of MAP-kinase-kinase by anthrax lethal factor. *Science*. 1998;280:734–737.

93. Bhatt RR, Ferrell Jr JE. Cloning and characterization of *Xenopus* Rsk2, the predominant p90 Rsk isozyme in oocytes and eggs. *J Biol Chem*. 2000;275:32983–32990.

94. Chen Z, Gibson TB, Robinson F, et al. MAP kinases. *Chem Rev*. 2001;101:2449–2476.

95. Rapp UR, Goldsborough MD, Mark GE, et al. Structure and biological activity of v-raf, a unique oncogene transduced by a retrovirus. *Proc Natl Acad Sci U S A*. 1983;80:4218–4222.

96. Moodie SA, Willumsen BM, Weber MJ, Wolfman A. Complexes of Ras. GTP with Raf-1 and mitogen-activated protein kinase kinase. *Science*. 1993;260:1658–1661.

97. Warne PH, Viciana PR, Downward J. Direct interaction of Ras and the amino-terminal region of Raf-1 in vitro. *Nature*. 1993;364:352–355.

98. Koide H, Satoh T, Nakafuku M, Kaziro Y. GTP-dependent association of Raf-1 with Ha-Ras: identification of Raf as a target downstream of Ras in mammalian cells. *Proc Natl Acad Sci U S A*. 1993;90:8683–8686.

99. Leevers SJ, Paterson HF, Marshall CJ. Requirement for Ras in Raf activation is overcome by targeting Raf to the plasma membrane. *Nature*. 1994;369:411–414.

100. Stokoe D, Macdonald SG, Cadwallader K, Symons M, Hancock JF. Activation of Raf as a result of recruitment to the plasma membrane. *Science*. 1994;264:1463–1467.

101. Mineo C, Anderson RG, White MA. Physical association with ras enhances activation of membrane-bound raf (RafCAAX). *J Biol Chem*. 1997;272:10345–10348.

102. Kolch W. Meaningful relationships: the regulation of the Ras/Raf/MEK/ERK pathway by protein interactions. *Biochem J*. 2000;351(Pt 2):289–305.

103. Ory S, Zhou M, Conrads TP, Veenstra TD, Morrison DK. Protein phosphatase 2A positively regulates Ras signaling by dephosphorylating KSR1 and Raf-1 on critical 14-3-3 binding sites. *Curr Biol*. 2003;13:1356–1364.

104. Hekman M, Wiese S, Metz R, et al. Dynamic changes in C-Raf phosphorylation and 14-3-3 protein binding in response to growth factor stimulation: differential roles of 14-3-3 protein binding sites. *J Biol Chem*. 2004;279:14074–14086.

105. Rapp UR, Goldsborough MD, Mark GE, et al. Structure and biological activity of v-raf, a unique oncogene transduced by a retrovirus. *Proc Natl Acad Sci U S A*. 1983;80:4218–4222.

106. Jansen HW, Ruckert B, Lurz R, Bister K. Two unrelated cell-derived sequences in the genome of avian leukemia and carcinoma inducing retrovirus MH2. *EMBO J*. 1983;2:1969–1975.

107. Wellbrock C, Karasarides M, Marais R. The RAF proteins take centre stage. *Nat Rev Mol Cell Biol*. 2004;5:875–885.

108. Davies H, Bignell GR, Cox C, et al. Mutations of the BRAF gene in human cancer. *Nature*. 2002;417:949–954.

109. Wan PT, Garnett MJ, Roe SM, et al. Mechanism of activation of the RAF-ERK signaling pathway by oncogenic mutations of B-RAF. *Cell*. 2004;116:855–867.

110. Kallunki T, Su B, Tsigelny I, et al. JNK2 contains a specificity-determining region responsible for efficient c-Jun binding and phosphorylation. *Genes Dev*. 1994;8:2996–3007.

111. Kallunki T, Deng T, Hibi M, Karin M. c-Jun can recruit JNK to phosphorylate dimerization partners via specific docking interactions. *Cell*. 1996;87:929–939.

112. Masuda K, Shima H, Katagiri C, Kikuchi K. Activation of ERK induces phosphorylation of MAPK phosphatase-7, a JNK specific phosphatase, at Ser-446. *J Biol Chem*. 2003;278:32448–32456.

113. Vinciguerra M, Vivacqua A, Fasanella G, et al. Differential phosphorylation of c-Jun and JunD in response to the epidermal growth factor is determined by the structure of MAPK targeting sequences. *J Biol Chem*. 2004;279:9634–9641.

114. Tanoue T, Adachi M, Moriguchi T, Nishida E. A conserved docking motif in MAP kinases common to substrates, activators and regulators. *Nat Cell Biol*. 2000;2:110–116.

115. Brunet A, Pouyssegur J. Identification of MAP kinase domains by redirecting stress signals into growth factor responses. *Science*. 1996;272:1652–1655.

116. Canagarajah BJ, Khokhlatchev A, Cobb MH, Goldsmith EJ. Activation mechanism of the MAP kinase ERK2 by dual phosphorylation. *Cell*. 1997;90:859–869.

117. Dimitri CA, Dowdle W, MacKeigan JP, Blenis J, Murphy LO. Spatially separate docking sites on ERK2 regulate distinct signaling events in vivo. *Curr Biol*. 2005;15:1319–1324.

118. Roux PP, Blenis J. ERK and p38 MAPK-activated protein kinases: a family of protein kinases with diverse biological functions. *Microbiol Mol Biol Rev*. 2004;68:320–344.

119. Hauge C, Frodin M. RSK and MSK in MAP kinase signalling. *J Cell Sci.* 2006;119:3021–3023.

120. Erikson E, Maller JL. A protein kinase from *Xenopus* eggs specific for ribosomal protein S6. *Proc Natl Acad Sci U S A.* 1985;82:742–746.

121. Waskiewicz AJ, Flynn A, Proud CG, Cooper JA. Mitogen-activated protein kinases activate the serine/threonine kinases Mnk1 and Mnk2. *EMBO J.* 1997;16:1909–1920.

122. Lin TA, Kong X, Haystead TA, Pause A, Belsham G, Sonenberg N, Lawrence-JC J. PHAS-I as a link between mitogen-activated protein kinase and translation initiation [see comments]. *Science.* 1994;266:653–656.

123. Sonenberg N, Gingras AC. The mRNA 5′ cap-binding protein eIF4E and control of cell growth. *Curr Opin Cell Biol.* 1998;10:268–275.

124. Cuesta R, Xi Q, Schneider RJ. Adenovirus-specific translation by displacement of kinase Mnk1 from cap-initiation complex eIF4F. *EMBO J.* 2000;19:3465–3474.

125. Kruijer W, Cooper JA, Hunter T, Verma IM. Platelet-derived growth factor induces rapid but transient expression of the c-fos gene and protein. *Nature.* 1984;312:711–716.

126. Adachi M, Fukuda M, Nishida E. Two co-existing mechanisms for nuclear import of MAP kinase: passive diffusion of a monomer and active transport of a dimmer. *EMBO J.* 1999;18:5347–5358.

127. Adachi M, Fukuda M, Nishida E. Nuclear export of MAP kinase (ERK) involves a MAP kinase kinase (MEK)-dependent active transport mechanism. *J Cell Biol.* 2000;148:849–856.

128. Treisman R. Regulation of transcription by MAP kinase cascades. *Curr Opin Cell Biol.* 1996;8:205–215.

129. Murphy LO, Smith S, Chen RH, Fingar DC, Blenis J. Molecular interpretation of ERK signal duration by immediate early gene products. *Nat Cell Biol.* 2002;4:556–564.

130. Murphy LO, MacKeigan JP, Blenis J. A network of immediate early gene products propagates subtle differences in mitogen-activated protein kinase signal amplitude and duration. *Mol Cell Biol.* 2004;24:144–153.

131. Leprince D, Gegonne A, Coll J, de Taisne C, Schneeberger A, Lagrou C, Stehelin D. A putative second cell-derived oncogene of the avian leukaemia retrovirus E26. *Nature.* 1983;306:395–397.

132. de Taisne C, Gegonne A, Stehelin D, Bernheim A, Berger R. Chromosomal localization of the human proto-oncogene c-ets. *Nature.* 1984;310:581–583.

133. Sternberg MJ, Gullick WJ. Neu receptor dimerization. *Nature.* 1990;339:587.

134. Pulverer BJ, Fisher C, Vousden K, Littlewood T, Evan G, Woodgett JR. Site-specific modulation of c-Myc cotransformation by residues phosphorylated in vivo. *Oncogene.* 1994;9:59–70.

135. Gupta S, Seth A, Davis RJ. Transactivation of gene expression by Myc is inhibited by mutation at the phosphorylation sites Thr-58 and Ser-62. *Proc Natl Acad Sci U S A.* 1993;90:3216–3220.

136. Oskarsson T, Trumpp A. The Myc trilogy: lord of RNA polymerases. *Nat Cell Biol.* 2005;7:215–217.

137. Whitmarsh AJ, Davis RJ. A central control for cell growth. *Nature.* 2000;403:255–256.

138. Treinies I, Paterson HF, Hooper S, Wilson R, Marshall CJ. Activated MEK stimulates expression of AP-1 components independently of phosphatidylinositol 3-kinase (PI3-kinase) but requires a PI3-kinase signal To stimulate DNA synthesis. *Mol Cell Biol.* 1999;19:321–329.

139. Bakiri L, Lallemand D, Bossy-Wetzel E, Yaniv M. Cell cycle-dependent variations in c-Jun and JunB phosphorylation: a role in the control of cyclin D1 expression. E. *EMBO J.* 2000;19:2056–2068.

140. Herber B, Truss M, Beato M, Muller R. Inducible regulatory elements in the human cyclin D1 promoter. *Oncogene.* 1994;9:2105–2107.

141. Ladha MH, Lee KY, Upton TM, Reed MF, Ewen ME. Regulation of exit from quiescence by p27 and cyclin D1-CDK4. *Mol Cell Biol.* 1998;18:6605–6615.

142. Suzuki T, Tsuzuku J, Ajima R, Nakamura T, Yoshida Y, Yamamoto T. Phosphorylation of three regulatory serines of Tob by Erk1 and Erk2 is required for Ras-mediated cell proliferation and transformation. *Genes Dev.* 2002;16:1356–1370.

143. Lee WH, Shew JY, Hong FD, Sery TW, Young LJ, Bookstein R, Lee EY. The retinoblastoma susceptibility gene encodes a nuclear phosphoprotein associated with DNA binding activity. *Nature.* 1987;329:642–645.

144. Graves LM, Guy HI, Kozlowsk P, et al. Regulation of carbamoyl phosphate synthetase by MAP kinase. *Nature.* 2000;403:328–332.

145. Pages G, Brunet A, L'Allemain G, Pouyssegur, J. Constitutive mutant and putative regulatory serine phosphorylation site of mammalian MAP kinase kinase (MEK1). *EMBO J.* 1994;13:3003–3010.

146. Kahan C, Seuwen K, Meloche S, Pouyssegur J. Coordinate, biphasic activation of p44 mitogen-activated protein kinase and S6 kinase by growth factors in hamster fibroblasts. Evidence for thrombin-induced signals different from phosphoinositide turnover and adenylylcyclase inhibition. *J Biol Chem.* 1992;267:13369–13375.

147. Aplin AE, Stewart SA, Assoian RK, Juliano RL. Integrin-mediated adhesion regulates ERK nuclear translocation and phosphorylation of Elk-1. *J Cell Biol.* 2001;153:273–282.

148. Choi KY, Satterberg B, Lyons DM, Elion EA. Ste5 tethers multiple protein kinases in the MAP kinase cascade required for mating in *S. cerevisiae.* *Cell.* 1994;78:499–512.

149. Elion EA. The Ste5p scaffold. *J Cell Sci.* 2001;114:3967–3978.

150. Lamson RE, Takahashi S, Winters MJ, Pryciak PM. Dual role for membrane localization in yeast MAP kinase cascade activation and its contribution to signaling fidelity. *Curr Biol.* 2006;16:618–623.

151. Bhattacharyya RP, Remenyi A, Good MC, Bashor CJ, Falick AM, Lim WA. The Ste5 scaffold allosterically modulates signaling output of the yeast mating pathway. *Science.* 2006;311:822–826.

152. Kolch W. Coordinating ERK/MAPK signalling through scaffolds and inhibitors. *Nat Rev Mol Cell Biol*. 2005;6:827–837.

153. Therrien M, Chang HC, Solomon NM, Karim FD, Wassarman DA, Rubin GM. KSR, a novel protein kinase required for RAS signal transduction. *Cell*. 1995;83:879–888.

154. Therrien M, Michaud NR, Rubin GM, Morrison DK. KSR modulates signal propagation within the MAPK cascade. *Genes Dev*. 1996;10:2684–2695.

155. Kornfeld K, Hom DB, Horvitz HR. The *ksr-1* gene encodes a novel protein kinase involved in Ras-mediated signaling in *C. elegans*. *Cell*. 1995;83:903–913.

156. Nguyen A, Burack WR, Stock JL, et al. Kinase suppressor of Ras (KSR) is a scaffold which facilitates mitogen-activated protein kinase activation in vivo. *Mol Cell Biol*. 2002;22:3035–3045.

157. Stewart S, Sundaram M, Zhang Y, Lee J, Han M, Guan KL. Kinase suppressor of Ras forms a multiprotein signaling complex and modulates MEK localization. *Mol Cell Biol*. 1999;19:5523–5534.

158. Morrison DK. KSR: a MAPK scaffold of the Ras pathway? *J Cell Sci* 2001;114:1609–1612.

159. Muller J, Ory S, Copeland T, Piwnica-Worms H, Morrison DK. C-TAK1 regulates Ras signaling by phosphorylating the MAPK scaffold, KSR1. *Mol Cell*. 2001;8:983–993.

160. Matheny SA, Chen C, Kortum RL, Razidlo GL, Lewis RE, White MA. Ras regulates assembly of mitogenic signalling complexes through the effector protein IMP. *Nature*. 2004;427:256–260.

161. Lozano J, Xing R, Cai Z, et al. Deficiency of kinase suppressor of Ras1 prevents oncogenic ras signaling in mice. *Cancer Res*. 2003;63:4232–4238.

162. Yeung K, Seitz T, Li S, et al. Suppression of Raf-1 kinase activity and MAP kinase signalling by RKIP. *Nature*. 1999;401:173–177.

163. Corbit KC, Trakul N, Eves EM, Diaz B, Marshall M, Rosner MR. Activation of Raf-1 signaling by protein kinase C through a mechanism involving Raf kinase inhibitory protein. *J Biol Chem*. 2003;278:13061–13068.

164. Luttrell LM, Lefkowitz RJ. The role of β-arrestins in the termination and transduction of G-protein-coupled receptor signals. *J Cell Sci*. 2002;115:455–465.

165. Wunderlich W, Fialka I, Teis D, et al. A novel 14-kilodalton protein interacts with the mitogen-activated protein kinase scaffold mp1 on a late endosomal/lysosomal compartment. *J Cell Biol*. 2001;152:765–776.

166. Teis D, Wunderlich W, Huber LA. Localization of the MP1-MAPK scaffold complex to endosomes is mediated by p14 and required for signal transduction. *Dev Cell*. 2002;3:803–814.

167. Wiley HS, Shvartsman SY, Lauffenburger DA. Computational modeling of the EGF-receptor system: a paradigm for systems biology. *Trends Cell Biol*. 2003;13:43–50.

168. Schoeberl B, Eichler-Jonsson C, Gilles ED, Muller G. Computational modeling of the dynamics of the MAP kinase cascade activated by surface and internalized EGF receptors. *Nat Biotechnol.* 2002;20:370–375.

169. Buday L, Warne PH, Downward J. Downregulation of the Ras activation pathway by MAP kinase phosphorylation of Sos. *Oncogene.* 1995;11:1327–1331.

170. Corbalan-Garcia S, Yang SS, Degenhardt KR, Bar-Sagi D. Identification of the mitogen-activated protein kinase phosphorylation sites on human Sos1 that regulate interaction with Grb2. *Mol Cell Biol.* 1996;16:5674–5682.

171. Wang Z, Moran MF. Requirement for the adapter protein GRB2 in EGF receptor endocytosis. *Science.* 1996;272:1935–1939.

172. Citri A, Yarden Y. EGF-ERBB signalling: towards the systems level. *Nat Rev Mol Cell Biol.* 2006;7:505–516.

173. Haglund K, Dikic I. Ubiquitylation and cell signaling. *EMBO J.* 2005;24:3353–3359.

174. Pulverer BJ, Kyriakis JM, Avruch J, Nikolakaki E, Woodgett JR. Phosphorylation of c-jun mediated by MAP kinases. *Nature.* 1991;353:670–674.

175. Hibi M, Lin A, Smeal T, Minden A, Karin M. Identification of an oncoprotein- and UV-responsive protein kinase that binds and potentiates the c-Jun activation domain. *Genes Dev.* 1993;7:2135–2148.

176. Kyriakis JM, Avruch J. Mammalian mitogen-activated protein kinase signal transduction pathways activated by stress and inflammation. *Physiol Rev.* 2001;81:807–869.

177. Kyriakis JM, Avruch J. pp54 microtubule-associated protein 2 kinase. A novel serine/threonine protein kinase regulated by phosphorylation and stimulated by poly-L-lysine. *J Biol Chem.* 1990;265:17355–17363.

178. Weston CR, Davis RJ. The JNK signal transduction pathway. *Curr Opin Genet Dev.* 2002;12:14–21.

179. Brewster JL, de Valoir T, Dwyer ND, Winter E, Gustin MC. An osmosensing signal transduction pathway in yeast. *Science.* 1993;259:1760–1763.

180. Albertyn J, Hohmann S, Thevelein JM, Prior BA. GPD1, which encodes glycerol-3-phosphate dehydrogenase, is essential for growth under osmotic stress in *Saccharomyces cerevisiae*, and its expression is regulated by the high-osmolarity glycerol response pathway. *Mol Cell Biol.* 1994;14:4135–4144.

181. Nishimoto S, Nishida E. MAPK signalling: ERK5 versus ERK1/2. *EMBO Rep.* 2006;7:782–786.

182. Shields JM, Pruitt K, McFall A, Shaub A, Der CJ. Understanding ras: 'It ain't over 'til it's over'. *Trends Cell Biol.* 2000;10:147–154.

183. Whitmarsh AJ, Davis RJ. Structural organization of MAP-kinase signaling modules by scaffold proteins in yeast and mammals. *Trends Biochem Sci.* 1998;23:481–486.

184. Vignais ML, Sadowski HB, Watling D, Rogers NC, Gilman M. Platelet-derived growth factor induces phosphorylation of multiple JAK family kinases and STAT proteins. *Mol Cell Biol*. 1996;16:1759–1769.

185. Zhong Z, Wen Z, Darnell JEJ. Stat3: a STAT family member activated by tyrosine phosphorylation in response to epidermal growth factor and interleukin-6. *Science*. 1994;264:95–98.

186. Wagner BJ, Hayes TE, Hoban CJ, Cochran BH, Doolittle RF. The SIF binding element confers sis/PDGF inducibility onto the c-fos promoter. *EMBO J*. 1990;9:4477–4484.

187. Doolittle RF, Hunkapiller MW, Hood LE, et al. Simian sarcoma virus oncogene, v-*sis*, is derived from the gene (or genes) encoding a platelet-derived growth factor. *Science*. 1983;221:275–277.

188. Blaukat A, Barac A, Cross MJ, Offermanns S, Dikic I. G protein-coupled receptor-mediated mitogen-activated protein kinase activation through cooperation of $G\alpha_q$ and $G\alpha_i$ signals. *Mol Cell Biol*. 2000;20:6837–6848.

189. Schmitt JM, Stork PJ. $G\alpha$ and $G\beta\gamma$ require distinct Src-dependent pathways to activate Rap1 and Ras. *J Biol Chem*. 2002;277:43024–43032.

190. Hall RA, Premont RT, Lefkowitz RA. Heptahelical receptor signaling: beyond the G protein paradigm. *J Cell Biol*. 1999;145:927–932.

191. Kranenburg O, Verlaan I, Hordijk PL, Moolenaar WH. G_i mediated activation of the Ras/MAP kinase pathway involves a 100 kDa tyrosine-phosphorylated Grb2 SH3 binding protein, but not Src nor Shc. *EMBO J*. 1997;16:3097–3105.

192. Daub H, Wallasch C, Lankenau A, Herrlich A, Ullrich A. Signal characteristics of G protein-transactivated EGF receptor. *EMBO J*. 1997;16:7032–7044.

193. Prenzel N, Zwick E, Daub H, Leserer M, Abraham R, Wallasch C, Ullrich A. EGF receptor transactivation by G-protein-coupled receptors requires metalloproteinase cleavage of proHB-EGF. *Nature*. 1999;402:884–888.

194. Zwick E, Daub H, Aoki N, Yamaguchi-Aoki Y, Tinhofer I, Maly K, Ullrich A. Critical role of calcium-dependent epidermal growth factor receptor transactivation in PC12 cell membrane depolarization and bradykinin signaling. *J Biol Chem*. 1997;272:24767–24770.

195. Pai R, Soreghan B, Szabo IL, Pavelka M, Baatar D, Tarnawski AS. Prostaglandin E2 transactivates EGF receptor: a novel mechanism for promoting colon cancer growth and gastrointestinal hypertrophy. *Nat Med*. 2002;8:289–293.

196. Regan JW. EP2 and EP4 prostanoid receptor signaling. *Life Sci*. 2003;74:143–153.

197. Buchanan FG, Gorden DL, Matta P, Shi Q, Matrisian LM, DuBois RN. Role of β-arrestin 1 in the metastatic progression of colorectal cancer. *Proc Natl Acad Sci U S A*. 2006;103:1492–1497.

198. Wang D, DuBois RN. Prostaglandins and cancer. *Gut*. 2006;55:115–122.

CHAPTER 13

Signal Transduction to and from Adhesion Molecules

Here we consider the adherence of cells to surfaces or to other cells and ask how this affects their responses to soluble agonists such as growth factors; also, how soluble agonists affect cellular adherence, itself a signalling event important in the maintenance of stem cell compartments and the epithelial mesenchymal transition (see page 432). The molecules that effect adhesion serve both as targets for signals that are generated within cells (*inside-out signalling*) and as receptors for extracellular signals (*outside-in signalling*). These two aspects are well exemplified in the regulation of survival, proliferation, and differentiation, and in leukocyte trafficking (also discussed in the following chapter). This list is far from complete. Adhesion molecules are of prime importance in the functioning of synapses, nerve cell differentiation, differentiation of keratinocytes, gene expression in epithelial cells, thymic selection, and T lymphocyte activation.

Adhesion molecules

Adhesion molecules were originally thought of merely as a sort of glue. We now recognize that they also act as signalling molecules and are properly described as receptors (Figure 13.1). However, the ligands that interact with adhesion molecules are generally insoluble, frequently adhesion molecules themselves. They are presented on adjacent cells or by the extracellular matrix on the surfaces of epithelia or endothelia, or by the interstitial matrix of connective tissue.

Adhesion molecules first came to light around 1970, as a result of investigations of brain development. It was realized that the very precise organization of neuronal cells in the central nervous system must require a dynamic process of cell guidance and cell adhesion. This would drive the direction-seeking processes of neurite outgrowth and synapse formation. Two main ideas were considered.[1] The first suggested that during development, in order to establish precise cell–cell contacts, the interacting cells must each present unique adhesion molecules that fit each other like a lock and key (*chemoaffinity hypothesis*). The second idea was that the set of adhesion molecules is limited, but their binding capacity could be modulated over time. For instance, developing neuronal cells would all offer the same molecule and during outgrowth, this would be in a low-affinity state. The cell might then convert the adhesion molecule into its high-affinity state, to promote binding to its counterpart on a nearby cell.

It now appears that there is truth in both these propositions. The number of adhesion molecules is certainly limited and their capacity to interact with counter receptors is regulated by their levels of expression (presence or absence) and also by their binding state (affinity). In the realm of immunology,

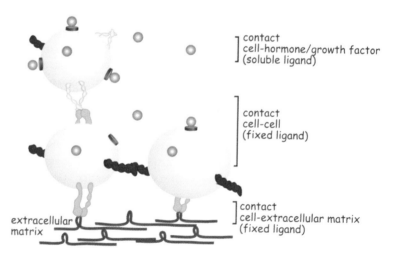

FIG 13.1 Three levels of communication. Cells are not only in contact with soluble ligands (e.g. hormones and cytokines), but also with each other and with the extracellular matrix. These contacts are mediated by adhesion molecules, which in many instances should be regarded as receptors that convey signals into the cells.

contact
cell-hormone/growth factor
(soluble ligand)

contact
cell-cell
(fixed ligand)

contact
cell-extracellular matrix
(fixed ligand)

extracellular matrix

the set of adhesion molecules expressed on a cell surface and their state of activation has been called the 'area code'.[2]

Clear evidence for specific adhesion interactions came from studies of the re-aggregation of disaggregated tissues. Tissue cells dispersed by treatment with trypsin, which strips proteins from the surface, only recover their adherent properties after a period in culture. The need for a recovery period suggested that new adhesion molecules must be expressed on the cell surface. The re-aggregation can be prevented by monovalent Fab_1 fragments prepared from antibodies that bind to cell surface epitopes (Figure 13.2). The possibility of weak non-specific interactions was precluded. It followed that there is a limited number of specialized molecules that determine cell–cell interactions. These are the adhesion molecules and the first to be discovered was the neuronal cell adhesion molecule (NCAM).[3]

Early attempts to identify individual adhesion molecules were hampered by the lack of specific antibodies. The key was provided by the advent of monoclonal antibodies and their application to questions regarding cell adhesion. This led to the identification of Mac-1, which is expressed on the surface of macrophages and plays a pivotal role in the binding of leukocytes to the vascular endothelium.[4] It also determines the binding of the serum complement factor iC3b, which, together with the Fc receptor, mediates activation of the respiratory burst. Later, a monoclonal antibody that recognizes LFA-1 (lymphocyte function-associated antigen-1) was used to show that the binding of cytotoxic T lymphocytes to their target cells is also mediated by specific adhesion molecules.[5]

The many abbreviations used in this chapter are collected together at the end of the chapter.

Complement is the term originally used to refer to the heat-labile factor in serum that causes immune cytolysis, the lysis of antibody-coated cells. It now refers to the entire functionally related system comprising at least 20 distinct serum proteins. As well as being the effector of cytolysis, it has other functions.

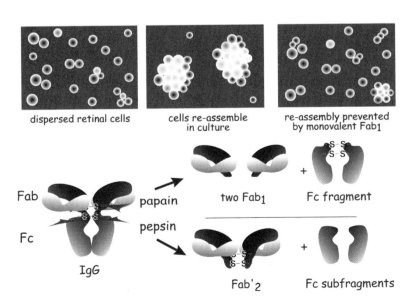

dispersed retinal cells

cells re-assemble in culture

re-assembly prevented by monovalent Fab₁

Fab

Fc

IgG

papain

pepsin

two Fab₁

Fc fragment

Fab'₂

Fc subfragments

FIG 13.2 Reassembly of retinal dispersed cells: discovery of NCAM. (a) The re-assembly of dispersed cells is prevented by antibodies that recognize cell membrane proteins. By this approach, the role of the adhesion molecule NCAM in the maintenance of tissue integrity was discovered. Since intact bivalent antibodies would have forced the cells to aggregate, it was essential to use monovalent Fab fragments that are unable to form cross-links.(drawing of image from Brackenbury et al.[3]). (b) Generation of the fragments Fc, Fab₁ (monovalent), and Fab₂ (divalent) by enzyme digestion of immunoglobulin G.

Adhesion molecules not only link cells to surfaces but may also make the connection between the extracellular matrix and the cytoskeleton. Because it was perceived that this class integrates intracellular and extracellular events, they were called integrins.[6] These proteins can bind to a wide range of extracellular matrix molecules, including fibronectin, fibrinogen, laminin, osteopontin, thrombospondin, vitronectin, and von Willebrand factor. A number of integrins bind these proteins by recognizing short sequences such as RGD or EILDV.[7] Mac-1 and LFA-1 share substantial sequence homology with the integrins and it was confirmed that they too are members of this family of adhesion molecules.

Naming names

Adhesion molecules have been given names that reflect their function (inter-cellular adhesion molecule-1, ICAM-1), their location and function (endothelium leukocyte adhesion molecule-1, ELAM-1), their need for Ca^{2+} (Ca^{2+}-dependent adherence, cadherins), their time of expression during T cell activation (very late antigen-4, VLA-4), or their integration of the extracellular matrix with the intracellular cytoskeleton (integrins). Of course, there are other names, often bestowed after cloning (αM integrin) or through recognition by specific monoclonal antibodies (CD11b, cluster of differentiation 11b). If this is not all perfectly clear, then one should appreciate that CD18/CD11b is synonymous with Mac-1, which is synonymous with integrin αMβ2, and that CD62E is synonymous with ELAM-1 which is synonymous with E-selectin. αLβ2 is synonymous with LFA-1. Such are the problems of nomenclature when molecular cloning rubs shoulders with immunology, pharmacology, and all the rest.

Immunoglobulin superfamily

NCAM is a member of a large family of cell surface proteins that express repeated immunoglobulin-like domains at their extracellular N-termini. Proteins of the immunoglobulin superfamily that play a role in adhesion are called Ig-cellular adhesion molecules (Ig-CAMs). Their Ig-like, globular, loop structures are stabilized by sulfydryl bridges and are resistant to proteases (Figure 13.3a). The Ig-like domains are then subclassified as C1-, C2-, V-, and I-set (or -type), based on their similarity to the variable and constant regions of antibodies.

Examples of C2-set Ig-CAMs are VCAM-1 (vascular cell adhesion molecule-1),[8] NCAM-1 and -2, PECAM (platelet endothelial cell adhesion molecule, CD31), and IGS4B (immunoglobulin superfamily member 4B) (Figure 13.3b). The latter contains two C2-set Ig-like domains and one V-set domain. There are two splice variants of VCAM-1 that express either five or seven Ig-like C2-set domains. VCAM-1 is highly expressed on endothelial cells and certain fibroblasts when present in inflamed tissues. VCAM-1 binds integrins α4β1 (VLA-4) and αDβ2. Many of the Ig-CAM subfamilies have not yet been fully explored with respect to their function and to their signalling pathways.

ICAM

Intercellular adhesion molecules, ICAM-1–5, are a subfamily of Ig-CAMs with varying numbers of Ig-like C2-set domains. They bind the integrin αLβ2 (LFA-1) and a diverse range of other ligands/counter-receptors. ICAM-1 and -2 are important for the recruitment of leukocytes into the tissues. Upon binding of LFA-1, they signal into the cell through members of the Rho family of GTPases. This mediates subsequent transendothelial migration through the loosening of the VE-cadherin cell–cell contact sites and by rearrangement of the actin cytoskeleton.[10,11] This will be further elaborated in Chapter 14.

The interaction with the actin cytoskeleton is mediated through members of the ERM family (ezrin/moesin/radixin). ICAM-2 on vascular endothelial cells also supports homophilic interactions that may be involved in vascular tube formation during the process of angiogenesis (formation of new vasculature from a pre-existing vascular bed).[12] ICAM-3 is restricted to leukocytes[13] and ICAM-4 to red blood cells, constituting the Landsteiner–Wiener blood group.[14] ICAM-5 is expressed in neurons[15] (see Figure 13.3).

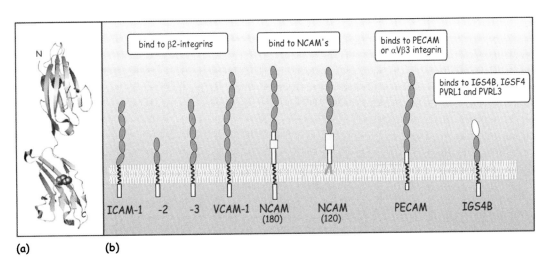

FIG 13.3 Adhesion molecules of the immunoglobulin superfamily. (a) Ig family adhesion molecules are characterized by repeated domain structures that are homologous to those in immunoglobulins (Ig domains). The disulfide bonds are indicated by red spheres. (b) There are several members of this family, all having a single membrane-spanning domain. They interact with different ligands (or counter-receptors). NCAM 120 is attached to the membrane by a glycosyl-phosphatidyl inositol anchor. (1epf[9]).

The word 'sialic' is derived from the Greek **sialos,** meaning saliva, where sialic acids were first discovered. They are nine-carbon sugars.

SIGLEC

The SIGLECs (sialic acid binding Ig-like lectin proteins) constitute another subgroup within the superfamily of immunoglobulins. Its members include sialoadhesin (SIGL1), CD22 (SIGL2), CD33 (SIGL3), MAG (SIGLET4), and SIGL5–11. They are characterized by the presence of a single N-terminal Ig-like V-set domain which has the characteristics of an I-type lectin[16] that binds sialic acid (see also below, page 392). Anywhere between 1 and 17 Ig-like C2-set domains lie between the sialic acid binding site and the plasma membrane (Figure 13.4). Each SIGLEC has a preference for a specific type of sialic acid and for a specific type of linkage to the subterminal sugar. The cytosolic tails vary in sequence and length but most have conserved tyrosine residues within immunoreceptor tyrosine-based inhibition motifs (ITIMs) that are implicated

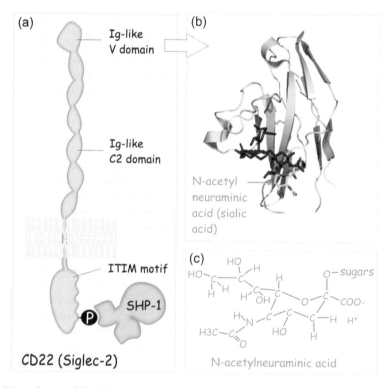

FIG 13.4 Structure of CD22/SIGLEC. (a) The extracellular domain of CD22 is composed of six Ig-like C2-set domains with an N-terminal Ig-like V-set domain that binds sialic acid. The cytoplasmic domain contains four ITIM motifs with tyrosine phosphorylation sites. One of these is recognized by one of the two SH2 domains of the phosphatase SHP-1. (b) Detail of the binding of N–acetylneuraminic (a sialic acid, green sticks) to the Ig-like V-set domain (mouse). (c) Structure of sialic acid. (2hrl[17]).

in signalling functions. SIGLEC adhesion molecules are present in myeloid cells (granulocytes, lymphocytes, monocytes), Schwann cells, and placental trophoblasts. They are not expressed in *Drosophila* or *Caenorhabditis elegans*, probably representing a later adaptation of an ancient protein binding Ig domain (see Figure 13.4 and Table 13.1).

Of the many SIGLECs, CD22, a single-span membrane protein which recognizes sialic acid linked to galactose and is expressed on cells of B-cell lineage, has attracted much attention. This is because it is an inhibitory coreceptor that down-modulates signalling through the B cell receptor (BCR). It does this by setting a threshold that prevents overstimulation, important in the maintenance of tolerance to some antigens. In mice, disruption of the CD22 gene gives rise to a hyper-responsive BCR and the animals exhibit an augmented immune response.[18]

CD22 binds to sialic acid residues that are attached to the B cell itself (*in cis*), as well as those that are attached to other cells that present antigens (*in trans*). Binding *in cis* keeps CD22 away from the BCR and enhances signalling.[19] When this is prevented, for instance through binding to sialic acid residues of an adjacent antigen-presenting cell or by binding directly to the engaged antigen receptor, it inhibits BCR signalling. The reason for this is that the activated BCR recruits Lyn kinase which phosphorylates the adjacent tyrosine residues in the three ITIM motifs of CD22.[20] This in turn engages the tyrosine

TABLE 13.1 Human Siglec adhesion molecules

Human Siglecs	Ig-C2 domains	Expression
Sialoadhesin/hSiglec-1	17	macrophage
CD22/Siglec-2	7	B cell
CD33/Siglec-3,	2	monocyte, myeloid progenitor
MAG/hSiglec-4,	5	oligodendrocyte, Schwann cell
hSiglec-5	4	monocyte, neutrophil, B cell
hSiglec-6	3	B cell, placental trophoblast
hSiglec-7	3	monocyte, NK cell
hSiglec-8	3	eosinophil, basophil, mast cell
hSiglec-9	3	monocyte, neutrophil, NK cell
hSiglec-10	5	monocyte, NK cell
hSiglec-11	5	macrophage

phosphatase SHP-1 which prevents BCR signalling by dephosphorylation of its own ITAM motifs (see also page 517).[21]

Junctional adhesion molecules (JAMs)

The JAMs constitute a further subgroup of the immunoglobulin superfamily. Four members, JAMs 1–3 and JAML1, are expressed in humans. They are single-span membrane proteins with two extracellular Ig-like domains, of which the N-terminal is a V-set and the C-terminal an I-set domain. They possess a PDZ binding motif which interacts with the adaptor protein ZO-1 (zonula occludens-1) (Figure 13.5). They contribute to the architecture of tight junctions in epithelial and endothelial cells and are also expressed on lymphocytes, megakaryocytes, platelets and red blood cells. Both homo- and heterophilic interactions have been reported. Importantly, they play a role in leukocyte transendothelial migration (see page 500).

Claudins

The claudins form a large group of adhesion molecules (21 members in human, designated Cldn-1–21). They have a molecular mass of 23 kDa, span the membrane four times, and have short cytosolic N- and C-terminal domains. They are members of an even larger family of adhesion molecules, the PMP22/EMP/MP20/claudin proteins, all having the same membrane topology and a characteristic signature motif in the first extracellular loop (Figure 13.5).

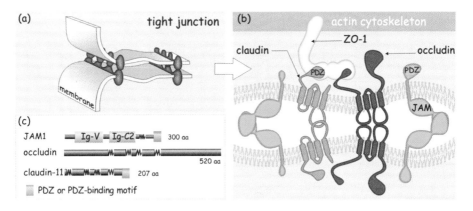

FIG 13.5 Tight junction proteins: JAM, occludin, and claudin.

(a) Diagram illustrating a tight junction as present in endothelial and epithelial tissues. (b) The junctions are composed of three types of protein. JAM is a member of the immunoglobulin family of adhesion molecules, but claudin and occludin are not. Both of these span the membrane four times and have very short extracellular domains. Their interaction ensures close apposition of the two membranes. There are many variants of the claudins. The combination expressed in tight junctions determines the accessibility of the paracellular space. They all bind to ZO-1 (or ZO-2, ZO-3) which connects the junctional proteins to the actin cytoskeleton. (c) Domain architecture of junctional proteins. Cyan indicates PDZ or PDZ binding motif.

The claudins contribute to the selectivity of paracellular transport (meaning transport between, not through, cells). Thus, they are the principle barrier-forming proteins of the tight junction and their expression pattern determines the effectiveness of barriers imposed by epithelial sheets. Epithelia may have high resistance, maintaining steep ionic gradients (such as in the distal nephron and urinary bladder) or they may be leaky, of low resistance, allowing the movement of large volumes of iso-osmotic fluid (as occurs in much of the gastrointestinal tract).[22,23] The C-terminus contains a PDZ motif which binds homologous motifs in adaptor proteins such as ZO-1, 2, and 3.[24]

Occludin

Occludin is a four-transmembrane-spanning protein of 60 kDa having a short N-terminal and a much longer C-terminal region both exposed in the cytosol. Only a single gene exists. It is a component of tight junctions in endothelial and epithelial cells (Figure 13.5). Its typical apical localization occurs through interaction with the ZO adaptor proteins, which are linked to the actin cytoskeleton, but unlike the claudins it does not act as a regulator of paracellular transport.

Integrins

The integrins are the most dynamic and versatile of the adhesion molecules. They are composed of two subunits (α and β), linked non-covalently. There are at least 18 α- and 8 β-subunits and these can form at least 25 different integrin heterodimers (Figure 13.6). Depending on the particular $\alpha\beta$ combination, they may bind to either ICAMs, VCAM-1, or MadCAM (on mucosal cells). They may also bind to components of the extracellular matrix such as collagen, fibronectin, laminin, or vitronectin and to the blood proteins fibrinogen, or von Willebrand factor.

The subunits of the integrins (α and β) are transmembrane glycoproteins consisting of a head, a stalk, and a short (40–60 amino acid) cytoplasmic region. Integrin $\beta4$, which is specialized to connect to the keratin cytoskeleton in hemidesmosomes (complexes at epithelial junctions with the extracellular matrix), has a more extended cytoplasmic region. The stalk is composed of a number of highly conserved domains as depicted in Figure 13.7). As will become evident, the positioning of the 'hybrid domain' within the stalk region of the β-integrin subunit has a role in regulating the affinity for ligand. The head regions of β-integrin subunits all have a $\beta1$ domain (also referred to as the I-like domain), while the head regions of α-integrin subunits vary according to integrin type. They all contain a β-propeller structure, but some have an additional αI domain; (these integrins are indicated with an asterisk in Figure 13.6). Two examples of the different α-subunits are presented in Figure 13.7

Claudins 3 and 4 act as receptors for the bacterial endotoxin CPE produced by *Clostridium perfringens.* This causes a profound change in the structure of the tight junctions of the enterocytes, followed by an increased paracellular leakage of water. The failure to reabsorb water in the colon results in diarrhoea.

Tight junctions are also referred to as **zonula occludens** (hence ZO). The Latin word occludens comes from **occludere,** meaning to lock up. Claudin is derived from the Latin **claudere,** meaning to close.

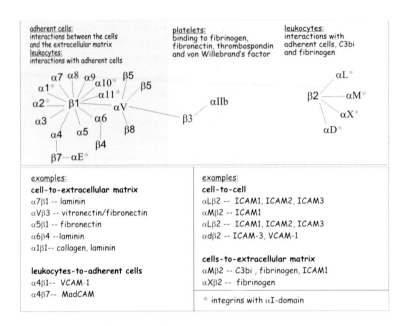

FIG 13.6 Mixing and matching integrin subunits.

Numerous combinations of α- and β-subunits create integrins with specific ligand-binding characteristics. The β2-integrins are restricted to blood cells, β1 are present on all cells. Platelets have the unique combination αIIbβ3, vital in blood clotting. Integrins bind to specific sequences in proteins of the extracellular matrix such as RGD or EILDV.

(comparing αV and αL). Importantly, αI- and β1-domains have similar folds and can both interact with ligand.

The β-propeller structure and the β1-domain are responsible for linking the α- and β-subunits.[26] This interaction resembles the linking of α- and β-subunits of heterotrimeric GTP binding proteins (see page 83). The β1-domain is a hotspot for mutations, which can result in a failure of association and can give rise to leukocyte adhesion deficiency (LAD), characterized by frequent unresolved infections (see Chapter 16).

Inactive to primed

Integrins can exist in low affinity (inactive), high affinity (primed), or ligand-bound (activated) states, determined by conformational changes in the head region.

The shift from low to high affinity is directed from inside the cell by association of proteins with the cytosolic region of the integrin α- and β-subunits (inside-out signalling). Association of these proteins disrupts the interaction between the juxtamembrane region of the α- and β-subunits

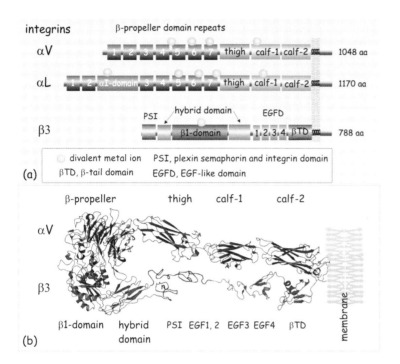

FIG 13.7 Domain architecture of integrins. (a) Integrins, involved in cell–cell and cell–extracellular matrix contact, are composed of two non-covalently bound subunits (α and β). Two examples of a subunit are shown. The stalk region of the α-subunit comprises two calf domains and a thigh domain. The protein articulates at the border of calf-1 and thigh, the PSI and hybrid domain of the β-subunit being pushed outward in a switchblade movement. This plays an important role in the activation process. The heads contain various subdomains of which the β-propeller (on the α-subunit) and β1 domain (on the β-subunit) assure the association of the two subunits. (b) Molecular structure of αV and β3. Depending on the type of integrin, ligand recognition either occurs through the β1 domain or through the α domain (in αL). Yellow spots indicate divalent metal ions; PSI, plexin 7 integrin domain; βTD, β-tail domain; EGFD, EGF-like domain (1jv2[25]).

which leads to their separation. This then alters the stalk region, which unfolds from a bent to an extended conformation, like an upward switchblade movement (Figure 13.8a). A possible candidate for the juxtamembrane positioning of the integrins is talin, which binds both integrins and components of the actin cytoskeleton. A mutation in the FERM domain (F3) of talin that prevents binding to integrin has a profound effect on the affinity of integrins for extracellular matrix in *Drosophila* but, surprisingly, does not perturb the organization of the actin cytoskeleton around focal adhesion complexes[27] (see page 398).

The switchblade movement of the stalk region and its effect on the head structure has been studied in great detail for integrin αIIbβ3, expressed on platelets.[28] As the stalk region straightens, there is a movement of the hybrid domain of the β-subunit which causes a shape change of the neighbouring β1 domain (see Figure 13.8a, b). This exposes the ligand binding site and converts the integrin into a primed, high-affinity, adhesion molecule. Similar changes occur on those integrins that carry an I domain (αLβ2) but here the change has to be propagated by means of an intrinsic ligand (Figure 13.8c).[29,30]

The integrins of circulating cells such as leukocytes are maintained in an inactive state and are stimulated by chemokines acting through G-protein-coupled 7TM receptors. In contrast, for tissue cells the integrins appear to be constitutively activated, though local inactivation can occur, as must be the case in migrating cells or cells that round up for division.

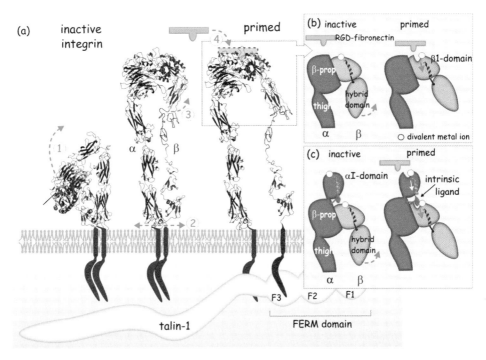

FIG 13.8 Mechanisms of integrin activation. (a) Integrin activation is depicted here as a three-step process. Inactive integrin is in a folded state, head down (1). Though the mechanism that separates the cytosolic domains is not fully understood, binding of talin-1, through the F3 region in the FERM domain, is important. The separation (2) causes an upward switch-blade movement, followed by a shift of the hybrid domain of the β-subunit, which moves to the outside position (3). The integrin is now primed and ready to bind ligand (4). (b) The outward movement of the hybrid domain changes the conformation of the β1 domain, creating a ligand binding site. (c) For those integrins that carry an α1 domain, the mechanism of activation is identical, but the change in conformation is indirect. The movement of the hybrid domain causes an effective intrinsic ligand to bind at a site which, when occupied, leads to a conformational change in the α1 domain, creating a binding site for fibronectin. Yellow spots indicate divalent metal ions. Information from Xiong et al.,[25] Shimaoka et al.,[26] Xiao et al.,[31] and Carman and Springer[32] (1jv2[25]).

Primed to active

Ligand binding causes further alterations in the relative positioning of the stalk regions[33] so that they recruit proteins that constitute an intracellular signalling complex. Some 50 proteins have been implicated in the localization (transient or stable) of these focal adhesion complexes.[34,35] It is not yet clear which proteins actually sense the ligand-bound conformational change and pass the message on into the cell. The clustering of integrins that follows ligand binding (formation of focal adhesion contacts) probably plays an important part in the formation of intracellular signalling complexes, though clustering alone fails to reproduce full outside-in signalling events.[36]

Finally, amongst the proteins that localize in the focal adhesion complex are ILK (integrin-linked kinase), α-actinin, talin, paxillin, and filamin. These are all multidomain proteins that connect integrins with components of the actin cytoskeleton.[37] Once again, integrin β4 present in hemidesmosomes is an exception (see above), being coupled to intermediate filaments through plectin.

Cadherins

In the formation of the early embryo, the compaction of the morula is mediated through Ca^{2+}-dependent adhesion molecules[38] (Figure 13.9). Uvomorulin, one of the first to be identified, is instrumental in the transition from a grape-like to a mulberry-like object. The cellular junctions thus formed

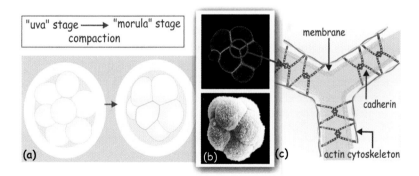

FIG 13.9 Role of cadherin in the compaction of the eight-cell stage mouse embryo. Scanning electron microscopy reveals that after three cell divisions (eight-cell stage) mouse embryos change from a uva (grape)-like to a morula (mulberry)-like aggregate. This process, called compaction, enables the cells to attach firmly to each other, manifesting the first signs of polarization. Their morphology now presents distinct basal (contact) and apical (peripheral) membrane surfaces. Cadherins play an important role in the compaction process, they are localized at cell–cell boundaries and their link to the actin cytoskeleton allows both the compaction of the embryo and the profound shape change of the cells. Images in centre panel courtesy of Dr Alexandre, Mons, Belgium.

A **morula** (from the Latin *morus*, a mulberry) is an embryo at an early stage of embryonic development, consisting of ~12–32 cells (called blastomeres). During compaction, the blastomeres change their shape and tightly align with each other to form a compact ball of cells. This process is mediated by adhesion molecules, cadherin being one of the most important.

are linked by a contractile network of actin filaments that by shortening, pulls the embryonic cells together in a manner similar to a purse-string. Compaction not only tightens the bonds between the cells, it also introduces, for the first time, morphological polarity. From here on, embryonic cells have a smooth basal surface on one aspect and an apical surface dotted with microvilli on the other.

Investigations using teratocarcinoma F9 embryonal carcinoma cells have revealed a range of Ca^{2+}-dependent adhesion molecules, collectively the cadherins.[39] They comprise a large family (at least 36 members in humans) mediating homotypic cell–cell adhesion, acting as both receptor and ligand. The first to be discovered were E-cadherin (CDH1, initially named L-CAM)[40] and uvomorulin.[41] Many cadherins are named after the tissue in which they were discovered (epithelial, placental, neuronal, etc.), but these labels have little meaning since most of them are more widely expressed. The individual cadherins do, however, show restricted and distinct expression patterns.

Cadherins are categorized on the basis of their conserved ectodomain modules (EC), each of about 110 amino acids (so-called cadherin repeats) (Figure 13.10b). These modules are numbered, starting with EC1 at the N-terminus. Some cadherins have as many as 34 of these repeats, though most have only 4 or 5. Today's subdivision of the cadherin family is set out in Table 13.2. Class I (classical) and class II (atypical) cadherins have very similar EC1 modules, but class I cadherins are distinguished by the presence of a His-Ala-Val (HAV) sequence. Two other subfamilies, the desmocollins and desmogleins (collectively, desmocadherins), are associated with intermediate filaments, not actin filaments as are most of the cadherins. Desmogleins differ from the desmocollins by virtue of their more extended cytoplasmic domains. Remaining are those cadherins having low (<44%) sequence similarity with E-cadherins and which are apparently not linked to the cytoskeleton. Among the cadherin-related proteins are the protocadherins, which show 30% sequence similarity with E-cadherin. On this basis and further detailed sequence analysis of 50 different EC1 domains, 6 major subfamilies are now recognized amongst different species, besides several solitary members (see Table 13.2).[42]

Most, if not all, of what we know about the cadherins concerns the classical or type-I cadherins and in particular its archetype, E-cadherin. Classical cadherins are transmembrane glycoproteins having 5 EC domains. Of these, EC1–4 are very similar while EC5 is more distant and therefore referred to as the membrane-proximal extracellular domain (EM). Interdomain stabilization is achieved through the binding of three calcium ions at the domain interfaces[42] (Figure 13.10b).

Cadherins generally mediate homotypic cell–cell adhesion, acting as both receptor and ligand (Figure 13.10b). Both types I and II cadherins exhibit a dimeric configuration with adhesive interfaces that are confined to the EC1

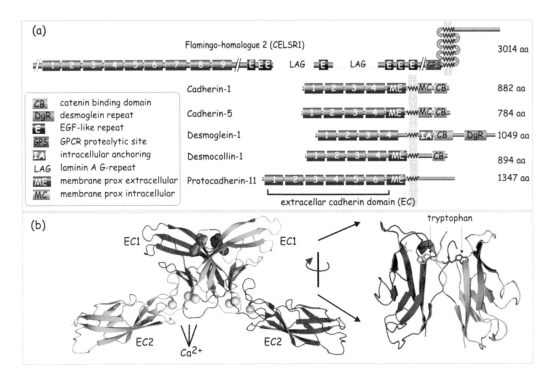

FIG 13.10 Cadherin domain structure. (a) Cadherins form a large family of adhesion molecules, characterized by extracellular cadherin (EC) domains and classified on the sequence of EC1. In many cases, the membrane-proximal EC domain is only distantly related and better regarded as the membrane-proximal extracellular domain. Many cadherins carry a catenin binding domain which links to the actin or intermediate filament cytoskeleton. (b) A ribbon representation of two cadherin repeats (EC1 and EC2 of cadherin-11) with Ca^{2+} binding sites at the domain interface. The interaction between the two cadherins occurs through the N-terminal EC1 domain and is based on a domain swap, in which tryptophan residues, (sticks) play an important role. Ca^{2+} ions indicated as yellow spots. (2a4e[43]).

domains.[43] Anchoring occurs through the insertion of an EC1 side chain into a complementary hydrophobic pocket in the partner molecule and vice versa. This so-called *strand exchange* exemplifies a more general domain-swapping strategy which enables homophilic interactions between proteins having low affinity yet high specificity.[44]

Full functionality of the classical cadherins requires structural linkage to the cytoskeleton. To this end, they are organized as a 'core complex'. This also includes β-catenin bound directly to the cytoplasmic domain of cadherin and α-catenin bound to the N-terminal region of β-catenin. α-Catenin is also bound to actin and to several actin binding proteins such as α-actinin, ZO-1, vinculin, and formin. All this suggests that the complex plays a role in

TABLE 13.2 Human cadherin subfamilies

1 Classical/type I cadherins	Epithelial (E-cadherin) (CDH1, cadherin-1)
	Neural (N-cadherin) (CDH2)
	Placental (P-cadherin) (CDH3)
	Retinal (R-cadherin) (CDH4)
	T-cadherin (truncated cadherin-GPI anchor) (CDH13)
	Muscle (M-cadherin) (CDH15) (Hs)
2 Atypical/type II cadherins	Vascular endothelial (VE-cadherin) (CDH5)
	Kidney (K-cadherin) (CDH6)
	Cadherin-7 (CDH7)
	Cadherin-8 (CDH8)
	Cadherin-9 (CDH9)
	Cadherin-10 (CDH10)
	Osteoblast (OB-cadherin) (CDH11)
	Brain (BR-cadherin) (CDH12)
	CDH18 type 2
	CDH19 type 2
	CDH20 type 2
3 Desmocollins (desmo-cadherins)	Desmocollin-1 (Dsc1)
	Desmocollin-2 (Dsc2)
	Desmocollin-3 (Dsc3)
4 Desmogleins (desmo-cadherins)	Desmoglein-1 (Dsg1)
	desmoglein-2 (Dsg2)

	desmoglein-3 (Dsg3)
5 Protocadherins	Kidney (Ksp-cadherin) (CDH16)
	Liver–intestine (LI-cadherin) (CDH17)
	Protocadherin α1 (Pdch-α1)
	Protocadherin γc3
	Protocadherin 7
	FAT
	Protocadherin 11
	Protocadherin β15
	Protocadherin 1
	Protocadherin γb1
	Protocadherin γa1
	Protocadherin 8
	Protocadherin β1
	Ret
	Protocadherin 68
6 Flamingo cadherins	CELSR1 (flamingo homolog 2)
(also qualify as 7TM receptors)	CELSR2 (flamingo 1)
	CELSR3 (flamingo homologue 1)
7 Solitary members (some non-human)	Muscle (M-cadherin) (CDH15)
	Ret (Hs)
	Cadherin 3 (Ce)
	Dachsous (Dm)

Adapted from Nollet et al.[42] and from PROSITE.

Catenins were discovered in a search for proteins that associate with uvomorulin. Three proteins of 102, 88, and 80 kDa were repeatedly coimmunoprecipitated and found to be present in mouse, chicken and human cells. The were named catenin (α, β, and γ respectively), from the Latin *catena*, meaning chain, since one major function was thought to be linking the adhesion molecule with the cytoskeleton.[47]

Selectin: homologous to the Ca^{2+}-dependent (C-type) lectin. Lectin is derived from the word 'select', and originally applied to plant proteins that bind to specific carbohydrate residues present on nitrogen-fixing bacteria. They have been used in cell biological work because they bind to specific glycosidic residues present in the Golgi system and on cell surfaces. Only later was it found that animal cells also possess similar proteins, generally on the surfaces of endothelial and myeloid cells.

the formation and maintenance of stable adhesions. Formin is involved in the polymerization of actin, which indicates that cadherin core complexes contribute to the elongation of the actin filaments to which they bind.[45] The juxtamembrane region of cadherin also binds p120ctn (or δ-catenin), involved in the recruitment of cytosolic protein tyrosine phosphatases of various sorts.[46]

In developmental processes, expression of each cadherin subclass is regulated both spatially and temporally and this is associated with individual morphogenic events. Thus, the tendency of cells to segregate or to aggregate correlates with the expression of particular cadherins. This is particularly apparent during gastrulation and the formation of the neural tube. In addition, regulation can also occur through activation, as occurs at the onset of compaction (see Figure 13.9) or in the process of axon guidance where local changes in cadherin adhesion steer the growth cone. Axon guidance cues, like Neurocan and Slit, rapidly inactivate cadherin through the phosphorylation of its intracellular partner β-catenin.[48,49] By eliminating cadherin-mediated traction, these axon guiding cues allow for growth cone extension. Many of the cadherins are implicated in long-term potentiation, memory formation, and spatial learning, most likely through their capacity to form synaptic contacts and thus spatially organize the brain cortical layer.[50]

In epithelial cells, cadherins are present in cell–cell junctions, adherens junctions, and desmosomes (Figure 13.11), being associated with bundles of actin filaments or intermediate filaments via the intermediates of β-catenins and α-catenin or plakoglobin and desmoplakin respectively. By forming cell–cell contacts, the cadherins appear to act as tumour suppressors and when these fail, as in metastasizing epithelial tumours, there is a loss of basolateral localization (see also Chapter 14).

Selectins

Selectins, present on the surface of white blood cells, platelets, and also the endothelial cells that line blood vessels, form a family of adhesion molecules (the selectin/LECAM family, leukocyte–endothelial cell adhesion molecules). They mediate the initial low-affinity adhesion sites for lymphocytes and for leukocytes such as neutrophils. This prepares the way for cell migration into the lymph nodes and tissues. At the N-terminus there is a C-type (Ca^{2+}-binding) lectin domain that enables binding to particular carbohydrate residues (galactose, *N*-acetyl glucosamine, and fucose) present on cell surface glycoproteins and glycolipids. They also contain a single EGF-like domain and a series of Sushi repeats (also referred to as complement regulatory proteins, CRP) (Figure 13.12).

Lymphocytes can be displaced from lymph nodes by monosaccharides (L-fucose and D-mannose) and by the polysaccharide fucoidin (rich in L-fucose). This gives an indication that carbohydrate residues may be involved in the

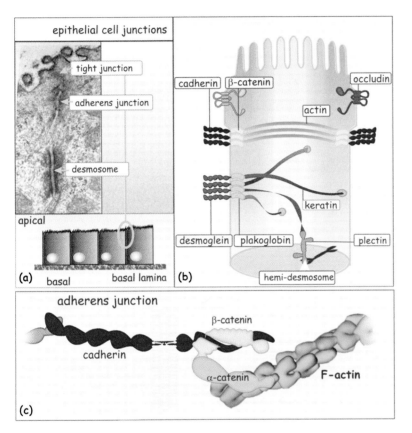

FIG 13.11 Epithelial adherent junctions. (a) Epithelial cells are firmly attached to each other by adherens junctions which play an important role in maintaining both cellular and tissue integrity. (b) Adherens junctions are formed by homophilic interactions between cadherins and they are connected to the cellular adhesion belt formed of actin filaments. Desmosomes are formed by homophilic interactions of desmogleins (or desmocollins), linked to intermediate filaments (keratin in the case of epithelial cells). At the basal membrane these filaments form a hemidesmosome, interacting with integrins via plectin. (c) The tight association between cadherin, β-catenin, α-catenin, and filamentous actin illustrates the physical stress-resisting quality of cell junctions.

homing of lymphocytes.[51] This was confirmed by the finding of a lymph-node-specific homing receptor which possesses a C-type lectin domain.[52] Various selectins, mediating intercellular interactions, have since been identified in vascular endothelial cells (E), platelets (P), and leukocytes (L) (Figure 13.12a). L-Selectin is responsible for lymphocyte homing and is expressed on all circulating leukocytes except for a subpopulation of memory cells. It recognizes CD34 (a heavily glycosylated mucin) on endothelial cells. E- and P-selectins are expressed on endothelial cells and recognize, respectively, sialyl Lewis^X and PSGL-1 on leukocytes. P-selectin and PSGL-1 interact in a dimeric mode; two P-selectins bind to one PSGL-1.

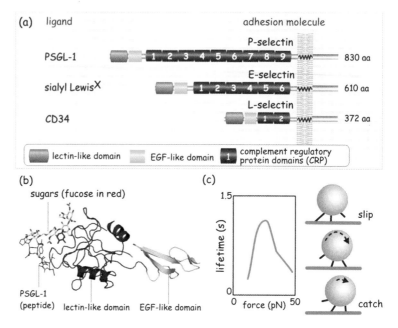

FIG 13.12 Domain structure of selectins. (a) The selectins are characterized by the N-terminal lectin-homology domain and a variable number of complement regulatory protein domains (CRPs). Through their lectin domain, selectins interact with sugar residues present in cell surface glycoproteins and glycolipids. (b) Ribbon representation of the lectin-like domain of E-selectin bound to PSGL-1. Sugars in stick representation; fucose is red. (c) The lifetime of the bonding of P-selectin to PSGL-1 increases with force, then decreases. This may explain the rolling of white blood cells (that express P-selectin) on the vascular endothelium (that expresses PSGL-1) under shear stress (blood flow). Image adapted from Yago et al.[53] and Marshall et al.[54] (1g1s[55]).

Selectins are recognized by their lectin-like domain (LD) which binds sugars (Figure 13.12b). This has the remarkable property that allows its affinity for ligands to be modulated by the imposition of shear forces.[56] The interaction between L-selectin and PSGL-1 shows a biphasic relationship by which low shear forces decrease off-rates, the bond lifetime being prolonged ('catch bonds'), while higher shear forces increase off-rates ('slip bonds'). It appears, although the molecular details remain uncertain, that shearing deforms the interacting molecules such that they lock more tightly.[54] This behaviour, catch and slip, might explain the rolling of leukocytes on the vascular endothelial layer under flow conditions (Figure 13.12).[57]

Little is known about the role of the selectins or the selectin-binding molecules in signal transduction. Several binding partners including α-actinin, calmodulin, and members of the ezrin/radixin/moesin (ERM) family of cytoskeletal proteins attach to the short cytoplasmic tail (17–35 residues) of selectins. With respect to extracellular ligands, PGSL-1 induces tyrosine phosphorylation by Syk, mediated by moesin. This pathway leads to transcriptional activation of the SRE element.[58]

Cartilage link proteins

The glycosaminoglycan hyaluronan is a high molecular weight polysaccharide present in the tissue matrix and body fluids of all vertebrates. It plays a fundamental role in regulating cell migration and differentiation.[59] The majority of hyaluronan-binding proteins belong to the cartilage link protein superfamily. All contain a conserved link module (~100 residues) that has a three-dimensional structure resembling the sugar-binding domain of E-selectin (Figure 13.13). They can be further subdivided into two groups, those that are true cellular adhesion molecules such as CD44[60,61] and LYVE-1,[62] and those that are components of the extracellular matrix such as cartilage link protein itself, aggrecan, versican, brevican, and TSG-6.[63] This second group plays an important role in the architecture of the extracellular matrix, bringing together the pressure-resistant qualities of glycosaminoglycans and the tension-resistant qualities of the extracellular matrix proteins (collagen, elastin), jointly, an essential quality of cartilage in articulated joints.

CD44, originally discovered as the homing receptor of lymphocytes, is the best studied of the cartilage link proteins. It is required when lymphocytes bind to high endothelial venules and exit the circulation, seeking lymph nodes.[66] In a manner reminiscent of the integrins expressed on circulating blood cells, CD44 is functionally inactive. It binds hyaluronan only after T cell receptor triggering or exposure of the lymphocytes to inflammatory cytokines such as TNF-α and IFN-γ. In this way, selected lymphocytes are directed to inflammatory sites by binding to hyaluronan present on the surface of the vascular endothelial cells.

CD44 is a single membrane-spanning protein encoded by 19 exons of which nine in the extracellular domain (v2–v10) are variably spliced (Figure 13.13). Because of the extensive glycosylation and the sequence variability, the molecular mass of CD44 can vary widely, anything between 85 and 200 kDa. The link module is implicated in the binding of hyaluronan but the splicing extends the range of binding capacities. For instance, the inclusion of exon v3 allows the attachment of heparan or chondroitin sulfate, enabling it to interact with fibroblast growth factor-4 or -8 (FGF-4 or -8) and with hepatocyte growth factor/scatter factor (HGF/SF). Other potential ligands are osteopontin, fibronectin, and ankyrin. Homophilic intercellular interactions with CD44 are also reported.

The splice variants stimulated great interest when it appeared that CD44v4–7 might determine metastasis in rat pancreatic tumour cells. In short, it seemed

Glycosaminoglycans and the extracellular matrix proteins, in particular collagen, share with wood and reinforced concrete the valuable qualities of resistance to the forces of both tension and compression. These are two important requirements for the construction not only of organisms but also of buildings.

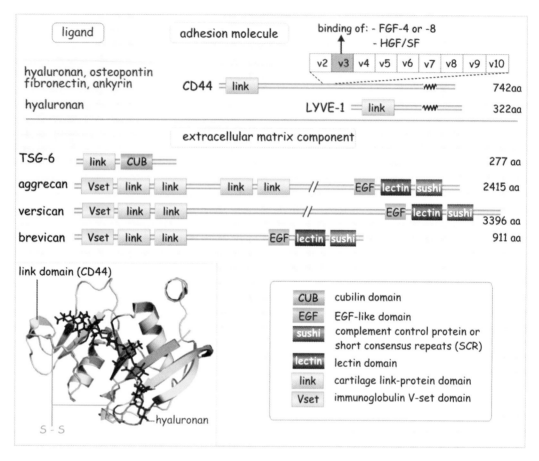

FIG 13.13 Domain structure of the cartilage link proteins. Cartilage link proteins come in two forms, those that are transmembrane proteins and act as receptors, and those that are extracellular matrix proteins and play an important role in connecting glycosaminoglycans with proteins. They are characterized by the link domains that bind hyaluronan. CD44 has the capacity to form many splice variants (inserted variable exons v2–v10), which provide new binding properties independent of the link domain. The inset shows a ribbon representation of the link domain, stabilized by two disulfide bonds, connecting CD44 to a short stretch of hyaluronan (2bvk[64] and 1poz[65]).

at first to have the quality of a metastasis gene product.[67] More than this, a monoclonal antibody raised against v6 prevents metastasis. Unfortunately however, although human tumours often express splice variants of CD44 and in certain instances this predicates poor prognosis, there seem to be no functional implications. The signal transduction pathway emanating from CD44 remains uncertain. A number of membrane proteins, including ezrin, radixin, and moesin, that associate with it, belong to the group of erythrocyte band 4.1-related proteins. These are proposed to function as links between the membrane and the cytoskeleton,[68] but it remains to be shown that they actually relay signals from CD44 into the cell. Another point of interest is of course, the capacity of CD44v3 to bind growth factors of the FGF and HGF/SF families. This may mean that it is involved in the recruitment of the respective receptors, thereby mediating intracellular signalling. In the following sections, signalling to and from CD44 will not be further discussed.

Integrins, cell survival, and cell proliferation

Inside-out signalling and the formation of integrin adhesion complexes

Cells organize their contacts with the extracellular matrix in the form of integrin clusters (see Figure 13.15a, b). These are composed of a number of proteins, some with structural and others with signalling roles. Formation of such focal adhesion sites involves activation of integrins, their binding to a component of the extracellular matrix (fibronectin, vitronectin, laminin, or collagen), the formation of actin filaments, and their attachment to these newly formed filaments. This process also requires the monomeric GTPases Rap1 and RhoA.

Rap1 (see page 251) contributes to integrin activation, and more generally it is a key regulator for the adhesion of blood borne cells and platelets.[69,70] The mechanism of activation is shown in Figure 13.14. Rap1 is activated by the guanine nucleotide exchange factor C3G[71] recruited to tyrosine-phosphorylated receptor tyrosine kinases (PDGFR or EGFR) via the SH2/SH3-adaptor protein Crk (see pages 325, 326). Fibroblasts deficient in C3G display impaired adhesion and accelerated migration.[72] Loading Rap1 with GTP allows an interaction with two effectors that play a role in integrin activation. One of these is RapL, which associates with the α-chain of integrins.[73] Mice lacking RapL show impaired transendothelial migration because their integrins fail to attach to the endothelial cells (see Chapter 16). The other effector is Riam which, in lymphocytes, localizes Rap1-GTP to adhesion structures,[74] through its interaction with $PI(4,5)P_2$. Riam then recruits VASP to the membrane, so setting the stage for actin polymerization.

RhoA mediates its effect in the assembly process through the activation of PI 5-kinase, generating $PI(4,5)P_2$ (Figure 13.14 and Figure 13.15). As a result, both Riam and vinculin attach to the plasma membrane. In the case of vinculin, this unmasks cryptic binding sites that allow it to form a cross-bridge between talin (which is linked to the β-subunit of integrins) and to α-actinin (which is linked to filamentous actin)[75] (Figure 13.15c). A solid structure, integrating extracellular matrix and actin cytoskeleton, is thus formed.

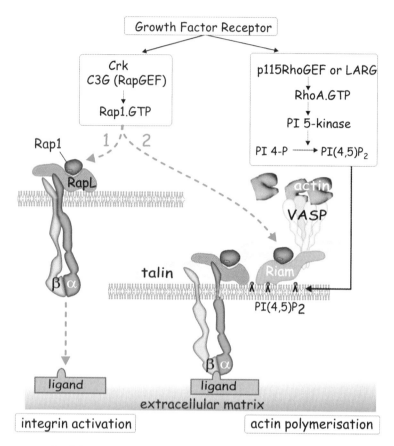

FIG 13.14 Growth-factor-mediated activation of integrins. Tyrosine phosphorylated growth factor receptors bind the adaptor Crk that links to the guanine nucleotide exchange factor C3G and activates Rap1. This addresses two effectors, RapL and Riam. RapL binds the cytosolic domain of the α-subunits of integrins and is implicated in their activation. Riam is recruited to the membrane by inositol lipids and binds VASP, leading to the initiation of actin polymerization. The growth factor also causes activation of the RhoA guanine exchange factors 115RhoGEF or LARG. These activate PI 5-kinase with production of PI(4,5)P$_2$.

Outside-in signalling from integrin adhesion complexes

Anchorage dependent growth and survival

An essential requirement for the proliferation of tissue cells driven by growth factors is their attachment to a suitable surface. The fact that cells must adhere to substrate provides a likely molecular explanation for the anchorage-dependent growth of cells that form part of a solid tissue. Adherence-dependent signalling targets have much in common with

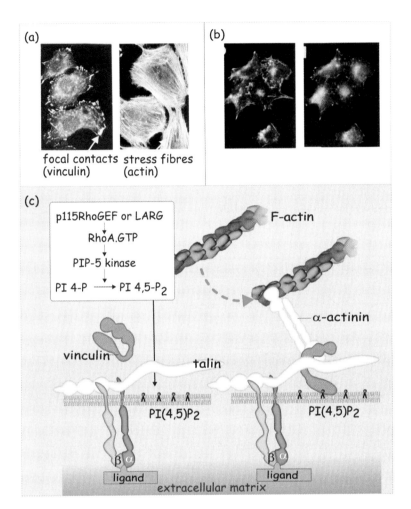

FIG 13.15 RhoA-mediated formation of focal contact sites. (a) The role of RhoA in focal contact formation is illustrated in micrographs of cells stained with fluorescent phalloidin (specifically binds to actin) or with antibodies against vinculin. In the presence of matrix proteins and serum, the fibroblasts spread out, forming stress fibres and focal adhesion sites. (b) However, when a non-functional RhoA is introduced, actin stress fibres are absent and strong focal contacts are not formed. Adapted from Clark et al.[76] (c) Growth factors play an essential role in the formation of focal adhesion sites through activation of the Rho nucleotide exchange factors p115RhoGEF or LARG. RhoA-GTP promotes formation of $PI(4,5)P_2$ through activation of PI 5-kinase. Vinculin binds to the $PI(4,5)P_2$, causing it to open up so that it attaches to both talin and α-actinin, making the link between the integrins and the actin cytoskeleton.

One might expect that circulating cells, such as leukocytes, must have specialized mechanisms to protect them from apoptosis. For sure, some white blood cells such as neutrophils are very short-lived (just a few hours in the blood), but others, such as the lymphocytes responsible for immunological memory, survive for years. Whether memory cells need to be continually reminded to survive by occasional encounters with other cells or an extracellular matrix remains to be established. Interestingly, lymphocytes employ an alternative pathway to activate PKB which involves the GTPase Rac.[79]

growth factor signalling targets (described in Chapters 12 and 18) and they are seemingly redundant. Signalling targets are evident in various stages of the Ras-MAPkinase pathway; upstream of Ras, between Ras and Raf or between Raf and MEK. Other targets are the JNK pathway, STAT-mediated regulation of transcription and the PKB pathway. For all of these, adherence and growth factor signalling act in synergy, the one amplifying the other, and this is necessary to engage the cell in a round of proliferation. The survival of endothelial and epithelial cells also depends critically upon the contacts they make with the extracellular matrix (and with each other). Without contact, they die through the controlled process of apoptosis, which in this particular context is referred to as 'anoikis' (meaning homeless).[77,78]

Cells have an intrinsic drive to self-destruct, but are prevented from doing so by signals emanating from specific rescue pathways. One such signal (outside-in) follows from the attachment of the integrins to the extracellular matrix. This mechanism may have evolved to protect the organism against stray cells colonizing inappropriate locations (metastasis, see page 307). The section below illustrates integrin-dependent signalling pathways that control cell survival and early events in the cell cycle (G1 phase).

Integrins and cell survival

The formation of an integrin signalling complex

Focal adhesion complexes form the sites of attachment for the tyrosine kinase FAK (focal adhesion kinase) (see Figure 13.16). Although a focal adhesion targeting domain has been described,[80] it is unclear which structural component acts as the docking site for FAK. Both paxillin and talin have been suggested. Attachment, resulting in autophosphorylation and activation, enables FAK to act as a docking site for Src (and also other members of the Src family of kinases such as Fyn and Yes). Src phosphorylates further tyrosine residues converting FAK into an SH2 domain docking protein. Besides binding SH2 domains, FAK also possesses two proline rich regions one of which interacts with the SH3 domains of CAS[81] (see Figure 13.16). CAS, with its multiple tyrosine phosphorylation sites, acts as a docking protein attracting numerous SH2-containing effectors and adaptors (a role similar to IRS-1/2 for the insulin receptor or LAT in the case of TCR signalling). A lack of CAS prevents integrin-mediated FAK signalling.[82]

The importance of FAK is underlined by the finding that cells expressing a constitutively active form survive in suspension even though they are effectively 'homeless'.[85] Here, FAK is active regardless of the failure to make contact with an extracellular matrix.

FAK-mediated activation of PKB

Amongst the proteins that attach to tyrosine phosphorylated FAK is the p85-regulatory subunit of PI 3-kinase (bound to the catalytic subunit p110)

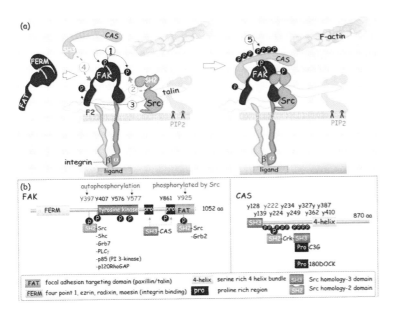

FIG 13.16 A two-step process to create an integrin signalling complex. (a) With the structural components of the focal adhesion site assembled, the focal adhesion kinase FAK associates with the F2 lobe of the FERM domain of talin; (Fak also has a FERM domain). Autophosphorylation FAK then generates a docking site (Y397) for the SH2 domain of Src which phosphorylates at Y925, in the FAT domain). Src and FAK next phosphorylate the FAK-associated docking protein CAS at multiple sites. An integrin-signalling complex is formed that acts in a manner similar to growth factor-receptor signalling complexes, i.e. attachment of adaptors and effectors and tyrosine phosphorylation of substrate. (b) Illustration of the main domains and essential phosphorylation sites of FAK and CAS.

The CAS protein family – CAS (Crk-associated substrate) and EFS (embryonal Fyn-associated substrate) – are SH3-containing docking proteins with multiple tyrosine phosphorylation sites that interact with SH2 domains of Crk (adaptor), Nck, and Abl (tyrosine protein kinases). In humans, CAS was identified as a protein whose over-expression confers resistance to anti-oestrogen treatment (tamoxifen) of breast cancer (**BCAR1**).[83] CAS is an important mediator of Src-mediated cell transformation. CAS phosphorylation correlates with inhibition of expression of Fhl1, a transcription factor. Loss of Fhl1 is a marker of anchorage-independent cell growth.[84]

(Figure 13.17). This generates 3-phosphorylated inositol lipids with the subsequent activation of the Ser/Thr kinase PKB (see Chapter 18).

PKB effects a number of phosphorylations[86–88] that promote rescue by at least four different mechanisms (see Figure 13.17). The first is by direct phosphorylation and inactivation of components of the apoptotic machinery, including Bad, Caspase-9, and XIAP.[89] Phosphorylation of Caspase-9 prevents its proteolytic activation while phosphorylation of Bad shifts its location from the mitochondrial membranes to the cytosol due to sequestration by 14-3-3 proteins. Since the phosphorylation site of human Caspase-9 is not conserved in mouse and rat it seems unlikely that this is a major control mechanism.

PKB also acts to control transcription of a number of genes that regulate apoptosis by phosphorylating the three members of the FoxO subfamily of Forkhead transcription factors (FoxO1, FoxO3, and FoxO4, previously known as FKHR, FKHRL1, and AFX) causing them to be retained in the cytosol,

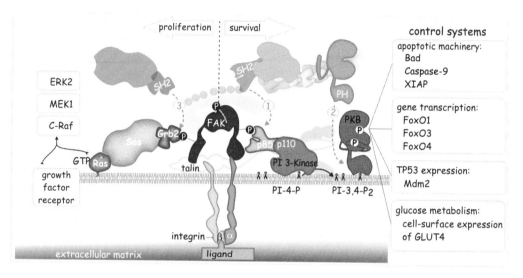

FIG 13.17 Survival and proliferation. The focal adhesion site promotes cell proliferation signals through activation of Ras. The Src phosphorylated Tyr925 acts as a binding site for the SH2 domain of Grb2. This interaction recruits the Ras guanine nucleotide exchange factor Sos to the membrane, leading to activation of Ras (1). Ras-GTP initiates the activation of the Raf-ERK pathway, necessary for initiation of the cell cycle. The focal adhesion site promotes cell survival signals through activation of the serine/threonine protein kinase-B (PKB). The phosphorylated Tyr residue of focal adhesion kinase (FAK) provides a binding site for the SH2 domain of the regulatory subunit (p85) of PI 3-kinase (2). Subsequent production of PI(3,4)P₃ provides a binding site for the PH domain of PKB (3). After its activation by phosphorylation, PKB phosphorylates a large number of proteins that directly or indirectly deal with cell death (see text for further detail).

again by 14-3-3 proteins. In this way, they are prevented from activating genes critical for induction of factors that promote cell death such as the Fas ligand,[90] TRAIL,[91] TRADD, and the pro-apoptotic protein Bim.[92] That this constitutes an essential survival mechanism is indicated by the finding that over-expression of FoxO3 triggers a programme of cell death in the interleukin-3-dependent haematopoietic cell line Ba/F3.[93] The death process is identical to that which occurs after withdrawal of the cytokine interleukin-3 (normally essential for these cells' survival).

Another substrate of PKB, Mdm2, is a component of an E3 ubiquitin ligase complex (see page 469) that ubiquitylates the tumour suppressor protein p53, preparing it for destruction by the proteasome. Mdm2-assisted destruction of p53 occurs in cells that have undergone DNA damage ('genotoxic stress'), but which seek to resist apoptosis. Phosphorylation of Mdm2 by PKB causes its nuclear localization and renders p53 ubiquitylation more efficient.[94] As with caspase-9, this mechanism of protection by PKB is operational in human cells, although the phosphorylation sites are not conserved throughout mammalian species.

PKB also protects against apoptosis by intervening in carbohydrate metabolism. When cells are deprived of growth factors, nutrient uptake is severely reduced and they become quiescent. This may reach such a point they fail to maintain their size and this inevitably results in a loss of viability.[95] The apoptotic process that ensues is due to the enhanced permeability of the mitochondrial outer membrane to cytochrome *c*. In cells starved of glucose, this is signalled by detachment of hexokinase from the mitochondrial outer membrane. Beyond its widely recognized role in phosphorylating glucose, hexokinase may also be involved in the linkage between the mitochondrial outer membrane pores and the adenine nucleotide transporter of the inner membrane, ensuring pore integrity and transport selectivity. By maintaining the cell surface expression of the glucose transporter GLUT4 (see page 558), sustained activation of PKB prevents apoptosis of cells even in the absence of growth factors. As a consequence, the activity and the mitochondrial association of hexokinase is maintained and leakage of cytochrome c is prevented.[96]

Finally, the action of PKB is certainly not limited to these survival processes. It also protects cyclin D1 against degradation and stimulates protein synthesis through activation of the kinase mTOR (see page 559).

Integrins and cell proliferation

FAK signalling reinforces the Ras–ERK pathway

As we have seen, clustered integrins bind FAK which undergoes autophosphorylation and then recruits Src to cause further phosphorylation and the formation of an activated tyrosine protein kinase complex. With most integrins, the phosphorylated FAK binds the adaptor complex Grb2/Sos and then activates the Ras–ERK pathway (Figure 13.17).[97] With other integrins ($\alpha1\beta1$, $\alpha5\beta1$, and $\alpha V\beta3$), activation of the ERK pathway requires the palmitoylated kinase Fyn that links to the α-subunit and, through SH3 linkage to Shc. Shc is then phosphorylated and this leads to recruitment of the Grb2/Sos complex.[98] This additional Ras–ERK signal is important for a sustained activation of ERK. If EGF or PDGF is provided to suspended fibroblasts, the activation of the ERK pathway is merely transient; there is no increase in the expression of cyclin D1 and expression of the cell cycle inhibitors p21[CIP/WAF] and p27[KIP] is fully maintained. The cells fail to proliferate and, sooner or later, they undergo apoptotic death.[99]

An alternative route for the stimulation of ERK is through phosphorylation of CAS by Src/FAK. Tyrosine phosphorylated CAS preferentially binds the SH2/SH3 adaptor protein Crk, which is either linked to the Rap1 guanine exchange factor C3G or to the Rac1 exchange factor 180DOCK[100,101] (Figure 13.18b). This results in activation of both Rap1 and Rac1, leading to activation of B-Raf, with prolonged activation of ERK, and leads to activation of JNK-1.[102,103] The sustained signal ensures progression from G0 to G1 and entry into the cell cycle.[104]

403

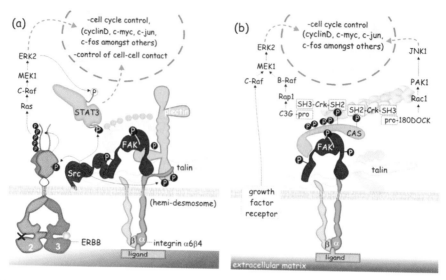

FIG 13.18 FAK-mediated signalling. (a) In epithelial cells, integrin α6β4 attached to the extracellular matrix forms a hemidesmosome, a specialized adhesion complex. These are linked to intermediate filaments via plectin (resembles plakoglobin). ErbB2/3 receptors are recruited, leading to phosphorylation of ErbB2 by Src bound to FAK. This promotes the activity of the ErbB2 kinase and enhances the growth factor receptor signalling output. (Note that ErbB3 is kinase dead and cannot phosphorylate ErbB2, see page 318). Src also phosphorylates STAT3, this signal being enforced by a second (serine) phosphorylation by ERK2. ERK2 and STAT3 cooperate in the regulation of cell–cell contacts and the cell cycle (cyclin D, c-Myc, c-Jun, c-Fos, and more). In breast tumour cells, this pathway promotes cellular invasion. Finally, the α4-integrin subunit is also a target of Src and this may affect its interaction with plectin. The ghosts decorating CAS indicate tyrosine phosphorylation sites. (b) Two examples of FAK signalling via the intermediate of CAS. The phosphotyrosines of CAS bind the SH2 domain of Crk. The proline-rich regions (PR) of C3G and 180DOCK (guanine nucleotide exchange factors) bind the SH3 domain of Crk. Their recruitment to the focal adhesion complex causes the activation of Rap1 and Rac1. Rap1 signals to B-Raf, which then phosphorylates MEK1, thus enforcing the growth factor receptor-stimulated Ras–ERK pathway, whereas Rac1 stimulates PAK1, which signals to JNK1. Both ERK and JNK stimulate expression of genes that initiate progression into the G1 phase of the cell cycle (cyclinD, c-myc, etc.).

FAK-mediated regulation of cell cycle inhibitors

Focal adhesion sites also regulate the expression of the cell cycle inhibitor p27^{KIP}, which plays a dual role in cell cycle regulation. At low levels it promotes progression through G1 because it stabilizes the complexes of cyclin D with CDK4 or CDK6, and it enhances nuclear retention, resulting in a more effective phosphorylation of the retinoblastoma protein. At high concentrations it blocks proliferation by inhibiting the activity of the cyclin complexes. Replacement of FAK with a kinase-dead mutant blocks DNA synthesis due to growth factors and this coincides with sustained elevated levels of p27^{KIP}.

FAK effectively controls the expression level of p27^{KIP} by stimulating its destruction. More directly, it acts to induce Skp2, the receptor component of one of the different SCF E3-ubiquitylation complexes. The elevated levels of

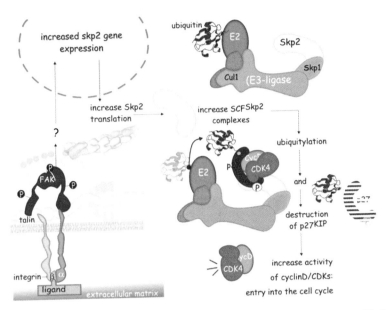

FIG 13.19 Regulation of destruction of p27^{KIP} by the integrin signalling complex. Increased expression of Skp2, due to activated FAK, enhances the formation of the ubiquitin ligase complex SCF^{Skp2}. Skp2 is the receptor component of this complex and it selectively binds (phosphorylated) p27^{KIP}, leading to its polyubiquitylation and destruction. The cells thus lose a cell cycle inhibitor, so favouring entry into G1 phase.

Skp2 augment the assembly of SCF^{Skp2} complexes and this targets p27^{KIP} for ubiquitylation and destruction by the proteasome[105] (see **Figure 13.19**).

FAK-mediated activation of growth factor receptors

Growth factor receptors certainly play a role in the activation of integrins, but the reverse is also true.[34,106] A good example is the interaction between the integrin αVβ3 and the epidermal growth factor receptor (EGFR).

Adherence to extracellular matrix causes phosphorylation of four tyrosine residues on the EGFR in a manner dependent on the adaptor protein CAS and the kinase Src. The number of receptors activated and the number of tyrosines phosphorylated is less than that induced by EGF, and there is a failure to achieve a full proliferative response. Addition of EGF then phosphorylates a further tyrosine and so it seems that the full response to low-level stimulation by EGF requires the additional integrin-mediated phosphorylation of the receptor.[107] The importance of this interplay between integrins and growth factor receptors is accentuated by the finding that integrin α6β4 amplifies the transformation of breast cancer cells that over-express the ErbB2 receptor. This amplification occurs through Src-mediated phosphorylation of ErbB2[108]

405

(see Figure 13.18a) though the tyrosine residue in question is not normally phosphorylated by the EGFR itself (see Figure 12.5 page 322, and page 323). Src also phosphorylates the transcription factor STAT3, a process that relates to enhanced invasiveness of epithelial cells.

List of Abbreviations

Abbreviation	Full name/description	SwissProt entry	Other names/OMIM
115RhoGEF	Rho guanine nucleotide exchange factor 115 kDa	Q92888	
actinin-α		P12814	F-actin cross-linking protein
Bad	Bcl2-associated death promotor	Q92934	Bcl2 antagonist of cell death
C3G	Crk SH3-binding guanine nucleotide release protein	Q13905	RapGEF1
Cadherin-1	Ca^{2+}-dependent adherence	P12830	E-cadherin, Uvomorulin
CAS	Crk-associated substrate	P56945	p130CAS
caspase-9	cystein-containing aspartate cleaving protease	P55211	ICE-like apoptotic protease 6
catenin-α		P35221	cadherin-associated protein
catenin-β		P35222	
catenin-δ		O60716	p120ctn
CD	cluster of differentiation		CD56, NCAM140
CD44	cluster of differentiation-44	P16070	Hermes antigen, Epican
CELSR1	Cadherin EGF LAG Seven-pass Receptor-1	Q9NYQ6	flamingo homologue 1
claudin-4		O14493	CPE-receptor
Crk	C10 regulator of kinase	P46108	p38
cyclinD1		P24385	PRAD1 and BCL-1 oncogene
desmocollin		Q08554	desmosomal glycoprotein-2/3
desmoglein	desmosomal glycoprotein-1	Q02413	
desmoplakin	desmosomal plaque protein	P15924	

Continued

Abbreviation	Full name/description	SwissProt entry	Other names/OMIM
ELAM-1	endothelium leukocyte adhesion molecule-1	P16581	CD62E, E-selectin
ERBB2	erythroblastoma-B2	P04626	Neu proto-oncogene, human EGF receptor-2 (Her2), MIM:164870
ERBB3	erythroblastoma-B3	P21860	human EGF receptor-3 (Her3), MIM:190151
FAK-1	focal adhesion kinase-1	Q05397	pp125FAK
FAT	focal adhesion targeting domain (in FAK)		
FERM	domain found in four point 1, ezrin, radixin and moesin		
Flamingo	see www.sdbonline.org/fly/dbzhnsky/starynt1.htm	Q9V5N8	starry night, stan
FoxO1	forkhead box protein 01	Q12778	FKHR
glut4	glucose transporter-4 (insulin responsive)	P14672	
Grb2	growth factor receptor bound protein-2	P62993	
ICAM-1	inter cellular adhesion molecule-1	P05362	CD54
IGS4B	immunoglobulin superfamily-4	Q8N126	
integrin αL		P20701	CD11a, LFA-1 α chain
integrin αV		P06756	CD51, vitronectin receptor
integrin β2		P05107	CD18, LFA-1, p150/95, C3
integrin β3		P05106	GPIIIa
JAM1	junctional adhesion molecule-1	Q9Y624	
LARG	leukemia-associated RhoGEF	Q9NZN5	
LFA-1	lymphocyte function antigen	P20701	CD11a, αLβ2
LYVE	lymphatic vessel endothelial hyaluronic acid receptor-1	Q9Y5Y7	
Mac-1	macrophage-1	P11578	αM(β2), CD11b(CD18)

Continued

Abbreviation	Full name/description	SwissProt entry	Other names/OMIM
Mdm2	mouse double minute-2 protein	Q00987	E3-ubiquitin protein ligase
NCAM-1	neural cell adhesion molecule-1	P13591	
p27KIP	(cyclin-dependent) kinase inhibitory protein of 27 kDa	P46527	
PAK1	p21-activated kinase-1	Q13153	
PECAM	platelet endothelial cell adhesion molecule, CD31	P16284	
PI 3-kinase p110	phosphatidylinositol 3-kinase protein 110 kDa	P42336	
PI 3-kinase p85	phosphatidylinositol 3-kinase regulatory subunit protein 85 kDa	Q92569	
PKBα	protein kinase B alpha	P31749	Akt1, C-Akt
plectin-1		Q15149	hemidesmosomal protein-1
Rap1A	ras-related protein-1A	P62834	Krev-1
RapL	regulator for cell adhesion and polarization in lymphoid tissues	QWWW0	ras association domain-containing protein 5
Rbx1	Ring Box protein-1	P62877	Roc1
RhoA	ras homologue-A	Q9QUI0	
Riam	rap1-GTP interacting adaptor molecule	Q7Z5R6	amyloid beta (A4) precursor protein-binding
Selectin-E		P16581	CD62E, ELAM
Selectin-L		P14151	CD62L, LECAM, LAM-1
Selectin-P		P16109	
SIGLEC-12	sialic acid binding Ig-like lectin proteins	Q96PQ1	
Skp2	S-phase kinase-associated protein-2	Q13309	F-box protein/LRR-repeat protein1
Sos1	son of sevenless-1	Q07889	RasGEF
Src	Sarcoma	P12931	p60Src, c-Src
STAT3	signal transducer and activator of transcription-3	P40763	

Continued

Abbreviation	Full name/description	SwissProt entry	Other names/OMIM
talin-1		Q9Y490	
TP53	tumour supressor protein 53 kDa	P04637	p53
Vasp	vasodilator-stimulated phosphoprotein	P50552	
VCAM-1	vascular cell adhesion molecule-1	P19320	CD106
vinculin		P18206	ʹ
VLA-4	very late antigen-4	P13612	CD49d
Xiap	X-linked inhibitor of apoptosis	P98170	
ZO-1	zonala occludens-1	Q07157	tight junction protein-1

References

1. Edelman GM. Cell adhesion molecules. *Science*. 1983;219:450–457.
2. Springer TA. Traffic signals for lymphocyte recirculation and leukocyte emigration: the multistep paradigm. *Cell*. 1994;76:301–314.
3. Brackenbury R, Thiery JP, Rutishauser U, Edelman GM. Adhesion among neural cells of the chick embryo. I. An immunological assay for molecules involved in cell–cell binding.. *J Biol Chem* 1977;252:6835–6840.
4. Springer T, Galfre G, Secher DS, Milstein C. Mac-1: a macrophage differentiation antigen identified by monoclonal antibody. *Eur J Immunol*. 1979;9:301–306.
5. Davignon D, Martz E, Reynolds T, Kurzinger K, Springer TA. Lymphocyte function-associated antigen 1 (LFA-1): a surface antigen distinct from Lyt-2,3 that participates in T lymphocyte-mediated killing. *Proc Natl Acad Sci U S A*. 1981;78:4535–4539.
6. Tamkun JW, DeSimone DW, Fonda D, et al. Structure of integrin, a glycoprotein involved in the transmembrane linkage between fibronectin and actin. *Cell*. 1986;46:271–282.
7. Ruoslahti E, Pierschbacher MD. New perspectives in cell adhesion: RGD and integrins. *Science*. 1987;238:491–497.
8. Osborn L, Hession C, Tizard R, et al. Direct expression cloning of vascular cell adhesion molecule 1, a cytokine-induced endothelial protein that binds to lymphocytes. *Cell*. 1989;59:1203–1211.
9. Kasper C, Rasmussen H, Kastrup JS, et al. Structural basis of cell–cell adhesion by NCAM. *Nat Struct Biol*. 2000;7:389–393.
10. Wojciak-Stothard B, Ridley AJ. Shear stress-induced endothelial cell polarization is mediated by Rho and Rac but not Cdc42 or PI 3-kinases. *J Cell Biol*. 2003;161:423–439.

11. Barreiro O, Yanez-Mo M, Serrado JM, et al. Dynamic interaction of VCAM-1 and ICAM-1 with moesin and ezrin in a novel endothelial docking structure for adherent leukocytes. *J Cell Biol*. 2002;157:1233–1245.

12. Huang MT, Mason JC, Birdsey GM, et al. Endothelial intercellular adhesion molecule (ICAM)-2 regulates angiogenesis. *Blood*. 2005;106:1636–1643.

13. de Fougerolles AR, Springer TA. Intercellular adhesion molecule 3, a third adhesion counter-receptor for lymphocyte function-associated molecule 1 on resting lymphocytes. *J Exp Med*. 1992;175:185–190.

14. Bailly P, Hermand P, Callebaut I, et al. The LW blood group glycoprotein is homologous to intercellular adhesion molecules. *Proc Natl Acad Sci U S A*. 1994;91:5306–5310.

15. Kilgannon P, Turner T, Meyer J, Wisdom W, Gallatin WM. Mapping of the *ICAM-5* (telencephalin) gene, a neuronal member of the ICAM family, to a location between *ICAM-1* and *ICAM-3* on human chromosome 19p13.2. *Genomics* 1998;54:328–330.

16. Powell LD, Varki A. I-type lectins. *J Biol Chem*. 1995;270:14243–14246.

17. Attrill H, Imamura A, Sharma RS, Kiso M, Crocker PR, van Aalten DM. Siglec-7 undergoes a major conformational change when complexed with the α(2,8)-disialylganglioside GT1b. *J Biol Chem*. 2006;281:32774–32783.

18. O'Keefe TL, Williams GT, Davies SL, Neuberger MS. Hyperresponsive B cells in CD22-deficient mice. *Science*. 1996;274:798–801.

19. Collins BE, Smith BA, Bengtson P, Paulson JC. Ablation of CD22 in ligand-deficient mice restores B cell receptor signaling. *Nat Immunol*. 2006;7:199–206.

20. Nishizumi H, Horikawa K, Mlinaric-Rascan I, Yamamoto T. A double-edged kinase Lyn: a positive and negative regulator for antigen receptor-mediated signals. *J Exp Med*. 1998;187:1343–1348.

21. Nitschke L. The role of CD22 and other inhibitory co-receptors in B-cell activation. *Curr Opin Immunol*. 2005;17:290–297.

22. Turksen K, Troy TC. Barriers built on claudins. *J Cell Sci*. 2004;117:2435–2447.

23. Van Itallie CM, Anderson JM. Claudins and epithelial paracellular transport. *Annu Rev Physiol*. 2006;68:403–429.

24. Itoh M, Furuse M, Morita K, Kubota K, Saitou M, Tsukita S. Direct binding of three tight junction-associated MAGUKs, ZO-1, ZO-2, and ZO-3, with the COOH termini of claudins. *J Cell Biol*. 1999;147:1351–1363.

25. Xiong JP, Stehle T, Diefenbach B, et al. Crystal structure of the extracellular segment of integrin αVβ3. *Science*. 2001;294:339–345.

26. Shimaoka M, Takagi J, Springer TA. Conformational regulation of integrin structure and function. *Annu Rev Biophys Biomol Struct*. 2002;31:485–516.

27. Tanentzapf G, Brown NH. An interaction between integrin and the talin FERM domain mediates integrin activation but not linkage to the cytoskeleton. *Nature Cell Biol*. 2006;8:601–606.

28. Xiao T, Takagi J, Coller BS, Wang JH, Springer TA. Structural basis for allostery in integrins and binding to fibrinogen-mimetic therapeutics. *Nature* 2004;432:59–67.

29. Law SK, Tan SM, Ranganathan S, Cheng M. The integrin αLβ2 hybrid domain serves as a link for the propagation of activation signal from its stalk regions to the I-like domain. *J Biol Chem.* 2004;279:54334–54339.

30. Carman CV, Springer TA. Integrin avidity regulation: are changes in affinity and conformation underemphasized. *Curr Opin Cell Biol.* 2003;15:547–556.

31. Xiao T, Takagi J, Coller BS, Wang JH, Springer TA. Structural basis for allostery in integrins and binding to fibrinogen-mimetic therapeutics. *Nature* 2004;432:59–67.

32. Carman CV, Springer TA. Integrin avidity regulation: are changes in affinity and conformation underemphasized? *Curr Opin Cell Biol.* 2003;15:547–556.

33. Mould AP, Humphries JH. Adhesion articulated. *Nature* 2004;432:27–28.

34. Miranti CK, Brugge JS. Sensing the environment: a historical perspective on integrin signal transduction. *Nat Cell Biol.* 2002;4:E83–E90.

35. Miyamoto S, Teramoto H, Coso OA, et al. Integrin function: molecular hierarchies of cytoskeletal and signaling molecules. *J Cell Biol.* 1995;131:791–805.

36. Hato T, Pampori N, Shattil SJ. Complementary roles for receptor clustering and conformational change in the adhesive and signaling functions of integrin αIIb β3. *J Cell Biol.* 1998;141:1685–1695.

37. Brakebusch C, Fassler R. The integrin-actin connection, an eternal love affair. *EMBO J.* 2003;22:2333.

38. Ducibella T, Anderson E. Cell shape and membrane changes in the eight-cell mouse embryo: prerequisites for morphogenesis of the blastocyst. *Dev Biol.* 1975;47:45–58.

39. Yoshida C, Takeichi M. Teratocarcinoma cell adhesion: identification of a cell-surface protein involved in calcium-dependent cell aggregation. *Cell.* 1982;28:217–224.

40. Gallin WJ, Edelman GM, Cunningham BA. Characterization of L-CAM, a major cell adhesion molecule from embryonic liver cells. *Proc Natl Acad Sci U S A.* 1983;80:1038–1042.

41. Boller K, Vestweber D, Kemler R. Cell-adhesion molecule uvomorulin is localized in the intermediate junctions of adult intestinal epithelial cells. *J Cell Biol.* 1985;100:327–332.

42. Nollet F, Kools P, van Roy F. Phylogenetic analysis of the cadherin superfamily allows identification of six major subfamilies besides several solitary members. *J Mol Biol.* 2000;299:551–572.

43. Patel SD, Ciatto C, Chen CP, et al. Type II cadherin ectodomain structures: implications for classical cadherin specificity. *Cell.* 2006;124:1255–1268.

44. Chen CP, Posy S, Ben-Shaul A, Shapiro L, Honig BH. Specificity of cell–cell adhesion by classical cadherins: Critical role for low-affinity dimerization through β-strand swapping. *Proc Natl Acad Sci U S A*. 2005;102:8531–8536.

45. Vavylonis D, Kovar DR, O'Shaughnessy B, Pollard TD. Model of formin-associated actin filament elongation. *Mol Cell*. 2006;21:455–466.

46. Lilien J, Balsamo J. The regulation of cadherin-mediated adhesion by tyrosine phosphorylation/dephosphorylation of β-catenin. *Curr Opin Cell Biol*. 2005;17:459–465.

47. Ozawa M, Baribault H, Kemler R. The cytoplasmic domain of the cell adhesion molecule uvomorulin associates with three independent proteins structurally related in different species. *EMBO J*. 1989;8:1711–1717.

48. Li H, Leung TC, Hoffman S, Balsamo J, Lilien J. Coordinate regulation of cadherin and integrin function by the chondroitin sulfate proteoglycan neurocan. *J Cell Biol*. 2000;149:1275–1288.

49. Rhee J, Mahfooz NS, Arregui C, Lilien J, Balsamo J, VanBerkum MF. Activation of the repulsive receptor Roundabout inhibits N-cadherin-mediated cell adhesion. *Nat Cell Biol*. 2002;4:798–805.

50. Curran T, D'Arcangelo G. Role of reelin in the control of brain development. *Brain Res Brain Res Rev*. 1998;26:285–294.

51. Stoolman LM, Rosen SD. Possible role for cell-surface carbohydrate-binding molecules in lymphocyte recirculation. *J Cell Biol*. 1983;96:722–729.

52. Lasky LA, Singer MS, Yednock TA, et al. Cloning of a lymphocyte homing receptor reveals a lectin domain. *Cell* 1989;56:1045–1055.

53. Yago T, Wu J, Wey CD, Klopocki AG, Zhu C, McEver RP. Catch bonds govern adhesion through L-selectin at threshold shear. *J Cell Biol*. 2004;166:913–923.

54. Marshall BT, Long M, Piper JW, Yago T, McEver RP, Zhu C. Direct observation of catch bonds involving cell-adhesion molecules. *Nature* 2003;423:190–193.

55. Somers WS, Tang J, Shaw GD, Camphausen RT. Insights into the molecular basis of leukocyte tethering and rolling revealed by structures of P- and E-selectin bound to SLe(X) and PSGL-1. *Cell* 2000;103:467–479.

56. Alon R, Chen S, Fuhlbrigge R, Puri KD, Springer TA. The kinetics and shear threshold of transient and rolling interactions of L-selectin with its ligand on leukocytes. *Proc Natl Acad Sci U S A*. 1998;95:11631–11636.

57. Ivetic A, Ridley AJ. The telling tail of L-selectin. *Biochem Soc Trans*. 2004;32:1118–1121.

58. Urzainqui A, Serrador JM, Viedma F, et al. ITAM-based interaction of ERM proteins with Syk mediates signaling by the leukocyte adhesion receptor PSGL-1. *Immunity* 2002;17:401–412.

59. Laurent TC, Fraser JR. Hyaluronan. *FASEB J*. 1992;6:2397–2404.

60. Aruffo A, Stamenkovic I, Milnick M, Underhill CB, Seed B. CD44 is the principal cell surface receptor for hyaluronate. *Cell* 1990;6:1303–1313.

61. Mytherye K, Blobe GC, Proteoglycan signaling co-receptors: Roles in cell adhesion, migration and invasion. *Cell Signal.* 2009 in press.

62. Banerji S, Ni J, Wang SX, et al. LYVE-1, a new homologue of the CD44 glycoprotein, is a lymph-specific receptor for hyaluronan. *J Cell Biol.* 1999;144:789–801.

63. Day AJ. The structure and regulation of hyaluronan-binding proteins. *Biochem Soc Trans.* 1999;27:115–121.

64. Almond A, Deangelis PL, Blundell CD. Hyaluronan: the local solution conformation determined by NMR and computer modeling is close to a contracted left-handed 4-fold helix. *J Mol Biol.* 2006;358:1256–1269.

65. Teriete P, Banerji S, Noble M, et al. Structure of the regulatory hyaluronan binding domain in the inflammatory leukocyte homing receptor CD44. *Mol Cell.* 2004;13:483–496.

66. Goldstein LA, Zhou DF, Picker LJ, et al. A human lymphocyte homing receptor, the hermes antigen, is related to cartilage proteoglycan core and link proteins. *Cell* 1989;56:1063–1072.

67. Gunthert U, Hofmann M, Rudy Y, et al. A new variant of glycoprotein CD44 confers metastatic potential to rat carcinoma cells. *Cell* 1991;65:13–24.

68. Tsukita S, Oishi K, Sato N, Sagara J, Kawai A. ERM family members as molecular linkers between the cell surface glycoprotein CD44 and actin-based cytoskeletons. *J Cell Biol.* 1994;126:391–401.

69. Reedquist KA, Ross E, Koop EA, et al. The small GTPase, Rap1, mediates CD31-induced integrin adhesion. *J Cell Biol.* 2000;148:1151–1158.

70. Franke B, van Triest M, de Bruijn KM, et al. Sequential regulation of the small GTPase Rap1 in human platelets. *Mol Cell Biol.* 2000;20:779–785.

71. Tanaka S, Morishita T, Hashimoto Y, et al. C3G, a guanine nucleotide-releasing protein expressed ubiquitously, binds to the Src homology 3 domains of CRK and GRB2/ASH proteins. *Proc Natl Acad Sci U S A.* 1994;91:3443–3447.

72. Ohba Y, Ikuta K, Ogura A, et al. Requirement for C3G-dependent Rap1 activation for cell adhesion and embryogenesis. *EMBO J.* 2001;20:333–3341.

73. Katagiri K, Maeda A, Shimonaka M, Kinashi T. RAPL, a Rap1-binding molecule that mediates Rap1-induced adhesion through spatial regulation of LFA-1. *Nat Immunol.* 2003;4:741–748.

74. Lafuente EM, van Puijenbroek AA, Krause M, et al. RIAM, an Ena/VASP and Profilin ligand, interacts with Rap1-GTP and mediates Rap1-induced adhesion. *Dev Cell.* 2004;7:585–595.

75. Gilmore AP, Burridge K. Regulation of vinculin binding to talin and actin by phosphatidyl-inositol-4–5-bisphosphate. *Nature* 1996;381:531–535.

76. Clark EA, King WG, Brugge JS, Symons M, Hynes RO. Integrin-mediated signals regulated by members of the rho family of GTPases. *J Cell Biol.* 1998;142:573–586.

77. Frisch SM, Francis H. Disruption of epithelial cell-matrix interactions induces apoptosis. *J Cell Biol.* 1994;124:619–626.

78. Raff MC. Cell suicide for beginners. *Nature* 1998;396:119–122.

79. Genot E, Arrieumerlou C, Ku G, Burgering BM, Weiss A, Kramer IM. The T-cell receptor regulates Akt (protein kinase B) via a pathway involving Rac1 and phosphatidylinositide 3-kinase. *Mol Cell Biol*. 2000;20:5469–5478.

80. Prutzman KC, Gao G, King ML, et al. The focal adhesion targeting domain of focal adhesion kinase contains a hinge region that modulates tyrosine 926 phosphorylation. *Structure* 2004;12:881–891.

81. Lim Y, Han I, Jeon J, Park H, Bahk YY, Oh ES. Phosphorylation of focal adhesion kinase at tyrosine 861 is crucial for Ras transformation of fibroblasts. *J Biol Chem*. 2004;279:29060–29065.

82. Iwahara T, Akagi T, Fujitsuka Y, Hanafusa H. CrkII regulates focal adhesion kinase activation by making a complex with Crk-associated substrate, p130Cas. *Proc Natl Acad Sci U S A*. 2004;101:17693–17698.

83. Brinkman A, van der Flier S, Kok EM, Dorssers LC. BCAR1, a human homologue of the adapter protein p130Cas, and antiestrogen resistance in breast cancer cells. *J Natl Cancer Inst*. 2000;92:112–120.

84. Shen Y, Jia Z, Nagele RG, Ichikawa H, Goldberg GS. SRC uses Cas to suppress Fhl1 in order to promote nonanchored growth and migration of tumor cells. *Cancer Res*. 2006;66:1543–1552.

85. Frisch SM, Vuori K, Ruoslahti E, Chan-Hui PY. Control of adhesion-dependent cell survival by focal adhesion kinase. *J Cell Biol*. 1996;134:793–799.

86. King WG, Mattaliano MD, Chan TO, Tsichlis PN, Brugge JS. Phosphatidylinositol 3-kinase is required for integrin-stimulated AKT and Raf-1/mitogen-activated protein kinase pathway activation. *Mol Cell Biol*. 1997;17:4406–4418.

87. Khwaja A, Rodriguez VP, Wennstrom S, Warne PH, Downward J. Matrix adhesion and Ras transformation both activate a phosphoinositide 3-OH kinase and protein kinase B/Akt cellular survival pathway. *EMBO J*. 1997;16:2783–2793.

88. Downward J. PI 3-kinase, Akt and cell survival. *Semin Cell Dev Biol*. 2004;15:177–182.

89. Downward J. Mechanisms and consequences of activation of protein kinase B/Akt. *Curr Opin Cell Biol*. 1998;10:262–267.

90. Brunet A, Bonni A, Zigmond MJ, et al. Akt promotes cell survival by phosphorylating and inhibiting a Forkhead transcription factor. *Cell* 1999;96:857–868.

91. Modur V, Nagarajan R, Evers BM, Milbrandt J. FOXO proteins regulate tumor necrosis factor-related apoptosis inducing ligand expression. Implications for PTEN mutation in prostate cancer. *J Biol Chem*. 2002;277:47928–47937.

92. Dijkers PF, Medema RH, Lammers JW, Koenderman L, Coffer PJ. Expression of the pro-apoptotic Bcl-2 family member Bim is regulated

by the forkhead transcription factor FKHR-L1. *Curr Biol*. 2000;10: 1201–1204.

93. Dijkers PF, Birkenkamp KU, Lam EW, et al. FKHR-L1 can act as a critical effector of cell death induced by cytokine withdrawal: protein kinase B-enhanced cell survival through maintenance of mitochondrial integrity. *J Cell Biol*. 2002;156:531–542.

94. Zhou BP, Liao Y, Xia W, Zou Y, Spohn B, Hung MC. HER-2/neu induces p53 ubiquitination via Akt-mediated MDM2 phosphorylation. *Nat Cell Biol*. 2001;3:973–982.

95. Plas DR, Rathmell JC, Thompson CB. Homeostatic control of lymphocyte survival: potential origins and implications. *Nat Immunol*. 2002;3:515–521.

96. Majewski N, Nogueira V, Bhaskar P, et al. Hexokinase-mitochondria interaction mediated by Akt is required to inhibit apoptosis in the presence or absence of Bax and Bak. *Mol Cell*. 2004;16:819–830.

97. Mitra SK, Mikolon D, Molina JE, et al. Intrinsic FAK activity and Y925 phosphorylation facilitate an angiogenic switch in tumors. *Oncogene*. 2006;25:5969–5984.

98. Wary KK, Mariotti A, Zurzolo C, Giancotti FG. A requirement for caveolin-1 and associated kinase Fyn in integrin signaling and anchorage-dependent cell growth. *Cell* 1998;94:625–634.

99. Assoian RK. Anchorage-dependent cell cycle progression. *J Cell Biol*. 1997;136:1–4.

100. Hasegawa H, Kiyokawa E, Tanaka S, et al. DOCK180, a major CRK-binding protein, alters cell morphology upon translocation to the cell membrane. *Mol Cell Biol*. 2006;16:1770–1776.

101. Kiyokawa E, Hashimoto Y, Kobayashi S, Sugimura H, Kurata T, Matsuda M. Activation of Rac1 by a Crk SH3-binding protein, DOCK180. *Genes Dev*. 1998;12:3331–3336.

102. Bottazzi ME, Zhu X, Bohmer RM, Assoian RK. Regulation of p21(cip1) expression by growth factors and the extracellular matrix reveals a role for transient ERK activity in G1 phase. *J Cell Biol*. 1999;146: 1255–1264.

103. Lenormand P, Sardet C, Pages G, L'Allemain G, Brunet A, Pouyssegur J. Growth factors induce nuclear translocation of MAP kinases (p42mapk and p44mapk) but not of their activator MAP kinase kinase (p45mapkk) in fibroblasts. *J Cell Biol*. 1993;122:1079–1088.

104. Aplin AE, Stewart SA, Assoian RK, Juliano RL. Integrin-mediated adhesion regulates ERK nuclear translocation and phosphorylation of Elk-1. *J Cell Biol*. 2001;153:273–282.

105. Bryant P, Zheng Q, Pumiglia K. Focal adhesion kinase controls cellular levels of p27/Kip1 and p21/Cip1 through Skp2-dependent and -independent mechanisms. *Mol Cell Biol*. 2006;26:4201–4213.

106. Wang R, Kobayashi R, Bishop JM. Cellular adherence elicits ligand-independent activation of the Met cell-surface receptor. *Proc Natl Acad Sci U S A*. 1996;93:8425–8430.

107. Giancotti FG, Tarone G. Positional control of cell fate through joint integrin/receptor protein kinase signaling. *Annu Rev Cell Dev Biol*. 2003;19:173–206.

108. Guo W, Pylayeva Y, Pepe A, Yoshioka T, Muller WJ, Inghirami G, Giancotti FG. β4 integrin amplifies ErbB2 signaling to promote mammary tumorigenesis. *Cell* 2006;126:489–502.

CHAPTER 14

Adhesion Molecules in the Regulation of Cell Differentiation: Mainly about Wnt

Here we consider the adhesion molecules as mediators of differentiation with regard to epithelial tissues. We discuss how cytokines affect adhesion, with special emphasis on Wnt signalling. Errors in these pathways cause a transition towards a loosely attached mesenchymal phenotype that gives rise to cancer. We also discuss the importance of Wnt family proteins in the maintenance of the self renewal capacity of stem cells in the crypts of the intestinal tract.

Destabilization of adherens junctions causes cellular de-differentiation

The surfaces that separate interior from exterior, such as the skin, the linings of the body cavities, and the gut are formed from tight sheets of epithelial cells. Similarly, the linings of the vasculature, including the heart, blood vessels and lymphatics, are formed from endothelial cells which are

The many abbreviations used in this chapter are collected together at the end of the chapter.

specialized epithelial cells. Unlike bone or interstitium, the extracellular matrix is minimal, comprising a thin sheet, the basal lamina. It is the cells themselves that bear the brunt of physical stresses such as the peristalsis of the gut and the pulsatile movement of the vessels. Tissue structure is maintained by an elaborate series of cellular attachments, tight junctions (also known as zonula occludens), adherens junctions and desmosomes (Latin *occludo,* to block up; *desmos,* meaning band or ligament) (see Figures 13.5 and 13.11).

If the adherens junctions stabilizing a cell layer are artificially destabilized, for instance by antibodies against cadherin, there is a loss of cell polarity and the spatial distribution of he membrane proteins becomes randomized. Insulin-like growth factor receptor type-I (IGFR, normally present on the basal surface) and the glucose transporters (SGLT, on the apical surface), are now distributed all over the surface of the cell. The tight junctions disassemble through detachment of the zonula adherens-associated proteins (ZO-1) and the link between the cadherins and the catenins is broken; ZO-proteins are scattered into the cytosol[1] and β-catenin is present in the nucleus[2] (Figure 14.1). The cells produce more fibronectin (extracellular matrix) and switch from the production of keratin to vimentin (both forming intermediate filaments). In

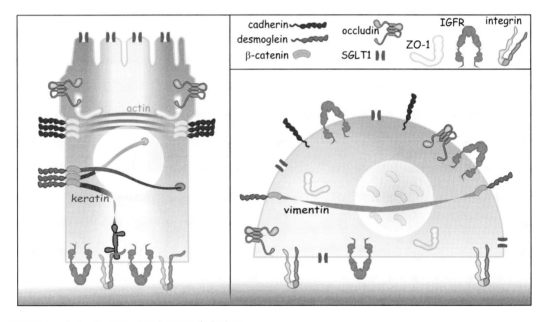

FIG 14.1 Loss of cell–cell contact and the dissipation of cell polarity.
The polarized state is well illustrated by the segregation of the components of the tight and adherent junctions on lateral surfaces: IGFRs at the basal surface and the Na^+/glucose symporter (SGLT1) at the apical surface. With the dissipation of the adherent junctions, this functional polarization is lost. The Na^+/glucose symporters and IGF receptors become randomly distributed and keratin is replaced by vimentin. The cells become de-differentiated.

short they begin to resemble fibroblasts (mesenchymal cells) for changes in gene expression, protein activity, and protein localization (see Table 14.1).

This disorientation, also referred to as the epithelial mesenchymal transformation (EMT), plays an important role in the reorganization of tissue in the developing embryo.[4] It allows tight sheets of cells to loosen up, to detach and then migrate to establish new tissues elsewhere (organ development). Typical examples are gastrulation and neural crest cell migration. A similar, but uncontrolled, transition occurs in the formation of cancer cells as a consequence of transformation of epithelial cells[3,5] and down-regulation of E-cadherin expression.[6] The importance of cell–cell contact in the maintenance of the differentiated phenotype was well demonstrated in experiments in which ectopic expression of E-cadherin suppressed the transformation of a colon carcinoma cell line (that normally contains very little E-cadherin)[7] (see Table 14.2).

In certain situations, tissue cells exist in a de-differentiated state that allows self renewal, albeit in a highly controlled fashion. Here we are referring to epithelial stem cells. After each cell division, one of the progeny retains the potential to renew itself, while the other becomes committed, after one or two more rounds of division, to differentiation into epithelial cells that replace damaged and dying cells.[8]

Signalling through the canonical WNT pathway

We consider the similarity of the elements that assure the maintenance of a stem cell compartment with those that are involved during development and

TABLE 14.1 Changes in epithelial cells marking the epithelial mesenchymal transition

Gain of expression	Fibronectin, vimentin, snail1, snail2 (slug), twist, goosecoid, foxC2, sox10, mmp-9, mmp-3, integrinαVβ6, EF1/ZEB1, SIP1/ZEB2, N-cadherin
Loss of expression	E-cadherin, desmoplakin, occludin, keratin (intermediate filament)
Increase in activity	Integrin-linked kinase (ILK), GSK3β, Snail2 (slug), Twist, EF1/ZEB1, SIP1/ZEB2
Accumulation in the nucleus	β-Catenin, Smad2/3, Snail1, Snail2 (Slug), Twist, EF1/ZEB1, SIP1/ZEB2
Functional markers	Increased migration, increased matrix invasion, increased scattering, elongation of cell shape, resistance to anoikis

From Lee et al.[3]

TABLE 14.2 Factors that determine cell differentiation or de-differentiation

Differentiation	Promoted by
Destruction of β-catenin (not linked to cadherin)	S/T phosphorylation of β-catenin by CK1, GSK3β etc.
Stabilization of E-cadherin core complex	S/T phosphorylation of E-cadherin by CK2 and GSK3β
Stabilization of E-cadherin core complex	dephosphorylation of β-catenin by tyrosine phosphatases (PTPμ, VEPTP, LAR, PTP1B, LMW-PTP, SHP2, DEP1)
De-differentiation	
Destabilization of E-cadherin core complex	EGF, FGF, HGF/SF or TGFβ (tyrosine phosphorylation by Src, Fer, Fyn, Abl, or c-Met)
Endocytosis of E-cadherin core complex	tyrosine phosphorylation on residues 754–755

in cell transformation (leading to cancer). Several growth factors including EGF, FGF, and TGFβ play a role. Now we focus on the Wnts and discuss in detail the so-called *canonical pathway*. This operates by switching the operation of β-catenin from its role in the maintenance of tissue integrity, to a transcription factor involved in the regulation of proliferation. The canonical Wnt pathway illustrates the intricate link between the action of cytokines and adhesion molecules.[9]

The discovery of the Wnt family of cytokines

The discovery of Wnt arose from several independent lines of research. Work with *Drosophila* brought to light a gene linked to wing development, which was given the name *wingless* (Wg).[10] Outright deletion of *wingless* is lethal, causing defects that are manifest in early development. This gene and its downstream signalling components are often referred to as 'segment polarity genes'.[11] We now know that the Wnt pathway is used in different stages of *Drosophila* development: in the formation of the segments of the head, thorax and trunk, in the development of appendages such as legs and antennae, and in certain aspects of oogenesis and neurogenesis.[12]

In an unrelated investigation it was found that mouse mammary tumours caused by the retrovirus MMTV result from the insertion of the viral DNA into the int-1 locus of the mouse genome.[13] Later, it transpired that int-1 and Wg are homologous genes[14–16] and a new name, Wnt, was proposed (an amalgamation of *Wg* and *int*-1).[17] Furthermore, a previously known mouse mutant, *swaying*, also turned out to be an allele of the Wnt-1 gene.[18] Due to a malformed cerebellum, these animals manifest a lack of muscular coordination (ataxia), swaying of the head being one of the symptoms. Here, the mutated Wnt-1 protein lacks the C-terminal half as a result of a single base pair mutation that creates a premature stop codon.

From mutation analysis in *Drosophila* and screening for homologous genes in mice, it emerged that Wnt is a member of a large family of genes (at least 19 in mammals), having different functions and different signal transduction pathways. As with the Ras-MAPkinase pathway (see page 327), the study of mutants in genetically accessible organisms greatly facilitated investigations of the role of Wnt in the regulation of proliferation, differentiation, and cell fate in mammalian cells.

Following its discovery, Wnt remained a gene that evaded proper definition as a protein for about 15 years. Its effects were studied through gene amplification and deletion, through injection of mRNA, or, latterly, through translation arrest by applied siRNA. Proteins of the Wnt family are now recognized as secreted palmitoylated glycoproteins.[20] The lipid attachment, essential for signalling, renders them insoluble and upon exocytosis they remain associated with the cell membrane. However, Wnt can diffuse into the tissues to a limited extent, over a distance of up to 30 cells, in this way forming morphogenic gradients.[21]

Wnt signals through β-catenin

Mutational analysis in *Drosophila* led to the discovery of a signal transduction pathway consisting of the ligand (wingless, Wg), its receptor (frizzled, Fz), and downstream components. By scoring for rescue or aggravation of the phenotype (so-called epistatic tests), these were found to act in the sequence:

dishevelled (Dsh) $\rightarrow$ shaggy (Sgg) $\rightarrow$ armadillo (Arm) $\rightarrow$ pangolin (Pan)

resulting in the expression of the transcription factor engrailed (en) (Figure 14.2)[11,22,23] (for review see Willert and Nusse[24]). The genes involved are named after the mutant phenotypes they give rise to, having either abnormal hair patterns on their body or wings (as in frizzled, shaggy, or dishevelled) or with abnormal larval cuticle formation (armadillo or pangolin).

Understanding these genes and their interrelationships, and how they are integrated into pathways that determine cell fate or polarity, has required a number of different experimental approaches. Sequence analysis and immunocytochemical experiments have provided evidence that armadillo is a homologue of the mammalian β-catenin and that it resembles plakoglobin, a component of desmosomes.[30,31] Pangolin encodes a protein homologous to the transcription factor, LEF-1,[32] that interacts with β-catenin. This interaction is essential for transmission of the Wnt signal into the nucleus.[33,34] Armadillo also enters the cell nucleus to form a binary transcription complex with a

Wnt alleles and disease.
Wnt may be regarded as a proto-oncogene (having the potential to cause cancer), but so far no human tumours have been correlated with a Wnt defect. However, certain components of the Wnt-signalling transduction pathway do qualify as true oncogenes. Dysfunctional Wnt alleles have been found in other human pathologies such as obesity, type 2 (late onset) diabetes, Müllerian duct regression, and the embryo-lethal phenotype tetra-amelia (lack of extremities).[19]

Wnt in development.
For an informative review of Wnt effects in *Drosophila* see Klingensmith and Nusse.[25] For a well illustrated description of Wnt's role in morphogenesis in *Drosophila* see Martinez-Arias.[26] For developmental disorders in mice Wnt-knockouts, see Logan and Nusse.[27] A continuing update of Wnt and its signal transduction pathways, including figures and gene tables, can be found on the Wnt homepage, maintained by the laboratory of Roel Nusse, http://www.stanford.edu/~rnusse/wntwindow.html

Regulation of gene expression. Regulation of gene expression is mediated by several mechanisms such as DNA methylation, ATP-dependent chromatin remodelling, and post-translational modifications of histones, which include acetylation/de-acetylation of lysine residues present in the tail of core histones. The histone acetyltransferases (HATs) act as transcriptional coactivators, and histone deacetylases (HDACs) form part of transcriptional corepressor complexes. Wnt uncouples Groucho from TCF and in so doing, it also uncouples the associated de-acetylation complex (containing HDAC2). This opens many gene loci for transcription. A mutation in HDAC2, which makes it insensitive to inhibitory factors, predisposes to colorectal cancer.[42]

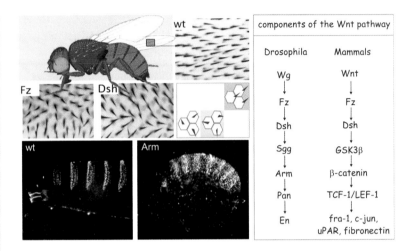

FIG 14.2 Examples of *Drosophila* phenotypes due to mutations in the Wnt pathway.

In the wild-type fly, actin, which forms a single hair, polymerizes in a defined position with uniform orientation within the plane of the epithelial wing cells. Mutations in Fz and Dsh cause a loss of planar polarity of these cells. Polymerization occurs at random points within the plane of the epithelium giving rise to a frizzled (curly) or dishevelled (disorganized) phenotype. Mutations in Arm give rise to a highly segmented embryo resembling the banded shell of an armadillo.

Image of hairs from Wong and Adler 1993,[28] originally published in the *Journal of Cell Biology* 123, 209–221.
Images of *Drosophila* larvae Wt and Arm are kindly provided by Dr Martinez-Arias, Cambridge, UK

Drosophila homologue of the mammalian TCF-1 gene (dTCF)[35] (for a review, see Clevers and van de Wetering[36]).

The closely related TCF-1 and LEF-1 were initially discovered as nuclear proteins involved in lymphocyte development and in the expression of T-cell-receptor complex components.[37,38] With the Wnt signal silenced, TCF-1 and LEF-1, together with Groucho (a nuclear protein that recruits histone deacetylase) repress DNA transcription. The ensuing de-acetylation of histone causes dense packing of the DNA in nucleosomes (chromatin condensation) making it inaccessible for RNA polymerases[39–41] (see Figure 14.3 and page 288). Binding of β-catenin (armadillo) to TCF-1 (or LEF-1) interrupts the interaction with Groucho and its repressor partners, and allows re-acetylation of the DNA. The TCF-1 (or LEF-1) associated with β-catenin then initiates transcription of genes leading to proliferation, differentiation or cell death (depending on cell type and context).

From all this it was understood that β-catenin is localized either at the cell membrane or in the nucleus and that its presence in the nucleus profoundly alters the course of gene expression. It remains to be seen how Wnt affects the translocation of β-catenin from the membrane to the nucleus.

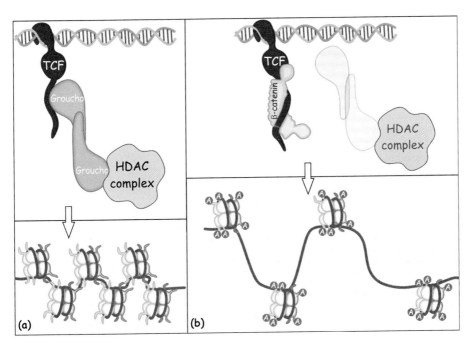

FIG 14.3 The Groucho/TCF-1 repressor complex.

(a) TCF-1 (or LEF-1), Groucho and HDAC form a repressor complex that renders DNA inaccessible to the transcription machinery (RNA polymerases).

(b) Binding of β-catenin to TCF-1 (or LEF-1) removes Groucho and with it the histone deacetylase. Acetylation proceeds, allowing the relaxation of the nucleosomes. Not all components of the deacetylation complex are shown.

Contribution of different species to the elucidation of the Wnt signal transduction pathway

We have limited the description of the discovery of Wnt and its downstream signalling components to the epidermal segment polarity of the larval cuticle in *Drosophila*. However, experiments using *Xenopus laevis* have also contributed to the elucidation of this pathway. The *Xenopus* model attracted much attention after it was found that injection of Wnt mRNA into developing embryos causes a new and complete dorsal–ventral axis, giving rise to two-headed tadpoles.[43] The local production of Wnt creates an independent Spemann organizer. With respect to molecular mechanisms of Wnt signalling, Xenopus has helped to elucidate the mechanism of β-catenin stabilization and subsequent nuclear localization, it has helped to define the Wnt coreceptors LRP5 and LRP6, the discovery that β-catenin binds TCF and the Wnt antagonists such as Dkk, sFRP, and Cerberus. Finally, Xenopus has played an important role in the discovery of non-canonical (β-catenin-independent) pathways (for a review, see Moon et al.;[44] for a note on the Spemann organizer, see page 618).

The TCF family of vertebrate transcription factors consists of four members: TCF-1–4. They all contain an HMG box, which is a DNA-binding

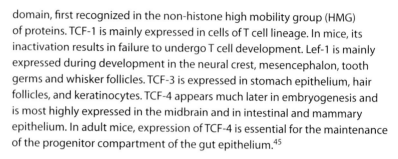

TCF, T cell factor, initially cloned as a lymphoid cell transcription factor. Members of the TCF family are now recognized as activators and repressors of genes implicated in regulation of the cell cycle in a number of cell types. The name T cell factor should not be confused with the ternary complex factor p62^{TCF}, a quite distinct entity, linked with the MAP kinases (see page 342).

domain, first recognized in the non-histone high mobility group (HMG) of proteins. TCF-1 is mainly expressed in cells of T cell lineage. In mice, its inactivation results in failure to undergo T cell development. Lef-1 is mainly expressed during development in the neural crest, mesencephalon, tooth germs and whisker follicles. TCF-3 is expressed in stomach epithelium, hair follicles, and keratinocytes. TCF-4 appears much later in embryogenesis and is most highly expressed in the midbrain and in intestinal and mammary epithelium. In adult mice, expression of TCF-4 is essential for the maintenance of the progenitor compartment of the gut epithelium.[45]

Dual regulation by TCF/LEF. Through their interaction with Groucho these proteins prevent access of the transcriptional machinery to DNA, an action similar to that of the retinoblastoma protein (see page 308). Association with β-catenin (or with plakoglobin) relieves the repression and turns TCF into a true transcription factor, initiating transcription of genes that affect cell fate, proliferation, adhesion, and extracellular matrix degradation.[46]

Groucho and chromatin condensation. Groucho is the name of a mutant that has an increased number of bristles around the eye, reminiscent of the bushy-browed Groucho Marx (Figure 14.4). It encodes a nuclear protein expressed ubiquitously in both embryos and imaginal discs, and acts as a transcriptional repressor of several genes in *Drosophila* development.[47] *Drosophila* Groucho is the archetype of a conserved family of transcriptional corepressors collectively named Groucho/TLE. It participates in a protein complex, enhancer of split ((spl)m9/m10) comprising at least five gene products. Together they repress neurogenesis in proneural clusters of the ventral neuroectoderm (to avoid an excess of brain formation). Sequence analysis indicates that one of the components contains WD domains (sequence terminating with Trp-Asp), also characteristic of the β-subunits of G proteins. The human variants of Groucho are therefore known as transducin-like enhancer-of-split (TLE). The murine and human genomes harbour four full-length homologues of Groucho, as well as a gene that encodes a truncated Groucho protein termed hAES (amino-terminal enhancer of split). In mice, the homologues are called Groucho-related genes, Grg (1–4). There is also a truncated version, Grg-5. Members of this family all share a N-terminal Q (glutamine-rich) domain, GP (glycine/proline-rich) domain, CCN domain (containing putative casein kinase 2/Cdc2 phosphorylation sites and nuclear localization signal), SP (serine/proline-rich) domain, and four WD40 repeats (protein interaction domain). Numerous transcription factors interact with Groucho.[45]

Adenomatous polyposis coli (APC) and the localization of β-catenin

Hints of the mechanism by which β-catenin is localized were provided by an inherited disorder, adenomatous polyposis coli (APC). This is characterized by

FIG 14.4 Groucho Marx.

early onset of numerous polyps in the colon, which progress to malignancy.[48] The polyps are a consequence of germline and (subsequent) somatic mutations that cause truncation of a protein, also termed APC[49] (for a review, see Polakis[50]) (Figure 14.5).

Intact APC forms part of a complex of proteins that interact with β-catenin and bring about its destruction, while stimulation of the Wnt pathway protects the β-catenin, allowing it to re-accumulate.[51] In *Drosophila*, genetic ablation of APC causes up-regulation of β-catenin signalling.[52] In humans, a lack of APC and the ensuing nuclear translocation of β-catenin are linked to cell transformation giving rise to colorectal polyps. When APC is over-expressed there is a reduction in the cellular content of β-catenin due to enhanced degradation.[53]

The importance of APC has also been established in a transgenic mouse model allowing inducible ablation of the APC gene. Here, Wnt signalling is acutely activated in intestinal epithelial cells. There is copious nuclear accumulation of β-catenin, and this coincides with transformation towards a 'crypt progenitor-like' phenotype.[54] Finally, over-expression of β-catenin

APC (adenomatous polyposis coli) is not to be confused with APC/C (anaphase promoting complex/cyclosome).

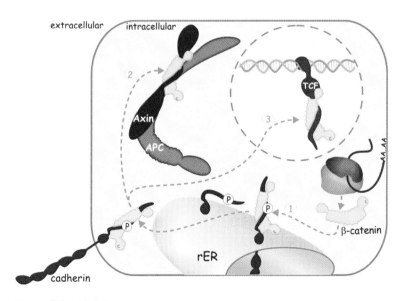

FIG 14.5 β-Catenin's choice.

Following synthesis on free ribosomes, β-catenin binds to the cytosolic segment of cadherin at the surface of the rough endoplasmic reticulum (RER) (1). This binding is rendered effective through prior phosphorylation of cadherin. The complex transfers to the plasma membrane where it forms a component of the adherent junctions. Destabilization of the junctions liberates β-catenin, which is then recognized by Axin and APC (2). This leads to its destruction. Any β-catenin that escapes enters the nucleus and binds transcription factors of the TCF/LEF family (3).

through gene transfection gives rise to a transformed phenotype as found in colorectal cancer.[55] Thus, Wnt signalling causes an elevated level of β-catenin by disabling the destruction-promoting activity of APC. It is the accumulation of cytosolic free β-catenin that favours its translocation into the nucleus.

Importantly, intact APC also forms a complex with glycogen synthase kinase-3β (GSK3β), the homologue of the *Drosophila* gene *shaggy*. This finding positions APC and GSK3β at the centre of the *wingless* signal transduction pathway.[56]

Take your partner: which way β-catenin?

β-Catenin, freshly released from the ribosome has three potential binding partners (see **Figure 14.5**). Two of these, the cadherins and Axin/APC, are located in the cytoplasm; the other, TCF/LEF, in the nucleus. The first choice of partner is cadherin and association occurs as the nascent cadherin appears on the rough endoplasmic reticulum. Binding of β-catenin first requires that this cytoplasmic tail is phosphorylated (by casein kinase 2 or GSK3β). The attachment of β-catenin then facilitates cadherin transport through the Golgi and prevents its degradation by masking the so-called PEST region (or destruction box) that would otherwise attract ubiquitylation[57] (see page 467). On reaching the plasma membrane, the two proteins contribute to the establishment of cell–cell contacts.

A shortage of cadherins or destabilization of the interaction between the two proteins directs β-catenin to its second cytoplasmic partner, the Axin/APC complex.[58] Here, it is first phosphorylated by casein kinase 1α (CK1α) and then by GSK3β, after which it dissociates to undergo multiple rounds of ubiquitylation by SCF[Tcrp] ubiquitin E3-ligase. It is then recognized by the regulatory PA700 and destroyed in the proteasome (see Figure 14.6). Activation of the Wnt pathway allows a stay of execution, causing disruption of the axin/APC/CK1α/GSK3β complex, so preventing phosphorylation. The free β-catenin can now enter the nucleus to unite with its third partner, the transcription factors of the TCF/LEF family.

The (β-catenin-dependent) canonical Wnt pathway

Our description of the continuing sequence of events refers to Wnts 1, 3a, and 8. These bind to two receptors, both discovered in genetic screens. Frizzled-1 (Fz-1) is a 7TM receptor (see page 55) having an extracellular cysteine-rich portion (CRD or Fz domain) thought to bind direct to Wnt. This group of receptors forms a large family, 10 members in humans (Figure 3.14, page 61).[27] The other receptor (or coreceptor) is LRP (low-density lipoprotein-related protein, the *arrow* gene in *Drosophila*), which has only a single membrane-spanning segment. The LRPs also comprise a large family of

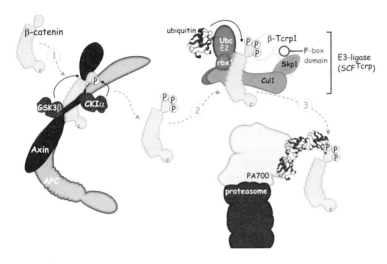

Interestingly, CK and GSK3β contribute both to the destruction of β-catenin (through its phosphorylation) and its protection (by phosphorylation of LRP6 in the presence of Wnt Figure 14.8), a subtle mechanism that genetic screening in *Drosophila* failed to illuminate.

FIG 14.6 Path to destruction.
Binding of β-catenin to axin and APC brings it into the vicinity of CK1α and GSK3β. These phosphorylate the N-terminal at numerous serines and one threonine residue. When dissociated from the complex, phosphorylated β-catenin is recognized by Tcrp1, the receptor component of an SCF E3-ubiquitin ligase complex (2). The polyubiquitylated protein is next recognized by the PA700 subunit of the proteasome (3). The protein is unfolded and split into small polypeptides.

which LRP5 and LRP6 are of particular relevance.[59] It is clear that both types of receptor are required for proper signalling, but how they interact with Wnt and with each other remains to be determined. The domain architecture of the main components of the Wnt pathway are illustrated in Figure 14.7.

Binding of Wnt activates a membrane-bound CK1γ (*Drosophila* homologue Gilgamesh, Gish), which with GSK3β causes phosphorylation of residues in the cytoplasmic tail of the LRP6 receptor (see Figures 14.8 and 14.9).[60,61] Phosphorylated LRP6 acts as a docking site for axin, which is normally associated with APC, CK1α, and GSK3β as part of the β-catenin destruction complex. The docking destabilizes the complex, liberating β-catenin and, in consequence, its cytosolic concentration gradually increases. Simultaneously, Dsh is recruited to the membrane by Fz where it becomes phosphorylated by (soluble) CK1ε (Figure 14.8). Fz, acting conventionally as a G-protein-linked receptor, also activates G_o. Although loss of Dsh and $G\alpha_o$ impairs Wnt-mediated nuclear localization of β-catenin, no effectors have so far been clearly defined (though possible candidates are CKII and Daam1).[59,62]

A failure to destroy β-catenin or the lack of cytoplasmic binding sites leads to its nuclear localization. Several modes of transport to the nucleus have been proposed, including free exchange, APC-mediated import, or Bcl-9-mediated import (also known as Legless).

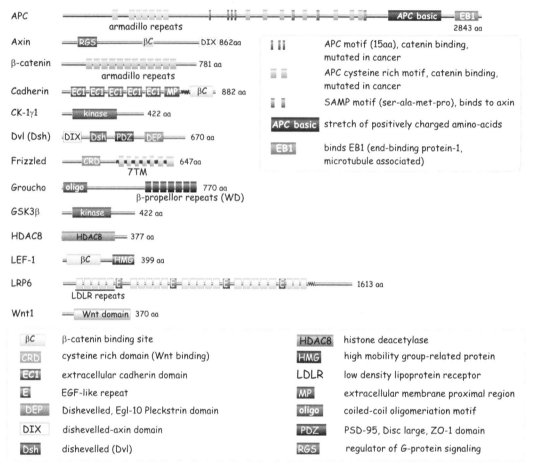

APC motif (15aa), catenin binding, mutated in cancer

APC cysteine rich motif, catenin binding, mutated in cancer

SAMP motif (ser-ala-met-pro), binds to axin

APC basic   stretch of positively charged amino-acids

EB1   binds EB1 (end-binding protein-1, microtubule associated)

βC	β-catenin binding site
CRD	cysteine rich domain (Wnt binding)
EC1	extracellular cadherin domain
E	EGF-like repeat
DEP	Dishevelled, Egl-10 Pleckstrin domain
DIX	dishevelled-axin domain
Dsh	dishevelled (Dvl)
HDAC8	histone deacetylase
HMG	high mobility group-related protein
LDLR	low density lipoprotein receptor
MP	extracellular membrane proximal region
oligo	coiled-coil oligomeriation motif
PDZ	PSD-95, Disc large, ZO-1 domain
RGS	regulator of G-protein signaling

FIG 14.7   Domain architecture of components of the Wnt pathway.

In the nucleus, β-catenin finds its third binding partner, the transcription factors of the TCF/LEF-1 family (Figure 14.10). Importantly, its binding to TCF/LEF-1 displaces Groucho (see **Figure 14.3**), thereby removing the transcription repression complex. Effective transcriptional activation requires the presence of many other factors, including Bcl-9, Pygopus (Pygo), CREB-binding protein (CBP), and Brg1 (the catalytic subunit of a chromatin-remodelling complex.[63–66]

Casein kinases are constitutively active, though not insensitive to allosteric modifications. Access to substrate is controlled by compartmentalization and

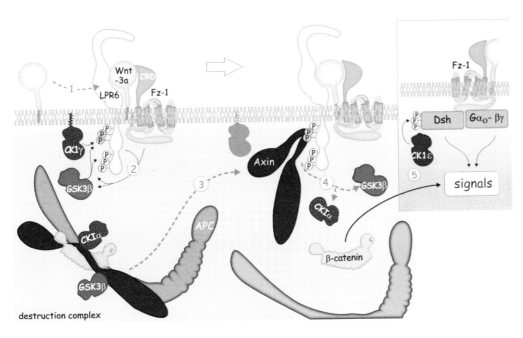

FIG 14.8 The canonical Wnt pathway.

Binding of Wnt3A to Fz-1 and LRP6 (1) causes activation of membrane-bound CK1-γ and the cytoplasmic GSK3β (2). These phosphorylate the cytoplasmic segment of LRP6 at numerous serine and threonine residues, allowing Axin to bind at the position normally occupied by CK1α (3). As a consequence, the destruction complex dissipates and β-catenin is liberated (4). Simultaneously (5), the occupied Fz-1 receptor recruits Dsh, which is then phosphorylated by CK1ε. It also recruits and activates G_o. All three components, β-catenin, Dsh, and Gα_o, play a role in the signalling process.

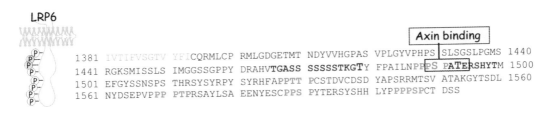

FIG 14.9 Numerous CK1-γ and GSK3β phosphorylation sites in the cytosolic sequence of LRP6.

LRP6 contains numerous phosphorylation sites. The axin binding site (in the motif PPPSPATER) is first phosphorylated on threonine by CK1-γ. This acts as a priming point for further phosphorylations by GSK3β. Downstream of the axin binding site, four GSK3β consensus sequences ((P)PPS/TP) are evident. CK1 phosphorylation site clusters are in red (only the bold Ts are confirmed). Potential GSK3β phosphorylation sites are in green (the bold S is confirmed).

association of regulatory subunits. Substrates are numerous and include cytoskeletal proteins, receptors, ribosomal proteins, translation and transcription factors, kinases, phosphatases, and metabolic enzymes. There are two main families, CK1 and CK2, and in addition, a related mammary gland enzyme.

429

The name **Casein kinases (CK)** is misleading because the majority of them have little or nothing to do with phosphorylation of casein (a highly phosphorylated protein present in milk). CK2 was discovered in a search for protein phosphorylating enzymes present in mitochondria.[67] In order to test the ability of mitochondrial enzyme systems to phosphorylate proteins other than those native to the organelle, various purified proteins were added to mitochondrial suspensions. Of the proteins tried, only casein showed a large and reproducible phosphorylation. One of its substrates, present in the inner mitochondrial membrane, is glycerol-3-phosphate acyltransferase, involved in de novo glycerolipid biosynthesis.[68]

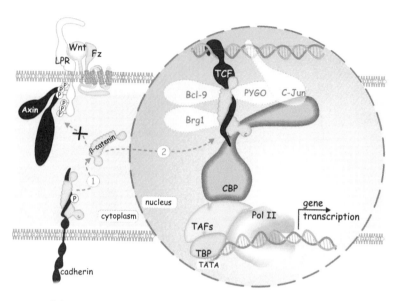

FIG 14.10 β-Catenin/TCF transcription complex.

Destabilization of the cadherin core complex results in liberation of β-catenin (1). In the presence of Wnt, binding to Axin/APC is prohibited and the β-catenin enters the nucleus, there to bind TCF/LEF-1 transcription factors (2). Numerous proteins associate with the complex, of which Bcl-9, Brg1, Pygo, and c-Jun are indicated. The DNA-bound complex interacts with CREB binding protein (CBP), which in turn plays an important role in positioning RNA polymerase at the transcription initiation site. It does this by interaction with general transcription factors such as TATA-binding proteins (TBP) and the TBP-associated factors (TAF).

The CK1 family are further subdivided into α, β, γ1, γ2, γ3, δ, and ε, of which the α and ε forms exhibit a number of splice variants. The CK1γ members are potentially membrane bound because they contain a consensus palmitoylation attachment site.[69]

CK2 is heterotetrameric with two different catalytic subunits α and α' associated with two regulatory β-subunits. It is not surprising that CK2 catalytic subunits are found in mitochondria, given their highly positively charged arginine-rich N-terminal region which serves as a 'destination signal' for soluble mitochondrial proteins. CK2 is involved in a number of processes, amongst which the regulation of cell polarity and morphology have attracted much attention.[70] Casein is an excellent substrate.

Wnt target genes with a TCF-binding element
Many genes are directly affected by Wnt signalling. Here we only consider those having TCF binding sites and those listed below should not be considered universal Wnt targets, because common transcriptional outputs are unlikely to exist. It is the cell and its context, more than the signal itself, which determines

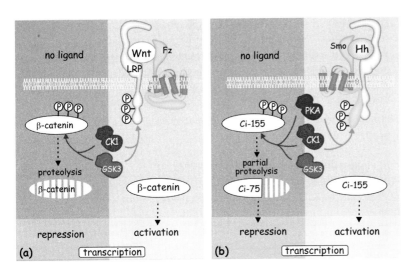

FIG 14.11 Parallels between the Wnt and Hedgehog pathways.
(a) Wnt and (b) Hh signalling pathways share components including GSK3β and members of the CK1 and Lrp families. They also have similar mechanisms. In the absence of ligands, transcription factors are phosphorylated (red arrows) and destroyed or partially degraded, resulting in repression of transcription. In the presence of their ligands, the receptors are phosphorylated (green arrows) and this prevents phosphorylation and destruction of the transcription factors, which now enter the nucleus to activate transcription (Smo, Smoothened).

the nature of the response. However, there are some themes that can be discerned in the types of target genes that are induced by Wnt:

- Genes that code for positive or negative regulators of the pathway. Amongst these are genes encoding TCF-1, LEF1, Axin-2, frizzled 7, Wnt-3a (mouse), Dickkopf
- Genes involved in proliferation and cell de-differentiation. Among these are Engrailed-2, Snail1, Snail2 (Slug), PPARδ, c-Jun, Fra-1, ITF-2, FGF4, FGF18, VEGF, Dpp (*Drosophila*, homologue of BMP), or c-Myc binding protein
- Genes involved in adhesion, E-cadherin
- Genes coding for components of the extracellular matrix and their modifying enzymes, fibronectin and MMP7.

For further details, see the Wnt homepage: http://www.stanford.edu/~rnusse/wntwindow.html

Wnt and Hedgehog

Hedgehog (Hh) is another secreted signalling protein that directs cell growth and patterning. The signalling pathways of Wnt and Hh have striking similarities (Figure 14.11). When Hh is absent from its receptor (Smoothened, panel b), PKA, CK1, and GSK3β sequentially phosphorylate the transcription

Extracellular inhibitors of Wnt and its receptors. A number of proteins regulate Wnt signalling negatively. They do this by binding either to Wnt itself or to the Wnt receptors.[27] Dickkopf (Dkk), one of the best characterized, down-regulates the LRP5/6 receptor. Dickkopf are secreted proteins, four members in human and mouse. Binding to LRP5/6 requires the presence of a member of the Kremen family of cell surface proteins (kringle-containing protein marking the eye and the nose). Thus Dkk2 binds to LRP6 and Krm2 in order to inhibit Wnt signalling in human fibroblasts and in *X. laevis* embryos, to prevent dorsalization.[71] Another inhibitor is Wise, that binds LRP6 and prevents Wnt8 binding.[72]

There are also soluble inhibitors that bind Wnts, preventing their binding to cellular receptors. Examples are WIF1, highly expressed in the unsegmented paraxial presomitic mesoderm in *X. laevis* embryos.[73] Also, there are the secreted Frizzled receptor-related proteins (SFRP), five members in human, that sequester Wnt.[74]

factor Ci-155 (red arrows). This causes it to be converted into a transcriptional repressor (Ci-75). In the presence of Hh, the cytoplasmic domain of the receptor is sequentially phosphorylated by PKA and CK1 (green arrows) preventing phosphorylation of Ci-155. Similar events occur in Wnt signalling (Figure 14.11a). Thus, phosphorylation of both Wnt and Hh receptors prevents transcriptional repression. This is in striking contrast to the tyrosine kinase containing cytokine receptors, in which phosphorylation invariably initiates a cascade of activation events.[75,76]

Wnt and the epithelial to mesenchymal transition

Wnt has a permissive role in β-catenin signalling. The cellular localization of β-catenin determines the fate of mammalian epithelial cells. At the membrane it promotes a differentiated epithelial cell having numerous cell to cell contacts, while in the nucleus it signals a loosely attached cell having a propensity to proliferate and degrade its extracellular matrix. Many of the Wnt-mediated effects on transcription resemble those that occur in the epithelial mesenchymal transformation. The cadherin/β-catenin complex may be regarded as the guardian of the differentiated epithelial phenotype.[77,78]

The factors that determine whether cells differentiate or de-differentiate are listed in Table 14.2 (page 420). Key to this is the formation of stable cell–cell adhesions which depends on the association of the cytoplasmic domains of type 1 cadherin with β-catenin, and of β-catenin with α-catenin (the core complex). The binding of β-catenin to these partners depends on its phosphorylation status. This is regulated by receptor tyrosine protein phosphatases (PTPμ, VEPTP, or LAR) and soluble tyrosine phosphatases (PTP1B, LMW-PTP, SHP2, and DEP1). Opposing effects have been reported for a number of cytokines. The receptors for EGF, FGF, HPG, and TGFβ, by activation of cytoplasmic tyrosine kinases such as Src, Fer, Fyn, or Abl directly or indirectly phosphorylate components of the cadherin core complex resulting in its disruption.[79] Moreover, phosphorylation of β-catenin by the EGFR increases binding to the TATA-box-binding protein (TBP), whereas phosphorylation by Fer, Fyn, or Met promotes binding to Bcl-9. Either way, these phosphorylations favour nuclear localization of β-catenin (Figure 14.12). Not only this, FGF, EGF, and TGFβ also induce expression of numerous transcription factors (E12, Twist, SIP1, δEF1, Snail1, and Slug) that repress transcription of E-cadherin. The cytokines also induce the pro-migratory adhesion molecule N-cadherin (figure 14.13).

Transfer of β-catenin to the nucleus is normally limited by its attachment to its second cytosolic partner, axin/APC (see **Figure 14.5**). This ensures its destruction. To be effective, β-catenin therefore requires a second signal to prevent this. Herein lies the crucial role of Wnt (see Figure 14.10). By disabling the destruction complex, entry of β-catenin into the nucleus is favoured, there to reinforce (or enable) the FGF, EGF, or TGFβ signal by removing the repressor machinery associated with Groucho (and leaving the chromatin in an open configuration).

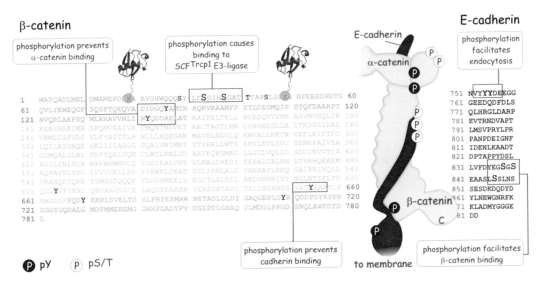

FIG 14.12 Determination of cell fate: phosphorylation and dephosphorylation of the cadherin core complex.

Phosphorylation of β-catenin plays an important role in the determination of cell fate. Serine/threonine phosphorylations (indicated in red) prepare the protein to bind to the SCF[Trcp1] E3-ubiquitin ligase leading to its destruction and favouring differentiation. Serine phosphorylation of E-cadherin stabilizes the core complex and promotes formation of adherent junctions. Numerous tyrosine phosphatases are also involved. These stabilize the cadherin core complex by maintaining β-catenin in a tyrosine-dephosphorylated state.

De-differentiation is favoured by tyrosine phosphorylations (dark blue) of β-catenin and E-cadherin. These destabilize the cadherin core complex by dissociating β-catenin from cadherin. It is of note that Src, a known oncogene, can phosphorylate β-catenin at Y654.

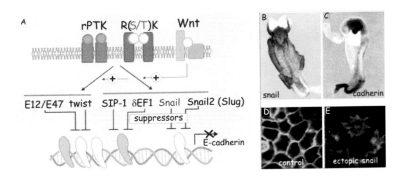

FIG 14.13 Suppression of E-cadherin expression by growth factors and Wnt.

(a) Activation of the receptors for growth factors such as EGF, FGF (rPTK), and TGFβ1 (a receptor serine/threonine kinase) induces suppressors of E-cadherin expression. The growth factor signals are strongly enhanced by the Wnt signal. (b) and (c) Expression of the transcriptional suppressor Snail is inversely related to that of E-cadherin in mouse embryos at 8.5 days of development (mRNA revealed by an *in situ* hybridization procedure using anti-sense DNA). (d) and (e) Ectopic expression of Snail in epidermal keratinocytes causes an epithelial mesenchymal transformation, indicated by the loss of adherent junctions and epithelial structure, and the appearance of detached, non-polarized cells. Expression of protein was detected by antibodies coupled to a fluorescent dye. Figures b–e from Cano et al.[80]

A similar role has been assigned to Wnt in *Drosophila* embryos, both at the level of segment polarity and of wing development. Here, unlike the more conventional inductive morphogens (such as Spatzle, BMP, Activin, Dpp, or Sonic Hedgehog), it acts to modulate the effects of inductive molecules.[26]

Wnt organizes the villous epithelium of the small intestine

Maintenance of the stem cell compartment requires Wnt signalling

The surface of the absorptive epithelium of the small intestine is lined with villi and with crypts of Lieberkühn (see Figure 14.14). Differentiated cells that

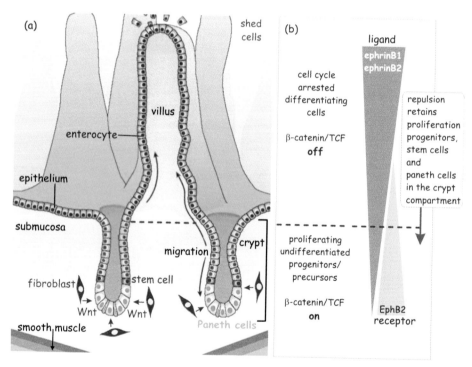

FIG 14.14 Migration of epithelial cells in intestinal crypts.

(a) The crypts of the small intestine contain stem cells that continuously produce committed progenitors. These move either downwards to transform into Paneth cells or up the villi to form the absorptive epithelium. (b) Wnt contributes to the maintenance of the stem cell compartment by ensuring the proliferation of the stem cells and their progenitors and also their localization within the crypt. It does this by induction of the ephrin receptor B2 and by inhibiting expression of the ephrin ligands -B1 and -B2. This creates opposing gradients of both receptor and ligand. Stem cells and progenitors expressing high levels of the receptor are prevented from migrating out of the crypt through the repulsive action of the ephrins. Only when committed progenitor cells have switched off the canonical Wnt pathway do they escape repulsion and move up to the villus.

Panel (a) adapted from Sancho et al.[81] Adapted and reprinted, with permission, from the *Annual Review of Cell and Developmental Biology* Volume 20 © 2004 by Annual Reviews (www.annualreviews.org).

reside on the villi are enterocytes, whose role is to transport metabolites towards the interior. Here, enteroendocrine cells secrete the hormones gastrin, secretin, and cholecystokinin in response to the intestinal content and its acidity, and goblet cells secrete the mucin that lubricates the gut lining. Paneth cells, functionally similar to neutrophils, contribute to the extrinsic gastrointestinal barrier by releasing α-defensins (cryptdins), lysozyme, and phospholipase A2. The crypts also harbour the stem and progenitor cell compartment, situated halfway between the bottom of the crypt and the start of the villi. The colonic epithelium has a flat surface pitted with crypts. The upper segment of the crypt contains goblet cells and mature enterocytes, whereas the lower segment is dominated by the stem cell/progenitor cell compartment.

The lifespan of intestinal epithelial cells is less than a week, during which time they migrate from the base of the crypts to the tips of the villi. Here they die and are shed into the faecal stream. Stem cells and Paneth cells escape migration. The stem cells cycle slowly, continuously producing progenitor cells which proliferate rapidly. As they attain the crypt–villus axes in the small intestine (or the upper crypt segments in the colon), they undergo cell cycle arrest and commence expressing differentiation markers.[82] Cells of the crypt epithelium possess nuclear β-catenin, the hallmark of Wnt signalling.

Wnt aligns committed progenitor cells along the crypt–villus axis

A screen for genes induced or repressed by β-catenin/TCF-4 in intestinal epithelial cells revealed both Ephrin-B receptors (EphB) and their ligands (B-type ephrins) amongst the targets.[83] Here, β-catenin/TCF exerts inverse control, enhancing expression of the receptors and suppressing the ligands. This creates opposed gradients, with high receptor expression at the base of the crypts and high ligand expression at the points of the villi (see **Figure 14.14**).

Expression of EphB receptors is essential for the correct positioning of epithelial cells along the crypt–villus axis. Although they maintain their intestinal villi, mice carrying EphB2/EphB3 double mutants lose the characteristic localization of the Paneth cells in the crypts and the characteristic compartmentalization of proliferating and differentiated cells along the axis (put simply, everything is all higgledy-piggledy).[85] EphB receptors, which are highly expressed at the bottom of the crypt, may thus act to restrict the upward migration of stem and Paneth cells.

Canonical Wnt signalling is vital for the maintenance of the epithelial stem cell compartment. Neonatal mice, deficient in TCF-4, lack the crypt progenitor compartment[86] and transgenic mice expressing Dickkopf, an inhibitor of Wnt signalling, also show a loss of crypts.[87,88]

Other examples of Wnt signalling include the self-renewing capacities of mammalian hair follicles [89] and of haematopoietic stem cells.[20,90] Of note,

Ephrin receptors represent the largest subfamily of receptor tyrosine kinases (14 members in human). Based on their ligand binding specificities, these receptors are grouped into two subclasses. EphA receptors bind A-type ephrins, EphB bind B-type ephrins. Ephrin ligands are GPI-anchored proteins and the interaction between Eph receptors and their ligands therefore involves direct cell–cell interactions. Eph receptors and their ligands are said to be involved in directed migration through repulsion. They constitute repulsive cues in processes such as axon path-finding, migration of neural crest cells and boundary formation between segmented structures such as rhombomeres.[84]

Paneth cells are insensitive to the Wnt signal and remain non-proliferative differentiated cells.

Wnt and the asymmetric division of stem cells

In order to maintain their capacity for self-renewal, stem cells must undergo asymmetric division. The segregating proteins and mRNA are distributed in a way that generates one daughter stem cell and one daughter progenitor cell. The progenitor may still divide a number of times but is committed to differentiate into a non-proliferating cell with a specialized function. The stem cell does not differentiate and retains its capacity to divide.

Molecular insight into the asymmetric distribution of cellular products during division has been provided by studies on brain neuroblasts. These cells achieve an unequal distribution of two proteins, brain tumour (Brat) and prospero (Pros), assisted by miranda (Mira, a scaffold protein). The daughter cell containing Brat and Pros becomes the precursor of ganglion mother cells, whereas its sister that lacks them, retains its identity as a neuroblast. Pros is a homeobox transcription factor that represses cyclins E and A as well as Cdc25, a dual-specificity phosphatase required for activation of cyclin-dependent kinases. Brat prevents translation of c-myc mRNA, a transcription factor involved in RNA-polymerase 1-dependent transcription of rRNA, thus a driver of ribosome production.[91,92]

To achieve the asymmetric distribution of cellular products, correct orientation of the microtubule spindle, prior to division, is essential (see Figure 14.15). Wnt and its 7TM Fz receptor are clearly implicated, but not the coreceptor LRP5/6.[93–95] The signal is transmitted to Dsh, but is without effect on axin binding, instead switching towards Daam1 which is associated with RhoA.[96] The RhoGEF involved remains to be identified.

Rho: regulator of the actin cytoskeleton

Rho is a key regulator in the organization of the actin cytoskeleton.[97] Through its effector mDia, RhoA facilitates nucleation and polymerization of actin into long filaments. Through another effector, RhoA coiled-coil kinase (ROCK), it converts these filaments into stress fibres, cross-linking them with actinin and activating the motor protein myosin II. Both ROCK and mDia also operate in centrosome positioning during prometaphase.[98] They control the connection of the astral microtubules with the cell cortex and hence they determine the orientation of the mitotic spindle.[99] Interestingly, APC, a component of the β-catenin destruction complex (see page 424), binds the microtubules and is involved in organizing their linkage to the cortical network (see also page 589).[100] It is not known what determines the site of fixation of the astral microtubules in epithelial stem cells. How this pathway signals to the ubiquitously expressed and evolutionarily conserved Par proteins that determine cell polarity is also unresolved.[101] The non-canonical pathway involved in the induction of cell polarity is referred to as the 'planar cell polarity' (PCP) pathway.

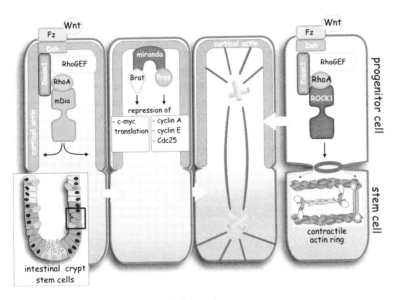

FIG 14.15 Wnt and asymmetric distribution of cellular products.

By activating the non-canonical pathway, through Dsh, Daam1, RhoGEF, and RhoA, Wnt is involved in asymmetric production of cortical actin. Downstream of RhoA, mDia assists the local polymerization of actin. One consequence is the asymmetric distribution of the scaffold protein Miranda which binds to cortical actin bringing with it, Brat and Pros. Brat suppresses translation of c-myc thereby limiting the protein synthetic capacity of the future committed progenitor cell. Pros suppresses the transcription of cyclins A and E and the dual specificity phosphatase Cdc25. This prevents repetitive cell division. The asymmetric distribution of cortical actin also determines the orientation of the mitotic spindle, through orientation of the astral microtubules. Attachment of these to the cell cortex is facilitated by the RhoA-controlled kinase ROCK, another component of the non-canonical pathway. Finally, this pathway also operates in the formation of the contractile actin ring, positioned perpendicular to the mitotic spindle, which ultimately divides the cell in two. This occurs through the action of ROCK1 which phosphorylates and activates myosin II. The events depicted represent a compilation of data from different species and cell types and in different contexts.

Different mutations in APC. The penetrance of germline mutations that increase the risk of colorectal cancer varies. The APC polymorphism I1307K has a penetrance of 20%, meaning that 1 in 5 persons who carry it will develop adenomatous polyposis coli. This mutation does not directly affect the activity of APC, but enhances the chance of further 'somatic' mutations. The reason for this is that in the mutated protein, the codon ATA (isoleucine) has been changed to AAA (leucine). This gives rise to a DNA sequence consisting of eight adenines in a row, which increases the risk of slippage during replication. It therefore increases the chance of further mutations of which some will eventually give rise to a frame shift and loss of APC function. The penetrance of truncation mutations approaches 100%.

Mutations of β-catenin, Axin, and APC in human cancers

Several mutations in components of the Wnt signal transduction pathway give rise to cancer.[50] Most of these result in the expression of non-functional proteins at both alleles. A prime example is the APC gene. Here, most of the mutations accumulate in the so-called *mutation cluster region* in the central part of the peptide chain. The truncated proteins that arise are unable to bind β-catenin and axin.[102] Cancer generating mutations also occur in axin and in β-catenin. β-Catenin is protected against degradation because the mutations either replace the phosphorylation sites for CK1α or GSK3β in the destruction

Several somatic mutations in β-catenin have been associated with colorectal, liver, and hair follicle cancer. These mutations often occur at sites subject to phosphorylation by CK1a and GSK3β. Mutations are also found in armadillo repeats (see **Figure 14.7**): mutations in the 3rd and 4th repeat affect binding to APC or axin, mutations in the 12th enhance transcriptional activity. All these are so-called *activation mutations* because they enhance the output of β-catenin, either through an increase in protein content or by an increase in its transcriptional activity.

box or they prevent its binding to the axin/APC complex. They are therefore considered as 'activating' mutations (even though they do not actually activate the protein). In some patients having familial gastric cancer, the gene for E-cadherin expresses a truncated product[103] which is unable to participate in junctional complexes. As a result, the adherent junctions are improperly formed so that over time, the cells detach and become invasive.

Non-canonical signal transduction pathways

A number of developmental processes require Wnt (in particular, Wnts 5a, 7a, or 11), Fz, and Dsh, but not β-catenin. They are responsible for cell polarization, e.g. during gastrulation of *X. laevis* embryos, cell movement during ovarian morphogenesis, and planar cell polarity during gastrulation of zebrafish. These processes involve the Rho-related GTPases, Rho, Rac, and Cdc42 and their effectors, amongst which, mDia, Rho kinase (ROCK), and the MAP kinase JNK (reviewed in Veeman et al.[29] and Strutt;[104] see Figure 14.15). A non-canonical Wnt pathway also operates in muscle development in chick and mouse embryos. This requires Wnt1, generated by the neural tube, and Wnt7A, generated by the surface ectoderm tissue. Wnt serves to differentiate mesoderm into muscle cells. Initiation of this process requires the Fz receptor, but in this case it is coupled to a heterotrimeric GTP-binding protein that activates adenylyl cyclase. A classical PKA to CREB signal transduction pathway ensues (see page 248) that leads to the induction of the differentiation markers such as Pax3, MyoD and Lyf5.[105]

A role for cadherin in contact inhibition

Contact inhibition is the term used to describe the abrupt arrest of the cell cycle that occurs in cultures of rapidly proliferating epithelial cells at the point when a confluent monolayer forms. The phenomenon has been recognized for many years, but even now a full description remains elusive. It was clear from the start that the arrest is not due to the accumulation of inhibitory factors, but is mediated by an adhesion molecule.[106] Cadherin plays an important role. As the density of the cells increases and multiple contacts are established, the forming adherent junctions sequester the free β-catenin and in this way reduce the proliferation promoting action of TCF/β-catenin.[107]

Regulation may also occur directly at the receptor level. Clustering of VE-cadherin at intercellular junctions blocks the proliferative response of endothelial cells to VEGF. This is reflected in the inhibition of tyrosine phosphorylation of the VEGF-R2.[108] Dephosphorylation of the VEGF receptor is by the receptor tyrosine phosphatase DEP1/CD148[109] which binds the cadherin associated protein p120-catenin (p120ctn).[110]

Other examples of signalling through adhesion molecules

Cadherin in the central nervous system

Apart from their roles in the formation and maintenance of neuronal synapses, the cadherins have signalling functions in nerves. These involve neither E-cadherin nor β-catenin but instead, N-cadherin and the γ-protocadherins. Through a pathway that is as yet unresolved, membrane depolarization or NMDA-R activation (see page 55) causes the activation of membrane proteases, among which are ADAM10 and γ-secretase. ADAM10 removes the extracellular N-terminus of N-cadherin, which is shed into the synaptic cleft, while γ-secretase cleaves the cytoplasmic C-terminus. The C-terminal fragment binds CBP (CREB binding protein: see page 430) and together they are destined for degradation by the proteasome. In this way N-cadherin disables CPB/CREB-mediated transcription. The C-terminal fragment of γ-protocadherin translocates to the nucleus, where it appears to have a transcriptional role. In this way cadherins also make a contribution to synaptic plasticity as regulators of transcription.

JAM and the regulation of differentiation

The abutment of the junctional adhesion molecule JAM with JAM on a neighbouring membrane initiates a series of events that contribute to the organization of polarized epithelial cells (see Figure 14.16). JAM recruits PARs 3 and 6 which in turn attract the GTPase Cdc42 causing activation of the atypical PKCζ (see page 585). This sets about a process that contributes to junctional assembly and inhibition of proliferation. More surprising, the structural component ZO-1, which binds to occludin, claudin, and JAM, also interacts with the transcription factor ZONAB, required for controlled proliferation rates.[111] Sequestration of ZONAB at the tight junction disables transcriptional activity, and cyclin D is amongst the genes that are no longer expressed. In addition, at the membrane ZONAB attracts CDK4 and, in so doing, halts the cell cycle in G1.[112]

Occludin interacts with the TGFβ type I receptor

In the presence of TGFβ1, its receptor (TGFβRI or TβRI) binds occludin, a major component of the tight junction. This interaction plays a role in the depolarization of epithelial cells during the epithelial to mesenchymal transition. Epithelial cells expressing an occludin that lacks the second (C-terminal) loop, retain their tight junctions upon addition of TGFβ1. This effect is entirely independent of Smad signalling[113] (normally linked to TβRI signalling, see page 600). Thus occludin recruits the activated TGFβ type-I

FIG 14.16 Components of the tight junction control junctional organization and cell proliferation. Opposing JAMs initiate a series of events that result in the organization of the tight junction and arrest of cell proliferation. For details see text.

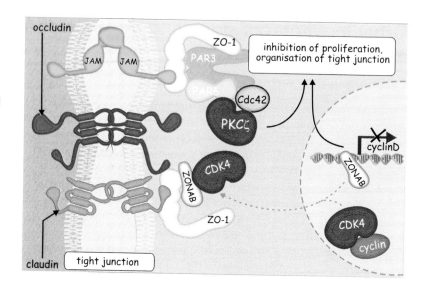

receptor to the tight junction and this determines their efficient dispersal during the epithelial to mesenchymal transition.

Occludin prevents Raf-1-mediated cell transformation

Transformation of epithelial cells with a constitutively active mutant of Raf-1 is associated with transcriptional down-regulation of occludin. Introduction of exogenous occludin into Raf-1 transformed cells rescues the epithelial phenotype and induces reassembly of functional tight junctions. Moreover, these cells also regain anchorage-dependent growth, indicating that occludin does more than merely restore cell–cell association.[114]

List of Abbreviations

Abbreviation	Full name/description	SwissProt entry	Other names/OMIM
APC	adenomatous polyposis coli	P25054	MIM: 175100
arm	armadillo (Drosophila)	P18824	
axin-1	axis inhibition	O15169	
Bcl-9	B cell lymphoma protein-9	O00512	legless (Lgs) homologue
Brg1	Brahma relate gene-1	P51532	SMARCA4
Cadherin-E	calcium-dependent adherence	P12830	uvomorulin

Continued

Abbreviation	Full name/description	SwissProt entry	Other names/OMIM
catenin-β		P35222	OMIM:116806
CBP	CREB binding protein	Q92793	
Ci	cubitus interruptus (*Drosophila*)	P19538	
CK1-ε	casein kinase-1 isoform epsilon	P49674	
CK1-γ1	casein kinase-1 isoform gamma-1	Q9HCP0	
DbpA	DNA binding protein A	P16989	
Dsh	dishevelled human	O14640	
Dsh	dishevelled *Drosophila*	P51140	
Dvl-1	segment polarity dishevelled homologue-1	O14649	
EB1	end binding protein-1	Q15691	APC binding protein, MAPRE
EphB2	ephrin type-B receptor-2	P29323	tyrosine-protein kinase receptor EPH-3
EphB3	ephrin type-B receptor-3	P54753	tyrosine protein kinase receptor HEK-2
ephrin B1		P98172	EPH-related receptor tyrosine kinase ligand-2
ephrin B2		P52799	EPH-related receptor tyrosine kinase ligand-5
Fz-1	Frizzled-1	Q9UP38	
Gα_o	G-protein subunit α_o	P09471	
Groucho		P16371	
GSK3β	glycogen synthase kinase-3 beta	P49841	
HDAC	histone deacetylase	Q13547	
HGF	hepatocyte growth factor	P14210	scatter factor
Hh	hedgehog *Drosophila*	Q02936	
LEF-1	lymphoid enhancer-binding factor-1	Q9UJU2	TCF-1α
Lgs	legless *Drosophila*	O00512	
LRP5	low density lipoprotein receptor protein	O75197	

Continued

Abbreviation	Full name/description	SwissProt entry	Other names/OMIM
LRP6	LDL-receptor related protein-6	O75581	
mDia1	mouse Diaphanous homologue-1	O08808	
Pan	pangolin (*Drosophila*)	P91943	TCF
PDZ	domain found in PSD95, DlgA and ZO-1		
PKC*ι*	protein kinase C-iota	P41743	atypical protein kinase C λ
plakoglobin	junctional plaque protein	P14923	desmoplakin-3
PSD95	post-synaptic density protein-95	P78352	disc large homolog-4
Pygo1	pygopus homologue-1	Q9Y3Y4	
ROCK	Rho-associated coiled-coil containing protein kinase-1	Q13464	
SCF	ubiquitin protein E3-ligase complex comprising Skp1, cullin-1 and F-box protein		
Sgg	shaggy (rough-haired) (*Drosophila*)	P18431	zeste-white (GSK3 kinase)
Smo	smoothened homologue	Q99835	Gx protein
Snail		O95863	SNAI1, SnaH
TAF1	TBP-associated factor-1	Q15573	RNA polymerase subunit A
TBP	TATA-box binding protein	P20226	TFIID TBP subunit
TCF-4	T cell specific transcription factor-4	Q9NQB0	TCF7L2, MIM: 602228
TLE-1	transducin-like enhancer protein-1	Q04724	
Wnt-3a	wingless/int-3a	P56704	
ZONAB	ZO-1 associated nucleic acid binding protein	Q9N1Q2	DbpA (DNA binding protein-A)

References

1. Fialka I, Schwarz H, Reichmann E, Oft M, Busslinger M, Beug H. The estrogen-dependent c-JunER protein causes a reversible loss of mammary epithelial cell polarity involving a destabilization of adherens junctions. *J Cell Biol*. 1996;132:1115–1132.
2. Eger A, Stockinger A, Schaffhauser B, Beug H, Foisner R. Epithelial mesenchymal transition by c-Fos estrogen receptor activation involves nuclear translocation of β-catenin and upregulation of β-catenin/

lymphoid enhancer binding factor-1 transcriptional activity. *J Cell Biol*. 2000;148:173–188.

3. Lee JM, Dedhar S, Kalluri R, Thompson EW. The epithelial-mesenchymal transition: new insights in signaling, development, and disease. *J Cell Biol*. 2006;172:973–981.

4. Hay ED. The mesenchymal cell, its role in the embryo, and the remarkable signaling mechanisms that create it. *Dev Dyn*. 2005;233:706–720.

5. Huber MA, Kraut N, Beug H. Molecular requirements for epithelial-mesenchymal transition during tumor progression. *Curr Opin Cell Biol*. 2005;17:548–558.

6. Berx G, van Roy F. The E-cadherin/catenin complex: an important gatekeeper in breast cancer tumorigenesis and malignant progression. *Breast Cancer Res*. 2001;3:289–293.

7. Gottardi CJ, Wong E, Gumbiner BM. E-cadherin suppresses cellular transformation by inhibiting β-catenin signaling in an adhesion-independent manner. *J Cell Biol*. 2001;153:1049–1060.

8. Reya T, Clevers H. Wnt signalling in stem cells and cancer. *Nature*. 2005;434:843–850.

9. Nelson WJ, Nusse R. Convergence of Wnt, β-catenin, and cadherin pathways. *Science*. 2004;303:1483–1487.

10. Sharma RP, Chopra VL. Effect of the Wingless (wg1) mutation on wing and haltere development in *Drosophila melanogaster*. *Dev Biol*. 1976;48: 461–465.

11. Nusslein-Volhard C, Wieschaus E. Mutations affecting segment number and polarity in *Drosophila*. *Nature*. 1980;287:795–801.

12. Gerhart J. Signaling pathways in development. *Teratology*. 1999;60: 226–239.

13. Nusse R, Varmus HE. Many tumors induced by the mouse mammary tumor virus contain a provirus integrated in the same region of the host genome. *Cell*. 1982;31:99–109.

14. Baker NE. Molecular cloning of sequences from wingless, a segment polarity gene in *Drosophila*: the spatial distribution of a transcript in embryos. *EMBO J*. 1987;6:1765–1773.

15. Cabrera CV, Alonso MC, Johnston P, Phillips RG, Lawrence PA. Phenocopies induced with antisense RNA identify the wingless gene. *Cell*. 1987;50: 659–663.

16. Rijsewijk F, Schuermann M, Wagenaar E, Parren P, Weigel D, Nusse R. The *Drosophila* homolog of the mouse mammary oncogene *int-1* is identical to the segment polarity gene *wingless*. *Cell*. 1987;50:649–657.

17. Nusse R, Brown A, Papkoff J, et al. A new nomenclature for *int-1* and related genes: the Wnt gene family. *Cell*. 1991;64:231.

18. Thomas KR, Musci TS, Neumann PE, Capecchi MR. *Swaying is a mutant allele of the proto-oncogene Wnt-1*. *Cell*. 1991;67:969–976.

19. Moon RT, Kohn AD, De Ferrari GV, Kaykas A. WNT and β-catenin signalling: diseases and therapies. *Nat Rev Genet*. 2004;5:691–701.

20. Willert K, Brown JD, Danenberg E, et al. Wnt proteins are lipid-modified and can act as stem cell growth factors. *Nature*. 2003;423:448–452.
21. Panakova D, Sprong H, Marois E, Thiele C, Eaton S. Lipoprotein particles are required for Hedgehog and Wingless signalling. *Nature*. 2005;435: 58–65.
22. Siegfried E, Chou TB, Perrimon N. *wingless* signaling acts through zeste-white 3, the *Drosophila* homolog of glycogen synthase kinase-3, to regulate engrailed and establish cell fate. *Cell*. 1992;24:1167–1179.
23. Noordermeer J, Klingensmith J, Perrimon N, Nusse R. *dishevelled* and *armadillo* act in the wingless signalling pathway in *Drosophila*. *Nature*. 1994;367:80–83.
24. Willert K, Nusse R. β-Catenin: a key mediator of Wnt signalling. *Curr Opin Genet Dev*. 1998;8:95–102.
25. Klingensmith J, Nusse R. Signaling by wingless in *Drosophila*. *Dev Biol*. 1994;166:396–414.
26. Martinez-Arias A. Wnts as morphogens? The view from the wing of *Drosophila*. *Nat Rev Mol Cell Biol*. 2003;4:321–325.
27. Logan CY, Nusse R. The Wnt signaling pathway in development and disease. *Annu Rev Cell Dev Biol*. 2004;20:781–810.
28. Wong LL, Adler PN. Tissue polarity genes of *Drosophila* regulate the subcellular location for prehair initiation in pupal wing cells. *J Cell Biol*. 1993;123:209–221.
29. Veeman MT, Axelrod JD, Moon RT. A second canon. Functions and mechanisms of β-catenin-independent Wnt signalling. *Dev Cell*. 2003;5:367–377.
30. Peifer M, Wieschaus E. The segment polarity gene *armadillo* encodes a functionally modular protein that is the *Drosophila* homolog of human plakoglobin. *Cell*. 1990;63:1167–1176.
31. Peifer M, McCrea PD, Green KJ, Wieschaus E, Gumbiner BM. The vertebrate adhesive junction proteins β-catenin and plakoglobin and the *Drosophila* segment polarity gene *armadillo* form a multigene family with properties similar. *J Cell Biol*. 1992;118:681–691.
32. Brunner E, Peter O, Schweizer L, Basler K. Pangolin encodes a Lef-1 homologue that acts downstream of Armadillo to transduce the Wingless signal in *Drosophila*. *Nature*. 1997;385:829–833.
33. Behrens J, von Kries JP, Kuhl M, Bruhn L, Wedlich D, Grosschedl R, Birchmeier W. Functional interaction of β-catenin with the transcription factor LEF-1. *Nature*. 1996;382:638–642.
34. Riese J, Yu X, Munnerlyn A, Eresh S, Hsu SC, Grosschedl R, Bienz M. LEF-1, a nuclear factor coordinating signaling inputs from *wingless* and *decapentaplegic*. *Cell*. 1997;88:777–787.
35. van de Wetering M, Cavallo R, Dooijes D, et al. Armadillo coactivates transcription driven by the product of the *Drosophila* segment polarity gene *dTCF*. *Cell*. 1997;88:789–799.

36. Clevers H, van de Wetering M. TCF/LEF factor earn their wings. *Trends Genet*. 1997;13:485–489.

37. van de Wetering M, Oosterwegel M, Dooijes D, Clevers H. Identification and cloning of TCF-1, a T lymphocyte-specific transcription factor containing a sequence-specific HMG box. *EMBO J*. 1991;10:123–132.

38. Travis A, Amsterdam A, Belanger C, Grosschedl R. *LEF-1*, a gene encoding a lymphoid-specific protein with an HMG domain, regulates T-cell receptor α enhancer function. *Genes Dev*. 1991;5:880–894.

39. Bienz M. TCF: transcriptional activator or repressor? *Curr Opin Cell Biol* 1998;10:366–372.

40. Cavallo RA, Cox RT, Moline MM, et al. *Drosophila* Tcf and Groucho interact to repress Wingless signalling activity. *Nature*. 1998;395:604–608.

41. Daniels DL, Weis WI. β-catenin directly displaces Groucho/TLE repressors from Tcf/Lef in Wnt-mediated transcription activation. *Nat Struct Mol Biol*. 2005;12:364–371.

42. Ropero S, Fraga MF, Ballestar E, et al. A truncating mutation of *HDAC2* in human cancers confers resistance to histone deacetylase inhibition. *Nat Genet*. 2006;38:566–569.

43. Sokol S, Christian JL, Moon RT, Melton DA. Injected Wnt RNA induces a complete body axis in *Xenopus* embryos. *Cell*. 1991;67:741–752.

44. Moon RT, Bowerman B, Boutros M, Perrimon N. The promise and perils of Wnt signaling through β-catenin. *Science*. 2002;296:1644–1646.

45. Brantjes H, Roose J, van de Wetering M, Clevers H. All Tcf HMG box transcription factors interact with Groucho-related co-repressors. *Nucleic Acids Res*. 2001;29:1410–1419.

46. Simcha I, Shtutman M, Salomon D, et al. Differential nuclear translocation and transactivation potential of β-catenin and plakoglobin. *J Cell Biol*. 1998;141:1433–1448.

47. Lindsley DL, Grell EH. *Genetic variations of Drosophila melanogaster* Publication 627. Washington DC: Carnegie Institute; 1968.

48. Lynch HT, de la Chapelle A. Hereditary colorectal cancer. *N Engl J Med*. 2003;348:919–932.

49. Groden J, Thliveris A, Samowitz W, et al. Identification and characterization of the familial adenomatous polyposis coli gene. *Cell*. 1991;66:589–600.

50. Polakis P. Wnt signaling and cancer. *Genes Dev*. 2000;14:1837–1851.

51. Hinck L, Nelson WJ, Papkoff J. Wnt-1 modulates cell–cell adhesion in mammalian cells by stabilizing β-catenin binding to the cell adhesion protein cadherin. *J Cell Biol*. 1994;124:729–741.

52. Ahmed Y, Hayashi S, Levine A, Wieschaus E. Regulation of *armadillo* by a *Drosophila* APC inhibits neuronal apoptosis during retinal development. *Cell*. 1998;93:1171–1182.

53. Munemitsu S, Albert I, Souza B, Rubinfeld B, Polakis P. Regulation of intracellular β-catenin levels by the adenomatous polyposis coli (APC) tumor-suppressor protein. *Proc Natl Acad Sci U S A*. 1995;92:3046–3050.

54. Sansom OJ, Reed KR, Hayes AJ, et al. Loss of Apc in vivo immediately perturbs Wnt signaling, differentiation, and migration. *Genes Dev*. 2004;18:1385–1390.

55. Orford K, Orford CC, Byers SW. Exogenous expression of β-catenin regulates contact inhibition. *anchorage-independent growth, anoikis, and radiation-induced cell cycle arrest, J Cell Biol*. 1999;146:855–868.

56. Rubinfeld B, Albert I, Porfiri E, Fiol C, Munemitsu S, Polakis P. Binding of GSK3β to the APC-β-catenin complex and regulation of complex assembly. *Science*. 1996;272:1023–1026.

57. Huber AH, Stewart DB, Laurents DV, Nelson WJ, Weis WI. The cadherin cytoplasmic domain is unstructured in the absence of β-catenin. A possible mechanism for regulating cadherin turnover. *J Biol Chem*. 2001;276:12301–12309.

58. Su LK, Vogelstein B, Kinzler KW. Association of the APC tumor suppressor protein with catenins. *Science*. 1993;262:1734–1737.

59. Cadigan KM, Liu YI. Wnt signaling: complexity at the surface. *J Cell Sci*. 2005;119:395–402.

60. Davidson G, Wu W, Shen J, et al. Casein kinase 1 γ couples Wnt receptor activation to cytoplasmic signal transduction. *Nature*. 2005;438:867–872.

61. Huang H, Tamai K, Doble B, et al. A dual-kinase mechanism for Wnt co-receptor phosphorylation and activation. *Nature*. 2005;438:873–877.

62. Gao Y, Wang HY. Casein kinase 2 is activated and essential for Wnt/β-catenin signalling. *J Biol Chem*. 2006;281:18394–18400.

63. Tsukiyama T. The in vivo functions of ATP-dependent chromatin-remodelling factors. *Nat Rev Mol Cell Biol*. 2002;3:422–429.

64. Hoffmans R, Stadeli R, Basler K. Pygopus and legless provide essential transcriptional co-activator functions to armadillo/β-catenin. *Curr Biol*. 2005;15:1207–1211.

65. Tolwinski NS, Wieschaus E. A nuclear escort for β-catenin. *Nat Cell Biol*. 2004;6:579–580.

66. Townsley FM, Cliffe A, Bienz M. Pygopus and Legless target Armadillo/β-catenin to the nucleus to enable its transcriptional co-activator function. *Nat Cell Biol*. 2004;6:626–633.

67. Burnett G, Kennedy EP. The enzymatic phosphoryation of proteins. *J Biol Chem*. 1954;211:969–980.

68. Onorato TM, Chakraborty S, Haldar D. Phosphorylation of rat liver mitochondrial glycerol-3-phosphate acyltransferase by casein kinase 2. *J Biol Chem*. 2005;280:19527–19534.

69. Knippschild U, Gocht A, Wolff S, Huber N, Lohler J, Stoter M. The casein kinase 1 family: participation in multiple cellular processes in eukaryotes. *Cell Signal*. 2005;17:675–689.

70. Canton DA, Litchfield DW. The shape of things to come: an emerging role for protein kinase CK2 in the regulation of cell morphology and the cytoskeleton. *Cell Signal*. 2006;18:267–275.

71. Mao B, Niehrs C. Kremen2 modulates Dickkopf2 activity during Wnt/LRP6 signaling. *Gene*. 2003;302:179–183.

72. Itasaki N, Jones CM, Mercurio S, et al. Wise, a context-dependent activator and inhibitor of Wnt signalling. *Development*. 2003;130:4295–4305.

73. Hsieh JC, Kodjabachian L, Rebbert ML, et al. A new secreted protein that binds to Wnt proteins and inhibits their activities. *Nature*. 1999;398:431–436.

74. Finch PW, He X, Kelley MJ, et al. Purification and molecular cloning of a secreted, Frizzled-related antagonist of Wnt action. *Proc Natl Acad Sci U S A*. 1997;94:6770–6775.

75. Jia J, Tong C, Wang B, Luo L, Jiang J. Hedgehog signalling activity of Smoothened requires phosphorylation by protein kinase A and casein kinase I. *Nature*. 2004;432:1045–1050.

76. Apionishev S, Katanayeva NM, Marks SA, Kalderon D, Tomlinson A. *Drosophila* Smoothened phosphorylation sites essential for Hedgehog signal transduction. *Nat Cell Biol*. 2004;7:86–92.

77. Christofori G. Changing neighbours, changing behaviour: cell adhesion molecule-mediated signalling during tumour progression. *EMBO J*. 2003;22:2318–2323.

78. Pagliarini RA, Xu T. A genetic screen in *Drosophila* for metastatic behavior. *Science*. 2003;302:1227–1231.

79. Lilien J, Balsamo J. The regulation of cadherin-mediated adhesion by tyrosine phosphorylation/dephosphorylation of β-catenin. *Curr Opin Cell Biol*. 2005;17:459–465.

80. Cano A, Perez-Moreno MA, Rodrigo I, et al. The transcription factor snail controls epithelial-mesenchymal transitions by repressing E-cadherin expression. *Nat Cell Biol*. 2000;2:76–83.

81. Sancho E, Batlle E, Clevers H. Signaling pathways in intestinal development and cancer. *Annu Rev Cell Dev Biol*. 2004;20:695–723.

82. Booth C, Potten CS. Gut instincts: thoughts on intestinal epithelial stem cells. *J Clin Invest*. 2000;105:1493–1499.

83. van de Wetering M, Sancho E, Verweij C, et al. The β-catenin/TCF-4 complex imposes a crypt progenitor phenotype on colorectal cancer cells. *Cell*. 2002;111:241–250.

84. Wilkinson DG. Multiple roles of EPH receptors and ephrins in neural development. *Nat Rev Neurosci*. 2001;2:156–164.

85. Beghtel H, Henderson JT, van den Born MM, et al. β-catenin and TCF mediate cell positioning in the intestinal epithelium by controlling the expression of EphB/ephrinB. *Cell*. 2002;111:251–263.

86. Korinek V, Barker N, Moerer P, et al. Depletion of epithelial stem-cell compartments in the small intestine of mice lacking Tcf-4. *Nat Genet*. 1998;19:379–383.

87. Pinto D, Gregorieff A, Begthel H, Clevers H. Canonical Wnt signals are essential for homeostasis of the intestinal epithelium. *Genes Dev*. 2003;17:1709–1713.

88. Kuhnert F, Davis CR, Wang HT, et al. Essential requirement for Wnt signaling in proliferation of adult small intestine and colon revealed by adenoviral expression of Dickkopf-1. *Proc Natl Acad Sci U S A*. 2006;101:266–271.

89. Alonso L, Fuchs E. Stem cells in the skin: waste not, Wnt not. *Genes Dev*. 2003;17:1189–1200.

90. Reya T, Duncan AW, Ailles L, et al. A role for Wnt signalling in self-renewal of haematopoietic stem cells. *Nature*. 2003;423:409–414.

91. Caussinus E, Gonzalez C. Induction of tumor growth by altered stem-cell asymmetric division in *Drosophila melanogaster*. *Nat Genet*. 2005;37:1125–1129.

92. Betschinger J, Mechtler K, Knoblich JA. Asymmetric segregation of the tumor suppressor brat regulates self-renewal in *Drosophila* neural stem cells. *Cell*. 2006;124:1241–1253.

93. Gho M, Schweisguth F. Frizzled signalling controls orientation of asymmetric sense organ precursor cell divisions in *Drosophila*. *Nature*. 1998;393:178–181.

94. Gong Y, Mo C, Fraser SE. Planar cell polarity signalling controls cell division orientation during zebrafish gastrulation. *Nature*. 2004;430:689–693.

95. Ahringer J. Control of cell polarity and mitotic spindle positioning in animal cells. *Curr Opin Cell Biol*. 2003;15:73–81.

96. Habas R, Kato Y, He X. Wnt/Frizzled activation of Rho regulates vertebrate gastrulation and requires a novel Formin homology protein Daam1. *Cell*. 2001;107:843–854.

97. Hall A, Nobes CD. Rho GTPases: molecular switches that control the organization and dynamics of the actin cytoskeleton. *Phil Trans R Soc Lond B Biol Sci*. 2000;355:965–970.

98. Narumiya S, Yasuda S. Rho GTPases in animal cell mitosis. *Curr Opin Cell Biol*. 2006;18:199–205.

99. Cowan CR, Hyman AA. Asymmetric cell division in *C. elegans*: cortical polarity and spindle positioning. *Annu Rev Cell Dev Biol*. 2004;20:427–453.

100. Reilein A, Nelson WJ. APC is a component of an organizing template for cortical microtubule networks. *Nat Cell Biol*. 2005;7:463–473.

101. Kemphues K. PARsing embryonic polarity. *Cell*. 2000;101:345–348.

102. Miyoshi Y, Nagase H, Ando H, et al. Somatic mutations of the APC gene in colorectal tumors: mutation cluster region in the APC gene. *Hum Mol Genet*. 1992;1:229–233.

103. Guilford P, Hopkins J, Harraway J, et al. E-cadherin germline mutations in familial gastric cancer. *Nature*. 1998;392:402–405.

104. Strutt D. Frizzled signalling and cell polarisation in *Drosophila* and vertebrates. *Development*. 2003;130:4501–4513.

105. Chen AE, Ginty DD, Fan CM. Protein kinase A signalling via CREB controls myogenesis induced by Wnt proteins. *Nature*. 2005;433: 317–322.

106. Schutz L, Mora PT. The need for direct cell contact in 'contact' inhibition of cell division in culture. *J Cell Physiol*. 1968;71:1–6.

107. St Croix B, Sheehan C, Rak JW, Florenes VA, Slingerland JM, Kerbel RS. E-Cadherin-dependent growth suppression is mediated by the cyclin-dependent kinase inhibitor p27(KIP1). *J Cell Biol*. 1998;142:557–571.

108. Grazia Lampugnani M, Zanetti A, Corada M, et al. Contact inhibition of VEGF-induced proliferation requires vascular endothelial cadherin, β-catenin, and the phosphatase DEP-1/CD148. *J Cell Biol*. 2003;161: 793–804.

109. Ostman A, Yang Q, Tonks NK. Expression of DEP-1, a receptor-like protein-tyrosine-phosphatase, is enhanced with increasing cell density. *Proc Natl Acad Sci U S A*. 1994;91:9680–9684.

110. Holsinger LJ, Ward K, Duffield B, Zachwieja J, Jallal B. The transmembrane receptor protein tyrosine phosphatase DEP1 interacts with p120(ctn). *Oncogene*. 2002;21:7067–7076.

111. Balda MS, Garrett MD, Matter K. The ZO-1-associated Y-box factor ZONAB regulates epithelial cell proliferation and cell density. *J Cell Biol*. 2003;160:423–432.

112. Sourisseau T, Georgiadis A, Tsapara A, et al. Regulation of PCNA and cyclin D1 expression and epithelial morphogenesis by the ZO-1-regulated transcription factor ZONAB/DbpA. *Mol Cell Biol*. 2006;26:2387–2398.

113. Barrios-Rodiles M, Brown KR, Ozdamar B, et al. High-throughput mapping of a dynamic signaling network in mammalian cells. *Science*. 2005;307:1621–1625.

114. Wang Z, Mandell KJ, Parkos CA, Mrsny RJ, Nusrat A. The second loop of occludin is required for suppression of Raf1-induced tumor growth. *Oncogene*. 2005;24:4412–4420.

CHAPTER 15

Activation of the Innate immune System: the Toll-like Receptor 4 and Signalling through Ubiquitylation

Host defence in mammals is provided by innate and adaptive (or acquired) mechanisms of immunity, both able to discriminate between the self and non-self. In innate immunity, invading microbes are sensed either directly, through the release of cellular components, or indirectly through the deposition of lectins, C-reactive protein, or complement (C3b) on their outer membranes.[1,2] This first line of defence, though not always fully protective, buys time and is required to establish the much more potent processes of adaptive immunity by which invading microbes are sensed by specific immunoglobulins (Ig) and then destroyed. This occurs directly by means of cytolytic antibodies that activate the complement system. Or it may occur indirectly by means of cytotoxic T cells that recognize virus-infected cells, or through stimulation of phagocytosis by granulocytes and macrophages.

In an evolutionary perspective, innate immunity predated adaptive immunity, dendritic cells constituting the primary cellular link between the two processes. The establishment of adaptive immunity depends on antigen presentation by dendritic cells, macrophages and B cells. There follows the expansion of antigen-specific clones of T cells (both CD4$^+$ T-helper cells and CD8$^+$ cytolytic T cells) and immunoglobulin-producing B cells. These processes are dependent on the creation of an inflammatory environment; effectively, the production of inflammatory mediators such as interleukin IL-1β, IL-6, IL-12, IFN-α/β/γ or TNF-α, and chemotactic factors such as RANTES, MIP1α/β, IL-8, or MCP-1. These play important roles in the recruitment of leukocytes by regulating the expression of adhesion molecules (see Chapters 13 and 16). They also participate in the maturation of dendritic cells, preparing them to initiate the differentiation of effector T cells and of memory cells. Further down the line, the inflammatory mediators also play a role in the maturation of B cells into plasma cells, and thus stimulate the production of microbe-specific immunoglobulins.

The many abbreviations used in this chapter are collected together at the end of the chapter.

Metchnikov and Ehrlich

The foundations for the understanding of immunity, both innate and adaptive, were laid by Ilya Metchnikov and Paul Ehrlich, who, in 1908, shared the Nobel prize 'in recognition of their work on immunity'. We can regard Metchnikov as an early developmental biologist. He detected a population of highly motile cells associated with the larvae of starfish (echinoderms) and postulated that these could somehow be implicated in defence. Putting this idea to the test, he added crusted thorns from a tangerine tree which had served as Christmas decorations for his children, and found the next morning that they were surrounded by these mobile cells. The scene reminded him of white blood cells accumulating at sites of infection and he suggested that the two phenomena represent the mobilization of specialized cells in order to destroy invaders. His friend Professor Claus suggested the name 'phagocytes' (from the Greek *phagein,* eating) to describe the mobile cells. His first paper on 'phagocytosis' appeared in 1884.[3]

Ehrlich, around 1895, was investigating the protective effect of serum passive immunization against bacterial toxins. He found that it contained 'antitoxins' that disabled the bacterial products. He was one of the first to recognize the importance of specific molecular interactions or, as he stated it, *'corpora non agunt nisi fixate'*, meaning, 'substances do not act if they are not bound'. The antitoxins we now recognize as antibodies (immunoglobulins). A similar principle of molecular recognition was also elaborated by John Newport Langley who, in 1906, postulated that 'receptive substances' transmit physiological effects (see page 16).

Sensing the microbial universe

To initiate an immunological response, the invading microbes must first be recognized. This occurs through a family of *pattern-recognition receptors* (PRRs). These have a limited diversity, are expressed in all cells,

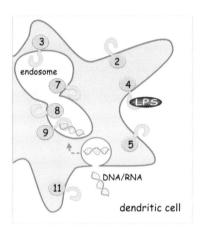

FIG 15.1 Subcellular localization of the TLR pattern-recognition receptor family in dendritic cells. TLR2, 4, 5, and 11 are cell surface receptors that recognize membrane components of invading microorganisms. TLR3, 7, 8, and 9 are present in the endosomal compartment, where they recognize different forms of nucleic acids taken up by the dendritic cell.[6,17] TLR1 and TLR6, involved in the recognition of diacyl/triacyl lipopeptides and lipoteichoic acid, are not shown.

and, importantly, function independently of immunological memory. They recognize highly conserved components that are released by invading microbes, collectively named the pathogen-associated molecular patterns (PAMS).[4] Among the PRRs are the Toll-like receptors that recognize essential components of microorganisms, such as membrane proteins, lipids, and nucleic acids[5] (**Figure 15.1** and Table 15.1). This chapter deals with the signalling pathway downstream of the Toll 4 (TLR4) receptor.

The toll receptor in *Drosophila*

Flies possess two strategies of defence against bacterial and fungal invasion: (1) a cellular pathway involving plasmatocytes (the major blood cell type) and (2) a humoral pathway that involves the production of bacteriolytic peptides (such as cecropins, diptericin, drosocin, and defensin), and antifungal peptides (drosomycin and metchnikowin).[7]

The events leading to the expression of the drosomycin gene were revealed by comparing signal transduction pathways in mammals and flies that have similar components but are instrumental in completely different processes.[8,9] Thus, there are striking parallels between the pathway operated by the morphogen 'dorsal' in embryonic flies, controlled by the Toll receptor and the mediation of the inflammatory response in mammals by IL-1.[10–12] Also, the promoter sequence of the diptericin gene has DNA motifs related to NF-κB binding sites.

Toll. Lack-of-function mutants produce dorsalized *Drosophila* embryos, dominant gain-of-function alleles result in ventralized embryos. The German word *toll* means 'amazing', 'fantastic'. According to Christiane Nüsslein-Volhard, 'When Eric Wieschaus and I first saw the mutant phenotype, we were amazed, because it was so novel and unexpected. I must have yelled "*toll*", and the name stuck'.

Beyond its role in development, Toll plays a key role in the host defence mechanisms of adult flies. While *Drosophila* Toll recognizes the cytokine spätzle, the mammalian TLRs (TLR1–12) recognize specific microbial components, ranging from single-stranded DNA to lipopolysaccharides[6] (see **Figure 15.4**).

TABLE 15.1 Ligands for Toll-like receptors

	Ligand	Source organisms	TLR
Viruses	DNA		9
	dsRNA		3
	ssRNA		7 & 8
	envelope proteins	RSV, MMTV	4
	haemagglutinin	measles	2
Bacteria	lipopolysaccacharides	gram-negative	4
	diacyl lipopeptides	mycoplasma	2 & 6
	triacyl lipopeptides	(myco)bacteria	1 & 2
	LTA	streptococcus B	2 & 6
	peptidoglycan	gram-positive	2
	porins	*Neisseria*	2
	flagellin	flagellated bacteria	5
	CpG-DNA	(myco)bacteria	9
Fungi	zymosan	*S. cerevisiae*	2 & 6
	phospholipomannan	*C. albicans*	2
	mannan	*C. albicans*	4
	glucuronoxylomannan	*Cryptococcus n.*	2 & 4
Protozoans	tGPI-mutin	*Trypanosoma*	2
	glycoinositol phopsholipids	*Trypanosoma*	4
	hemozoin	*Plasmodium*	9
	profilin-like molecule	*Toxoplasma gondii*	11
Host	Hsp 60		4
	Hsp 90		4
	fibrinogen		4

The idea that inflammatory responses may be conserved throughout species, and that the dorsal pathway may be replayed by adult flies to control expression of antimicrobial peptides, acted as a spur to investigate the role of the 'dorsal/IL-1β' pathway in the regulation of *Drosophila* host defence.[7] This yielded dividends when it was found that expression of drosomycin occurs

TABLE 15.2 Toll and IL-1 pathways in *Drosophila* and mammals

Function	*Drosophila*	Mammal
Ligand	Spätzle	IL-1β
Receptor (TIR domain)	Toll	IL-1R
Adaptor (TIR/death domain)	Tube	MyD88
Serine/threonine protein kinase (death domain)	Pelle	IRAK-4
Inhibitory protein	Cactus	IkB
Transcription factor	Dorsal	NF-κB
Gene expression	drosomycin, metchinikowin, cecropin, attacin, defensin	numerous, see Figures 15.5 and 16.4

through the dorsal pathway. Ablation of the Toll receptors results in flies that are susceptible to microbial infection, particularly by fungi. The Toll–dorsal pathway is a costimulant for the expression of antimicrobial peptides.[13]

Spätzle and IL-1β are, respectively, the ligands of Toll and IL-1 receptors.[14] Although these ligands have no sequence homology, spätzle being closer to the mammalian nerve growth factors (NGF), the processing of their inactive precursors is controlled, in part, by immune challenge. During development, spätzle is processed by three extracellular serine proteases ('easter', 'gastrulation defective', and 'snake'). An alternative proteolytic cascade, activated by an immune challenge, may constitute the activation mechanism in host defence.[15] IL-1β is processed by an intracellular protease (interleukin-1β-converting enzyme, ICE, also called caspase-1). In monocytes the activity of this enzyme is controlled by, among other factors, lipopolysaccharides.[16]

Although the ligands are dissimilar, the receptors, Toll and IL-1R share a common intracellular protein–protein interacting TIR domain, through which they bind the adaptors TIRAP, TRIF, TRAM, and MyD88 (in mammalian cells) and MyD88 (in *Drosophila*)[19,20] (see Table 15.2 and **Figure 15.2**). The extracellular domains are completely different. The adaptors Tube and MyD88 both contain a death domain (DD) and are involved in the activation of the serine/threonine kinases pelle and IRAK-4 respectively. The protein interaction domain DD is also present in the C-terminal region of the TNF receptor (TNFR-1) and involved in TNF-mediated signalling for apoptosis (see page 489). The DD mediates self-association of these receptors, so directing the signal to downstream events.

The downstream targets of these two pathways are Dorsal and NF-κB, both of which contain a REL-homology domain (RHD) that binds DNA at the κB sequence motif (**Figure 15.3**) (first discerned in the kappa light chain enhancer region of immunoglobulin[24]). Classical NF-κB is a transcription

The TIR domain is the target of an escape strategy of vaccinia virus. It encodes two TIR-domain proteins which block receptor signalling by sequestering MyD88 and TRIF.[21]

Proteins carrying a death domain are not necessarily involved in the process of death signalling (apoptosis). Examples of death domain proteins are IRAK, pelle, tube, ankyrin (cytoskeleton), MyD88, RIP (apoptosis), and TRADD.

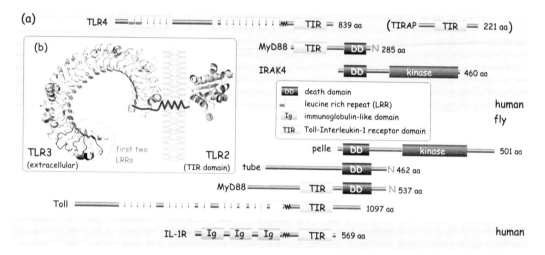

FIG 15.2 Comparison of domain the architectures of receptors, adaptors and effectors involved in signalling through the Toll and Toll-like receptors. Toll and TLR4 both have leucine-rich repeats in their extracellular segment (two of which, at the N-terminus, are highlighted in the inset panel) and an intracellular TIR domain. A signalling complex is recruited to the occupied receptor through adaptors containing TIR domains. Those that possess only a TIR domain (such as TIRAP) recruit other TIR-containing adaptors. Adaptors having an additional death domain (DD) recruit the DD-containing serine/threonine kinase IRAK-4 (or pelle in the case of Toll). Note that the IL-1R also possesses a TIR domain and therefore resembles both Toll and TLR4, but in addition it has three Ig-like domains in the extracellular segment. The N-terminal is indicated at the left, unless otherwise indicated (1ziw,[22] 1fyw[23]).

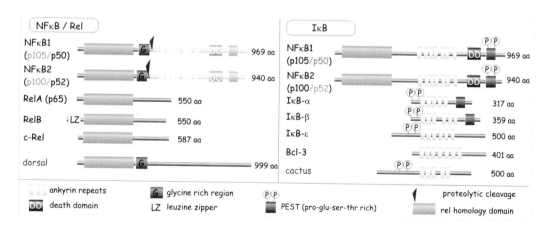

FIG 15.3 Domain architecture of the NF-κB/Rel and IκB proteins.
The Toll and IL-1R pathways control transcription factors of the NK-κB/Rel family which share a Rel-homology domain that determines nuclear localization, DNA binding, and subunit dimerization. Members of this family are held in check by inhibitors, members of the IκB family, characterized by numerous ankyrin repeats. (Truncated versions, p50 and p52, possess only the Rel-homology domain.) IκB proteins bind NF-κB/Rel and prevent both recognition of their nuclear localization signal (NLS) and binding to DNA. Partial proteolysis of the ankyrin repeat relieves this constraint. Phosphorylation of these proteins signals their destruction and this allows activation of NF-κB. Adapted from Beinke and Ley.[25]

factor complex comprising RelA (p65) and NF-κB1 (p50) (**Figure 15.3**). The founding member of this family of transcription factors is v-rel (p58), a viral gene responsible for reticuloendotheliosis in chickens. Mammalian REL-containing proteins include RelB (p60), c-Rel (p66) (involved in lymphopoiesis), and NF-κB2 (p100/p52).

Dorsal and NF-κB also share the characteristic of being sequestered in the cytoplasm by inhibitory proteins, cactus in *Drosophila* and IκB in mammals. These inhibitors are destroyed by an ubiquitylation-based process that permits their translocation to the nucleus where they bind to the κB sequence motif (see page 493).

Lipopolysaccharides: shield and signal

Bacteria with no outer membrane have cell walls composed of a thick layer of peptidoglycan, lipoteichoic acid and S-layer proteins. Bacteria with a second, outer membrane have a thin peptidoglycan layer between the two (Fig 15.4). The outermost membrane surface is almost entirely lipopolysaccharide (LPS). The thicker peptidoglycan layers may be detected by the Gram test (staining with crystal violet and iodine), initially devised to discriminate between pneumococci (gram positive) and *Klebsiella pneumoniae* (gram negative).[26] Mycobacteria form a separate group. They have a cell wall made of glycolipids and mycolic acid and only a thin peptidoglycan layer. They are able to survive because, inhabiting the intracellular environment, they are subject to little osmotic stress.

Because the innate immune system lacks any means of instigating receptor diversity, all its targets are highly conserved structural components of micro-organisms. One such is lipopolysaccharide.[27] LPS, shed from the outer membrane of gram-negative bacteria (Table 15.3), are detected at picomolar concentrations and serve to alert the host (**Figure 15.4**). Of the Toll-like receptors, TLR4 has emerged as a specific conduit for the innate immune response to bacterial LPS.[28,29] We focus on LPS-mediated signalling because it is one of the most potent immunostimulants.

LPS and endotoxin. At the start of the 20th century, long before its chemical composition was realized, Richard Pfeiffer identified LPS as endotoxin. It represents the major fever-evoking component of gram-negative bacteria and, when applied systemically, it causes shock.[29] We now recognize the lipid A-KDO2 moiety of LPS as being responsible for most of the toxic effects of gram-negative organisms (see Figure 15.4). The immediate pyrogenic effects of endotoxin are due to signalling through TLR-4 that elevates the expression of PLA2 and cyclo-oxygenase2 (COX2) and the production of prostaglandin E2 (PGE2).[31]

The cross-linking of the peptidoglycan is targeted by penicillin, which causes rapid lysis of microbes.

Virulence of *Yersinia pestis* and acylation of LPS. On entering a host having a body temperature of 37°C, the plague bacterium *Yersinia pestis* switches production of LPS to a form that lacks secondary acylations. This fails to activate TLR4 and antagonizes the binding of highly acylated LPS. Forced expression of an *E. coli* gene responsible for secondary acylations renders *Y. pestis* completely avirulent. Resistance to disease requires TLR4 and the adaptor protein MyD88. Both innate and adaptive responses are required for immunity against the modified strain. By evading TLR4 activation in this way, the altered LPS may contribute to the virulence of various gram-negative bacteria.[30]

The outer leaflet of the outer membrane of gram-negative bacteria is made up almost entirely of LPS. These have a basic structure composed of lipid A, consisting of 4–6 acyl chains covalently bound to two diaminodideoxy-d-glucose residues, and a hydrophilic heteropolysaccharide component. The inner component of the polysaccharide, the inner core, is composed of heptose and keto-3-deoxyoctonic acid (KDO). To this, a long chain of polysaccharides is attached, the outer core, of which, the O-antigen, is repeated many times. The long polysaccharide chain is not essential for bacterial survival, but the lipid-A moiety is vital. The polysaccharides play a role in pathogenesis and serve as an antigen for the generation of protective antibodies.

TABLE 15.3 Examples of Gram-negative bacteria

Bacteria	Diseases caused
Cyanobacteria	photosynthesis (3.8 billion years old)
Enterobacter cloacae	urinary and pulmonary tract infections
Escherichia coli	commensal in intestine, urinary tract infection, septicemia
Green sulfur and green non-sulfur bacteria	
Helicobacter pylori	peptic ulcer
Haemophilus influenzae (Pfeiffer baccillus)	lung infections and meningitis
Klebsiella pneumoniae	pneumonia, hospital-acquired urinary tract and wound infections
Legionells pneumophila	legionellosis (Legionnaires' disease)
Leptospiractero haemorrhagiae	Weil's disease
Moraxella catarrhalis	bronchitis, sinusitis, laryngitis and otitis media
Neisseria gonorrhoeae	gonorrhoea
Neisseria meningitidis	meningitis
Proteus mirabilis	urinary tract infections
Pseudomonas aeruginosa	pulmonary tract, urinary tract, burns, wounds infections
Salmonell enteridis	gastroenteritis
Salmonella typhi	typhoid fever
Serratia marcescens	conjunctivitis, keratitis endophthalmitis

Signalling through the TLR4 receptor

The TLR4 differs from other TLRs because it requires the collaboration of other receptors, most notably CD14 and MD-2, though how these interact with LPS and how they interact with each other remains uncertain. LPS, bound to LPS-binding protein (LBP, a trace 'acute phase' plasma protein involved in the early stages of septic shock)[32] interacts with CD14.[33] This presents LPS to MD-2, linked to the membrane via TLR4 (**Figure 15.5**).[34] Two major signalling pathways are activated. A signalling complex is formed with a

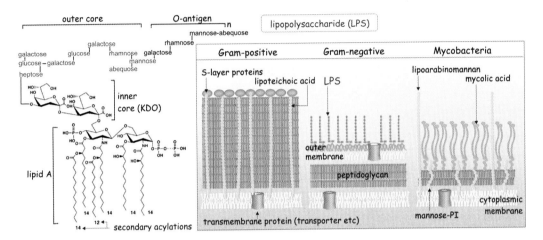

FIG 15.4 Composition of the cell wall of bacteria and molecular detail of LPS.

Bacteria are protected from the environment by a cell wall. In the case of gram-positive bacteria, this is composed of a lipid membrane surrounded by a thick layer composed of peptidoglycan, lipoteichoic acid, and a coating of S-layer proteins. Gram-negative bacteria are surrounded by two membranes separated by the peptidoglycan. Mycobacteria have a single membrane, a relatively thin peptidoglycan, and a thick outer layer composed of mannose-phosphatidylinositol, lipoarabinomannan, and mycolic acid. Adapted from Akira et al.[6]

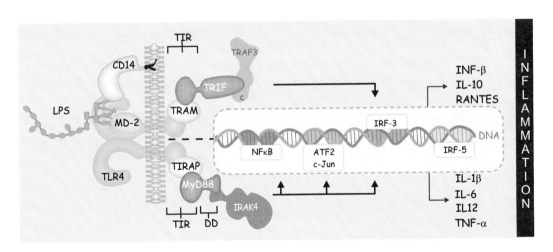

FIG 15.5 Signalling from the TLR4 receptor.

LPS binds different receptors, first CD14, then MD-2, and finally TLR4. Ligand-mediated dimerization (or multimerization) of TLR4 initiates the assembly of receptor signalling complexes through the intermediate of adaptors that harbour the TIR protein-interaction domain. One complex is formed by association of TRAM with TRIF which, through a death domain (DD), recruits the adaptor/effector TRAF3 (ubiquitin ligase). This causes activation of the transcription factor IRF3 and results in the expression of inflammatory cytokines IFN-β, IL-10, and RANTES. The other complex is formed by association of the adaptors TIRAP and MyD88 which, through a death domain, recruit the kinase IRAK4. This pathway reaches the transcription factors NF-κB, ATF2, c-Jun, and IRF5 and plays an important role in the expression of IL-1β, IL-6, IL-12, and TNF-α.

Discovery of TLR4, the LPS receptor. The earliest indication that LPS might interact with specific receptors came from the discovery of LPS-resistant mice.[35] The search for the receptor then commenced in earnest with the realization of two substrains both having defective responses to LPS. Four years of fine mapping of the Lps locus by several research groups culminated in the discovery of LTR4 as the sole candidate.[29] Although LPS interacts directly with TLR4, the interaction also requires one or more coreceptors, among which CD14, a glycosylphosphoinositol-anchored membrane protein expressed on leukocytes.[18,33]

LPS and mucosa. The form of LPS having secondary acylations (**Figure 15.5**) allows bacteria to persist in the host mucosa.[36–38] They are tolerated in this external environment but as soon as they traverse the epithelial barrier, as occurs in trauma, they are effectively eliminated.

number of adaptors and assembly then proceeds through the association of proteins via the TIR domain. Many components of the pathway also operate downstream of the TNF, Toll, or TGFβ-receptors (which explains why so many of the acronyms begin with the letter T). One downstream pathway of TLR4 requires the adaptors TIRAP and MyD88 and culminates in the production of pro-inflammatory cytokines (IL-1β, IL-12, TNF-α, IL-6). These signal to the transcription factor complexes NF-κB, ATF2/c-Jun (AP-1), and IRF-5 (Fig15.5 and pages 464 and 525). The other pathway requires the adaptors TRAM and TRIF and culminates in the production of IFN-β, IL-10, and RANTES. It signals to the transcription factor IRF-3 and acts in cooperation with NK-κB.

The TIRAP/MyD88 pathway

Oligomerization of the TLR4 receptor provokes the formation of an adaptor complex through the assembly of components TIRAP and MyD88, carrying TIR domains. How the oligomerization creates homophilic TIR-TIR binding sites remains to be elucidated. The death domain of MyD88 recruits the kinase IRAK4 which phosphorylates IRAK1[39,40] (see **Figure 15.6**). This draws TRAF6 to the receptor complex which now acts as a docking platform on to which another large signalling complex develops. The new docking sites are created, not by phosphorylation as is the case for growth factor receptors, but by polyubiquitylation of TRAF6, in the so-called K63-type reaction (i.e. they are interconnected through lysine-63 (see **Figure 15.13** and page 468). TRAF6 is an E3-ubiquitin ligase that by association with other factors (Ubc13/Mms2) constitutes the conjugation system that supplies activated ubiquitin monomers. Importantly, TRAF6, decorated with its polyubiquitin chains, brings together two kinase complexes that act in series to ensure phosphorylation of IκBα (inhibitor of NF-κB) and activation of the JNK and p38 pathways.

The significance of IRAK4 is clearly illustrated in patients who either fail to express it, or have a mutation in its kinase domain which renders the enzyme inactive. Because they have ineffective innate immunity, these people are particularly susceptible to infection with *Streptococcus pneumoniae*, suffering recurrent pyrogenic infections and bacteraemia.[41]

From TRAF6 to activation of NF-κB

The first complex recruited by TRAF6 is composed of the serine/threonine kinase TAK1 with its associated adaptor TAB2 (or TAB3) and its regulatory subunit TAB1 (both TAK1/TAB1/TAB2 and TAK1/TAB1/TAB3 complexes have been detected).[42,43] It is the TAB2 (or TAB3) that binds K63-ubiquitylated TRAF6 through a CUE domain[44] (**Figure 15.7** illustrates the domain architecture of these proteins). This activates TAK1 through autophosphorylation.[45] The

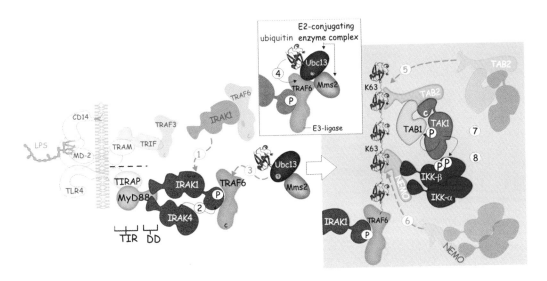

FIG 15.6 TRAF6-mediated signalling complex formation.
Recruitment of IRAK4 brings IRAK1 and TRAF6 to the receptor signalling complex (1). IRAK1 is phosphorylated by IRAK4 (2) and this induces catalytic activity the complex comprising TRAF6 (E3–ubiquitin ligase), Ubc13 and Mms2 (3). Autoubiquitylation of TRAF6 follows (4). The K63-linked ubiquitin chain acts as a docking site for two kinase complexes. One of these comprises TAB2 coupled to the serine/threonine kinase TAK1 with its regulatory subunit TAB1 (5). The other comprises NEMO attached to IκB-kinases-α and -β (IKK-α and -β). Recruitment of TAB2/TAB1/TAK1 results in autophosphorylation of the kinase (7) and subsequent phosphorylation and activation of IKK-β (8).

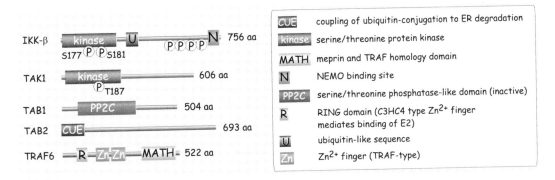

FIG 15.7 Domain architecture of proteins involved the TRAF6-mediated activation of IKKβ.
The phosphorylation sites in the kinase domain of IKKβ and TAK1 are situated in the activation segment. The TAK1 regulatory subunit TAB1 has a phosphatase fold similar to PP2C but is inactive. TAB2 binds K63-linked ubiquitin through its CUE domain. TRAF6 possesses an N-terminal RING domain with which it binds to the E2-conjugation complex (Ubc13/Mms2).

second complex is composed of the serine/threonine kinases IKKα and IKKβ, linked to the ubiquitin-binding protein that acts as the sensor for K63-linked polyubiquitylated TRAF6 (sometimes referred to as IKKγ, IKK signifying IκB

FIG 15.8 IKKβ-mediated activation of the RelA/NF-κB complex. Phosphorylated/activated IKKβ phosphorylates IκB on two serine residues in its N-terminal domain (1, 2) allowing its recognition by the SCFβ^Tcrp ubiquitin-ligase complex (3). Following (K48-type) ubiquitylation (4), IκBα is marked for destruction by the proteasome (5), liberating RelA (p65) and NF-κB1 (p50) and exposing their nuclear localization signal (NLS). They enter the nucleus (6), there to bind DNA at the κB element, driving expression of inflammatory cytokines.

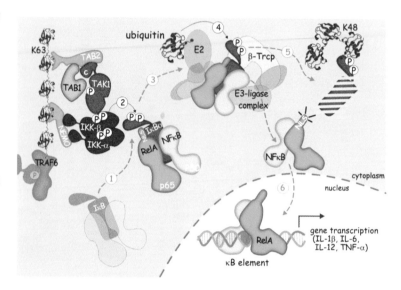

kinase).[46, 47] Upon binding to TRAF6, TAK1 phosphorylates and activates IKKβ which then phosphorylates IκBα in the 'destruction box' (Figure 15.8). The phosphorylated IκBα is detected by the receptor component (β-TrCP) of an E3-ligase complex and subsequent polyubiquitylation (K48-type) allows detection by the proteasome followed by its degradation. Finally, IKKα is also activated and this plays a role in the initiation of the 'non-canonical' pathway. Finally, NEMO too is ubiquitylated by TRAF6 and serves to recruit more TAB2/3-TAB1-TAK1 into the complex and enhance the output of the signalling pathway (not shown in Figure 15.6).

On destruction of IκBα, the NF-κB dimer, (comprising proteins of the Rel-family p65 and p50) is recognized by the α- and β-importins at the nuclear pore and enters the nucleus.[48] This pathway of activation of NK-κB, through destruction of IκB, is referred to as the 'canonical' NF-κB pathway. We have given here the example IκBα, but the other inhibitor proteins, IκBβ and IκBε, can also bind NF-κB thereby blocking the nuclear localization signal and thus transcriptional activity.[25]

Alternative activation pathways for NF-κB

These pathways involve unprocessed (precursor) members of the Rel family, NF-κB1 and NF-κB2. To complicate matters further, these proteins qualify as members of both the Rel and IκB families. They are composed of two regions, the one resembling Rel and the other IκB (ankyrin repeats). They bind to other members of the Rel family and, having an intrinsic inhibitory region that masks the nuclear localization signal, they render the complexes inactive with respect to their transcriptional function[25] (see Figure 15.3).

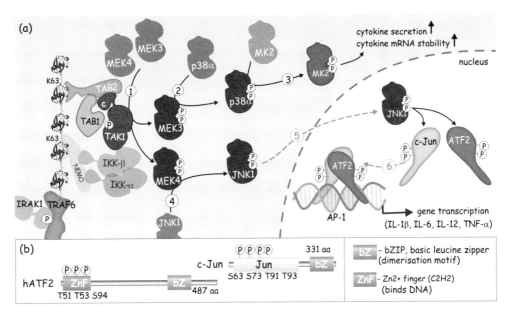

FIG 15.9 TAK1-mediated activation of p38 and JNK1.

(a) TAK1 phosphorylates MEK3 and MEK4 (MAP2K) (1). These in turn phosphorylate and activate p38α (2) and JNK1 respectively (4). P38α phosphorylates and activates MK2 (3) which enhances expression of inflammatory cytokines through stabilization of their mRNAs. It also promotes secretion. JNK1 enters the nucleus and phosphorylates the transcription factors ATF2 and c-Jun (5). These form heterodimers that bind the AP-1 response element (6) and drive expression of numerous inflammatory cytokines. (b) Domain architecture of ATF2 (human) and c-Jun. The N-terminal DNA binding regions of the proteins are phosphorylated by JNK1. They dimerize through a basic leucine zipper (bZIP) in the c-terminal segment.

The p105 pathway involves the unprocessed NF-κB1 (p105) which preferentially binds to a homodimer of processed NF-κB1. The complex is inactive and ubiquitylation of p105 followed by total destruction is required for liberation of the processed NF-κB dimer.

In the non-canonical (or alternative) pathway, NF-κB2 is bound to RelB and is phosphorylated in its IκB-like segment by IKKα. Poly-ubiquitylation is then followed by only partial proteolysis through the proteasome. This processed (or mature) form of NF-κB2 translocates, in association with Rel-B, to the nucleus. This pathway operates mainly in B cells in response to a subset of TNF receptors, including BAFF, LTα, LTβ and CD40.[49]

From TRAF to activation of p38α and JNK1

In parallel with the activation of NF-κB, two other pathways contribute to the transcriptional regulation of inflammatory mediators.[50] TAK1 phosphorylates the activation segments of MEK3 (or 6) and MEK4 (or 7) which then phosphorylate and activate p38α and JNK1 (respectively) (Figure 15.9). These members of the Big MAP kinase family are involved in the responses to cellular stress and inflammation (see page 349 and Figure 22, page 351).

Scaffold proteins implicated in the p38/JNK pathway.

P38 and JNK kinases are bound to the scaffold protein JIP (four variants, JIP1–4).[52] JIP1–3 act to potentiate JNK signalling through the assembly of a signalling cassette comprising members of the mixed-lineage protein kinase group MAP3K, MKK 7 (MAP2K), and JNK.[53,54] They also attach to the TPR domain of the light chain of the microtubule motor protein kinesin and through this they localize the JNK signalling cassette to growth cones of neurons.[55,56] JIP4 is somewhat different. It shares with JIP1–3 its interaction with JNK and with kinesin, but it has a role as an activator of p38-mitogen activated protein kinase by a mechanism that requires MKK3 and MKK6.[57] The precise role of JIP in the TAK1-mediated activation of MEK3 and MEK4 remains to be solved.

JNK1 (c-Jun N-terminal kinase) enters the nucleus to phosphorylate the transcription factors ATF2 (a member of the ATF/CREB family of transcription factors)[51] and c-Jun. Together they constitute the AP-1 complex (activator protein-1, which is comprised of different combinations of the transcription factors ATF, c-fos, fra1, fra2, c-Jun, or junB). Numerous genes carry AP-1 responsive elements in their promoter regions. P38 remains in the cytosol and acts to enhance inflammatory cytokine production through phosphorylation and activation of MK2 (a member of the MAP kinase-activated protein kinases) (Figure 15, page 340) that acts to enhance the expression and secretion of inflammatory cytokines through stabilization of the relevant mRNAs.

From TRAF6 to activation of IRF-3

The TRAF6 signalling complex also causes dimerization of the interferon-regulatory factor IRF-3 which is then transferred into the nucleus.

The TRAM, TRIF pathway

Here we concentrate on the role of TRAF3 as an adaptor that recruits a kinase complex to the TLR4 receptor (**Figure 15.10**). First it binds TANK which is associated with the kinases IKKε and TBK1, homologues of IKKα.[59] Recruitment of TBK1 causes phosphorylation of IFR3 which, as a homodimer (or heterodimer IRF3/IRF7), translocates to the nucleus to bind DNA at the ISRE-responsive element (see page 525 onwards). TBK1 also associates with CBP, inducing expression of IFN-β. Transcriptional activity of IFR3 requires the cooperation of NF-κB. Other genes expressed through this pathway are IL-10 and the chemokine RANTES. Absence of IFR3 prevents the production of IFN-β and IL-10 without reducing the expression of the pro-inflammatory cytokines TNF-α, IL-1β, or IL-6. Indeed, their production is actually enhanced, possibly due to failure of the negative control normally exerted by IL-10.[60]

The IRF family of transcription factors

The interferon regulatory factors (IRFs), first identified in the promoter region of the human interferon-β gene,[61] operate in both innate and in adaptive immunity. They recognize a consensus DNA sequence known as the IFN-stimulated response element (ISRE) and are functionally active as homo- or heterodimers. Actually, their application is widespread, with many genes containing the ISRE consensus sequence and they are activated by several receptors, including TLR4.[62] The mammalian IRF family comprises nine members (IRF1–9). Except for IRF1 and IRF2, their C-termini (IAD domain) resemble the Smad family of transcription factors (figure 8, page 609), suggesting interactions with other regulatory transcription factors or with the general machinery of

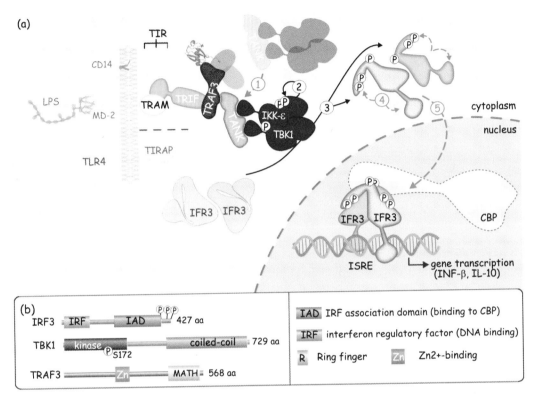

FIG 15.10 TRAM/TRIF-mediated receptor signal complex formation and the activation of IRF3.

(a) TLR4 recruits TRAM and TRIF through their TIR domains. TRIF is bound to TRAF3 (an E3-ligase) which in turn, binds the adaptor TANK coupled to IKKε and TBK1 (1). The formed complex causes auto-phosphorylation of IKKε, followed by phosphorylation and activation of TBK1 (2). Activated TBK1 phosphorylates IFR3 (3) which now unfolds, thereby exposing its DNA binding domain and its dimerization and transactivation domains (4). Dimerized IFR3 enters the nucleus, binds to ISRE and to CBP (5). This leads to induction of expression of IFN-β and IL-10. (b) Domain architecture of components of the IFR3 activation pathway. Activation phosphorylation sites of IRF are at the C-terminal in and just outside the IRF association domain that binds the transcription factor CBP. TBK1 is different from IKKε with respect to its activation segment, containing only a single serine phosphorylation site. TRAF3, like TRAF6, is an E3-ligase, with a RING domain (R) that binds the E2-conjugating enzyme. As illustrated here, it merely acts as an adaptor. Its role as ubiquitylating enzyme in the TLR4 pathway remains unclear.

gene transcription (see **Figure 15.10**).[63] Different IRFs play different roles, well exemplified by the phenotypes of knockout mice. Loss of IRF7 renders mice vulnerable to viral infection, correlating with a decrease in serum interferon levels.[64] Loss of IRF5 has little effect on IFN-α production but the animals are resistant to the lethal effect of systemically applied non-methylated DNA or LPS and this correlates with a reduction of serum IL-6, TNF-α, and IL-12.[58]

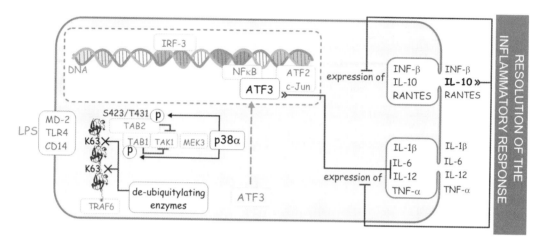

FIG 15.11 Negative feedback of the TLR4 signal transduction pathways: resolution of the inflammatory response.
Among the genes regulated by the different transcription factors are proteases that cleave the ubiquitin isopeptide bonds. Their action causes disassembly of the TRAF6-mediated signalling complexes. Activated p38 phosphorylates the ubiquitin binding protein TAB2 as well as the regulatory subunit TAB1. Both phosphorylations cause inactivation of TAK1. ATF3 is also one of the targets of the TLR4 signal pathways, acting to suppress transcription of IL-6 and IL-12 when bound to NF-κB and ATF2/c-Jun. Finally, IL-10 has a general inhibitory effect on expression of pro-inflammatory cytokines. Collectively, these negative feedback pathways act to resolve the inflammatory response. Compare with Figure 15.5 (1aar[70]).

Negative feedback control of the TLR4 pathway

We present four ways by which the TLR4 pathway is held in check. These either involve an activated protein kinase that feeds back on upstream components or involve gene products that are induced by the TLR4 pathway (see **Figure 15.11**).

1. The transcription factor ATF3 participates in the same transcription complex as NF-κB and ATF2, acting as a negative regulator of IL-6 and IL-12. It does this by altering the chromatin structure, thereby restricting access to the other transcription factors.[65]
2. Inhibition of TAK1 by phosphorylation of its regulators TAB1 and TAB2 (or TAB3) by p38α (see **Figure 15.6**).[66]
3. Expression of de-ubiquitylating enzymes which break the K63-peptide bond in TRAF6.[67,68]
4. Expression of the cytokine IL-10 which terminates the inflammatory response by deactivating macrophages and effector T cells and silencing the synthesis of TNF-α, IL-6, IL-1α, and an array of chemokines.[69]

Some consequences of TLR4-induced gene transcription

Below we list some of the mediators generated following provocation by foreign organisms and activation of pattern recognition receptors (PRR, see page 452). Although many cells contribute to mediator production and therefore contribute to the establishment of the adaptive immune response, the dendritic cell must be regarded as the primary cellular bridge linking innate and adaptive immunity. As the major antigen-presenting cell, it signals the T-effector cells and thereby triggers the production of antigen-specific immunoglobulins.

- IFN-β is necessary for the maturation of antigen-presenting dendritic cells: increased expression of MHCII and costimulatory proteins including CD40, CD80 and CD86 (Chapter 17).
- IFN-γ and IL-12 are necessary for formation of Th1 cells, important in antigen-specific activation of B cells, leading to the production of specific antibodies.
- IL-6 stimulates the maturation of B cells towards high-level immunoglobulin-producing plasma cells.
- TNF-α and IL-1β are essential for the increased expression of adhesion molecules on endothelial cells, leading to enhanced extravasation of leukocytes and dendritic cells at sites of infection (Chapter 16).
- Several chemokines (RANTES, MIP-1α, IL-8, CCL5, CXCL9, etc.) cause activation of integrins on leukocytes, including dendritic cells. This leads to their enhanced extravasation and tissue infiltration in order to combat the infecting agent.

In Chapter 16 we show how the TLR4 response leads to recruitment of leukocytes at sites of infection. In Chapter 17 we explain how the innate immune response establishes an adaptive immune response, involving the antigen presenting molecule MHCII and signalling downstream of the T cell receptor (TCR).

Essay: ubiquitylation and SUMOylation

Ubiquitylation

Ubiquitylation is the process by which the peptide ubiquitin is coupled to the ε-amino groups of lysines present in the target protein (Figure 15.12). It is self-evident that it must specify its targets with great accuracy; if it did not, protein degradation would be uncontrolled. The reaction is reversible.[71]

The attachment of one or more ubiquitins does more than just marking proteins for destruction by the proteasome complex. It also serves as a signal for endocytosis of cell surface receptors and subsequent selection for lysosomal destruction. It is involved in the regulation of protein kinases, it regulates gene transcription and it acts as a protein–protein interaction motif involved in the recruitment of signalling complexes.[72]

Ubiquitylation: a process involving three activities (but not necessarily three proteins)

Ubiquitin conjugation is catalyzed in three steps. Similar to adenylation, ubiquitin is first 'activated', bound to E1-cysteine through a glycine–sulfhydryl

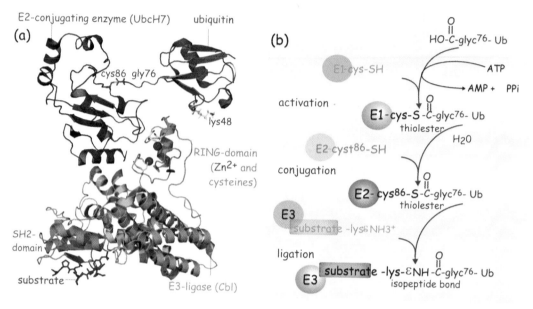

FIG 15.12 The ubiquitylation reaction.

(a) Structure of the Cbl E3-ligase complex. Cbl has three roles. It selects substrate, in this example the phosphorylated kinase ZAP-70 bound by the SH2 domain of Cbl. It recruits the E2-conjugating enzyme (UbcH7, loaded with ubiquitin) through its RING domain and contributes to the transfer of the glycine-76 of ubiquitin to a lysine in the substrate (not shown). Zn^{2+} ions indicated as red spheres. (b) The ubiquitylation reaction occurs in three steps. Following adenylation ('activation'), ubiquitin is bound to E1-cysteine through a thioester bond. This allows transfer of the ubiquitin to E2-cysteine (residue 86 in this example). With substrate bound to E3, the ubiquitin transfers to a lysine ε-amino group to create a stable isopeptide bond. This reaction can be repeated by addition of further ubiquitins. Only polyubiquitylated proteins are recognized for destruction. (1aar[70] and 1fbv[73]).

thioester link. It is then transiently coupled to a cysteine on the E2-conjugating enzyme. The substrate (target protein) and E2-ubiquitin are brought in close proximity through a third component, E3-ligase, which determines the specificity of the conjugation reaction (**Figure 15.12**). Ubiquitin detaches from E2 to form an isopeptide bond linking the α-carboxyl group of its terminal glycine with an ε-amino group of a lysine residue in the substrate.

63K or 48K conjugation

Repeated conjugation results in the formation of polyubiquitin chains. Ubiquitin itself possesses six lysines, all of which can make the link to the N-terminal glycine, but two of these, Lys48 and Lys63, predominate (48K or 63K type). Recognition of target proteins by the proteasome complex requires four or more linked 48K-type ubiquitins (**Figure 15.13**). The 63K-type polyubiquitin chain, though not excluded, is not necessarily involved in

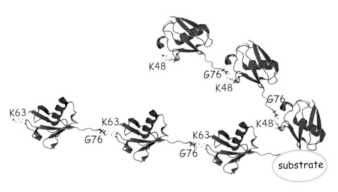

FIG 15.13 K48- or K63-linked
ubiquitin chains.
Ubiquitins are coupled to each
other through isopeptide linkages
between either Lys48 or Lys63 and
the C-terminal Gly76. K48-type
polyubiquitin chains are recognized
by ubiquitin binding proteins that
communicate with the proteasome.
K63-linked ubiquitins play a role in the
recruitment of signalling complexes
(1aar[70]).

recognition by the proteasome, serving more readily as a docking site
for other ubiquitin-binding proteins. It also plays a role in cell signalling,
importantly in TNF-α-mediated activation of NF-κB.[74] Two examples of K63-
type polyubiquitin-binding proteins relevant for this discussion are NEMO
and TAB2 but how these make the distinction between 48K- and 63K-type
ubiquitylation is not known.

Two classes of E3-ubiquitin ligases

There are two classes of E3-ubiquitin ligases; RING- and HECT-type
(Figure 15.14).

1. The RING type are abundant. The RING domain comprises a series of zinc
   finger motifs that act as a protein–protein interaction domain. Some RING-
   domain proteins interact with E2-ubiquitin conjugating enzymes. This class
   of E3-ubiquitin ligases comes in two flavours:

   - *Single protein*. Here the E3-ligase acts both as a binding site for E2
     and as a receptor for substrates. An example is Mdm2 involved in the
     ubiquitylation of p53, Cbl, involved in the ubiquitylation of the EGFR
     (Figure 21, page 350) and the kinase ZAP-70 (that acts downstream of
     the T cell receptor) (Figure 17.3, page 518).
   - *Multiprotein complex*. The RING domain protein binds the E2-ubiquitin
     conjugating enzyme, but the E3-ligase and the receptor are kept apart
     by scaffold proteins. Examples are SCF and APC/C, both involved in
     destruction of components of the cell cycle machinery (anaphase
     promoting complex/cyclosome).
2. The HECT type E3-ligase harbours both E2- and E3-activity.[75] An example
   is Nedd4 which operates in the regulation of the activity and plasma
   membrane expression of ENaC, an epithelial cell-surface Na$^+$ channel. We
   return to HECT type E3-ligases in relation to signalling by TGFβ (Chapter
   20). A variant, Nedd4-2, plays a role in the regulation of destruction of PTEN
   (page 668 onwards).

The **ENaC Na$^+$** channel
is involved in fluid
resorption in kidney, lung,
and colon. Mutated ENaC,
resistant to the attention
of Nedd4, results in its
enhanced expression at
the cell surface. This leads
to excessive re-uptake of
Na$^+$ in nephrons, giving
rise to hypertension
(Liddle's syndrome).[76]

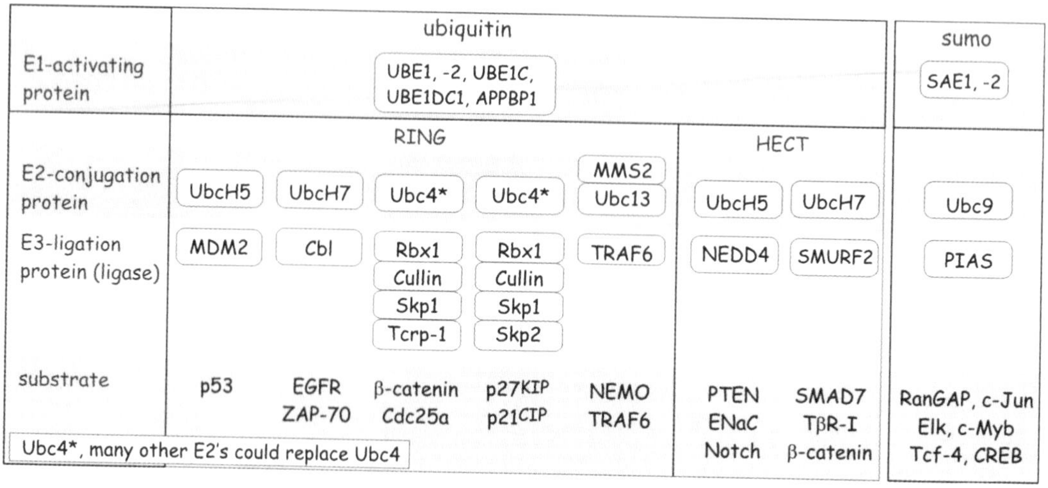

	ubiquitin							sumo
E1-activating protein	UBE1, -2, UBE1C, UBE1DC1, APPBP1							SAE1, -2
	RING					HECT		
E2-conjugation protein	UbcH5	UbcH7	Ubc4*	Ubc4*	MMS2 Ubc13	UbcH5	UbcH7	Ubc9
E3-ligation protein (ligase)	MDM2	Cbl	Rbx1 Cullin Skp1 Tcrp-1	Rbx1 Cullin Skp1 Skp2	TRAF6	NEDD4	SMURF2	PIAS
substrate	p53	EGFR ZAP-70	β-catenin Cdc25a	p27KIP p21CIP	NEMO TRAF6	PTEN ENaC Notch	SMAD7 TβR-I β-catenin	RanGAP, c-Jun Elk, c-Myb Tcf-4, CREB
Ubc4*, many other E2's could replace Ubc4								

FIG 15.14 Examples of ubiquitin- and sumo-ligase complexes.

The human genome contains five genes that code for E1-ubiquitin activating proteins. These transfer ubiquitin to a much wider range of E2-conjugating enzymes (37 genes). Relevant examples are UbcH5, UbcH7, UbcH4, and Ubc13 (other Ubcs can replace Ubc4). Ubc13 is only active when bound to MMS2. In the case of the RING domain-containing ligase complexes, the E2-conjugating enzyme transfers the ubiquitin chain to substrate which is bound to a separate E3-ligase. With Mdm2, Cbl, and TRAF6, the E3-ligase constitutes a single protein; in other cases it forms a multiprotein complex comprising a RING domain protein (Rbx1), scaffold proteins (Cullin and Skp1), and a receptor (β-Tcrp-1 or Skp2). In the case of HECT-type E3-ligases, both activities (E3 and E2) are harboured within one protein (NEDD4, for instance). Proteins can also be modified by ubiquitin-like SUMO proteins. Specialized E1-SUMO activating proteins are SAE1 and -2. These transfer the SUMO to Ubc9, which, in the presence of PIAS, is transferred to substrate.

Ubiquitin-binding proteins

The first entity that binds ubiquitin with high specificity to be identified was the regulatory proteasome particle S5a (RPN10).[77] However, since deletion of the homologous gene in yeast has somewhat modest phenotypic effects, a search for additional proteins was instigated. We have already mentioned NEMO and TAB2/3 but there are many others, of which a large proportion is involved in targeting proteins to the proteasome.[78] Representative examples and their domains are shown in Figure 15.15.

SUMO and sumoylation

Overall, SUMO (small ubiquitin-like modifier) and ubiquitin share only 18% sequence identity, but the folded structures of their C-termini are almost superimposable. In vertebrates, there are three SUMO-related peptides. SUMO is implicated in a number of regulatory events such as enhancement of protein stability, protein–protein interactions, subcellular translocation, and transcriptional control.[80–82] Because ubiquitin and SUMO compete for the

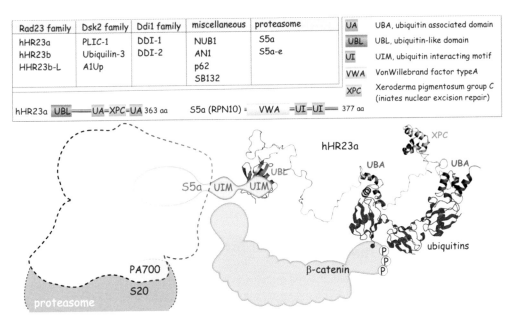

Rad23 family	Dsk2 family	Ddi1 family	miscellaneous	proteasome		UA	UBA, ubiquitin associated domain
hHR23a	PLIC-1	DDI-1	NUB1	S5a		UBL	UBL, ubiquitin-like domain
hHR23b	Ubiquilin-3	DDI-2	AN1	S5a-e		UI	UIM, ubiquitin interacting motif
HHR23b-L	A1Up		p62			VWA	VonWillebrand factor typeA
			SB132			XPC	Xeroderma pigmentosum group C (iniates nuclear excision repair)

hHR23a [UBL]━━━[UA]=[XPC]-[UA] 363 aa   S5a (RPN10) = [VWA] =[UI]=[UI]━━ 377 aa

FIG 15.15 Ubiquitin-binding proteins involved in substrate presentation to the proteasome.

There are numerous proteins that bind ubiquitins and which are involved in the presentation of substrate to the proteasome. They contain ubiquitin-associated (UBA) or ubiquitin-interacting motifs (UIM) (among others). The domain organization of two examples are illustrated. hHR23a is composed of two UBA sites and one ubiquitin-like domain (UBL). It binds ubiquitylated substrate at both UBA sites, and this may be one mechanism by which only poly-ubiquitylated proteins are selected for destruction. With its UBL domain it binds the ubiquitin-interacting motif from S5a, an integral component of the PA700 particle (1oqy,[79] 1aar[70]).

same target lysine residues, sumoylation can prevent ubiquitylation and can thus serve to protect proteins from degradation at the proteasome.[81]

The PIAS family of E3-ligases involved in the conjugation of SUMO represent a variant on RING-type E3 ligases.[81] They are able to interact with some E2-ubiquitin conjugating enzymes but not the E2-SUMO conjugation enzyme, Ubc9 being preferred (Figure 15.14). Sumoylation of c-Jun, Elk, c-Myb, and the androgen receptors by PIAS suppresses their activity.[83] Sumoylation can also act as an activator, TCF-4 and CREB being good examples.[84,85]

Essay: the proteasome complex

The continuous recycling of cellular proteins requires a coherent process of synthesis and degradation. The numbers are striking. Of the total protein body mass of healthy human adults, 2%, about 50 g, is renewed every day. About 75% of the amino acid nitrogen is reused for *de novo* protein synthesis, the remaining 25% being eliminated as urea. It follows that about 200 g (wet weight!) of meat, eggs, or milk products are needed to compensate the loss.

Only the nitrogen content is eliminated as urea or uric acid. The keto acids enter the Krebs cycle as pyruvate, which may be used to make fat or metabolized to release energy.

Dairy products, fish, meat, pulses, fruit, vegetables, nuts, all contain protein. Vegetarians have to eat more than carnivores to satisfy their needs. This applies to all animals. Grazers have to eat almost continuously, while meat-eating predators like lions and tigers eat infrequently and have the luxury of free time. Humans didn't have much free time until they became large-scale meat eaters.

The rate of protein turnover in the liver is much greater, $\sim$40% per day. The enormous disparity is due to the difference between proteins that constitute the extracellular mass, with half-lives of days to months, and cellular proteins that exist for only a few hours.

Disposal of proteins occurs by either ubiquitylation and degradation in the proteasome or by degradation of proteins (or whole organelles) by lysosomes, or by autodestruction of cells by caspases (apoptosis). All these are subject to regulation. Many proteases are synthesized as inactive precursors, requiring partial proteolysis that occurs in specialized organelles. Access of substrate to the proteolytic activity may be dependent on post-translational modifications (as is the case of the proteasome which only recognizes ubiquitylated substrates).

The proteasome

Polyubiquitylated proteins, in particular the 48K-type, are recognized by receptors on the proteasome, effectively a large complex of proteases. The proteasome can be regarded as the functional opposite of the ribosome, which it also rivals with respect to size ($\sim$30 nm) and complexity. It is composed of two major subunits, a 20S particle and a regulatory 11S (PA28) or 19S (PA700) particle (proteasome activators) (**Figure 15.16**). The 20S particle is highly conserved, being present in archaea, eubacteria, and all eukaryotes. The proteasome occurs in different configurations; single 20S, singly and doubly PA28-capped, singly PA700-capped, PA700 and PA28 capped (hybrid), and doubly PA700-capped. All these species are highly dynamic, with rapid recruitment and exchange of regulatory caps. It is suggested that the association of different caps signifies different stages of the degradation process.[86]

20S particle

The 20S particle, of which numerous copies are present in both cytosol and nucleus, has a cylindrical form (15$\times$11 nm). It is composed of 28 subunits; 2 copies each of subunits α1–7 and β1–7 (**Figure 15.16**). The proteolytic activity is situated on the inside of the cylinder and is associated with three different subunits, each targeting different amino acids. The caspase-like β1 subunit cuts after acidic residues, the trypsin-like β2 cuts after basic residues and the chymotrypsin-like β5 cuts after hydrophobic residues. Together they assure the degradation of a large range of proteins into fragments of 3–25 residues.[88] (**Figure 15.17**).

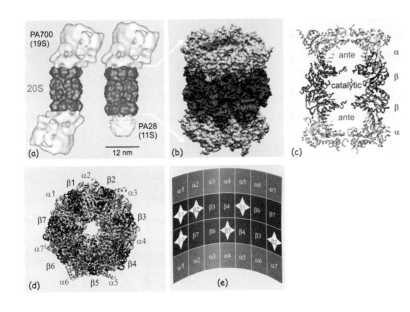

FIG 15.16 Proteasome structure.
(a) The proteasome is composed to two functional entities, the actual protease complex (20S particle) and the activating subunits, the caps. Of these two variants exist, PA700 (19S particle) and PA28 (11S particle). Different combinations of 20S and the caps exist. They play a role in binding, unfolding, and transport of the substrate. (b–d) The 20S particle is composed of α- and β-subunits, the α-subunits forming the antechamber, the β-subunits the catalytic chamber. (e) Not all β-subunits are functional proteases; this activity is restricted to β1, β2 and β5. Image in (a), courtesy of B. Baumeister, Martinsried, Germany. Image in (e), courtesy of K. Hendil, Copenhagen, Denmark (1ryp[87]).

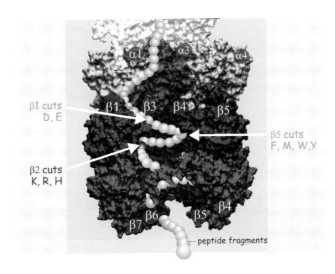

FIG 15.17 Proteasome chain degradation.
Within the catalytic chamber, the unfolded substrate slides along the active domains of the proteases. β1 cuts after acidic residues, β2 after basic residues, and β5 after large hydrophobic residues. The polypeptide chain is cut up into small chunks of 3–25 amino acids (1fnt[89]).

FIG 15.18 Activation mechanism of the 20S proteasome.
(a) Entry into the catalytic chamber is blocked by the α-subunits that form the lining of the antechamber of the 20S particle. (b) Binding of the activating particle (red) changes their conformation, opening the entry pore. In this way selectivity of protein degradation is assured because only capped 20S particles are accessible to substrate (1fnt,[89] 1ryp[87]).

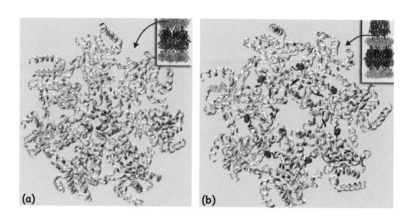

(a)   (b)

Proteasome activator (PA) subunits

The regulatory PA700 subunit is composed of a minimum of 20 proteins, referred to as Rpn and Rpt.[88] Together they organize (1) the fixation of substrate (with the help of an adaptor), (2) unfolding and transport of the polypeptide chain towards the proteasome interior,[90] (3) removal of the ubiquitin chains, which are then recycled, and (4) the opening of the entry pore of the proteasome 20S particle (Figure 15.18).[87,89] The composition of PA28 is less clear.

List of Abbreviations

Abbreviation	Full name/description	SwissProt entry	Other names/OMIM
ATF2	activating transcription factor-2	P15336	CRE binding protein (CRE-BP1)
cactus		Q03017	
Cbl	casitas B-lineage lymphoma proto-oncogene	P22681	
CBP	CREB binding protein	Q92793	
CD14	cluster of differentiation-14	P08571	
cJun	avian sarcoma virus 17 oncogene homologue (junana=17 in Japanese)	P05412	
CUE	coupling of ubiquitin conjugation to endoplasmic reticulum degradation		ubiquitin binding domain

Continued

Abbreviation	Full name/description	SwissProt entry	Other names/OMIM
dorsal		P15330	
HECT	homologous to E6AP C-terminus		
hHR23a	human homologue of radiation repair protein 23a	P54725	RAD23A
IκBα	inhibitor of NF-κB type alpha	P25963	
IKK-α	Inhibitor of NF-κB alpha (IkB)-kinase type alpha		
IKK-β	inhibitor of NF-κB alpha (IkB)-kinase type beta	O14920	
IKK-ε	Inhibitor of NF-κB alpha (IkB)- kinase type epsilon	Q14164	
IL-1R1	interleukin-1 alpha receptor-1	P14778	
IRAK1	interleukin-1 receptor-associated kinase-1	P51617	
IRAK4	interleukin-1 receptor associated kinase-4	Q9NWZ3	
IRF3	interferon regulatory factor-3	Q14653	
IRSE	IFN-stimulated response element		
JNK1	c-Jun N-terminal kinase 1	P45983	MAPK8, MK08
MD-2	myeloid differentation gene-2	Q9Y6Y9	LPS co-receptor, lymphocyte antigen 96
MEK3	MAP-kinase ERK-activating kinase 3	P46734	MKK3, MAP2K3
MEK6	MAP-kinase ERK-activating kinase 6	P52564	MKK6, SAPKK3
MK2	MAP-kinase activated protein kinase-2	P49137	MAPKAP K2
MMS2	mutant sensitive to methanesulfonate-2 (in *Saccharomyces cerevisiae*)	Q15819	Uev1, UB2V2
MyD88	myeloid differentiation primary response protein 88	Q99836	
NEDD4	neuronal precursor cell expressed developmentally down regulated-4	P46934	E3 ubiquitin protein ligase
NEMO	NFkB essential modulator	Q9Y6K9	IKKg
NF-κB1	nuclear factor kappa-B1	P19838	p105/p50 truncated
NF-κB2	nuclear factor kappa-B2	Q00653	p100/p52 truncated
p38α	mitogen activation protein kinase p38a	Q16539	SAPK21, MK14

Continued

Abbreviation	Full name/description	SwissProt entry	Other names/OMIM
pelle		Q05652	
PIAS1	protein inhibitor of activated STAT	O75925	RNA helicase II-binding protein
proteasome subunit-β1		P20618	PSMB1, component C5
proteasome subunit-β2		P49721	PSMB2, component C7-1
proteasome subunit-β5		P28074	PSMB5, epsilon chain, chain 6
REL-A	reticuloendotheliosis virus like protein A	Q04206	NF-k-B p65 subunit
RING	really interesting new gene		
S5a	human proteasome subunit, Svedberg-sedimentation constant 5a	P55036	Rpn10, PSMD4
SCF	ubiquitin E-3 ligase complex comprising Skp, Cullin-1 and F-box protein		
Skp2	S-phase kinase-associated protein-2	Q13309	F-box/LRR-repeat protein
SUMO-1	small ubiquitin-related modifier-1	P63165	RanGAP modifying protein
TAB1	TAK1-binding-1	Q15750	MAP3K7IP1
TAB2	TAK1-binding-2	Q9NYJ8	MAP3K7IP2
TAB3	TAK-1-binding-3	Q8N5C8	MAP3KIP3
TAK1	transforming growth factor-B-activted kinase-1	O43318	MAP3K7
TANK	TRAF-family member associated with NFkB activator	Q92844	ITRAF
TBK1	TANK-binding kinase 1	Q9UHD2	T2K or NAK
TIR	toll/interleukin-1 receptor (protein–protein interaction domain)		
TIRAP	toll-interleukin-1-receptor (TIR) domain containing adaptor protein	P58753	Mal
TLR4	toll-like receptor-4	O00206	

Continued

Abbreviation	Full name/description	SwissProt entry	Other names/OMIM
Toll	refers to its remarkable effect on dorsal ventral-polarity of *drosophila* embryos	P08953	
TRAF3	TNF-receptor-associated factor-3	Q13114	
TRAF6	TNF-receptor-associated factor-6	Q9Y4K3	
TRAM	TRIF-related adaptor molecule	Q86XR7	TICAM2 Toll/IL-1R-domain containing adaptor molecule-2
Trcp	Transducin-repeat containing protein	Q9Y297	F-box/WD repeat protein
TRIF	TIR domain-containing adaptor inducing IFNβ	Q8IUC6	TICAM1 Toll/IL-1R-domain containing adaptor molecule-1
Tube		P22812	
Ubc13	ubiquitin carrier-13	P61088	UBE2N (E2-conjugating enzyme)
Ubc4	ubiquitin conjugating enzyme	P62837	UBE2D2 (E2-conjugating enzyme)
ubiquitin	ubiquitously expressed	P62988	

References

1. Kang YS, Do Y, Lee HK, et al. A dominant complement fixation pathway for pneumococcal polysaccharides initiated by SIGN-R1 interacting with C1q. *Cell*. 2006;125:47–58.
2. Medzhitov Jr R, Preston-Hurlburt P, Janeway CA. A human homologue of the *Drosophila* toll protein signals activation of adaptive immunity. *Nature*. 1997;388:394–397.
3. Metchnikoff E. Über eine sprosspilzkrankheit der daphnien. Beitrag zur lehre über den kampf der phagocyten gegen krankheitserrenger. *Arch Pathol Anat Physiol Klin Med*. 1884;96:177–195.
4. Janeway CA. The immune system evolved to discriminate infectious nonself from noninfectious self. *Immunol Today*. 1992;13:11–16.
5. Rock FL, Hardiman G, Timans JC, Kastelein RA, Bazan JF. A family of human receptors structurally related to *Drosophila* toll. *Proc Natl Acad Sci U S A*. 1998;95:588–593.
6. Akira S, Uematsu S, Takeuchi O. Pathogen recognition and innate immunity. *Cell*. 2006;124:783–801.

7. Ferrandon D, Imler JL, Hoffmann JA. Sensing infection in *Drosophila*: toll and beyond. *Semin Immunol*. 2004;16:43–53.

8. Lemaitre B, Kromer-Metzger E, Michaut L, et al. A recessive mutation, immune deficiency (imd), defines two distinct control pathways in the *Drosophila* host defense. *Proc Natl Acad Sci U S A*. 1995;92: 9465–9469.

9. Belvin MP, Anderson KV. A conserved signaling pathway: the *Drosophila* toll-dorsal pathway. *Annu Rev Cell Dev Biol*. 1996;12:393–416.

10. Nusslein-Volhard C, Lohs-Schardin M, Sander K, Cremer C. A dorso-ventral shift of embryonic primordia in a new maternal-effect mutant of *Drosophila*. *Nature*. 1980;283:474–476.

11. Morisato D, Anderson KV. Signaling pathways that establish the dorsal-ventral pattern of the *Drosophila* embryo. *Annu Rev Genet*. 1995;29: 371–399.

12. Mercurio F, Manning AM. Multiple signals converging on NF-κB. *Curr Opin Cell Biol*. 1999;11:226–232.

13. Lemaitre B, Nicolas E, Michaut L, Reichhart JM, Hoffmann JA. The dorsoventral regulatory gene cassette *spatzle/toll/cactus* controls the potent antifungal response in *Drosophila* adults. *Cell*. 1996;86:973–983.

14. Weber AN, Tauszig-Delamasure S, Hoffmann JA, et al. Binding of the *Drosophila* cytokine spatzle to toll is direct and establishes signaling. *Nat Immunol*. 2003;4:794–800.

15. Levashina EA, Langley E, Green C, et al. Constitutive activation of toll-mediated antifungal defense in serpin-deficient *Drosophila*. *Science*. 1999;285:1917–1919.

16. Schumann RR, Belka C, Reuter D, et al. Lipopolysaccharide activates caspase-1 (interleukin-1-converting enzyme) in cultured monocytic and endothelial cells. *Blood*. 1998;91:577–584.

17. Wagner H, Bauer S. All is not Toll: new pathways in DNA recognition. *J Exp Med*. 2006;203:265–268.

18. Dauphinee SM, Karsan A. Lipopolysaccharide signaling in endothelial cells. *Lab Invest*. 2006;86:9–22.

19. Sims JE, Acres RB, Grubin CE, et al. Cloning the interleukin 1 receptor from human T cells. *Proc Natl Acad Sci U S A*. 1989;86:8946–8950.

20. Heguy A, Baldari CT, Macchia G, Telford JL, Melli M. Amino acids conserved in interleukin-1 receptors (IL-1Rs) and the *Drosophila* Toll protein are essential for IL-1R signal transduction. *J Biol Chem*. 1992;267:2605–2609.

21. Stack J, Haga IR, Schroder M, et al. Vaccinia virus protein A46R targets multiple toll-like-interleukin-1 receptor adaptors and contributes to virulence. *J Exp Med*. 2005;201:1007–1018.

22. Choe J, Kelker MS, Wilson IA. Crystal structure of human toll-like receptor 3 (TLR3) ectodomain. *Science*. 2005;309:581–585.

23. Xu Y, Tao X, Shen B, et al. Structural basis for signal transduction by the toll/interleukin-1 receptor domains. *Nature*. 2000;408:111–115.

24. Sen R, Baltimore D. Multiple nuclear factors interact with the immunoglobulin enhancer sequences. *Cell*. 1986;705–716.

25. Beinke S, Ley SC. Functions of NF-κB1 and NF-κB2 in immune cell biology. *Biochem J*. 2004;382:393–409.

26. Gram H. Über die isolierte färbung der schizomyceten in schnitt- und trockenpräparaten. *Fortschritte Der Medizin*. 1884;2:185–189.

27. Galloway SM, Raetz CR. A mutant of *Escherichia coli* defective in the first step of endotoxin biosynthesis. *J Biol Chem*. 1990;265:6394–6402.

28. Poltorak A, He X, Smirnova I, et al. Defective LPS signaling in C3H/HeJ and C57BL/10ScCr mice: mutations in *Tlr4* gene. *Science*. 1998;282:2085–2088.

29. Beutler B, Poltorak A. The sole gateway to endotoxin response: how LPS was identified as Tlr4, and its role in innate immunity. *Drug Metab Dispos*. 2001;29:474–478.

30. Montminy SW, Khan N, McGrath S, et al. Virulence factors of *Yersinia pestis* are overcome by a strong lipopolysaccharide response. *Nat Immunol*. 2006;7:1066–1073.

31. Steiner AA, Chakravarty S, Rudaya AY, Herkenham M, Romanovsky AA. Bacterial lipopolysaccharide fever is initiated via toll-like receptor 4 on hematopoietic cells. *Blood*. 2006;107:4000–4002.

32. Schroder NW, Schumann RR. Non-LPS targets and actions of LPS binding protein (LBP). *J Endotoxin Res*. 2005;11:237–242.

33. Wright SD, Ramos RA, Tobias PS, Ulevitch RJ, Mathison JC. CD14, a receptor for complexes of lipopolysaccharide (LPS) and LPS binding protein. *Science*. 1990;249:1431–1433.

34. Gioannini TL, Teghanemt A, Zhang D, et al. Isolation of an endotoxin-MD-2 complex that produces toll-like receptor 4-dependent cell activation at picomolar concentrations. *Proc Natl Acad Sci U S A*. 2004;101:4186–4191.

35. Heppner G, Weiss DW. High susceptibility of strain a mice to endotoxin and endotoxin-red blood cell mixtures. *J Bacteriol*. 1965;90:696–703.

36. Guo L, Lim KB, Poduje CM, et al. Lipid a acylation and bacterial resistance against vertebrate antimicrobial peptides. *Cell*. 1998;95:189–198.

37. Lu M, Zhang M, Takashima A, et al. Lipopolysaccharide deacylation by an endogenous lipase controls innate antibody responses to gram-negative bacteria. *Nat Immunol*. 2005;6:989–994.

38. Munford RS, Varley AW. Shield as signal: lipopolysaccharides and the evolution of immunity to gram-negative bacteria. *PLoS Pathog*. 2006;2:e67.

39. Li S, Strelow A, Fontana EJ, Wesche H. IRAK-4: a novel member of the IRAK family with the properties of an IRAK-kinase. *Proc Natl Acad Sci U S A*. 2002;99:5567–5572.

40. Suzuki N, Suzuki S, Duncan GS, et al. Severe impairment of interleukin-1 and toll-like receptor signalling in mice lacking IRAK-4. *Nature*. 2002;416:750–756.

41. Picard C, Puel A, Bonnet M, et al. Pyogenic bacterial infections in humans with IRAK-4 deficiency. *Science*. 2003;299:2076–2079.

42. Takaesu G, Kishida S, Hiyama A, et al. TAB2, a novel adaptor protein, mediates activation of TAK1 MAPKKK by linking TAK1 to TRAF6 in the IL-1 signal transduction pathway. *Mol Cell*. 2000;5:649–658.

43. Wang C, Deng L, Hong M, Akkaraju GR, Inoue J, Chen ZJ. TAK1 is a ubiquitin-dependent kinase of MKK and IKK. *Nature*. 2001;412:346–351.

44. Kanayama A, Seth RB, Sun L, et al. TAB2 and TAB3 activate the NF-κB pathway through binding to polyubiquitin chains. *Mol Cell*. 2004;15: 535–548.

45. Kishimoto K, Matsumoto K, Ninomiya-Tsuji J. TAK1 mitogen-activated protein kinase kinase kinase is activated by autophosphorylation within its activation loop. *J Biol Chem*. 2000;275:7359–7364.

46. DiDonato JA, Hayakawa M, Rothwarf DM, Zandi E, Karin M. A cytokine-responsive IκB kinase that activates the transcription factor NF-κB. *Nature*. 1997;388:548–554.

47. Wu CJ, Conze DB, Li T, Srinivasula SM, Ashwell JD. Sensing of Lys 63-linked polyubiquitination by NEMO is a key event in NF-κB activation. *Nat Cell Biol*. 2006;8:398–406.

48. Chen ZJ. Ubiquitin signalling in the NF-κB pathway. *Nat Cell Biol*. 2005;7:758–765.

49. Xiao G, Harhaj EW, Sun SC. NF-κB-inducing kinase regulates the processing of NF-κB2 p100. *Mol Cell*. 2001;7:401–409.

50. Ninomiya-Tsuji J, Kishimoto K, Hiyama A, Inoue J, Cao Z, Matsumoto K. The kinase TAK1 can activate the NIK-I κB as well as the MAP kinase cascade in the IL-1 signalling pathway. *Nature*. 1999;398:252–256.

51. Sano Y, Harada J, Tashiro S, Gotoh-Mandeville R, Maekawa T, Ishii S. ATF-2 is a common nuclear target of Smad and TAK1 pathways in transforming growth factor-β signaling. *J Biol Chem*. 1999;274:8949–8957.

52. Dickens M, Rogers JS, Cavanagh J, et al. A cytoplasmic inhibitor of the JNK signal transduction pathway. *Science*. 1997;277:693–696.

53. Ito M, Yoshioka K, Akechi M, et al. JSAP1, a novel jun N-terminal protein kinase (JNK)-binding protein that functions as a scaffold factor in the JNK signaling pathway. *Mol Cell Biol*. 1999;19:7539–7548.

54. Dziarski R. Deadly plague versus mild-mannered TLR4. *Nat Immunol*. 2006;7:1017–1019.

55. Verhey KJ, Meyer D, Deehan R, et al. Cargo of kinesin identified as JIP scaffolding proteins and associated signaling molecules. *J Cell Biol*. 2001;152:959–970.

56. Wilkinson KD, Urban MK, Haas AL. Ubiquitin is the ATP-dependent proteolysis factor I of rabbit reticulocytes. *J Biol Chem*. 1980;256:7529–7532.

57. Kelkar N, Standen CL, Davis RJ. Role of the JIP4 scaffold protein in the regulation of mitogen-activated protein kinase signaling pathways. *Mol Cell Biol*. 2005;25:2733–2743.

58. Takaoka A, Yanai H, Kondo S, et al. Integral role of IRF-5 in the gene induction programme activated by Toll-like receptors. *Nature*. 2005;434:243–249.

59. Chariot A, Leonardi A, Muller J, Bonif M, Brown K, Siebenlist U. Association of the adaptor TANK with the I κ B kinase (IKK) regulator NEMO connects IKK complexes with IKK ε and TBK1 kinases. *J Biol Chem*. 2002;277:37029–37036.

60. Hacker H, Redecke V, Blagoev B, et al. Specificity in Toll-like receptor signalling through distinct effector functions of TRAF3 and TRAF6. *Nature*. 2006;439:204–207.

61. Miyamoto M, Fujita T, Kimura Y, et al. Regulated expression of a gene encoding a nuclear factor, IRF-1, that specifically binds to IFN-β gene regulatory elements. *Cell*. 1988;54:903–913.

62. Honda K, Taniguchi T. IRFs: master regulators of signalling by toll-like receptors and cytosolic pattern-recognition receptors. *Nat Rev Immunol*. 2006;6:644–658.

63. Lohoff M, Mak TW. Roles of interferon-regulatory factors in T-helper-cell differentiation. *Nat Rev Immunol*. 2005;5:125–135.

64. Honda K, Yanai H, Negishi H, et al. IRF-7 is the master regulator of type-I interferon-dependent immune responses. *Nature*. 2005;434:772–777.

65. Gilchrist M, Thorsson V, Li B, et al. Systems biology approaches identify ATF3 as a negative regulator of toll-like receptor 4. *Nature*. 2006;441:173–178.

66. Cheung PC, Campbell DG, Nebreda AR, Cohen P. Feedback control of the protein kinase TAK1 by SAPK2a/p38α. *EMBO J*. 2003;22:5793–5805.

67. Brummelkamp TR, Nijman SM, Dirac AM, Bernards R. Loss of the cylindromatosis tumour suppressor inhibits apoptosis by activating NF-κB. *Nature*. 2003;424:797–801.

68. Kovalenko A, Chable-Bessia C, Cantarella G, Israel A, Wallach D, Courtois G. The tumour suppressor CYLD negatively regulates NF-κB signalling by deubiquitination. *Nature*. 2003;424:801–805.

69. Moore KW, de Waal MR, Coffman RL, O'Garra A. Interleukin-10 and the interleukin-10 receptor. *Annu Rev Immunol*. 2001;19:683–765.

70. Cook WJ, Jeffrey LC, Carson M, Chen Z, Pickart CM. Structure of a diubiquitin conjugate and a model for interaction with ubiquitin conjugating enzyme (E2). *J Biol Chem*. 1992;267:16467–16471.

71. Nijman SM, Luna-Vargas MP, Velds A, et al. A genomic and functional inventory of deubiquitinating enzymes. *Cell*. 2005;123:773–786.

72. Liu YC. Ubiquitin ligases and the immune response. *Annu Rev Immunol*. 2004;22:81–127.

73. Zheng N, Wang P, Jeffrey PD, Pavletich NP. Structure of a c-Cbl-UbcH7 complex: RING domain function in ubiquitin-protein ligases. *Cell.* 2000;102:533–539.

74. Chen ZJ, Parent L, Maniatis T. Site-specific phosphorylation of IκBαby a novel ubiquitination-dependent protein kinase activity. *Cell.* 1996;84:853–862.

75. Huibregtse JM, Scheffner M, Beaudenon S, Howley PM. A family of proteins structurally and functionally related to the E6-AP ubiquitin-protein ligase. *Proc Natl Acad Sci U S A.* 1995;92:2563–2567.

76. Abriel H, Staub O. Ubiquitylation of ion channels. *Physiology (Bethesda).* 2005;20:398–407.

77. Deveraux Q, Ustrell V, Pickart C, Rechsteiner M. A 26S protease subunit that binds ubiquitin conjugates. *J Biol Chem.* 1994;269:7059–7061.

78. Elsasser S, Finley D. Delivery of ubiquitinated substrates to protein-unfolding machines. *Nat Cell Biol.* 2005;7:742–749.

79. Walters KJ, Lech PJ, Goh AM, Wang Q, Howley PM. DNA-repair protein hHR23a alters its protein structure upon binding proteasomal subunit S5a. *Proc Natl Acad Sci U S A.* 2003;100:12694–12699.

80. Muller S, Hoege C, Pyrowolakis G, Jentsch S. SUMO, ubiquitin's mysterious cousin. *Nat Rev Mol Cell Biol.* 2001;2:202–210.

81. Johnson ES. Protein modification by SUMO. *Annu Rev Biochem.* 2004;73:355–382.

82. Yang XJ, Gregoire S. A recurrent phospho-sumoyl switch in transcriptional repression and beyond. *Mol Cell.* 2006;23:779–786.

83. Girdwood DW, Tatham MH, Hay RT. SUMO and transcriptional regulation. *Semin Cell Dev Biol.* 2004;15:201–210.

84. Yamamoto H, Ihara M, Matsuura Y, Kikuchi A. Sumoylation is involved in β-catenin-dependent activation of Tcf-4. *EMBO J.* 2003;22:2047–2059.

85. Comerford KM, Leonard MO, Karhausen J, Carey R, Colgan SP, Taylor CT. Small ubiquitin-related modifier-1 modification mediates resolution of CREB-dependent responses to hypoxia. *Proc Natl Acad Sci U S A.* 2003;100:986–991.

86. Shibatani T, Carlson EJ, Larabee F, McCormack AL, Fruh K, Skach WR. Global organization and function of mammalian cytosolic proteasome pools: Implications for PA28 and 19S regulatory complexes. *Mol Biol Cell.* 2006;17:4962–4971.

87. Groll M, Bajorek M, Kohler A, et al. A gated channel into the proteasome core particle. *Nat Struct Biol.* 2000;7:1062–1067.

88. Hendild KB, Hartmann-Petersen R. Proteasomes: a complex story. *Curr Protein Pept Sci.* 2004;5:135–151.

89. Whitby FG, Masters EI, Kramer L, et al. Structural basis for the activation of 20S proteasomes by 11S regulators. *Nature.* 2000;408:115–120.

90. Braun BC, Glickman M, Kraft R, et al. The base of the proteasome regulatory particle exhibits chaperone-like activity. *Nat Cell Biol.* 1999;1:221–226.

Traffic of White Blood Cells

Inflammation and leukocytes

The first sightings of white cells adhering to the walls of the finer vessels and then emigrating into the tissues under conditions of inflammation were recorded more than 150 years ago (Figure 16.1).[1–3] Cohnheim's detailed histological description of 1882[4,5] remained the basis of most textbook accounts of inflammation for a further 80 years. Later, the first electron microscopic investigation of leukocyte adherence and migration provided much-needed detail but failed to reveal how cells adhere to the inflamed endothelium.[6] Nor were any ideas forthcoming about how the cells penetrate an apparently coherent endothelial cell layer.

Inflammatory mediators

It is essential that migratory cells such as leukocytes can tolerate changes in their environments. They must be able to move from the bone marrow (where they are attached) into the blood (where they float free) and then

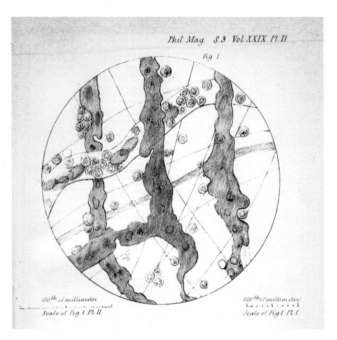

FIG 16.1 Augustus Waller's microscopic examination of cell adherence and extravasation from the vessels of an inflamed frog tongue.

'The blood, as we are aware, consists of a transparent fluid holding in suspension numerous particles, most of which are red and of a flattened shape, while a few others are colourless, and spherical in form ... The peculiar manner in which the lymph-globules, or corpuscles, conduct themselves when in the capillaries, when in an organ in a state of irritation, has of late engaged much attention. The experiments of Mr W. Addison of Malvern, have greatly contributed to show these important functions in inflammation. In the tongue of the frog and toad they may be frequently seen circulating with the red particles in the vessels, down to the minutest capillaries. As it has already been pointed out, these spherules are generally found, when they come into contact with the parieties of the vessels, to retain their adherence with greater force than is manifested by the red particles in the like circumstances; as in the figure, where the current was observed to continue for many minutes without displacing the globules near the sides of the vessel. Thus we frequently see a lymph-globule remain in the same place, notwithstanding the current of red particles sweeping and pushing by it. The appearance in the larger vessels of these spherules, adherent to their inner surface, has been very aptly compared to so many pebbles or marbles over which a stream runs without disturbing them ... The corpuscles, which are transparent, are occasionally seen to be granulated ... Let us now examine the admirable manner in which nature has solved the apparent paradox, of eliminating, from a fluid circulating in closed tubes, certain particles floating in it, without causing any rupture or perforation in the tubes, or allowing the escape of the red particles, which are frequently the smaller of the two, or that of the fluid part of the blood itself ... After the observation had continued for half an hour, numerous corpuscles were seen outside the vessels, together with a very few blood discs in the proportion of about one to ten of the former. No appearance of rupture could be seen in any of the vessels. The corpuscles were generally distant about 0^{mm} 03 from their parieties. After the experiment had lasted about two hours, thousands of these corpuscles were seen scattered over the membrane, with scarcely any blood discs ... No trace of the corpuscular extravasation could be seen, except the presence of the corpuscles themselves. . . . . . . I consider therefore as established, 1st , the passage of these corpuscles 'de toute pièce' through the capillaries; 2ndly, the restorative power in the blood, which immediately closes the aperture thus formed ... In endeavouring to account for the fact of the passage of the corpuscles through the vessels we find considerable difficulties. It cannot be referred to the influence of vitality, as it is observed likewise to take place after death. It may be surmised, either that the corpuscle, after remaining a certain time in contact with the vessel, gives off by exudation from within itself some substance possessing a solvent power over the vessel, or that the solution of the vessel takes place in virtue of some of those molecular actions which arise from the contact of two bodies; actions which are now known as exerting such extensive influence in digestion, and are referred to what is termed the catalytic power'. Paraphrased from Waller.[2,3]

into the tissues, particularly the lymph nodes (where they become attached again). To do all this they must be able to switch their integrins off and on. If the integrins on leukocytes were permanently switched on, the cells could never leave the bone marrow. Once in the blood, the integrin function must be switched off. Leukocytes leave the circulation when they encounter an appropriate signal of inflammation or infection, or more generally in the continuous process of immune surveillance. For monocytes and neutrophils, departure from the circulation is a one-way journey to their destiny with death in the tissues. By contrast, lymphocytes move from the blood into the tissues, then through the lymphatic system and then back into blood. We focus here on the regulation of leukocyte adhesion and extravasation under inflammatory conditions (**Figure 16.2**).

The many abbreviations used in this chapter are collected together at the end of the chapter.

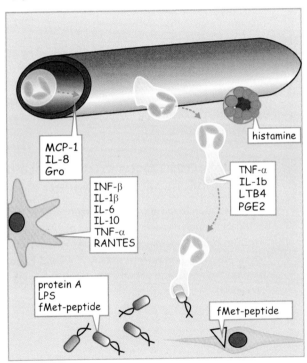

FIG 16.2 Generation of inflammatory mediators at sites of infection.

Tissue damage and bacterial infection induce the release of inflammatory mediators from various cell types. Bacterial protein A binds and activates the TNF-α receptor. Bacterial lipopolysaccharide (LPS), which acts on local fibroblasts, dendritic cells, vascular endothelial cells, mast cells and resident leukocytes induces release of Gro, histamine, IFN-β, IL-1β, IL-6, IL-8, MCP-1, RANTES, and TNF-α. Collectively, these mediators are responsible for the up-regulation of adhesion molecules on vascular endothelial cells and the activation of integrins on leukocytes. This results in their extravasation followed by migration to sites of damage/infection. Formylmethionyl peptides released from bacteria (as well as from the damaged mitochondria of host cells) facilitate the migratory response (chemotaxis). These processes allow the rapid accumulation of leukocytes, in particular neutrophils, at sites of infection as a first line of defence. It also initiates the process of tissue remodelling and repair (wound healing).

In bacteria, formylmethionine (fMet) is the initiating amino acid in protein synthesis; in eukaryotes, it is methionine. Synthetic peptides containing the sequence fMetXY (where X and Y are hydrophobic amino acids) act as powerful stimulants for neutrophil activation; among the most potent is fMetLeuPhe.[7] fMet peptides isolated from bacterial cultures also activate neutrophils.[8] Collectively, these have been called chemotactic peptides, although they act to stimulate all the multiple functions of these cells such as the respiratory burst and secretion of lysosomal enzymes in addition to chemotaxis.

At sites of inflammation (caused by injury or infection) mediators are released that affect the expression and affinity of adhesion molecules and affect the release of chemokines from endothelial cells. LPS, acting through its receptor TLR4 (Toll-like receptor: see Chapter 15) induces the expression of inflammatory mediators including the cytokines and chemokines. Other bacterial and cellular components (listed in Figure 16.2 and Table 16.1), can be added to the list of 'patterns' that provoke the release of inflammatory mediators. One of these is protein A, a staphylococcal surface protein that binds to the TNF-α receptor. Another is the class of formylmethionyl peptides, generated as the initiation sequences of prokaryotic proteins that are released by microbes. Histamine, released from mast cells and basophils, plays an important role in locally widening the vasculature, therefore allowing accumulation of blood and thus the accumulation of leukocytes at the site of inflammation.

In the following sections we describe the signalling pathway that emanates from the TNF-α receptor (TNFR1) in vascular endothelial cells. This results in elevated cell surface expression of adhesion molecules and the release of yet more inflammatory mediators. It also plays a crucial role in regulating the extravasation of circulating leukocytes (also known as transendothelial cell migration or diapedesis).

Tumour necrosis factor-α, potential anti-tumour agent or inflammatory cytokine?

The observation that the size of a tumour occasionally diminishes after bacterial infection has a long history. As early as 1848, Legrand noted two cases of long-standing scrofulous lymphoma which appeared to regress after infection with erysipelas.[13] In 1888, P. Bruns reported on the spontaneous regression of human tumours following infection with *S. erysipelas*.[14]

Protein A. Derived from *Staphylococcus aureus*, protein A binds with very high affinity to both the constant (Fcγ) and variable (Fab) regions of IgG and IgM antibodies.[9] Protein A contributes to the pathology of *Staphylococcus aureus* by directly binding to TNF-α receptors on the respiratory epithelium. This plays an important role in the onset of inflammation of the lower respiratory tract, leading to severe pneumonia associated with tissue damage and sometimes sepsis.[10] Protein A can also bind some membrane-associated B cell antigen receptors and promote antibody production.[11] This plays a role in the defence against the pathogens but is also associated with autoimmune disorders such as rheumatoid arthritis.

'Die öfters beobachtete Thatsache, dass Neubildungen, namentlich malingner Natur, durch ein interkurrentes Erysipel zur Verkleinerung oder zum Verschwinden gebracht werden, ist von hervorragendem theoretischem und praktischem Interesse. In ersterer Hinsicht verdient diese Thatsache gerade gegenwärtig besondere Beachtung, wo die Untersuchung der Aetiologie der malignen Neubildungen auf der Tagesordnung steht, deren Ergebnisse vielleicht geeignet sind, auch über die Art und Weise der salutären Wirkung des Erysipels Licht zu verbreiten. In praktischer Beziehung haben jene Beobachtungen zu Versuchen mit der kunstlichen Erzeugung eines kurativen Erysipels Veranlassung gegeben, um inoperable bösartige Neubildungen zur Heilung zu bringen.

Allein die Zulässichkeit.solcher Versuche ist noch eine offene Frage …':

'The frequently observable fact that new formations, particularly those of a malign nature may be caused to regress or to disappear by a concurrent erysipelas is of very great interest both in theory and practice. In respect of the former this fact merits particular attention, especially at the present time, given the current preoccupation with the study of the aetiology of new formations, the results of which may well be capable of shedding new light upon the beneficial effects of erysipelas. On the practical front these observations have given rise to experiments in which curative erysipelas is artificially induced in order to bring about the healing of inoperable new formations. However the admissibility of such experiments remains an open question …'

TABLE 16.1 Sources of inflammatory mediators

Source	Examples
Release from invading microorganisms	Lipopolysaccharide (LPS or endotoxin), shed from the surface. Also N-formylmethionyl peptides (fMet-peptides) derived from the initiator sequence of bacterial protein synthesis and protein A, a cell surface protein of *Staphylococcus aureus* that binds to and activates the TNF receptor-1
Release from resident cells	TNF-α and IL-1β from activated macrophages or fibroblasts, histamine from mast cells and chemokines (MCP-1, IL-8, Gro) from endothelial cells or resident leukocytes
Release from the mitochondria of damaged cells[12]	Peptides derived from the N-terminus of proteins (such as NAD^+ dehydrogenase) coded by the mitochondrial genome. These share with eubacteria the characteristic N-formylmethionyl initiator sequence
Components of complement and of fibrinolysis	C5a and thrombin respectively

The principle of bacterial infection as a component in the armory of anti-tumour medicine persists today. Lovaxin C and other *Listeria*-based vaccines currently under development, at the stage of phase I/II clinical trials, target cervical, breast, ovarian, and lung cancers. In animal models of breast cancer, the Lovaxin B vaccine, composed of *Listeria* with the Her2/neu antigen, appears to stop tumour growth.[28]

W. B. Coley[15] reported several cases of tumour regression and even disappearance following repeated inoculation of erysipelas. Speculating on possible mechanisms, he considered

> 1) that the erysipelas coccus has a direct destructive action upon the cell elements of the new growth; 2) that the high temperature (induced by the infection) alone is sufficient; 3) that sarcoma and carcinoma are both of bacterial origin and the erysipelas germ has a direct antagonistic effect upon the cancer bacillus … it seems in the light of present knowledge not improbable that all of the three theories may contain an element of truth, and that a larger theory combining all these elements is necessary to explain the curative action of erysipelas.

In the firm belief that the cure of cancer was at hand, Coley and others applied cell-free filtrates of streptococci as 'Coley's toxins' over a period of about 40 years, apparently with some success.[15–17] The effect of these bacterial products is associated with severe haemorrhagic reaction mainly confined to the core of the tumour which rapidly sloughs off. However, it generally leaves a ring of viable tissue which, unfortunately, continues to grow. Not surprisingly, such treatments also tend to cause widely disseminated systemic effects and all too frequently, these can lead to circulatory collapse and death. Early attempts to separate the haemorrhagic from the cytotoxic components in filtrates of cultured *Serratia marcescens* led to the isolation of LPS, lipopolysaccharide (or endotoxin, see page 457).[18,19] Among its many biological activities, this could elicit haemorrhagic necrosis of both experimental and primary subcutaneous tumours.[20]

All this provoked the hunt for immune mediators, cytotoxic for transformed cells but not causing septic shock.[25] The first of these, the tumour necrosis factors TNF-α and TNF-β, attracted immense interest.[26,27] However, TNF was also found to possess general pro-inflammatory properties and to provoke severe systemic toxicity. Thus the focus of interest shifted from tumour necrosis to mechanisms of inflammation.

TNF-α as target for anti-inflammatory drugs. An excess of TNF-α is associated with chronic inflammatory disorders. The best studied in this respect are Crohn's disease (destructive inflammation of the bowel) and rheumatoid arthritis (destructive inflammation of the synovial joints). TNF-α is the key cytokine regulating expression of all other inflammatory cytokines. It was concluded that by blocking its action it would be possible to reduce the general state of inflammation.[21] Two approaches to lower the TNF-α concentration have been exploited successfully. One uses a humanized antibody against TNF-α (infliximab), the other an extracellular fragment of the TNFR2 coupled to IgG (etanercept). Both these treatments provide considerable relief.[22,23] More recently, the pre-ligand assembly domains of the TNFR1 and -R2 have been targeted by a soluble cysteine-rich region to prevent homophilic trimerization of the receptors. This too reduces the symptoms of arthritis in animal models.[24]

The family of TNF proteins and receptors

TNF-α and -β are members of a large family of cytokines (**Figure 16.3**).[29] They are initially generated in precursor form, as transmembrane proteins arranged as stable homotrimers. From these, soluble homotrimeric cytokines are released by the action membrane bound metalloproteases (so-called furins). With one or two exceptions, these ligands bind their receptors both in a membrane bound and soluble state. One example of these exceptions is FASL (CD95L). As a membrane protein this causes apoptosis of cells bearing the FAS (CD95), whereas when soluble, it prevents apoptosis.[30] Another example is TNF-α, which only effectively activates the TNFR2 in its membrane form (memTNF-α) but not in its soluble form (sTNF-α).[31]

The TNF ligands signal through 29 different receptors, members of the TNFR family (Figure 16.3). These receptors can be divided into two groups: those having a death domain (DD), e.g. the ubiquitously expressed TNFR1, and those without, e.g. TNFR2, which is typically found in cells of the immune system and whose expression level is context dependent. Members of the first group interact with DD-possessing proteins and couple to a caspase-activating pathway leading to cell death. The second group has the capacity to interact with the effector TRAF. The possession of a DD, however, does not restrict the receptor from signalling to pathways that have little to do with cell death. In fact, one of these pathways mediates activation of NF-κB and constitutes an important cell survival signal.[32] Conversely, death can be induced by receptors of the TNFR2 family that do not contain death domains.

Although some TNF-receptor signalling components have macabre names, TNF-α should be considered an inflammatory cytokine with an optional ability to cause cell death.[33] Generally, TNFR1 is the key mediator of TNF-α signalling.

TNF-α and regulation of adhesion molecule expression in endothelial cells

Receptor activation
In the absence of ligand, the N-terminal cysteine-rich domains (PLAD) of TNFR-1 and -2 interact with each other, maintaining the silent state (**Figures 16.4 and 16.5**). Binding to the trimeric ligand TNF-α either induces an activating conformational change or allows the formation of higher-order complexes that bestow signal competence.[33] Non-engaged receptors interact with a signal silencing protein, SODD (silencer of DD). When over-expressed, this blocks the NF-κB pathway as described below.[34]

In endothelial cells, binding of TNF-α causes enhanced expression of VCAM-1 and the ICAMs.[36] It causes release of chemokines (IL-8, MCP-3, MIP-1a) and cytokines (IL-1β, IL-6, TNF-α, GM-CSF) and the expression of the enzymes

TNF-β was initially named lymphotoxin-α.

Furin is also exploited by a number of pathogenic microorganisms. Thus, the envelope proteins of viruses such as influenza, HIV, and dengue fever must be processed by furin in order to become functional. Likewise, the toxins from anthrax and *Pseudomonas* must be cleaved in order to become membrane permeant. Compounds capable of acting as inhibitors of furin are currently being considered as therapeutic agents for the treatment of anthrax infection.

FIG 16.3 Domain architecture of the ligands and receptors of the TNF superfamily.

All ligands, except TNF-β and VEGI, are transmembrane proteins and have local effects. Under pathological conditions they can be liberated by a variety of proteases, some of which are indicated: furin, operating in the Golgi, and ADAM matrilysin at the plasma membrane. The C-terminal segments, carrying the TNF homology domain, have 20–30% sequence identity and bind the receptor. The receptors are transmembrane proteins, except for DCRs 1 and 3 and OPG (which possess inositol-lipid anchors). NGFR is a low affinity nerve growth factor receptor that is structurally related to the TNF receptors but has quite distinct actions. Figure adapted from Aggarwal.[29]

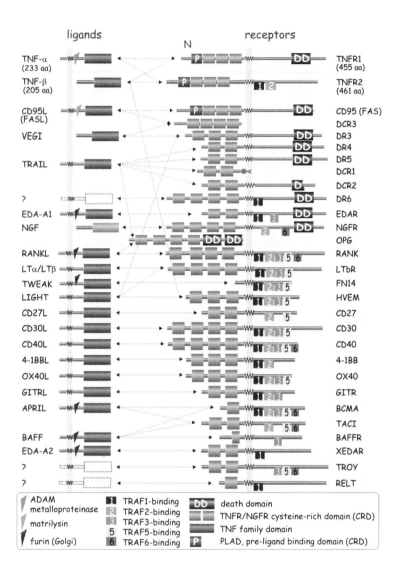

NF-κB: nuclear factor κB, a protein complex that attaches to the immunoglobulin κ light-chain gene.[38]

iNOS, cyclo-oxygenase-2, and cytosolic PLA2. As a result of all this, the adherence of leukocytes is greatly enhanced.[37]

All these effects of TNF-α and of inflammatory stimuli are mediated through the transcription factors NF-κB (Figure 16.6) and AP-1(ATF2, C-Jun).

Signalling downstream of TNFR1

Activation of the TNF1R (receptor for TNF-α) causes SODD to dissociate and initiates the formation of two complexes on the basis of homophilic

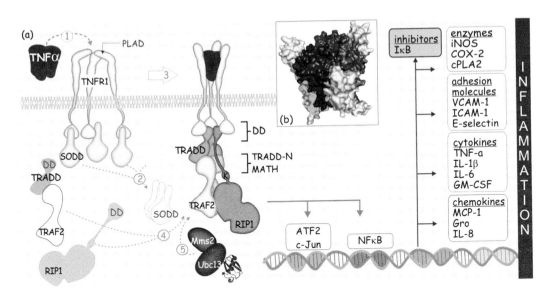

FIG 16.4 NFR activation.

(a) In the absence of ligand, TNF-α receptors are associated through their N-terminal pre-ligand assembly domain (PLAD). The intracellular segment is bound to SODD. PLAD and SODD control the silencing of the unoccupied receptor. Binding of ligand induces trimerization of the receptors and this reveals death domain docking sites. TRADD and RIP1 associate with the receptors. TRAF2 is recruited to the protein complex through a TRADD N-terminal interaction with the MATH domain of TRAF2. (b) Structure DR5 (member of the TNF receptor family) bound to TRAIL (member of the TNF family). Three receptor molecules bind a trimeric ligand (1du3[35]).

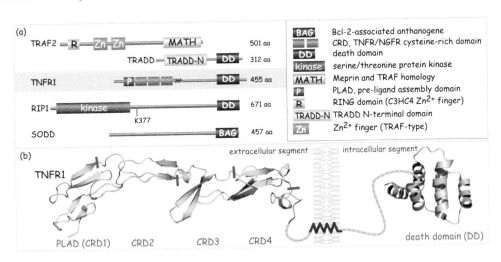

FIG 16.5 Domain architecture of proteins that associate with TNFR1.

(a) The death domain of TNFR1 interacts with the death domain of RIP1 and TRADD. TRADD in turn binds with its N-terminal domain to the MATH domain of TRAF2. SODD binds to inactive receptors. K377 indicates the ubiquitylation site of RIP1. (b) Molecular structure of TNFR1. The four cysteine-rich domains are indicated by green bars. CRD1 acts as a pre-ligand assembly domain. CRD2 and CDR3 are involved in TNF-α binding.

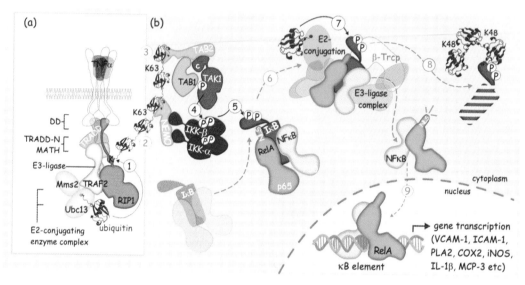

FIG 16.6 TNFR1-mediated activation of NF-κB.

(a) TRAF2, associated with MMS2 and Ubc13, causes polyubiquitylaion of RIP1 (at K377) in a K63-fashion (1). (b) The polyubiquitin K63 chain acts as a docking site for the adaptor proteins NEMO (2) and TAB2 (3), each bringing along specific proteins kinases. NEMO recruits IKK-α and -β whereas TAB2 associates with TAK1 and its regulatory subunit TAB1. A series of phosphorylations ensues: autophosphorylation of TAK1 and phosphorylation of IKK-α and -β (4) by TAK1. Activated IKK-β causes the phosphorylation of IκBα (5) and this signals its recognition by the E3-ligase complex (through the receptor β-Trcp) (6). Polyubiquitylated IκBα (7) is degraded by the proteasome (8). This unmasks the nuclear localization signal of the NF-κB transcription complex (comprising RelA and NFkB) which now migrates into the nucleus (9), there to bind to its DNA-response element (κB). Numerous genes are transcribed.

DD interactions. The DD was first identified as a stretch of 80 amino acids in FAS and TNFR1 necessary for the induction of programmed cell death (apoptosis).[39] Through the DD domain the TNFR1 recruits:

In the Prosite database, the TRAF domains are referred to as MATH domains because of their resemblance to Meprin (meprin and traf homology).

- TRADD, an adaptor that is complexed with TRAF2.[40] This multifunctional protein comprises two interaction domains with which, besides TRADD, it binds to non-DD-containing TNF receptors (such as TNFR2) and to downstream signalling components such as RIP or the MAP3kinases, ASK1 and MEKK1. It is also contains a RING zinc finger and multiple TRAF zinc finger motifs, so qualifying it as an E3-ubiquitin ligase (see page 467). It complexes with the E2-conjugating enzyme complex Ubc13/MMS2 responsible for a K63-type polyubiquitylation reaction (see **Figure 16.6**, also page 468).
- RIP1. This has a DD at the C-terminus, an intermediate region which contains an important ubiquitylation site (K377), and an N-terminal serine/threonine kinase domain that has no apparent function in TNF-α signalling.[41] Binding of TRADD and RIP1 is mutually exclusive so that they associate with different receptors (Figures 16.4 and 16.5).

Signalling via NF-κB

The assembly of the two complexes at the TNF1R results in polyubiquitylation of RIP1 by TRAF2 (**Figure 16.4**). The polyubiquitin chains (K63 type) thus formed serve as docking sites for the TAB2/TAB3/TAK1 complex and for NEMO/IKKα/IKKβ. From here on the series of events is similar to that described for the activation of NF-κB by TLR4 in response to LPS (page 460). TAK1 phosphorylates IKKβ causing its activation and subsequent phosphorylation of IκBα, which after ubiquitylation is degraded by the proteasome. The dissociation of IκB reveals a targeting signal on the NF-κB complex that directs it to the nucleus.[42] Here it promotes transcription of the genes for VCAM-1, ICAM, etc, (see also figure 15.3, page 456).

TNF-α (but also histamine, IL-1β, LPS, thrombin, etc.), induces a short-lived up-regulation of VCAM-1 and ICAM-1 and also of E- and P-selectins on vascular endothelial cells. P-selectin, stored in specialized secretory granules, is rapidly transferred to the plasma membrane. Expression of E-selectin, ICAM-1, and VCAM-1 occurs more slowly and then persists for a few days. The increased density of these adhesion molecules on the surface of endothelial cells enhances their 'avidity' (the product, density × affinity) for leukocytes.

The pathway triggered by TNF-α is self-limiting because NF-κB also enhances the expression of IκB which initiates a negative feedback loop that attenuates the TNF signal (Figure 16.4). This secures the return to the basal expression of VCAM-1 and ICAM-1. Their appearance at the cell surface is therefore transient, allowing termination of the inflammatory response. Without this, the influx of leukocytes would be prolonged, causing chronic tissue damage, as occurs in the formation of ulcers.

Control of inflammation by apoptotic neutrophils and T cells. The final resolution of the inflammatory response not only requires that transcription of inflammatory mediators ceases: they must also be eliminated from the tissues. Massive apoptosis of neutrophils and T cells is a hallmark of the termination of inflammation. Whilst dying, and before being eliminated by macrophages or other tissue cells, these leukocytes express high levels of the chemokine receptor CCR5. This also acts as a sink that sequesters the chemokines CCL3, -4, and 5. Mice lacking CCR5 manifest a prolonged presence of inflammatory mediators in their tissues.[45]

Signalling via p38 and JNK

TRAF2 and RIP serve as docking platforms for a number of MAP3Kinases (NIK, MEKK1, ASK1 for RAF2 and MEKK3, TAK1 for RIP) all capable of activating the stress kinases p38 and JNK. In turn these phosphorylate transcription factors, thereby activating c-Jun and ATF2 (AP-1 complex).

TNF-α and cachexia.
Prolonged exposure to low concentrations of TNF-α causes the wasting syndrome cachexia (from the Greek *kakos*, bad, and *hexis*, constitution). Indeed, cachectin, the mediator responsible for cachexia in mice exposed to trypanosome infection, has been identified as TNF-α.[43] Cancer patients suffer from cachexia, manifested as anorexia, early satiety, muscle wasting, weight loss, fatigue, and impaired immune response. Other inflammatory mediators that are involved in cachexia are IL-1, IL-6, and IFN-γ. These are either secreted by the tumour alone or in concert with host-derived factors.[44] They exert their effect directly on muscle metabolism, causing wasting of tissue, and indirectly by modifying neurotransmitter release in the hypothalamus (changing appetite, taste, and social behaviour).

Chemokines became a focus of attention when it was shown that the envelope glycoprotein of human immunodeficiency virus (HIV) binds competitively at chemokine receptors. Together with CD4, these receptors are the port of entry of HIV-1 into T lymphocytes. Occupation of chemokine receptors, for instance of CXCR-4, prevents the penetration of cells by the HIV-1.[47] In some rare individuals, due to a homozygous mutation, the cytokine receptor CCR-5 is not exposed at the cell surface. The virus cannot gain entry and these fortunate people appear to be immune to infection by HIV.[48] The mutation does not appear to induce any deleterious phenotype which indicates that there is some redundancy among these receptors. Essential functions are well covered in the absence of CCR-5.

Chemokines and activation of integrins on leukocytes

A family of chemokines

Though technically cytokines, the chemokines form a family of small, structurally related proteins (8–14 kDa) originally identified and characterized as chemotactic agents. Leukocytes, when attached to substrate, are capable of sensing concentration differences of chemokines as small as 1% over a distance of 40 μm, approximating the linear dimensions of a cell. They respond by moving in the direction of the source, in the process of *chemotaxis*.[46]

With respect to transendothelial migration, it is not chemotaxis that directs the movement, because the cells are floating free and buffeted by the blood. In this context, chemokines 'attract' leukocytes by stimulating their attachment to and subsequent migration across the endothelial barrier. It is believed that the chemokines, generated by inflamed tissues, are presented on to the surface of the endothelial cell layer. This has been demonstrated for IL-8.[49,50] The detection of the chemokines activates adhesion molecules, members of the β1 and β2-integrins on the leukocytes, initiating a series of events that leads to their arrest, flattening and transendothelial migration (see Figure 16.13). By attracting leukocytes to sites of infection/inflammation, chemokines mediate an essential first step in host defence, tissue repair, and healing.

About 40 human chemokines have been identified (Table 16.2). They mainly act on neutrophils, monocytes, lymphocytes, eosinophils, endothelial cells, and their (stem cell) precursors.[52,53] The chemokines are small proteins having four conserved cysteines that form two essential intramolecular disulfide bonds (Cys1 with Cys3, and Cys2 with Cys4: Figure 16.7). They are classified as CC, CXC, and CX3C chemokines, according to the spacing of the first two cysteines in the N-terminal segment. As an exception to the rule, two members of this family only have one cysteine.[51]

The chemokine receptors are coupled to G proteins

Of the 17 human receptors that have been characterized, most recognize more than one chemokine (Table 16.2). They are linked to both pertussis-toxin sensitive (G_i/G_o) and insensitive G proteins (G_{12}/G_{13}). It is predominantly the Gβγ subunits that carry the signal into cell,[55,56] but in the case of the bacterial chemokine, fMLP, activation also occurs through $G\alpha_{12}$ and $G\alpha_{13}$.[57] The chemokines elicit a number of signal transduction events having different biological effects, which are illustrated in Figure 16.8.[58]

Activation of integrins

Although the activation of the β1- and β2-integrins can be elicited by over-expression of different PKC isoforms or by phorbol esters[59,60] there is the

TABLE 16.2 Chemokines

Systemic name	Human ligand	Mouse ligand	Receptor
C chemokine/receptor family			
XCL1	Lymphotactin/SC-1α	lymphotactin	XCR1
XCL2	SCM-β	unknown	XCR1
CC chemokine/receptor family			
CCL1	I-309	TCA-3, P500	CCR8
CCL2	MCP-1/MCAF	JE?	CCR2
CCL3	MIP-1α/LD78a	MIP-1α	CCR1, CCR5
CCL4	MIP-1β	MIP-1β	CCR5
CCL5	RANTES	RANTES	CCR1, CCR3, CCR5
(CCL6)	Unknown	C10, MRP-1	Unknown
CCL7	MCP-3	MARC?	CCR1, CCR2, CCR3
CCL8	MCP-2	MCP-2	CCR3
(CCL9/10)	Unknown	MRP-2, CCF18/MIP-1γ	Unknown
CCL11	Eotaxin	Eotaxin	CCR3
(CCL12)	unknown	MCP-5	CCR2
CCL13	MCP-4	Unknown	CCR2, CCR3
CCL14	HCC-1	Unknown	CCR1
CCL15	HCC-2/Lkn-1/MIP-1δ	Unknown	CCR1, CCR3
CCL16	HCC-4/LEC	LCC-1	CCR1
CCL17	TARC	TARC	CCR4
CCL18	DC-CK1/PARC AMAC-1	Unknown	Unknown
CCL19	MIP-3β/ELC/exodus-3	MIP-3β/ELC/exodus-3	CCR7
CCL20	MIP-3α/SLC/exodus-1	MIP-3α/LARC/exodus-1	CCR6
CCL21	6Ckine/SLC/exodus-2	6Ckine/SLC/exodus-2	CCR7
CCL22	MDC/STCP-1	ABCD-1	CCR4
CCL23	MPOF-1	Unknown	CCR1

Continued

TABLE 16.2 Continued

Systemic name	Human ligand	Mouse ligand	Receptor
CCL24	MPIF-2/Eotaxin-2	Unknown	CCR3
CCL25	TECK	TECK	CCR9
CCL26	Eotaxin-3	unknown	CCR3
CCL27	CTACK/ILC	ALP/CTACK/ILC ESkine	CCR10
CXC chemokine/receptor family			
CXCL1	GROα/MGSA-α	GRO/KC?	CXCR2 > CXCR1
CXCL2	GROβ/MGSA-β	GRO/KC?	CXCR2
CXCL3	GROg/MGSA-γ	GRO/KC?	CXCR2
CXCL4	PF4	PF4	unknown
CXCL5	ENA-78	LIX?	CXCR2
CXCL6	GCP-2	CKa-3	CXCR1, CXCR2
CXCL7	NAP-2	Unknown	CXCR2
CXCL8	IL-8	Unknown	CXCR1, CXCR2
CXCL9	MIg	Mig	CXCR3
CXCL10	IP-10	IP-10	CXCR3
CXCL11	I-TAC	Unknown	CXCR3
CXCL12	SDF-1αβ	SDF-1	CXCR4
CXCL13	BLC/BCA-1	BLC/BCA-1	CXCR5
CXCL14	BRAK/bolekine	BRAK	Unknown
CXCL15	unknown	Lungkine	Unknown
CX3C chemokine/receptor family			
CX3CL1	fractalkine	neurotactin	CX3CR1

Adapted from Zlotnik and Yoshie.[51]

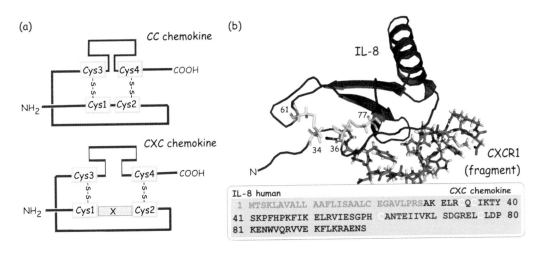

FIG 16.7 Chemokine structure.
(a) Representation of intramolecular disulfide bonding in CC and CXC cytokines. (b) Molecular structure of IL-8 bound to a short fragment of the CXCR1 receptor (1ilp[54]).

	enzyme	products	biological effect	
	phospholipase A2	-lysophospholipid -free fatty acid (arachidonate)	leukotrienes ↑ [lipoxygenase] ↑ arachidonate ↓ [cyclo-oxygenase] ↓ prostaglandins	LTB4 - chemotactic agent PGE2 - relax vasc smooth muscle (redness) - permeability endothelium (swelling) - pain
	phospholipase Cβ	-IP₃ -diacylglycerol	-elevation of intracellular free Ca²⁺ - exocytosis -activation of integrins via CalDAG-GEF1/Rap1	
	phospholipase D	-phosphatidic acid -choline	-activation of the respiratory burst (NADPH oxidase)	
	PI 3-kinase	-PI(3,4,5)P₃ -PI(3,4)P₂	-regulation of actin cytoskeleton-cell spreading and migration	

FIG 16.8 Signals emanating from chemokine receptors and some of their biological effects in leukocytes.
Chemokine receptors activate PI 3-kinase and a number of phospholipases. The products of phospholipase A2 are released as messengers that play an important role in the manifestation of inflammation (redness, swelling, and pain). Phospholipase Cβ induces production of IP₃ and diacylglycerol which, among many possible actions, cause activation of integrins and liberation of granules. Phospholipase D produces phosphatidic acid which activates the respiratory burst (production of reactive oxygen species). Finally, PI 3-kinase produces 3-phosphorylated phosphoinositides which play a role in cell spreading and cell migration.

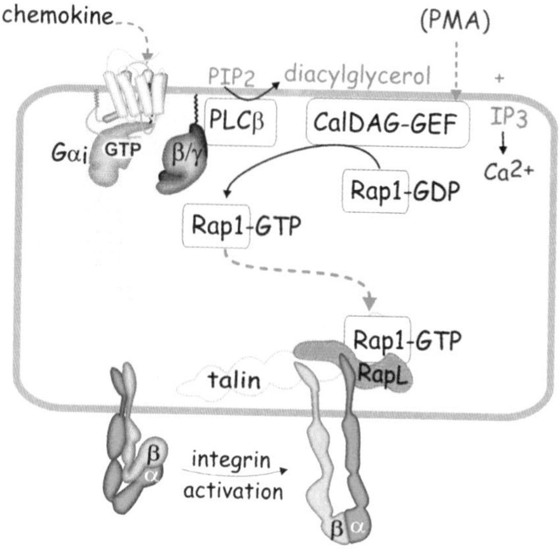

FIG 16.9 Chemokine-mediated activation of integrins in leukocytes. Binding of the chemokines causes activation of G_i. The β-γ subunits activate PLCβ giving rise to diacylglycerol and IP_3, both of which are involved in the activation of CalDAG-GEF, which activates Rap1. This binds its effector RapL responsible for activation of integrins (for instance αLβ2 (LFA1)).

CalDAG-GEF, also known as RapGRP, is a Ca^{2+} and diacylglycerol activated guanine nucleotide exchange factor. It contains a C1 domain, first identified as a conserved region in members of the PKC family, through which it binds diacylglycerol. It is worth noting that there are numerous proteins containing C1 domains and so potential targets of the action of diacylglycerol or phorbol esters (see page 234). It also contains an EF-hand motif, which confers Ca^{2+} sensitivity (see page 779).

Depending on the type of leukocyte, different integrins participate in the firm arrest to endothelium. For instance, in monocytes αMβ2 is predominantly involved in arrest, whereas in neutrophils this role is taken by αLβ2. Lymphocytes rely mainly on α4β1 (see Figure 13.6, page 384).

problem that chemokine-induced integrin activation is not prevented when PKC is inhibited.[61] The stimulatory effect of phorbol ester can also be explained through its direct effect on the exchange factor CalDAG-GEF causing activation of the GTPase Rap1b[62] (**Figure 16.9**). Rap1,[63] although initially considered as an antagonist of Ras[64,65] (see page 251), gained further physiological significance with the realization that it activates integrins in a number of blood-borne cells, including platelets.[66–69] Although the pathway has yet to be fully elucidated, activated Rap1 interacts either with RapL or Riam[70] and this causes both activation and clustering of integrins, thereby augmenting their affinity and avidity for ligands.[71,72] Riam interacts with the cytoskeletal protein talin, and this in turn binds integrins and thus modifies their configuration (see also Figure 13.14, page 398).[73,74] Integrin activation causes arrest of the leukocytes on endothelial cells that possess the ligands VCAM and ICAM-1, both of which are highly expressed due to the presence of TNF-α (and other inflammatory cytokines). In this way, floating leukocytes take on the characteristics of adherent cells (**Figure 16.13**).

Transendothelial migration

The leukocytes flatten, then migrate to the endothelial cellular junctions in a complex process involving reorganization of the cytoskeleton. β-γ subunits derived from G_i interact with the p110-γ subunit of class-I PI 3-kinase inducing the generation of $PI(3,4,5)P_3$ (see page 546). This engages GEFs that activate both Rac and Cdc42 (Rho family of GTPases)[75] (**Figure 16.11**). In particular, Rac1 is a key player in the initiation of a branched network of actin filaments and in orchestrating the formation of lamellipodia (protrusions) which cause

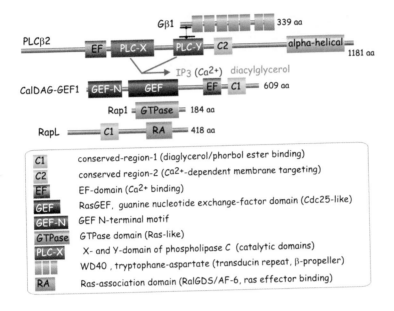

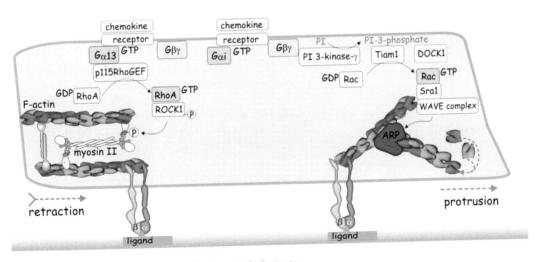

FIG 16.10 Domain architecture of components of the integrin activation pathway in leukocytes. The N-terminal segment of the Gβ1 subunit binds PLCβ2, leading to production of diacylglycerol and IP$_3$. IP$_3$ liberates Ca^{2+} from intracellular stores. Together, Ca^{2+} and diacylglycerol recruit CalDAG-GEF to the membrane and cause its activation, leading to GTP-loading of Rap1. Rap1-GTP binds to the Ras-association domain of RapL and this plays a role in the activation process of integrins.

FIG 16.11 Members of the Rho family are instrumental in the control of cell migration.
Chemokine receptors are linked to heterotrimeric GTP-binding proteins. With respect to the formation of protrusions (at the leading edge), it is the β-γ-subunits that are of importance. These bind PI 3-kinase-γ resulting in the formation of PI(3)P. This attracts and activates the Rac1-guanine nucleotide exchange factor Tiam. GTP-loaded Rac1 binds Sra1, a component of the WAVE2 complex, and this interacts with ARP2/3. This binds first to existing actin filaments, followed by the nucleation of new filaments. The elongating actin filaments push the membrane forward. With respect to retraction at the rear of the cell, chemokine receptors mediate their effect through Gα$_{13}$ which binds p115RhoGEF. GTP-loaded RhoA interacts with the serine/threonine kinase ROCK1 which, in turn, phosphorylates MLC. This enables ATP hydrolysis at the head of myosin II, resulting in movement and sliding of the actin filaments (contracting stress fibres).

cell spreading.[76] Rac1 is activated by a number of GEFs of which two, Tiam1 and DOCK1, qualify in the context of leukocyte migration. Activated Rac1 binds to Sra1, part of the WAVE2 complex[77] that is enabled to bind to ARP2/3. This complex of actin-like proteins binds to existing actin filaments and catalyses the further polymerization of actin, thus initiating new filaments that press the membrane forward.[76]

Migration within the tissue

Having traversed the endothelial intercellular junction and then the basal membrane, leukocytes search for sites of infection, migrating up the gradient of chemokines (such as formylmethionyl peptides) released by bacteria. This movement is characterized by the assembly of actin fibres into protrusions at the leading edge of the cell, and by formation of contractile actin–myosin complexes at the rear and along the sides. Both these processes are initiated following receptor activation of the heterotrimeric G proteins, G_i controlling forwardness and $G_{12/13}$ controlling backwardness.[84]

Directionality is not necessary for leukocyte binding to endothelial cells or for transendothelial cell migration. What matters is that the activated integrins are guided to the intercellular junctions. This occurs through binding of all three types of integrin involved in the arrest of leukocytes to the adhesion molecules (JAMs) that constitute the tight junctions. LFA1 ($\alpha L\beta 2$) shifts its attention from ICAM-1 towards JAM-A,[80] VLA-4 ($\alpha 4\beta 1$) shifts its attention from VCAM1 towards JAM-B,[81] and Mac-1 ($\alpha M\beta 2$) shifts its attention from ICAM1 or -2 towards JAM-C.[82,83] Integrin binding occurs at the membrane-proximal immunoglobin domain of the JAMs. This is thought to guide the leukocyte across the intercelluar junctions. It may also destabilize the numerous cadherin junctions that form the zonula adherens. Blocking of JAMs prevents leukocyte recruitment in cerebrospinal fluid.

As already related, $\beta\gamma$ subunits derived from G_i induce formation of $PI(3,4,5)P_3$. This occurs at the point of the cell closest to the source of the chemokine, in this way positioning the leading edge (**Figures 16.12 and 16.13**). The GTPase Rac1 focuses membrane protrusions uniquely at the leading edge, while the role of Cdc42 is to ensure that there is no formation of secondary protrusions elsewhere.[57] Retraction of the cell body at the rear is mediated through activation of the α-subunits of G_{12} or G_{13}.[85] This occurs through the formation of actin stress fibres formed by antiparallel filaments of actin intercalated with myosin II (see Figure 16.11). Activation of the myosin II occurs through phosphorylation of the myosin light chain (MLC) in a pathway involving p115RhoGEF, RhoA and ROCK.

In the tissue, the activated integrins augment neutrophil responses such as phagocytosis, exocytosis, and generation of reactive oxygen metabolites, necessary for the killing and clearing of microorganisms. They also determine the lifespan of the leukocyte in the tissues.[86]

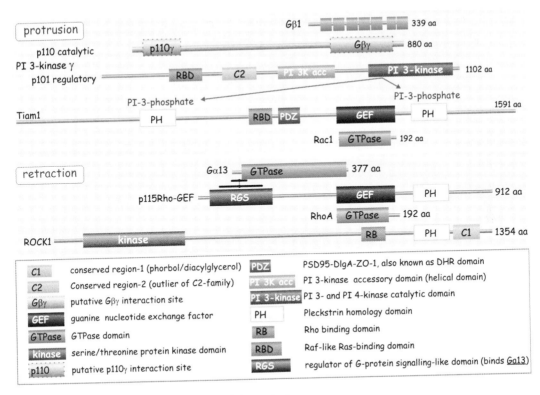

FIG 16.12 Domain architecture of proteins involved in signalling leukocyte migration.

Protrusion: β~γ subunits (derived from G$_i$) interact with p101 causing activation of the catalytic p110~γ. Tiam1 binds to PI(3)P through its PH domains. Rac1 interacts with the GEF domain. *Retraction:* The RGS domain of p115RhoGEF binds to the N-terminal of Gα$_{13}$. RhoA is loaded with GTP and interacts with the Rho-binding domain of its effector ROCK1.

The three-step process of leukocyte adhesion to endothelial cells

All these processes form part of a more general sequence of events that is now referred to as the three-step process of leukocyte adhesion to endothelial cells. Leukocytes make contact with the vascular endothelium from time to time, by the presentation of surface ligands (L-selectin, Sialyl Lewis-X, or PSGL-1: see page 392) to receptors exposed on the endothelial cells (CD34, E- or P-selectin). These interactions tether the cells in a manner by which they can attach and then roll on the surfaces of the endothelial cells and finally detach (Figure 16.13a). Their close contact with the endothelial surface allows them to verify the presence of membrane-bound chemokines. Their integrins become

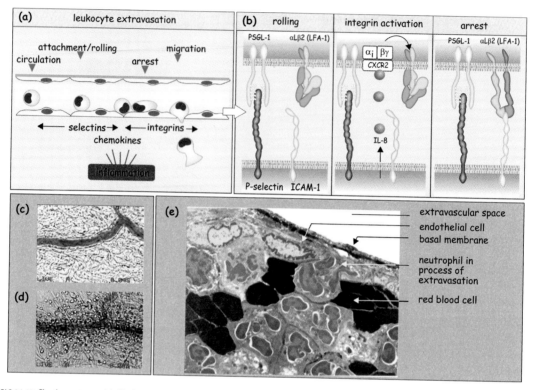

FIG 16.13 The three-step model of leukocyte transendothelial migration.
(a) Representation of the three-step model of leucocyte transendothelial migration through the wall of a venule. This comprises (1) selectin-mediated rolling of leukocytes on the vascular endothelial cell surface, then (2) chemokine-mediated activation of integrins on the leukocytes, resulting in (3) arrest and migration through the vascular endothelial cell layer. Most leukocytes are mildly attached and just roll along on the endothelial surface. Migration is triggered by firm arrest due to local inflammation as a consequence of the release of chemokines and enhanced expression of the selectins, VCAM-1 and ICAM. (b) A molecular representation of the same three steps, with mauve (at the top) representing the leukocyte and orange (at the bottom) representing the endothelial cell. *Left panel:* Initial contact between the two cells is made through P-selectin interacting with the PSGL-1. This allows the cells to roll. *Centre panel:* The close proximity to the endothelial cells allows detection of IL-8 by its receptor CXCR2. This causes activation of GTP-binding proteins that in turn activate the integrin αLβ2. *Right panel:* The activated integrin now binds ICAM-1 causing firm attachment of the leukocyte to the endothelial cell. This is a prerequisite for further flattening and then migration through the endothelial cell layer (see page 384 onwards). (c) Video-micrograph illustrating a rat mesenteric venule ($\sim$40 μm diameter). The blurring within the venule is due to the flow of blood. (d) Video-micrograph illustrating leukocyte accumulation into the mesenteric tissue after upstream administration of the chemokine LTB4. (e) Thin-section electron micrograph, showing a neutrophil leukocyte (multi-lobed nucleus) migrating through a layer of endothelial cells (blue) into rat mesenteric tissue. Images from Dr Sussan Nourshargh, London, UK.

activated so that they bind to ICAM-1 or VCAM-1 present at elevated levels due to the presence of inflammatory mediators so causing their arrest. They flatten and become competent to pass through the tight endothelial layer.[87,88] Not all bound cells transmigrate. Some merely perch on the endothelial cell surface and then detach. Those that do succeed, pass through the basal membrane and make it right through into the tissues.

List of Abbreviations

Abbreviation	Full name/description	SwissProt entry	Other names
ATF2	activating transcription factor-2	P15336	CRE binding protein (CRE-BP1)
CalDAG-GEF	calcium/diaglycerol-sensitive guanine nucleotide exchange factor	Q7LDG7	
Cdc42	Cell division cycle-42	P60953	
c-Jun	avian sarcoma virus 17, oncogene homologue (junana = 17 in Japanese)	P05412	
DcR2	decoy receptor-2	Q9UBN6	TRAIL-R4
DD	death domain		
DOCK1	dedicator of cytokinesis protein-1	Q14185	
FAS	FS-7 fibroblast cell surface antigen	P25445	CD95, Apo-1
FASL	FS-7 fibroblast cell surface antigen ligand	P48023	CD95L
Gα13	guanine nucleotide binding protein α-13 subunit	Q14344	
Gβ1	guanine nucleotide binding protein β-1	P62873	
ICAM-1	intercellular adhesion molecule-1	P05362	CD54
IKK-α	inhibitor of κB kinase-α	O14920	
IKK-β	inhibitor of κB kinase-β	Q14164	
IL-8	interleukin-8	P10145	CXCL8
NEMO	NF-κB essential modulator	Q9Y6K9	IKKγ
NF-κB1	nuclear factor kappa-B1	P19838	p105/p50 truncated
p115RhoGEF	RhoA guanine exchange factor protein of 115 kDa	Q92888	LSC, Lbc's second cousin
PI 3-kinase-γ p101	phosphatidylinositol-4,5-bisphosphate 3-kinase-γ regulatory subunit p101	Q8WYR1	

Continued

Abbreviation	Full name/description	SwissProt entry	Other names
PI 3-kinase-γ p110	phosphatidylinositol-4,5-bisphosphate 3-kinase-γ catalytic subunit p110	P48736	
PLCβ2	phospholipase Cβ2	Q00722	
PLAD	N-terminal pre-ligand assembly domain in TNF-α receptor		
P-selectin		P16109	GMP-140, CD62P
PSGL-1	P-selectin glycoprotein ligand-1	Q14242	CD162
Rac1	Ras-related C3-botulinum toxin substrate	P63000	
RANTES	regulated upon activation, normal T cell expressed and secreted	P13501	
Rap1b	Ras-like protein1-β	P621224	TC25
RapL	Ras-like protein ligand	Q8WWW0	
Rel-A	reticuloendotheliosis virus-like protein A	Q04206	NF-k p65 subunit
RhoA	Ras-homology protein	Q2LJ65	
RIP1	receptor interacting protein-1	Q13546	
ROCK1	Rho-associated coiled-coil containing protein kinase-1	Q13464	
SODD	silencer of DD	O95429	BAG-4
Sra-140	specifically Rac-1-associated protein of 140 kDa	Q14467	FMR-1/CYFIP1
TAB2	TAK1-binding-1	Q15750	MAP3K7IP1
TAB3	TAK1-binding-2	Q9NYJ8	MAP3K7IP2
TAK1	transforming growth factor-β-activated kinase-1	O43318	MAP3K7
Tiam1	T-lymphoma invasion and metastasis protein-1	Q13009	

Continued

Abbreviation	Full name/description	SwissProt entry	Other names
TNF-α	tumour necrosis factor-alpha	P01375	
TNF-β	tumour necrosis factor-beta	P01374	LT-α
TNFR1	tumour necrosis factor receptor-1	P19438	CD120a
TNFR2	tumour necrosis factor receptor-2	P20333	CD120b, TNFR1B
TRADD	TNF-receptor associated death domain protein	Q15628	
TRAF2	TNF-receptor associated factor-2	Q12933	
VCAM-1	vascular cellular adhesion molecule-1	P19320	CD106
WAVE	Wasp-verproline homology domain-containing protein	Q2558	

References

1. Addison W. Experimental and practical researches on the structure and function of blood corpuscles; on inflammation; and on the origin and nature of tubercles in the lungs. *Trans Provinc Med Surg Assoc*. 1843;11:233–305.
2. Waller A. Microscopic examination of some of the principal tissues of the animal frame, as observed in the tongue of the living frog, toad etc. *Phil Mag*. 1846;29:271–287.
3. Waller A. Microscopic observations on the perforation of the capillaries by the corpuscles of the blood, and on the origin of mucus and pus-globules. *Phil Mag*. 1846;29:397–405.
4. Cohnheim J. *Lectures on General Pathology*. London: The New Sydenham Society; 1882.
5. Malkin HM. Julius Cohnheim (1839-1884). His life and contributions to pathology. *Ann Clin Lab Sci*. 1984; 14: 335–342.
6. Marchesi VT, Florey HW. Electron micrographic observations on the emigration of leucocytes. *Quart J Exp Physiol*. 1960;45:343–348.
7. Toniolo C, Bonora GM, Showell H, Freer RJ, Becker EL. Structural requirements for formyl homooligopeptide chemoattractants. *Biochemistry*. 1984;23:698–704.

8. Bennett JP, Hirth KP, Fuchs E, Sarvas M, Warren GB. The bacterial factors which stimulate neutrophils may be derived from procaryote signal peptides. *FEBS Lett*. 2006;116:57–61.

9. Graille M, Stura EA, Corper AL, et al. Crystal structure of a *Staphylococcus aureus* protein A domain complexed with the Fab fragment of a human IgM antibody: structural basis for recognition of B-cell receptors and superantigen activity. *Proc Natl Acad Sci U S A*. 2000;97:5399–5404.

10. Gomez MI, Lee A, Reddy B, et al. *Staphylococcus aureus* protein A induces airway epithelial inflammatory responses by activating TNFR1. *Nat Med*. 2004;10:842–848.

11. Kristiansen SV, Pascual V, Lipsky PE. Staphylococcal protein A induces biased production of Ig by VH3-expressing B lymphocytes. *J Immunol*. 1994;153:2974–2982.

12. Shawar SM, Rich RR, Becker EL. Peptides from the amino-terminus of mouse mitochondrially encoded NADH dehydrogenase subunit 1 are potent chemoattractants. *Biochem Biophys Res Commun*. 1995;211: 812–818.

13. LeGrand A. De l'analogie et des différences entire les tubercules et les scrofuels: Influence des maladies éruptives sur le développement et la marche des scrofules et les tubercules. *Rev. Méd. francaise et étrangère*. 1848;2:392–148.

14. Bruns P. Die Heilwirkung des Erysipels auf Geschwülste [The healing effect of erysipelas on tumours]. *Beitr Klin Chir*. 1888;3:443–466.

15. Coley WB. Contribution to the knowledge of sarcoma. *Ann Surg*. 1891;14:199–220.

16. Pearl R. Cancer and tuberculosis. *Am J Hyg*. 1929;9:97–159.

17. Nauts HC, Fowler GA, Bogatko FH. A review of the influence of bacterial infection and of bacterial products (Coley's toxins) on malignant tumors in man. *Acta Med Scan suppl*. 1953;275:5–103.

18. Shear MJ, Andervont HB. Chemical treatment of tumors. III. Separation of hemorrhage-producing fration of *B. coli* filtrate. *Proc Soc Exp Biol Med*. 1936;34:323–325.

19. Shear MJ, Turner FC. Chemical treatment of tumors. V. Isolation of the hemorrhage-producing fraction from *Serratia marcescens (Bacillus prodigiosus)* culture filtrate. *J Natl Cancer Inst*. 1943;4:81–97.

20. Anderson BF, Legallaies FY. Vascular reactions of normal and malignant tissues in vivo: the role of hypotension in the action of a bacterial polysaccharide on tumors. *J Natl Cancer Inst*. 1952;12:1279–1295.

21. Feldmann M, Brennan FM, Williams RO, et al. Evaluation of the role of cytokines in autoimmune disease: the importance of TNF-α in rheumatoid arthritis. *Prog Growth Factor Res*. 1992;4:247–255.

22. Feldmann M, Maini RN. Anti-TNF-α therapy of rheumatoid arthritis: what have we learned? *Annu Rev Immunol* 2001;19:163–196.

23. Ostermann G, Weber KS, Zernecke A, Schroder A, Weber C. JAM-1 is a ligand of the β_2 integrin LFA-1 involved in transendothelial migration of leukocytes. *Nat Immunol.* 2002;3:151–158.

24. Deng GM, Zheng L, Chan FK, Lenardo M. Amelioration of inflammatory arthritis by targeting the pre-ligand assembly domain of tumor necrosis factor receptors. *Nat Med.* 2005;11:1066–1072.

25. Carswell EA, Old LJ, Kassel RL, Green S, Fiore N, Williamson B. An endotoxin-induced serum factor that causes necrosis of tumors. *Proc Natl Acad Sci U S A.* 1975;72:3666–3670.

26. Kolb WP, Granger GA. Lymphocyte in vitro cytotoxicity: characterization of human lymphotoxin. *Proc Natl Acad Sci U S A.* 1968;61:1250–1255.

27. Ruddle NH, Waksman BH. Cytotoxicity mediated by soluble antigen and lymphocytes in delayed hypersensitivity. 3. Analysis of mechanism. *J Exp Med.* 1968;128:1267–1279.

28. Paterson Y, Maciag PC. Listeria-based vaccines for cancer treatment. *Curr Opin Mol Ther.* 2005;7:454–460.

29. Aggarwal BB. Signalling pathways of the TNF superfamily: a double-edged sword. *Nat Rev Immunol.* 2003;3:745–756.

30. Suda T, Hashimoto H, Tanaka M, Ochi T, Nagata S. Membrane Fas ligand kills human peripheral blood T lymphocytes, and soluble Fas ligand blocks the killing. *J Exp Med.* 1997;186:2045–2050.

31. Grell M, Douni E, Wajant H, et al. The transmembrane form of tumor necrosis factor is the prime activating ligand of the 80 kDa tumor necrosis factor receptor. *Cell.* 1995;83:793–802.

32. Tamatani M, Che YH, Matsuzaki H, et al. Tumor necrosis factor induces Bcl-2 and Bcl-x expression through NFκB activation in primary hippocampal neurons. *J Biol Chem.* 1999;274:8531–8538.

33. Wajant H, Pfizenmaier K, Scheurich P. Tumor necrosis factor signaling. *Cell Death Differ.* 2003;10:45–65.

34. Takada H, Chen NJ, Mirtsos C, et al. Role of SODD in regulation of tumor necrosis factor responses. *Mol Cell Biol.* 2003;23:4026–4033.

35. Cha SS, Sung BJ, Kim YA, et al. Crystal structure of TRAIL-DR5 complex identifies a critical role of the unique frame insertion in conferring recognition specificity. *J Biol Chem.* 2000;275:31171–31177.

36. Osborn L, Hession C, Tizard R, et al. Direct expression cloning of vascular cell adhesion molecule 1, a cytokine-induced endothelial protein that binds to lymphocytes. *Cell.* 1989;59:1203–1211.

37. Pohlman TH, Stanness KA, Beatty PG, Ochs HD, Harlan JM. An endothelial cell surface factor(s) induced in vitro by lipopolysaccharide, interleukin 1, and tumor necrosis factor-α increases neutrophil adherence by a CDw18-dependent mechanism. *J Immunol.* 1986;136:4548–4553.

38. Sen R, Baltimore D. Multiple nuclear factors interact with the immunoglobulin enhancer sequences. *Cell.* 1986:705–716.

39. Itoh N, Nagata S. A novel protein domain required for apoptosis. Mutational analysis of human Fas antigen. *J Biol Chem*. 1993;268: 10932–10937.

40. Hsu H, Xiong J, Goeddel DV. The TNF-receptor 1-associated protein TRADD signals cell death and NF-κB activation. *Cell*. 1995;81:495–504.

41. Ea CK, Deng L, Xia ZP, Pineda G, Chen ZJ. Activation of IKK by TNGα requires site-specific ubiquitination of RIP1 and polyubiquitin binding by NEMO. *Mol Cell*. 2006;22:245–257.

42. Verstrepen L, Bekaert T, Chau TL, Tavernier J, Chariot A, Beyaert R. TLR-4, IL-1R and TNF-R signaling to NF-kB: variations on a common theme. *Cell Mol Life Sci*. 2008;65:2964–78.

43. Beutler B, Greenwald D, Hulmes JD, et al. Identity of tumour necrosis factor and the macrophage-secreted factor cachectin. *Nature*. 1985;316:552–554.

44. Esper DH, Harb WA. The cancer cachexia syndrome: a review of metabolic and clinical manifestations. *Nutr Clin Pract*. 2005;20:369–376.

45. Ariel A, Fredman G, Sun YP, et al. Apoptotic neutrophils and T cells sequester chemokines during immune response resolution through modulation of CCR5 expression. *Nat Immunol*. 2006.

46. Zigmond SH. Ability of polymorphonuclear leukocytes to orient in gradients of chemotactic factors. *J Cell Biol*. 1977;75:606–616.

47. Oberlin E, Amara A, Bachelerie F, et al. The CXC chemokine SDF-1 is the ligand for LESTR/fusin and prevents infection by T-cell-line-adapted HIV-1. *Nature*. 1996;382:833–835.

48. Liu R, Paxton WA, Choe S, et al. Homozygous defect in HIV-1 coreceptor accounts for resistance of some multiply-exposed individuals to HIV-1 infection. *Cell*. 1996;86:367–377.

49. Middleton J, Neil S, Wintle J, et al. Transcytosis and surface presentation of IL-8 by venular endothelial cells. *Cell*. 1997;91:385–395.

50. Proudfoot AE. The biological relevance of chemokine-proteoglycan interactions. *Biochem Soc Trans*. 2006;34:422–426.

51. Zlotnik A, Yoshie O. Chemokines: a new classification system and their role in immunity. *Immunity*. 2000;12:121–127.

52. Lapidot T, Dar A, Kollet O. How do stem cells find their way home? *Blood* 2005;106:1901–1910.

53. Dar A, Goichberg P, Shinder V, et al. Chemokine receptor CXCR4-dependent internalization and resecretion of functional chemokine SDF-1 by bone marrow endothelial and stromal cells. *Nat Immunol*. 2005;6:1038–1046.

54. Skelton NJ, Quan C, Reilly D, Lowman H. Structure of a CXC chemokine-receptor fragment in complex with interleukin-8. *Structure*. 1999;7: 157–168.

55. Arai H, Tsou CL, Charo IF. Chemotaxis in a lymphocyte cell line transfected with C-C chemokine receptor 2B: evidence that directed migration is mediated by βγ dimers released by activation of Gαᵢ-coupled receptors. *Proc Natl Acad Sci U S A*. 1997;94:14495–14499.

56. Neptune ER, Bourne HR. Receptors induce chemotaxis by releasing the βγ subunit of Gi, not by activating Gq or Gs. *Proc Natl Acad Sci U S A*. 1997;94:14489–14494.

57. Van Keymeulen A, Wong K, Knight ZA, et al. To stabilize neutrophil polarity, PIP3 and Cdc42 augment RhoA activity at the back as well as signals at the front. *J Cell Biol*. 2006;174:437–445.

58. Baggiolini M. Chemokines and leukocyte traffic. *Nature*. 1998;392: 565–568.

59. Wright SD, Silverstein SC. Tumor-promoting phorbol esters stimulate C3b and C3b′ receptor-mediated phagocytosis in cultured human monocytes. *J Exp Med*. 1982;156:1149–1164.

60. Hogg N, Laschinger M, Giles K, McDowall A. T-cell integrins: more than just sticking points. *J Cell Sci*. 2003;116:4695–4705.

61. Laudanna C, Campbell JJ, Butcher EC. Role of Rho in chemoattractant-activated leukocyte adhesion through integrins. *Science*. 1996;271: 981–983.

62. Crittenden JR, Bergmeier W, Zhang Y, et al. CalDAG-GEFI integrates signaling for platelet aggregation and thrombus formation. *Nat Med*. 2004;10:982–986.

63. Pizon V, Lerosey I, Chardin P, Tavitian A. Nucleotide sequence of a human cDNA encoding a ras-related protein (rap1B). *Nucleic Acids Res*. 1988;16:7719.

64. Kitayama H, Sugimoto Y, Matsuzaki T, Ikawa Y, Noda M. A ras-related gene with transformation suppressor activity. *Cell*. 1989;56:77–84.

65. Zwartkruis FJ, Bos JL. Ras and Rap1: two highly related small GTPases with distinct function. *Exp Cell Res*. 2000;1253:157–165.

66. Franke B, Akkerman JW, Bos JL. Rapid Ca^{2+}-mediated activation of Rap1 in human platelets. *EMBO J*. 1997;16:252–259.

67. Reedquist KA, Ross E, Koop EA, et al. The small GTPase, Rap1, mediates CD31-induced integrin adhesion. *J Cell Biol*. 2000;148:1151–1158.

68. Liu L, Schwartz BR, Tupper J, Lin N, Winn RK, Harlan JM. The GTPase Rap1 regulates phorbol 12-myristate 13-acetate-stimulated but not ligand-induced β₁ integrin-dependent leukocyte adhesion. *J Biol Chem*. 2002;277:40893–40900.

69. Shimonaka M, Katagiri K, Nakayama T, Fujita N, Tsuruo T, Yoshie O, Kinashi T. Rap1 translates chemokine signals to integrin activation, cell polarization, and motility across vascular endothelium under flow. *J Cell Biol*. 2003;161:417–427.

70. Lafuente EM, van Puijenbroek AA, Krause M, et al. RIAM, an Ena/VASP and Profilin ligand, interacts with Rap1-GTP and mediates Rap1-induced adhesion. *Dev Cell.* 2004;7:585–595.

71. Katagiri K, Maeda A, Shimonaka M, Kinashi T. RAPL, a Rap1-binding molecule that mediates Rap1-induced adhesion through spatial regulation of LFA-1. *Nat Immunol.* 2003;4:741–748.

72. Bos JL. Linking Rap to cell adhesion. *Curr Opin Cell Biol.* 2005;17:123–128.

73. Han J, Lim CJ, Watanabe N, et al. Reconstructing and deconstructing agonist-induced activation of integrin $\alpha II\beta_3$. *Curr Biol.* 2006;16:1796–1806.

74. Campbell ID, Ginsberg MH. The talin-tail interaction places integrin activation on FERM ground. *Trends Biochem Sci.* 2004;29:429–435.

75. Fleming IN, Batty IH, Prescott AR, et al. Inositol phospholipids regulate the guanine-nucleotide-exchange factor Tiam1 by facilitating its binding to the plasma membrane and regulating GDP/GTP exchange on Rac1. *Biochem J.* 2004;382:857–865.

76. Pollard TD, Borisy GG. Cellular motility driven by assembly and disassembly of actin filaments. *Cell.* 2003;112:453–465.

77. Suetsugu S, Kurisu S, Oikawa T, Yamazaki D, Oda A, Takenawa T. Optimization of WAVE2 complex-induced actin polymerization by membrane-bound IRSp53, PIP(3), and Rac. *J Cell Biol.* 2006;173:571–585.

78. Nagasawa T, Hirota S, Tachibana K, et al. Defects of B-cell lymphopoiesis and bone-marrow myelopoiesis in mice lacking the CXC chemokine PBSF/SDF-1. *Nature.* 1996;382:635–638.

79. Gunn MD, Kyuwa S, Tam C, Kakiuchi T, Matsuzawa A, Williams LT, Nakano H. Mice lacking expression of secondary lymphoid organ chemokine have defects in lymphocyte homing and dendritic cell localization. *J Exp Med.* 1999;189:451–460.

80. Martin-Padura I, Lostaglio S, Schneemann M, et al. Junctional adhesion molecule, a novel member of the immunoglobulin superfamily that distributes at intercellular junctions and modulates monocyte transmigration. *J Cell Biol.* 1998;142:117–127.

81. Cunningham SA, Rodriguez JM, Arrate MP, Tran TM, Brock TA. JAM2 interacts with $\alpha 4\beta 1$. Facilitation by JAM3. *J Biol Chem.* 2002;277: 27589–27592.

82. Arrate MP, Rodriguez JM, Tran TM, Brock TA, Cunningham SA. Cloning of human junctional adhesion molecule 3 (JAM3) and its identification as the JAM2 counter-receptor. *J Biol Chem.* 2001;276:45826–45832.

83. Santoso S, Sachs UJ, Kroll H, et al. The junctional adhesion molecule 3 (JAM-3) on human platelets is a counterreceptor for the leukocyte integrin Mac-1. *J Exp Med.* 2002;196:679–691.

84. Thomas KM, Pyun HY, Navarro J. Molecular cloning of the fMet-Leu-Phe receptor from neutrophils. *J Biol Chem.* 1990;265:20061–20064.

85. Alblas J, Ulfman L, Hordijk P, Koenderman L. Activation of Rhoa and ROCK are essential for detachment of migrating leukocytes. *Mol Biol Cell*. 2001;12:2137–2145.

86. Mayadas TN, Cullere X. Neutrophil β_2 integrins: moderators of life or death decisions. *Trends Immunol*. 2005;26:388–395.

87. Butcher EC. Leukocyte-endothelial cell recognition: three (or more) steps to specificity and diversity. *Cell*. 1991;67:1033–1036.

88. Springer TA. Traffic signals for lymphocyte recirculation and leukocyte emigration: the multistep paradigm. *Cell*. 1994;76:301–314.

Tyrosine Protein Kinases and Adaptive Immunity : TCR, BCR, Soluble Tyrosine Kinases and NFAT

The family of non-receptor protein tyrosine kinases

There is an important family of receptors that, while possessing no intrinsic catalytic activity, induce responses similar to those of the receptor tyrosine kinases. They operate by recruiting tyrosine kinases that are either soluble components in the cytosol or are membrane-associated. These are the non-receptor protein tyrosine kinases (nrPTKs). They can be divided into families based on the sequence of their kinase domains or on the presence of other protein interaction domains (PH, SH3, etc.) (**Figure 17.1**).[1] Recruitment of nrPTKs and the tyrosine phosphorylations that ensue are usually the first steps in the assembly of substantial signalling complexes, consisting of a dozen or more components that bind and interact with each other.

The many abbreviations used in this chapter are collected together at the end of the chapter.

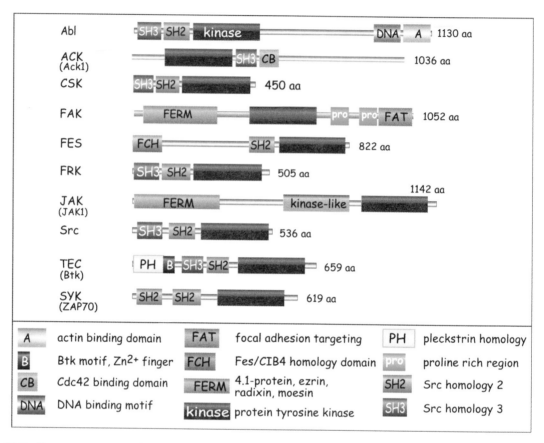

FIG 17.1 Non-receptor protein tyrosine kinase families.

These are subdivided into 10 families most of which contain SH2 and SH3 domains (Src-A and Src-B families are shown as one). Several were discovered as transforming genes of viral genomes, hence names like Src or Abl, derived from Rous sarcoma virus or Abelson murine leukaemia virus. Adapted from Robinson et al.[1]

The receptors that recruit nrPTKs include those that mediate immune responses:

- **The T lymphocyte receptor** (TCR). Involved in the detection of processed antigens, presented in the context of the major histocompatibility complex (MHC). Subsequently it regulates the clonal expansion of T cells.[2]
- **The B lymphocyte receptor (BCR)**. Involved in the detection of non-processed antigens. Important for the elaboration of antigen-specific immunoglobulins necessary for defence against infection by microorganisms.[3]
- **The interleukin-2 receptor (IL-2R)**. The cytokine IL-2, secreted by a subset of T-helper cells, enhances the proliferation of activated T and B cells, increases the cytolytic activity of natural killer (NK) cells and the secretion of IgG.

- **The high-affinity receptor for IgE** (IgE-R). Present on mast cells and blood-borne basophils,[4] this plays an important role in hypersensitivity and in the initiation of acute inflammatory responses.
- **Erythropoietin receptors**. The cytokine erythropoietin (EPO) plays an important role in the final stage of maturation of erythroid progenitor cells into mature red blood cells. The receptor is also present on other cell types suggesting that erythropoietin may have effects on other tissues. EPO has been abused by athletes aiming to boost their performance in endurance sports such as cycling.
- **Prolactin receptors**. Primarily involved in the regulation of lactation. In addition, prolactin has been implicated in modulation of immune responses. For instance, it regulates the level of NK-mediated cytotoxicity. It has also gained attention as a potential male contraceptive.
- **Other non-receptor PTKs** are discussed in Chapter 13, where they are shown to play pivotal roles in cell survival and proliferation.

T-cell receptor signalling

T lymphocytes are central role to cell-mediated immunity. When activated, they proliferate and differentiate, becoming cytotoxic T cells, helper T cells, or long-lasting memory cells. Cytotoxic T cells kill specific targets, most commonly virus-infected cells; helper T cells assist other cells of the immune system, mainly B lymphocytes. Effective T cell activation requires the coordinated attentions of several different ligands with their receptors. This is not the only unusual aspect. The primary ligands that activate T cells are not classical first messengers (i.e. blood-borne soluble molecules), but proteins presented on the membranes of neighbouring cells. For example, the stimulation of T lymphocytes that are naive to any previous form of activation involves the T cell receptor (Figure 17.2a). This detects foreign proteins (antigens) in the form of short peptides that are proffered on the surface of the target cells. An infected cell, expressing viral proteins, presents fragments fixed in a groove on its MHC class I molecules. These are expressed on virtually all nucleated cells and are characteristic of the host so that they are recognized by the immune system as 'self'. The task of the T cell is to kill the virus-infected cell. Circulating antigens are processed by specialized cells such as dendritic cells, B lymphocytes, and macrophages which then present protein fragments on their surfaces, again attached to an antigen-presenting molecule, this time MHC class II (Figure 17.2b). Here, the main function of the T cell is to help the target cell, the B lymphocyte, to make antibodies.

More than one lymphocyte receptor must be engaged to ensure activation

The cell–cell interaction necessary for T cell activation requires intimate contact, calling on specialized adhesion molecules, as well as three quite

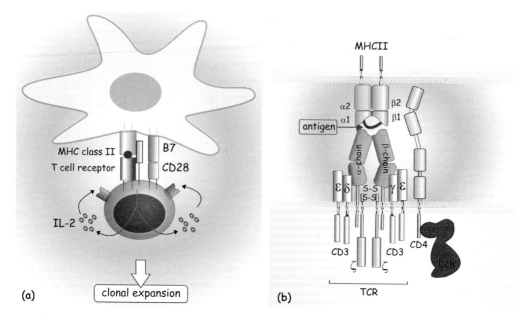

FIG 17.2 Clonal expansion of naive T lymphocytes.

(a) Clonal expansion is activated through binding of an antigen, presented in the groove of MHC II on an antigen-presenting dendritic cell (above), to the receptor (TCR) on the naive T cell (below). There are also essential costimulatory interactions involving CD4, B7, and CD28. The result is the production of IL-2 together with its receptor on the T cell. This induces a proliferation signal that is responsible for the expansion of a clone of T cells that exclusively recognize this particular antigen. (b) The TCR is a disulfide-linked heterodimer of α and β chains. These have hypervariable regions that detect the antigen, presented as a short peptide in the groove of the MHC molecule. This heterodimer forms a complex with six polypeptides, two ζ chains and the δ, ε, and γ chains of the CD3 molecule. A CD4 or CD8 molecule is also associated with the TCR in helper or cytotoxic T cells respectively. Importantly, attachment of CD4 (or CD8) to the MHC chain brings the non-receptor PTK Lck, into the vicinity of the T cell receptor.

distinct ligand–receptor interactions (**Figure 17.2a**). It is the combination of antigen and MHC that initiates the T cell response, but there are two possible outcomes. Only in the presence of a second, 'costimulatory' signal do the cells become fully activated. The full response comprises transcription of early response genes, followed by synthesis and release of the cytokine IL-2, entry into the cell cycle, and differentiation into an effector or memory cell. In the absence of a costimulatory signal, the T cell becomes unresponsive or anergic.

Initial cell–cell contact involves low-specificity interaction between B7 (on the antigen-presenting cell) and CD28 (on the T cell). The α and β chains of the T cell receptor (TCR) may then bind to the peptide presented by the antigen-presenting cell. The TCR is associated with CD3 as components of a complex of six polypeptides, all of which span the membrane (**Figure 17.2b**). The TCR/CD3 bound to MHCII assemble as microclusters, comprising 11–150 copies of each that converge to form the so-called immunological synapse.[5] By the recruitment of non-receptor PTKs, components of the TCR/CD3 complex are rapidly phosphorylated. In addition, there is activation of PLC-γ1 with production of IP$_3$ and diacylglycerol and elevation of cytosol Ca^{2+}. Thus, the consequences of receptor ligation are not dissimilar from those induced by the receptors for EGF or PDGF.

An early candidate explaining the induction of tyrosine kinase activity emerged with the discovery of Lck (p56^{lck}). This T cell-specific kinase of the Src family[6,7] (see **Figure 17.1** and Table 17.4) is associated with the membrane and with the cytosolic tail of CD4 (in helper T cells) or CD8 (in cytotoxic T cells)[8] (**Figure 17.2b**). The extracellular domains of CD4 and CD8 attach to the MHC, strengthening the rather weak interaction established between the TCR and antigen. They also bring CD4/CD8 into the vicinity of the TCR complex, leading Lck to its targets on their ζ and other CD3 chains.[9]

T cell receptor signal-complex formation

In resting cells, Lck exists in a primed state. This has an open conformation, accessible to substrate, but not phosphorylated at the activating residue (Y394) in its activation segment. The open state of Lck associated with CD4/CD8 is maintained by CD45, a transmembrane phosphatase that targets pY505 in the C-terminal tail. Its function is described more fully in Chapter 21. As with c-Src, Csk is thought to be the kinase that drives Lck into its inactive state[10] (see **Figure 17.13**).

Activation of primed Lck is by transphosphorylation by another Lck molecule present in the TCR microcluster. It can then phosphorylate the γ, δ, ε and ζ chains of CD3.[11] The target tyrosines are confined to so-called ITAM motifs. Phosphorylation of ITAMs provides multiple docking sites for SH2 domain-bearing molecules, amongst them ZAP-70, another kinase which binds to the ζ chain of CD3.[12] (**Figure 17.3**). ZAP-70 is now phosphorylated by Lck and thereby activated, causing further phosphorylation of multiple substrates. The sequence of events then follows a pattern in which phosphotyrosines bind SH2 or PTB domain-containing proteins that may themselves be PTKs and that can phosphorylate and recruit yet more proteins in succession. At each stage there is the opportunity for the pathways to diverge, involving a range of effectors.[9]

One important branch point is offered by the integral membrane protein LAT, which presents no fewer than nine substrate tyrosine residues.[13] When phosphorylated, these recruit a broad range of signalling molecules that include Grb2, SLP76, PLC γ1, PI 3-kinase (through its p85 regulatory subunit) and the guanine nucleotide exchange factors Dbl and Vav (see inset in **Figure 17.3**).

The formation of a signalling complex around the TCR and the branching pathways that emanate from it resemble the mechanisms used by the growth factors. However, not all the destinations of these pathways are clear. We outline two, one involving calcineurin, the other PKC θ (**Figure 17.4**). An important starting point is the production of diacylglycerol and IP$_3$ by PLC γ1.

TCR activation and Fyn. The ITAM motifs (immunoreceptor tyrosine-based activation motif) in the δ, γ, ε, and ζ chains of CD3 are also targets of Fyn (p59^{fyn}) another member of the large Src kinase family. Fyn is also activated by phosphorylation but its mode of recruitment to the TCR microclusters is not yet understood. Fyn is required for efficient TCR signalling during antigen presentation but it plays a more important role in the survival of naïve T cells.[9]

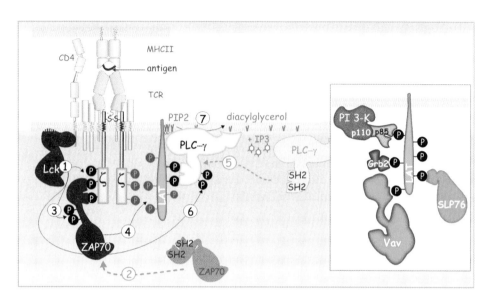

FIG 17.3 Signalling complex formation at the TCR complex.

The antigen/MHC-bound TCR activates Lck that phosphorylates the ITAM motifs present in the ζ chains of CD3 (1). The resulting pY residues form a docking site for the SH2 domains of ZAP70 (2), another cytosolic PTK. ZAP70 is phosphorylated by Lck in the linker region between the SH2 domains and the catalytic domain (3). The activated ZAP70 then phosphorylates several (maximally nine) tyrosine residues on the transmembrane adaptor LAT (4). Various proteins attach. These include PLC ‑γ (5), which upon attachment becomes phosphorylated and activated by Lck (6). Other molecules include the guanine nucleotide exchange factor Vav, the adaptors Grb2 and SLP76, and the p85 regulatory subunit of P I3-kinase. All play important roles in the activation of the IL-2 gene. Generation of diacylglycerol and IP_3 (7) by PLC ‑γ is a starting point for the two signalling pathways described in this section. The main components of the TCR are indicated in Figure 17.2.

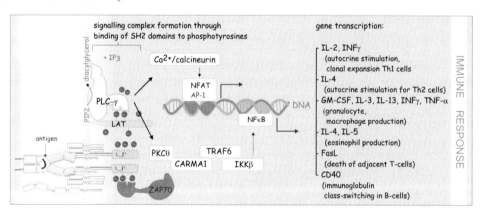

FIG 17.4 Two signalling pathways downstream of PLC ‑γ.

The second messengers diacylglycerol and IP_3 activate two signalling pathways. One involves IP_3-released Ca^{2+}, which results in the activation of the serine/threonine phosphatase calcineurin and leads to activation of NFAT. The other involves the diacylglycerol-mediated activation of PKC θ which phosphorylates the adaptor CARMA1. The unfolded protein then acts as a docking site for the assembly of a TRAF6-ubiquitin ligase complex that causes activation of IKKβ and then nuclear translocation of NF-κB. The genes regulated by NFAT and NF-κB are involved in various aspects of the immune response.

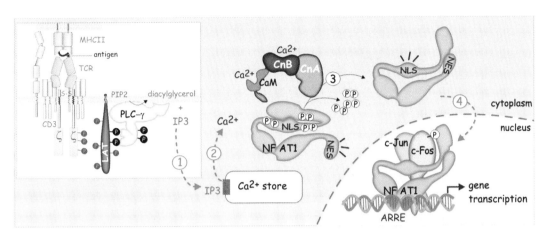

FIG 17.5 Calcineurin-mediated activation of NFAT.
Ca^{2+} binds to calmodulin (CaM) and calcineurin B (CnB), leading to the activation of the serine/threonine phosphatase calcineurin. Dephosphorylation of NFAT1 ensues, exposing its nuclear localization sequence (NLS) and masking the nuclear export sequence (NES). NFAT enters the nucleus and associates with c-Jun and c-Fos (AP-1 complex) to drive gene expression.

PLC-γ1 to NFAT

Phosphorylated LAT recruits PLC-γ1 to the TCR/CD3 complex.[14] This is facilitated by the concerted actions of an array of other proteins such as Vav1, SLP-76, Itk, and c-Cbl. In the assembled complex, PLC-γ1 is activated through phosphorylation by Lck in the linker region between SH2 and SH3 (see **Figure 17.3**). The consequent production of DAG and IP_3 and elevation of Ca^{2+} leads to activation of calcineurin (PP2B, see page 682). This (serine/threonine) phosphatase utilizes two Ca^{2+} sensors, calmodulin and calcineurin B (**Figure 17.5**). Calcineurin dephosphorylates multiple pS residues in the N-terminal segment of the transcription factor NFAT, thereby exposing a nuclear localization sequence (NLS).[15]

> **Multiple protein kinases and multiple phosphorylation sites on NFAT.**
> The N-terminal segment of NFAT contains numerous serine phosphorylation sites (about 14) which reside in short conserved motifs, serine-rich regions (SRR), and serine/proline motifs (SP1, SP2, and SP3) (see figure 17.6). These serine residues surround the NLS. Phosphorylation occurs in the nucleus. DYRKA initially phosphorylates the SP3 site and this opens the way to further phosphorylations by casein kinase and glycogen synthase kinase β. Phosphorylation leads to the masking of the NLS and unmasking of the nuclear export signal NES.[17] (see **Figure 17.5**). NFAT1 is actively transported out of the nucleus by exportin-1/RanGTP.

> **Two calcium sources.**
> An initial IP_3-induced release of Ca^{2+} from intracellular stores is augmented by cADPR-induced release, and then by influx of extracellular calcium ions through plasma membrane channels (CRAC; see page 210). This source of Ca^{2+} is required for expression of NFAT-controlled genes. Loss of function of CRAC causes immune deficiency.[16]

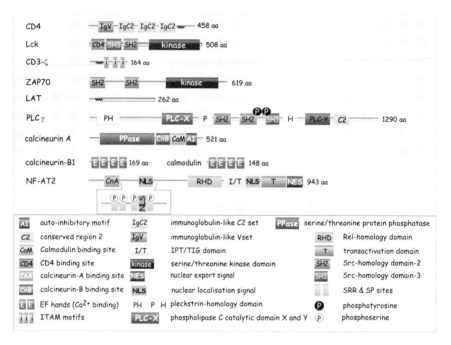

FIG 17.6 Domain architecture of proteins involved in TCR-mediated activation of NFAT.

A variety of proteins associates with the TCR. They are brought together through a cascade of phosphotyrosine-SH2 interactions. An important signal is transmitted via phospholipase C-γ which, through activation of calcineurin, results in the nuclear translocation of NFAT2. The domain organization of PLC γ is illustrated in Figure 5.11, page 151.

Nuclear translocation of NFAT is essential for clonal expansion of T cells because of its pivotal role in the induction of IL-2 and IFN-γ (for Th1 cells) and IL-4 (for Th2 cells). Full transcriptional activity of NFAT only occurs after phosphorylation by ERK1/2 or p38.[18] In order to drive full expression of IL-2, NFAT must associate with the AP-1 complex and with NF-κB (see below). The domain architecture of proteins involved in this pathway is illustrated in **Figure 17.6.**

Calcineurin, NFAT, and islet development in the pancreas. Ciclosporin, a cyclic peptide of 11 residues, of fungal origin, has been applied following transplantation surgery to suppress tissue rejection. However, it was found that it can raise the concentration of blood glucose, effectively causing diabetes mellitus, due to the loss of the insulin-producing β-cells in the pancreatic islets of Langerhans.[19] This raised the possibility that calcineurin (PP2B), being the target of ciclosporin, might play a normal role in adaptive islet responses. Mice deficient in the calcineurin regulatory subunit (CnB1) develop age-related diabetes due to loss of β-cells. Conditional expression of active NFAT1 in these β-cells rescues the defect and prevents diabetes. NFAT1 regulates transcription of genes critical for β-cell endocrine function and which are linked to type 2 (late onset) diabetes.[20]

The PLC γ1 to NF-κB pathway

In essence this process involves the formation of an E3 ubiquitin–ligase complex followed by K63-type ubiquitylation and recruitment of components that result in the activation of IKK-α and -β (see page 460 onwards)· This leads to destruction of the inhibitor IκB and hence nuclear translocation of NF-κB. The assembly process involves the CARD domain, first discerned in the recruitment and activation of caspases (proteases involved in apoptosis). All of this depends on the activation of PKC θ (see Fig 9.8, page 254) by diacylglycerol in the lipid raft supporting the TCR-CD3 complex (see page 522). Activated PKC-θ phosphorylates the membrane bound adaptor CARMA1 which causes it to unfold, resulting in the attachment of the adaptor Cbl-10, through a CARD-CARD domain interaction.

Cbl-10 is associated with the adaptor Malt1, which in turn binds to the TRAF6 E3-ubiquitin ligase complex (comprising TRAF6, Ubc13, and Mms2) (**Figure 17.7**). K63-type ubiquitylation of TRAF6 ensues and this recruits NEMO/IKKα/IKKβ, as well as TAB1, -2, -3/TAK1[23] (see page 460). The recruitment of TAK1 allows phosphorylation and activation of IKK-β and this phosphorylates IκB rendering the inhibitor susceptible to ubiquitylation (K48 type) by the SCF E3-ubiquitin ligase complex. Ubiquitylated IκB is destroyed by the proteasome (see Figure 15.8, page 462). NEMO is also ubiquitylated, allowing further recruitment of TAB/TAK1 complexes and enhancing the signal.[24] The liberated NF-κB now exposes its nuclear localization signal and translocates to the nucleus where it participates in the induction of interleukins together with NF-AT and AP-1.

As already mentioned, full T cell activation requires a second, costimulatory signal communicated through a separate receptor. For some helper T cells, this is provided by CD28 which binds to the B7 molecule on antigen-presenting cells. The signalling pathway activated by CD28 involves PI 3-kinase and mTOR, and it acts to regulate the expression of cyclin E necessary for T cell proliferation. In fact, prolonged occupation of both TCR and CD28 reduces the requirement for an autocrine IL-2 signal.[25] The involvement of mTOR in the costimulatory signal explains why rapamycin, an inhibitor of mTOR, acts as a potent inhibitor of T cell function (immunosuppressant)[26] (see page 565).

Down-regulation of the TCR response

The strength of the TCR signal depends on the strength of the interaction between receptors and the antigen in the context of the MHC class II. It also depends on the number of TCRs that contribute to the formation of microclusters. This requires efficient recycling because as much as 1% of cell surface TCR-CD3 complexes are internalized every minute. Binding of the adaptor SLAP prevents recycling. It binds to the ζ chain of the TCR and recruits

The Cbl family. Cbl was discovered as an oncogenic protein v-Cbl, carried by the Cas NS-1 retrovirus, cause of B cell lymphomas in mice.[21] It acts as an E3-ubiquitin ligase and plays a role in the down-regulation of tyrosine kinase-induced signalling pathways. It works by ubiquitylating activated kinases. In the case of the EGFR and the TCR, this has the effect of directing the receptor to late endosomes so preventing their recycling to the outer membrane. Cbls also down-regulates other effectors such as tyrosine phosphorylated Vav (a RhoGEF), tyrosine phosphorylated STAT5, or the FcεRI in mast cells. The oncogenic form, v-Cbl, may act as an inhibitor of endogenous Cbl activity. Cbl members also act as adaptors for numerous proteins such as PI 3-kinase, c-Src, tubulin, or yet another adaptor protein SLAP.[22]

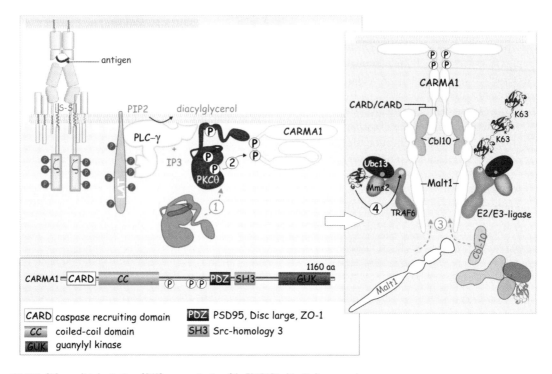

FIG 17.7 PLC-γ-mediated activation of PKCθ causes activation of the TRAF6 E3-ubiquitin ligase complex.
Diacylglycerol causes membrane attachment with multiple phosphorylations and activation of PKCθ (1). This phosphorylates and unfolds the adaptor CARMA1 (2). Its CARD domain attaches to the CARD domain of Cbl-10 and this recruits Malt1, TRAF6 (E3-ligase) and Ubc13/Mms2 (E2-conjugating enzyme) into the complex (3). As a result, TRAF6 is polyubiquitylated (4). The K63-type ubiquitin chain acts as a docking site for NEMO/IKKα/IKKβ and also the TAB1-2/TAK1 protein kinase complex (not shown). The domain organization of CARMA1 is shown in the lower panel; nPKCθ is illustrated in Figure 9.8, page 254.

Cbl into the complex. Subsequent polyubiquitylation (K48 type) retains the TCR/CD3 in the endosomal compartment, promoting its degradation in lysosomes.[27] A set of reactions regulating the recycling of the EFGR is illustrated in Figure 12.21, page 350.

The lipid raft hypothesis

A feature of cells that are stimulated through immune recognition receptors is the involvement of heterogeneous microdomains in the plasma membrane. These are known as lipid rafts. The enrichment of particular sets of membrane-associated or transmembrane proteins in such regions is, however, a widespread phenomenon and likely to be a key feature of many signalling mechanisms. Because they involve lipids that have a propensity for rapid lateral diffusion, rafts tend to be dynamic features and are in many respects poorly defined.

Initially, the discovery of detergent-resistant fractions in membranes obtained from epithelial cells[28] led to the postulation of membrane regions enriched in glycosphingolipids and cholesterol. These were also referred to as detergent-resistant (or detergent-insoluble) membrane domains, or more simply as lipid rafts. They are thought to be phase-separated regions, which, because of the saturated nature of the acyl chains of the component sphingolipids and the presence of cholesterol, possess a higher order of rigidity than the remainder of the membrane. Direct visualization of lipid microdomains in a smooth muscle cell by 'single molecule' fluorescence microscopy indicates dimensions ranging between 0.2 and 2 μm at ambient temperatures, but at 37 °C they may be smaller.[29] Both their size and density are likely to vary with cell type.

In the context of this chapter, signalling molecules that appear to be confined to lipid rafts include CD4/CD8 in T cells and the dual acylated (N-myristoylated and palmitoylated) non-receptor tyrosine kinases of the Src family. The non-acylated Src kinase Lck is also raft-associated, presumably through association with CD4/CD8. Other important signalling components located in rafts are LAT and CARMA1 (see above). Membrane proteins such as the transferrin receptor and the tyrosine phosphatase CD45 are generally absent. The primary receptors – the TCR, the BCR, and the IgE-R – were thought to exist outside rafts until cross-linked during activation, but it now appears that a proportion of these receptors is associated with rafts under resting conditions and that this increases following stimulation. Exactly how lipid rafts function as platforms for the assembly of signalling complexes remains unclear.

Signalling through interferon receptors

Classification of the interferons (IFN), cytokines originally recognized for their antiviral properties,[30] has been difficult. The main classes and subtypes, and their receptors are listed in Table 17.1 and illustrated in **Figure 17.8.**

The interferons are classified as types I, II, and III. The type I subfamily contains 8 members that all bind to the same receptor heterodimer composed of IFNAR-1 and -2. The type II interferon, represented uniquely by IFN-γ, interacts with IFNGR-1 and -2. Type III interferons comprise three variants of IFN-λ, each binding to the interleukin 28Ra/10R2 receptor dimer.

Hours to days before the adaptive immune response gets under way to combat viral infection, IFN-α, -β, and -γ are already at work.[33] Mice that lack the receptors for IFN-α, -β, and -γ exhibit increased susceptibility to viral infection.[34] The interferons induce the expression of protein kinase-R (dsRNA-dependent protein kinase), 2'–5' oligoadenylate synthetases (OAS) and methyl transferases. These (respectively) bring about (either directly or indirectly) arrest of protein synthesis, degradation of single-stranded RNA, and inhibition of transcription, thus hindering the propagation of the virus. In some cases, interferons cause

TABLE 17.1 Classification of interferons and their receptors

Class	Ligand	Receptor	Distribution
I	IFN-α	IFNAR-1/IFNAR-2	ubiquitous
	IFN-β	IFNAR-1/IFNAR-2	ubiquitous
	IFN-δ	IFNAR-1/IFNAR-2	trophoblasts
	IFN-ε	IFNAR-1/IFNAR-2	uterus/ovary
	IFN-ω	IFNAR-1/IFNAR-2	leukocytes
	IFN-κ	IFNAR-1/IFNAR-2	epidermal keratinocytes
	IFN-τ	IFNAR-1/IFNAR-2	trophoblasts
	IFN-ζ	IFNAR-1/IFNAR-2	spleen/ thymus/ lymph nodes
II	IFN-γ	IFNGR-1/IFNGR-2	activated T cells, macrophages, NK cells
III	IFN-λ1,2,3	L-28Ra/IL-10R2	ubiquitous

Adapted from Takaoka and Yanai.[31]

FIG 17.8 Interferons and their receptors.
The figure illustrates the organization of IFNAR1 and IFNAR2 bound to IFN-α. (1n6v[32]).
Adapted from Chill et al.[32]

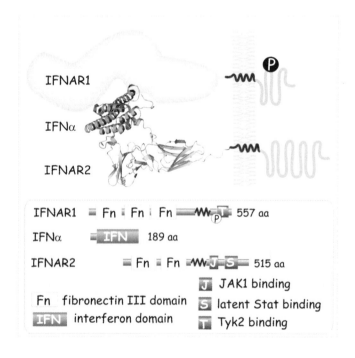

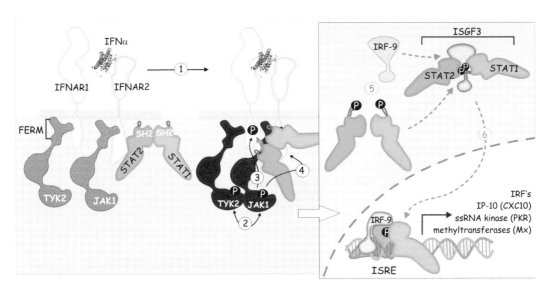

FIG 17.9 Activation of STATS by the interferon receptor.

Binding of IFN-α causes dimerization of the IFN-α receptor (IFNARs 1 and 2) (1). This brings together the associated cytosolic kinases TYK2 and JAK1 which phosphorylate each other (2) and then IFNAR1 (3). STAT2, which in its latent form is bound to IFNAR2, associates with the tyrosine phosphorylated IFNAR1 through its SH2 domain and in so doing it too becomes phosphorylated (4). The phosphorylated STAT2 catalyses the phosphorylation of STAT1 and these detach from the receptor complex. The dimer is accompanied (5) by an interferon regulatory factor IRF-9. The ternary complex, ISGF3, enters the nucleus (6) and binds to DNA at the interferon stimulated response element (ISRE). The complex induces expression of genes such as interferon regulatory factors (IRF), ssRNA kinase, methyltransferases (Mx), and the chemokine IP-10.

cell death through apoptosis.[33] IFN-α, -β, and -γ also play essential roles in immune modulation. They promote maturation of dendritic cells by increasing surface expression of costimulatory molecules and MHC class I and II. As a result, the ability of dendritic cells to stimulate the clonal expansion of effector T cells (e.g. Th1) is enhanced. In certain cases they may induce tolerance.[35]

Besides their roles in viral defence, IFN-α and -β are important in the control of differentiation and growth. Generated in the bone marrow, they affect the differentiation of haematopoietic B cells, T cells, osteoclasts, and myeloid dendritic cells.[36] IFN-α2 is used in adjuvant therapy as an anticancer drug. In particular, it has been applied in the treatment of 'hairy cell leukaemia' (a chronic B cell leukaemia[37]), chronic myelogenous leukaemia,[38] and to treat some metastasizing cancers, such as renal carcinoma and Kaposi's sarcoma, which affects AIDS sufferers.

Interferon-α receptor and STAT proteins

The interferon-α (type 1) receptor is made up of two subunits, IFNAR1 and IFNAR2, coupled to the FERM domain of TYK2 and JAK1 respectively[39] (**Figure 17.9**). Both are members of the Janus kinase family. IFNAR2 also binds the latent forms of the cytosol proteins STAT1 or STAT2. To date, seven members of the mammalian family of STATs, all substrates of the JAKs, have been characterized (see also page 353).

Binding of IFN-α2 causes receptor subunit dimerization followed by mutual activation of their associated PTKs (Figure 17.9) which then phosphorylate the IFNAR1 receptor subunit. Binding of STAT2 brings the transcription factor into the vicinity of the activated kinases. Phosphorylated STAT2 promotes phosphorylation of STAT1 and these, attached through their SH2 domains to the opposing phosphotyrosine residues, then dissociate from the receptor as a heterodimer. In the cytoplasm they combine with IRF-9. This trimeric complex, ISGF3, translocates to the nucleus and binds to DNA at the interferon-stimulated response element (ISRE).[40] Amongst the genes that contain the ISRE are interferon-regulated factors which amplify the production of type-1 interferons, the chemokine IP-10 (CXC10), and also genes coding for the ssRNA-dependent protein kinase (PKR), OAS, and the methyltransferases (Mx), which act collectively to keep the viral infection in check.

An alternative complex, less abundant and composed of STAT1 dimers, can also result from IFNAR1/IFNAR2 signalling. This complex binds to the GAS element and induces expression of IRFs. For the IFN-γ receptors, STAT1 homodimers predominate (see Table 17.2).

The STATs also convey signals issuing from other interferon receptors. The specificity of the intracellular signal is determined by the particular combinations of STAT that are phosphorylated and activated.

Alternative signalling pathways

The terms **PKB** and **Akt** both occur in the literature. Akt1, 2, and 3 are the same as PKB α, β, and γ. The PKB/Akt kinases are assembled in the SwissProt database under the 'RAC' subfamily of serine/threonine protein kinases because the mammalian sequence was discovered as a kinase related to PKA and PKC.[43]

The pathway described above does not account for all the biological effects of type 1 interferons. Activation of IFNAR1 and 2 also causes phosphorylation of the insulin receptor substrates IRS-1 and -2. These then act as docking proteins for SH2 domain-bearing proteins such as the p85/p110 PI 3-kinase complex. Membrane localization of this complex results in the production of PI(3,4)P$_2$ with subsequent activation of the PKB signal pathway (see page 550). This has been described as being both dependent and independent of JAK1/TYK2,[39] depending on the particular cell type. STAT1 can also be phosphorylated by the serine/threonine kinases ERK2, p38, casein kinase 2 (CK2), MEK1 (or 2), or protein kinase C-δ (for an example, see Figure 13.18, page 404). Serine phosphorylation of STAT1 plays a role in the survival of transformed stem cells in Wilms (kidney) tumours.[42]

TABLE 17.2 Diversity in cytokine-induced signalling: The spectrum of downstream STATs and STAT-associated proteins determine the outcome of the cellular response to a particular member of the interferon family. A thick red arrow represents the predominant pathway, thin blue arrows signify alternative pathways.

Ligand	IFN-α/β		IFN-γ	IFN-λ
Receptor	IFNAR1 IFRAR2		IFNGR-1 IFNGR-2	IL-28 IL-29
Transcription factor complex	CrkL STAT5	STAT4 STAT4	STAT1 STAT2 IRF-9 (ISGF3)	STAT1 STAT1 (AAF/GAF)
DNA binding element	GAS	STAT	ISRE	GAS
Gene expression	IRF-1 IRF-2 IRF-8 IRF-9	IFN-γ	ISG15 IP-10 IRF-5 IRF-7 2',5' OAS PKR	IRF-1 IRF-2 IRF-8 IRF-9

Information from Takaoka and Yanai.[31] Adapted from Ihle.[41]

Down-regulation of the JAK-STAT pathway

Nuclear dephosphorylation and recycling of STATs

Dimerized and phosphorylated STATs are rapidly dephosphorylated in the nucleus.[44] This is thought to be caused by interactions that force a rearrangement of the dimer partners, so breaking the SH2 domain phosphotyrosine links. This exposes the phosphotyrosines to nuclear phosphatases (**Figure 17.10**).[45] The enzymes involved are not clearly established. In order to maintain the system of signalling, the dephosphorylated STATs exit the nucleus to be re-phosphorylated at the membrane and return as active dimers. When large numbers of cytokine receptors are active, the STAT proteins recirculate rapidly and in consequence, a large proportion is retained as active dimers in the nucleus, stimulating gene transcription. When only few receptors are active, the majority accumulate in the cytoplasm in their inactive state. STAT proteins appear to act as a remote sensors, communicating between the cell surface and the nucleus, though, as described below, the situation is actually more complex.[46]

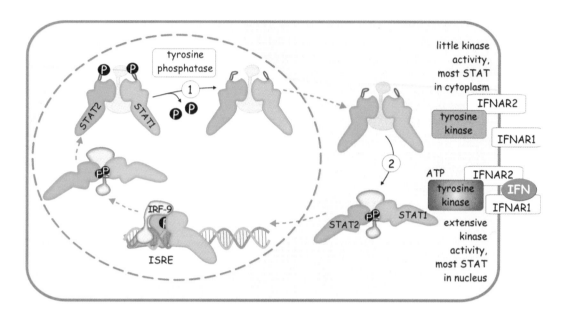

FIG 17.10 STAT proteins act as remote sensors.

STAT proteins that dissociate from the DNA are dephosphorylated by tyrosine phosphatases (1). They return to the cytoplasm and only re-enter the nucleus if they are rephosphorylated by (active) IFN receptors (2). With a large proportion of receptors occupied, the cycle is rapid and STAT accumulates in the nucleus. With few receptors occupied, the cycle is slow, resulting in an accumulation of STAT in the cytoplasm. Effectively, STAT acts as a remote sensor of receptor occupation.

Dephosphorylation of the IFN-α receptor-1

The dual specificity phosphatases SHP-1 and SHP-2 (see page 653) bind phosphorylated IFNAR1, thereby competing with and dephosphorylating STAT and the tyrosine kinases. The pathway is effectively silenced (**Figure 17.11a**). SHP-1 is mainly expressed in haematopoietic cells, SHP-2 ubiquitously. SHP-2 may also act in the nucleus and participate in the dephosphorylation of nuclear STATs.[47] Loss of SHP-1 results in hyperphosphorylation of JAK1 and is the cause of immunological and haematopoietic dysfunctions (mice exhibit the 'motheaten' phenotype[48]). Lack of SHP-2 is embryonic lethal. CD45 is also involved, since its absence also causes hyperactivation of JAK1. Other cytosolic phosphatases, PTP1B and T Cell-PTP, have the capacity to dephosphorylate both JAK2 and TYK2. This results in reduced phosphorylation of IFNAR1 and thus an arrest of STAT signalling (**Figure 17.11b**). This subject is further considered in Chapter 21.

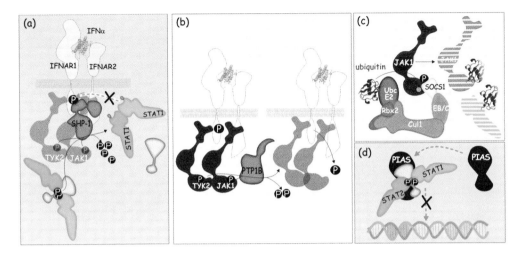

FIG 17.11 Inactivation of the IFN pathway.

Down-regulation of the IFN pathway occurs at different levels. (a) Tyrosine phosphatase SHP-1 binds the phosphorylated IFNAR1 receptor, so preventing association of STAT2. It dephosphorylates and de-activates STAT1/2 complexes and also the tyrosine protein kinases associated with the IFN receptors. (b) PTP1B deactivates TYK2 and JAK1, so preventing further activation of STATs-1 and -2. (c) The SH2 domain of SOCS recognizes tyrosine-phosphorylated IFNAR1 and JAK1. Binding prevents further access of STAT2 to the receptor and inhibits JAK1 activity. SOCS proteins associate with components of a big E3-ligase complex causing polyubiquitylation (K48) of IFNAR1, JAK1 and STAT proteins. Polyubiquitylated proteins are destroyed by the proteasome. (d) PIAS proteins associate with STAT1/2 and IRF-9 and prevent their association with DNA.

Cytokines, including IL-2, IL-3, and EPO, induce the 'suppressors of cytokine signalling' (SOCS1, 2, and 3). These all carry an SH2 domain with which they bind to the tyrosine phosphorylated IFNAR1 and in the case of SOCS1, to tyrosine phosphorylated JAK1 (**Figure 17.11c**). The prevailing hypothesis for their function is that they inhibit both by competing with STAT binding and by linking the components of the receptor–JAK–STAT pathway to an E3-ubiquitin ligase complex. This then leads to ubiquitylation (K48 type) and destruction of JAK1, STAT, or the IFNAR1 receptor. Loss or mutation of SOCS is associated with primary mediastinal B cell lymphoma.[49]

STAT signalling without phosphorylation

STAT signalling is also regulated independently of protein phosphorylation. The family of protein inhibitors of activated STATs (PIAS) form complexes with selected STATs and prevent their stimulation of transcription (**Figure 17.11d**).[50,51] PIASx and PIASy also bind to histone deacetylases and contribute to the repression of transcription, by rendering the DNA less accessible to RNA polymerases (see Table 17.3). Loss of PIAS3 and PIASy are associated with anaplastic T cell lymphoma, which is characterized by high levels of nuclear STAT3.

Mediastinal large B cell lymphoma accounts for about 1 in 50 of all cases of non-Hodgkin's lymphoma. It is a diffuse, large B cell lymphoma deriving from a rare type of B cell present in the thymus. It occurs most commonly in people between 25 and 40 and is twice as common in women as in men.

TABLE 17.3 Protein inhibitors of activated STATs

Inhibitor of activated STAT	STAT isotype
PIAS1	STAT1
PIAS3	STAT3
PIASx (recruits HDAC)	STAT4
PIASy (recruits HDAC)	STAT1

Oncogenes, malignancy, and signal transduction

Viral oncogenes

Infection by viruses carrying oncogenes can cause malignant cell growth. Although first recognized as causative agents in avian cancers just over 100 years ago (page 298), for much of the 20th century there was doubt that any human cancers were initiated in this way. Even now, almost all the information in this area refers to non-human animals. There are a number of problems here. First of all, as was already apparent in the first decade of the century,[52-54] demonstration of a viral mode of transmission depends on the induction of disease by transfer of tissue filtrates from animal to animal. Some viruses only become oncogenic as a consequence of multiple passages and through different animal species. Secondly, while there are many human cancers that are certainly associated with viral infection, in most cases it is far from certain whether it is the virus that actually initiates the condition or whether it is merely permissive of induction by another agent, such as a chemical carcinogen.

An exception is the human papilloma virus (HPV), the causative agent of skin warts[55] and epithelial (cervical) cancer. In general, the transforming products of the viral oncogenes behave as persistently activated mutants of (hijacked) endogenous cellular proteins having key regulatory roles in mitogenesis.

Figure 17.12 is an illustration of how the avian Rous sarcoma oncogene v-Src product (RSVH1) has arisen from a change of the amino acid composition of the C-terminus of Src. Most significantly, this involves the loss of the most C-terminal tyrosine residue with the result that the mutated protein, v-Src is unable to fold into its inactive state. Alternatively, viral oncogene products are authentic viral proteins that interfere with the functioning of nuclear proteins such as the tumour suppressors p53 and Rb. Examples are the transforming

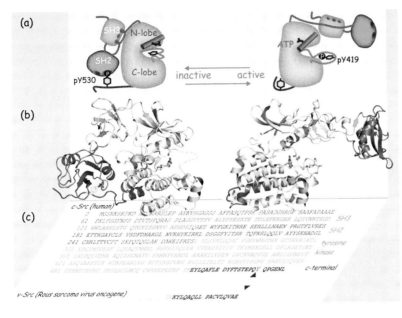

(a)

SH3 N-lobe

SH2 C-lobe

pY530 inactive active

ATP pY419

(b)

(c)

c-Src (human)

    0 MGSNKSKPKD SKRRSLEP AENVHGAGGG AFPASQTPSK PASADGHRGP SAAFAPAAAE
   61 PKLFGGFNSS DTVTSPQRAG PLAGGVTTFV ALYDYESRTE TDLSFKKGER LQIVNNTEGD
  121 WWLAHSLSTG QTGYIPSNYV APSDSIQAEE WYFGKITRRE SERLLLNAEN PRGTFLVRES
  181 ETTKGAYCLS VSDFDNAKGL NVKHYKIRKL DSGGFYITSR TQFNSLQQLV AYYSKHADGL
  241 CHRLTTVCPT SKPQTQGLAK DAWEIPRES. RLEVKLGQGC FGEVWMGTWN GTTRVAIKTL
  301 KPGTMSPEAF LQEAQVMKKL RHEKLVQLYA VVSEEPIYIV TEYMSKGSLL DFLKGETGKY
  361 LRLPQLVDMA AQIASGMAYV ERMNYVHRDL RAANILVGEN LVCKVADFGL ARLIEDNEYT
  421 ARQGAKFPIK WTAPEAALYG RFTIKSDVWS FGILLTELTT KGRVPYPGMV NREVLDQVER
  481 GYRMPCPPEC PESLHDLMCQ CWRKEPEERP TFEYLQAFLE DYFTSTEPQY QPGENL c-terminal

tyrosine

kinase

SH3

SH2

v-Src (Rous sarcoma virus oncogene) ...KYLQAQLL PACVLQVAE

FIG 17.12 From c-Src to the deregulated kinase v-Src.
(a) c-Src exists in two states. In its inactive form its SH2 domain folds back and binds to pY530 in the C-terminal segment. Dephosphorylation of pY530 or displacement by a higher affinity phosphotyrosine releases the SH2 domain, allowing Src to adopt its open form and become phosphorylated in its activation segment (Y419). (b) Ribbon diagram of c-Src in its inactive and active states. (c) v-Src lacks C-terminal residues that include Y530. It cannot adopt the closed conformation and is constitutively active.[57] (2src[58], 1y57[59]).

Soon after its discovery in 1960, SV40 was found as a contaminant in poliovirus vaccine. Clearly, this was a cause of concern. Already, more than 98 million Americans (and countless more throughout the world) had received one or more doses of the vaccine during the period 1955–63. Could immunization cause cancer? As luck turned out, nothing came of it. No widespread incidence of SV40 infection in the population, no increase of tumours, nor any direct role for SV40 in human cancer.[63]

antigens (T-antigens) of DNA papovaviruses such as simian vacuolating virus SV-40 (primate), BK (human), or JC (human).[56,67]

Non-viral oncogenes

Tumours that are not caused by viral infection (such as those caused by chemical carcinogens or radiation, or simply due to the multiple errors that accumulate as a consequence of ageing) also express persistently activated gene products. As an example of the role of oncogenes in cell transformation, mutated forms of Ras are found in 40% of all human cancers[60] and in more than 90% of pancreatic carcinomas. These oncogenes are gain-of-function mutants of the wild-type proteins. They include receptor tyrosine kinases, adaptor proteins, and guanine nucleotide exchange factors. Serine and threonine protein kinases can also act as oncogenes, but in comparison with tyrosine protein kinases, their contribution is relatively modest.[61,62]

A number of these mutated proteins operate in the early stages of tyrosine kinase signal transduction pathways. Cells may be transformed as a consequence of hypersecretion of growth factors, expression of variant forms of PTKs, over-expression of SH2/SH3-containing adaptor proteins, over-expression of serine/threonine protein kinases, or expression of variants of the small GTPases or their accessory proteins. At the downstream end of the signal transduction pathway, variants of transcription factors also act as potent cell transformers. Although tyrosine kinase phosphorylation accounts for only about 5% of total cellular phosphorylation activity, it has a key position in many signal transduction pathways and it is probably for this reason that the incidence of these genes in malignancy is so high (see Table 17.5).

A good example of cell transformation related to aberrant PTK activity is the role of JAK2 in acute lymphoblastic leukaemia (ALL). This is the most common childhood cancer of pre-B cell origin. These cells rely on paracrine or autocrine cytokine stimulation in order to proliferate and survive. Forms of this disease that are more resistant to chemotherapy express a constitutively activated mutant of JAK2.[64] Apparently, this serves as a survival signal for these cells making them insensitive to the damage inflicted on DNA by chemotherapy which would normally induce apoptosis.[65] Another example is Bcr-Abl, the cause of chronic myelogenous leukaemia (CML). This is a fusion protein caused by somatic reciprocal translocation between (paternal) chromosome 9 and (maternal) chromosome 22. The kinase activity of the fusion product, carried by Abl, is deregulated; we return to this topic in Chapter 23.

Essay: Non-receptor protein tyrosine kinases and their regulation

The nrPTKs comprise a large family with diverse roles in the control of cell proliferation, differentiation and death. Some are widely expressed; others are

TABLE 17.4 Families of non-receptor protein tyrosine kinases

ABL	ACK	CSK	FAK	FES	FRK	JAK	SRC-A	SRC-B	TEC	SYK
ABL1	ACK1	CSK	FAK	FER	BRK	JAK1	FGR	BLK	BMX	SYK
ARG	TNK1	MATK	PYK2	FES	FRK	JAK2	FYN	HCK	BTK	ZAP70
					SRMS	JAK3	SRC	LCK	ITK	
						TYK2	YES1	LYN	TEC	
									TXK	

restricted to particular tissues. Their early classification was dominated by the discovery of pp60^{src}, to the extent that the major group of kinases was simply known as the Src family. There are at least 10 known subfamilies of non-receptor PTKs (see **Figure 17.1** and Table 17.4).

The Src family kinases all share a similar structure. A unique chain at the N-terminus is followed by SH3 and SH2 domains (actually, the very archetypes of these domains that are so widely expressed) followed by a kinase domain and short tail at the C-terminus. Many Src kinases function by association with macromolecular signalling complexes assembled at membrane sites. Membrane targeting may be promoted by the unique N-terminal domain. Within the Src family, Src itself (pp60^{c-src}), Fyn, Lyn, and Yes are myristoylated at the N-terminus. This provides the opportunity for membrane attachment that may be strengthened by palmitoylation at a nearby cysteine. Similarly,

TABLE 17.5 Components of tyrosine kinase signal transduction cascades manifested as cellular or viral oncogenes: These oncogenes are gain-of-function mutants of the wild-type proteins, such as receptor tyrosine kinases, adaptor proteins, or guanine exchange factors. Serine or threonine protein kinases can also acts as oncogenes, but in comparison with tyrosine protein kinases their contribution is relatively modest

Receptor tyrosine protein kinase	Non-receptor tyrosine protein kinase	Serine/threonine protein kinase	SH2/SH3 adaptor	Nucleotide exchange factor	GTPase
Bek	Abl	Akt/PKB	Crk	Bcr	H-Ras
Eck	Blk	Cot	Nck	Dbl	K-Ras
Elk	Fgr	Mos		Ost	N-Ras
Eph	Fsp	Pim		Tiam	
ErbB	Fyn	Raf		Vav	
Flg	Hck				
Fms	Jak				
Kit	Lck				
Met	Lyn				
Neu	Src				
Ret	Yes				
TrkA					
TrkB					
TrkC					

members of the Btk/Tec family may become membrane-associated through their PH domains which can bind polyphosphoinositide lipids. Other non-receptor PTKs are recruited to their sites of action through the association of their SH2 domains with phosphotyrosine residues present in their targets. Regardless of their location, most Src family kinases are held in an inactive state by an intramolecular interaction that links a phosphorylated tyrosine in the C-terminus with the N-terminal SH2 domain.

The three-dimensional structure of human c-Src in its inactive and active forms is shown in **Figure 17.12**. The catalytic machinery is contained within the kinase domain, while control is exerted through its interaction with the adjacent structures. The elements that direct the regulation are the SH2 and SH3 domains. pY530 provides an intramolecular binding site for the SH2 domain. In its normal inactive conformation, the SH2 domain binds to pY530 so allowing Src to adopt a compact, closed conformation in which the SH3 and SH2 domains are packed against the kinase domain. This is achieved through the interaction of the SH3 domain with the chain that links the SH2 domain and the small kinase lobe. This linker adopts a left-handed helical conformation (a type II polyproline helix, although there is only one proline residue) that binds the SH3 domain on the one side and through hydrophobic interactions, the small lobe on the other. Thus the SH2 and SH3 domains pack against the kinase domain. This distorts the small lobe, impedes access to the cleft, and causes an outward rotation of the C-helix in the small lobe.

Activation follows events that destabilize the compact conformation. Lacking an isoleucine at pY+3, the sequence motif surrounding pY530 is not optimal for SH2 binding. This makes pY530 more accessible to tyrosine phosphatases as well as to competing SH2 domains of other proteins. Both events may lead to activation of Src, because it removes the inhibitory clamp of the SH2 and SH3 domains. These two domains, together with the linker, then flip out, allowing the kinase domain to relax from its distorted form and for autophosphorylation at Y419 in the activation segment, rendering the protein kinase catalytically competent.

For historical reasons the residues of human c-Src were numbered according to the sequence of chicken c-Src. Thus residues 83–533 (chicken numbering) correspond to 86–536 in humans. We have used human numbering here.

The major suppressor of Src activation is Csk (c-Src kinase), itself a member of the Src family. Csk has no C-terminal tyrosine residue and no lipid modification and in consequence it is both constitutively active and confined to the cytosol. Its target is the Src C-terminal tyrosine (Y530). How does this cytosolic enzyme interact effectively with the plasma membrane-associated Src? It seems likely that a transmembrane docking protein Cbp[66] recruits Csk through an interaction between the SH2 domain of Csk and a phosphotyrosine on Cbp (see **Figure 17.13**). Cbp and Src are both

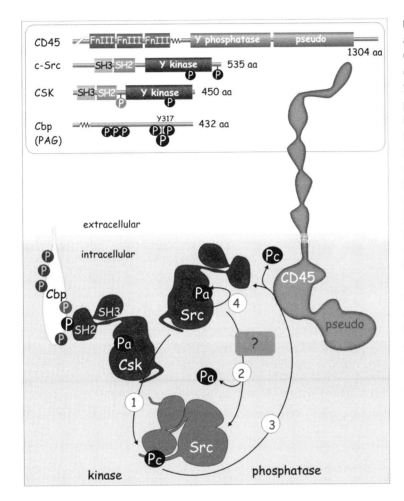

FIG 17.13 Regulation of Src kinase activity by Csk and CD45. Csk is recruited in the TRC signalling complex through interaction of its SH2 domain with the transmembrane phosphoprotein Cbp. Src is inactivated by Csk through tyrosine phosphorylation of its C-terminal tail (1). A tyrosine phosphatase removes the phosphate from the tyrosine in the activation segment (2). Inactive Src can be reactivated by CD45 which removes the phosphate in the C-terminus (Pc) (3). Autophosphorylation of the activation segment completes the process (P$_A$) (4). This mode of regulation is a good example of how retro-control can limit activity. Active Src phosphorylates Cbp causing de-activation. Loss of Src activity and subsequent dephosphorylation of Cbp re-initiates a cycle of Src activation. Many oncogenic tyrosine kinases have the effect of obliterating such oscillations because they are unable to adopt their inactive state.

localized in lipid rafts. The inhibitory action is linked to the removal of the phosphate from the activation segment of Src by a phosphatase of unknown identity. Activation of Src might then initially involve the dephosphorylation of Cbp by a tyrosine phosphatase, amongst which CD45 is a good candidate. CD45 also removes the phosphate from the pY530, thus allowing auto-activation of Src.

Abbreviation	Full name/description	SwissProt entry	Other names, OMIM links
Abl-c	Abelson murine leukaemia viral oncogene homologue	P00519	TNK1 (tyrosine non-receptor kinase)
ACK1	activated Cdc42 protein kinase	Q07912	
BTK	Bruton tyrosine kinase (Bruton's agammaglobulinaemia)	Q06187	
calcineurin A		Q08209	PP2B, PPP3 catalytic subunit a
calcineurin-B1		P63098	PP2B, PPP3 regulatory subunit 1
Calmodulin		P62158	CaM
CARD	caspase recruiting domain		
CARMA1	CARD-containing MAGUK protein-1	Q9BXL7	CARD11
Cbp	Csk-binding protein	Q9NWQ8	PAG1
CD3 ε-chain	cluster of differentiation-3 epsilon chain (TCR)	P07766	
CD3 ζ-chain	cluster of differentiation-3 zeta chain (TCR)	P20963	CD247, TCR ζ-chain
CD4	cluster of differentiation-4	P01730	
CD45	cluster of differentation-45	P08575	OMIM:608971
CIB4	calcium and integrin binding family member 4	A0PJX0	
CK1	casein kinase-1	P48729	
CSK	c-Src kinase	P41240	cyl
CSK	C-terminal Src kinase	P41240	
DYRK1A	dual specificity tyrosine-phosphorylation regulated kinase 1A	Q13627	
FERM	Four-one (band 4.1), Ezrin, Radixin, Moesin domain		
FES	feline sarcoma (Snyder–Theilen feline sarcoma virus)	P07332	FPS

Continued

Abbreviation	Full name/description	SwissProt entry	Other names, OMIM links
FRK	Fyn-related protein kinase	P42685	nuclear kinase Rak
GAS	gamma-interferon activated sequence		
GSK3β	glycogen synthase kinase-3 beta	P49841	
IFN-α	interferon alpha	P01562	
IFNAR1	interferon alpha receptor-1	P17181	
IFNAR2	interferon alpha receptor-2	P48551	
IFN-γ	interferon gamma	P01579	
IFNGR1	interferon gamma receptor1	P15260	
IFNGR2	interferon gamma receptor2	P38484	
IFN-λ	interferon lambda	Q8IU54	IL-29
IFR	interferon regulatory factors		
IFR-9	interferon regulatory factor-9	Q00978	
ISGF3	interferon-stimulated gene factor-3		
JAK1	janus kinase-1	P23458	
LAT	linker of activated T cells	O43561	pp36
Lck	lymphocyte cell-specific protein kinase	P06239	
MAGUK	membrane associated guanylate kinase domain		
NFAT1	nuclear factor of activated T cells cytosolic component	Q13469	NFATC2
NFAT2	nuclear factor of activated T cells nuclear component	O05644	NFATC1
PIAS-1	protein inhibitor of activated STAT protein-1	O75925	
PKCθ	protein kinase C theta	Q04759	
PLC-γ	phospholipase C gamma	P19174	

Continued

537

Abbreviation	Full name/description	SwissProt entry	Other names, OMIM links
PTP-1B	protein tyrosine-phosphatase-1	P18031	
SHP-1	SH2-containing tyrosine phosphatase-1	P29350	PTPN6, PTP1C
SHP-2	SH2-containing tyrosine phosphatases-2	Q06124	MIM:151100
SLP76	SH2-domain containing leukocyte protein 76 kDa	Q13094	
SOCS1	suppressor of cytokine signalling-1	O15524	Tec-interacting protein
Src	Sarcoma		
Src-c	cellular sarcoma protein	P12931	p60-Src
Src-v	viral sarcoma protein (transforming protein from strain H-19)	P25020	
STAT1	signal transducer and activator of transcription-1	P42224	
STAT2	signal transducer and activator of transcription-2	P52630	
SYK	spleen tyrosine protein kinase	P43405	
TCR	T cell receptor		
TEC	tyrosine kinase expressed in hepatocellular carcinoma	P42680	
TYK2	tyrosine kinase-2	P29597	
ZAP70	zeta-chain associated protein kinase 70 kDa	P43403	

References

1. Robinson DR, Wu YM, Lin SF. The protein tyrosine kinase family of the human genome. *Oncogene*. 2000;19:5548–5557.
2. Cantrell DA. T cell receptor signal transduction pathways. *Annu Rev Immunol*. 1996;14:259–274.
3. Burg DL, Furlong MT, Harrison ML, Geahlen RL. Interactions of Lyn with the antigen receptor during B cell activation. *J Biol Chem*. 1994;269:28136–28142.
4. Metzger H. The receptor with high affinity for IgE. *Immunol Revs*. 1992;125:37–48.

5. Dustin ML, Tseng SY, Varma R, Campi G. T cell-dendritic cell immunological synapses. *Curr Opin Immunol*. 2006;18:512–516.

6. Gacon G, Fagard R, Boissel JP, et al. Identification of a 58,000 daltons phosphoprotein with tyrosine protein kinase activity in a murine lymphoma cell line. *Biochem Biophys Res Commun*. 1984;122: 563–570.

7. Marth JD, Peet R, Krebs EG, Perlmutter RM. A lymphocyte-specific protein-tyrosine kinase gene is rearranged and overexpressed in the murine T cell lymphoma LSTRA. *Cell*. 1985;43:393–404.

8. Barber EK, Dasgupta JD, Schlossman SF, Trevillyan JM, Rudd CE. The CD4 and CD8 antigens are coupled to a protein-tyrosine kinase (p56[lck]) that phosphorylates the CD3 complex. *Proc Natl Acad Sci USA*. 1989;86: 3277–3281.

9. Palacios EH, Weiss A. Function of the Src-family kinases, Lck and Fyn, in T-cell development and activation. *Oncogene*. 2004;23:7990–8000.

10. Hermiston ML, Xu Z, Weiss A. CD45: a critical regulator of signaling thresholds in immune cells. *Annu Rev Immunol*. 2003;21:107–137.

11. Cantrell D. T cell antigen receptor signal transduction pathways. *Annu Rev Immunol*. 1996;14:259–274.

12. Neumeister EN, Zhu Y, Richard S, Terhorst C, Chan AC, Shaw AS. Binding of ZAP-70 to phosphorylated T-cell receptor ζ and η enhances its autophosphorylation and generates specific binding sites for SH2 domain-containing proteins. *Mol Cell Biol*. 1995;15:3171–3178.

13. Zhang W, Sloan-Lancaster J, Kitchen J, Trible RP, Samelson LE. LAT: the ZAP-70 tyrosine kinase substrate that links T cell receptor to cellular activation. *Cell*. 1998;92:83–92.

14. Braiman A, Barda-Saad M, Sommers CL, Samelson LE. Recruitment and activation of PLCγ1 in T cells: a new insight into old domains. *EMBO J*. 2006;25:774–784.

15. Okamura H, Aramburu J, Garcia-Rodriguez C, et al. Concerted dephosphorylation of the transcription factor NFAT1 induces a conformational switch that regulates transcriptional activity. *Mol Cell*. 2000;6:539–550.

16. Feske S, Gwack Y, Prakriya M, et al. A mutation in *Orai1* causes immune deficiency by abrogating CRAC channel function. *Nature*. 2006;441: 179–185.

17. Gwack Y, Sharma S, Nardone J, et al. A genome-wide *Drosophila* RNAi screen identifies DYRK-family kinases as regulators of NFAT. *Nature*. 2006;441:646–650.

18. Avots A, Buttmann M, Chuvpilo S, et al. CBP/p300 integrates Raf/Rac-signaling pathways in the transcriptional induction of NF-ATc during T cell activation. *Immunity*. 1999;10:515–524.

19. Weir MR, Fink JC. Risk for posttransplant diabetes mellitus with current immunosuppressive medications. *Am J Kidney Dis*. 1999;34:1–13.

20. Heit JJ, Apelqvist AA, Gu X, et al. Calcineurin/NFAT signalling regulates pancreatic β-cell growth and function. *Nature*. 2006;443:345–349.

21. Langdon WY, Hartley JW, Klinken SP, Ruscetti SK, Morse HC. v-*cbl*, an oncogene from a dual-recombinant murine retrovirus that induces early B-lineage lymphomas. *Proc Natl Acad Sci U S A*. 1989;86:1168–1172.

22. Swaminathan G, Tsygankov AY. The Cbl family proteins: ring leaders in regulation of cell signaling. *J Cell Physiol*. 2006;209:21–43.

23. Sun L, Deng L, Ea CK, Xia ZP, Chen ZJ. The TRAF6 ubiquitin ligase and TAK1 kinase mediate IKK activation by BCL10 and MALT1 in T lymphocytes. *Mol Cell*. 2004;14:289–301.

24. Liu YC, Penninger J, Karin M. Immunity by ubiquitylation: a reversible process of modification. *Nat Rev Immunol*. 2005;5:941–952.

25. Colombetti S, Basso V, Mueller DL, Mondino A. Prolonged TCR/CD28 engagement drives IL-2-independent T cell clonal expansion through signaling mediated by the mammalian target of rapamycin. *J Immunol*. 2006;176:2730–2738.

26. Morris RE. Rapamycin: FK506's fraternal twin or distant cousin? *Immunol Today* 1991;12:137–140.

27. Myers MD, Dragone LL, Weiss A. Src-like adaptor protein down-regulates T cell receptor (TCR)-CD3 expression by targeting TCRζ for degradation. *J Cell Biol*. 2005;170:285–294.

28. Simons K, Ikonen E. Functional rafts in cell membranes. *Nature*. 1997;387:569–572.

29. Schutz GJ, Kada G, Pastushenko VP, Schindler H. Properties of lipid microdomains in a muscle cell membrane visualized by single molecule microscopy. *EMBO J*. 2000;19:892–901.

30. Isaacs A, Lindemann J. Virus interference. I. The interferon. *Proc R Soc Lond B Biol Sci*. 1957;147:258–267.

31. Takaoka A, Yanai H. Interferon signalling network in innate defence. *Cell Microbiol*. 2006;8:907–922.

32. Chill JH, Quadt SR, Levy R, Schreiber G, Anglister J. The human type I interferon receptor: NMR structure reveals the molecular basis of ligand binding. *Structure*. 2003;11:791–802.

33. Stark GR, Kerr IM, Williams BR, Silverman RH, Schreiber RD. How cells respond to interferons. *Annu Rev Biochem*. 1998;67:227–264.

34. van den Broek MF, Muller U, Huang S, Aguet M, Zinkernagel RM. Antiviral defense in mice lacking both α/β and γ interferon receptors. *J Virol*. 1995;69:4792–4796.

35. Reis e Sousa C. Dendritic cells in a mature age. *Nat Rev Immunol*. 2006;6:476–483.

36. Honda K, Yanai H, Takaoka A, Taniguchi T. Regulation of the type I IFN induction: a current view. *Int Immunol*. 2005;17:1367–1378.

37. Genot E. Interferon α and intracytoplasmic free calcium in hairy cell leukemia cells. *Leuk-Lymphoma*. 1994;12:373–381.

38. Kantarjian HM, Talpaz M, O'Brien S, et al. Survival benefit with imatinib mesylate versus interferon-α-based regimens in newly diagnosed chronic-phase chronic myelogenous leukemia. *Blood*. 2006;108: 1835–1840.

39. van Boxel-Dezaire AH, Rani MR, Stark GR. Complex modulation of cell type-specific signaling in response to type I interferons. *Immunity*. 2006;25:361–372.

40. Shual K, Ziemiecki A, Wilks AF, et al. Polypeptide signalling to the nucleus through tyrosine phosphorylation of Jak and Stat proteins. *Nature*. 1993;366:580–583.

41. Ihle JN. Cytokine receptor signalling. *Nature*. 1995;377:591–594.

42. Timofeeva OA, Plisov S, Evseev AA, et al. Serine-phosphorylated STAT1 is a prosurvival factor in Wilms' tumor pathogenesis. *Oncogene*. 2006.

43. Jones PF, Jakubowicz T, Pitossi FJ, Maurer F, Hemmings BA. Molecular cloning and identification of a serine/threonine protein kinase of the second-messenger subfamily. *Proc Natl Acad Sci U S A*. 1991;88:4171–4175.

44. Haspel RL, Darnell JE. A nuclear protein tyrosine phosphatase is required for the inactivation of Stat1. *Proc Natl Acad Sci U S A*. 1999;96:10188–10193.

45. Zhong M, Henriksen MA, Takeuchi K, et al. Implications of an antiparallel dimeric structure of nonphosphorylated STAT1 for the activation-inactivation cycle. *Proc Natl Acad Sci U S A*. 2005;102:3966–3971.

46. Swameye I, Muller TG, Timmer J, Sandra O, Klingmuller U. Identification of nucleocytoplasmic cycling as a remote sensor in cellular signaling by databased modeling. *Proc Natl Acad Sci USA*. 2003;100:1028–1033.

47. Wu TR, Hong YK, Wang XD, et al. SHP-2 is a dual-specificity phosphatase involved in Stat1 dephosphorylation at both tyrosine and serine residues in nuclei. *J Biol Chem*. 2002;277:47572–47580.

48. Tsui HW, Siminovitch KA, de Souza L, Tsui FWL. Motheaten and viable motheaten mice have mutations in the haematopoietic cell phosphatase gene. *Nature*. 1993;4:124–129.

49. Valentino L, Pierre J. JAK/STAT signal transduction: regulators and implication in hematological malignancies. *Biochem Pharmacol*. 2006;71:713–721.

50. Liu B, Liao J, Rao X, et al. Inhibition of Stat1-mediated gene activation by PIAS1. *Proc Natl Acad Sci USA*. 1998;95:10626–10631.

51. Shuai K, Liu B. Regulation of gene-activation pathways by PIAS proteins in the immune system. *Nat Rev Immunol*. 2005;5:593–605.

52. Warthin AS. Leukemia of the common fowl. *J Infect Diseases*. 1907;4:369–380.

53. Warthin AS. *Trans Assoc Am Physicians*. 1904;19:421.

54. Ellermann V, Bang O. Experimentelle leukämie bei hünern. *Zentr Bakteriol Parasitenk*. 1908;4:595–609.

55. Kidd JG, Beard JW, Rous P. Serological reactions with a virus causing rabbit papllomas which become cancerous. *J Exp Med*. 1936;64:63–77.

56. Doherty J, Freund R. Polyomavirus large T antigen overcomes p53 dependent growth arrest. *Oncogene*. 1997;14:1923–1931.

57. Wilkerson VW, Bryant DL, Parsons JT. Rous sarcoma virus variants that encode src proteins with an altered carboxy terminus are defective for cellular transformation. *J Virol*. 1985;55:314–321.

58. Xu W, Doshi A, Lei M, Eck MJ, Harrison SC. Crystal structures of c-Src reveal features of its autoinhibitory mechanism. *Mol Cell*. 1999;3:629–638.

59. Cowan-Jacob SW, Fendrich G, Manley PW, et al. The crystal structure of a c-Src complex in an active conformation suggests possible steps in c-Src activation. *Structure*. 2005;13:861–871.

60. Bos JL. ras oncogenes in human cancer: a review. *Cancer Res*. 1989;49:4682–4689.

61. Spector DH, Varmus HE, Bishop JM. Nucleotide sequences related to the transforming gene of avian sarcoma virus are present in DNA of uninfected vertebrates. *Proc Natl Acad Sci U S A*. 1978;75:4102–4106.

62. Bishop JM, Baker B, Fujita D, et al. Genesis of a virus-transforming gene. *Natl Cancer Inst Monogr*. 1978:219–223.

63. Poulin DL, DeCaprio JA. Is there a role for SV40 in human cancer? *J Clin Oncol* 2006;24:4356–4365.

64. Lacronique V, Boureux A, Valle VD, et al. A TEL-JAK2 fusion protein with constitutive kinase activity in human leukemia. *Science*. 1997;278:1309–1312.

65. Meydan N, Grunberger T, Shahar M, et al. Inhibition of acute lymphoblastic leukaemia by a Jak-2 inhibitor. *Nature*. 1996;379:645–648.

66. Kawabuchi M, Satomi Y, Takao T, et al. Transmembrane phosphoprotein Cbp regulates the activities of Src-family tyrosine kinases. *Nature*. 2000;404:999–1003.

67. De Luca A, Baldi A, Esposito V, et al. The retinoblastoma gene family pRb/p105, p107, pRb2/p130 and simian virus-40 large T-antigen in human mesotheliomas. *Nat Med*. 1997;3:913–916.

CHAPTER 18

Phosphoinositide 3-Kinases, Protein Kinase B, and Signalling through the Insulin Receptor

In a *viva voce* examination at one of the more august universities, a student was being questioned about insulin and how it works. After some embarrassed hesitation, he assured the examiners that he did know how it works, but unfortunately he had now forgotten. 'What a pity', came the response. 'That means that now, nobody knows.'

Insulin receptor signalling; it took a little time to work out the details

As should be abundantly clear from Chapter 1, a lack of knowledge about how a particular drug or therapy works has never been an impediment to doctors working in the clinic. Indeed, for most practitioners, it is generally a very secondary consideration. A well-known case in point is aspirin, first introduced by the Bayer Company in 1898. With the wholehearted encouragement of the medical profession, it has been consumed by the

The many abbreviations used in this chapter are collected together at the end of the chapter.

tonne, and with good effect too. Yet it was more than 80 years before its most sensitive target, cyclo-oxygenase, was identified as responsible for the synthesis of prostaglandins.[1] Another example is the treatment of childhood (type I) diabetes with insulin.[2,3]

Several glycogenolytic hormones (adrenaline, glucagon, vasopressin, growth hormone) act to mobilize the metabolic stores of carbohydrate and fat, to maintain the concentration of blood glucose, but insulin uniquely has the reverse effect. It increases the net uptake of glucose from the blood into the tissues and increases its conversion to glycogen and triglyceride, at the same time inhibiting their breakdown. Since the acute effects of the glycogenolytic hormones all require elevations of cAMP or Ca^{2+}, it follows that the actions of insulin must be mediated through a third route, one that escapes the attentions of either PKA or PKC. Only in 1980 did it become apparent that the mechanism involves autophosphorylation of the receptor.[4] The discovery of protein kinase B (also known as Akt) and the 3-phosphorylated inositol lipids eventually provided the key.[4]

The association of diabetes mellitus with the pancreas was established by the work of Oscar Minkowski.[5] A dog from which he had removed the pancreas began to suffer an uncontrollable polyuria. The following day, there being no cage large enough to accommodate the animal, which had been well house-trained, it was kept tied up in the laboratory. According to the story (subsequently denied by Minkowski), the assistant noticed that flies settled wherever it had passed urine. Regardless of this, the animal passed 12% of sugar and was suffering from diabetes mellitus.

It is customary to credit the Canadians Banting and Best with the discovery of insulin, their colleagues McLeod and Collip somehow standing close by (or not quite so close by in the case of McLeod, who may have spent some of the critical months on a fishing holiday on the Isle of Skye). For sure, were it not for Banting's certainty and indeed obsession, it is unlikely that patients would have been successfully treated within a few months of the commencement of the experimental trials. However, it is legitimate to ask whether the Canadians were actually the first discoverers of insulin. Here, we are not considering the widely canvassed claim of Nicholas Paulescu,[2,3] who was certainly in the race at about the same time, but that of Eugene Gley who was working 25 years earlier. Sadly, his right to scientific immortality fails on two counts. First, he was never in a position to apply his preparation in the treatment of patients. This is the importance of Banting and Best's contribution. Whether he had the means to make the repeated and rapid analyses of blood glucose concentration, necessary to monitor its antiglycosuric effect, is uncertain. More likely, he just lacked the audacity to inject his preparation into people, a restraint to which Banting was quite insensitive. Anyway, Gley was working in 1895 and all this had to wait for another quarter century, but we do know from his report that there can be little doubt that he had successfully isolated an antiglycosuric agent from extracts of pancreas. He provides sufficient detail, rare in his day, for the modern reader to have some confidence that his preparation was as good as he claimed.

But where was Gley's report all this time? Instead of going public, he sealed it in an envelope which he placed into the hands of the secretary of the Société de Biologie de Paris, with the firm instruction that it was not to be opened until directed by him.[6] With this bizarre act he waived all claims to be credited as the discoverer of insulin, and so it lay hidden until word came through from Canada in 1921. The application of insulin was certainly one of the key milestones of modern medical practice, yet its mechanism of action as a glucose-lowering hormone has been widely regarded as a 'mystery', even into the last decade of the 20th century.[7,8]

Signalling through phosphoinositides

In Chapter 5 we discussed the effector, phospholipase C, that utilizes the minority phospholipid, phosphatidylinositol 4,5 bisphosphate ($PI(4,5)P_2$), as its substrate. Cleavage of the phospholipid head group produces the soluble second messenger inositol trisphosphate (IP_3), which releases Ca^{2+} from intracellular stores into the cytosol. The other product is the lipid, diacylglycerol (DAG) which activates PKC. $PI(4,5)P_2$ is present in unstimulated cells and its level in cell plasma membranes was originally considered to be under homeostatic control; it was therefore not classed as a signalling molecule. However, this is an oversimplification. Quite apart from being a substrate for PLC, the head group of $PI(4,5)P_2$ provides an important anchoring point on membranes for many cytosolic signalling proteins. (Principally those with phosphoinositide-binding domains; see page 774.) Furthermore, this phospholipid also exists in other cellular membranes, such as those of the Golgi apparatus, and even in the plasma membrane it is distributed unevenly, tending to segregate into dynamic microdomains rich in cholesterol and sphingolipids, commonly referred to as lipid rafts (see page 522). Exactly how the levels of $PI(4,5)P_2$ in these different pools are regulated is unclear, but there is evidence for local control through the phosphatidylinositol phosphate kinases (PIPkins) that convert phosphatidylinositol-4-phosphate and phosphatidylinositol-5-phosphate (PI(4)P and PI(5)P) to $PI(4,5)P_2$. For example, PIPKinI (also known as PI(4)P 5-kinase) functions downstream of the small GTPases Rho[9], Rac,[10] and Arf.[11]

In addition to all this, $PI(4,5)P_2$ is just one of many phosphoinositides that exist in cell membranes. They are formed by a series of reversible phosphorylations and dephosphorylations of the head group. An array of kinases and phosphatases regulate their interconversion. The metabolism of the seven principal phosphoinositides is outlined in Figure 18.1. Although

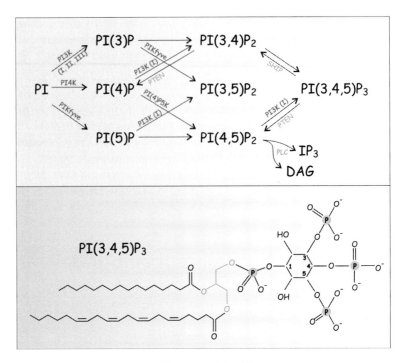

FIG 18.1 Phosphoinositide metabolism and the structure of PI(3,4,5)P$_3$.
A reaction scheme outlining the pathways of formation of the seven principal phosphoinositides. Selected kinases are shown in red. PTEN and SHIP are phosphatases. The PI 3-kinases phosphorylate the inositol headgroup at the 3′ position. Other kinases include the phosphatidylinositol phosphate kinases (PIPKins) of which PI(4)P 5-kinase is a PIPKinI and PIKfyve a PIPKinIII.

this scheme may seem complicated, it is important to bear in mind that the kinases are restricted to particular membranes so that the different phospholipids are enriched at specific locations. Thus PI(4)P is the predominant phosphoinositide on cytosolic surface of the Golgi apparatus, while PI(3)P, PI(4)P and PI(3,5)P$_2$ are enriched on the surfaces of the vesicular compartments that characterize the endocytic pathway, such as endosomes. PI(4,5)P$_2$ and PI(3,4)P$_2$ are predominant on the plasma membrane. The head groups of each of these phosphoinositides then provide specific tethers for the relevant proteins that recruit the machinery of endocytosis, exocytosis and intracellular vesicular transport (see also Chapter 24).

PI 3-kinase, PI(3,4)P$_2$ and PI(3,4,5)P$_3$

While PI(4)P and PI(4,5)P$_2$ are present in resting cells, the 3-phosphoinositides PI(3,4)P$_2$ and PI(3,4,5)P$_3$ are only formed upon the activation of PI 3-kinase

(Figure 18.1). PI(3,4,5)P$_3$ was first detected in human neutrophils after stimulation with formylmethionyl peptides.[12] It was initially thought to be another substrate for PLC, and thus the source of a water-soluble second messenger (inositol-1,3,4,5-tetrakisphosphate or IP$_4$), but it soon became clear that the lipid itself is a signalling entity.

Phosphatidylinositol occupies a special place among phospholipids, because its head group offers a multiplicity of phosphorylation sites. Although inositol lipids have been detected in analyses of bacteria, there is no indication that any of them plays a regulatory role. However, it is clear that some bacteria rely on the inositide metabolism of host cells to regulate their invasiveness.[13] PI is present in all eukaryotic cells, though its metabolism in unicellular organisms such as yeast is restricted, since they lack the means to generate either PI(4)P or PI(4,5)P$_2$. The substrate specificity of the yeast PI 3-kinase (coded by the gene VPS34) is appropriately limited to PI and therefore it only generates the monophosphorylated derivative PI(3)P (which can then be converted to PI(3,5)P$_2$, essential for Golgi membrane recycling[14,15]). The PI 3-kinases catalysing the formation of PI(3,4)P$_2$ and PI(3,4,5)P$_3$ probably evolved with the need for more complex forms of metabolic regulation following the emergence of the metazoans.

Inositol phospholipids having a phosphate group at the 3′ position do not serve directly as substrates for phospholipase C. Instead, they are metabolized by the hydrolysis of the phosphate groups at the 3′- and 5′- positions by the phosphatases PTEN and SHIP respectively (**Figure 18.1**). Both the 3-kinases and the 3-phosphatases are regulated by receptor-mediated processes.

A family of PI 3-kinases

Cloning and screening strategies have revealed a family of PI 3-kinases consisting of three types that have distinct substrates and various forms of regulation (Figure 18.2). They all have four homologous regions, the kinase domain being the most conserved[16].

Type I PI 3-kinases

Type I PI 3-kinases phosphorylate PI, PI(4)P and PI(4,5)P$_2$ (the preferred substrate). They are heterodimers *in vivo* comprising a regulatory (p55 or p85) subunit that maintains the enzyme in a low activity state and a catalytic subunit (p110). Each exists in various forms (Figure 18.2). Their multi-domain structure (in particular p85) enables them to interact with other signalling proteins allowing activation.[17] The SH2 domains enable them to bind to phosphotyrosine residues. Similarly, the SH3 domains allow interaction with proline-rich sequences, present for instance in the focal adhesion kinase FAK, the adaptor molecule Shc, the GTPase-activating protein Cdc42GAP, or the regulator of TCR signalling, Cbl.[18] In addition, the p85-subunit contains a

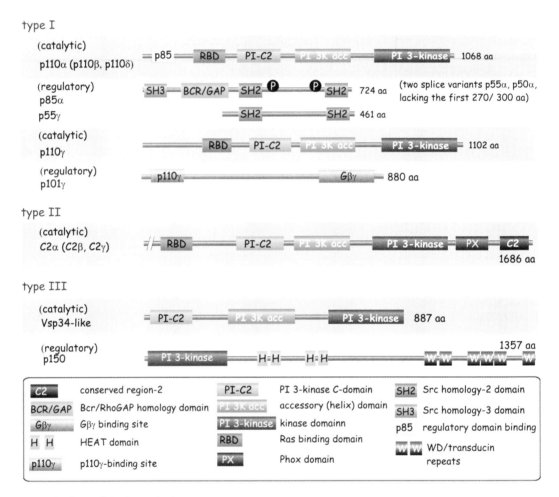

FIG 18.2 Classification of phosphoinositide 3-kinases.

This is based on the structure of the kinase domain and there are three main types. Type I enzymes are heterodimeric and are subdivided into two groups. Group A comprises either p110α, -β, or -δ with the regulatory subunit p85α (or its splice variants p55α and p50α) or p55γ. Group B has one member, p110γ associated with the p101γ regulatory subunit. This is involved in Gβγ-mediated activation of the kinase. Type II comprises three isoforms, C2α, -β, and -γ, none of which requires a regulatory subunit for its activity. Type III has only one member, PI 3-kinase-III (also known as Vsp34-like) which has a p150 regulatory subunit. The mechanisms regulating the type II and III kinases are not yet established.

BCR/GAP homology domain[19] that interacts with Rac and Cdc42, members of the Rho family of GTPases, providing yet further opportunities for regulation[20] (Figure 18.4).

There are four isoforms of the p110 catalytic-subunit (α, β, δ, and γ) all of which contain a kinase domain and a Ras binding site.[21] In addition, the α, β, and δ isoforms possess interaction sites for the p85-subunit. The type I enzymes can be further subdivided. **Type IA** enzymes (α, β, and δ) interact through their SH2 domains with phosphotyrosines present on either protein tyrosine kinases or to docking proteins such as insulin receptor substrates (IRS, see below) or LAT (see page 517). Uniquely, the class IA enzymes activate protein kinase B (PKB, also called Akt, see below). **Type IB** consists of a single member, p110γ linked to a unique regulatory subunit, p101γ, that has no apparent sequence homology with other proteins. Importantly, p110γ can be regulated by $\beta\gamma$-subunits.[22,23]

The type I PI 3-kinase enzymes respond to different upstream signals and have different functions. This is apparent in macrophages in which distinct isotypes modulate separate cellular responses. Here, mitogenic signalling induced by colony stimulating factor-1 is mediated by p110α, whereas actin organization and cell migration require the β or δ isoforms.[24]

In mammalian cells the class I PI 3-kinases have roles in the modification of intracellular membranes, affecting the recruitment of the fission and fusion machinery necessary for intracellular vesicle traffic. The class I PI 3-kinases are also involved in the formation of phagosomes.[25]

Type II PI 3-kinases

The three members of this group, α, -β, and -γ, have substrate specificity for PI and PI(4)P. They are all monomeric and possess a C-terminal C2 domain. Their mode of activation is unclear.

Type III PI 3-kinases

These are represented by the human homologues of the yeast gene product VPS34. They only phosphorylate PI to form PI(3)P. Like the yeast enzyme, the human form is tightly coupled to a regulatory subunit (p150, homologue of VPS15), an serine/threonine kinase which both phosphorylates and recruits the catalytic unit to membranes. Under conditions of starvation, the activity of hVps34 is inhibited, causing restriction of the synthesis of enzymes involved in a pathway that acts via the kinase complex mTOR/Raptor (see below). How hVps34 regulates mTOR/Raptor activity is not known, but it is clear that it is a key component of the nutrient sensing system. In yeast, hVsp34 is involved in autophagy, a rescue mechanism that allows renewal of cellular components in times of prolonged starvation.[26]

Studying the role of PI 3-kinase

Wortmannin, an inhibitor of PI 3-kinase

It might be thought that the availability of an inhibitor of the PI 3-kinases, in this case wortmannin, would have provided an unambiguous key to the understanding of the pathways and cellular functions that they control. This antifungal antibiotic, isolated from *Penicillium wortmannii*, was originally

Unfortunately, it has all too readily been assumed that the actions of **wortmannin** are specific to the 3-kinases, even though it was in use as an inhibitor of myosin light chain kinase for at least a couple of years before its first application as an inhibitor of PI 3-kinases.[30] At concentrations above 10^{-7} M, wortmannin also inhibits a form of PI 4-kinase and then, in the micromolar range, in addition to its effects on myosin light chain kinase, it possibly inhibits other protein kinases as well. As a result of its uncritical use, the 3-phosphorylated lipids were considered to be implicated in several processes that are quite innocent of any such relationship.

This could be taken as an object lesson in the use of inhibitors in general. Experience advises that the terms 'potent' and 'highly specific', frequently used to promote pharmacological agents on their first outing, may wear a bit thin after a year or two. As the evidence of side effects accumulates, new 'potent' and 'highly specific' agents come to take their place. The application of 'potent' and 'highly specific' inhibitors of calmodulin and later of protein kinase C muddled progress in an earlier generation.

identified as a toxic agent causing acute necrosis of lymphoid tissues, severe myocardial haemorrhage and haemoglobinuria.[27] At an appropriate dose, it is also a powerful anti-inflammatory agent.[28]

At concentrations above 10^{-9} M, wortmannin associates covalently with the p110 catalytic subunit of PI 3-kinase[29] (Figure 18.3). Used judiciously, and in parallel with other inhibitors (e.g. Lilly compound LY294002) and independent approaches, wortmannin certainly has its place in the battery of techniques for investigating signal transduction. In the quest for inhibitory effects, however, it has often been applied at high concentrations, leading to false positive reports of inhibition. Furthermore, due to the existence of multiple isoforms of PI 3-kinase, not all of which are targets of this compound, a negative result does not necessarily rule out a role for the products of 3-phosphorylation.

Pathways of activation for PI 3-kinase

The process by which receptors for growth factors activate the type IA PI 3-kinases through interaction with the SH2 domain of the p85 regulatory subunits is well established (see Figure 18.4). PI 3-kinase can also be recruited and activated by non-receptor tyrosine kinases such as those of the Src family through the same mechanism.[32] In platelets that have been stimulated by thrombin, it appears that activation of PI 3-kinase is linked to FAK. This kinase, associated with integrin signalling and cytoskeletal organization at focal adhesion sites, contains proline-rich regions that interact with the SH3 domain on p85.[33] In addition, phosphorylation of FAK allows interaction with the SH2 domain of p85 and this appears to be the route of activation following attachment of cells to solid substrates. Apparently, there are two routes to the activation of PI 3-kinase. One of these is direct, the other indirect through the involvement of focal adhesion sites (for further discussion concerning the link between FAK and the cytoskeleton (see page 400). PI 3-kinase is also activated by Ras which interacts with directly with the catalytic p110α subunit (Figure 18.4).[21] Other GTPases, particularly those of the Rho family, are also involved as regulators (and downfield effectors) in pathways regulated by PI 3-kinase.[23]

Protein kinase B and activation through PI(3,4)P$_2$

The viral oncogene v-akt (acutely transforming retrovirus AKT8) encodes a fusion product of a cellular serine/threonine kinase and the viral structural component Gag. This kinase, PKB, is similar to both PKCε (73% identity to the catalytic domain) and PKA (68%). Unlike the other kinases, it contains a PH domain that enables it to bind to polyphosphoinositide head groups. There are three subtypes, all of which show a broad tissue distribution (Figure 18.5).

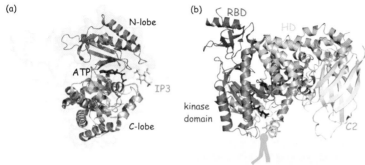

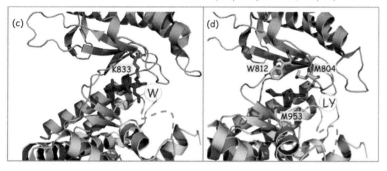

FIG 18.3 Inhibition of PI 3-kinase-γ by wortmannin and LY294002.
(a) The structure of the catalytic domain of PI 3-kinase resembles that of the classical protein kinases (Chapter 24). It is composed of a smaller N-lobe and a larger C-lobe and between them the catalytic cleft that harbours ATP and forms the substrate entry point. The activation segment (dotted orange line) determines the affinity of the kinase for phosphatidylinositol lipids. (b) Structure of PI 3-kinase showing the Ras-binding domain (RBD) (blue), C2 domain (yellow), helical domain (green), and catalytic domain (pink). The RBD contacts the N-lobe and to a lesser degree the C-lobe, suggesting that Ras binding will have an allosteric (stimulatory) effect on the catalytic domain. The C2 domain participates in membrane interaction. The helical domain is common to the PI 3-kinase and PI 4-kinase families and provides the framework for the attachment of the other domains. (c) Wortmannin (W) binds covalently at K833 involved in coordination of the α-phosphate of ATP. It irreversibly blocks access of ATP and thus kinase activity. (d) LY294002 (LY) also occupies the ATP binding pocket. Its predominant sites of interaction are at M804, Y812, and M953. Only the catalytic cleft of the kinase domain is shown and the structure lacks the N-terminal regions that interact with the regulatory subunit (p85 or p55) (1e8x,[16] 1e7u,[16] 1e7v[31]).

With respect to their structural organization and mode of activation, they belong to the AGC subfamily of serine/threonine protein kinases.

Shortly after its discovery, it was found that PKB is activated by PI(3,4)P$_2$ but the actual mechanism of activation turns out to be far from simple. The phospholipid plays two distinct roles, acting both as recruiting sergeant and then as an activating signal. The first is direct, the lipid head group binding to

The terms PKB and Akt both occur in the literature: see text box on page 526. The PKB/Akt kinases are assembled in the SwissProt database under the 'RAC' subfamily of serine/threonine protein kinases because the mammalian sequence was discovered as a kinase related to PKA and PKC.[34]

The **AGC family** of protein kinases comprises serine/threonine kinases having structural homology with PKA, PKG, and PKC, hence the name AGC. This classification is not based on sequence homology. They contain a C-terminal hydrophobic motif (FxxF[S/T]Y) (except for PDK1). They also share a mode of activation in which (for most of them), PDK1 acts as the master-switch. Examples are members of the S6Kinase subfamily (S6K1, RSK and MSK1), members of protein kinase C family (PKCα, β1, β2, δ, γ, ξ, and PRK), Serum-glucocorticoid responsive kinase (SGK1) PDK1 and PKBα, β, γ[35] (for domain architecture of some of these protein kinases see Figures 18.7).

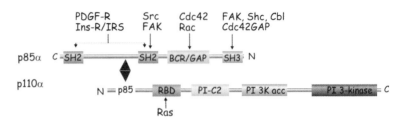

FIG 18.4 Multiple pathways to activate PI 3-kinase class 1 A.

Several proteins interact with the p85 regulatory subunit of PI 3-kinase. These include members of Rac and Cdc42 (Rho-related GTPases) that interact at the BCR/GAP domain. The receptors for PDGF and insulin, insulin receptor substrate (IRS), and also FAK and Src all interact at the SH2 domains. A proline-rich region present in FAK, Shc, Cbl, and Cdc42GAP interacts with the SH3 domain. The catalytic subunit p110α can interact directly with Ras. Domains denoted as in Figure 18.2.

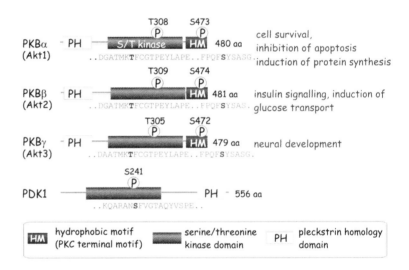

FIG 18.5 Classification of the protein kinases B/Akt.

Originally identified as the oncogene in the transforming retrovirus AKT8, there are multiple isoforms of PKB/Akt (α, β, and γ). They all contain a PH domain and a hydrophobic motif at the C-terminus but differ slightly in the locations of their regulatory phosphorylation sites PKD1 is not a member of the PKB family but plays an important role in their activation.

the PH domain in the N-terminal segment of PKB.[36] The other interaction is indirect, involving the soluble kinase PDK1, also endowed with a PH domain.[37] Binding of PI(3,4)P$_2$ is crucial since it enables PDK1 and PKB to embrace, an encounter that, as shown below, proves beneficial to both.

Two changes are required in order to render PKB catalytically competent. Firstly, the αC-helix becomes structured through the interaction with the serine phosphorylated C-terminal hydrophobic motif (HM) (Figure 18.6a, b).

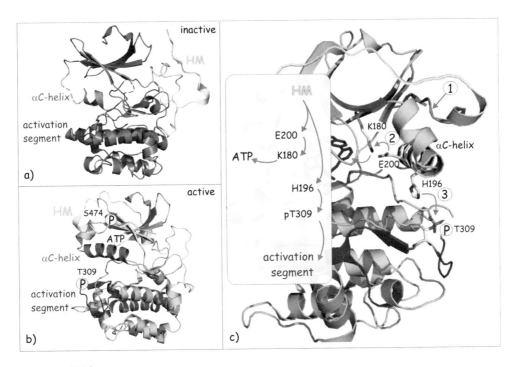

FIG 18.6 Activation of PKBβ.
(a) In the inactive PKB structure, the various regions of the kinase domains comprising the αC-helix of the N-lobe and the activation segment are disordered. Substrate and ATP do not bind. The organization of the C-terminal segment with its hydrophobic motif (HM, shown in yellow) is indicated approximately. (1gzk[40]). (b) Binding of the C-terminal hydrophobic motif (HM, yellow) with the αC-helix is facilitated by phosphorylation of S474. This induces reorganization of the αC-helix (pink) and a second phosphorylation in the activation segment (T309) organizes the activation segment (red). Binding of ATP and substrate ensues (2jdr[41]). (c). Binding of the hydrophobic motif to the αC-helix (1) leads to structural rearrangements in which E200 engages and correctly positions K180 (2) that coordinates the binding of ATP in the catalytic cleft. The H196 of the ordered αC-helix also engages the (PDK) phosphorylated T309 resulting in a reorientation of the activation segment (3). The kinase now binds both ATP and substrate and is fully competent to phosphorylate substrate.(2jdr[41]).

Then, again following phosphorylation, the activation segment is reorganized (Figure 18.6b,c). Two separate kinases are required. Unlike most kinases, PDK1 is constitutively (but only partially) active. Also, unlike the other kinases of this class, having no intrinsic hydrophobic (HM) motif, the activating impetus is provided by its substrates, which, in return, become phosphorylated (Figure 18.7). The full activation signal requires a priming phosphorylation in the C-terminal domain catalysed by the tentatively named PDK2. Its identity remains elusive. The kinase complex mTOR/Rictor (mTORC2) is among the favourite candidates. Unless both buttons are pressed on PKB, little happens.[38] The need for 3-phosphoinositides to activate PKB may go beyond that of colocalization with PDK1 since binding of its PH domain to PI(3,4)P$_2$ induces a conformational change that facilitates phosphorylation of the activation segment by PDK1.[39]

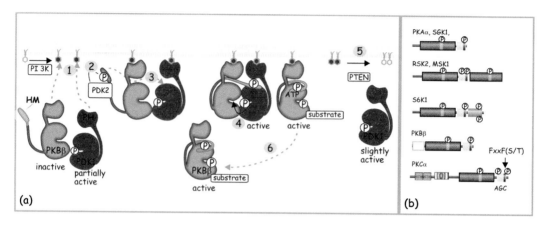

FIG 18.7 Activation of PKB.

(a) PI(3,4)P$_2$ serves as a membrane recruitment signal for both PKB and PDK1 (1). A first phosphorylation of PKB (by PDK2) occurs in the C-terminal hydrophobic motif (2). Close apposition of PKD1 causes binding of the phosphorylated HM to the αC-helix of PDK1 (3), which, now fully competent, phosphorylates the activation segment of PKB so that full kinase activity is achieved (4). Detachment of activated PKB from the membrane may occur after dephosphorylation of PI(3,4)P$_2$ by PTEN (5 and 6). (b) Domain architecture of the AGC kinases. The activities are controlled by paired phosphorylation sites, one in a hydrophobic motif, sometimes part of a larger autoregulatory domain, the other in the activation segment. For these kinases, PDK1 acts as the master switch by phosphorylation of the activation segment. AGC signifies AGC kinase C-terminal domain; other domains as in Figure 18.2.

PDK1, an AGC kinase, has an activation segment that is stabilized by autophosphorylation. It therefore only requires stabilization of the αC-helix to express full activity. This occurs through interaction with substrates that possess a phosphorylated hydrophobic motif such as other members of the AGC family of protein kinases. PDK1 may therefore be considered as a constitutively active protein kinase.

Activated PKB phosphorylates its substrates either at the membrane, or, following its detachment, in the cytosol. The viral oncogene product, v-Akt, has a lipid anchor (myristoyl group) and its attachment to the membrane may facilitate its activation or prevent its deactivation by phosphatases. It may also prevent phosphorylations that ensure negative feedback, thus perpetuating the signal.

Insulin: the role of IRS, PI 3-kinase, and PKB in the regulation of glycogen synthesis

From the insulin receptor to PKB

Insulin signalling pathways all commence at the insulin receptor, INSR. The product of a single gene, this is post-translationally cleaved into an extracellular α-subunit (containing a ligand binding site) and a transmembrane β-subunit (which has a tyrosine protein kinase catalytic domain) (Figure 18.8). The mature functional receptor comprises four components, two α and two β, held together by disulfide bonds.

In common with the related IGF-1R and INSRR (an insulin-related orphan receptor), the insulin receptor is dimeric in its inactive state. In this respect, it differs from all other tyrosine kinase-containing receptors. Binding of the

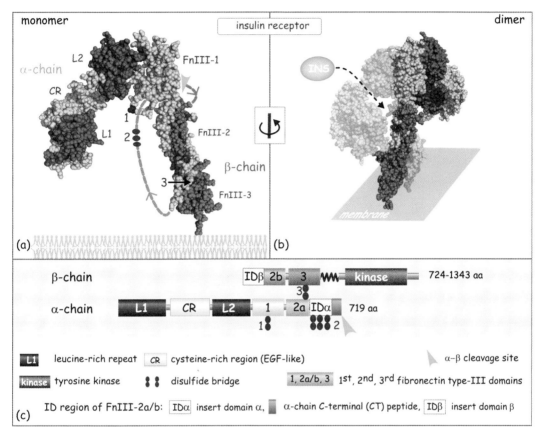

FIG 18.8 Insulin receptor structure.
(a) The insulin receptor is a homodimer. In the mature form, each component is present as two chains, α and β. Each monomer possesses seven domains, L1 (leucine rich region-1), CR (cysteine rich region), L2, FnIII-1 (fibronectin-III-like domain 1), FnIII-2, FNIII-3, and the intracellular tyrosine kinase domain (structure not shown). Both the α and β chains contribute to the FnIII-2 domain. The long insert region (shown as a dotted line) separating FnIII-2a from -2b contains three disulfide bridges (2) that link the monomers. FNIII-1 provides a fourth disulfide bridge (1). b) A possible arrangement of the two peptides is shown. The insulin-binding space is between the central β-sheet of L1 and the bottom loops of the FnIII-1 domain. (c) Domain architecture showing locations of the positions numbered in (a).
From McKern et al.[42] (2dtg[42]).

ligand, insulin or IGF-1, must alter the receptor conformation or the relative position of the receptor subunits in order to allow the intracellular catalytic domains to approach each other (or to remove an inhibitory constraint). Three transphosphorylations in each activation segment ensue in orderly sequence. All are required for maximal activation.[43] Further phosphorylations then occur in positions beyond the catalytic site (see Figure 18.9).

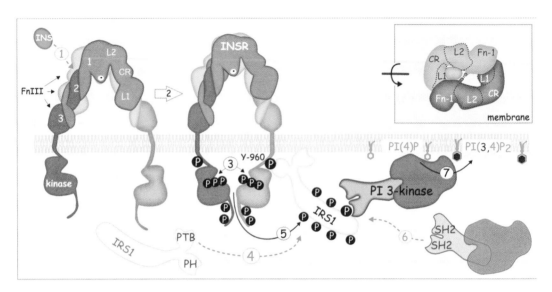

FIG 18.9 Activation of PI 3-kinase by the insulin receptor.
Insulin binding to the receptor dimer (1) induces a conformational change (2) that causes transphosphorylation of the activation segments at three tyrosine residues (3). Further phosphorylation follows on both sides of the catalytic domain. The IRS-1 binds pY960 with its PTB domain (4). Phosphorylation on further tyrosine residues follows (5) generating a docking site for the SH2 domains of the p85 regulatory subunit and activation of PI 3-kinase (6) leading to the phosphorylation, at the 3-position, of inositide lipids (7).

A second peculiarity of the insulin receptor is that it signals through the intervention of large docking proteins. These insulin receptor substrates (IRS-1–4) all possess PH and PTB domains. The PTB domains bind directly to the tyrosine phosphorylated region of the receptor, which corresponds to pY960 as illustrated in Figure 18.9, and results in phosphorylation of the docking protein. Of the four docking proteins, IRS-1 and -2 are essential for insulin signalling; lack of IRS-1 is linked to insulin resistance in muscle and adipose tissue. In mice, absence of IRS-2 causes insulin resistance in the liver (together with many other developmental defects).

The p85 regulatory subunit of PI 3-kinase (type IA) binds phosphorylated IRS-1/2 through its SH2 domain[44] (Figure 18.9), so negating the inhibitory constraint imposed on the catalytic subunit and bringing the kinase into the vicinity of its substrate. Prolonged activation of PI 3-kinase and the production of 3-phosphorylated polyphosphoinositides encourage a number of serine/threonine kinases to associate with the plasma membrane. Important among these is PKB.

The intervention of PI 3-kinase in the activation of PKB is well supported.[45] It is quiescent in serum-starved fibroblasts but becomes active shortly after the addition of insulin or platelet-derived growth factor (PDGF). This fails in cells that contain a mutant form of PKB lacking the PH domain and depends on the presence of phosphotyrosines on IRS-1 or on the PDGF receptor. Note that $PI(3,4)P_2$ rather than $PI(3,4,5)P_3$ appears to activate PKB and that this correlates with the ability of both the intact kinase and its isolated PH domain to bind to this particular phospholipid.[46] Of the three members of the PKB family, PKBβ plays the major role. PKBα is not involved in insulin signalling.[47]

Similar to insulin, PDGF is able to activate PKB but is without immediate effect on glucose metabolism. Other factors must be in place to give direction to the signal transduction pathways downstream of these receptors. For instance, the proximity of phosphorylated IRS-1 to the insulin receptor relays the activation of PKB in the sense of glucose metabolism. This connection is absent in the case of the PDGF receptor.

Lastly, note that the insulin receptor does not signal exclusively through the IRS, because it can also bind the adaptor APS. This regulates membrane expression of the glucose transporter GLUT-4.[48] Moreover, p85a (regulatory subunit of group IA PI 3-kinases) can bind directly to the C-terminal end of INSR.

From PKB to glycogen synthase

The minimum sequence motif required for efficient phosphorylation of small peptides by PKB is RxRxx(S/T)(F/L) and there are several substrates in which this is present. With respect to insulin signalling, one such is glycogen synthase kinase-3, GSK3β. This has two unusual features. First, the activation segment is fully structured when dephosphorylated so that it needs no further modification to become active. Second, it recognizes only those substrates that offer an array of serine phosphorylation sites, of which the most C-terminal is already phosphorylated ('primed substrate'). (We considered a similar situation in relation to the phosphorylation of β-catenin – see page 432). Phosphorylation of the most N-terminal serine by PKB then exposes a 10-residue pseudo-substrate sequence that neatly fits the catalytic cleft of GSK3β (see Figure 18.10b). Having a proline, not serine, four residues upstream, it blocks rather than activates kinase activity.[32,49] Importantly, inactivation of GSK3β necessarily reduces phosphorylation at regulatory serine residues on glycogen synthase, so that it becomes activated.

This alone, however, is insufficient to allow an abrupt onset of glycogen synthesis. In order for this to occur, it is also necessary to activate the protein phosphatase-1G (PP1G) which removes phosphate groups from glycogen synthase (Figure 18.10a). This is mediated through the action of the insulin-stimulated protein kinase ISPK (not shown). The insulin-stimulated cell is now fully engaged in the synthesis of glycogen.

Diabetes mellitus affects 1–2% of people in Western societies, being especially prevalent among Hispanic and native American peoples. Type 1 (or early onset) diabetes is due to autoimmune destruction of the insulin-producing β cells in the islets of Langerhans in the pancreas, and is the more severe form of the disease. It accounts for most instances of blindness in Western populations. Multiple genes are involved that determine a susceptibility, but not an inevitability, to express the condition that might be precipitated by a viral infection.

The more common type II (or late onset) form of diabetes appears to be a correlate of good living. It is associated with obesity, though genetic factors undoubtedly play a part. It is characterized by resistance to insulin and indeed, the early stages are characterized by very high levels of the hormone which are, however, insufficient to overcome the resistance.

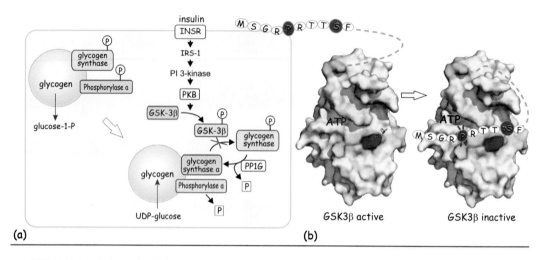

(a)

(b)

GSK-3β N-terminal pseudo substrate

GSK-3β phosphorylation sites on glycogen synthase

(c)

FIG 18.10 Insulin-mediated activation of glycogen synthase.

(a) Binding of insulin to its receptors leads to the activation of PKB. One of the substrates of PKB is glycogen synthase kinase-3β, which becomes inactivated and so unable to phosphorylate and inactivate glycogen synthase. Also, the phosphatase PP1G ensures rapid dephosphorylation and activation of glycogen synthase allowing glycogen synthesis to recommence. In addition, phosphorylase a is inactivated by dephosphorylation. Thus glycogenolysis is arrested. (b) Detail of the inhibition of GSK3β by PKB. The N-terminal (pseudosubstrate) segment of GSK3β is most likely disordered, a wagging tail projecting out of the kinase. When phosphorylated (at S9) it takes on the characteristics of a primed substrate and binds at the catalytic cleft. However, lacking a second serine at a position four amino acids upstream, the catalytic action is hindered and the peptide remains firmly attached. Access of other substrates is prevented. (c) Comparison of the N-terminal sequence of GSK3β with the C-terminal sequence of glycogen synthase. The lack of a second serine renders the phosphorylated N-terminal sequence a pseudo-substrate.

Insulin induces translocation of the GLUT4 transporter

Insulin is instrumental in the control of glucose uptake by the tissues, in particular muscle, fat, and those areas of the brain that regulate metabolic homeostasis.[50] It does this by the recruitment of GLUT4 transporters to the plasma membrane from intracellular vesicular stores.

The mechanism by which this occurs is not entirely clear. The insulin acts at two levels, activating the Rap GTPase on the vesicles and then orchestrating the assembly of the exocyst complex at the plasma membrane. Activation of Rap occurs through the PI 3-kinase/PKB pathway and it involves phosphorylation and deactivation of RapGAP, so promoting the GTP-bound state. In this form it interacts with the docking proteins of the exocyst. Which of the many Raps initiates the docking process is not known. Insulin is also involved in exocyst assemblage in a process that is independent of PI 3-kinase and involves insulin receptors that are recovered from lipid rafts and which then associate with the adaptor protein APS.[52]

The role of PI 3-kinase in activation of protein synthesis

Insulin is a growth factor. Again, operating through the PI 3-kinase pathway, it increases protein synthesis by a mechanism that involves the GTPase Rheb and that acts upon the initiator factor-4E (eIF-4E) and the ribosomal p70 S6-kinase-1 (S6K1). We return to the role of Rheb below.

eIF-4E is the limiting ribosomal initiation factor in most cells and it plays a principal role in determining global translation rates. It is regulated by phosphorylation, for instance through the ERK pathway (see page 341), but more importantly by binding to the translational repressor 4E-BP. eIF-4E binds to the 7-methylguanine (m7G) cap at the 5'-end of mRNA and enables the recruitment of other initiation factors. These, together with the 40S ribosomal subunit, form the pre-initiation complex which promotes the search for the start codon (AUG) on the mRNA (see Figure 18.12). Progression is facilitated by the helicase complex eIF4A/eIF4B which breaks the intramolecular base-pairing in the mRNA. Recognition of the AUG codon requires tRNA[Met], which contains the anticodon 3'-UAC-5' and comes to the ribosomal pre-initiation complex bound to eIF-2γ.GTP. As the AUG codon is recognized, GTP is hydrolysed and the ribosomal 60S particle, carrying the peptidyl transferase activity, joins the initiation complex, so initiating the elongation phase of protein synthesis. Binding of 4E-BP to eIF-4E prevents all this.

In response to glucagon and to numerous growth factors, the p70 S6-kinase1 associated with eIF-3 (a large complex of small initiation factors), acts to phosphorylate rbS6, a component of the 40S ribosomal subunit.[53,54] Cells lacking rbS6 are smaller in size, but how it regulates cell size remains an enigma. Cellular localization of rbS6 is not restricted to the 40S ribosomal particle, so it may control cell size independently of protein synthesis at the ribosomal level[55]. p70 S6-kinase1 also phosphorylates the regulatory domain of eIF-4 A (helicase).

Rheb and TSC

TSC1 and TSC2, are downstream components of the PI 3-kinase pathway, discovered as loss-of-function mutants in patients suffering from tuberous sclerosis.[56,57]

The TSC1/TSC2 complex constitutes a GAP that holds Rheb in its GDP-bound inactive state.[62] Phosphorylation of TSC2 by PKB inhibits this activity, so allowing accumulation of Rheb.GTP which activates mTOR (Figure 18.11).[63,64] The kinase mTOR when complexed with Raptor and GβL binds to eIF-3

A missense point mutation present in the kinase domain of PKBγ is present in members of a family showing autosomal dominant inheritance of severe insulin resistance. This finding provides additional evidence for the important role for PKB in insulin signal transduction.[51]

The **S6** protein kinases form a subfamily of serine/threonine protein kinases with members KS6A1-A6, KS6B1, and KS6B2. The S6 kinase discussed here is KS6B1. It phosphorylates the ribosomal S6 protein (rbS6) at six serine residues. Two of these are also phosphorylated by RSK, a protein kinase downstream of the Ras/MAPK pathway.

Tuberous sclerosis (TSC). First described by Désiré-Magloire Bourneville in 1880 (and called Bourneville's disease in France),[58,59] TSC is a rare (30–100 per million, worldwide), multisystem genetic disease characterized by the growth of benign tumours in the brain and on other vital organs including the kidneys, heart, eyes, lungs, and skin. It is caused by mutations on either of two TSC genes, 1 or 2, which encode for the proteins hamartin and tuberin respectively. These act as tumour suppressors, regulating cell proliferation and differentiation. PTEN, which acts in the same signal transduction pathway, is also a tumour suppressor but loss of its function is linked to Cowden disease in which patients have a predisposition to malignancies such as prostate cancer, glioblastoma, endometrial tumours, and small-cell lung carcinoma.[60,61]

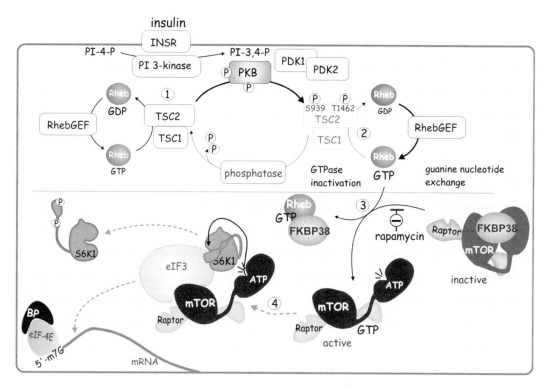

FIG 18.11 Regulation of mTOR by PKB-mediated phosphorylation of TSC2.

The complex TSC1/TSC2 acts as a GAP, maintaining Rheb in its (GDP) inactive state and holding protein synthesis in check (1). Insulin augments the rate of protein synthesis through activation of Rheb. This occurs through phosphorylation and inactivation of TSC2 by PKB (2). Rheb then accumulates in its GTP bound state (probably involving an as yet unidentified GEF) and interacts with the kinase mTOR (3), complexed to Raptor, FKBP38 (and GβL, not shown). Rheb bound to GTP sequesters the inhibitor FKBP38 and this unveils the kinase activity of mTOR and causes the complex to associate with eIF3 (4).

and coordinates the assembly of the initiation complex through a series of ordered phosphorylation events. First, mTOR/Raptor phosphorylates the eIF-4E binding protein, 4E-BP, at multiple sites, causing its detachment[65] and allowing the association of eIF-4G with the initiation complex (Figure 18.12). Importantly, binding of eIF-4G allows interaction of the 5′-mRNA initiation complex with the 3′-mRNA poly A binding proteins (PAPB) (Figure 18.12). This drives translation and ensures that only intact mRNA is translated.

mTOR/Raptor also phosphorylates S6K1,[66] which then dissociates from eIF-3 and associates with PDK1, which phosphorylates the activation segment of S6K1 causing full activation. (This mechanism resembles the phosphorylation/activation of PKB by PDK1: see Figure 18.07.) It phosphorylates the S6 ribosomal

When first discovered, the G-protein β-like subunit, GβL, was thought to be a novel β subunit, the sixth in the human genome. Although it has WD repeats, characteristic of β subunits, it is otherwise distinct and has now been classified as member of a tiny WD repeat subfamily, WD repeat Lst8.

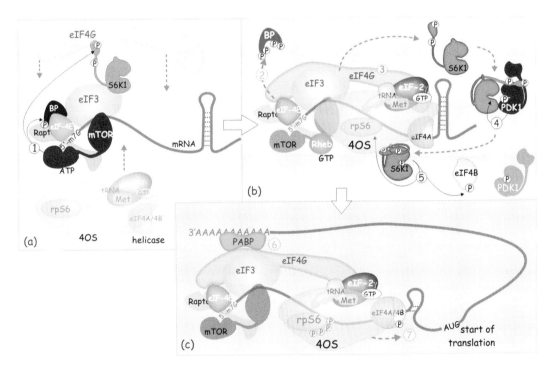

FIG 18.12 A series of ordered phosphorylations facilitates assembly of the translation initiation complex on mRNA. (a) Activation and association of mTOR with the pre-initiation complex causes phosphorylation of 4E-BP (BP) (1) and of S6K1. Phosphorylated 4E-BP detaches from eIF-4E and phosphorylated S6K1 detaches from eIF-3. (eIF-3 and the 40S ribosomal particle are represented as a single entities). (b) Dissociation of 4E-BP (2) and S6K1 permits the association of the large initiation factor eIF-4G (3) that interacts with eIF-4E, eIF-3, and eIF-4. S6K1, phosphorylated in its regulatory domain (including the hydrophobic motif) encounters PDK1 (4) causing phosphorylation of its activation segment. S6K1 now phosphorylates the ribosomal proteins S6 and eIF-4B, the regulatory subunit of the DNA helicase eIF-4 A (5). (c) The conditions now favour binding of poly-A binding protein (PABP) which is attached to the 3′-poly A tail of the mRNA (6). The pre-initiation complex moves towards the start AUG codon (7) where it is joined with the 60S ribosomal particle.

protein and also eIF-4B, a regulatory subunit of the eIF-4A helicase.[67] As eIF-2γ.GTP and tRNA^{Met} combine with the pre-initiation complex, the small ribosomal subunit starts its progression towards the 3′ end. It halts when it attains the AUG codon, binding to the anticodon of tRNA^{Met}. Protein translation commences. The domain architectures of the components of the TSC/mTOR pathway are shown in Figure 18.13.

Although insulin-mediated activation of ERK is without significant effect on glucose transport or glycogen synthesis, it does have a significant effect on

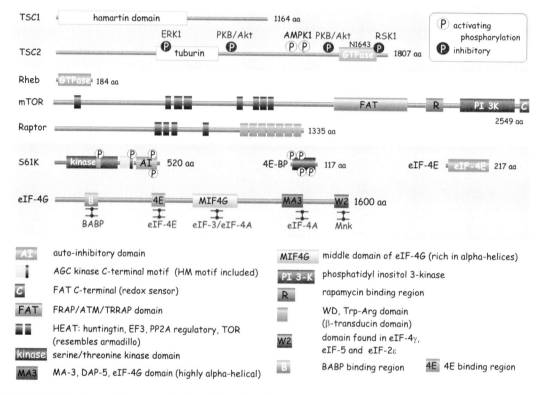

FIG 18.13 Domain architecture of the components of the mTOR signalling pathway.
Many of the components of the mTOR pathway are large proteins, some having still poorly defined domains. TSC2 carries a GAP domain responsible for activation of hydrolysis of GTP by Rheb. The numerous phosphorylation sites have either stimulatory effects (sites phosphorylated by AMPK1), or inhibitory effects (red). mTOR bears the signature sequence of a PI 3-kinase but operates as a protein kinase. S6K1 has an autoinhibitory and an AGC kinase C-terminal domain with hydrophobic motif, both of which have to be phosphorylated before phosphorylation and activation of the activation segment. 4E-BP has to be phosphorylated at four residues before it effectively detaches from eIF-4E. mTOR binds directly at the C-terminus or indirectly via Raptor. mTOR phosphorylates at least four residues, one of which is not sensitive to rapamycin and can thus be phosphorylated by another kinase. eIF-4G has a number of domains and regions that interact with other components of the initiation complex. Note the interaction with Mnk, a member of MAPkinase family (see Figure 12.22, page 351 and 12.15, page 340).

protein synthesis.[68,69] This occurs through ERK1/2-mediated phosphorylation of TSC2 at sites distinct from those targeted by PKB and it also causes inhibition of the GAP activity.[70] This occurs through RSK1, one of the kinases downstream of ERK that phosphorylates TSC2.[71]

Control of translation and transcription

Both the Ras/MAPkinase and PKB pathways have profound and selective effects on the pattern of expression of individual proteins that occurs quite independently of their effects on gene transcription. Thus, after ectopic expression of constitutively activated forms of K-Ras or PKBβ in mouse glial-progenitor cells, polyribosome complexes engaged in the translation of mRNAs that encode proteins involved in the regulation of growth, transcription, and cell–cell interactions, become enriched. How the mRNAs are selected remains to be determined.[72] It seems that protein translation, in addition to gene transcription, is also an important target of transforming proteins.

In order to give a further boost to protein synthesis, mTOR/Raptor also controls the transcription of the different forms of RNA required for ribosomal biogenesis and formation of the translation machinery. It stimulates RNA polymerase-I leading to the transcription of ribosomal RNA, RNA polymerase-II leading to the transcription of mRNA coding for ribosomal proteins (approximately 82), and RNA polymerase-III, which transcribes the 32 different transfer RNAs.[73]

Integration of growth factor and nutrient signalling

A lack of amino acids or the presence of 5′-AMP (see page 243 and Figure 9.1, page 244) must intervene with the growth factor action of insulin in order to prevent the production of unwanted or unfinished proteins.[74] Although the mechanism remains elusive, amino acids regulate mTOR through the intervention of Rheb, but independently of TSC1/2.[75] Regulation by 5′-AMP occurs at the level of AMP-activated kinase (AMPK) (Figure 18.14) leading to phosphorylation of the TSC1/2 complex on residues distinct from those phosphorylated by PKB (compare **Figure 18.14** with **Figure 18.11**, also see **Figure 18.13**). This prevents its inactivation by PKB. Consequently, the Rheb-GAP activity of TSC1/2 remains elevated and protein synthesis is inhibited through failure to stimulate mTOR.

5′-AMP is not a direct activator of AMPK in the way that cyclic AMP activates PKA. Instead, similar to the activation of PKB by PI(3,4)P$_2$, it acts as a recruiting agent. By binding to the γ-regulatory subunit of AMPK, 5′-AMP induces a conformational change that allows the interaction of the enzyme with LKB1, a constitutively activated kinase. This results in phosphorylation and activation of AMPK. The importance of this pathway in the regulation of protein synthesis is illustrated by the finding that lack of LKB1 activity predisposes to cancer

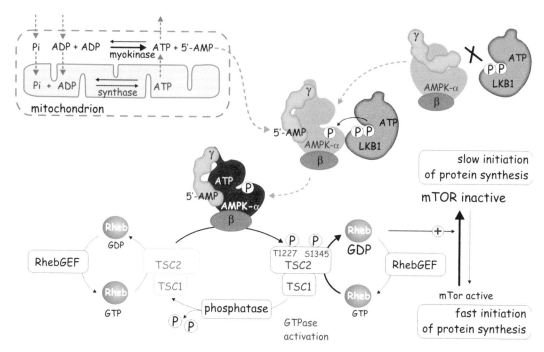

FIG 18.14 Signalling through AMP kinase.

Hints of ATP depletion are manifested by the leakage of 5′-AMP from mitochondria which binds the regulatory subunit of AMPK. Subsequent phosphorylation by LKB1 leads to its activation. AMPK phosphorylates TSC2 thereby preventing the inhibitory control by PKB. The active TSC1/2 complex maintains Rheb in its GDP-bound state thus preventing activation of mTOR. Protein synthesis is suppressed.

(Peutz–Jeghers syndrome). LKB1 thus qualifies as a tumour suppressor. It prevents excessive stimulation of the mTOR pathway and in this way it maintains strict control over protein synthesis.[76] Further evidence for a tumour suppressor role of LKB1 comes from the finding that its activation, through the induction of its pseudokinase partner STRAD causes repolarization of transformed epithelial cells with the reformation of tight junctions. LKB1 thus appears to classify as both a polarity and a tumour suppressor gene.[77]

PI 3-kinase: regulator of cell size, proliferation, and transformation

The (p110) catalytic subunit of PI 3-K class I A is the only component of this family that has been implicated in cancer. Activating mutations or mutations that cause its over-expression are particularly frequent in colorectal cancers and in glioblastomas.[78] Indeed, after K-Ras, it is the most mutated gene implicated in cancer. On the other hand, activating mutations in PKB are not associated with human cancers though over-expression has been detected (particularly PKBβ), again predominantly in colorectal cancers.

The first indication that PI 3-kinase might act as a regulator of cell size and proliferation came from studies of *Drosophila*. Over-expression of Dp110

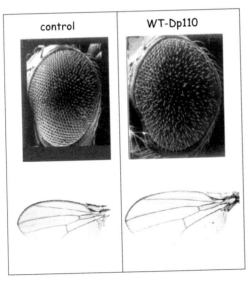

control	WT-Dp110

FIG 18.15 Overexpression of *Drosophila* type I PI 3-kinase (Dp110) yields flies with giant eyes (and wings). From Leevers et al.[85]

(catalytic subunit of PI 3-kinase) results in enlarged wings or eyes. Conversely, kinase-defective mutants have very small organs[79] (Figure 18.15). The PI 3-phosphatase, PTEN, correspondingly reduces both cell size and number.[80] PKB stimulates those pathways that specifically regulate cell and compartment size independently of cell number.[81]

Why are PTEN inactivation mutants so often linked to malignancy, but TSC1/2 mutants only rarely so? PTEN has many roles. Not only does it regulate cell size, but it is also a determinant of cell survival and it regulates the cell cycle through expression of cyclin D (acting in the G1 phase). Indeed, this may explain why its loss of function gives rise to a 'higher level' of cell transformation. A second explanation has emerged from a recent study in which mice lacking TSC2 manifest deficient signalling of the PKB pathway, which may limit the transforming capacity. However, loss of a single copy of the PTEN gene, so-called *haplodeficiency*, restores the activation level of PKB and with it, its transforming consequences.[82]

Over-expression of eIF-4E can induce full tumorigenic transformation.[83] It also leads to enlarged cells and increased proliferation rates and it can rescue protein synthesis even when mTOR is inhibited by rapamycin.[84] Not surprisingly, the PI 3-kinase/mTOR pathway is regarded as a promising target for therapeutic intervention and many derivatives of rapamycin are now being tested both in animal models and in clinical trials. Oncogenes and tumour suppressors linked to the PI 3-kinase/mTOR pathway are listed in Table 18.1.

Rapamycin is a triene macrolide antibiotic obtained from *Streptomyces hygroscopius*, first dug up out of the earth of Rapa Nui (Easter Island). It binds to a cellular receptor FKBP12 and, together, they form a complex with mTOR and Raptor (mTORC1). This prevents interaction with substrates. Such inhibitory complex formation does not occur when mTOR is associated with Rictor (mTORC2).

565

TABLE 18.1 Protooncogenes and tumour suppressors associated with the mTOR/Raptor (mTORC1) signalling pathway

Protooncogenes functionally linked to mTORC1 signalling	
PI 3-K	elevated activity in cancers
PKB/Akt	amplified gene expression in cancers
Rheb	elevated expression, possible target of farnesyl transferase inhibitors
Ras	mutations resulting in hyperactivity
eiF-4E	over expressed in many cancers
S6K1	elevated expression in breast cancer
Tumour suppressors functionally linked to mTORC1 signalling	
PTEN	loss-of-function in numerous cancers. Cowden disease
TSC1/TSC1	loss-of-function-linked hamartomas in several organs
NF1	mutations in neurofibromatosis type-1
LKB1	mutations in Peutz–Jeghers syndrome
4E-BP1	overexpression blocks c-Myc induced transformation

That the understanding of how tumours arise is impressive yet still limited is given by the observation that mice lacking PTP1B exhibit no predisposition to cancer. This, despite the fact that this phosphatase has a central place in the deactivation of the insulin receptor and that when absent, the PKB signalling pathway is strongly enhanced. These mice are highly responsive to insulin and show little propensity to obesity (see page 648).[86]

Other processes mediated by the 3-phosphorylated inositol phospholipids

A number of seemingly disparate functions are modified in cells and organisms in which the synthesis of the 3-phosphorylated lipids is chronically altered (see **Figure 18.1**). These include

- A constitutively activated retrovirus-encoded PI 3-kinase induces transformation of fibroblasts. In chickens, formation of haemangiomas.
- Inhibition of PI 3-kinase prevents T cell activation by preventing the translocation of NFAT.
- In vascular endothelial cells subjected to shear stress, nitric oxide synthase is activated by PKB. The released NO relaxes the vascular smooth muscle.[87]

- Rac and Cdc42 cooperate with PI(4)P 5-kinase and PI 3-kinase in the assembly of the submembranous actin filament system, leading to the formation of the membrane protrusions necessary for neutrophil migration.[88] Conversely, membrane retraction occurs through localized activation of PTEN at the trailing edge of migrating neutrophils. This pathway involves RhoA-mediated activation of ROCK, which in turn phosphorylates and activates PTEN.[89,90]
- In *C. elegans* , PKB phosphorylates and activates a transcription factor of the forkhead/winged-helix family.[91] This has a function in resisting apoptosis and may regulate lifespan. Mutants with reduced activity of the insulin/IGF-1-receptor homologue DAF-2, and therefore unable to activate PI 3-kinase, live twice as long as normal.
- In epithelial cells, activation of the apoptosis pathway is suppressed as a consequence of PI 3-kinase activation (see page 400).

So, who did discover insulin?

Probably nobody discovered insulin. It existed first as an idea (Minkowski), then as a proven hypothesis (Gley and others), and finally as a practical way of alleviating diabetes (Paulescu and Banting) (adapted from Henderson[3]).

List of Abbreviations

Abbreviation	Full name/description	SwissProt entry	Other names, OMIM links
4E-BP	eIF4E-binding protein	Q13541	
AMPK-α1	5'-AMPK activated protein kinase subunit-α1	Q13131	
AMPK-β1	5'-AMPK activated protein kinase subunit-β1	Q9Y478	
AMPK-γ1	5'-AMPK activated protein kinase subunit-γ1	P54619	
diabetes type I	early onset (childhood) diabetes		MIM: 222100
diabetes type II	late onset diabetes		MIM: 125853
eIF-4E	eukaryotic initiation factor-4E	P06730	PHAS-1
eIF-4G	eukaryotic initiation factor-4G	Q04637	
FAT	domain found in FRAP, ATM, TRAPP		
FRAP	FKBP-12-rapamycin complex-associated protein	P42345	mTOR

Continued

567

glycogen synthase		P54840	MIM: 240600
GSK3β	glycogen synthase-3β	P49841	
HEAT	domain found in huntingtin, EF3, PP2A regulatory, TOR		
IGF-1R	insulin-like growth factor receptor-1	P08069	
INSR1	insulin receptor I	P06213	MIM: 147670 ; MIM: 262190 ; MIM: 246200;
INSRR	insulin receptor related protein	P14616	orphan receptor
insulin	from insula (referring to islets of Langerhans)	P01308	MIM: 176730
IRS1	insulin receptor substrate	P35568	
LKB1	liver kinase type 1B	Q15831	STK11, MIM: 175200
mTOR	mammalian target of rapamycin (see FRAP)	P42345	FRAP
PDK1	3-phosphoinositide-dependent protein kinase-1	O15530	PDPK1
PI 3-kinase C2α	catalytic type II α	O00443	
PI 3-kinase C2β	catalytic type II β	O00750	
PI 3-kinase C2γ	catalytic type II γ	O75747	
PI 3-kinase III	catalytic type III	Q8NEB9	Vsp34p-like
PI 3-kinase p101γ	regulatory	Q8WYR1	
PI 3-kinase p110α	phoshatidylinositol 3-kinase catalytic subunit type I α	P42336	MIM: 171834
PI 3-kinase p110β	phoshatidylinositol 3-kinase catalytic subunit type I β	P42338	
PI 3-kinase p110γ	phoshatidylinositol 3-kinase catalytic subunit type Iγ	P48736	
PI 3-kinase p110δ	phoshatidylinositol 3-kinase catalytic subunit type I δ	O00329	
PI 3-kinase p150	phosphatidylinositol 3-kinase regulatory subunit-4	Q99570	
PI 3-kinase p85α	phosphatidylinositol 3-kinase regulatory subunit-α	P27986	MIM: 171833
PI 3-kinase p85β	phosphatidylinositol 3-kinase regulatory subunit-b	O00459	
PI 3-kinase p55γ	phosphatidylinositol 3-kinase regulatory subunit-γ	Q92569	

Continued

PKBα	protein kinase B α	P31749	Ak1
PKBβ	protein kinase B β	P31751	Ak2
PKBγ	protein kinase B γ	Q9Y243	Akt3
Raptor	regulatory associated protein of mTOR	Q8N122	
Rheb	Ras homologue enriched in brain	Q15382	
Rictor	rapamycin insensitive companion of mTOR	Q6R327	
rpS6 protein	ribosomal protein S6 (S for Svedberg constant)	P62753	component of the S40 ribosomal subunit
S6K1	S6 ribosomal particle protein kinase-1	P23443	
TSC1	tuberous sclerosis complex protein-1	Q92574	hamartin, MIM: 191100, MIM: 607341
TSC2	tuberous sclerosis complex protein-2	P49815	tuberin, MIM: 606690, MIM: 191100
VPS34	(yeast) vacuolar protein sorting-associated	P22543	

References

1. Vane JR. Inhibition of prostaglandin synthesis as a mechanism of action for aspirin-like drugs. *Nat New Biol*. 1971;231:232–235.
2. Murray I. Paulescu and the isolation of insulin. *J Hist Med*. 1971;26:150–157.
3. Henderson JR. Who really discovered insulin? *Guy's Hospital Gazette* 1971;85:314–318.
4. Kasuga M, Karlsson FA, Kahn CR. Insulin stimulates the phosphorylation of the 95,000-dalton subunit of its own receptor. *Science*. 1982;215:185–187.
5. von Mehring J, Minkowski O. Diabetes mellitus nach Pankreasexstirpation. *Arch Exp Pathol Pharmakol*. 1890;26:371–387.
6. Gley E. Action des extraits de pancréas sclérosé sur des chiens diabétiqués (par extirpation du pancréas). *C R Soc Biol (Paris)*. 1922;87:1322–1325.
7. Freychet P. Pancreatic hormones. In: Baulieu E-E, Kelly PA, eds. *Hormones: From Molecules to Disease*. London: Chapman & Hall; 1990:490–532.
8. Foster DW, McGarry JD. Glucose, lipid and protein metabolism. In: Griffin JE, Ojeda SR, eds. *Textbook of Endocrine Physiology*. New York: Oxford University Press; 2006:393–419.

569

9. Chong LD, Traynor-Kaplan A, Bokoch GM, Schwartz MA. The small GTP-binding protein Rho regulates a phosphatidylinositol 4-phosphate 5-kinase in mammalian cells. *Cell.* 1994;79:507–513.

10. Tolias KF, Hartwig JH, Ishihara H, Shibasaki Y, Cantley LC, Carpenter CL. Type Iα phosphatidylinositol-4-phosphate 5-kinase mediates Rac-dependent actin assembly. *Curr Biol.* 2000;10:153–156.

11. Honda A, Nogami M, Yokozeki T, et al. Phosphatidylinositol 4-phosphate 5-kinase α is a downstream effector of the small G protein ARF6 in membrane ruffle formation. *Cell.* 1999;99:521–532.

12. Traynor Kaplan AE, Harris AL, Thompson BL, Taylor P, Sklar LA. An inositol tetrakisphosphate-containing phospholipid in activated neutrophils. *Nature.* 1988;334:353–356.

13. Ireton K, Payrastre B, Chap H, et al. A role for phosphoinositide 3-kinase in bacterial invasion. *Science.* 1996;274:780–782.

14. Gary JD, Wurmser LS, Emr SD. Fap1p is essential for PtdIns(3)P 5-kinase activity and the maintenance of vacuolar size and membrane homeostasis. *J Cell Biol.* 1998;143:65–79.

15. Herman PK, Stack JH, DeModena JA, Emr SD. A novel protein kinase homolog essential for protein sorting to the yeast lysosome-like vacuole. *Cell.* 1991;64:425–437.

16. Walker EH, Perisic O, Ried C, Stephens L, Williams RL. Structural insights into phosphoinositide 3-kinase catalysis and signaling. *Nature.* 1999;402:313–320.

17. Yu J, Zhang Y, McIlroy J, Rordorf-Nikolic T, Orr GA, Backer JM. Regulation of the p85/p110 phosphatidylinositol 3′-kinase: stabilization and inhibition of the p110 α catalytic subunit by the p85 regulatory subunit. *Mol Cell Biol.* 1998;18:1379–1387.

18. Rudd E, Schneider H. Lymphocyte signaling: Cbl sets the threshold for autoimmunity. *Curr Biol.* 2000;10:R344.

19. Musacchio A, Cantley LC, Harrison SC. Crystal structure of the breakpoint cluster region-homology domain from phosphoinositide 3-kinase p85 α subunit. *Proc Natl Acad Sci U S A.* 1996;93:14373–14378.

20. Diekmann D, Brill S, Garrett MD, et al. Bcr encodes a GTPase-activating protein for p21rac. *Nature.* 1991;351:400–402.

21. Rodriguez-Viciana P, Warne PH, Dhand R, et al. Phosphatidylinositol-3-OH kinase as a direct target of Ras. *Nature.* 1994;370:527–532.

22. Voigt P, Brock C, Nurnberg B, Schaefer M. Assigning functional domains within the p101 regulatory subunit of phosphoinositide 3-kinase γ. *J Biol Chem.* 2005;280:5121–5127.

23. Wymann MP, Pirola L, Pirola L. Structure and function of phosphoinositide 3-kinases. *Biochim Biophys Acta.* 1998;1436:127–150.

24. Van Haesebroeck B, Jones GE, Allen WE, et al. Distinct PI(3)Ks mediate mitogenic signalling and cell migration in macrophages. *Nature Cell Biol.* 1999;1:69–71.

25. Foster FM, Traer CJ, Abraham SM, Fry MJ. The phosphoinositide (PI) 3-kinase family. *J Cell Sci*. 2003;116:3037–3040.

26. Dann SG, Thomas G. The amino acid sensitive TOR pathway from yeast to mammals. *FEBS Lett*. 2006;580:2821–2829.

27. Gunther R, Abbas HK, Mirocha CJ. Acute pathological effects on rats of orally administered wortmannin-containing preparations and purified wortmannin from Fusarium oxysporum. *Food Chem Toxicol*. 1989;27:173–179.

28. Closse A, Haefliger W, Hauser D, Gubler HU, Dewald B, Baggiolini M. 2,3-Dihydrobenzofuran-2-ones: a new class of highly potent antiinflammatory agents. *J Med Chem*. 1981;24:1465–1471.

29. Wymann MP, Bulgarelli-Leva G, Zvelebil MJ, et al. Wortmannin inactivates phosphoinositide 3-kinase by covalent modification of Lys-802, a residue involved in the phosphate transfer reaction. *Mol Cell Biol*. 1996;16:1722–1733.

30. Nakanishi S, Kakita S, Takahashi I, et al. Wortmannin a microbial product inhibitor of myosin light chain kinase, *J Biol Chem*. 1992;267:2157–2164.

31. Walker EH, Pacold ME, Perisic O, et al. Structural determinants of phosphoinositide 3-kinase inhibition by wortmannin, LY294002, quercetin myricetin, and staurosporine. *Mol Cell*. 2000;6:909–919.

32. Cross DA, Alessi DR, Cohen P, Andjelkovich M, Hemmings BA. Inhibition of glycogen synthase kinase-3 by insulin mediated by protein kinase B. *Nature*. 1995;378:785–789.

33. Guinebault C, Payrastre B, Racaud SC, et al. Integrin-dependent translocation of phosphoinositide 3-kinase to the cytoskeleton of thrombin-activated platelets involves specific interactions of p85 α with actin filaments and focal adhesion kinase. *J Cell Biol*. 1995;129:831–842.

34. Jones PF, Jakubowicz T, Pitossi FJ, Maurer F, Hemmings BA. Molecular cloning and identification of a serine/threonine protein kinase of the second-messenger subfamily. *Proc Natl Acad Sci USA*. 1991;88:4171–4175.

35. Frodin M, Antal TL, Dummler BA, et al. A phosphoserine/threonine-binding pocket in AGC kinases and PDK1 mediates activation by hydrophobic motif phosphorylation. *EMBO J*. 2002;21:5396–5407.

36. Bellacosa A, Testa JR, Staal SP, Tsichlis PN. A retroviral oncogene, akt, encoding a serine-threonine kinase containing an SH2-like region. *Science*. 1991;254:274–277.

37. Corvera S, Czech MP. Direct targets of phosphoinositide 3-kinase products in membrane traffic and signal transduction. *Trends Neurosci*. 1998;8:442–447.

38. Yang J, Cron P, Good VM, Thompson V, Hemmings BA, Barford D. Crystal structure of an activated Akt/protein kinase B ternary complex with GSK3-peptide and AMP-PNP. *Nat Struct Biol*. 2002;9:940–944.

39. Mora A, Komander D, van Aalten DM, Alessi DR. PDK1, the master regulator of AGC kinase signal transduction. *Semin Cell Dev Biol*. 2004;15:161–170.

40. Yang J, Cron P, Thompson V, Good VM, Hess D, Hemmings BA, Barford D. Molecular mechanism for the regulation of protein kinase B/Akt by hydrophobic motif phosphorylation. *Mol Cell*. 2002;9:1227–1240.

41. Davies TG, Verdonk ML, Graham B, et al. A structural comparison of inhibitor binding to PKB, PKA and PKA-PKB chimera. *J Mol Biol*. 2007;367:882–894.

42. McKern NM, Lawrence MC, Streltsov VA, et al. Structure of the insulin receptor ectodomain reveals a folded-over conformation. *Nature*. 2006;443:218–221.

43. White MF, Shoelson SE, Keutmann H, Kahn CR. A cascade of tyrosine autophosphorylation in the β-subunit activates the phosphotransferase of the insulin receptor. *J Biol Chem*. 1988;263:2969–2980.

44. Valverde AM, Lorenzo M, Pons S, White MF, Benito M. Insulin receptor substrate (IRS) proteins IRS-1 and IRS-2 differential signaling in the insulin/insulin-like growth factor-I pathways in fetal brown adipocytes. *Mol Endocrinol*. 1998;12:688–697.

45. Franke TF, Yang SI, Chan TO, et al. The protein kinase encoded by the Akt proto-oncogene is a target of the PDGF-activated phosphatidylinositol 3-kinase. *Cell*. 1995;81:727–736.

46. Franke TF, Kaplan DR, Cantley LC, Toker A. Direct regulation of the Akt proto-oncogene product by phosphatidylinositol-3,4-bisphosphate. *Science*. 1997;275:665–668.

47. Jiang ZY, Zhou QL, Coleman KA, Chouinard M, Boese Q, Czech MP. Insulin signaling through Akt/protein kinase B analyzed by small interfering RNA-mediated gene silencing. *Proc Natl Acad Sci U S A*. 2003;100:7569–7574.

48. Hu J. Liu.J, Ghirlando R, Saltiel AR, and Hubbard SR. Structural basis for recruitment of the adaptor protein APS to the activated insulin receptor. *Mol Cell*. 2006;12:1379–1389.

49. Frame S, Cohen P, Biondi RM. A common phosphate binding site explains the unique substrate specificity of GSK3 and its inactivation by phosphorylation. *Mol Cell*. 2001;7:1321–1327.

50. Benomar Y, Naour N, Aubourg A, et al. Insulin and leptin induce Glut4 plasma membrane translocation and glucose uptake in a human neuronal cell line by a phosphatidylinositol 3-kinase- dependent mechanism. *Endocrinology*. 2006;147:2550–2556.

51. Wilson JC, Rochford JJ, Wolfrum C, et al. A family with severe insulin resistance and diabetes due to a mutation in AKT2. *Science*. 2004;304:1325–1328.

52. Chang L, Chiang S, Saltiel AR. Insulin signaling and the regulation of glucose transport. *Mol Med*. 2006;10:65–71.

53. Novak-Hofer I, Thomas G. Epidermal growth factor-mediated activation of an S6 kinase in Swiss mouse 3T3 cells. *J Biol Chem*. 1985;260:10314–10319.

54. Smith CJ, Rubin CS, Rosen OM. Insulin-treated 3T3-L1 adipocytes and cell-free extracts derived from them incorporate 32P into ribosomal protein S6. *Proc Natl Acad Sci U S A*. 1980;77:2641–2645.

55. Ruvinsky I, Meyuhas O. Ribosomal protein S6 phosphorylation: from protein synthesis to cell size. *Trends Biochem Sci*. 2006;31:342–348.

56. van Slegtenhorst M, de Hoogt R, Hermans C, Nellist M, Janssen B, et al. Identification of the tuberous sclerosis gene TSC1 on chromosome 9q34. *Science*. 1997;277:805–808.

57. The European Chromosome 16 Tuberous Sclerosis Consortium. Identification and characterization of the tuberous sclerosis gene on chromosome 16. *Cell*. 1993;75:1305–1315.

58. Bourneville DM. Sclérose tubéreuse des circonvolutions cérébrales. *Arch Neurol (Paris)*. 1880;1:81–89.

59. Bourneville DM, Brissaud E. Encéphalite ou sclérose tubéreuse des circonvolutions cérébrales. *Arch Neurol (Paris)*. 2008;1:390–410.

60. Cheadle JP, Reeve MP, Sampson JR, Kwiatkowski DJ. Molecular genetic advances in tuberous sclerosis. *Hum Genet*. 2000;107:97–114.

61. Cantley LC, Neel BG. New insights into tumor suppression: PTEN suppresses tumor formation by restraining the phosphoinositide 3-kinase/AKT pathway. *Proc Natl Acad Sci U S A*. 1999;96:4240–4245.

62. Inoki K, Li Y, Xu T, Guan KL. Rheb GTPase is a direct target of TSC2 GAP activity and regulates mTOR signaling. *Genes Dev*. 2003;17:1829–1834.

63. Inoki K, Li Y, Wu J, Guan KL. TSC2 is phosphorylated and inhibited by Akt and suppresses mTOR signaling. *Nat Cell Biol*. 2002;4:648–657.

64. Dan HC, Sun M, Yang L, et al. Phosphatidylinositol 3-kinase/Akt pathway regulates tuberous sclerosis tumor suppressor complex by phosphorylation of tuberin. *J Biol Chem*. 2002;277:35364–35370.

65. Gingras AC, Raught B, Sonenberg N. Regulation of translation initiation by FRAP/mTOR. *Genes Dev*. 2001;15:807–826.

66. Burnett PE, Barrow RK, Cohen NA, Snyder SH, Sabatini DM. RAFT1 phosphorylation of the translational regulators p70 S6 kinase and 4E-BP1. *Proc Natl Acad Sci USA*. 1998;95:1432–1437.

67. Holz MK, Ballif BA, Gygi SP, Blenis J. mTOR and S6K1 mediate assembly of the translation preinitiation complex through dynamic protein interchange and ordered phosphorylation events. *Cell*. 2005;123:569–580.

68. Yamamoto-Honda R, Tobe K, Kaburagi Y, et al. Upstream mechanisms of glycogen synthase activation by insulin and insulin-like growth factor-I. Glycogen synthase activation is antagonized by wortmannin or LY294002 but not by rapamycin or by inhibiting p21ras. *J Biol Chem*. 1995;270:2724–2729.

69. Dorrestijn J, Ouwens DM, Van den Berghe N, Bos JL, Maassen JA. Expression of a dominant-negative Ras mutant does not affect stimulation of glucose uptake and glycogen synthesis by insulin. *Diabetologia*. 1996:558–563.

70. Ma L, Chen Z, Erdjument-Bromage H, Tempst P, Pandolfi PP. Phosphorylation and functional inactivation of TSC2 by Erk implications for tuberous sclerosis and cancer pathogenesis. *Cell*. 2005;121:179–193.

71. Roux PP, Ballif BA, Anjum R, Gygi SP, Blenis J. Tumor-promoting phorbol esters and activated Ras inactivate the tuberous sclerosis tumor suppressor complex via p90 ribosomal S6 kinase. *Proc Natl Acad Sci U S A*. 2004;101:13489–13494.

72. Rajasekhar VK, Viale A, Socci ND, Wiedmann M, Hu X, Holland EC. Oncogenic Ras and Akt signaling contribute to glioblastoma formation by differential recruitment of existing mRNAs to polysomes. *Mol Cell*. 2003;122:889–901.

73. Martin DE, Hall MN. The expanding TOR signaling network. *Curr Opin Cell Biol*. 2005;17:158–166.

74. Shamji AF, Nghiem P, Schreiber SL. Integration of growth factor and nutrient signaling: implications for cancer biology. *Mol Cell*. 2003;12:271–280.

75. Roccio M, Bos JL, Zwartkruis FJ. Regulation of the small GTPase Rheb by amino acids. *Oncogene*. 2006;25:657–664.

76. Alessi DR, Sakamoto K, Bayascas JR. LKB1-dependent signaling pathways. *Annu Rev Biochem*. 2006;75:137–163.

77. Baas AF, Kuipers J, van der Wel NN, et al. Complete polarization of single intestinal epithelial cells upon activation of LKB1 by STRAD. *Cell*. 2004;116:457–466.

78. Samuels Y, Velculescu VE. Oncogenic mutations of *PIK3CA* in human cancers. *Cell Cycle*. 2004;3:1221–1224.

79. Leevers SJ, Weinkove D, MacDougall LK, Hafen E, Waterfield MD. The *Drosophila* phosphoinositide 3-kinase Dp110 promotes cell growth. *EMBO J*. 1996;15:6584–6594.

80. Goberdhan DC, Paricio N, Goodman EC, Mlodzik M, Wilson C. *Drosophila* tumor suppressor PTEN controls cell size and number by antagonizing the Chico/PI3-kinase signaling pathway. *Genes Dev*. 1999;13:3244–3258.

81. Verdu J, Buratovich MA, Wilder EL, Birnbaum MJ. Cell-autonomous regulation of cell and organ growth in *Drosophila* by Akt/PKB. *Nat Cell Biol*. 1999;1:500–506.

82. Manning BD, Logsdon MN, Lipovsky AI, Abbott D, Kwiatkowski DJ, Cantley LC. Feedback inhibition of Akt signaling limits the growth of tumors lacking Tsc2. *Genes Dev*. 2005;19:1773–1778.

83. Lazaris-Karatzas A, Montine KS, Sonenberg N. Malignant transformation by a eukaryotic initiation factor subunit that binds to mRNA 5′ cap. *Nature*. 1990;345:544–547.

84. Fingar DC, Salama S, Tsou C, Harlow E, Blenis J, Mammalian cell size is controlled by mTOR and its downstream targets S6K1 and 4EBP1/eIF4E. *Genes Dev*. 202;16:1472–1487.

85. Leevers SJ, Weinkove D, MacDougall LK, Hafen E, Waterfield MD. The *Drosophila* phosphoinositide 3-kinase Dp110 promotes cell growth. *EMBO J*. 1996;15:6584–6594.

86. Elchebly M, Payette P, Michaliszyn E, Cromlish W, Collins S, et al. Increased insulin sensitivity and obesity resistance in mice lacking the protein tyrosine phosphatase-1B gene. *Science*. 1999;283:1544–1548.

87. Dimmeler S, Fleming I, Fisslthaler B, Hermann C, Busse R, Zeiher AM. Activation of nitric oxide synthase in endothelial cells by Akt-dependent phosphorylation. *Nature*. 1999;399:601–605.

88. Kwiatkowska K, Sobotka A. Signaling pathways in phagocytosis. *Bioessays*. 1999;21:422–431.

89. Xu J, Wang F, Van Keymeulen A, et al. Divergent signals and cytoskeletal assemblies regulate self-organizing polarity in neutrophils. *Cell*. 2003;114:201–214.

90. Li Z, Hannigan M, Mo Z, et al. Directional sensing requires G$\beta\gamma$-mediated PAK1 and PIXα-dependent activation of Cdc42. *Cell*. 2003;114:215–227.

91. Kops GJ, de Ruiter ND, de Vries-Smits AM, Powell DR, Bos JL, Burgering BM. Direct control of the Forkhead transcription factor AFX by protein kinase B. *Nature*. 1999;398:630–634.

CHAPTER 19

Protein Kinase C Revisited

PKC in cell transformation

Long before their association with PKC became apparent, it was evident that extracts of croton oil can act as cocarcinogens, enhancing the tumorigenic activity of other substances such as benzo[a]pyrene.[1] In some culture systems the phorbol ester PMA increases the number of mitoses,[2] but in the absence of a tumour promoter it fails to effect full cell transformation. One might think that PKC should be implicated in all this, maybe acting to phosphorylate a transcription factor to induce specific gene expression. However, the matter remains unresolved and PKC isoforms responsive to phorbol ester have failed to qualify as true oncogenes.

Despite a voluminous literature, understanding of the relationship between PKC and tumour formation remains somewhat hazy. At best, it seems to have accessory roles, boosting the activity of the real transformers such as Ras and Raf. Since more and more information focuses on the action of atypical

The five-ring polycyclic aromatic hydrocarbon **benzo[a]pyrene** is highly carcinogenic. It is the component of coal tar responsible for the cancers (sooty warts, cancers of the scrotum) suffered by chimney sweeps in 18th century England.[3] To this day it presents a health hazard as a carcinogen present in cigarette and marijuana smoke, and possibly in barbecued meat. As with many other chemical carcinogens, benzopyrene is better regarded as a procarcinogen, requiring conversion to an active derivative, benzo[a]pyrene diol epoxide. This slots into the major groove of DNA,[4] specifically targeting the p53 gene.[5]

The many abbreviations used in this chapter are collected together at the end of the chapter.

PKC(ι, ζ) in the regulation of cell polarity, we concentrate on this aspect of PKC signalling.

With the aim of understanding the mechanisms underlying tumour promotion by phorbol esters, two independent strategies have been applied. By searching for transcriptional control elements that mediate the phorbol-ester-induced alterations in gene expression, it should be possible to work backwards, identifying first the transcription factor(s) that bind these elements and then the signal transduction pathway that regulates their activation.[6] The alternative approach has been to over-express various isoforms of PKC and to study changes in cell phenotype.[7]

The search for transcription factors that mediate phorbol ester effects

From TRE to SRE

Analysis of the promoter regions of a number of genes (for instance, collagenase, metallothionein IIA, and stromelysin) induced by phorbol ester reveals a conserved 7 bp palindromic motif, the TPA-responsive element (TRE). This is recognized by AP-1, understood to be at the receiving end of a complex pathway that transmits the effects of phorbol ester tumour promoters from the plasma membrane to the transcriptional machinery, possibly involving protein kinase C.[8] AP-1 collectively describes a group of structurally and functionally related members of the Jun and Fos protein family, together with some members of the ATF and MAF families (in combination with Jun proteins)[9] and linked by a leucine zipper protein–protein interaction motif (see Figure 19.1 and Tables 19.1 and 19.2).[10] These transcription factors are similar in structure to CREB (see page 248).[11–14] CREB can also bind to the TRE[13] and so inhibit activation through c-Jun.

AP-1

The activator protein-1 transcription factor complexes owed their initial definition to a fortuitous convergence of tumour virology and transcription factor research. AP-1 had been described as a TPA-inducible (or PMA-inducible) activity that addresses specific sequences in the enhancer of the metallothionein gene and in the 72 bp repeat of the simian virus 40 enhancer-region. The oncoproteins v-Fos and v-Jun were identified as transforming proteins in the Finkel–Biskis–Jinkins murine osteosarcoma virus and in avian sarcoma virus-17, respectively. v-Jun was linked with AP-1 because of its homology with the yeast transcriptional regulator GCN4, which binds an identical DNA sequence. Indeed, the DNA-binding domains of Jun and of GCN4 are exchangeable. The cellular homologue of v-Fos-associates with a protein of 39 kDa, recognized as c-Jun. This finding identified Fos and Jun as partners, together forming the AP-1 complex that binds DNA.[15]

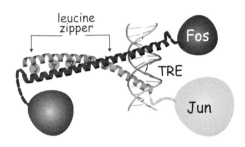

FIG 19.1 Activator protein-1 (AP-1) complexes.

Jun and Fos, members of AP-1 complex proteins, are connected through numerous leucine residues, the leucine zipper. Here, they bind DNA at the palindromic -TGACTACA-/-ACTGAGT- sequence (TRE)Image adapted from Eferl and Wagner.[16]

TABLE 19.1 Examples of palindromic DNA sequences recognized by AP1 factors

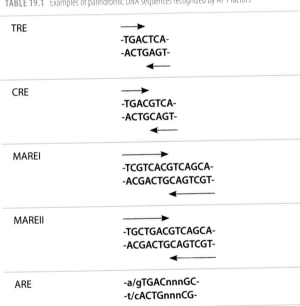

TRE	→ -TGACTCA- -ACTGAGT- ←
CRE	→ -TGACGTCA- -ACTGCAGT- ←
MAREI	→ -TCGTCACGTCAGCA- -ACGACTGCAGTCGT- ←
MAREII	→ -TGCTGACGTCAGCA- -ACGACTGCAGTCGT- ←
ARE	-a/gTGACnnnGC- -t/cACTGnnnCG-

v-Fos and v-Jun are transforming proteins by virtue of their prolonged half-lives, enhanced transcriptional activity and a change in the spectrum of target genes (compared to their cellular counterparts).[11,12] It became apparent that cellular AP-1 is composed of different combinations of proteins, always containing either c-Jun or c-Fos, but not necessarily binding to the TRE. Other response elements recognized by AP-1 are CRE, MARE I, MARE II, and ARE. Possible combinations, with their preferred DNA response element, are listed in Table 19.1.

TABLE 19.2 Partners of Jun and Fos in AP-1 complexes

Jun		Fos	
c-Jun	TRE > CRE	c-Jun	TRE > CRE
JunB	TRE > CRE	JunB	TRE > CRE
junD	TRE > CRE	JunC	TRE > CRE
FosB	TRE > CRE		
Fra1	TRE > CRE		
FRA2	TRE ≈ CRE		
ATFa	CRE > TRE	ATFa	Non-binding
ATF2	CRE > TRE	ATF2	CRE > TRE
ATF4	(CRE)	ATF4	CRE
B-ATF	TRE > CRE		
c-MAF	(MAREI/II)	c-MAF	(MAREI/II)
		MAFA	(MAREI/II)
		MAFB	(MAREI/II)
NRL	(TRE-related)	NRL	(TRE-related)
MAFF/G/K	(MAREI/II)	MAFF/G/K	(MAREI/II)
NRF1	(ARE)		
NRF2	(ARE)	NRF2	(ARE)
NFIL-6	(ARE)	NFIL6	(ARE)

Although they were originally discovered in the context of tumorigenesis, not all forms of AP-1 cause cell transformation and their effect is dependent on context. Roughly speaking, c-Fos, FosB, and c-Jun are considered oncogenes although they have not been found associated with cancer (no mutants). Fra1 and Fra2 are considered weak oncogenes whereas JunB and junD have tumour suppressor qualities.[4,5,16]

Regulation of the AP-1 complex

The upstream signal transduction pathway that regulates AP-1 activity proved a hard nut to crack. Initially it appeared that activation of PKC causes *de*phosphorylation of c-Jun, just in the basic region where it binds DNA.[17] Phosphorylation of this segment can also be achieved (in the test tube) by glycogen synthase kinase-3β (GSK3β) and so it was postulated that PKC stimulates the binding of c-Jun DNA through inhibition of GSK3β, which would result in dephosphorylation of the basic region. Consistent with this is that activation of PKC (α, β1, β2, and γ) causes phosphorylation and thus deactivation of GSK3β.[17,18] However the interaction between GSK3β and c-Jun needed verification and the phosphatase that strips the phosphates remained enigmatic. That this could not be the whole story became clear when it was found that phosphorylation of four residues at the N-terminus is far more important for c-Jun-mediated transcriptional activity and cell transformation.[19,20]

The discovery of a Jun N-terminal protein kinase, JNK-1, which phosphorylates c-Jun,[21,22] drew attention away from PKC and on to the newly emerging family of MAP kinases and their effects on the serum response element, SRE (see page 342).

PKCε activates Raf through phosphorylation of RKIP

If PKC is not the initiator of TRE- and SRE-mediated gene transcription, it certainly acts to reinforce the input derived from growth factors and stress conditions (including inflammatory factors). PKCα activates the Ras-activated kinase c-Raf, an oncogene that cooperates in the transformation of NIH3T3 fibroblasts[23] (see page 327). In rat embryo fibroblasts, activation of Raf-1 is also essential for the transforming effect of PKCε.[24]

Since all growth factors induce the generation of diacylglycerol and hence activate PKC, it follows that PKC reinforces the Ras-mediated growth factor signal, either at the level of Ras itself or at the level of Raf-1. Indeed, PKC phosphorylates the inhibitor of c-Raf, RKIP, causing its detachment from c-Raf and so reinstates the Ras–ERK pathway.[25,26] In this way PKC contributes to gene transcription through the SRE (Figure 19.2).[27,28]

Phorbol ester also activates Ras directly, through membrane binding of the guanine nucleotide exchange factor RasGRP. Again, this pathway is reinforced by PKC through phosphorylation and activation of RasGRP. This feedforward mechanism plays an important role in effective B and T cell signalling.

PKC rewires the ERK pathway: role of RACK1 and JIP

A large proportion of human melanomas harbour a highly active Ras–ERK pathway due to gain-of-function mutations in the genes coding for B-Raf and/or N-Ras (see page 327 onwards). CyclinD, which drives progression of the cell division cycle in G1, is also up-regulated. The transcription factors c-Jun and ATF2, forming an AP-1 complex, are regulators of cyclinD expression. c-Jun must be phosphorylated in order to be transcriptionally active and this could occur through at least two pathways, the one comprising the Ras–Raf–MEK–ERK cascade,[29] the other comprising MKK4 or -7 and JNK. Studies with kinase

With the discovery of the Jun kinases, PKC lost any claim or right of ownership over specific pathways related to cell transformation. This is a point of interest. The Raf–ERK system could equally well have been described as a PKC pathway had the series EGFR–Ras–Raf–ERK (as described in Chapter 12) not been discovered first. PKC is now better regarded as a booster of the Ras–Raf–ERK pathway, but it could also have been the other way round, Ras being the booster of PKC–Raf–ERK.

FIG 19.2 PKC causes activation of SRE via RKIP/RAF.

PKC and growth factors were initially thought to activate distinct signal transduction pathways resulting in the activation of TRE and SRE respectively. However, PKC can activate SRE independently through phosphorylation of RKIP (1), an inhibitor of Raf. Phosphorylated RKIP loses its grip on Raf, which now associates with the Ras, MEK, ERK signalling cassette (3). Diacylglycerol also affects SRE through direct activation of the Ras guanine exchange factor, RasGRP. This activation is reinforced by PKCε-mediated phosphorylation (2), thus boosting the Ras-MEK-ERK pathway.

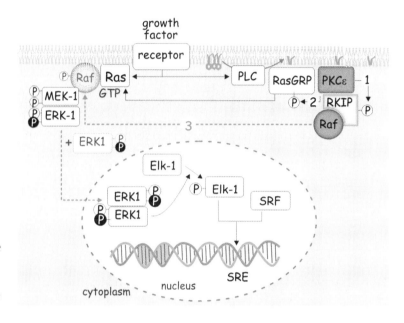

The S129 phosphorylation site on JNK. The S129 phosphorylation site is only slightly exposed and would seem unable to interact directly with the upstream kinases MKK-4 or -7. Nor does it appear to be able to affect the catalytic activity of JNK (according to the conformation observed in the non-phosphorylated JNK structure). However, binding of the scaffold protein JIP1 controls kinase activity and it is thus possible that pS129 changes this interaction and that the loss of JIP1 binding allows for more efficient MKK4 or MKK7-mediated phosphorylation of JNK[34] (see page 463).

inhibitors showed that ERK and JNK are equal partners in maintaining cyclinD at an elevated level.[30] However, the JNK pathway is not generally activated by mitogenic signalling, responding mainly to stress situations (see page 349).[31] In some way, ERK provokes a strong JNK signal and this is where PKC comes into play.

Although the mitogenic pathway does not generally cause activation of c-Jun, it does induce its expression and stabilization. Enhancement of c-Jun is caused by phosphorylation and activation of CREB (see table 12.3, page 339). This involves the nuclear kinase MSK, a substrate of ERK1. Protection from the ubiquitylation/proteasome machinery is due to inactivation of GSK3β which otherwise phosphorylates c-Jun, making it a good substrate for the APC-ubiquitin ligase (see page 467). Despite low levels of activation, increasing amounts of c-Jun lead to complex formation with ATF2 (forming an AP-1 complex) and together they bind the TRE-responsive element causing expression of RACK1. This binds to activated PKC and together they associate with JNK. The consequence is phosphorylation of JNK1. PKCα, β, and ε are likely candidates for this phosphorylation reaction, though other isoforms are not excluded. The phosphorylation enables subsequent phosphorylation and activation of JNK by the upstream kinases MKK4 or 7 to render JNK1 fully competent (Figure 19.3). Now things start to happen. The enhanced activity of JNK1 together with increased expression of c-Jun leads to highly active c-Jun, which in turn boosts expression of yet more RACK1. A feed-forward amplification loop is established.

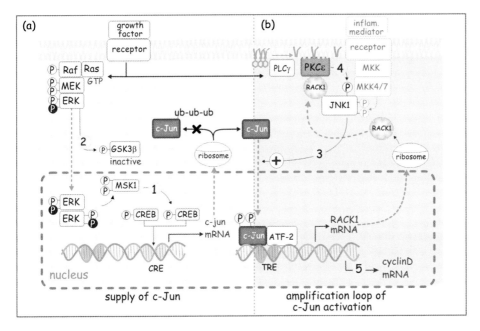

FIG 19.3 PKC rewires the ERK pathway into an amplified JNK pathway.

(a) Growth factors activate the Ras–ERK pathway which leads to enhanced expression of c-Jun in two ways, through phosphorylation and activation of CREB (1) which enhances c-Jun transcription and through phosphorylation/inactivation of GSK3β (2). As a result, c-Jun is no longer phosphorylated and is protected against ubiquitylation.(b) Accumulation of c-Jun and residual activation levels of MKK4 or -7 lead to its activation (3) and, after having joining up with ATF2, to increased expression of Rack1 mRNA. After translation, this attaches to PKCε and together they associate with JNK1 leading to its phosphorylation (4). This renders the MKK-JNK1 signalling cassette much more efficient. An amplification loop results in an elevated c-Jun/ATF2-directed transcription of cyclinD (5) and shortens the G1 phase of the cell cycle.

Constant provision of c-Jun, due to a highly active Ras–ERK system, is a prerequisite for this pathway, as witnessed by its slow decline when cells are treated with inhibitors of the Ras–ERK pathway. Indeed, in melanoma cells this mode of amplification is only detectable when there are high levels of ERK activity. As a consequence, elevated levels of c-Jun lead to elevated expression of cyclinD, again through a c-Jun/ATF2 complex which binds to its promoter. The c-Jun amplification loop explains, amongst many other things, how melanoma cells maintain such a high rate of proliferation. Not only this, elevated levels of RACK1 also protect against apoptosis[32] and conversely, depletion of RACK1 sensitizes melanoma cells to UV-induced apoptosis and reduces their tumorigenicity. Alterations in RACK1 have been reported in different types of human tumours, with high levels of expression in non-small cell lung carcinoma and in colon carcinomas.[33]

Over-expression of PKC isoforms and cellular transformation

Inflammatory mediators and cell transformation. We assume that both growth factor and inflammatory mediators are ever-present in the cellular environment. This is certainly the case for cancer cells. They often generate their own autocrine growth factors and these also elicit an inflammatory response with concomitant release of inflammatory mediators by surrounding cells and infiltrating blood-borne cells. This enhances their growth and dissemination because of the increased vascular permeability and general vascular remodelling around inflamed tissue.[42]

Over-expression of PKCβ1 in rat embryo (R6) fibroblasts induces the characteristic transformed cell phenotype (see page 306). The cells have a reduced requirement for growth factors and generate an autocrine mitogenic factor that possibly induces their own transformation.[35] However, when injected into nude mice, these cells induce tumours with lower frequency and with a longer latent period than in cells transformed by the H-Ras oncogene. This should not be entirely unexpected, as phorbol esters act as tumour promoters or cocarcinogens, not carcinogens in their own right. Coexpression of H-Ras and PKCβ1 certainly results in greatly enhanced transformation.[36] In contrast, for R6 cells over-expressing PKCα there is no tendency to transform.[35] Instead, growth is retarded and the cells achieve lower saturation densities than normal. In glioma cells, PKCα has a growth-promoting role, suppressing the expression of the cell cycle inhibitor p21$^{WAF1/CIP1}$.[37] A summary of phenotypic changes caused by ectopic expression of members of the PKC family is presented in Table 19.3.

PKCζ has tumour suppressor qualities, actually causing v-Raf transformation of NIH-3T3 cells to revert.[38] Likewise, PKCδ inhibits proliferation of vascular smooth muscle cells by preventing expression of cyclins D and E.[39] It appears that, depending on the cells and on the circumstances, different isoforms of PKC, when over-expressed, can either induce or suppress the formation of the transformed cell phenotype.

Unfortunately, it has not been possible to draw any strong generalizations from these experiments. For sure, the tumour-promoting tendency of phorbol esters cannot be explained by a simple mechanism involving the activation of PKC. In addition, because prolonged treatment with phorbol esters has the effect of down-regulating the expression of conventional and novel PKCs (a phenomenon often used to dissect their role in signal transduction), it

TABLE 19.3 Phenotypic changes caused by ectopic expression of different isoforms of PKC

PKCα	PKCα	PKCβ1	PKCδ	PKCε	PKCζ
rat colonic epithelial cells	glioma cells	rat embryo fibroblasts (R6)	vascular smooth muscle cells	rat fibroblasts and colonic epithelial cells	v-Raf transformed NIH-3T3
tumour suppression	tumour promotion	tumour promotion	tumour suppression	tumour promotion	tumour suppression
slower growth low densities	suppression of p21$^{WAF/CIP}$	growth in nude mice anchorage independence synergy with Ha-Ras	slower growth reduced expression of cyclin D and E	growth in nude mice anchorage independence synergy with Ras	reversion of transformed phenotype

is possible in some instances that the proliferative response may be due to absence of PKC rather than its activation. A case in point is PKCδ. This normally conveys an anti-proliferative and pro-apoptotic signal which is lost when its expression is down-regulated by prolonged treatment with PMA.[40] Aberrant expression of PKC frequently occurs in cancers, with a general tendency for up-regulation of PKCε and down-regulation of PKCα and PKCδ. In some cancers the correlation between expression of PKC isotypes and disease progression is striking, suggesting their potential uses as prognostic markers.[41]

Strong evidence against a crucial role of PKC in human tumorigenesis comes from the finding that few natural tumours have been found that express PKC mutations. There is just one lone example, a point mutation in PKCα in a sub-population of highly invasive pituitary tumours and in thyroid follicular adenomas and carcinomas.[43] This mutation alters the hinge region that links the regulatory and catalytic domains and, among other things, it causes a selective loss of substrate recognition and an altered subcellular localization.[44] How this explains aberrant growth regulation is not known.

Importantly, as already indicated, there are many other proteins that possess C1 domains, like those in the PKCs, that may also be regulated by phorbol esters. One of these, CalDAG-GEF (see page 234), is a guanine nucleotide exchange factor for Ras (and Rab), which promotes malignant transformation in fibroblasts when induced by phorbol ester.[45]

Regulation of cell polarity

Role of atypical PKC

The Par proteins: from *Caenorhabditis elegans* to mammals

The *C. elegans par* genes (*par-1–6*) were first identified in a genetic screen for maternal-effect mutations affecting the unequal partitioning of polar granules to the posterior cell, during the asymmetric division of the single-cell embryo.[46] Two of these genes, *par-3* and *par-6*, encode PDZ domain proteins that colocalize at the anterior cortex by the end of prophase. At this point, Par-2, a RING finger protein, and Par-1, an serine/threonine kinase, localize in a reciprocal manner to the posterior cortex (Figure 19.4). Par-4, another kinase, and Par-5, a 14-3-3 protein, are uniformly distributed throughout the cortex and in the cytoplasm. The posterior Par-3–Par-6 complex is joined by the atypical PKC-3 and together, by guiding astral microtubules to the posterior site, they regulate the positioning of the mitotic spindle. A similar process occurs at the anterior site.[47] The anterior and posterior are determined by the site of sperm entry. The microtubule organizing centre causes disruption of the actin–myosin network and Par-3, Par-6, and PKC-3 remain associated and are excluded from the entry point. This is the start of a series of events that enforces mutual exclusion of localization of the polarity complexes Par-3/6 and Par-1/2.[48]

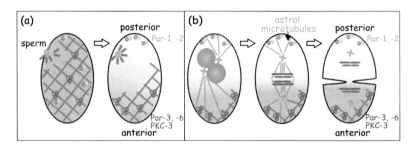

FIG 19.4 Par in *C. elegans*.

(a) The sperm microtubule organizing centre arranges the disassembly of the actin cytoskeleton. This leads to a redistribution of Par proteins. Par-3 and Par-6 remain associated with the cytoskeleton and dominate the future anterior site. Par-1 and Par-2 remain 'active' at the future posterior site. Each Par complex has its specific kinase activity, Par-1 at the posterior and PKC-3 at the anterior site. This leads to phosphorylation of other polarity-determining proteins and results in an asymmetric distribution of mRNA and proteins. (b) Then, following replication of DNA and during attachment of the chromosomes to the microtubule cytoskeleton, the Par complexes ensure the right orientation of the mitotic spindle by correctly positioning the astral microtubules.

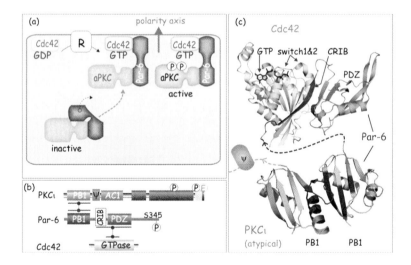

FIG 19.5 Activation of a PKC-Par-6 by Cdc42.

(a) Activation of Cdc42, by a receptor binding an external polarity cue, leads to binding of Par-6/aPKC. This acts to relieve inhibition and allows phosphorylation and activation of the atypical PKC. The active complex now phosphorylates substrates that operate in the determination or maintenance of cell polarity. (b) Domain architecture of the Par-6 polarity complex. Domain interactions are indicated by red lines. (c) Structure of the domains involved in formation of the Par-6 polarity complex. The switch 1 and 2 regions of GTP-bound Cdc42 interact with the partial CRIB and with the PDZ domains of Par-6. Par-6 and PKCι are linked through their PB1 domains (phox-bem domains, 1 and 2). Interacting surfaces are highlighted.

For more information about *C. elegans*, asymmetric cell division and axis formation in the embryo, refer to: http://www.wormbook.org/chapters/www_asymcelldiv/asymcelldiv.html

Homologues of the *C. elegans* proteins are also regulators of cell polarity in other organisms, ranging from flies to vertebrates. In mammals, homologues of Par-3, Par-6 (Par-6 A–D), and atypical PKCs ι and ζ form a complex which localizes at tight junctions and contributes to apical–basal polarity.[49] Beyond its roles in the determination of asymmetric division during development and in the organization of epithelial sheets, polarity also provides directionality to migrating cells, and determines the site of axon-outgrowth and the positioning of the hairs on the epithelial cells of *Drosophila* wings (planar polarity) (see also Figure 14.2, page 422).

Activation of atypical PKC by Cdc42

The atypical PKC proteins are attached to Par-6 through their PB1 domains, the complex maintaining them in an inactive state. Par-6 also contains a Cdc42 binding site composed of a partial-CRIB domain complemented by an adjacent PDZ domain[50,51] (see Figure 19.7). The GTPase Cdc42, in its active GTP-bound state, relieves the inhibitory constraint, opening the way to phosphorylation and activation of atypical PKCs. Where exactly this takes place is determined by: (1) the site of activation of Cdc42 (extrinsic polarity cue), (2) the subcellular compartment in which it accumulates after activation, or (3) the subcellular localization of the PKC–Par-6 complex (intrinsic polarity cues).

Spatial restriction of inositol lipids may also play a role in subcellular localization of polarity complexes. Polarizing or polarized cells exhibit segregation of $PI(3,4,5)P_3$ and $PI(4,5)P_2$. In particular, epithelial cells concentrate $PI(4,5)P_2$ at the apical surface and this is sufficient to recruit both the PKC–Par-6 complex as well as Cdc42.[52] Moreover, once recruited at the apical membrane, atypical PKC interacts with Par-3, which itself is attached to the adhesion molecule JAM-1. This cascade of interactions puts PKC–Par-6 at the site of the engaging cells, leading to the organization of the tight junction (see page 382).[53] Cdc42 also has a key role in the regulation of the actin cytoskeleton, a process not necessarily linked with activation of atypical PKC, and this too affects cell polarity.[54]

Polarity in migrating astrocytes

Astrocytes are the major source of glial cells in the brain. They operate in the differentiation and functioning of neurons, not only as supporting structures but also in the regulation of synaptic transmission and thus the organization of neuronal circuitry. When a scratch is made in a near-confluent culture of astrocytes, the surrounding cells present new tips that grow into the empty space.

At the site of the scratch, the astrocytes accumulate Scrib and βPIX[57] that cause the activation of Cdc42. From here two parallel polarizing processes

Although his body was cremated, Albert Einstein's brain was preserved for pathologists and posterity. A recent report indicates that his cerebral cortex contained a considerable excess of astrocytic tissue (in comparison with four others).[55] Of course, this news was promptly reported in the Sunday papers[56] but the authors of the original report wisely claimed no particular significance for their observation in terms of the great man's cognitive ability.

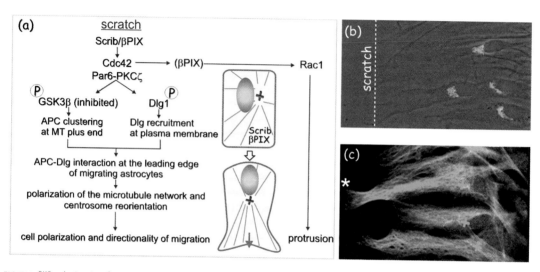

FIG 19.6 PKC and migration of astrocytes.

(a) A scratch in a near-confluent culture of astrocytes provokes reorientation and migration of the adjacent cells to fill the gap. This is guided by the accumulation of the scaffold protein Scrib, bound to βPIX (RhoGEF). Ensuing activation of the Par-6 polarity complex causes protrusion formation, involving the GTPase Rac1, and reorientation of the microtubule cytoskeleton. In order to achieve this, atypical PKCζ phosphorylates and inactivates GSK3β, causing clustering of APC at the microtubule plus ends. It also phosphorylates and activates Dlg, which is now recruited to the membrane. Binding of Dlg and APC assures anchorage of the microtubules to the protruding membrane to provide directionality to the migration process. (b) Immunocytochemical staining of the Golgi apparatus (green), the centrosome (red) and the nucleus (blue) reveals the orientation of the astrocytes perpendicular to the scratch line. (c) Immunocytochemical staining of APC (red and indicated by an asterisk) at the plus end of the microtubules (green). Locations of the centromeres are indicated by blue crosses and the nuclei are red.

Images b and c courtesy of Dr Etienne-Manneville, Institut Pasteur, Paris, France.

Scrib is the vertebrate homologue of *Drosophila* Scribble, a scaffolding protein originally discovered as an epithelial polarity protein.

βPIX is a Dbl-homology domain containing GTP exchange factor for members of the Rho family of GTPases (also known as Cool-1).

are set in train. First, Cdc42 orchestrates the formation of a membrane protrusion through activation of Rac1 which organizes the actin fibres into a gel-like network. The recruitment and activation of Rac1 again requires βPIX.[58] Secondly, Cdc42 reorients the centrosome-attached microtubule network perpendicular to the direction of the scratch, so that the Golgi apparatus is directed perpendicularly to the newly formed microtubule axis. Here, atypical PKC phosphorylates and inhibits GSK3β so allowing dephosphorylation of APC (adenomatous polyposis coli protein, see page 424). This now binds to the plus-ends of centromere-attached microtubules[59] (Figure 19.6).

The second substrate of PKC is Dlg-1 which on phosphorylation localizes to the plasma membrane. Interaction between Dlg1 and APC, in addition to the action of tubulin-bound motor proteins such as dynein, enable the microtubules to reorient the centromere. Cdc42-mediated activation of atypical PKC thus provides both the means to migrate (protrusion formation) and the necessary directionally for the migrating cell (reorientation of centrosome-attached microtubules).[60] The domain architecture of the components involved in astrocyte migration are shown in Figure 19.7.

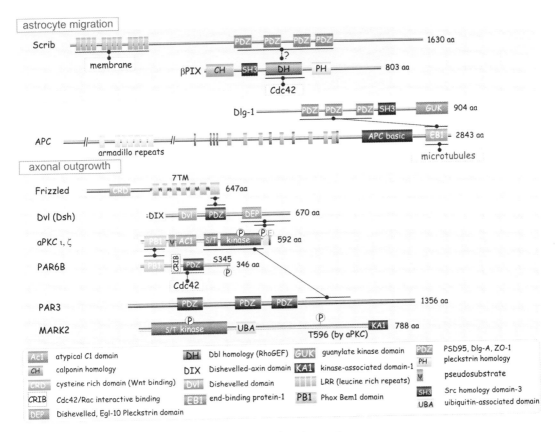

FIG 19.7 Domain architecture of proteins involved in astrocyte migration and axonal outgrowth. See also Figure 14, page 434.

Regulation of atypical PKC by Dishevelled and its role in axon outgrowth

Neurons transmit their signal via *axons*. These are typically long, thin processes of uniform width. Each cell produces a single axon, although near to its end the axon may branch to form one or more presynaptic terminals. Neurons receive inputs via other processes called *dendrites* which also extend from the cell body. They are relatively short, but close to the cell body they are thick and with increasing distance they become thinner, forming numerous Y-shaped branches. (Figure 19.8). These two cellular structures, with the synaptic contacts they make with other cells, are the basic means by which nerve cells receive, process, and transmit signals. Axon formation has been extensively studied in cultured hippocampal neurons. These cells initially form multiple projections called *neurites* which extend and retract, until, at a moment,

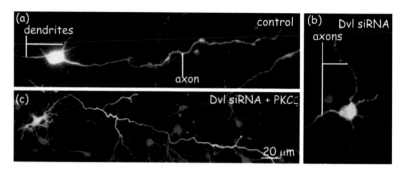

FIG 19.8 Axon outgrowth.

(a) Morphology of hippocampal neurons in culture. Note the numerous short dendrites and the single long axon.

(b) Loss of expression of Dvl, by siRNA knockdown, leads to loss of polarity and numerous short axons are produced.

(c) Injection of PKC-ζ into Dvl-depleted neurons restores polarity with the formation of a single long axon.
Adapted from Arimura and Kaibuchi.[61]

just one of them stretches a little beyond the others to become the unique axon. Its formation then prevents the extension of other neurites, most likely through depletion of Cdc42. In consequence, the remaining neurites become dendrites.

The tip of the growing axon accumulates dishevelled (Dvl is the human orthologue of *Drosophila* Dsh), associated with Frizzled (Fz), a receptor for the family of Wnt ligands (see page 421). Both Dsh/Dvl and Fz were discovered in *Drosophila* mutants in which the epithelial cells had lost the polarized positioning of the wing hairs (see Figure 14.2, page 422). Dvl binds the aPKC–Par-6–Par-3 polarity complex and this leads to stabilization and activation. Although Cdc42 also accumulates in the growing axon,[62] its precise role in the activation of Dvl-associated PKC remains to be determined.

An important substrate of the atypical PKC is the serine/threonine kinase MARK2 (Figure 19.9) which, on phosphorylation is inactivated and detaches from the membrane.[63] This removes an inhibitory constraint for a number of components that normally interact with microtubules. Among these are Tau and MAP1B. The other substrate is GSK3β which also becomes inactivated, so enabling the association of APC and CRMP2 with microtubules. These proteins stabilize microtubules and also facilitate their attachment to the plasma membrane. In so doing, they provide important support for the growing axon. The domain architecture of proteins involved in axon outgrowth is summarized in Figure 19.7.

The Rac1 guanine exchange protein Tiam1/STEF lies downstream of the aPKC–Par-6–Par-3 polarity complex. It therefore activates Rac1 which leads to cytoskeletal reorganization which also contributes to neuronal polarity.[64]

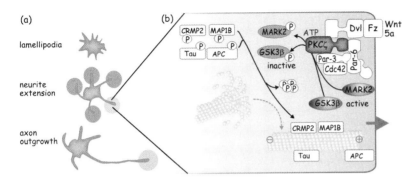

FIG 19.9 Dv1, PKCζ, Par-6, and polarity of axonal outgrowth.
(a) Hippocampal neurons form multiple neurites of which only one becomes an axon. (b) Axon outgrowth, stimulated by Wnt at the Fz receptor is initiated by the binding of a PKCζ–Par-3–Par-6 polarity complex to Dvl. Activated PKCζ phosphorylates and inactivates both MARK2 and GSK3β. Loss of kinase activity leads to a dephosphorylation of the microtubule binding proteins CRMP2, MAP1B, Tau, and APC. These now bind to microtubules, preventing depolymerization and allowing their interaction with the plasma membrane.

List of Abbreviations

Abbreviation	Full name/description	SwissProt entry	Other names
AC1	atypical C1 domain (not related to adenylyl cyclase 1)		
AP-1	activator protein-1		
APC	adenomatous polyposis coli protein	P25054	
ATF2	activating transcription factor-2	P15336	cAMP response element binding protein
PB1	Phox Bem1 domain		
Cdc42	cell division cycle protein 42	P60953	
CREB1	cyclic AMP responsive element binding protein-1	P16220	
CRMP2	collapsin response mediator protein-2	Q16555	DPYSL2
cyclinD		P24385	CCND1, BCL-1 oncogene
DGK-α	diacylglycerol kinase-alpha	P23743	
Dlg1	Disc large-1	Q12959	

Continued

Abbreviation	Full name/description	SwissProt entry	Other names
Dvl	dishevelled	P54792	
dPKC1	Drosophila PKC-1 (D. melanogaster)	P05130	PKC53E
EB1	end binding protein-1	Q15691	APC binding protein, MAPRE
ERK1	extracellular signal regulated kinase-1	P27361	p44 MAPK, MAPK3
Fos-c	feline osteosarcoma cellular homologue	P01100	
Fz-3	Frizzled-3	Q9NPG1	
GAP-43	growth cone associated protein of 43 kDa	P17677	neuronal phosphoprotein B-50, neuromodulin
GSK3β	glycogen synthase kinase-3 beta	P49841	
InaC	inactivation no after-potential C (*D. melanogaster*)	P13677	eye-PKC
InaD	inactivation no after-potential D (*D. melanagaster*)	Q24008	
JIP1	JNK-interacting protein-1	Q9UQF2	
JNK1	c-Jun N-terminal kinase-1	P45983	MAPK8
Jun-c	ju-nana, meaning 17 in Japanese (avian sarcoma virus 17)	P05412	
MAP1B	microtube-associated protein-1B	P46821	
MARCKS	myristoylated alanine-rich C-kinase substrate	P29966	p80
MARK2	MAP/microtubule affinity-regulated kinase-2	Q7KZ17	Par-1 homologue
MEK1	MAPK ERK activating kinase-1	Q02750	MAP2K1
MKK4	MAP kinase kinase-4	P45985	MAP2K4, JNK-activating kinase-1
MKK7	MAP kinase kinase-7	O14733	MAP2K7, JNK-activating kinase-2
NorpA	no receptor-potential (*D. melanogaster*)	P13217	PLC
P62	protein of 62 kDa	Q13501	ubiquitin binding protein, sequestosome-1
Par-3	partioning defective protein-3	Q8TEW0	
Par-6	partitioning defective protein-6	Q9BYG5	

Continued

Abbreviation	Full name/description	SwissProt entry	Other names
PDK1	3-phosphoinositide-dependent protein kinase-1	O15530	
PIX-b	PAK-interacting exchange factor-b	Q14155	Cool-1
PKC1	protein kinase C-1 (*C. elegans*)	P34885	
PKCα	protein kinase C alpha	P17252	
PKCβ1	protein kinase C beta-1	P05771	splice variant PKCb2
PKCδ	protein kinase C delta	Q05655	
PKCε	protein kinase C epsilon	Q02156	
PKCγ	protein kinase C gamma	P05129	
PKCη	protein kinase C eta	P24723	PKCL
PKCι	protein kinase C lambda/iota	P41743	
PKCθ	protein kinase C theta	Q04759	
PKCζ	protein kinase C zeta	Q05513	
PMA	phorbol myristate acetate		
PRK1	PKC-related kinase-1	Q16512	PKN
Rac1	Ras-related C3 botulinum toxin substrate-1	P63000	
RACK1	receptor for activated C-kinase-1	P63244	GNB2
RACK2	receptor for activated C-kinase-2	P35606	coatomer protein β (β-COP)
Raf-C	rat fibrosarcoma	P04049	
Ras-H	harvey rat sarcoma	P01112	
RKIP	Raf-kinase inhibitory protein	P30086	phosphatidylethanolamine-binding protein (PEBP)
Scrib	scribble homologue	Q14160	
Tau	(micro)tubule assembly unit	P10636	neurofibrillary tangle protein
Tiam1	T-lymphoma invasion and metastasis inducing protein-1	Q13009	STEF

Continued

Abbreviation	Full name/description	SwissProt entry	Other names
TPA	12-O-tetradecanoylphorbol-13-acetate		
TPA-1	transient (*C. elegans*)		
Trp	transient receptor-potential (D. melanogaster)	P19334	
UBA	ubiquitin associated domain		
Wnt5a	wingless (Wg) & insert (int) amalgamation	P41221	

References

1. Berenblum I. Cocarcinogenic action of croton resin. *Cancer Res.* 1941;1:48.
2. Sivak A, Van Duuren BL. Phenotypic expression of transformation: induction in cell culture by a phorbol ester. *Science.* 1967;157:1443–1444.
3. Cook J, Hewett C, Hieger I. The isolation of a cancer-producing hydrocarbon from coal tar. *J Chem Soc.* 1933:395–405.
4. Ling H, Sayer JM, Plosky BS, et al. Crystal structure of a benzo[*a*]pyrene diol epoxide adduct in a ternary complex with a DNA polymerase. *Proc Natl Acad Sci U S A.* 2004;101:2269.
5. Pfeifer GP, Denissenko MF, Olivier M, Tretyakova N, Hecht SS, Hainaut P. Tobacco smoke carcinogens, DNA damage and p53 mutations in smoking-associated cancers. *Oncogene.* 2002;21:7435–7451.
6. Imbra RJ, Karin M. Phorbol ester induces the transcriptional stimulatory activity of the SV40 enhancer. *Nature.* 1986;323:555–558.
7. Housey GM, Johnson MD, Hsiao WL, et al. Overproduction of protein kinase C causes disordered growth control in rat fibroblasts. *Cell.* 1998;52:343–345.
8. Angel P, Imagawa M, Chiu R, et al. Phorbol ester-inducible genes contain a common cis element recognized by a TPA-modulated trans-acting factor. *Cell.* 1987;49:729–739.
9. Shaulian E, Karin M. AP-1 in cell proliferation and survival. *Oncogene.* 2001;20:2390–2400.
10. Sassone-Corsi P, Ransone LJ, Lamph WW, Verma IM. Direct interaction between fos and jun nuclear oncoproteins: role of the 'leucine zipper' domain. *Nature.* 1988;336:692–695.
11. Karin M, Liu Z, Zandi E. AP-1 function and regulation. *Cell.* 1997;9:240–246.
12. Bohmann D, Tjian R. Biochemical analysis of transcriptional activation by Jun: differential activity of c- and v-Jun. *Cell.* 1989;59:709–717.
13. Kramer IM, Koornneef I, de Laat SW, van den Eijnden-van Raaij AJ. TGF-β1 induces phosphorylation of the cyclic AMP responsive element binding protein in ML-CCl64 cells. *EMBO J.* 1991;10:1083–1089.

14. Masquilier D, Sassone CP. Transcriptional cross-talk: nuclear factors CREM and CREB bind to AP-1 sites and inhibit activation by Jun. *J Biol Chem*. 1992;267:22460–22466.
15. Vogt PK. Jun, the oncoprotein. *Oncogene*. 2001;20:2365–2377.
16. Eferl R, Wagner EF. AP-1: a double-edged sword in tumorigenesis. *Nat Rev Cancer*. 2003;3:859–868.
17. Boyle WJ, Smeal T, Defize LH, et al. Activation of protein kinase C decreases phosphorylation of c-Jun at sites that negatively regulate its DNA-binding activity. *Cell*. 1991;64:573–584.
18. Goode N, Hughes K, Woodgett JR, Parker PJ. Differential regulation of glycogen synthase kinase-3 β by protein kinase C isotypes. *J Biol Chem*. 1992;267:16878–16882.
19. Smeal T, Binetruy B, Mercola DA, Birrer M, Karin M. Oncogenic and transcriptional cooperation with Ha-Ras requires phosphorylation of c-Jun on serines 63 and 73. *Nature*. 2000;354:494–496.
20. Behrens A, Jochum W, Sibilia M, Wagner EF. Oncogenic transformation by ras and fos is mediated by c-Jun N-terminal phosphorylation. *Oncogene*. 2000;19:2657–2663.
21. Derijard B, Hibi M, Wu IH, et al. JNK1: a protein kinase stimulated by UV light and Ha-Ras that binds and phosphorylates the c-Jun activation domain. *Cell*. 1994;76:1025–1037.
22. Kallunki T, Deng T, Hibi M, Karin M. c-Jun can recruit JNK to phosphorylate dimerization partners via specific docking interactions. *Cell*. 1996;87: 929–939.
23. Kolch W, Heidecker G, Kochs G, et al. Protein kinase Cα activates RAF-1 by direct phosphorylation. *Nature*. 1993;364:249–252.
24. Cacace AM, Ueffing M, Philipp A, et al. PKCε functions as an oncogene by enhancing activation of the Raf kinase. *Oncogene*. 1996;13:2517–2526.
25. Yeung K, Seitz T, Li S, et al. Suppression of Raf-1 kinase activity and MAP kinase signalling by RKIP. *Nature*. 1999;401:173–177.
26. Corbit KC, Trakul N, Eves EM, Diaz B, Marshall M, Rosner MR. Activation of Raf-1 signaling by protein kinase C through a mechanism involving Raf kinase inhibitory protein. *J Biol Chem*. 2003;278:13061–13068.
27. Schonwasser DC, Marais RM, Marshall CJ, Parker PJ. Activation of the mitogen-activated protein kinase/extracellular signal-regulated kinase pathway by conventional, novel, and atypical protein kinase C isotypes. *Mol Cell Biol*. 1998;18:790–798.
28. Whitmarsh AJ, Shore P, Sharrocks AD, Davis RJ. Integration of MAP kinase signal transduction pathways at the serum response element. *Science*. 1995;269:403–407.
29. Gille H, Downward J. Multiple ras effector pathways contribute to G(1) cell cycle progression. *J Biol Chem*. 1999;274:22033–22040.
30. Lopez-Bergami P, Huang C, Goydos JS, et al. Rewired ERK-JNK signaling pathways in melanoma. *Cancer Cell*. 2007;11:447–460.

31. Weston CR, Davis RJ. The JNK signal transduction pathway. *Curr Opin Genet Dev*. 2002;12:14–21.

32. Lopez-Bergami P, Habelhah H, Bhoumik A, Zhang W, Wang LH, Ronai Z. RACK1 mediates activation of JNK by protein kinase C. *Mol Cell* 2005;19:309–320.

33. Berns H, Humar R, Hengerer B, Kiefer FN, Battegay EJ. RACK1 is up-regulated in angiogenesis and human carcinomas. *FASEB J*. 2000;14:2549–2558.

34. Heo YS, Kim SK, Seo CI, et al. Structural basis for the selective inhibition of JNK1 by the scaffolding protein JIP1 and SP600125. *EMBO J*. 2004;23:2185–2195.

35. Borner C, Ueffing M, Jaken S, Parker PJ, Weinstein IB. Two closely related isoforms of protein kinase C produce reciprocal effects on the growth of rat fibroblasts Possible molecular mechanisms. *J Biol Chem*. 1995;270:78–86.

36. Hsiao WL, Housey GM, Johnson MD, Weinstein IB. Cells that overproduce protein kinase C are more susceptible to transformation by an activated H-ras oncogene. *Mol Cell Biol*. 1989;9:2641–2647.

37. Besson A, Yong VW. Involvement of p21(Waf1/Cip1) in protein kinase C α-induced cell cycle progression. *Mol Cell Biol*. 2000;20:4580–4590.

38. Kieser A, Sietz T, Adler HS, et al. Protein kinase Cζ reverts v-raf transformation of NIH-3T3 cells. *Genes Dev*. 1966;10:1455–1466.

39. Fukumoto S, Nishizawa Y, Hosoi M, et al. Protein kinase Cδ inhibits the proliferation of vascular smooth muscle cells by suppressing G1 cyclin expression. *J Biol Chem*. 1997;272:13816–13822.

40. Jackson DN, Foster DA. The enigmatic protein kinase Cδ: complex roles in cell proliferation and survival. *FASEB J*. 2004;18:627–636.

41. Griner EM, Kazanietz MG. Protein kinase C and other diacylglycerol effectors in cancer. *Nat Rev Cancer*. 2007;7:281–294.

42. Ruegg C. Leukocytes, inflammation, and angiogenesis in cancer: fatal attractions. *J Leukoc Biol*. 2006;80:682–684.

43. Prevostel C, Martin A, Alvaro V, Jaffiol C, Joubert D. Protein kinase Cα and tumorigenesis of the endocrine gland. *Horm Res*. 1997;47:140–144.

44. Prevostel C, Alvaro V, Vallentin A, Martin A, Jaken S, Joubert D. Selective loss of substrate recognition induced by the tumour-associated D294G point mutation in protein kinase Cα. *Biochem J*. 1998;334:393–397.

45. Tognon CE, Kirk HE, Passmore LA, Whitehead IP, Der CJ, Kay RJ. Regulation of RasGRP via a phorbol ester-responsive C1 domain. *Mol Cell Biol*. 1998;18:6995–7008.

46. Kemphues KJ, Priess JR, Morton DG, Cheng NS. Identification of genes required for cytoplasmic localization in early *C. elegans* embryos. *Cell*. 1988;52:311–320.

47. Henrique D, Schweisguth F. Cell polarity: the ups and downs of the Par6/aPKC complex. *Curr Opin Genet Dev*. 2003;13:341–350.

48. Munro E, Nance J, Priess JR. Cortical flows powered by asymmetrical contraction transport PAR proteins to establish and maintain anterior-posterior polarity in the early *C. elegans* embryo. *Dev Cell*. 2004;7:413–424.

49. Joberty G, Petersen C, Gao L, Macara IG. The cell-polarity protein Par6 links Par3 and atypical protein kinase C to Cdc42. *Nat Cell Biol.* 2000;2:531–539.

50. Etienne-Manneville S, Hall A. Integrin-mediated activation of Cdc42 controls cell polarity in migrating astrocytes through PKCζ. *Cell.* 2001;106:489–498.

51. Garrard SM, Capaldo CT, Gao L, Rosen MK, Macara IG, Tomchick DR. Structure of Cdc42 in a complex with the GTPase-binding domain of the cell polarity protein, Par6. *EMBO J.* 2003;22:1125–1133.

52. Martin-Belmonte F, Gassama A, Datta A, et al. PTEN-mediated apical segregation of phosphoinositides controls epithelial morphogenesis through Cdc42. *Cell.* 2007;128:383–397.

53. Ebnet K, Suzuki A, Ohno S, Vestweber D. Junctional adhesion molecules (JAMs): more molecules with dual functions? *J Cell Sci* 2004;117:19–29.

54. Wedlich-Soldner R, Altschuler S, Wu L, Li R. Spontaneous cell polarization through actomyosin-based delivery of the Cdc42 GTPase. *Science.* 2003;299:1231–1235.

55. Colombo JA, Reisin HD, Miguel-Hidalgo JJ, Rajkowska G. Cerebral cortex astroglia and the brain of a genius: a propos of A. Einstein's. *Brain Res Rev.* 2006;52:257–263.

56. Dobson R. Inside Einstein's Brain. *Independent on Sunday.* 2006;4.

57. Osmani N, Vitale N, Borg JP, Etienne-Manneville S. Scrib controls Cdc42 localization and activity to promote cell polarization during astrocyte migration. *Curr Biol.* 2006;16:2395–2405.

58. ten Klooster JP, Jaffer ZM, Chernoff J, Hordijk PL. Targeting and activation of Rac1 are mediated by the exchange factor βPix. *J Cell Biol.* 2006;172:759–769.

59. Mogensen MM, Tucker JB, Mackie JB, Prescott AR, Nathke IS. The adenomatous polyposis coli protein unambiguously localizes to microtubule plus ends and is involved in establishing parallel arrays of microtubule bundles in highly polarized epithelial cells. *J Cell Biol.* 2002;157:1041–1048.

60. Etienne-Manneville S, Manneville JB, Nicholls S, Ferenczi MA, Hall A. Cdc42 and Par6-PKCζ regulate the spatially localized association of Dlg1 and APC to control cell polarization. *J Cell Biol.* 2005;170:895–901.

61. Arimura N, Kaibuchi K. Neuronal polarity: from extracellular signals to intracellular mechanisms. *Nat Rev Neurosci.* 2007;8:194–205.

62. Schwamborn JC, Puschel AW. The sequential activity of the GTPases Rap1B and Cdc42 determines neuronal polarity. *Nat Neurosci.* 2004;7:923–929.

63. Hurov JB, Watkins JL, Piwnica-Worms H. Atypical PKC phosphorylates PAR-1 kinases to regulate localization and activity. *Curr Biol.* 2004;14:736–741.

64. Nishimura T, Yamaguchi T, Kato K, et al. PAR-6-PAR-3 mediates Cdc42-induced Rac activation through the Rac GEFs STEF/Tiam1. *Nat Cell Biol.* 2005;7:270–277.

CHAPTER 20

Signalling Through Receptor Serine/Threonine Kinases

The TGFβ family of growth factors

The receptor serine/threonine protein kinases[1] are all dedicated to relaying signals deriving from members of the transforming growth factor β family (TGFβ). The first of these to be identified, TGFβ1, emerged as a transforming factor for mesenchymal cells (see page 304). Related proteins were revealed through loss-of-function mutation studies in *Drosophila* (*Dpp* gene) and *Xenopus* (*Vg1* gene).

We now know that TGFβ is a member of a family of structurally related proteins, most of which have little or nothing to do with cell transformation. With 42 members in the human genome, 7 in *Drosophila*, and 4 in *Caenorhabditis elegans*, the TGFβ family is one of the most prominent families of first messengers.[2] The mammalian TGFβ family can be divided into a number of subfamilies: TGFβ itself, BMP, GDF, activin, inhibin, nodal, myostatin, and AMH. These factors may be produced by many cell types (TGFβ), or just a few (myostatin). They may be active from the earliest stages of embryo development through adulthood (BMP), or for only very limited periods (AMH).[3]

The many abbreviations used in this chapter are collected together at the end of the chapter.

Myostatin and big burgers

Myostatin, a member of the TGFβ family, is a negative regulator of muscle growth. It was first identified in null-mutant knockout mice that exhibit a widespread increase in muscle mass due to hyperplasia (fibre number) and hypertrophy (fibre thickness). An 11 bp deletion in the C-terminal coding region of the myostatin gene (Mstn) in the Belgian Blue cattle breed and a missense mutation in the Piedmontese breed are responsible for their double-muscled phenotypes[4] (Figure 20.1). The animals are apparently born with a normal phenotype, the so-called double muscling only appearing at about 4–6 weeks. It is not uncommon for the bulls to achieve a weight exceeding 1300 kg. The animals are docile and the beef is said to be delicious. An inherited mutation in the myostatin gene, boosting muscle growth and reducing fat, also occurs in humans. It may come as no surprise that products that claim to regulate myostatin are already used by many athletes and bodybuilders.

FIG 20.1 Big beef.
Springhill Wizzard, owned by Martin Brothers, Newtownards, was the best Belgian Blue Junior Bull and Reserve Male Champion at the Balmoral Show 2007. Thomas and James Martin are pictured exhibiting the prizewinner. Photograph by kind courtesy of Columba O'Hara and the British Blue Cattle Society.

Several members of the TGFβ family are released from cells as dimers still attached (non-covalently) to their propeptides (Figure 20.2a). These 'latent' complexes are often linked to one of the several TGFβ-associated proteins, also termed latent.[5] Release of the ligand occurs through proteolytic activity (by plasmin or metalloprotease MMP-2 or -9) or by interaction with thrombospondin or the integrins αVβ6 and αVβ8 (which modify the propeptide by binding to an RGD sequence). The importance of integrin-mediated activation of TGFβ is illustrated by the finding that lack of the RGD motif, and thus failure of the interaction with integrins, results in organ wasting similar to that observed in mice that do not express TGFβ1.[6,7] The general idea is the existence of a matrix-associated ligand-reservoir, ready to be activated under favourable conditions.

Sequestration of TGFβ family members also occurs through ligand traps such as chordin, DAN/Cerberus, decorin, follistatin, noggin, or α2M (see Figure 20.9). We will return to chordin and noggin, which, by preventing BMP signalling, play an important role in the formation of neuroectoderm in Xenopus laevis.

TGFβ receptors, type I and type II

TGFβ1 came to notice in a search for transforming growth factors and initially, it was anticipated that its receptor would be linked, either directly

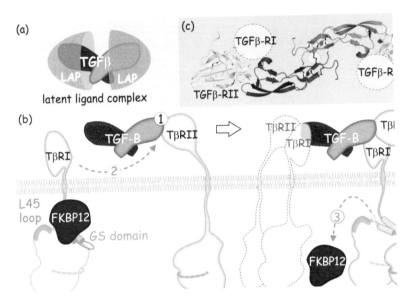

FIG 20.2 TGFβ induces receptor dimerization.

(a) The TGFβ1-ligand constitutes a dimer associated with its propeptides (LAP). To bind its receptor, the ligand detaches from LAP. Different modes of activation include degradation by proteases such plasmin, MMP-2 or -9 but also through contact with thrombospondin or the integrins αvβ8 or αvβ8 (vitronectin receptor). (b) TGFβ1 binds to the type II receptor (1) and this allows it to be recognized by the type I receptor (2). The inhibitory immunophilin FKBP12 detaches (3) and the type I receptor is phosphorylated by the type II (4). The activated type I receptor signals onward into the cell. Because TGFβ1 is itself a dimer, two receptor complexes can form around the growth factor. (c) Molecular structure of TGFβ3 associated with two type II receptors. The presumed orientation of the type I receptor is indicated. The cysteine knot composed of numerous intramolecular disulfide bonds is indicated in yellow (1 ktz[13]).

or indirectly, to tyrosine phosphorylations. When it was later found that it inhibits proliferation of epithelial cells, interest shifted to the possibility that it prevents growth factor signalling. Neither of these ideas proved very fruitful. TGFβ1 has no effect on tyrosine phosphorylation, nor (at least when measured on a time scale of minutes) does it affect the early events in EGF or PDGF receptor signalling. So it remained until 1990, when a receptor for activin was cloned and found to contain a putative transmembrane serine/threonine protein kinase domain.[8] Similar domains were then found in two of the TGFβ1 receptors.[9–11]

Based on their structural and functional properties, the receptors for this family of ligands are divided into subfamilies types I and II (TβR-I and TβR-II) (Figures 20.2b and 20.3).[12] They are very similar. Both receptors are glycoproteins. Both have a single membrane span and an intrinsic serine/threonine kinase domain in the intracellular C-terminal segment. What distinguishes the type I receptor is a highly conserved stretch of 30 residues, the GS domain,

FIG 20.3 Domain architecture of TGFβ, its receptors and associated proteins.

The types I and II receptors contain a serine/threonine protein kinase domain and a cysteine-rich TGFβ-binding domain. The type I receptor contains an additional GS domain and a number of phosphorylation sites that are involved in its activation by the type II receptor (which is constitutively active). Other abbreviations may be found in the list at the end of the chapter.

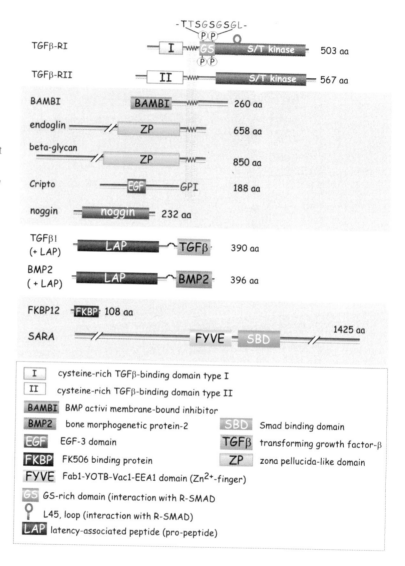

	cysteine-rich TGFβ-binding domain type I
I	cysteine-rich TGFβ-binding domain type I
II	cysteine-rich TGFβ-binding domain type II
BAMBI	BMP activi membrane-bound inhibitor
BMP2	bone morphogenetic protein-2
EGF	EGF-3 domain
FKBP	FK506 binding protein
FYVE	Fab1-YOTB-Vac1-EEA1 domain (Zn²⁺-finger)
GS	GS-rich domain (interaction with R-SMAD
L45, loop (interaction with R-SMAD)
LAP	latency-associated peptide (pro-peptide)
SBD	Smad binding domain
TGFβ	transforming growth factor-β
ZP	zona pellucida-like domain

immediately preceding the kinase domain that regulates TβR-I kinase activity. This stretch is attached to the kinase domain but is rather flexible and it can fluctuate between attached and detached states. This would allow the receptor to switch between inactive and active conformations. This is prevented by the inhibitory immunophilin FKBP12 which holds the GS domain firmly in place. (Figures 20.2b and 20.3). Phosphorylation of the GS domain by the type II receptor breaks the inhibitory spell.

As it has no GS domain, the type II receptor escapes inhibitory control and is therefore constitutively active, always in wait for the approach of the type I receptor.

The mammalian receptors are homologous with the punt (*put*), thickveins (*tkv*), and saxophone (*sax*) genes of *Drosophila* and with the dauer phenotypes 1 and 4 (*daf-1* and *daf-4*) genes of *C. elegans*. The downstream signalling pathway from the TGFβ1-receptors was revealed by searching for mammalian homologues of their counterparts in these organisms. It became apparent that mammalian TGFβ1 receptors transmit signals into cells through the Smad proteins, a unique set of transcription factors (see below).[14]

Here we outline some of the mechanisms by which members of the TGFβ family of receptors elicit their effects on target cells. We concentrate on the pathway activated by TGFβ1 through the TβR-I and TβR-II receptors and in particular on the pivotal role of the Smads in relaying signals to the nucleus. It will become apparent that these pathways are rather similar to those already described for the activation of the STATs through the tyrosine kinase-containing receptors such as those for EGF, PDGF, and interferon (see page 353). The main theme is that a receptor complex first recruits and then phosphorylates a transcription factor. This forms an oligomeric complex that translocates to the nucleus to interact with DNA-response elements in promoter regions of genes.

TGFβ-mediated receptor activation

TGFβ1 is a disulfide-linked homodimer. It gathers pairs of type I and II receptors to form heterotetrameric receptor complexes (Figure 20.2). Ligand-independent homo-oligomers may exist, but do not transmit signals. TGFβ1 can bind to TβR-II in the absence of TβR-I but not vice versa. TGFβ1, however, cannot signal into the cell in the absence of TβR-I. All this indicates that the most likely sequence of events is that TGFβ1 first binds to TβR-II, altering its conformation so that it can then be recognized by TβR-I. (In contrast, BMPs -2 and -4 first bind the type I receptors and then the type II). When the ligand brings the two receptors in close proximity, the type II receptor phosphorylates the serines and threonines in the sequence TTSGSGSGL of the GS domain of TβR-I (see Figures 20.2b and 20.5).

As with the tyrosine kinase-containing receptors, oligomerization and phosphorylation together constitute the signal for recruitment of effector proteins. However, the receptor serine/threonine kinases employ a quite different approach in the control of kinase activity. Here phosphorylation alters catalytic competence by removing an inhibitory constraint (the GS domain that wedges into the N-lobe) as well as by preventing binding of the inhibitor FKBP12 (Figure 20.4).[15] Moreover, the detached phosphorylated GS domain now acts as a docking site for Smad proteins, the substrates of the type I receptors (Figure 20.5). This general mechanism applies to all TGFβ and BMP receptors, from *Drosophila* to mammals.

603

FIG 20.4 Activation of the kinase domain of the TGFβ receptor. The structure of the TGFβ type I receptor reveals a kinase domain. The activation segment (red) is structured but the presence of the GS domain (light blue), wedged into the N-lobe, causes a rotation of the αC-helix and thus a separation between K232 and E245. This prevents the correct positioning of ATP in the cleft. The threonine and three serine residues phosphorylated in the GS domain are shown as sticks in an orange ellipse. In order to achieve catalytic competence the GS domain detaches partially (1), so allowing K232 and E245 to approach each other (2), allowing ATP to locate correctly. The L45 loop, which interacts with the L3 loop of Smad proteins is purple.

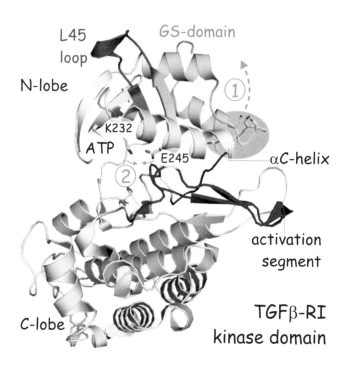

FIG 20.5 Schematic view of TGFβ receptor activation.
The TGFβ type I receptor exists in three different states. *Inactive* has GS wedged into the N-lobe (held in place by FKBP12); *intermediate* allows oscillation between the attached and partly-detached positions of GS; in the *active* state the phosphorylated GS cannot wedge into the N-lobe. Phosphorylation is induced by the close proximity of the types I and II receptors. The phosphorylated GS, together with the L45 loop, form the binding site of receptor-regulated Smad proteins (R-Smad) (1), which are subsequently phosphorylated on two serine residues at their C-termini (2). SARA facilitates the interaction between R-Smads and the type I receptor. Once phosphorylated, R-Smad proteins detach from the receptor (3). FYVE, MH1 and MH2 indicate protein domains: see text.

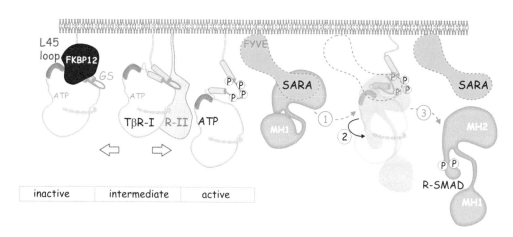

Accessory and pseudo receptors: betaglycan, endoglin, cripto, and BAMBI

The quest for cell surface TGFβ-binding proteins revealed a third set of receptors, quite distinct from the TβRs I and II. Their intracellular domains are devoid of any sequence motif that could be involved in signal transduction. Betaglycan, endoglin, and cripto are coreceptors that support signalling. BAMBI is a pseudo-receptor that inhibits signalling.

Betaglycan

Betaglycan is a transmembrane proteoglycan that binds isoforms of TGFβ.[16] This is vital in the case of TGFβ2, as it facilitates its binding to the receptor, leading to the formation of a ternary complex comprising TGFβ2, TβRII, and betaglycan.[17] TβRI then binds to the complex, displacing the betaglycan. Expression of betaglycan in cells normally lacking this coreceptor causes an increase in TGFβ2 binding and a concomitant increase in sensitivity to the ligand (Figure 20.6a).[18]

Inhibins and activins were first identified as factors that respectively suppress or stimulate secretion of FSH from pituitary gonadotropes.[19] This functional antagonism is explained by the finding that they both bind the activin receptor ActR-II. Binding of activin promotes the recruitment of the activin type I receptor, ActR-IB, to initiate the signal. Inhibin recruits betaglycan to form a stable complex so that there is a loss of the activin signal due to depletion of available ActR-II. This puts a block on further proceedings[20] (Figure 20.6b). Expression of betaglycan in cells that normally respond poorly enhances the sensitivity to inhibin.

Endoglin

Endoglin, originally discovered as a glycoprotein antigen highly expressed on human endothelial cells,[23] facilitates the interaction of TGFβ1 already bound to TβRII with ALK-1 (type -I receptor) in endothelial cells[24] (Figure 20.6c). Thus it determines the balance between ALK-1 signals, leading to proliferation and migration of endothelial cells, and TβRI (also referred to as ALK-5) signals that inhibit proliferation and migration. Since both receptors are expressed on endothelial cells, the presence of endoglin decides the outcome of the response to TGFβ, favouring an ALK-1-mediated proliferative and invasive response. It facilitates the formation of new blood vessels (angiogenesis).[25,26] Inherited absence of functional endoglin is associated with the disorder haemorrhagic telangiectasia type I (HHT1), characterized by a leaky and poorly developed vascular bed.[27]

Cripto

Cripto is a small extrinsic membrane protein of the EGF–CFC gene family, members of which take part in embryonic anterior-posterior axis

In vivo, the functional role of **betaglycan as a coreceptor** is vital. Mice lacking betaglycan die during embryogenesis with heart and liver defects,[21] and targeting betaglycan during development disrupts mesenchyme formation in the heart and branching morphogenesis in the lung.[22] In the adult, loss of betaglycan is associated with tumour formation.

The receptor **TβRI** was first identified biochemically, as a component necessary for the transmission of TGFβ signals, but molecular characterization was spearheaded by the activin receptors. Sequences with high similarity were named activin-like kinases or ALKs. This explains the double nomenclature of TβRs and ALKs.

development.[28] It binds to nodal, promoting the formation of active receptor dimers (ActRIB with ActRIIB) and also binds to activin, when attached to ActRII, but here it impedes access to the type I receptor (Figure 20.6d). Its tumour promoting action may in part be explained by its capacity to block activin-mediated growth inhibition of epithelial cells.[29]

FIG 20.6 Accessory and pseudo-receptors.
Several membrane proteins bind the TGFβ family of ligands. None of these has intracellular signalling motifs; instead, they reinforce or inhibit signalling by affecting the interaction between the ligand and its receptors. (a) Betaglycan binds TGFβ and facilitates its access to TβR-II. Once a complex is formed with the type I receptor (TbR-I), it detaches. Betaglycan is indispensable for TGFβ2-mediated signalling. (b) Inhibin binds betaglycan and ActR-IIB, and prevents the interaction with ActR-1B (Alk-4) resulting in formation of non-functional type II receptors. (c) Endoglin facilitates the formation of TβR-II/ALK-1 complexes and in endothelial cells it shifts the balance towards an ALK-1 response (rather than an TbR-I response). (d) Cripto is bound to the membrane by a glycosylphosphatidylinositol anchor. It binds activin attached to ActRII and blocks signal propagation. (e) Expression of BAMBI is induced by TGFβ signalling. It forms a complex with TβR-I and prevents its phosphorylation by TβR-II, thereby blocking the TGFβ response. Red spots represent TGFβ2, activin, or TGFβ in general; the dark blue spot is inhibin.

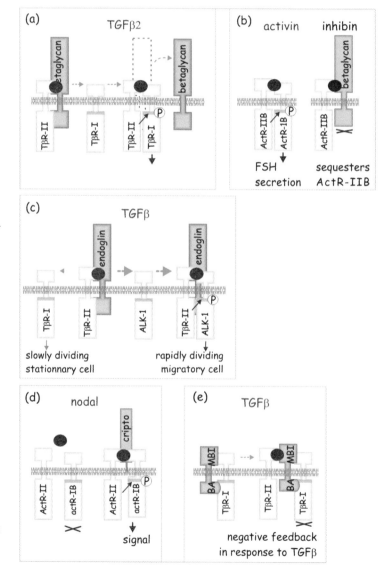

606

BAMBI

BAMBI associates with the cytoplasmic domain of TβRI receptors in the absence of ligand (Figure 20.6e). In so doing, it prevents phosphorylation of TβRI and inhibits signalling from activin, BMP, and TGFβ.[30]

Downstream signalling: *Drosophila*, *Caenorhabitidis,* and Smad

Genetic screens of accessible organisms such as *Drosophila* and *C. elegans* have provided pointers to the mechanism of TGFβ signalling in mammalian cells. The gene *decapentaplegic (Dpp),* when mutated, causes pattern deficiencies and duplications in structures derived from one or more of the 15 major imaginal disks in *Drosophila*.[31] *Dpp,* equivalent to BMPs-2 and 4 of mammalian cells, is responsible for dorsal/ventral polarity in early development. Later, with the appearance of the segments, *Dpp* functions in the definition of boundaries between the segmental compartments. As part of this process, it defines the position of the future limbs, wings, legs, and antennae. It also has a role in the structuring of the mesoderm (see web page http://www.sdbonline.org/fly/aimain/1aahome.htm)

The Dpp gene product acts through three receptors that are homologous to the family of mammalian TGFβ receptors (thickveins, saxophone (TβRI), and punt (TβRII)). Genetic screening has revealed a mutation that, when combined with *Dpp* mutations associated with a feeble phenotype, generates one that is more severely affected. The mutant gene so obtained, *Mad,* yields flies that exhibit defects resembling those due to mutated *Dpp*.[32] However, since wild-type *Dpp* cannot restore the defects induced by mutations in *Mad, Dpp* must lie upstream of *Mad* (Figure 20.7).[33]

Decapentaplegic, paralysed at 15 sites.

Imaginal, from *imago,* the final and perfect stage or form of an insect after its metamorphoses.

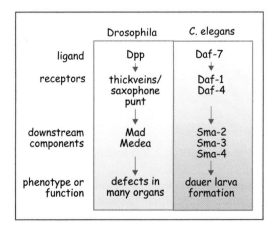

FIG 20.7 Signal transduction pathways downstream of serine/threonine kinase receptors in two phyla.
Homologies between the genes implicated in these pathways has allowed the elucidation of the sequence of events downstream of the TGFβ receptor in mammals.

The nematode *C. elegans* responds to conditions of overcrowding and starvation by developmental arrest as a dauer (resilient, durable) larva. Screening of mutants having this phenotype revealed a number of genes (*daf-1–4*, etc.) of which *daf-1* and *daf-4* code for serine/threonine receptor protein kinases.[34,35] *daf-4* mutants are dauer-constitutive and the larvae are smaller than the wild types. Screening for mutants with similar phenotypes has revealed three more genes, *sma-2–4*,[36] which act downstream of *daf-4*. Mad (fly) and Sma (worm) proteins are homologous and with the sequences to hand, eight human homologues coding for Smad proteins have been identified.[14,37]

Smad proteins have multiple roles in signal transduction

The Smad proteins have two regions of homology, MH1 and MH2, connected by a more divergent linker segment (Figure 20.8a). The N-terminal MH1, highly conserved in all Smads (except Smads 6 and 7), binds to DNA at the Smad binding element (though, curiously, the most abundant splice form of Smad2 carries an insert that blocks binding to DNA). MH1 also interacts with transcription factors such as Jun, ATF3, Sp1, and Runx. The linker constitutes a flexible segment that is a hotspot of phosphorylation sites and operates in the integration of several signalling pathways. It also contains a PPxY motif that acts as an E3-ubiquitin ligase binding site. In Smad4 the linker contains a nuclear export signal (NES) (Figure 20.8b).

The MH2 domain, a particularly versatile protein-interacting module, is conserved in all Smad proteins. A set of adjoining hydrophobic patches (the hydrophobic corridor) mediates interactions with cytoplasmic retention proteins such as SARA, nuclear pore proteins (NUP214) and with a number of transcription cofactors. The MH2 domains of Smads 1, 2, 3, 5 and 8 carry a conserved C-terminal SxS motif, the substrate of TβR-I. When phosphorylated, this interacts with a basic pocket in MH2 to provoke the assembly of hetero-oligomeric complexes (see below).

Smad proteins have different functions, controlled through their selective interactions with TβR-I and with each other[38] (Figures 20.9 and 20.10). On the basis of their sequences and functions they are divided into three groups: (i) receptor-regulated Smads 1, 2, 3, 5, and 8; (ii) common mediator Smads 4 and 3; inhibitory Smads (6 and 7). Below we elaborate further on their structure and function.

Receptor-regulated Smads 1, 2, 3, 5 and 8: receptor recognition

The receptor-regulated Smads are phosphorylated by the activated TβRI receptors. Short structural elements in TβRI and in the Smads determine the specificity of interaction: the exposed L45 loop in the type I receptor kinase

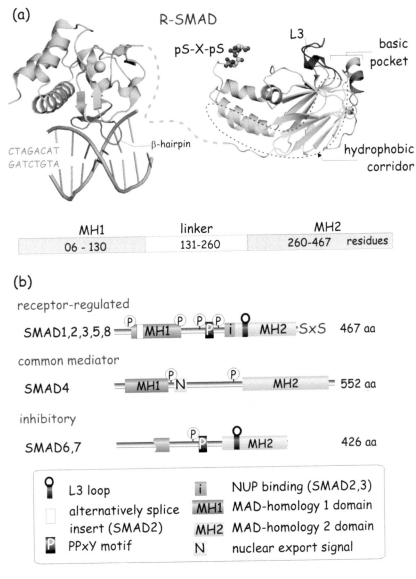

FIG 20.8 Molecular structure and domain architecture of Smad proteins.

(a) Receptor-regulated Smad proteins have three distinct segments, a conserved MH1 domain, a more divergent linker region, and a highly conserved MH2 domain. MH1 is stabilized by Zn^{2+} (yellow sphere) and contains a β-hairpin that binds the Smad-binding element (SBE). The MH2 domain contains a C-terminal SxS motif that is phosphorylated by the type I receptor. It also has an L3 loop and basic pocket which interact with L45 and the phosphorylated GS domain of the receptor. It exposes a hydrophobic corridor which is the site of attachment of numerous proteins among which is SARA. It also interacts with a large number of other components including TβR-1, nuclear pore proteins, coactivators and repressors, and DNA binding cofactors. It is involved in Smad oligomerization (1ozj,[39] 1 khx[40]). (b) Smad transcription factors are subdivided in three groups based on functional and structural criteria. Receptor-regulated Smad proteins contain the SxS motif which is phosphorylated by type I receptors. They contain a NUP-binding motif as well as a PPxY motif with which they interact with the E3-ubiquitin ligase Smurf. The common mediator Smad4 forms complexes with all receptor-regulated Smad proteins interacting with their phosphorylated SxS motif. It lacks the L3 loop and does not interact with type 1 receptors. Smads 6 and 7 lack a functional MH1 domain as well as the SxS motif. They prevent TGFβ signalling. They too bind Smurf through a PPxY motif in the linker region. All Smad proteins contain numerous sites that are phosphorylated by serine/threonine protein kinases (just a few are indicated).

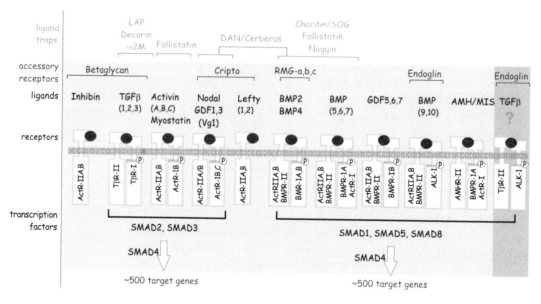

FIG 20.9 Ligands and their traps of the TGFβ family.

Traps are proteins that trap ligand or block its access to the receptor. Accessory receptors facilitate ligand binding and receptor complex formation. The different receptor combinations can be divided into two groups: the TGFβ/activin group recruits Smad-2 and 3, whereas the BMP group recruits Smad 1, 5, and 8. The type II receptors phosphorylate and activate the type I receptors. Unfortunately receptor nomenclature remains unsettled; ActR-1 = ALK-2, ActR-1B = ALK-4, ActR-1C = ALK-7. The combination in the shaded column on the right is not firmly established.

The **MH2** domain qualifies as a phosphoserine-binding domain and offers yet another means by which receptor signalling complexes are formed. Structurally, it shares striking homology with the forkhead-associated domain (FHA) but not with 14-3-3, both of which also recognize phosphoserine/threonine residues.[43]

domain interacting with the L3 loop in the MH2 domain of the Smads (see Figures 20.3 and 20.8). By exchanging selected residues in these domains it is possible to switch the signalling specificity of the TGFβ and BMP pathways.[41,42] Thus, the TβR-I (ALK-5) and the nodal/activin type I receptors (ALK-4 and ALK-7) recognize Smads 2 and 3, whereas ALK-1, -2, -3 and -6 recognize Smads 1, 5 and 8 (Figure 20.9). In this way, different ligands, binding different receptor combinations, provide different signals through the recruitment of specific combinations of Smad proteins.

Whereas the interaction between loops L45 and L3 determines specificity, phosphorylation of the GS domain acts as the on/off signal, roughly equivalent to phosphotyrosine binding to SH2- or PTB domain containing proteins (see page 768). The phosphorylated GS domain is recognized by a basic pocket in the Smad proteins, adjacent to the L3 loop (Figure 20.8b). Serine phosphorylation of the SxS motif in the MH2 domain causes a structural alteration that decreases affinity of the Smads for cytoplasmic anchors, increases their affinity for nuclear pore proteins and allows them to bind each other and to Smad4 (Figure 20.10).

Cytoplasmic retention of receptor-regulated Smad proteins

In their basal steady state, the receptor-regulated Smads are mainly present in the cytoplasm.[44] They are prevented from moving to the nucleus by a number of proteins that function as anchors. Examples are Disabled-2, Axin, cPML, SARA (all of which bind Smads 2 and 3), and ELF (Smads 3 and 4). TRAP-1 and TLP bind to Smad4, but only after treatment with TGFβ, and it remains uncertain whether they have a role in cytosolic retention in 'resting' cells.[3] The best characterized of these is SARA.[45]

SARA operates in two ways. First, it binds the MH2 domain in the hydrophobic corridor which also binds the nuclear transport proteins[46] (see Figure 20.8). Phosphorylation of the SxS motif disrupts the binding with SARA and this may constitute the nuclear translocation signal. Secondly, SARA has a FYVE domain that can bind to PI(3)P, abundant on early endosomal vesicles.[47,48] Indeed, it has been suggested that effective phosphorylation of Smads 2 and 3 only occurs when the receptors are present in the early endosomal membrane. Despite all this, there is no genetic evidence to support the idea that SARA is directly involved in signalling through Smads 2 or 3,[3] and importantly, there is no interaction between Smad1 and SARA, so it is not clear if a similar protein is involved in signalling from BMP receptors.

Common mediator Smad4

Smad4 does not interact with the receptors and it lacks the C-terminal SxS phosphorylation motif, though it does form hetero-oligomeric complexes with the receptor-regulated Smads (Figure 20.10). In mammalian cells it binds to phosphorylated Smads 1, 2, 3, 5, and 8, forming complexes that transfer to the nucleus, there to act as transcription factors. It plays an important role in the interaction with cotranscription factors such as CPB or P300.

It is not certain whether Smad4 links with other Smads in the cytoplasm or the nucleus. For sure it does not interact with SARA, and no cytosolic retention proteins have been detected in resting cells, but it does carry a nuclear export signal (NES) which interacts with CRM1, a nuclear export receptor. In this way Smad4 is retained in the cytosol unless it teams up with phosphorylated receptor-regulated Smads or with other DNA-binding transcription factors that mask the nuclear export signal.[49,50]

Hetero-oligomeric complex formation

Smad–Smad complexes

The C-terminal SxS phosphorylation motif of the receptor-regulated Smads resembles the phosphorylation motif of the GS domain in the TβRI receptors. Following phosphorylation, the basic pocket in the MH2 domain turns its attention to fellow Smads to form hetero-oligomeric complexes (Figure 20.10).[40,52] The stoichiometry of the hetero-oligomers remains a matter of some debate, but

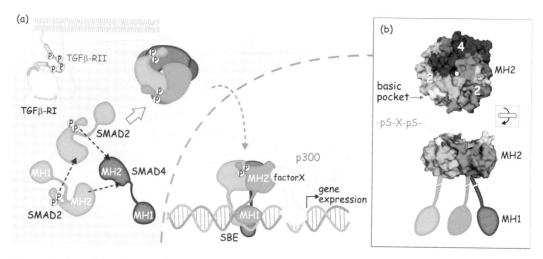

FIG 20.10 Smad activation and nuclear translocation.

(a) On phosphorylation of the C-terminal SxS motif, receptor-regulated Smads complex with each other and then with Smad4. The phosphoserines bind the basic pocket in the MH2 domain. The trimers enter the nucleus to bind DNA at the SBE. The complexes also bind other DNA-binding proteins and transcriptional cofactors (for instance p300). (b) Detail of a Smad2–Smad4 complex showing the interaction of the phosphoserines in the C-terminus with the basic pocket of the MH2 domain. The Smad proteins are viewed from two different angles to show the presumed orientation of the MH2 domain relative to the DNA-binding MH1 domain (1u7v[51]).

heterotrimers made up of two phospho-R-Smads and one Smad4 seem to be prevalent.[51]

Individual Smads also complex with TCF/β-catenin or TIF1γ

Both Smads 2/3 and Smad4 also combine to form complexes with other DNA-binding proteins. Quite independently of TGFβ signalling, Smad4 forms a complex with TCF and β-catenin to induce expression of c-myc and the tight junction protein claudin-1 (Figure 20.11).[53,54] Smads 2 and 3 bind TIF1γ in competition with Smad4.[55] This alternative complex comes into action during TGFβ-mediated differentiation of haematopoietic progenitors into red blood cells, acting in parallel with the canonical TGFβ pathway, giving rise to a complex of Smads 2/3 with Smad4, which inhibits the proliferation of the progenitor cells.

Nuclear import and export

Migration of the Smads into the nucleus occurs both with and without the intervention of nuclear importins. The hydrophobic corridor of the MH2 domain interacts directly with the FG repeat region on nucleoporins (see Figure 20.13), in this way obviating the nuclear transport receptors (importins α and -β).[56] The Smads are then retained within the nucleus through the formation of hetero-oligomeric complexes that bind DNA and/or transcription

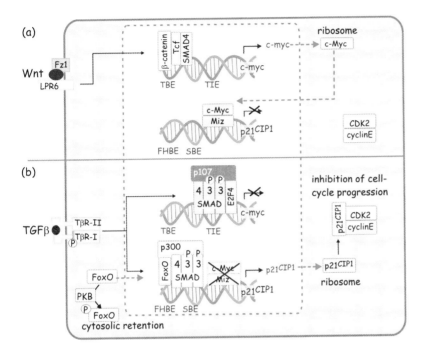

FIG 20.11 A self-enabling response regulates expression of p21CIP1.

(a) Expression of the cyclin-dependent kinase inhibitor p21^{CIP1} is repressed by the transcription factors c-Myc and Miz, both expressed through the influence of β-catenin/TCF complexed with Smad4. (b) TGFβ inverses the situation. Smad3/Smad4 bind the TGFβ-inhibitory element in the c-myc promoter. With E2F4 (or -F5) and p107 they repress expression of c-*myc*. The c-Myc/Miz-mediated repression is lifted and a new transcriptional complex, comprising Smad3/Smad4, FoxO, and p300 drive expression of p21^{CIP1}. After translation the inhibitor binds to CDK4–CyclinD and halts cell cycle progression. Regulation of the inhibitor p15^{INK4B} is similar.

cofactors. Dephosphorylation of the SxS motif leads to dissociation and allows export back to the cytosol (see Figure 20.17).

SMAD transcriptional complexes

Smad hetero-trimers bind to DNA through their MH1 domains, but due to the weak interaction, they interact with other DNA-binding factors to achieve high affinity and selectivity.[57] Furthermore, the sequence of the β-hairpin that interacts with the base pairs that constitute the SBE is highly conserved (Figure 20.8). This suggests that it is not the DNA contact that provides selectivity for gene targeting between the different Smad proteins.[58] Rather, the combinations of the Smads with their partners provide the necessary levels of specificity. Several cofactors have been identified, belonging to different families of DNA-binding proteins, that account for the great breadth of TGFβ transcriptional responses. Examples of complexes are presented in

Table 20.1. Important in all this is the recruitment of additional cofactors that either render the DNA accessible or inaccessible to RNA polymerase.

Activation of gene expression

In addition to all that has just been related, the heterotrimeric Smad complexes also interact with other DNA-binding proteins and transcriptional cofactors (see below) to form even bigger aggregates. These recruit histone acetyltransferases (HATs) such as p300 and CBP (see Figure 14.3, page 423)[67,68] and, through interaction with the MH2 domains of Smads1–4, render the DNA accessible to RNA polymerases.

Repression of gene expression

About a quarter of all TGFβ responses in mammalian cells involve gene repression.[69] Repressive complexes act in different ways. For instance, p107, a protein that resembles the retinoblastoma tumour suppressor, recruits histone methyltransferases to the promoter region to create silent heterochromatic regions.[70] TGIF recruits, via the intermediate of CTBP1, the histone deacetylation complex mSin3/HDAC and this leads to tightly wrapped DNA that is inaccessible to RNA polymerases.[71]

A self-enabling response: repression of myc is prerequisite for expression of cell cycle inhibitors

A common response to TGFβ in epithelial cells, inhibition of the cell division cycle, occurs by enhancing the expression of inhibitors of cyclin-dependent protein kinases. This is mediated through a self-enabling process in which Smad proteins first repress the expression of c-myc and then upregulate expression of the inhibitors. This has been studied in detail for p21^{Cip1} and p15^{INK4B}.[57,72] The c-myc promoter contains two TCF-binding elements (TBE) that bind TCF/LEF, β-catenin, and Smad4 and drive gene expression. The c-myc promoter region also contains a degenerate Smad-binding sequence, TGFβ-inhibitory element (TIE) (Figure 20.11).[73] In response to TGFβ, this binds Smads 3 and 4 combined with E2F4 (or E2F5). This complex then recruits the corepressor p107, which leads to effective silencing of c-myc.[65] Since the promoter region of the two cell cycle inhibitors contains a c-Myc/Miz binding element that represses its transcription, loss of c-myc abrogates this negative control and leaves the way open for Smad-mediated activation of transcription. This occurs through two promoter elements, SBE binding Smads 3 and 4 and FHBE which binds the forkhead transcription factor FoxO.[53] Together these recruit p300 which opens the way for the RNA polymerase. Elevated levels of p21^{CIP1} and p15^{INK4b} then inhibit the cell cycle through binding and inhibiting the kinase complex cyclinD–CDK4 that facilitates progression through the G1 phase.

TABLE 20.1 Transcription factors that interact with Smad heterotrimeric complexes

Protein family	Members	Interaction	Gene and function
Forkhead family (winged helix proteins)	FoxH1 (FAST1)	Smad2 Smad4	induction of expression of Mix, leading to mesoderm induction in *Xenopus* in response to activin or nodal-like signals[59]
	FoxO1, FoxO2, FoxO3	Smad3 Smad4	induction of expression of p21^{CIP1} in epithelial cells, leading to inhibition of the cell division cycle[57]
RUNX	Runx1, -2, -3 (also known as Cbfa)	Smad2 or -3 Smad4	induction of expression of the constant region of Igα in B cells[60]
			repression of expression of osteocalcin, leading to the inhibition osteoblast differentiation[61]
Mix Homeodomain	Mixer, Milk	Smad3 Smad4	restriction of Goosecoid expression to the dorsal marginal zone of the early *Xenopus* gastrula embryo[62]
Zinc finger proteins	OAZ (ZNF423) (Olf-1 associated zinc finger)	Smad1 Smad4	induction of expression of vent2, a gene that orchestrates ventral mesoderm induction in *Xenopus* in response to BMP[63]
AP1	c-Jun, c-fos	Smad3 Smad4	induction of expression of JunB, c-Jun, collagenase-1 (mmp1), interleukin-11 and plasminogen activator inhibitor-1[64]
E2F/DP	E2F4, E2F5	Smad3 Smad4	repression of expression of c-myc by TGFβ in epithelial cells, leading to cell cycle inhibition (through release of c-Myc-mediated repression of cell cycle inhibitors)[65]
ATF/CREB	ATF3	Smad3 Smad4	repression of expression of Id1, leading to cell cycle inhibition of epithelial cells[66]

Smad proteins as integrators of signal pathways

Through their cooperativity and the multitude of their interactions, the Smads provide a versatile platform that provides many forms of cross-talk between signalling pathways. These are summarized in Figure 20.12 and in the following paragraphs.

- The activation of STAT3 (page 525) by leukaemia inhibitor factor (LIF) enforces BMP-induced expression of GFAP. This way, STAT3 and Smad1 work together to drive differentiation of neural stem cells into astrocytes.
- p38 and JNK (page 350) contribute to TGFβ signalling through activation of ATF3 and c-Jun, both of which interact with Smad proteins. ATF3 and

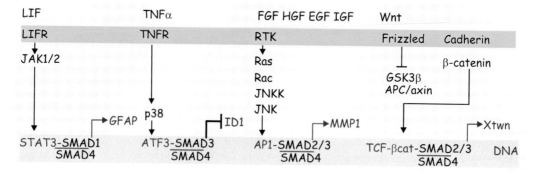

FIG 20.12 Signal integration through different transcriptional partners.
Through interaction with different DNA-binding transcription factors, Smad proteins form nodes of integration that produce different signalling outputs.
Image adapted from Massagué et al.[3]

Smad3 act together to repress expression of ID1. In epithelial cells this is necessary for effective arrest of the cell cycle. Thus TGFβ and TNF-α act in concert to slow down the cell cycle, reducing DNA replication and therefore protecting the organism against DNA damaging agents.

- c-Jun, in complex with c-Fos (AP1 complex, see Figure 19.1, page 579) aids Smads 2 and 3 in the induction of extracellular matrix proteins and proteases in response to TGFβ.[74]
- Smads also cooperate with the Wnt pathway, in both a TGFβ-independent and a dependent manner. We have already mentioned that in the absence of a TGFβ signal, Smad4 interacts with β-catenin/TCF and this complex activates expression of c-myc (Figure 20.11).[53] Cooperation between Smads and Wnt in the presence of TGFβ occurs in expression of the *Xenopus* homeobox gene twin (Xtwn). Here Smads 3 and 4 team up with β-catenin/Lcf in the transformation of dorsal mesoderm to form the Spemann organizer.[75]
- Inflammatory mediators, such as TNF-α, IFN-γ, and IL-1β, activate the transcription factors NF-κB/RelA and IRF3, which combine with Smads 2 and 4, leading to the induction of the inhibitory Smad7. In this way, they abrogate the response to BMP or TGFβ (see below and Figure 20.16).
- Coactivators such as CBP or p300 also serve as platforms for pathway cross-talk. An example is the collaboration between BMP2 and LIF, leading (E3) to activation of Smad1/Smad4 and STAT3 respectively, each binding to different sites in p300 and cooperatively activating the glial fibrillary acidic protein promoter.[76]

The Smad linker region: hotspot for kinases and an E3-ligase
The linker region of the receptor-regulated Smads contains numerous phosphorylation sites which are targeted by various kinases, including

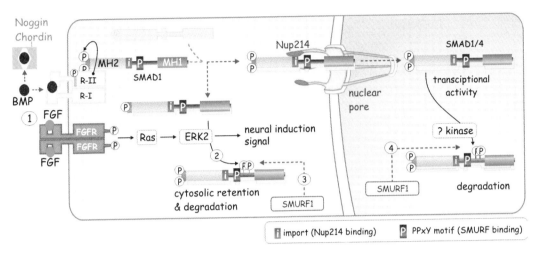

FIG 20.13 Opposing effects of BMP and FGF signalling on the translocation of Smad complexes into the nucleus.
The linker region of Smad proteins contains several phosphorylation sites, targets of CDK2/4, ERKs1 and 2 and GSK3β. The PPxY motif binds the E3-
ubiquitin ligases Smurf. Phosphorylation by ERK facilitates binding of Smurf1 (an ubiquitin E3 ligase). In embryonic development, the BMP signal is
counteracted by secretion of trapping proteins such as noggin and chordin and also by FGF (1). This, through the Ras–ERK pathway causes phosphorylation
of serine residues in the linker region (2). ERK-mediated phosphorylation of Smad1 facilitates the binding of Smurf1 to PPxY (3). The Smurf/Smad
complex cannot bind the nuclear pore protein Nup214 and remains in the cytoplasm. Moreover, Smurf1 causes destruction of Smad. In the nucleus, linker
phosphorylation of Smads and Smurf-mediated destruction also occurs (4) but the kinase remains unknown.

CDK2 and CDK4, GSK3β, and MAPkinases (ERK, JNK, and p38) (Figure 20.13).
Phosphorylation of Smad3 by CDK4, and of Smad1 by MAPkinases, causes
cytosolic retention and subsequent degradation.[77–79] The functional
consequences of phosphorylation by GSK3β are not known. Phosphorylation
of Smad1 by ERK2 has been studied in more detail and we treat this topic in
the context of neural development in *Xenopus laevis*.[80]

In the early gastrula stage, the dorsal ectoderm transforms into
neuroectoderm, which, at a later point provides the neural plate from which
the notochord develops.[81] The ectoderm develops into dermis under the
influence of BMP, but at the dorsal site of the embryo the signal is opposed by
FGF, migrating from the mesodermal marginal zone,[82,83] and by chordin and
noggin (and other factors) that trap the BMP ligand. These diffuse from the
Spemann organizer at the onset of gastrulation (Figure 20.15). FGF opposes
the action of BMP and it also presents a neuroinductive signal. The opposing
action of FGF is mediated through its receptor tyrosine kinase (FGFR),
activating the Ras–ERK pathway and resulting in phosphorylation of the linker
region of Smad1. This favours the interaction of Smad1 with Smurf1, an
E3-ubiquitin ligase, thereby blocking the nuclear import binding site located

ID1 is required for
progression through G1,
one of its functions being
to disable the action
of Rb. ID proteins are
helix–loop–helix proteins
that lack a DNA binding
domain. They bind to
other helix–loop–helix
proteins and, in so doing,
prevent their access to
DNA. In this way they
prevent the action of,
for instance, MyoD. Their
expression is blocked in
senescent fibroblasts.

in the MH2 domain. Smad1 fails to enter the nucleus and is degraded following ubiquitylation (figure 20.13).[78]

There are other instances in which the activities of the FGF and BMP pathways have opposing effects. These include the pairs FGF4/BMP2 in limb bud formation, FGF10/BMP4 in lung morphogenesis, FGF2/BMP4 in cranial suture fusion and FGF8/BMP4 in the initiation of tooth development.[80]
These processes may also be coordinated by signal integration at the level of Smad1.

Hans Spemann and the organizer

'How does that harmonious interlocking of separate processes come about which makes up the complete process of development? Do they go on side by side independently of each other (by 'self-differentiation', Roux), but from the very beginning so in equilibrium that they form the highly complicated end product of the complete organism, or is their influence on each other one of mutual stimulation, advancement or limitation?'

The question was put by Hans Spemann, professor of zoology at the University of Freiburg im Breisgau.[84] With his graduate student, Hilde Mangold, he examined the transplantation of structures from the giant embryos of amphibians. Portions of surface cells (presumptive ectoderm) were removed and then implanted into an intact embryo. In many instances the implanted cells adopted the fate of the cells surrounding the new insertion site, but cells from a limited area, namely the region of the upper and lateral blastopore lip failed to do so. These gave rise to the formation of a nearly complete secondary dorsal axis, commencing with the formation of a second medullary plate (*Medullarplatte*) and giving rise to the formation of a secondary: neural tube (*Neuralrohr*), notochord (*chorda*), prevertebral discs (*Urwirbel*), kidneys (*Niere*), and even gut (*Darmlumen*). This area was named 'organizer' and we now know it as the 'Spemann organizer' (see Figures 20.14 and 20.15).

Of several hundred chimeric embryos, only five survived, and these formed the basis for the famous paper that was published in 1924.[85] Spemann and Mangold were the first to prove the reality of 'induction', by which one group of cells influences the developmental fate of the other. The identity of the inducing signal long remained the holy grail of developmental biology, but now, after about 80 years, we begin to realize that it constitutes a number of secreted factors, amongst which the antagonists of BMP play an important role.

In 1935 Spemann was awarded the Nobel Prize in Physiology or Medicine 'for his discovery of the organizer effect in embryonic development'. He was relieved of his professorial position in 1937, his place being taken by Otto Mangold, more acceptable to the regime and previously the husband of Hilde who had been killed in a kitchen explosion 13 years earlier.

The linker region of Smad3 is also phosphorylated by the cell-division cycle kinases CDK2 and CDK4. As their phosphorylation sites overlap with those of ERK, the consequences are similar and Smad3 remains sequestered in the cytoplasm to face destruction by Smurf. Tumour cells that over-express these kinases may therefore be less sensitive to the growth inhibitory effect of TGFβ.[77]

The transcriptional activity of the Smads is also inhibited by PKC-mediated phosphorylation of the DNA-binding MH1 domain and enhanced by binding of Ca^{2+}–calmodulin.[87] For example, Ca^{2+}–calmodulin enhances the ventral mesoderm-forming activity of Smad1 (BMP) and blocks the activity

The **BMP** proteins are involved in the development of the craniofacial complex (teeth, peridontium, and jaws). The discovery of stem cells in dental pulp now offers the possibility of regenerative therapy of the craniofacial complex. These rather fancy words disguise the idea that this approach may ultimately offer the possibility of regenerating whole teeth after extraction.[86]

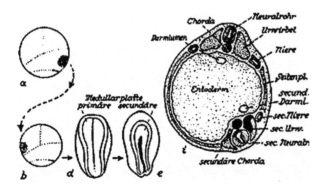

FIG 20.14 Secondary dorsal axis formation after implanting "the organizer".

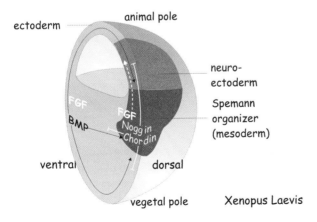

FIG 20.15 Neuroectoderm induction through FGF and the release of the BMP-traps noggin and chordin. Neuroectoderm formation in *Xenopus laevis* occurs through inhibition of the BMP signal by release of noggin and chordin from the Spemann organizer and through diffusion of FGF from the mesoderm marginal zone into dorsal ectoderm. Neuroectoderm develops into notochord and gives rise to the brain.

of Smad2 (activin/TGFβ). Moreover, calmodulin prevents ERK2-mediated phosphorylation of Smads 1 and 2 and, conversely, phosphorylation by Erk2 inhibits the binding of calmodulin.[88]

Smurf-mediated Smad degradation

Interaction of receptor-regulated Smads with the E3-ubiquitin ligases Smurfs1 and 2 leads to their degradation by the proteasome (see page 472). Smurf1 binds Smads 1 and 5, whereas Smurf2 has broad specificity. Matters are more complicated for Smad3 which first binds the transcription factor SnoN, priming it for ubiquitylation by Smurf2.[89]

Holding the TGFβ pathway in check

Inhibitory Smads

Smads 6 and 7 lack the C-terminal SxS phosphorylation site and have MH1 domains that are only distantly related to those of the other Smads (see Figure 20.8). In the absence of TGFβ, they are retained in the nucleus. Both their expression and then their translocation into the cytoplasm are induced by TGFβ, activin, and BMP. Here, they act as inhibitors (Figure 20.16a).[90,91]

In the cytoplasm, Smad6 forms complexes with phosphorylated Smad1, thereby preventing its interaction with Smad4 and in this way, inhibiting BMP signalling.[92] For Smad7, the mechanism is much more elaborate. It is bound to Smurf2 and, for reasons not fully understood, is protected against ubiquitylation in the nucleus. On addition of TGFβ, the Smad–Smurf complex moves into the cytosol to associate with the membrane through the C2 domain in Smurf2. Here, Smad7 binds to the receptor TβR-I. Ubiquitylation ensues and the complex is recognized by the proteasome for partial degradation (Figure 20.16a).[93] The expression of Smad7 is also induced by inflammatory mediators (such as TNF-α, IFN-γ, or IL-1β) via the intermediate of NF-κB/RelA and IRF3 transcription factors. This may be one mechanism by which these mediators lift the generally immunosuppressive influence of TGFβ.[7,94]

It seems that Smad7 originated as a transcription factor and then evolved to become a structural component of the Smurf ubiquitylation complex. Here it acts as a substrate receptor and also enables the association of the E2-ubiquitin conjugation protein (UbcH7) (Figure 20.16b). In view of this, it is possible that other Smad-interacting proteins are targets of ubiquitin-mediated destruction. The finding that Smad7/Smurf2 also binds β-catenin, causing its degradation, provides evidence for yet another mechanism of pathway cross-talk.[95]

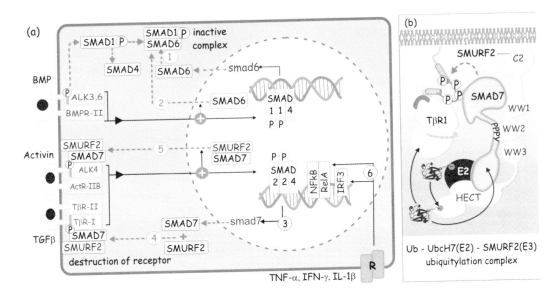

FIG 20.16 Induction of inhibitory Smad proteins.

(a) Transcripts induced by members of the TGFβ family, include those coding for Smads 6 and 7. The newly expressed Smad6 associates with phospho-Smad1, preventing the formation of functional Smad1/Smad4 complexes (1). BMP also causes the nuclear export of existing Smad6 (2) leading to the formation of the same non-functional complexes. The newly synthesized Smad7 protein (3) associates with Smurf2 and then with the type I receptor (4). It induces degradation of the Smad–receptor complex. TGFβ or activin also causes nuclear export of existing Smad7–Smurf complexes (5). Expression of Smad7 is also induced by inflammatory mediators such as TNF-α, IFN-γ, or IL-1β and these may act to attenuate the TGFβ response (6). (b) Smad7 is an essential component of the Smurf2 ubiquitylation complex. It brings together Smurf2 (E3) and the E2-conjugating protein UbcH7, from which ubiquitin is transferred to the HECT domain of Smurf2 and then on to its substrate. Smad7 thus acts as a receptor, bringing Smurf2 and UbcH7 to the relevant substrates. This is not restricted to the type I TGFβ receptors, other proteins interacting with Smad7 are potential targets.

BAMBI, a signal inhibitory pseudo receptor

BMP, activin, and TGFβ induce the expression of BAMBI. Its expression correlates inversely with the metastatic potential of human melanoma cell lines. [96] It is a truncated type I receptor (closely resembling BMP-RI) that binds ligand but lacks GS and serine/threonine kinase domains. However, the cytosolic segment binds TβRI receptors and, in so doing, prevents phosphorylation by type II receptors. The effect is reduced signalling output from TGFβ-receptors (see Figure 20.6e).[30] It thus acts as a pseudo-receptor.

Smad phosphatases

A steady-state level of phosphorylation of Smads2 and 3 is achieved in cells within 15–30 min of adding TGFβ, and this can last for several hours. Addition of a receptor kinase inhibitor, to interrupt the TβR-I signal, then allows rapid (30 min) dephosphorylation.[49]

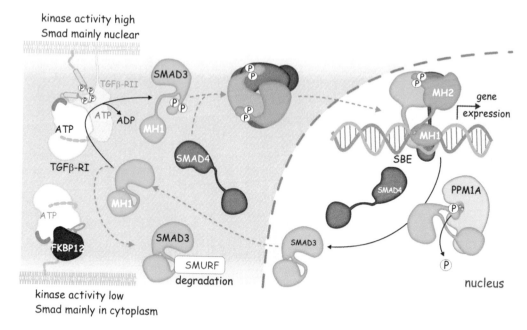

kinase activity high
Smad mainly nuclear

nucleus

kinase activity low
Smad mainly in cytoplasm

FIG 20.17 Signal attenuation by nuclear phosphatase PPM1A.

Nuclear Smad transcription complexes, dephosphorylated by PPM1A in the nucleus, disaggregate and transfer as monomers to the cytoplasm. Under conditions of low receptor activity, they remain here for some time and risk recognition by Smurf ubiquitin ligases. With high receptor activity, they are immediately rephosphorylated and return to the nucleus as hetero-oligomeric transcription factor complexes.

SCP phosphatases were originally found to control the phosphorylation state of the C-terminal domain of RNA polymerase II. This determines the choice of proteins that associate with the polymerase. These play important roles in processing the primary RNA transcript, such as 5'-capping, splicing and 3'-polyadenylation. SCP2 is amplified in sarcomas.[101]

The Smad-specific serine/threonine phosphatase PPM1A, member of the PP2C subfamily, is present in both in the cytosol and in the nucleus.[97] Of 39 phosphatases tested, only PPM1A is reactive against Smads2 and 3. It causes dephosphorylation of the C-terminal SxS motif, allowing dissociation of the Smad complex and export of the individual components from the nucleus (Figure 20.17).

PPM1A (Figure 20.18) also affects transcription and other responses. For instance, in keratinocytes depleted of PPM1A, repression of c-myc is increased and expression of the cell cycle inhibitors p15^{INK4b} and p21^{CIP1} in response to TGFβ is enhanced.[98] It is likely that PPM1A is constitutively active so that the level of nuclear Smad is determined by receptor activity. The picture that emerges for Smad signalling is thus similar to that described for the STATs, in which the transcription factors act as remote sensors of receptor activity (see Figure 17.10, page 528).

PDP (pyruvate dehydrogenase phosphatase, essentially a mitochondrial enzyme but also present in the nucleus), is reactive against the SxS motif of Smad1 (BMP signalling) but not of Smads 2 or 3 (TGFβ/activin signalling).[99]

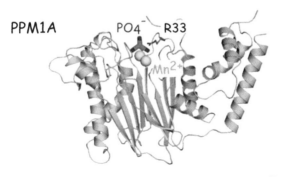

FIG 20.18 Structure of PPM1A. R33 is a key residue for coordination of the phosphate. (1a6q[97]).

Finally, SCPs 1, 2, and 3 dephosphorylate the linker region of Smads 2 and 3 and the linker and the SxS motif of Smad1.[100]

TGFβ: tumour suppressor and metastatic promoter?

Given the importance of TGFβ in suppressing cell proliferation, terminal differentiation of haematopoietic cells, and the activation of cell death mechanisms, it is perhaps not surprising that mutations in the components of the signal transduction pathway can increase susceptibility to transformation. Together with other mutations that favour proliferation, these may ultimately result in the formation of tumours. The common mediator Smad4 was initially identified as one of the mutated or deleted genes linked to pancreatic, colorectal or other carcinomas.[102] Most of these mutations are present in the MH2 domain, which contains the SxS phosphorylation site. They act to prevent complex formation with other Smad proteins, decrease the stability of the protein, or prevent its interaction with non-Smad transcriptional partners.

The importance of Smad4 in colon cancer has been extensively studied in mice. Those that lack a single allele develop gastric/intestinal tumours within 1 year. These reveal a slow but continuous progression from initial hyperplastic lesions to more advanced stages with clear dysplasia, similar to human colorectal cancer.[103] In more advanced stages, the cells also lose the second (functional) Smad4 allele.[104] Mice lacking one allele of both APC/C and Smad4 develop tumours earlier and they die earlier, suggesting that TGFβ-signalling is a component of APC/C-driven progression towards malignancy.

There are also hereditary and somatic forms of colorectal cancer in which the TβR-II receptor is mutated, so that the cells lack their normal growth-inhibitory signalling mechanism. Paradoxically, properly functional Smad4 is also required for TGFβ-induced epithelial–mesenchymal transition (EMT), a process generally considered to be tumorigenic. In breast cancer, this transition is linked with metastasis.[105] Pancreatic tumour cells that are still sensitive to TGFβ-mediated growth inhibition become highly metastatic with over-expression of the TβR-I receptor.[106] The general view is that TGFβ acts as a tumour suppressor by holding proliferation in check, thus reducing the chance of transforming mutations (fewer replication cycles, fewer errors). On the other hand, for cells already partly transformed (resistant to apoptosis and having reduced requirement of growth factors for proliferation), TGFβ may act as a metastasis promoter due to its capacity to induce the EMT.[107]

High levels of TGFβ also enable transformed cells to escape immune surveillance. There is much evidence for a role of the immune system in the elimination of cells in their early stages of transformation, and mice that lack essential components of the innate or adaptive immune system are prone to develop tumours.[108] Transformed cells are recognized by cytotoxic T lymphocytes (CTL) (and other immune cells) and by tumour-specific antibodies, leading to complement or NK-cell-dependent lysis. Activated CTLs recognize antigens on the tumour cell membrane so that they merge to form an immune synapse in which the CTLs release perforin and granzymes (serine

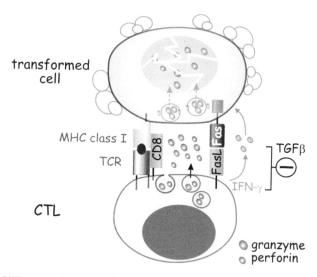

FIG 20.19 TGFβ suppresses immune surveillance of transformed cells.
Cytotoxic T lymphocytes (CTLs) recognize transformed cell antigens presented by MHC class I, through their T cell receptors and CD8 molecules. Immune synapses are formed into which granzymes (serine proteases) and perforin are secreted. Perforin facilitates the uptake of granzymes into large endocytic vesicles in the target cell. Later the granzymes escape and enter the cell nucleus, where they destroy proteins. The CTLs also present FasL which binds to CD95/Fas (not shown; see page 489). Both the receptor engagement and the action of granzymes lead to target cell death. Released IFN-γ also has a cytolytic effect on transformed cells. TGFβ prevents expression of these cytolytic agents and therefore protects the transformed cell.

proteases) (Figure 20.19). Perforin triggers the rapid uptake of the granzymes into enormous endosomal vesicles. The granzymes then escape and induce apoptosis by cleaving nuclear proteins. The cytotoxic T cell also exposes TRAIL and FAS-ligands, which, by binding to their receptors on the target cells, cause the recruitment and activation of caspases which cleave numerous cellular proteins and induce apoptosis. Finally, cytotoxic T cells release IFN-γ which has cytolytic effects on transformed cells (Figure 20.19).

On average, by making serial contacts with multiple targets, a cytotoxic immune cell can kill four tumour cells, mainly by induction of apoptosis. However, serial killing is associated with a loss of their own content of perforin and granzymes, so the exhausted cells must replenish themselves in order to go on killing.[109] This is where TGFβ protects tumour cells from being eliminated: it prevents replenishment of the granules by repressing transcription of the cytotoxic agents.[110] Alternatively, the effect is indirect. Neutralization of TGFβ with antibodies or soluble TβRII receptors restores T-cell-mediated tumour clearance.

TGFβ and the epithelial–mesenchymal transition

The EMT is central to the morphogenic processes that occur in early embryonic development,[111] and it also operates in the orchestration of metastasis in transformed cells. It involves the disaggregation of epithelial cell layers through dissolution of tight junctions, modulation of adherens junctions, reorganization of the actin cytoskeleton, and loss of cell polarity (see Figure 14.1, page 418). TGFβ is one of the factors that drive the EMT.

Among the many genes induced by TGFβ1 are Snail2, Fz-1 and PP2A.[112] Snail2 is a transcription factor[113] that represses expression of E-Cadherin leading to the loss of adherens junctions. Fz-1 is a receptor for Wnt that re-enforces the TGFβ signal by amplifying the gene responses that determine the mesenchymal phenotype (see also Figure 14.13, page 433). The phosphatase PP2A opposes the action of CK1α.[114] It maintains axin in a dephosphorylated state, so preventing binding of GSK3β (see also Figure 14.7, page 429). As a result, β-catenin is not phosphorylated and is not recognized by the ubiquitylation apparatus. In short the Wnt signal is enhanced (Figure 20.20).

Other genes, in particular those involved in matrix remodelling and cell motility, are also induced by TGFβ1, but their expression requires the MEK–ERK pathway and they are thus indirect targets. Among these are metalloproteases (MMP1, MMP12); matrix components such as fibronectin, laminin, and thrombospondin; and the integrin adhesion molecules. How the two pathways are connected has not been fully resolved.

In the EMT, TGFβ provokes the loss of tight junctions without involvement of the Smads. The TβR-I receptors, retained at tight-junctions by occludin, bind to Par-6, a scaffold protein that operates in the assembly of components that determine

Snail2 is Slug. Snail, so named because early *Drosophila* embryos that carry this lethal mutant resemble a snail. Slug, a vertebrate homologue of Snail, has roles in the formation of mesoderm during gastrulation and in the migration of neural crest cells. When the nomenclature was revised, Slug became Snail2. For more on slugs and snails, see Barrallo-Gimeno and Nieto.[115]

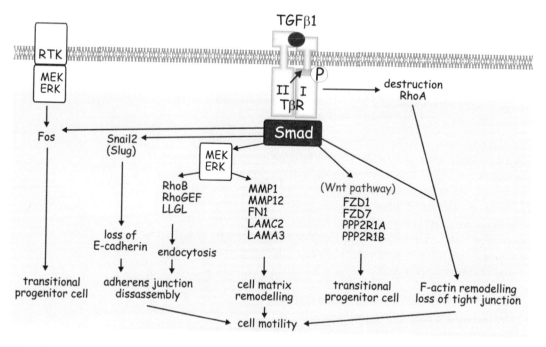

FIG 20.20 Genetic programme underlying TGFβ-mediated EMT.
In the process of EMT, there are 728 genes regulated by TGFβ. They can be grouped on the basis of the different signalling pathways which operate them and their different phenotypic outcomes. Numerous genes require a contribution from the MEK–ERK pathway. A selection of genes, those that have been discussed in this book, are presented here.
Figure adapted from Zavadil et al.[112]

polarity and which is bound to the junctional ZO-1 (Figure 20.21, see also page 439).[116] TβR-II receptors are redistributed to the tight junctions on addition of TGFβ1, causing phosphorylation and activation of the TβR-I receptors and phosphorylation of Par-6 which now recruits the E3-ubiquitin ligase Smurf1. This leads to destruction of RhoA and consequent dissolution of the tight junctions. The tissue dissociates, transforming into a mass of motile fibroblast-like cells. Loss of tight junctions also enables the expression of cyclin D, and by returning CDK4 to the nucleus, sets up a programme that initiates the cell cycle.

We conclude this chapter by pointing out that the Smads do not possess the exclusive rights to the relaying of TGFβ signals into the cell. In addition to the Par-6/RhoA pathway, TGFβ receptors also recruit PP2A (β-regulatory subunit), STRAP/PDK1, and eIF-2α, all of which affect protein synthesis. They also recruit FTα, which affects Ras (and possibly ERK) signalling, and they recruit TAK1, which initiates the p38 and JNK pathway (see Figure 15.9, page 463).[74]

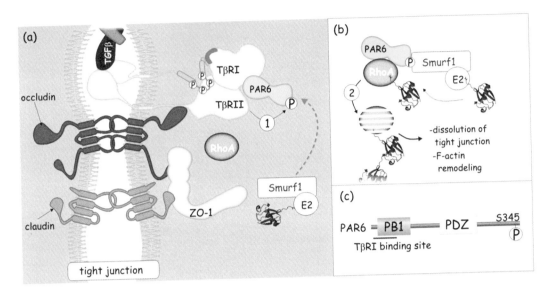

FIG 20.21 TGFβ-mediated dissolution of the tight junction.
(a) TβR-I is localized to the tight junction by association with occludin. Here it binds the cell polarity protein Par-6. Assembly of the tight junction relies on RhoA that dictates the configuration of the actin cytoskeleton. Addition of TGFβ leads to recruitment of TβR-II into the tight junction to activate TβRII/TβRI and Par-6. (b) Phosphorylated Par-6 recruits Smurf1 (and its associated E2-ubiquitin conjugating protein) causing destruction of RhoA with consequent remodelling of the actin cytoskeleton and dissolution of the tight junction (c) Domain architecture of Par 6.

List of Abbreviations

Abbreviation	Full name/description	SwissProt Entry	Comments, OMIM links
α2M	alpha(2)-macroglobulin	P01023	inhibitor of plasmin
activin-A	activates FSH secretion in pituitary	P08476	homodimer of β-A inhibin chains.
ALK-5	activin-receptor like kinase-5	P36897	TGFR-1, TβR-II
AMP	anti-Mullerian hormone	P03971	Muellerian-inhibitory substance (MIS)
AP-1	activator protein-1 (complex of two transcription factors)		
ATF3	activating transcription factor-3	P18847	binds the cAMP response element (CRE)
Axin1	axis-inhibiting protein-1 (mutant causes axial duplication in mice)	O15169	

Continued

627

Abbreviation	Full name/description	SwissProt Entry	Comments, OMIM links
BAMBI	BMP, activin membrane-bound inhibitor homologue	Q13145	non-metastatic gene A protein (NMA)
β-catenin	from catena (L. chain)	P35222	
betaglycan	TGFβ-binding proteoglycan	Q03167	TGFβ receptor type III
BMP2	bone morphogenetic protein-2	P12643	
CBP	CREB binding protein	Q92793	
CDK4	cyclin dependent protein kinase-4	P11802	
cerberus	multihead *Xenopus* phenotye, three-headed dog in Greek mythology	O95813	cDNA induces ectopic heads in *Xenopus*, DAN4
chordin	strongly expressed in notochord in *Xenopus laevis* embryos	Q9H2X0	
c-Jun	homologue of sarcoma virus-17 oncogene, junana=17	P05412	
c-myc	cellular homologue of avian myelocytomatosis MC29 oncogene	P01106	p64
cPML	cytosolic promyelocytic leukemia protein	P29590	cytosolic isoform is truncated version of PML
Cripto	mysterious lack to known proteins and signalling pathways	P13385	
CRM1	chromosome region maintenance-1 protein homology	14980	exportin
CTBP1	C-terminus binding protein (binds C-terminus of adenovirus protein E1A)	Q13363	
Daf	dauer formation (TGFβ type-1 receptor in *C. elegans*)		
DAN4	differential screening-selected gene aberrative in neuroblastoma	O95813	cerberus
disabled	disables axonal connections in *Drosophila* (ablates action of Abl)	O75553	
DPP	decapentaplegic (TGFβ- and BMP-like morphogen in *Drosophila*)	P07713	
E2F4	adenovirus E2A-promoter binding factor (E2F family)	Q16254	

Continued

Abbreviation	Full name/description	SwissProt Entry	Comments, OMIM links
eIF2α	eukaryotic initiation factor-2 alpha	P05198	
ELF	embryonic liver fodrin	Q62261	spectrin β-chain
endoglin	endothelium-specific glycoprotein	P17813	CD105, MIM:187300
FHBE	forkhead transcription factor DNA binding element		
FKBP12	FK506-binding protein of 12 kDa	P62942	FKBP1A
follistatin	follicle stimulating hormone inhibitor (inhibits biosynthesis and secretion)	P19883	activin-binding protein, FST
FoxH1	forkhead activin signal transducer H1	O75593	FAST-2
Foxo1	forkhead box protein-1	Q12778	FKHR
FTα	farnesyl-protein transferase-alpha	P49354	
Fz-1	frizzled-1 (messed up hair-alignment on wings of mutant *Drosophila*)	Q9UP38	
GDF8	growth and differentiation factor-8	O14793	myostatin
GFAP	glial fibrillary acidic protein	P14136	class-III intermediate filament, MIM:203450
granzyme A	granule proteolytic enzyme-A	P12544	CTL tryptase, Hannukah factor
HDAC	histone deacetylation complex	Q13547	
ID1	inhibitor of DNA-binding	P41134	
IFN-γ	interferon gamma	P01579	
inhibin-A	Inhibits FSH secretion in pituitary	P05111 and P08476	dimer of α and β-A inhibin chains
IRF3	interferon regulatory factor-3	Q14653	
LAP	latency-associated peptide	P01137	also contains TGFβ1
LIF	leukocyte inhibitory factor	P15018	
MAD	Mothers against decapentaplegic (transcription factors in Dpp pathway)		

Continued

Abbreviation	Full name/description	SwissProt Entry	Comments, OMIM links
mixer	MIX-like endodermal regulator (*Xenopus laevis*)	O73867	
mSin3A	mammalian homologue of yeast Sin3 (switch-independent-3 mutant)	Q96ST3	HDAC complex subunit, paired amphiphatic helix protein Sin3a
NF-κB1	nuclear factor kappa B-1	P19838	p105/p50 subunit
Nodal	localized in the node at the anterior of the primitive streak in mice embryos	Q96S42	
Noggin	phenotype of dorsalized embryo; noggin being slang for a head	Q13253	MIM:185800, MIM:196500
NUP153	nucleoporin 153 (nuclear pore complex protein)	P49790	
OAZ	Olf-1 associated zinc finger protein	Q2M1K9	ZNF423, EBFAZ, ROAZ
p107	protein of 107 kDa	P28749	retinoblastoma-like protein (RBL1)
p15INK4B	15 kDa inhibitor of CDK4	P42772	CDKN2B
p21CIP	21 kDa CDK-inhibitory protein	P38936	CDKN1A
p300	protein of 300 kDa	Q09472	E1A-associated protein p300
Par-6	partitioning defective 6 homologue	Q9NPB6	
PDK1	3-phosphoinositide-dependent protein kinase-1	O15530	
PDP1	pyruvate dehydrogenase phosphatase-1	Q9P0J1	
perforin	Ca^{2+}-dependent perforation of plasma membrane	P1422	
PP2A	protein phosphatase-2A catalytic subunit	P30153	PP2R1A (member of the PP2C family)
PPM1A	protein phosphatase Mn^{2+}-dependent	P35813	PP2Cα (member of the PP2C family)
RhoA	Ras homologue A	P61586	
Runx1	Runt-related transcription factor-1	Q01196	PEBP2, CBF-α2
SARA	Sm anchor for receptor activation	O95405	Zinc finger FYVE domain containing protein-9

Continued

Abbreviation	Full name/description	SwissProt Entry	Comments, OMIM links
SCP1	small C-terminal domain phosphatase-1	Q9GZU7	nuclear LIM interactor-interacting factor 3
Ski	Sloan-Kettering Institute virus gene product	P12755	
Sma	small larvae (transcription factors in Daf pathway)		
Smad2	small/mothers against Decapentaplegic homologue-2	Q15796	
Smad4	small/mothers against Decapentaplegic homologue-4	Q13485	deletion target in pancreatic carcinoma-4 (DPC-4), MIM:260350, MI:174900
Smad7	small/mothers against Decapentaplegic homologue-7	O15105	
Smurf1	Smad ubiquitylation regulatory factor-1	Q9HCE7	Smad E3-ubiquitin ligase
Smurf2	Smad ubiquitylation regulatory factor-2	Q9HAU4	Smad E3-ubiquitin ligase
Snail2	snail phenotype of *drosophila* embryo	O43623	Slug (snail-like)
SnoN	Ski novel related gene product non-Alu containing	?	
Sp1	SV40-promoter specific protein-1 (from HeLa cell extracts)	P08047	
STAT3	signal transducer and activator of transcription-3	P40763	
STRAP	serine-threonine kinase receptor-associated protein	Q9Y3F4	
TAK1	TGFbeta-activated kinase	O43318	MAP3K7
TCF1	T cell factor-1	P36402	lymphocyte enhancer-binding factor (LEF)
TGFβ1	transforming growth factor-β1	P01137	also contains LAP
TGFβ-R1	transforming growth factor-β type-I receptor	P36897	TβR-I, ALK-5, MIM:609192, MIM:610168, MIM:610380, MIM:608967
TGFβ-RII	transforming growth factor-b type II receptor	P37173	TβR-II, MIM:190182, MIM:154705, MIM:33239

Continued

631

Abbreviation	Full name/description	SwissProt Entry	Comments, OMIM links
TGIF1	5′-TG-3′-interacting factor-1	Q15583	homeobox protein, MIM:142946, MIM:236100
TIF1γ	transcription intermediate factor 1-gamma	Q9UPN9	tripartite motif containing protein 33 (TRIM33)
TLP	TRAP-like protein	?	
TRAP-1	TGF-beta receptor associated protein-1	genbank AF022795	
UbcH7	ubiquitin conjugation factor H7	P68036	ubiquitin-protein ligase L3, E2-F1
ZO-1	zona occludens protein-1	Q07157	tight-junction protein-1

References

1. Manning G, Whyte DB, Martinez R, Hunter T, Sudarsanam S. The protein kinase complement of the human genome. *Science*. 2002;298:1912–1934.
2. Shi Y, Massagué J. Mechanisms of TGFβ signaling from cell membrane to the nucleus. *Cell*. 2003;113:685–700.
3. Massagué J, Seoane J, Wotton D. Smad transcription factors. *Genes Dev*. 2005;19:2783–2810.
4. McPherron AC, Lee SJ. Double muscling in cattle due to mutations in the myostatin gene. *Proc Natl Acad Sci USA*. 1997;94:12457–12461.
5. Oklu R, Hesketh R. The latent transforming growth factor β binding protein (LTBP) family. *Biochem J*. 2000;352(pt 3):601–610.
6. Yang Z, Mu Z, Dabovic B, et al. Absence of integrin-mediated TGFβ1 activation in vivo recapitulates the phenotype of TGFβ1-null mice. *J Cell Biol*. 2007;176:787–793.
7. Shull MM, Ormsby I, Kier AB, et al. Targeted disruption of the mouse transforming growth factor-β1 gene results in multifocal inflammatory disease. *Nature*. 1992;359:693–699.
8. Mathews LS, Vale WW. Expression cloning of an activin receptor, a predicted transmembrane serine kinase. *Cell*. 1991;65:973–982.
9. Bassing CH, Yingling JM, Howe DJ, et al. A transforming growth factor β type I receptor that signals to activate gene expression. *Science*. 1994;263:87–89.
10. Lin HY, Wang X-F, Ng-Eaton E, Weinberg RA, Lodish HF. Expression cloning of the TGF-β type II receptor, a functional transmembrane serine/threonine kinase. *Cell*. 1992;68:775–785.

11. ten-Dijke P, Yamashita H, Ichijo H, et al. Characterization of type I receptors for transforming growth factor-β and activin. *Science*. 1994;264:101–104.

12. Luo K, Lodish HF. Signaling by chimeric erythropoietin-TGF-β receptors: homodimerization of the cytoplasmic domain of the type I TGF-β receptor and heterodimerization with the type II receptor are both required for intracellular signal transduction. *EMBO J*. 1996;15:4485–4496.

13. Hart PJ, Deep S, Taylor AB, Shu Z, Hinck CS, Hinck AP. Crystal structure of the human TβR2 ectodomain–TGF-β3 complex. *Nat Struct Biol*. 2002;9:203–208.

14. Liu F, Hata A, Baker JC, et al. A human Mad protein acting as a BMP-regulated transcriptional activator. *Nature*. 1996;381:620–623.

15. Huse M, Muir TW, Xu L, Chen YG, Kuriyan J, Massagué J. The TGFβ receptor activation process: an inhibitor- to substrate-binding switch. *Mol Cell*. 2001;8:671–682.

16. Wang X-F, Lin HY, Ng EE, Downward J, Lodish HF, Weinberg RA. Expression cloning and characterization of the TGF-β type III receptor. *Cell*. 1991;67:797–805.

17. Lopez-Casillas F, Cheifetz S, Doody J, Andres JL, Lane WS, Massagué J. Structure and expression of the membrane proteoglycan betaglycan, a component of the TGF-β receptor system. *Cell*. 1991;67:785–795.

18. Lopez-Casillas F, Wrana JL, Massagué J. Betaglycan presents ligand to the TGF-β signaling receptor. *Cell*. 1993;73:1435–1444.

19. Ling N, Ying SY, Ueno N, et al. Pituitary FSH is released by a heterodimer of the β-subunits from the two forms of inhibin. *Nature*. 1986;321:779–782.

20. Lewis KA, Gray PC, Blount AL, et al. Betaglycan binds inhibin and can mediate functional antagonism of activin signalling. *Nature*. 2000;404:411–414.

21. Wiater E, Harrison CA, Lewis KA, Gray PC, Vale WW. Identification of distinct inhibin and transforming growth factor β-binding sites on betaglycan: functional separation of betaglycan co-receptor actions. *J Biol Chem*. 1999;281:17011–17022.

22. Brown CB, Boyer AS, Runyan RB, Barnett JV. Requirement of type III TGF-β receptor for endocardial cell transformation in the heart. *Science*. 1999;283:2080–2082.

23. Gougos A, Letarte M. Identification of a human endothelial cell antigen with monoclonal antibody 44G4 produced against a pre-B leukemic cell line. *J Immunol*. 1988;141:1925–1933.

24. Cheifetz S, Bellon T, Cales C, et al. Endoglin is a component of the transforming growth factor-β receptor system in human endothelial cells. *J Biol Chem*. 1992;267:19027–19030.

25. Goumans MJ, Valdimarsdottir G, Itoh S, et al. Activin receptor-like kinase (ALK)1 is an antagonistic mediator of lateral TGFβ/ALK5 signaling. *Mol Cell*. 2003;12:817–828.

26. Lebrin F, Goumans MJ, Jonker L, et al. Endoglin promotes endothelial cell proliferation and TGFβ/ALK1 signal transduction. *EMBO J.* 2004;23:4018–4028.

27. McAllister KA, Grogg KM, Johnson DW, et al. Endoglin, a TGFβ binding protein of endothelial cells, is the gene for hereditary haemorrhagic telangiectasia type 1. *Nat Genet.* 1994;8:345–351.

28. Shen MM, Schier AF. The EGF-CFC gene family in vertebrate development. *Trends Genet.* 2000;16:303–309.

29. Gray PC, Shani G, Aung K, Kelber J, Vale W. Cripto binds transforming growth factor β (TGFβ) and inhibits TGFβ signaling. *Mol Cell Biol.* 2006;26:9268–9278.

30. Onichtchouk D, Chen YG, Dosch R, et al. Silencing of TGFβ signalling by the pseudoreceptor BAMBI. *Nature.* 1999;401:480–485.

31. Spencer FA, Hoffmann FM, Gelbart WM. Decapentaplegic: a gene complex affecting morphogenesis in *Drosophila melanogaster*. *Cell.* 1982;28:451–461.

32. Raftery LA, Twombly V, Wharton K, Gelbart WM. Genetic screens to identify elements of the decapentaplegic signaling pathway in *Drosophila*. *Genetics.* 1995;139:241–254.

33. Newfeld SJ, Chartoff EH, Graff JM, Melton DA, Gelbart WM. Mothers against dpp encodes a conserved cytoplasmic protein required in DPP/TGF-β responsive cells. *Development.* 1996;122:2099–2108.

34. Estevez M, Attisano L, Wrana JL, et al. The *daf-4* gene encodes a bone morphogenetic protein receptor controlling *C. elegans* dauer larva development. *Nature.* 1993;365:644–649.

35. Georgi LL, Albert PS, Riddle DL. *daf-1*, a *C. elegans* gene controlling dauer larva development, encodes a novel receptor protein kinase. *Cell.* 1990;61:635–645.

36. Savage C, Das P, Finelli AL, et al. *Caenorhabditis elegans* genes sma-2, sma-3, and sma-4 define a conserved family of transforming growth factor β pathway components. *Proc Natl Acad Sci USA.* 1996;93:790–794.

37. Heldin CH, Miyazono K, ten-Dijke P. TGF-β signalling from cell membrane to nucleus through SMAD proteins. *Nature.* 1997;390:465–471.

38. Kretzschmar M, Massagué J. SMADs: mediators and regulators of TGF-β signaling. *Curr Opin Genet Dev.* 1998;8:103–111.

39. Chai J, Wu JW, Yan N, Massagué J, Pavletich NP, Shi Y. Features of a Smad3 MH1-DNA complex. Roles of water and zinc in DNA binding. *J Biol Chem.* 2003;278:20327–20331.

40. Wu JW, Hu M, Chai J, et al. Crystal structure of a phosphorylated Smad2. Recognition of phosphoserine by the MH2 domain and insights on Smad function in TGFβ signaling. *Mol Cell.* 2001;8:1277–1289.

41. Chen YG, Hata A, Lo RS, et al. Determinants of specificity in TGF-β signal transduction. *Genes Dev.* 1998;12:2144–2152.

42. Feng XH, Derynck R. A kinase subdomain of transforming growth factor-beta (TGF-β) type I receptor determines the TGF-β intracellular signaling specificity. *EMBO J.* 1997;16:3912–3923.

43. Durocher D, Taylor IA, Sarbassova D, et al. The molecular basis of FHA domain:phosphopeptide binding specificity and implications for phospho-dependent signaling mechanisms. *Mol Cell.* 2000;6:1169–1182.

44. Pierreux CE, Nicolas FJ, Hill CS. Transforming growth factor β-independent shuttling of Smad4 between the cytoplasm and nucleus. *Mol Cell Biol.* 2000;20:9041–9054.

45. Tsukazaki T, Chiang TA, Davison AF, Attisano L, Wrana JL. SARA, a FYVE domain protein that recruits Smad2 to the TGFβ receptor. *Cell.* 1998;95:779–791.

46. Wu G, Chen YG, Ozdamar B, et al. Structural basis of Smad2 recognition by the Smad anchor for receptor activation. *Science.* 2000;287:92–97.

47. Misra S, Hurley JH. Crystal structure of a phosphatidylinositol 3-phosphate-specific membrane-targeting motif, the FYVE domain of Vps27p. *Cell.* 1999;97:657–666.

48. Di Guglielmo GM, Le RC, Goodfellow AF, Wrana JL. Distinct endocytic pathways regulate TGF-β receptor signalling and turnover. *Nat Cell Biol.* 2003;5:410–421.

49. Inman GJ, Nicolas FJ, Hill CS. Nucleocytoplasmic shuttling of Smads 2, 3, and 4 permits sensing of TGFβ receptor activity. *Mol Cell.* 2002;10:283–294.

50. Xu L, Alarcon C, Col S, Massagué J. Distinct domain utilization by Smad3 and Smad4 for nucleoporin interaction and nuclear import. *J Biol Chem.* 2003;278:42569–42577.

51. Chacko BM, Qin BY, Tiwari A, et al. Structural basis of heteromeric smad protein assembly in TGFβ signaling. *Mol Cell.* 2004;15:813–823.

52. Qin BY, Chacko BM, Lam SS, de Caestecker MP, Correia JJ, Lin K. Structural basis of Smad1 activation by receptor kinase phosphorylation. *Mol Cell.* 2001;8:1303–1312.

53. Lim SK, Hoffmann FM. Smad4 cooperates with lymphoid enhancer-binding factor 1/T cell-specific factor to increase c-myc expression in the absence of TGFβ signaling. *Proc Natl Acad Sci USA.* 2006;103:18580–18585.

54. Shiou SR, Singh AB, Moorthy K, et al. Smad4 regulates claudin-1 expression in a transforming growth factor-β-independent manner in colon cancer cells. *Cancer Res.* 2007;67:1571–1579.

55. He W, Dorn DC, Erdjument-Bromage H, Tempst P, Moore MA, Massagué J. Hematopoiesis controlled by distinct TIF1γ and Smad4 branches of the TGFβ pathway. *Cell.* 2006;125:929–941.

56. Xu L, Kang Y, Col S, Massagué J. Smad2 nucleocytoplasmic shuttling by nucleoporins CAN/Nup214 and Nup153 feeds TGFβ signaling complexes in the cytoplasm and nucleus. *Mol Cell.* 2002;10:271–282.

57. Seoane J, Le HV, Shen L, Anderson SA, Massagué J. Integration of Smad and forkhead pathways in the control of neuroepithelial and glioblastoma cell proliferation. *Cell*. 2004;117:211–223.

58. Shi Y, Wang YF, Jayaraman L, Yang H, Massagué J, Pavletich NP. Crystal structure of a Smad MH1 domain bound to DNA: insights on DNA binding in TGFβ signaling. *Cell*. 1998;94:585–594.

59. Chen X, Rubock MJ, Whitman M. A transcriptional partner for MAD proteins in TGF-β signalling. *Nature*. 1996;383:691–696.

60. Hanai J, Chen LF, Kanno T, et al. Interaction and functional cooperation of PEBP2/CBF with Smads. Synergistic induction of the immunoglobulin germline Cα promoter. *J Biol Chem*. 1999;274:31577–31582.

61. Alliston T, Choy L, Ducy P, Karsenty G, Derynck R. TGFβ-induced repression of CBFA1 by Smad3 decreases cbfa1 and osteocalcin expression and inhibits osteoblast differentiation. *EMBO J*. 2001;20:2254–2272.

62. Germain S, Howell M, Esslemont GM, Hill CS. Homeodomain and winged-helix transcription factors recruit activated Smads to distinct promoter elements via a common Smad interaction motif. *Genes Dev*. 2000;14:435–451.

63. Hata A, Seoane J, Lagna G, Montalvo E, Hemmati-Brivanlou A, Massagué J. OAZ uses distinct DNA- and protein-binding zinc fingers in separate BMP-Smad and Olf signaling pathways. *Cell*. 2000;100:229–240.

64. Zhang Y, Feng XH, Derynck R. Smad3 and Smad4 cooperate with c-Jun/c-Fos to mediate TGF-β-induced transcription. *Nature*. 1998;394:909–913.

65. Chen CR, Kang Y, Siegel PM, Massagué J. E2F4/5 and p107 as Smad cofactors linking the TGFβ receptor to c-myc repression. *Cell*. 2002;110:19–32.

66. Kang Y, Chen CR, Massagué J. A self-enabling TGFβ response coupled to stress signaling: Smad engages stress response factor ATF3 for Id1 repression in epithelial cells. *Mol Cell*. 2003;11:915–926.

67. Ogryzko VV, Schiltz RL, Russanova V, Howard BH, Nakatani Y. The transcriptional coactivators p300 and CBP are histone acetyltransferases. *Cell*. 1996;87:953–959.

68. Janknecht R, Wells NJ, Hunter T. TGFβ-stimulated cooperation of smad proteins with the coactivators CBP/p300. *Genes Dev*. 1998;12:2114–2119.

69. Valcourt U, Kowanetz M, Niimi H, Heldin CH, Moustakas A. TGFβ and the Smad signaling pathway support transcriptomic reprogramming during epithelial-mesenchymal cell transition. *Mol Biol Cell*. 2005;16:1987–2002.

70. Nicolas E, Roumillac C, Trouche D. Balance between acetylation and methylation of histone H3 lysine 9 on the E2F-responsive dihydrofolate reductase promoter. *Mol Cell Biol*. 2003;23:1614–1622.

71. Melhuish TA, Wotton D. The interaction of the carboxyl terminus-binding protein with the Smad corepressor TGIF is disrupted by a holoprosencephaly mutation in TGIF. *J Biol Chem*. 2000;275:39762–39766.

72. Seoane J, Pouponnot C, Staller P, Schader M, Eilers M, Massagué J. TGFβ influences Myc, Miz-1 and Smad to control the CDK inhibitor p15INK4b. *Nat Cell Biol*. 2001;3:400–408.

73. Frederick JP, Liberati NT, Waddell DS, Shi Y, Wang XF. Transforming growth factor β-mediated transcriptional repression of c-myc is dependent on direct binding of Smad3 to a novel repressive Smad binding element. *Mol Cell Biol*. 2004;24:2546–2559.

74. Derynck R, Zhang YE. Smad-dependent and Smad-independent pathways in TGFβ family signalling. *Nature*. 2003;425:577–584.

75. Nishita M, Hashimoto MK, Ogata S, et al. Interaction between Wnt and TGFβ signalling pathways during formation of Spemann's organizer. *Nature*. 2000;403:781–785.

76. Nakashima K, Yanagisawa M, Arakawa H, et al. Synergistic signaling in fetal brain by STAT3-Smad1 complex bridged by p300. *Science*. 1999;284:479–482.

77. Matsuura I, Denissova NG, Wang G, He D, Long J, Liu F. Cyclin-dependent kinases regulate the antiproliferative function of Smads. *Nature*. 2004;430:226–231.

78. Sapkota G, Alarcon C, Spagnoli FM, Brivanlou AH, Massagué J. Balancing BMP signaling through integrated inputs into the Smad1 linker. *Mol Cell*. 2007;25:441–454.

79. Kretzschmar M, Doody J, Massagué J. Opposing BMP and EGF signalling pathways converge on the TGFβ family mediator Smad1. *Nature*. 1997;389:618–622.

80. Pera EM, Ikeda A, Eivers E, De Robertis EM. Integration of IGF. FGF, and anti-BMP signals via Smad1 phosphorylation in neural induction. *Genes Dev*. 2003;17:3023–3028.

81. De Robertis EM, Larrain J, Oelgeschlager M, Wessely O. The establishment of Spemann's organizer and patterning of the vertebrate embryo. *Nat Rev Genet*. 2000;1:171–181.

82. Streit A, Berliner AJ, Papanayotou C, Sirulnik A, Stern CD. Initiation of neural induction by FGF signalling before gastrulation. *Nature*. 2000;406:74–78.

83. Delaune E, Lemaire P, Kodjabachian L. Neural induction in *Xenopus* requires early FGF signalling in addition to BMP inhibition. *Development*. 2005;132:299–310.

84. Spemann H. http://nobelprizeorg/nobel_prizes/medicine/laureates/1935/ 11935.

85. Spemann H, Mangold H. Über induktion von embryonalanlagen durch implantation artfremder organisatoren. *Wilhelm Roux Arch Entw Mech Org*. 1924;100:599–638.

86. Nakashima N, Reddi AH. The application of bone morphogenetic proteins to dental tissue engineering. *Nature Biotech*. 2003;21:1025–1032.

87. Yakymovych I, ten Dijke P, Heldin CH, Souchelnytskyi S. Regulation of Smad signaling by protein kinase C. *FASEB J*. 2001;15:553–555.

88. Scherer A, Graff JM. Calmodulin differentially modulates Smad1 and Smad2 signaling. *J Biol Chem*. 2000;275:41430–41438.

89. Bonni S, Wang HR, Causing CG, et al. TGFβ induces assembly of a Smad2-Smurf2 ubiquitin ligase complex that targets SnoN for degradation. *Nat Cell Biol*. 2001;3:587–595.

90. Nakao A, Afrakhte M, Moren A, et al. Identification of Smad7, a TGFβ-inducible antagonist of TGF-β signalling. *Nature*. 1997;389:631–635.

91. Afrakhte M, Moren A, Jossan S, et al. Induction of inhibitory Smad6 and Smad7 mRNA by TGFβ family members. *Biochem Biophys Res Commun*. 1998;249:505–511.

92. Hata A, Lagna G, Massagué J, Hemmati-Brivanlou A. Smad6 inhibits BMP/Smad1 signaling by specifically competing with the Smad4 tumor suppressor. *Genes Dev*. 1998;12:186–197.

93. Ogunjimi AA, Briant DJ, Pece-Barbara N, et al. Regulation of Smurf2 ubiquitin ligase activity by anchoring the E2 to the HECT domain. *Mol Cell*. 2005;19:297–308.

94. Wahl SM, Wen J, Moutsopoulos N. TGFβ: a mobile purveyor of immune privilege. *Immunol Rev*. 2006;213:213–227.

95. Han G, Li AG, Liang YY, et al. Smad7-induced β-catenin degradation alters epidermal appendage development. *Dev Cell*. 2006;11:301–312.

96. Degen WG, Weterman MA, van Groningen JJ, et al. Expression of nma, a novel gene, inversely correlates with the metastatic potential of human melanoma cell lines and xenografts. *Int J Cancer*. 1996;65:460–465.

97. Das AK, Helps NR, Cohen PT, Barford D. Crystal structure of the protein serine/threonine phosphatase 2C at 2.0 A resolution. *EMBO J*. 1996;15:6798–6809.

98. Lin X, Duan X, Liang YY, et al. PPM1A functions as a Smad phosphatase to terminate TGFβ signaling. *Cell*. 2006;125:915–928.

99. Chen HB, Shen J, Ip YT, Xu L. Identification of phosphatases for Smad in the BMP/DPP pathway. *Genes Dev*. 2006;20:648–653.

100. Sapkota G, Knockaert M, Alarcon C, Montalvo E, Brivanlou AH, Massagué J. Dephosphorylation of the linker regions of Smad1 and Smad2/3 by small C-terminal domain phosphatases has distinct outcomes for bone morphogenetic protein and transforming growth factor-β pathways. *J Biol Chem*. 2006;281:40412–40419.

101. Meinhart A, Kamenski T, Hoeppner S, Baumli S, Cramer P. A structural perspective of CTD function. *Genes Dev*. 2005;19:1401–1415.

102. Hahn SA, Schutte M, Hoque AT, et al. DPC4, a candidate tumor suppressor gene at human chromosome 18q21. *Science*. 1996;271:350–353.

103. Miyaki M, Iijima T, Konishi M, et al. Higher frequency of Smad4 gene mutation in human colorectal cancer with distant metastasis. *Oncogene*. 1999;18:3098–3103.

104. Alberici P, Jagmohan-Changur S, De Pater E, Van Der Valk M, Smits R, Hohenstein P, Fodde R. Smad4 haploinsufficiency in mouse models for intestinal cancer. *Oncogene*. 2006;25:1841–1851.

105. Deckers M, van Dinther M, Buijs J, et al. The tumor suppressor Smad4 is required for transforming growth factor β-induced epithelial to mesenchymal transition and bone metastasis of breast cancer cells. *Cancer Res*. 2006;66:2202–2209.

106. Schniewind B, Groth S, Sebens MS, et al. Dissecting the role of TGFβ type I receptor/ALK5 in pancreatic ductal adenocarcinoma: Smad activation is crucial for both the tumor suppressive and prometastatic function. *Oncogene*. 2007;26:4850–4862.

107. Oft M, Akhurst RJ, Balmain A. Metastasis is driven by sequential elevation of H-ras and Smad2 levels. *Nat Cell Biol*. 2002;4:487–494.

108. Zitvogel L, Tesniere A, Kroemer G. Cancer despite immunosurveillance: immunoselection and immunosubversion. *Nat Rev Immunol*. 2006;6:715–727.

109. Bhat R, Watzl C. Serial killing of tumor cells by human natural killer cells – enhancement by therapeutic antibodies. *PLoS ONE*. 2007;2:e326.

110. Thomas DA, Massagué J. TGFβ directly targets cytotoxic T cell functions during tumor evasion of immune surveillance. *Cancer Cell*. 2005;8:369–380.

111. Shook D, Keller R. Mechanisms, mechanics and function of epithelial-mesenchymal transitions in early development. *Mech Dev*. 2003;120:1351–1383.

112. Zavadil J, Bitzer M, Liang D, et al. Genetic programs of epithelial cell plasticity directed by transforming growth factor-β. *Proc Natl Acad Sci USA*. 2001;98:6686–6691.

113. Nieto MA, Sargent MG, Wilkinson DG, Cooke J. Control of cell behavior during vertebrate development by Slug, a zinc finger gene. *Science*. 1994;264:835–839.

114. Luo W, Peterson A, Garcia BA, et al. Protein phosphatase 1 regulates assembly and function of the β-catenin degradation complex. *EMBO J*. 2007;26:1511–1521.

115. Barrallo-Gimeno A, Nieto MA. The Snail genes as inducers of cell movement and survival: implications in development and cancer. *Development*. 2005;132:3151–3161.

116. Ozdamar B, Bose R, Barrios-Rodiles M, Wang HR, Zhang Y, Wrana JL. Regulation of the polarity protein Par6 by TGFβ receptors controls epithelial cell plasticity. *Science*. 2005;307:1603–1609.

CHAPTER 21

Protein Dephosphorylation and Protein Phosphorylation

Protein phosphorylation serves multiple roles in the regulation of cell function. But this is only half the story. If the transfer of phosphate groups to proteins is to serve as a precise and sensitive signalling mechanism, then necessarily it must operate against a low background. Dephosphorylation is therefore as important as phosphorylation, and it follows that the phosphoprotein phosphatases are integral components of the signalling systems operated by the protein kinases.[1] In a number of cases dephosphorylation serves as a true reset button, bringing proteins back to their inactive state.

A good example of this is the role of the serine/threonine phosphatase PP1c which dephosphorylates glycogen phosphorylase a, thereby terminating the breakdown of glycogen. There are other proteins (for example, glycogen synthase, GSK3β, Src, Lck, c-Jun, and NFAT) that are inactive when phosphorylated, only becoming active as a consequence of dephosphorylation (Figure 21.1) In the case of Lck, a tyrosine protein kinase, activation of the enzyme requires dephosphorylation of a C-terminal tyrosine

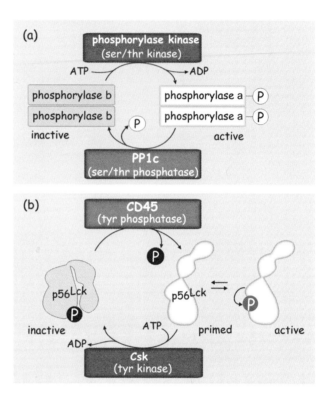

FIG 21.1 Phosphatases in the generation of signals for both deactivation and activation.
(a) A protein phosphatase as a reset button. The serine phosphatase PP1c dephosphorylates and deactivates
phosphorylase a. (b) A protein phosphatase as activator. CD45, a receptor-like tyrosine protein phosphatase,
primes p56Lck for activation.

residue and then phosphorylation of another tyrosine in the activation
segment, through an autophosphorylation mechanism (see pages 532
and 659). To become fully active, the transcription factor c-Jun undergoes
dephosphorylation of serine and threonine residues near the DNA binding
site and phosphorylation of serines near the N-terminus.

Protein tyrosine phosphatases

A soluble protein phosphatase specific for phosphotyrosines (PTP1B) was
first isolated from human placenta.[2] Its amino acid sequence has stretches
homologous with the tandem repeat domains present in the cytoplasmic
portion of CD45 (leukocyte common antigen),[3] a receptor-like protein

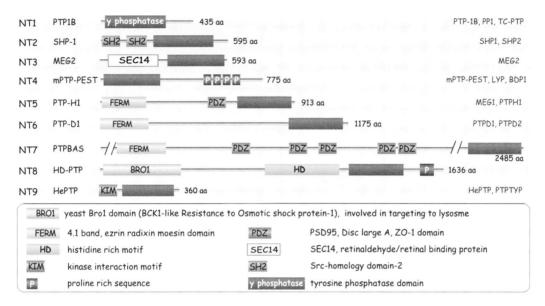

NT1	PTP1B	γ phosphatase	435 aa	PTP-1B, PP1, TC-PTP
NT2	SHP-1	SH2–SH2	595 aa	SHP1, SHP2
NT3	MEG2	SEC14	593 aa	MEG2
NT4	mPTP-PEST	P·P·P·P	775 aa	mPTP-PEST, LYP, BDP1
NT5	PTP-H1	FERM ··· PDZ	913 aa	MEG1, PTPH1
NT6	PTP-D1	FERM	1175 aa	PTPD1, PTPD2
NT7	PTPBAS	//FERM ··· PDZ PDZ–PDZ PDZ·PDZ //	2485 aa	
NT8	HD-PTP	BRO1 ··· HD ··· P	1636 aa	
NT9	HePTP	KIM	360 aa	HePTP, PTPTYP

BRO1	yeast Bro1 domain (BCK1-like Resistance to Osmotic shock protein-1), involved in targeting to lysosme			
FERM	4.1 band, ezrin radixin moesin domain	PDZ	PSD95, Disc large A, ZO-1 domain	
HD	histidine rich motif	SEC14	SEC14, retinaldehyde/retinal binding protein	
KIM	kinase interaction motif	SH2	Src-homology domain-2	
P	proline rich sequence	γ phosphatase	tyrosine phosphatase domain	

FIG 21.2 Domain architecture of cytosolic tyrosine phosphatase families. Nine distinct families of cytosolic tyrosine phosphatases are recognized on the basis of sequence differences in the phosphatase domain and of the presence of additional domains. Information from Andersen et al.[4]

expressed on cells of haematopoietic lineage. Using the DNA sequence that codes for the catalytic domain as a probe (the conserved 'signature motif'), some 113 distinct vertebrate PTP catalytic domains have been identified, with ~37 distinct *PTP* genes in the human genome.

Sequence comparison of the catalytic domains of the vertebrate genes reveals 17 subtypes that comprise both cytosolic (Figure 21.2) and transmembrane phosphatases (Receptor-PTP) (Figure 21.3). The diversity of tyrosine protein phosphatases is greatly enlarged through alternative splicing, alternative promoter usage, and post-translational modification. Notably, as a result of alternative splicing, some of the receptor-PTP subtypes also contain cytosolic variants (for example PTPε from the R4 group).[4]

For more information, refer to the tyrosine phosphatase database http://ptp.cshl.edu or the parallel site http://science.novonordisk.com/ptp).

The many abbreviations used in this chapter are collected together at the end of the chapter.

In contrast to the serine/threonine and tyrosine kinases, which share a common ancestry, the tyrosine phosphatases and the serine/threonine phosphatases bear no relation to each other. Unlike the serine/threonine phosphatases, in which substrate specificity is determined by associated targeting subunits, the tyrosine phosphatases are all monomeric enzymes. The various domains that flank the catalytic domain act as targeting sequences and play important roles in determining the subcellular localization and the control of activity. How the structural diversity reflects differences in substrate recognition remains unclear.

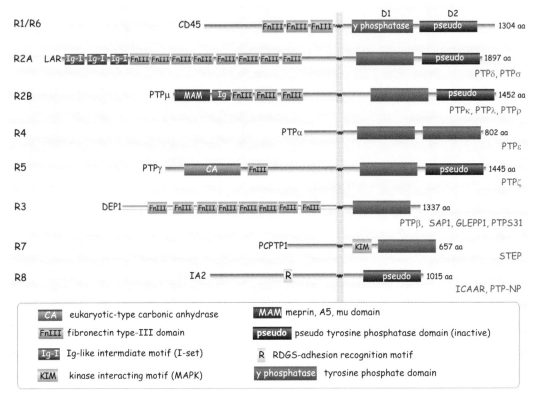

FIG 21.3 Domain architecture of receptor tyrosine phosphatases. All contain a tyrosine protein phosphatase motif, often two, D1 and D2, in tandem, and are distinguished by their extracellular domains. Some have elaborate extracellular structures resembling adhesion molecules or growth factor receptors, others appear rudimentary and ligand association is hard to imagine. No ligands have been identified for most of these 'receptor-like' phosphatases.

Cytosolic PTPs

The cytosolic PTPs are classified according to sequence homology of their catalytic domains. There are nine subclasses, each having different additional domains. An important subclass, comprised of SHP-1 and SHP-2, possesses SH2 domains, whereas others are characterized by the presence of proline-rich sequences in the vicinity of the C-terminus. Again, others have FERM and PDZ domains. We will return later to PTP1B, SHP-1, and SHP-2.

Included within the cytosolic PTP family are the dual-specificity phosphatases, which are active at both tyrosine and serine/threonine residues. Since the first of these to be identified was VH1, encoded by vaccinia virus, they are also referred to as VH1-like phosphatases.[5] The dual-specificity phosphatases also have homology with Cdc25, a regulator of mitosis in fission yeast (*Schizosaccharomyces pombe*). This activates cyclin-dependent kinase-2 by dephosphorylation of adjacent threonine and tyrosine residues.[6] Dual-specificity phosphatases share the general signature motif but otherwise display little similarity with the phosphotyrosine-specific phosphatases (and are therefore not included in Figure 21.2). We return to dual specificity phosphatases on page 661.

Transmembrane receptor-like PTPs

Nearly all the transmembrane PTPs contain the tandem repeats D1 and D2, both of which express the catalytic signature motif. However, with the exception of PTPα, only the membrane-proximal D1 domains are catalytically active. The inactivity of the D2 domains can be ascribed to the absence of invariant amino acids (tyrosine and asparagine, see next section) that converge around the active site. The preservation of the D2 domain among species and within nearly all receptor-like PTPs indicates that it probably serves an important physiological function. Its structural integrity is important for stability of the RPTP as a whole. Also, there is the possibility that D2 might be involved in the regulation of D1, or that it could assist in substrate recognition.

The transmembrane PTPs were originally classified on the basis of their extracellular ectodomains.[7] Happily, when they are sorted according to the sequence homology of their catalytic domains, the categorization remains almost the same. (The initial classification has therefore been retained.) The ectodomains range from very short chains, having no apparent function, to extended structures with putative ligand-binding domains similar to those present in adhesion molecules (fibronectin repeats or immunoglobulin repeats, Figure 21.3). Such diversity suggests a wide range of biological functions which have, however, proved hard to pin down. This is largely because identification of their physiological substrates has been hampered by the non-specificity of these enzymes when assayed for activity *in vitro*. Moreover, although some are predicted to be receptors, their ligands have so far been elusive. PTPμ and PTPκ may take part in homotypic adhesion through their MAM domains.[8,9] When expressed in insect cells, they induce Ca^{2+}-independent cell aggregation, but even here the downstream pathways remain unclear. We will return later to CD45 and DEP1.

Tyrosine specificity and catalytic mechanism

The protein tyrosine phosphatases are without effect on phosphoserine or phosphothreonine residues, but there are a large number of phosphotyrosine proteins that act as substrates. This suggests that the overall structure of the substrate is not the main determinant of selectivity; instead, substrate recognition is determined primarily by the presence of the pY residue in the context of its peptide environment. In particular, specificity results from the depth of the catalytic cleft (Figure 21.4b, c).

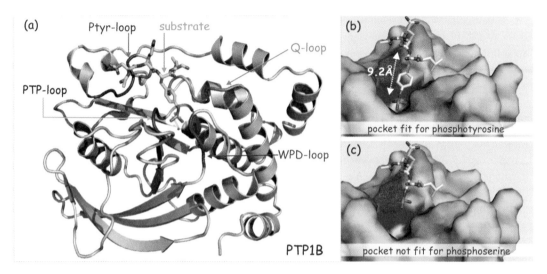

FIG 21.4 The arrangement at the active site of PTP1B. (a) The sides of the catalytic cleft of tyrosine phosphatases are characterized by three motifs: the WPD loop, containing an invariant Asp, the Q-loop, containing an invariant Gln and the phosphotyrosine-binding loop (Ptyr), containing an invariant Tyr (1ptt[10]). (b) The invariant tyrosine determines the depth of the catalytic cleft, ~10 Å, and allows the target phosphotyrosine to make contact with the catalytic cysteine residue (yellow). (c) The pocket is too deep for phosphoserine or phosphothreonine to contact the catalytic cysteine residue (2hnp[11]).

Unlike the kinases (see page 782), the catalytic domains of tyrosine protein phosphatases do not require post-translational modifications in order to become catalytically competent. The binding of substrate induces a conformational change in the protein in which the WPD loop closes around the substrate to create the recognition pocket, generating the catalytically active form of the enzyme (Figure 21.4a). Protein tyrosine phosphatases are, however, subject to regulation, by either intramolecular interactions, phosphorylation of non-catalytic domains, association of regulatory subunits, or spatial separation from substrate.

The catalytic mechanism of the tyrosine phosphatases is quite different from that of the serine/threonine phosphatases. These are metalloenzymes that dephosphorylate their substrates in a single reaction step involving a metal-activated nucleophilic water molecule. In contrast, the PTPs catalyse dephosphorylation through a cysteinyl-phosphate enzyme intermediate (reviewed in Tonks[12]) (Figure 21.5).

PTPs in signal transduction

At first, interest in protein tyrosine phosphatases was driven by the perception that they might be antitumorigenic. They appeared to offer the possibility

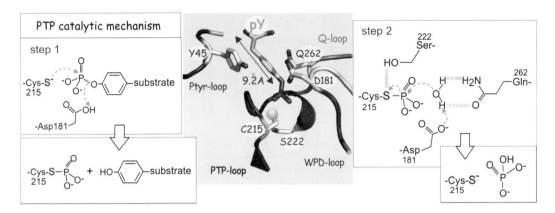

FIG 21.5 Catalytic mechanism of tyrosine phosphatases (PTP1B). In the first step, there is transfer of an electron from C215 to the substrate phosphate. The catalytic cysteine has a low pK_a (5.4) so that the sulfydryl group acts as a nucleophile (−S⁻). This is coupled with protonation of the tyrosyl-leaving group by the side chain of the conserved D181, leading to the formation of a cysteinyl-phosphate intermediate. In the second step involving Q262 and D181 there is hydrolysis of the catalytic intermediate with release of phosphate. The placement of the essential residues at the active site is depicted on the right. Note that the phosphotyrosine substrate is aligned and sandwiched between Y45 and F180. Y45 determines the depth of the catalytic cleft and is the major determinant of specificity for phosphotyrosine (1ptt[10]).

of counteracting the transforming effects of mutated, and therefore constitutively activated, PTKs and were therefore considered as possible tumour suppressors. Thus it came as quite a surprise when it was found that some PTPs, rather than opposing the actions of the kinases, actually cooperate with them to reinforce their signals. As an example, over-expression of PTPα in rat embryo fibroblasts causes persistent activation of Src with concomitant cell transformation.[13] In accordance with this, increased PTPα mRNA levels have been detected in late-stage colorectal tumours and enhanced levels of the enzyme occur in one third of breast carcinomas. However, over-expression of PTPα in breast carcinoma cells actually reduces tumour aggressiveness. The functional significance of PTPα thus depends on the cellular context and the type of tumour. 19 PTP genes that map to chromosomal regions frequently deleted in human cancers have been identified, and also 4 PTP genes that map to regions frequently amplified in human cancers (see http://ptp.cshl.edu or http://science.novonordisk.com/ptp).

We now exemplify the different roles of tyrosine phosphatases.

PTP1B, diabetes, and obesity

Perhaps the most spectacular example of a link between a PTP and human disease is type 2 (mature onset, insulin-resistant) diabetes and obesity. Both

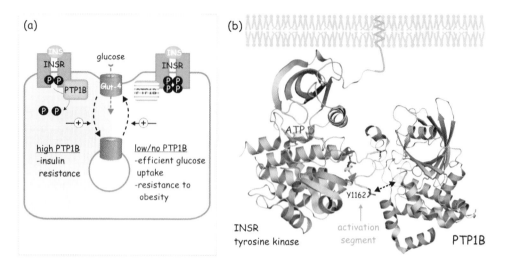

FIG 21.6 PTPB1 down-regulates insulin signalling by dephosphorylation and inactivation of the receptor and its substrates (IRS). (a) PTP1B inhibits the action of insulin by dephosphorylating the cytosolic segment of the insulin receptor. Lack of PTP1B activity promotes glucose uptake due to elevated cell surface expression of its transporter, but (in rats at least) it also protects against the effects of insulin resistance and obesity induced by a high-fat diet. (b) PTP1B binds preferentially to the bis- or tris-phosphorylated activation segment of the insulin receptor.[19] pY-1162 is most readily recognized by the catalytic cleft of the phosphatase. Dephosphorylation renders the insulin receptor catalytically inactive (see page 554). Sequence comparison with other PTPs suggests that the features that confer the specificity of this reaction are unique to PTP1B and its close relative TCPTP (1ptt[10] and 1irk[20]).

genetic and biochemical studies provide good evidence of a role of tyrosine phosphatase in the signalling events downstream of the insulin and leptin receptors.[14]

Dephosphorylation of the insulin receptor by PTPs is critical in the control of the cellular response to insulin. Numerous studies have demonstrated that in humans, and in animal models, the resistance to insulin in type 2 diabetes and obesity is accompanied by increases in PTP activity and increases in the level of expression of defined members of the PTP family (LAR and PTP1B).[15] However, disruption of the LAR gene in mice yields a complex phenotype. There is a post-receptor defect in insulin signalling but, surprisingly, it is associated with *impaired* activation of downstream signals, such as PI 3-kinase, rather than the opposite. Matters are somewhat clearer for PTP1B (Figure 21.6 and Table 21.1). Here, for *Xenopus* oocytes injected with the purified enzyme, insulin-induced activation of S6 kinase and transition through the G2/M checkpoint of the cell cycle is inhibited.[16,17] When over-expressed in Rat1 fibroblasts, PTP1B reduces insulin-induced phosphorylation of its own receptor and the downstream phosphorylation of components of the signalling cascade. It also reduces the translocation of GLUT-4 (glucose

TABLE 21.1 Consequences of the absence of PTP1B

Dephosphorylation of INSR	Reduced phosphorylation of IRS-1 and -2
	Glut-4 fails to translocate
	Low glycogen synthase activity
	Inhibition of G2/M transition (*X. laevis* oocytes)
Enhanced phosphorylation of INSR	Enhanced phosphorylation of IRS-1
	Enhanced expression of IRS-2
	Increased phosphorylation of PKB
	Increased glucose uptake
	Increased expression of RasGAP and p62DOK
	Reduced expression of genes involved in lipogenesis
	Resistance to obesity
	Protection against high fat diet-induced insulin resistance

transporter) to the membrane and glycogen synthesis.[18] Conversely, loading of hepatoma cells with neutralizing antibodies to PTP1B enhances insulin-induced phosphorylation of its receptor and of the receptor substrate IRS-1.

An unequivocal link between insulin signalling and the tyrosine phosphatase was established through the use of PTP1B knockout mice; these animals remained healthy and displayed an enhanced sensitivity to insulin.[21] Moreover, when fed a high-fat diet, they failed to become obese and retained their normal sensitivity to insulin. By contrast, their wild-type (PTP1B$^{+/+}$) litter mates suffered rapid weight gain and onset of insulin resistance, which coincided with a reduced level of tyrosine phosphorylation of the insulin receptor.[22] All this suggests that the insulin receptor β-subunit acts as a substrate of PTP1B. Surprisingly, the knockout mice did not show any predisposition to cancer despite the potential of PTP1B to regulate growth factor receptor tyrosine kinase signalling and to counteract the transforming effects of the kinases Src and Neu.[23,24] The activity of ERK present in fibroblasts obtained from these mice was only slightly enhanced and there was no effect on the activation level of PKB. A possible explanation comes from the finding that, due to a lack of PTP1B, expression of RasGAP and phosphorylation of p62Dok are elevated and that this leads to attenuation of the Ras–MAPK

The **leptin receptor** is a cytokine receptor. It resembles the granulocyte–monocyte colony stimulating factor (GM-CSF) receptor and signals mainly through the JAK2/STAT3 pathway. It also activates the IRS2–PI 3-kinase and Ras–MAPK pathways.

Insulin and **leptin** act on neurons in the hypothalamic arcuate nucleus to regulate the desire to eat. The two hormones signify positive energy balance and satiety, manifested by (among other changes) increased release of pro-opiomelanocortin (POMC) peptides and reduced release of NPY and AgRP. The consequence is suppression of appetite and a greater propensity to undertake physical activity. In contrast, depletion of fat stores gives rise to low basal insulin and low leptin levels, which favour food intake and economy of energy expenditure. Disruption of insulin and/or leptin signalling encourages food intake, reduces energy expenditure, and leads to an increase in body mass. Stimulating insulin and/or leptin signalling does the reverse.

pathway.[25,26] This is a good example of how feedback mechanisms prevent excess signalling (often referred to as the *robustness* of the system).

Introduction of antisense oligonucleotides to PTP1B into mice has the effect of suppressing *PTP1B* gene expression in the liver and adipose tissue, but not in muscle. Obese (*ob/ob*, leptin-deficient) and late onset diabetic (*db/db*, leptin receptor deficient) mice, treated in this way, exhibit enhanced sensitivity to insulin but maintain normal levels of blood glucose.[27] This, and the resistance to obesity, has made PTP1B an attractive target for treating type 2 diabetes as well as obesity in general. The quest for inhibitors of PTP1B has attracted much attention and more than 40 crystal structures of complexes of PTP1B with inhibitors or substrates have been deposited in the Protein Data Bank.

PTP1B as a possible therapeutic target for the treatment of type 2 diabetes and obesity

Optimal substrate recognition by PTP1B requires the amino acid motif D/E-pY-pY-R/K, also present in the kinases Trk, FGF-R, NGF-R, and Axl, and of particular relevance here, members of the JAK family.[28] With respect to obesity, the JAKs are instrumental in signalling by the satiety hormone leptin via its receptor OB-R_L (for a review, see Tartaglia[29]). JAK2, a physiological substrate of PTP1B, phosphorylates two residues on human OB-R_L and probably acts to regulate leptin signalling (blocking the satiety signal). Hypothalamic neurons derived from PTP1B$^{-/-}$ knockout mice display markedly increased leptin-induced STAT3 phosphorylation (STAT3 is the transcription factor involved in the leptin response) (Figure 21.7). Since leptin-activated JAK2 is a substrate of PTP1B but not STAT3 or the leptin receptor itself,[30] it follows that PTP1B negatively regulates leptin signalling. This provides a mechanism by which it may regulate obesity.[31,32]

Redox regulation of PTP1B: reactive oxygen species as second messengers

Numerous cells, including phagocytes, epidermal cells, fibroblasts, vascular smooth muscle cells, and osteoblasts, generate reactive oxygen species (ROS) in response to environmental cues.

ROS originate from two different sources. The first is production by the NADPH-oxidase complex (NOX). In response to the GTPase Rac1, the subunit NOX1 transfers electrons across the plasma membrane delivering them to the acceptor O_2, generating superoxide (O_2^-) (Figure 21.8). TGFβ1, IL-1, TNF-α, insulin, PDGF, EGF, angiotensin, thrombin, and lysophosphatidic acid are amongst the cytokines that activate the NOX pathway[33–35] (for a review see Chiarugi and Cirri[36]).

A second source of ROS, operational in all cell types, is the mitochondrial electron transport chain. Here, superoxide is released at the level of

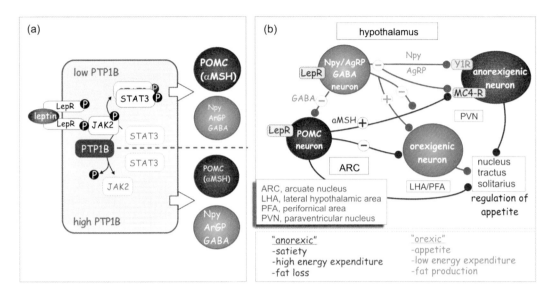

FIG 21.7 Role of PTP1B in satiety signalling. (a) High fat (or low fat turnover) releases leptin, a satiety signal, from adipose tissue. This binds to its receptor, LepRR, in neurons of the arcuate nucleus of the hypothalamus, and activates JAK2. This phosphorylates STAT3, which dimerizes and enters the nucleus, there to alter gene expression leading to an enhanced expression of POMC products (αMSH, CART, and CRH), and reduced expression of Npy, AgRP and GABA. (b) This translates into a predominantly anorexigenic signal (red neurons), leading to a feeling of satiety, high energy expenditure and loss of fat. Deletion of PTP1B renders mice resistant to obesity.

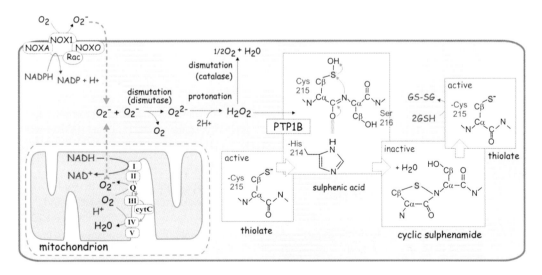

FIG 21.8 Generation of superoxide (O_2^-) by NADPH oxidase and the mitochondrial electron transport chain. Superoxide (O_2^-) is produced by NADPH oxidase, a complex comprising NOXA (NOX activator), NOXO (NOX organizer), Rac, and NOX1. It is also released from mitochondria. It is converted to oxygen and peroxide (O_2^{2-}), to become H_2O_2, which then transiently converts the active site cysteine thiolate into a catalytically inactive cyclic sulfenamide. Glutathione is required to restore enzyme activity. This reaction scheme applies for cysteines with a low pK_a.

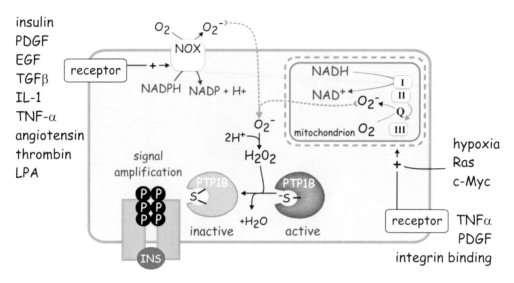

FIG 21.9 Stimulation of production of ROS by growth factors, hypoxia, oncogenes, and integrin binding. Various growth factors, cytokines, oncogenes, and integrins, as well as hypoxia, enhance production of ROS. Hydrogen peroxide causes inactivation of the active site cysteine of PTP1B. The transient inhibition enhances and prolongs phosphorylation of receptor tyrosine kinases and amplifies the growth factor signal. Other targets of H_2O_2 are proteins having an acidic cysteine residue (pK_a5.4) such as other tyrosine phosphatases (PTP) and the transcription factors NF-κB, AP1, and HIF.

ubiquinone (which normally shuttles electrons from complex II to III) (Figure 21.8). This plays a role in apoptosis, but transient production also occurs in response to other cues under conditions where apoptosis does not occur. Examples are hypoxia, the expression of oncogenic Ras or c-Myc, the presence of cytokines such as TNF-α and PDGF, or the interaction of integrins with the extracellular matrix (Figure 21.9). The mechanism by which the ubiquinone pathway is activated remains to be elucidated.[37,38]

Superoxide is rapidly converted (dismutated) into molecular oxygen (O_2) and peroxide (O_2^{2-}), which as H_2O_2, is converted by catalase into oxygen and water (Figure 21.8). The ROS, in particular H_2O_2, survive long enough to modify the active-site cysteine residue of tyrosine phosphatases.

There have been a number of claims for a role of ROS, in particular H_2O_2, as intracellular messengers that regulate protein phosphorylation and gene expression. An early hint was provided by the observation that H_2O_2 can mimic the stimulatory effect of insulin on glucose transport and lipid synthesis in adipocytes.[39,40] Later it was found that O_2^- can stimulate members of the MAP kinase family[41] and that blunting the spike of H_2O_2 (using catalase or the antioxidant N-acetylcysteine) obscures a number of signal transduction events such as phosphorylation and activation of MAP kinase in response to PDGF. The cells also fail to enter the S-phase of the cell cycle and lose their response to chemotactic stimuli.[34]

Further evidence for a second messenger role of ROS came from experiments that showed that EGF-mediated autophosphorylation of its receptor and the subsequent phosphorylation of receptor substrates can be can be up-regulated by the addition of H_2O_2 to the culture medium. The enhanced phosphorylation can be prevented by intracellular microinjection of catalase. Several studies have demonstrated an inhibitory effect of H_2O_2 (at micromolar concentrations) on various tyrosine phosphatases, both membrane bound and cytosolic.[42–45]

Striking evidence for the role of NADPH oxidase in cellular signalling came from the finding that over-expression of NOX1 in NIH3T3 cells produces particularly aggressive tumours following injection into nude (athymic) mice. Sustained and elevated levels of hydrogen peroxide are required to maintain the transformed state.[46,47]

It is now widely accepted that ROS that are transiently generated in cells in response to extracellular stimuli, and which act on effector systems, can be regarded as second messengers. Oxidative modification of catalytic-site cysteine residues resembles the mode of action of phosphorylation (Figure 21.8). rPTK-induced generation of ROS causes prolonged inactivation of tyrosine phosphatases and so maintains the phosphorylated (and activated) state of the receptor (positive feedback loop) (Figure 21.9). Their brief generation time and the presence of glutathione (GSH) ensure that over a period of about a few minutes the phosphatases return to their reduced state, capable once again of dephosphorylating (and inactivating) the receptor.[48–50]

SHP-1 and SHP-2

The SHP phosphatases comprise a subfamily of cytosolic PTPs having two N-terminal SH2 domains. They bind to tyrosine phosphorylated growth factor receptors (EGFR, FGFR, HGFR, PDGFR), cytokine receptors (IFNAR1, IL-R, TNFR1), scaffolding adaptors (IRS, DOS/Gab, FRS proteins) and immune inhibitory receptors (such as FcγRIIB (CD32) and KIR) that contain an ITIM motif).[51] The C-termini of SHP-1 and -2 contain two phosphorylation sites which are targets of the PDGF receptor or cytoplasmic kinases such as JAK or TYK2. When phosphorylated these tyrosines act as docking sites for other SH2 or PTB domain containing proteins (e.g. Grb2, see Figure 21.11).

There are two vertebrate SHP genes, SHP-1 and SHP-2, and invertebrate orthologues such as Csw (corkscrew) in *Drosophila* and Ptp-2 in *C. elegans*.[52] SHP-1 is mainly (but not exclusively) expressed in haematopoietic cells; SHP-2 is ubiquitously expressed. Despite the general similarity of SHP-1 and SHP-2, their functions are quite distinct. Whereas SHP-1 is generally inhibitory, SHP-2 acts as a stimulator of the Ras–MAPK pathway.

Regulation of SHP-1 and -2

Structural analysis of SHP-1 has revealed that part of the N-terminal SH2 domain is wedged into the catalytic domain (Figure 21.10). This simultaneously obstructs the catalytic cleft and distorts the phosphotyrosine binding site. In this state activity is low. Binding of this SH2 domain to a pY-containing peptide can relieve some of the intramolecular inhibition, but full activity is only achieved when both SH2 domains are engaged.[53] This occurs when SHP-1 (or-2) binds to tandem phosphorylation sites on phosphorylated tyrosine kinase receptors, some of their substrates or tyrosine phosphorylated cytokine receptors[54] (Figure 21.10).

SHP-1, JAKs, and STAT5

A role for SHP-1 was first indicated in mice having a somewhat motheaten appearance with patches of inflamed skin and loss of hair.[56] This phenotype, arising from the absence of SHP-1, is due to a systemic autoimmune condition caused by abnormalities in cells of haematopoietic lineage and it is marked by a general overexpansion of cell numbers. The differentiation and functions of natural killer (NK) cells and erythroid cells are also adversely affected. There is an accumulation of macrophages and neutrophils in the lungs and the mice die within weeks of birth. Two naturally occurring point mutations are at the origin of the motheaten phenotype. The motheaten allele (*me*) generates an

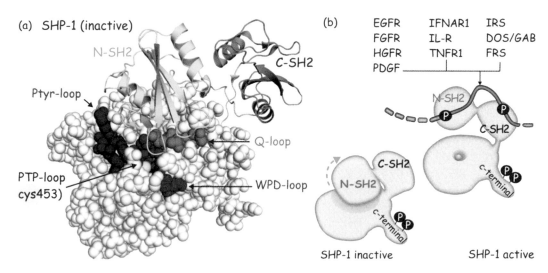

FIG 21.10 Autoinhibition of SHP tyrosine phosphatases. (a) The most N-terminal SH2 domain of SHP-1 wedges into the catalytic cleft of the tyrosine phosphatase domain. One of the loops of N-SH2 contacts the PTP-loop region to obscure the pocket that leads to the catalytic Cys residue (yellow). (b) When the SH2 domains are bound to tyrosine phosphorylated proteins, the inhibitory restraint on SHP is relieved. Engagement of both SH2 domains induces maximal activity. Examples include receptors (EGFR, FGFR, HGFR, PDGFR, ILR, TNFR1) and the adaptors Dos/GAB and IRS (2b3o[55]).

early frameshift and consequently homozygous *me/me* mice are protein-null (lethal after weeks). The viable motheaten allele (*me*^{v}) encodes two aberrant SHP-1 proteins that possess 20% of the normal enzyme activity (lethal after months) (reviewed in Neel et al.[52]). On this basis, SHP-1 can be regarded as a negative regulator of signalling.

Biochemical and functional studies established this role in a wide range of signalling pathways downstream of receptors for ligands that include EPO, prolactin, CSF-1, EGF, BCR, and TCR (reviewed in Neel et al.[52] and Zhang et al.[57]). There are multiple ways in which SHP-1 interferes, but the common theme is that the phosphatase is invariably recruited into signalling complexes through its SH2 domains. This brings the SHP-1 into the proximity of the phosphotyrosyl groups of receptors, recruited cytoplasmic kinases, or multiphosphorylated docking proteins. Moreover, SHP-1 itself can be phosphorylated on its C-terminal tyrosine residues, which converts it into an adaptor. Figure 21.11 illustrates the diverse actions of SHP-1.

SHP-1 also enhances the sensitivity of cells to apoptosis (Figure 21.12). Neutrophils from mice possessing just one *me* allele (*me*$^{+/-}$) are somewhat resistant to apoptosis and this might contribute to the increased number of white blood cells in the motheaten phenotype. SHP-1 binds to the receptors Fas and TNF-α, both of which contain a conserved cytoplasmic phosphorylation motif that is recognized by SHP-1, causing its activation. Activated SHP-1 then sensitizes the cell for apoptosis, possibly by counteracting the protective activity of the kinase Lyn or TYK2.[58]

Lastly, β1- and β2-integrin-mediated adherence to the extracellular matrix is enhanced in bone-marrow macrophages deficient in SHP-1. These cells also show enhanced uptake of opsonized red blood cells, which in part involves the β2-integrins (Mac-1), and it is therefore suggested that SHP-1 negatively regulates 'inside-out' and 'outside-in' integrin signalling. Again, the targets of SHP-1 remain unclear.

SHP-2 and the Ras–MAP kinase pathway

Targeted disruption of the gene encoding SHP-2 in mice causes early embryonic death, but other than concluding that it is important, little can be learned from this. Early indications about a possible role of SHP-2 in cell signalling came from three lines of evidence.

First, the gene is homologous to *Drosophila* Corkscrew (Csw). The Corkscrew pathway directs the differentiation of the terminal, non-segmented regions of the fly embryo.[59] Csw is involved in activation of the serine/threonine kinase Draf (equivalent to mammalian Raf). Draf is activated through the Dras pathway and this implies that the phosphatase provides a positive signal downstream of receptor tyrosine kinase activation. Molecular and genetic analyses indicate that the Dos protein is a likely substrate.[60] This contains a

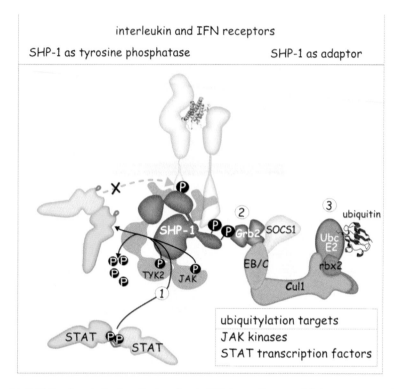

FIG 21.11 SHP-1 mediated inhibition of receptor functions. Direct recruitment of SHP-1 to cytokine or growth factor receptors causes dephosphorylation in the activation segment and inactivation of JAK. This is followed by dephosphorylation of STAT transcription factors and of the receptor (1). Tyrosine phosphorylated SHP-1 can also act as an adaptor to recruit Grb2/SOCS1 into the receptor signalling complex (2). SOCS1, in turn, recruits an E3-ubiquitin ligase complex to the cytokine receptor (3) and sets in train the ubiquitylation and subsequent destruction of JAK and STAT.

FIG 21.12 SHP-1 in sensitization of neutrophils to apoptosis. SHP-1 binds to tyrosine phosphorylated Fas or TNF-α receptors and then interferes with Lyn- and TYK2-mediated tyrosine phosphorylation of downstream effectors. In this way it blocks survival signals and renders neutrophils more sensitive to apoptosis.

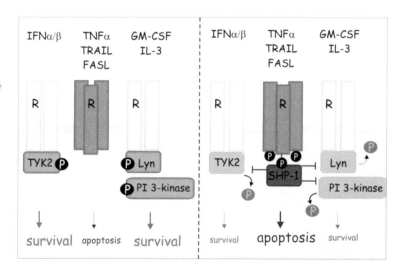

PH domain which permits membrane association, and numerous tyrosine phosphorylation sites through which it could act as a membrane-bound docking protein involved in relaying signals from Sevenless (an insulin-like rPTK) to Ras1 (see page 327).

Gab1, a human protein that resembles Dos, is also associated with SHP-2. In SHP-2$^{-/-}$ embryonic stem cells there is an enhancing effect of SHP-2 on EGF signalling which requires the concerted action of Gab1 and an unidentified tyrosine phosphorylated protein p90 (Figure 21.13).[61]

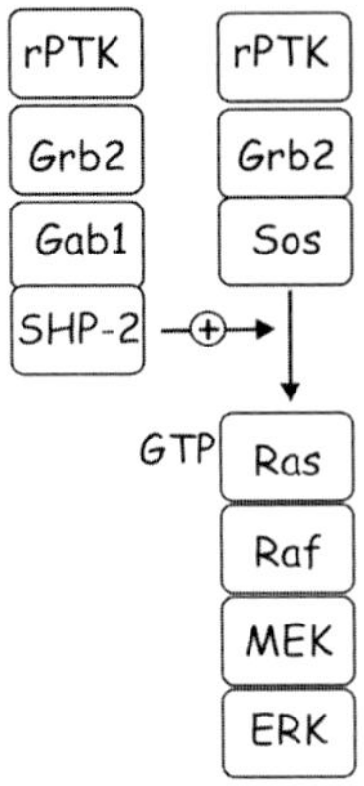

Secondly, expression of a mutant form of SHP-2 lacking the catalytic domain in *Xenopus* embryos causes tail truncations and prevents animal cap elongation (*short toad*, composed mainly of head structures).[62] These processes are determined in part by the receptor for FGF which also signals through the MAP kinase pathway and so the phosphatase SHP-2 must act between the FGF-R and the activation of MAP kinase. Similarly, fibroblasts expressing the catalytically inactive form of SHP-2 fail to activate MAP kinase in response to FGF, PDGF or insulin-like growth factor (IGF).[63]

Thirdly, in mice there is an essential role for SHP-2 in limb development and in formation of the branchial (pharyngeal) arch, two other pathways controlled by FGFR signalling.[64]

FIG 21.13 SHP-2, its role in activation of the Ras–MAP kinase pathway. Genetic and biochemical analyses place SHP-2 between Sos and Ras. How it stimulates GTP loading of Ras and so the MAP kinase pathway remains unknown.

Insight through the Noonan syndrome

Missense mutations in the SHP-2 gene have been identified as the underlying cause of Noonan syndrome. Most of the altered amino acid residues are located in or around the interacting surfaces of the N-terminal SH2 domain and the catalytic domain, and are predicted to relieve the intramolecular inhibition caused by binding of the SH2 domain to the catalytic domain.[65] In some cases there are mutations in residues present in the C-terminal SH2 domain and in the peptide linking the two SH2 domains. However, there is more to the Noonan story because there is no absolute correlation between basal activities of the SHP-2-mutants and the disease they induce. Some mutations, even those associated with severe symptoms, are without effect on the basal activity or on phosphotyrosine-mediated activation.[66]

Gain-of-function mutations in SHP-2 are linked to some leukaemias.[70] This is perhaps not surprising, given its stimulatory role in the Ras–MAP kinase pathway. Despite some uncertainty with respect to the exact mechanism of cell transformation, SHP-2 mutants provided the first example of a gain-of-function mutation in a PTP that is the underlying cause of a human disease.

Density enhanced PTP (DEP1)

For DEP1, matters are just a bit clearer. This phosphatase was originally discovered by probing a HeLa cell cDNA library with a PCR fragment encoding the catalytic domain of a putative PTP. The receptor-type PTP isolated is

Noonan syndrome is an autosomal dominant disorder characterized by short stature and multiple developmental abnormalities including congenital heart and skeletal malformations.[67,68] The incidence of this syndrome is high – 1 in 1000–2500 live births – and in a survey of more than 100 unrelated cases about 50% displayed a mutation in SHP-2. There is an even higher incidence of mutations among familial cases.[69] All these are missense mutations.

weakly expressed in non-confluent cultures of lung or foreskin fibroblasts, but is highly expressed in dense cultures.[71] Later, genome analysis revealed that one of the mouse genes for susceptibility to colorectal cancer encodes a receptor tyrosine phosphatase, PTPRJ. Loss of heterozygosity of the human gene (DEP1) was noted in 19 out of 39 colorectal adenocarcinomas.[72] DEP1 was the first tyrosine-specific PTP to be assigned a convincing role as a tumour suppressor relevant to the development of human cancers (the dual specificity phosphatase PTEN is also a tumour suppressor: see page 668). The receptor-PTK Met, which is up-regulated in several human tumours, is a substrate of DEP1, so raising the possibility of a functional interaction between this phosphatase–kinase pairing in the progression of cancer.

CD45 and the regulation of immune cell function

The CD45 tyrosine phosphatase, expressed in all nucleated cells of haematopoietic origin, first alerted the attention of immunologists because of its great abundance (5–10% of membrane proteins). It is a transmembrane tyrosine protein phosphatase having a receptor-like extracellular domain,[73] but as with many other receptor-like PTPs, no specific activating ligand has been identified. Its importance became evident with the discovery that T lymphocytes lacking this antigen fail to become activated through the TCR, although they respond quite normally to stimulation through the IL-2 receptor.[74–76] A total absence of surface CD45 causes severe combined immunodeficiency disease (SCID).[77] Conversely, transgenic mice bearing an activating mutation in CD45 display lymphoproliferation, autoantibody production (and severe nephritis).[78]

Normally, engagement of the TCR by an antigen-presenting MHC results in phosphorylation of the three ITAM motifs in the ζ chains of CD3 (part of the TCR complex). This occurs through the intervention of the kinase Lck, bound to either CD4 or CD8 (Figure 21.14 and Figure 17.2, page 516; also see page 515). The fully phosphorylated ζ chains then serve as docking sites for the formation of a signalling complex that transmits the signal into the cell. In cells lacking active CD45, Lck remains inactive and the subsequent cascade of tyrosine phosphorylations does not occur.

So where does the phosphatase activity of CD45 come into all this? Normally Lck activity is suppressed due to autoinhibition arising from the intramolecular interaction of its own SH2 domain with a phosphotyrosine residue situated in its C-terminus ("Pc" in Figure 21.14, the 'closed conformation'). Lck is activated first through dephosphorylation by CD45 which exposes the kinase active site, and then through autophosphorylation in the activation segment.[79] The fully active Lck phosphorylates the ITAM motifs of the ζ chains but the process only becomes efficient when engagement of the TCR by MHC peptides brings the catalytically competent Lck, bound to CD4 or CD8, into closer juxtaposition with the CD3 ζ chains.

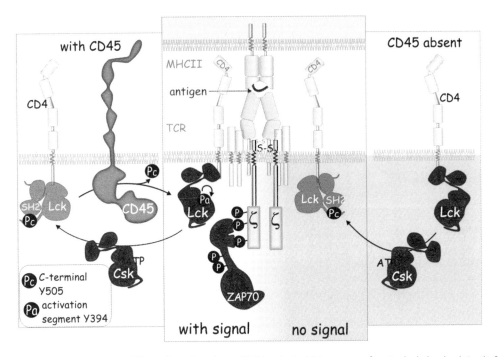

FIG 21.14 Role of CD45 in T cell activation. TCR signalling is dependent on CD45. It maintains Lck in an open configuration by dephosphorylating the C-terminal tyrosine residue (Pc). Subsequent autophosphorylation in the activation segment (Pa) renders Lck catalytically competent. The kinase Csk does the opposite. It phosphorylates Lck at Pc which makes contact with the SH2 domain, causing it to assume a compact configuration. In the absence of CD45, TCR signalling is prevented. For details of the TCR, see Figure 17.2, page 516.

Similar mechanisms of activation through dephosphorylation apply in the action of Fyn (associated with the ε-chain of CD3) and Src.[80] From this, it appears that CD45 maintains a pool of Lck, associated with CD4/CD8, in a competent state (but away from substrate). In this way, engagement of the MHC-peptide with the TCR leads to Lck-mediated phosphorylation of the CD3/TCR-ζ chains and thereby initiates TCR signal transduction.[81] The localization of CD45 with respect to the major transmembrane signalling proteins in an immunological synapse has been visualized by laser confocal microscopy, and some images are shown in Figure 21.15.

In addition to Fyn and Lck, the Janus kinases (JAK) are also substrates of CD45.[83] The JAKs are phosphorylated and activated following cytokine receptor ligation (IFN-α, IL-3) and in turn phosphorylate the cytoplasmic tail of these receptors to create docking sites for proteins bearing an SH2 domain. Among these are the transcription factor STAT and Src family kinases. Tyrosine phosphorylation of STAT causes homo- or heterodimerization which induces translocation to the nucleus. Bone marrow-derived mast cells

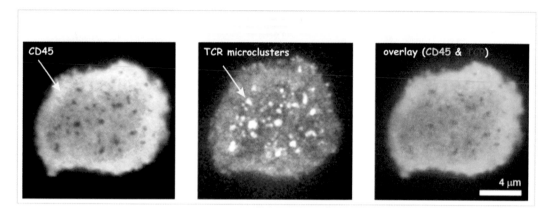

FIG 21.15 Localization of CD45 and T cell receptors in the regulation of T cell activation. The fluorescent images are of T cells that have been placed on a lipid monolayer doped with MHCII antigen and ICAM-1. They form adhesive structures called microclusters. While CD45 is necessary for the supply of primed Lck (Figure 21.14), it is excluded from the TCR microclusters. This avoids down-regulation of the early tyrosine phosphorylation-based signalling events. The lack of overlap in the fluorescence signals from CD45 (green) and the TCRs (red) indicates their segregation. Only later (after ~20 minutes), when so-called central supramolecular activation clusters are formed, is CD45 allowed to participate again and to terminate the activation process. From Varma et al.[82]

lacking CD45 respond more vigorously to IL-3 with respect to JAK2 activation, phosphorylation of STAT3 and 5 and proliferation.

Even now, the regulation of CD45 itself remains enigmatic. No ligand has been identified and structural analysis has failed to indicate how the phosphatase activity might be regulated

Regulating receptor PTPs

In direct contrast to the kinases, it appears that dimerization may the off-switch for the receptor tyrosine phosphatases. In order to test this idea, cells have been generated that express a chimaeric CD45 molecule that comprises its intracellular phosphatase-containing segment coupled to the extracellular domain of the EGF receptor. Application of EGF, expected to dimerize the receptor, is then found to switch off the phosphatase activity.[84] By analogy to RPTPα[85] and LAR (PTPRF), it was proposed that CD45 might form a symmetrical protein dimer in which the catalytic site of one component is blocked by a wedge structure formed by its partner (Figure 21.16). Separation of the two, exposing the catalytic site, would constitute the activation mechanism. In favour of this idea, the wedge region in CD45 is conserved at the sequence level and 'knock-in' mice carrying a glutamate to arginine mutation in the wedge are subject to enhanced lymphoproliferation.

Against this however, a recent structural analysis of CD45 has provided no evidence of dimerization, nor did it reveal the presence of the proposed

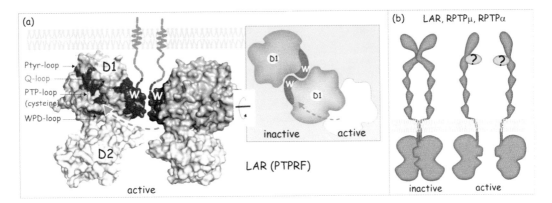

FIG 21.16 Regulation of receptor tyrosine phosphatases. Dimerization is the turn-off signal. Inhibition is thought to occur through reciprocal interaction of the two wedge structures (in red) with the catalytic site (blue/yellow) which impedes substrate access. It is unclear whether or not ligands are involved in all this, and if they are it is unclear how they act to keep the receptors apart. (a) The structure of the D1 and D2 domains of LAR (PTPRF). D1 is catalytically active, D2 is a pseudo phosphatase domain (inactive). Characteristic features of tyrosine phosphatases are indicated (Ptyr-loop, Q-loop, WPD-loop and the PTP-loop, which harbours the catalytic cysteine residue (in yellow). (b) Possible action of a hypothetical ligand of the LAR/RPTP phosphatases.(1lar[86]).

wedge. The catalytic site of D1 (membrane proximal catalytic domain, see page 644) appears to be unobstructed. This study did, however, indicate the importance of the catalytically inactive D2 domain (highly conserved within the tyrosine phosphatase family and across species). It concerns two flexible loop structures, acid and basic. Removal of the acidic loop diminishes CD3-mediated activation of T cells. Moreover, phosphorylation of serine residues in the acidic loop by casein kinase 2 appears to play a role in CD45 function. From this we conclude that D2 serves an important purpose, though what it might be remains far from clear.

Dual specificity phosphatases

The dual-specificity phosphatases (DSP) form a subfamily of the tyrosine protein phosphatases. Under *in vitro* conditions, these enzymes catalyse the dephosphorylation of all three phosphoamino acids, pT, pS, and pY. Examples include MKPs, VHR, KAP, CDC25, and PTEN. Although sharing no overall sequence similarity between themselves, nor with the tyrosine-specific phosphatases, the presence of the conserved PTP signature motif (I/V)HC-x_5-R) indicates that they are members of this wider family. The catalytic domains of VHR and KAP possess core structural features similar to the PTPs but appear to be truncated versions. Importantly, they lack the phosphotyrosine-binding loop and consequently the active site cleft has a depth sufficient

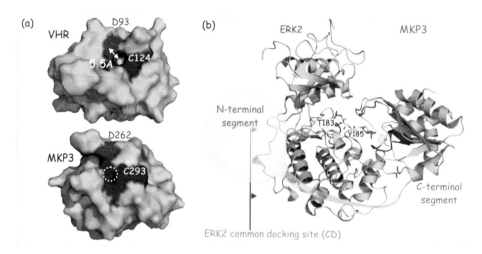

FIG 21.17 Structure of the dual specificity phosphatases VHR and MKP3. (a) Surface representation of VHR and MKP3. The PTP loop is dark blue and the aspartates D93 and D262, important for the dephosphorylation reaction, are red. The catalytic pocket measures 5.5 Å. The catalytic site cysteine in VHR is yellow. In MKP3 it is not exposed to the surface (no yellow spot visible). It appears that the binding of ERK alters the active site of MKP3 so as to expose the catalytic cysteine (C293). See also **Figure 21.5**. (b) Model illustrating the interaction of ERK2 with MKP3. The N-terminal segment of the phosphatase binds to the 'back' of ERK2 whereas the C-terminal segment binds to Y185 in the activation segment. The two segments are connected by a hinge region that folds around the C-lobe of the kinase. (1vhr,[87] 2erk,[88] and 1mkp[89]).

to accommodate both pY and pT (Figure 21.17). Numerous dual specificity phosphatases have been characterized, they are subdivided in four different groups, each comprising various subfamilies (Figure 21.18).

Regulation of MAP kinases by dual-specificity protein phosphatases (DS-MKP)

Members of the MAP kinase family are phosphorylated in the TEY sequence by dual-specific kinases such as the MEKs (see page 334). They are dephosphorylated by the MAP kinase phosphatases (MKP), of which some are serine/threonine phosphatases (PP2A and PP2C) while others are tyrosine phosphatases (PCPTP1, STEP and HePTP).[91,92] Yet others are DSPs (DS-MKPs). Ascertaining the roles of the specific MKPs in vertebrates has been hindered by the great diversity of phosphatase enzymes that inactivate MAP kinases and also their functional redundancy. Rapid inactivation of ERK1 in PC12 cells reflects, in part, dephosphorylation of phosphothreonine,[93] whereas in other systems, such as rat-1 fibroblasts, a dominant role for members of the PTP family is evident.[94]

To date, 13 human MKPs have been identified[95] (Table 21.2). They are structurally and functionally distinct and are grouped into four categories.

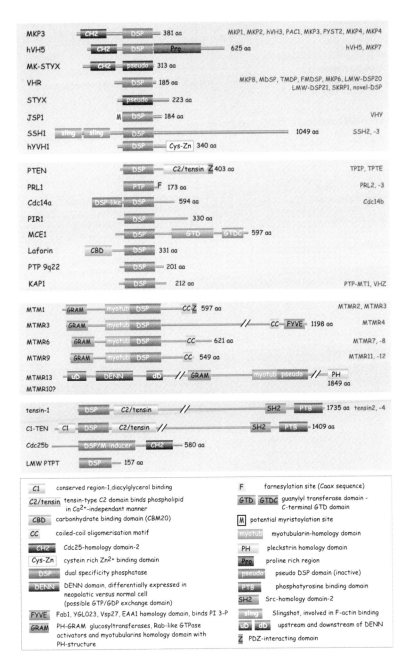

FIG 21.18 Domain architecture of the dual specificity phosphatases. Although characterized by the presence of a common signature motif, the DSPs are more diverse than the specific PTPs. As a consequence, numerous small subfamilies exist that possess a truncated version of the catalytic domain. Among other differences, this results in a shallower catalytic pocket. Both catalytically competent and pseudo phosphatases have been identified. Some accessory domains as well as interaction sites or post-translational modification sites are highlighted. Protein names and gene symbols do not always match. From Tonks.[90]

TABLE 21.2 MAP kinase phosphatases: Four categories of MAP kinase phosphatases are recognized. The first two comprise dual-specificity phosphatases of which members of the first group are directly induced by MAP kinases (immediate early genes) and are localized mainly in the nucleus. The second group are not induced by MAP kinases and act in the cytoplasm. The third category comprises members of the tyrosine phosphatase family (PTP) and the fourth comprises the S/T phosphatases PP2A and PP2C. Not all members of the MAP kinase family are equally susceptible as substrates; preferences are indicated for some.

Dual specificity phosphatases	
MKP-1 (CL100)	nuclear
MKP-2 (hVH2)	immediate early genes
hVH3 (B23)	growth factor or stress induced
PAC1	growth factor or stress induced
MKP-3 (Pyst1)	cytosolic
MKP-X (Pyst2)	non-immediate early genes
MKP-4	
MKP-5	
Tyrosine phosphatases	
PCPTP1	neuronal
STEP	targets ERK 1/2 & p38, not JNK
HePTP	lymphoid
	targets ERK 1/2 & p38, not JNK
Serine/threonine phosphatases	
PP2A	targets all MAP kinases
PP2Cα	p38 & JNK

Dephosphorylation of both pY and pT is believed to occur by a similar mechanism as that described for PTP1B (page 647).

A remarkable degree of selectivity between substrate and phosphatase is achieved through the interaction of MKP-3 to a common docking domain (CD) adjacent to the catalytic domain on ERK2 (see Figure 21.17). Binding of MKP-3 then causes rearrangement and activation of the catalytic site.[96] Importantly, catalytic activation of MKP-3 mirrors the selectivity of the docking domain, as the different members, JNK and p38, are neither able to bind nor to increase its catalytic activity. The docking site on ERK2 is also its point of interaction with the upstream kinase MEK and the downstream transcription factor Elk. Thus, all these components of the MAPK pathway compete for the same binding site.[97] It is not certain whether this mechanism of activation applies to all DS-MKPs.

Physiological role of the dual-specificity MAP kinase phosphatases

The archetypal dual-specificity phosphatase is the *VH1* gene of vaccinia virus.[5] A mammalian homologue of *VH1* was discovered as a product of a growth factor-induced early response gene.[98,99] Biochemical analysis revealed it to be a MAP kinase phosphatase (MKP-1).

Addition of serum to quiescent cells induces rapid but transient phosphorylation and activation of ERK2 (MAPK-1). This is sufficient to induce gene transcription and to promote entry into the G1 phase of the cell cycle.[100] The activity of MKP-1 is expressed within 20 min and coincides with the dephosphorylation of the pTEpY motif in the activation segment followed by the deactivation of the kinase. If synthesis of MKP-1 is prevented, or if its catalytic activity is suppressed by mutation, the duration of the activated state of ERK2 is greatly extended. Conversely, expression of a constitutively activated form of MKP-1 blocks G1 specific transcription and entry into S-phase.[101,102] Thus MKP-1 acts as a negative regulator of ERK2, serving to attenuate the growth factor signal (Figure 21.19). In addition, MKP-1 is a phosphorylation substrate of ERK2. When phosphorylated, it becomes less sensitive to ubiquitin-directed proteolysis and as a result, it accumulates.[103]

Despite all these promising results, this is certainly not the only pathway by which ERK2 is controlled because none of the mutations in MKP-1 predispose to tumour development. If this pathway were absolutely essential, then one might expect that a tumorigenic mutation would have emerged by now. Moreover, MKP-1-deficient mice appear to develop normally. Their fibroblasts respond normally with respect to the extent and timing of the expression of c-fos, indicating that the control of MAP kinase is unperturbed.[105] We now know that MKP-1 is more active against members of the p38 and JNK kinases[95] and that the ERKs are the preferred targets of MKP-3.[106]

Dual-specificity phosphatases in development

Most investigations of the dual-specificity MAP kinase phosphatases have focused on their roles in development. A particularly informative experiment, providing a direct demonstration of a physiological role for one of them in the regulation of MAP kinase activity, came from studies of dorsal closure during embryogenesis in *Drosophila*. In this process, lateral epithelial cells undergo cytoskeletal changes which enable them to spread out, providing

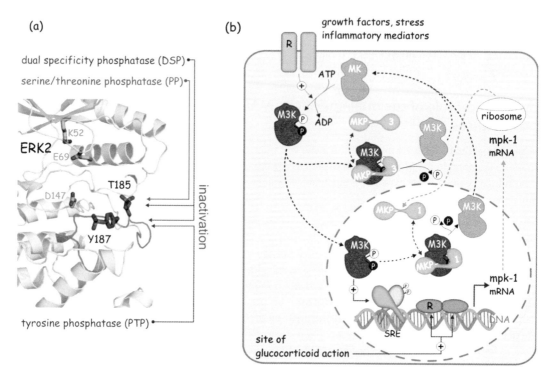

FIG 21.19 The role of MKP-3 in the regulation of receptor signalling. (a) All the phosphatases cause inactivation through dephosphorylation of one or both of the phosphoamino acid residues in the activation segment. Information from Farooq and Zhou.[95] (2erk[88]). (b) Another example of phosphatases acting as a reset button. The scheme shows how dual specificity phosphatases operate in receptor signalling. Receptors activate the MAP kinase pathway, resulting in the activation of M3K (members of the ERK, JNK, or p38 subfamilies). A portion of these operate in the cytoplasm until they are dephosphorylated and inactivated by MKP-3. Others transfer to the nucleus to induce gene expression through phosphorylation of transcription factors. The product of one of the immediate early genes, MKP-1, enters the nucleus to inactivate M3K. MKP-1 can be induced by the synthetic glucocorticoid dexamethasone. Dephosphorylated and inactivated M3K accumulates in the cytoplasm were it is 'recharged' by occupied cell surface receptors. Information from Camps et al.[104]

a cover for the dorsal region of the embryo[107] (Figure 21.20b). The *Drosophila* homologue of the mammalian JNK, Bsk, induces the expression of *puc*, which encodes a dual-specificity phosphatase and negatively regulates activity of Bsk. The activity of Puc must be regulated precisely. Lack of Puc results in a disordered border between the dorsal epithelial cells ('puckered' phenotype) (Figure 21.20a). In excess it induces a phenotype characterized by patches of uncovered dorsal openings.[108] Figure 21.20c shows the pathways that control dorsal closure.

The ERK-specific MKP-3 plays a similar role in the regulation of chicken limb outgrowth. This is initiated by FGF8 released from the epithelial cells

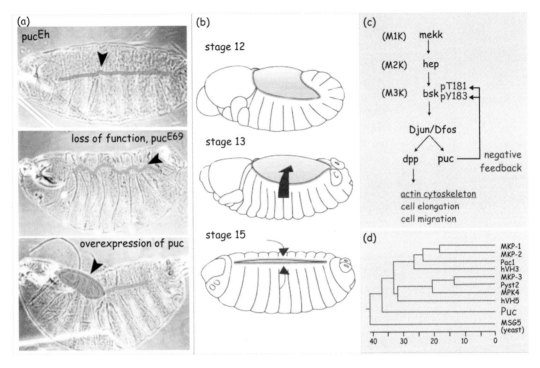

FIG 21.20 Dorsal closure in *Drosophila* and the role of the dual specificity phosphatase *puc*. (a) Phenotypes of *Drosophila* embryos arising as a consequence of altered or enhanced expression of the *puc* gene. The weak mutant *puc*^Eh yields an almost perfect seal between the two sides of the ectoderm. Loss of function (*puc*E69) causes a puckered phenotype, in which the cells of the two borders penetrate each other's territory. Over-expression prevents the leading edges reaching each other and the embryo reveals dorsal openings. (b) Stages in dorsal closure during *Drosophila* embryogenesis. The embryos are shown dorsal side up, with the anterior (future head region) to the left. At stage 12, a large part of the embryo is still covered by amnioserosa (pink). At stage 13, the ectoderm cell sheet extends towards the dorsal midline (red arrow). This movement takes about 2 h and is the consequence of the progressive flattening of cells which extend a leading edge towards the midline. At stage 15, the leading edges of both sides converge. The process of flattening and migration ceases. (c) The signal transduction pathway that controls dorsal closure. Positive control of flattening and migration occurs through induction of dpp (decapentaplegic, member of the TGFβ family of growth factors, see page 607) by a pathway involving *mekk* (M1K), *hep* (M2K), and *bsk* (M3K). When cells make contact, negative feedback, due to Puc-mediated dephosphorylation and inactivation of Bsk, arrests flattening and migration. (d) Phylogenetic tree analysis of *puc*, yeast MSG5, and human-dual specificity phosphatases. Image courtesy of Dr Martinez-Arias, Cambridge, UK. Embryonic stages according to Campos-Ortega and Hartenstein.[109]

that form the apical epidermal ridge. FGF8 causes proliferation of the underlying mesenchymal cells to form buds that extend into the limbs with differentiation of the mesenchymal cells into cartilage and then bone. MKP-3 is one of the genes up-regulated in mesenchymal cells by FGF8. Its inhibition of expression by siRNA, results in cell death in the mesenchyme.[110] In this system, MKP-3 expression is regulated by the PI 3-kinase–PKB pathway.

PTEN, a dual-specificity phosphatase for phosphatidyl inositol lipids

As indicated at the start of this chapter, interest in protein tyrosine phosphatases was sparked by the hope that they might act as tumour suppressors. However, of the many phosphatases investigated, only a few seem to exert such a role; PTEN is one of them. It provides the best example of a gene coding for a PTP in which loss of function is strongly linked with tumour development.[111,112] Somatic mutations, resulting in its inactivation, occur in multiple sporadic cancer types (figure 21.21c). Germline mutations are linked to inherited hamartoma and Cowden disease (see page 560). Thus it came as a surprise to find that it does so, not by controlling the level of tyrosine phosphorylation, but through its action as a phosphatase for 3-phosphoinositides ($PI(3,4)P_2$ and $PI(3,4,5)P_3$).[113] Although protein substrates are not totally excluded, (e.g. FAK and the adaptor protein Shc), the tumour-suppressive function of PTEN can be ascribed to its lipid phosphatase activity.[114]

The amino acid sequence of PTEN contains a tyrosine phosphatase signature motif and it displays the structural features of the tyrosine protein phosphatases (Figure 21.21a). In spite of this, it is generally a poor protein phosphatase. It catalyses the dephosphorylation of the phosphate from the 3-position of the phosphoinositide $PI(3,4,5)P_3$ with far greater alacrity.[113] Unlike the dedicated protein phosphatases, PTEN possesses positively charged amino acids (-CKAGKR-) in the PTP loop that may explain its ability to target

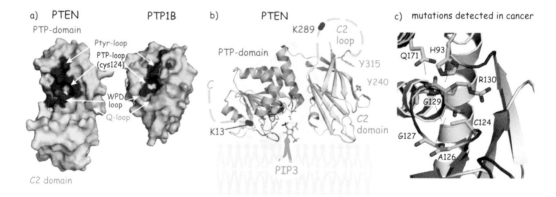

FIG 21.21 PTEN, a tyrosine phosphatase lookalike. (a) Molecular surface representations of PTEN and PTP1B reveal the characteristic features of the tyrosine phosphatases, with a PTP loop, containing the catalytic cysteine residue (yellow) and the pY-, WPD-, and Q-loops. PTEN contains an additional C2 domain. (b) PTEN attached to a phospholipid bilayer. The $PI(3,4,5)P_3$ head group protrudes into the catalytic pocket, close to the catalytic cysteine. The C2 domain (yellow) facilitates binding to the membrane. It has two tyrosine residues, Y240 and Y315, which are phosphorylated by Src. It also contains an unstructured loop of about 23 residues (hatched yellow line) harbouring lysine K289, which is ubiquitylated by NEDD4. Another lysine, K13, at the C-terminus of the protein is also ubiquitylated by this E3-ligase. (1d5r,[115] 1ptt[10]). (c) Inactivation mutations detected in cancer.

3-phosphoinositides [115]. The phosphatase and C2 domain associate across an extensive interface, suggesting that the C2 domain may serve to locate the catalytic domain at appropriate sites on the membrane (Figure 21.21b).

With a 3-phosphoinositide as its preferred substrate, PTEN is at the head of two well-characterized signalling pathways, the one determining cell survival (signals deriving from focal adhesion complexes), the other leading to cell proliferation (signals from receptor protein tyrosine kinases) (see pages 400 and 554). In essence, PTEN serves to keep 3-phosphoinositides at a low level. Unlike other players in the PI 3-kinase pathway, there is a lack of functional redundancy for PTEN and this may be the underlying reason for its prominence as a tumour suppressor. As a result of PTEN activity, the recruitment and activation of PH domain-containing enzymes such as PKB and PDK1 are held in check.[116]

Multiple factors have been found to influence the activity of PTEN in experimental *in vitro* systems, but just how they affect the tumour-suppressive activity of PTEN remains unclear.[117] Worth mentioning is the regulation by Src which reduces the half-life and the membrane association of the enzyme, and it is possible that this may be involved in Src-mediated cell transformation (Figure 21.22). Its relevance was demonstrated by the observation that herceptin, a humanized monoclonal antibody that targets the ErbB2 receptor and which can suppress some malignant breast tumours (see page 743) reduces Src-mediated PTEN phosphorylation so enhancing its membrane association and phosphatase activity.[118]

Another pertinent example is the regulation of PTEN by *DJ-1*, a gene, that emerged from a gain-of-function screen of PTEN-signalling in *Drosophila*.[119] The mechanism remains to be elucidated (Figure 21.22). Defects in *DJ-1* have also been noted in autosomal recessive early-onset Parkinson's disease. *DJ-1* also plays a role in fertilization, the regulation of androgen receptor signalling, and cell handling of oxidative stress. It is thus possible that it facilitates the redox regulation of PTEN (see page 650).

With respect to the suppression of cell transformation, the action of PTEN at the cell membrane is perhaps only half the story. Recent studies have demonstrated a role for PTEN in the nucleus (Figure 21.23). Nuclear translocation requires ubiquitylation in a process driven by NEDD4-1, a HECT-type E3-ubiquitin ligase (see Figures 15.12, page 468 and 15.14, page 470). Monoubiquitylated PTEN is recognized by the nuclear import machinery (comprising importin and RanGTP) and enters the nucleus[121] where it affects expression of *RAD51*, involved in the prevention of DNA double-strand breaks. Also it associates directly with CENP-C, an integral component of the kinetochore, and helps to stabilize centromeres, necessary for accurate segregation of chromosomes during mitosis. Together these actions are said to safeguard chromosome integrity and thus prevent genetic anomalies.[122] Whether or not PTEN also

DJ-1 (named after two students, Daisuke Nagakubo and Junko Maeda), was originally cloned as a putative oncogene capable of transforming NIH-3T3 fibroblasts in cooperation with *ras*.[120] Several lines of evidence indicate that *DJ-1* plays a role in human tumorigenesis: breast cancer patients have elevated levels of circulating DJ-1, cellular protein levels are increased in primary non-small cell lung carcinoma samples, and its expression is decreased in human lung cancer cells treated with the chemotherapy drugs paclitaxel (metaphase inhibitor) and U0216 (MEK inhibitor).

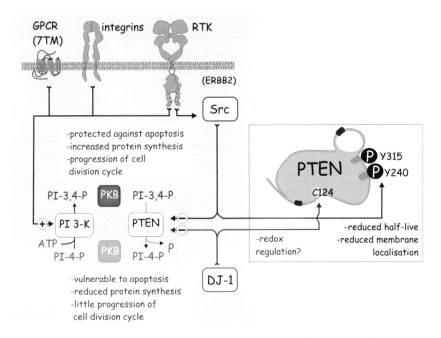

FIG 21.22 Regulation of PTEN. This scheme represents a simplified version of the life of a cell in which its fate is determined by PKB. High activity protects against apoptosis, promotes protein synthesis and progression of the cell cycle. Low activity prevents growth and renders the cell vulnerable to apoptosis. Signalling from 7TM receptors, tyrosine kinase receptors, and integrins results in activation of PI 3-kinase with activation of PKB. ErbB2 (EGF receptor) causes inhibition of PTEN, in a pathway that involves Src, promoting its destruction thereby reinforcing the PI 3-kinase signal. PTEN can also be inhibited by DJ-1 (possibly through redox regulation).

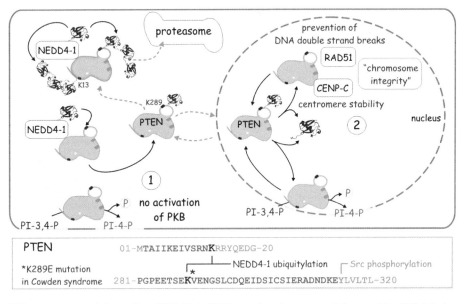

FIG 21.23 PTEN: action at two sites. In the cytoplasm, PTEN holds the PI 3-kinase pathway in check, so regulating the activity of PKB. Following mono-ubiquitylation by the HECT-type E3-ligase NEDD4-1, PTEN enters the nucleus and then after de-ubiquitylation it acts to up-regulate the expression of RAD51 which prevents double-strand breaks of DNA. It also associates with CENP-C, a kinetochore component, and together they assure centromere stability. It is not clear if PTEN affects the activation state of PKB in the nucleus. Deletion of lys289, as in Cowden disease, prevents nuclear localization.

regulates the activity of PKB in the nucleus is not yet clear. The importance of all this has been highlighted by the finding that the mutation K289E occurs in some cases of Cowden syndrome, and as a consequence PTEN fails to enter the nucleus. Not only this, but failure of nuclear localization also renders PTEN more vulnerable to polyubiquitylation, resulting in its recognition by the proteasome and leading to its destruction (Figure 21.23).

Serine/threonine phosphatases

While all the PTPs are monomeric and their differences are set by their domain structures, the majority of the serine/threonine phosphatases are oligomers and characterized by their association with targeting subunits. These direct them to particular locations, so restricting their action to a limited range of substrates. The first serine/threonine phosphatase recognized, PP1c, inactivates glycogen phosphorylase.[123,124] Other serine/threonine phosphatases, initially classified into four groups PP1, PP2A, PP2B, and PP2C, have since been identified (Table 21.3). Because they are relatively non-specific in their action, it was thought that this small number of enzymes would be sufficient to balance the effects of the numerous kinases.[125] Some functional diversity then became apparent with the discovery of inhibitors

TABLE 21.3 Classification of the S/T phosphatase catalytic (C) subunits

Name	Gene name	
PPP family		
PP1α, -β, -γ, -δ	*PPP1CA, PPP1CB, PPP1CC, PPP1CD*	PP1c is the catalytic subunit of PP1
PP2Aα, -β	*PPP2CA, PPP2CB*	plus subunit A-a or A-b
PP2Bα,-β, -γ	*PPP3CA, PPP3CB, PPP3CC*	plus subunit calcineurin B (and calmodulin)
PP4	*PPP4C*	PPP4R1
PP5 (Ppt1)	*PPP5C*	
PP6	*PPP6C*	
PPM family		
PP2Cα,-β	*PPM1A, PPM1B, PPM1D, PPM1F, PPM1G, PPM1K, PPM1L, PPM1M, ILKAP*	monomer, Mg^{2+}-dependent
PDP	*PPM2C*	mitochondrial pyruvate dehydrogenase phosphatase

The three-letter abbreviations adhere to the conventions of the human genome nomenclature in the specification of a family. PPP indicates PhosphoProtein Phosphatase, PPM activation by magnesium.

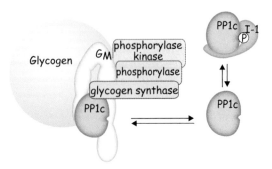

FIG 21.24 Serine/threonine phosphatases, targeting subunits, and inhibitor proteins. PP1c, involved in glycogen metabolism, never exists in a free soluble state. It is bound either to its regulatory subunit G_M, associated with glycogen (forming the enzyme complex PP1G), or it is bound to the inhibitory subunit I-1, which prevents uncontrolled dephosphorylation of other substrates.

that affect a restricted range of phosphatases. For example, ciclosporin A, widely used as an immunosuppressive agent after organ transplantation, is a selective inhibitor of PP2B (calcineurin) (page 683), and okadaic acid, a tumour promoter, is a potent inhibitor of PP1 and PP2A (see page 679).

Real diversity came to light when it was realized that the purified activities represent only the catalytic subunits, but that in the cellular context they are coupled with targeting or regulatory subunits. The subcellular distributions, substrate selectivities, and catalytic activities are largely determined by these regulatory subunits. For example, PP1 is active against a wide range of peptide and protein substrates and so its activity must be carefully limited. Under normal conditions, it is coupled with a G subunit (G_M in the case of muscle) which has very high affinity for glycogen (K_{app} ~6 nM) (Figure 21.24). This ensures that its free form is kept at vanishingly low levels. In the situation of severe glycogen depletion or upon detachment due to phosphorylation of G_M, soluble inhibitor proteins (inhibitor-1) ensure that its concentration in the cytosol remains low. In these ways, PP1 is prevented from acting as a loose cannon, randomly dephosphorylating any phosphoprotein that might come in range. The association with particular targeting proteins narrows the range of available substrates. The other enzymes of glycogen metabolism – phosphorylase kinase, phosphorylase, and glycogen synthase – are also tightly bound to glycogen and all three are good substrates for dephosphorylation by PP1c/G_M.

Classification of the serine/threonine phosphatases

The serine/threonine phosphatases are classified in two superfamilies, PPP and PPM, listed in Table 21.3.[126,127] The domain architectures of some representative members of these families are shown in Figure 21.25. The PPM

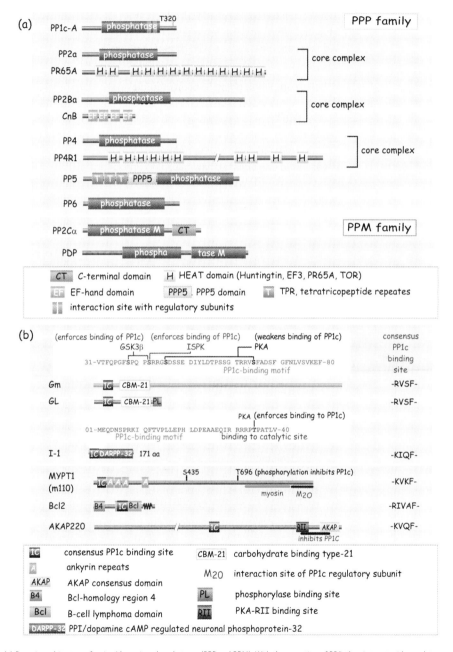

FIG 21.25 (a) Domain architecture of serine/threonine phosphatases (PPP and PPM). With the exception of PP5, they interact with regulatory or targeting subunits. (b) Domain architecture of regulatory subunits discussed in this chapter.

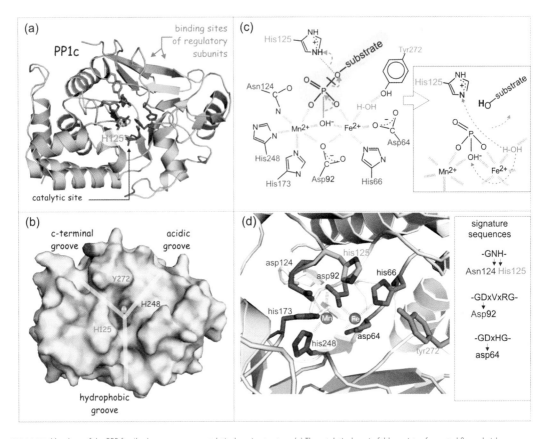

FIG 21.26 Members of the PPP family share a common catalytic domain structure. (a) The catalytic domain fold consists of a central β-sandwich surrounded on one side by seven α-helices and on the other by a subdomain comprising three α-helices and a three-stranded β-sheet. (b) The catalytic cleft has a Y-shape, with three branches commonly referred to as hydrophobic, acidic and C-terminal grooves. (c) Crystallographic data provide compelling evidence for the role of two metal ions in the catalytic reaction. Most data are consistent with a single step mechanism, employing a metal-activated nucleophilic water molecule or hydroxide ion. (d) The metal coordinating residues (asparagines, aspartates and histidines) are invariant amongst the PPP family members. The position of some of the amino-acids of the signature sequences are shown. (1jk7[134]).

family includes the Mg^{2+}-dependent PP2C and mitochondrial pyruvate dehydrogenase phosphatase (PDP). The PPP family is characterized by the presence of three invariant amino acid motifs in the catalytic domain and two metal ions required for their activity, though their identity remains controversial, possibly $Mn^{2+} + Mn^{2+}$ or $Mn^{2+} + Fe^{2+}$ (Figure 21.26). The human genome contains 40 genes encoding the catalytic subunits of the PPP family. Functional diversity is expanded by the expression of splice variants and the existence of a large number of targeting subunits. For instance, PP1c Table 21.4, Figure 21.25 forms complexes with 45 regulatory subunits (*bona fide* and putative).

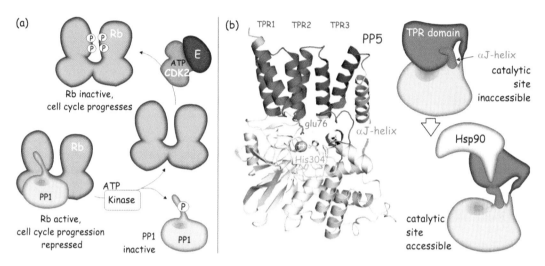

FIG 21.27 Regulation of serine/threonine phosphatases through phosphorylation or intramolecular domain interaction. (a) Phosphorylation of the C-terminal region of PP1c hinders its catalytic activity towards the retinoblastoma protein (Rb). It is suggested that phosphorylation breaches the interaction between the two proteins though other mechanisms could apply. Lack of phosphatase action leads to effective phosphorylation of Rb by CDK2/CyclinE and allows the cell cycle to proceed. (b) PP5 is one of the serine/threonine phosphatases that is not subject to regulation by subunits. Instead, its inactive state is maintained by its TPR domain that wedges into the catalytic site. Glu76 (E76) and the hydrophobic αJ-helix are highlighted because they stabilize the autoinhibited state. The catalytic His304, adjacent to the Mn^{2+}/Fe^{2+} binding site, is coloured yellow. Proteins that interact with the TPR domain, such as Hsp90, breach the interaction with the αJ-helix and expose the catalytic site.(1wao[130]).

Regulation of PPPs

Phosphorylation of the catalytic subunits

The C-terminal regions of the catalytic domains are critical for the communication of regulatory signals to the active site. For instance, phosphorylation by CDK2 of a C-terminal threonine residue in PP1 inhibits the activity in a manner dependent on the point in the cell cycle. This prevents the reversal of CyclinE/CDK2-mediated phosphorylation of the retinoblastoma protein (Rb) and therefore facilitates cell cycle progression towards the S-phase[128,129] (Figure 21.27).

Regulation by intramolecular domain interaction

The widely expressed PP5 participates in several stress-activated cellular signalling pathways involving the kinases p38 and JNK (see pages 351 and 463). It also binds to $G\alpha_{12/13}$, which gives a clue that it may be involved in responses to 7TM receptors.[131]

The *in vitro* basal activity of PP5 is extremely low. Moreover, whereas PP1, PP2A, and PP4 exist as dimers or trimers, with the catalytic subunit bound to a regulatory subunit (next section), PP5 is a momomer and regulation is due to the presence of a tetratricopeptide repeat (TPR, 34 residues) domain present at the N-terminus. The TPR blocks substrate access to the catalytic cleft and this inhibited conformation is stabilized by the αJ-helical domain at the C-terminus[130] (Figure 21.27b). The inhibition can be annulled by Gα subunits, Hsp90, or arachidonic acid, which reorient the TPR and disrupt the contact with the catalytic domain.

Regulatory subunits of PP1

PP1 is widely distributed and regulates a broad range of cellular functions. These include glycogen metabolism, muscle relaxation (both smooth and skeletal muscle), and cell-cycle progression. The substrate specificities and the regulatory characteristics of muscle and liver type-1 phosphatases are

TABLE 21.4 Cell regulation by PP1: Examples of the many aspects of signalling controlled by the regulatory subunits of PP1

Glycogen targeting	Gm (PPPIR3A), glycogen metabolism, muscle
	GL (PPP1R3B), glycogen metabolism, liver
	R5 (PPPIR3C), glycogen metabolism, liver/muscle
	R6 (PPPIR3D), glycogen metabolism, ubiquitous
Myosin/actin targeting	M110 (PPP1R12A), smooth muscle, relaxation
	MYPT2 (PPP1R12B), skeletal muscle, contraction
	p85 (PPP1R12C), actin cytoskeleton, ubiquitous
Spliceosome/RNA targeting	NIPP1 (PPP1R8), pre-mRNA splicing, nucleus
	PSF1 (−), pre-mRNA splicing, nucleus
	p99 (PPP1R10), RNA processing, nucleus
	Hoxl1 (−), cell cycle checkpoint, nucleus
	HCF (−), transcription, cell cycle, nucleus
Proteasome targeting	Sds22 (PPP1R7), exit from mitosis
Nuclear membrane targeting	AKAP149 (−), B-type lamin dephosphorylation
Plasma membrane and cytoskeleton targeting	neurabin I (PPPIR9A), neurite outgrowth
	spinophilin (PPPIR9B), glutamatergic signalling (GluR1)
	NF-L (−), synaptic transmission?
	AKAP220 (−), coordination of PKA/PP1 signalling

Continued

TABLE 21.4 continued

	Yotiao (−), glutamatergic signalling (NMDA-R)
	Ryanodyne receptor (−), calcium ion channel activity?
	NKCC1 (−), chloride ion transport, epithelium
Endoplasmic reticulum targeting	L5 (−), ribosomal protein, protein synthesis?
	RIPP1 (−), ribosomal inhibitor, protein sythesis?
	GADD34 (PPP1R15A), protein synthesis
Centrosome targeting	AKAP350 (−), centrosomal function?
	Nek2 (−), centrosome separation
Microtubule targeting	Tau (−), microtubule stability, neurons
Mitochondrial targeting	Bcl2 (−), dephosphorylation of Bad
Targeting to specific substrates	54BP2 (PPP1R13A), TP53 binding, cell cycle checkpoint?
	Rb (−), cell cycle progression
	PRIP-1 (−), phospholipase C, IP$_3$ signalling?
	PFK (−), glycolysis?
Activity modulators/chaperones	I-1 (PPP1R1A), inhibition PP1c
	DARPP-32 (PPP1R1B), inhibition PP1c, brain, kidney
	I-2 (PPP1R2), chaperone (folding), inhibitor PP1c
	I-3 (PPP1R11), inhibition of PP1c?
	CPI-17 (PPP1R14A), inhibition of PP1c, smooth muscle
	PHI-2 (PPP1R14A), inhibition of holoenzymes
	I1-PP2A (−), stimulation PP1c, activator PP2A?
	I2-PP2A (−), stimulation PP1c, activation PP2A?
	G-substrate (−), inhibition PP1c, brain
	Grp78 (−), chaperone, stress inducible

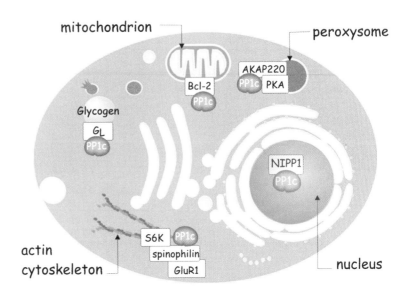

FIG 21.28 Regulatory subunits of PP1. Regulatory subunits engage PP1 in diverse metabolic events and localize the phosphatase to distinct subcellular compartments. Some examples of how regulatory subunits are involved in signalling, complex formation, and the gathering of substrates, kinase and phosphatase, are illustrated. A more comprehensive listing may be found in Table 21.4.

quite distinct, yet the proteins are identical, their differences arising from association with different regulatory subunits.[132] About 45 PP1-binding proteins have been identified (Table 21.4). They take part in different processes and they localize the catalytic subunit, PP1c, to different subcellular compartments (Figure 21.28). Although the regulatory subunits are very dissimilar, most, but not all, share a common PP1 binding sequence -K/R-V/I-x-F- (see Figure 21.25b).

In smooth muscle, the regulatory subunit MYPT1 (M_{110}), targets PP1c to myosin and renders it more active against the myosin regulatory light chain. The main outcome of the interaction of MYPT1 with PP1c is the formation of an extended acidic groove, well adapted to accommodate the basic N-terminal sequence of myosin and making it less attractive for other substrates.[133] This is illustrated in Figure 21.29a.

Besides defining substrate specificity, regulatory subunits also allow the activity of PP1c to be modulated by phosphorylation or by second messengers. For instance, Rho-mediated regulation of smooth muscle contraction and the formation of focal adhesion contacts are exerted in part by the Rho-regulated kinase ROCK. This phosphorylates the MYPT1 subunit, so suppressing PP1 activity towards the myosin regulatory light chain.

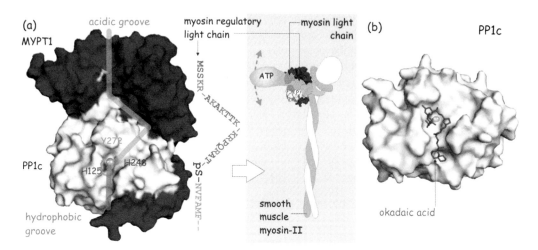

FIG 21.29 Regulation of PP1c by MYPT1 and inhibition by okadaic acid. (a) Attachment of the regulatory subunit MYPT1 to PP1c (the PP1δ variant) creates an extended acidic groove on its surface which forms a perfect fit for the N-terminal sequence of the regulatory light chain of myosin II, which in turn has many basic residues (K or R in orange). At its other end, the groove is hydrophobic enabling it to accommodate a stretch of hydrophobic residues (blue) in the regulatory light chain. This causes it to align so that the phosphoserine lies above the catalytic pocket (indicated by ++). Thus MYPT1 increases the affinity of PP1δ some 10-fold for this particular substrate. Dephosphorylation of myosin regulatory light chain inactivates the ATPase activity of myosin-II and hence prevents movement of its head, resulting in muscle relaxation. (b) Okadaic acid, a causative toxin of diarrhetic shellfish poisoning in Europe, inhibits PP1c by occupying the catalytic cleft, preventing access of substrate. (1jk7,[134] 1s70[133]).

Inhibitors of PP1, PP2A, PP4, and PP5

PP1, PP2A, PP4, and PP5 are potently inhibited by the tumour promoters okadaic acid (figure 21.29b) and microcystin LR[135]. PP2B (calcineurin) is inhibited by the complex tacrolimus (FK506) + FKBP12 (see page 683). Neither of these is capable of associating with the phosphatase alone, but together they form a stable system that inhibits its activity towards protein substrates while having no effect on the dephosphorylation of small model substrates (such as *p*-nitrophenol phosphate).[136] In brain, PP2B forms a complex with AKAP-79, an A-kinase anchoring protein that is a non-competitive inhibitor of PP2B.[137]

PP1 in the regulation of glycogen metabolism

Site-specific phosphorylation of the glycogen targeting subunit (G_M) of PP1c in muscle enables it to generate appropriate responses for adrenaline and Ca^{2+} on the one hand (glycogenolysis), and for insulin on the other (glycogen synthesis).

Regulation of glycogen metabolism: muscle

In skeletal muscle, glycogenolysis is activated by adrenaline. It causes the production of cAMP leading to the activation of PKA (Figure 21.30a) and sets in

Myosin light chain kinase and myosin phosphatase regulate smooth muscle contraction and relaxation. Phosphorylation of the myosin regulatory light chain turns on the ATPase activity of myosin and this initiates the movement of actin filaments by the myosin motor proteins.

train a number of phosphorylation reactions. First it phosphorylates G_M which destabilizes the complex by a factor of 10^4, allowing the dissociation of PP1c.[138] PKA also phosphorylates the inhibitory subunit of PP1, I-1, and through this it creates a pseudosubstrate binding site. As a result, the liberated PP1c is hidden, unable to dephosphorylate proteins associated with glycogen. Finally, PKA phosphorylates and stimulates glycogen phosphorylase kinase. Because of the lack of PP1 activity, kinase activity now predominates. Phosphorylase kinase effectively phosphorylates and activates glycogen phosphorylase, while simultaneously GSK3β phosphorylates and inactivates glycogen synthase. The consequence of all this is glycogenolysis, the net breakdown of glycogen into numerous units of glucose-1-P.

Insulin brings about the reverse reaction (Figure 21.30b). It causes activation of PI 3-K, which in turn leads to activation of PKB. This inactivates GSK3β (see Figure 18.10, page 558). PKB also phosphorylates and activates cyclic

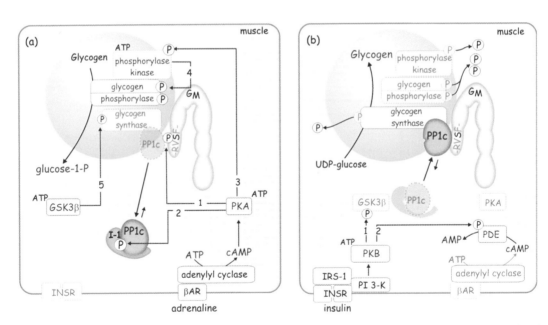

FIG 21.30 Decision-making in glycogen synthesis and breakdown: opposing effects of adrenaline and insulin. (a) Adrenaline drives glycogenolysis through the activation of PKA. This leads to phosphorylation of G_M in the PP1c binding motif (-RVSF-, see **Figure 21.25**), causing them to dissociate (1). It also leads to phosphorylation of I-1 which, through the creation of a pseudo-substrate site, more effectively sequesters free PP1c (2). Finally PKA phosphorylates and activates glycogen phosphorylase kinase (3) which phosphorylates and activates glycogen phosphorylase (4). With the phosphorylation and inhibition of glycogen synthase by GSK3β (5), the balance shifts towards glycogenolysis with liberation of glucose-1-P. (b) Insulin stimulates glycogen synthesis. This is achieved by silencing the protein kinases above. PKB is activated and this phosphorylates and inactivates GSK3β (1). It also phosphorylates and activates phosphodiesterase (PDE) so depleting cAMP (2). With both GSK3β and PKA turned off, PP1c now dominates the scene. It inactivates both phosphorylase kinase and glycogen phosphorylase, and it activates glycogen synthase. The balance shifts towards glycogen synthesis (incorporation of UDP-glucose).

nucleotide phosphodiesterase (PDE). As a result, the level of cAMP falls and PKA activity is suppressed. With the lack of both GSK3β and PKA activity, the phosphatase action of PPIc predominates, phosphorylase kinase and glycogen phosphorylase are inactivated, and glycogen synthase is activated. The balance shifts towards glycogen synthesis.

In order to switch from glycolysis to glycogen synthesis, PP1c is recruited to its regulatory G_M subunit. This requires the action of PP2B or PP2A to dephosphorylate both G_M and I-1 (Figure 21.31). This is another example of phosphatases acting as true reset buttons.[139]

Regulation of glycogen metabolism: liver
The mechanisms that control phosphatase (PP1) activity in liver and muscle are different. In liver, the activity of the G_L subunit is not controlled by phosphorylation. Instead, inhibition of phosphatase activity, necessary to suppress glycogen synthase activity, is exerted by the phosphorylated (active) form of the target enzyme glycogen phosphorylase[139] (Figure 21.32). This acts as an allosteric inhibitor of PP1c at extremely low concentrations. The interaction between the two is made possible because G_L binds both enzymes, PP1c and

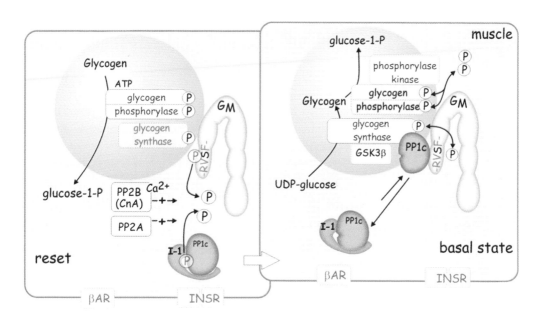

FIG 21.31 Rebalancing glycogen breakdown (PP2A and PP2B). When adrenaline is removed, glycogen metabolism is reset through dephosphorylation of the G_M-regulatory subunit. This is mediated by PP2A and Ca^{2+}-dependent PP2B (calcineurin). PP1c rejoins G_M, and glycogen metabolism again becomes sensitive to adrenaline and insulin. Under these 'resting' conditions, it is likely that PP1c shuttles between G_M and I-1, allowing a low basal release of glucose-1-phosphate.

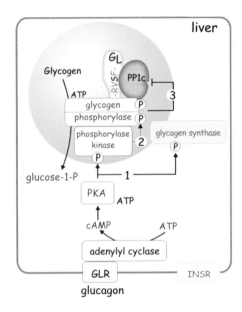

FIG 21.32 Regulation of glycogenolysis in liver: allosteric inhibition of PP1c by activated phosphorylase. In liver, both glycogen phosphorylase and PP1c bind the regulatory subunit G_L (distinct from G_M in muscle). PKA phosphorylates and activates phosphorylase kinase, and phosphorylates and inactivates glycogen synthase (1). Phosphorylase kinase then phosphorylates and activates glycogen phosphorylase (2), which then sterically hinders the adjacent PP1 (3). Glycogen breakdown ensues.

glycogen phosphorylase. (This gathering together of both enzymes does not operate for the much larger G_M.) Again, this provides an effective means for coupling the activation of glycogenolysis to the inhibition of glycogen synthesis, and vice versa.

PP2B (calcineurin)

Although we now recognize that PP2B has a wide tissue distribution,[140] it is identical to a Ca^{2+}-binding protein that was originally identified in neural tissue and named calcineurin.[141] Some time after its discovery, it was realized that calcineurin possesses phosphatase activity and that one of its regulatory subunits is the Ca^{2+}-binding protein calmodulin. The functional protein actually consists of three subunits: calcineurin A (CnA, the catalytic subunit), calcineurin B (CnB, a Ca^{2+}-binding, calmodulin-like subunit), and calmodulin. The principal role of CnB appears to be structural. The phosphatase activity of CnA is activated by Ca^{2+} (see page 230).

Dephosphorylation of NFAT: immunophilins show the way

Functional roles of calcineurin have come to light through the use of drugs that bind to cytosolic proteins called immunophilins. This name reflects the circumstances of their discovery. In 1972, cyclosporin A (now ciclosporin), a fungal product, was found to have remarkable immunosuppressive properties and few side effects.[142] It is effective in preventing transplant rejection and graft versus host disease. Indeed, since its discovery, the use of ciclosporin A has enormously increased the life expectancy of patients receiving kidney, heart, and liver transplants. It also finds application in the treatment of autoimmune diseases and asthma.[143] It was nearly 10 years before the cellular target of ciclosporin A was identified as the cytosolic protein cyclophilin A (CypA). Later, another more potent immunosuppressive drug of fungal origin, tacrolimus or FK506, was discovered. It targets the cytosolic binding protein FKBP12. CypA and FKBP12 are called immunophilins and are members of large protein families. The complexes of both tacrolimus and ciclosporin A with their respective target proteins inhibit calcineurin by hindering substrate access[144] (Figure 21.33 and Figure 8.4, page 230).

Tacrolimus (FK506), a macrolide derived from *Streptomyces,* inhibits peptidyl-prolyl isomerase activity by binding to the immunophilin **FKBP-12** (FK506-binding protein). The FKBP12–tacrolimus complex interacts with and inhibits calcineurin, thus inhibiting both T lymphocyte signal transduction and IL-2 transcription. Tacrolimus is widely used in the prevention of organ rejection after transplant surgery. Its activity is similar to that of ciclosporin A.

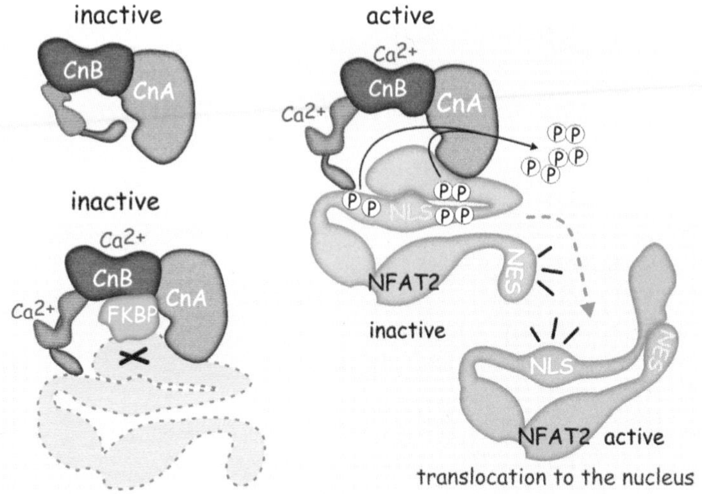

FIG 21.33 Immunophilins bound to their ligand inhibit the phosphatase activity of calcineurin. Under resting conditions calcineurin is in an autoinhibited state in which the C-terminus interferes with the access of substrate. Antigen stimulation causes an increase in cytosolic free Ca^{2+} which binds to the regulatory domains CnB and calmodulin (pink), thereby making the catalytic site available for substrate. In the context of T cell activation, an important substrate is the transcription factor NFAT2, from which numerous phosphates are removed. This exposes the nuclear localization signal. Tacrolimus (not shown) allows the attachment of FKBP12 to the calcineurin complex and this blocks phosphatase activity. See also Figure 8.4, page 230.

One important consequence of the inhibition of calcineurin is the failure of T cell activation after engagement of the TCR with antigen, presented in MHC-context (see Figure 17.5, page 519). This arises from the failure of calcineurin to dephosphorylate NFAT transcription factors in preparation for transfer into the nucleus. Inhibition by ciclosporin A or tacrolimus causes abrupt arrest of the TCR signal[145] and suppression of the immune response.

List of Abbreviations

Abbreviation	Full name/description	SwissProt entry	Other names/OMIM
AKAP220	A-kinase anchoring protein 220 kDa	Q9UKA4	AKAP11
Bsk	basket (*Drosophila*, MAPkinase involved in dorsal closure)	P92208	
Bcl-2	B cell lymphoma protein-2	P10415	
C1-TEN	C1 domain containing tensin-like protein	Q8NFF9	(splice variant a)
CD-3γ	cluster of differentiation-3 gamma chain	P20963	
CD45	cluster of differentiation-45	P08575	PTPRC
Cdc14A	cell division cycle-14 homologue A	Q9UNH5	
cdc25b	cell cycle division-25b	P30305	
CnA	calcineurin A	Q08209	PP2B, PPP3CA
CnB	calcineurin B	P63098	PPP3R1
Cytochrome b	glycoprotein 91 kDa of phagocyte NADPH-oxidase	P04839	NADPH oxidase, cytochrome b245, NOX2
DEP1	density enhanced tyrosine phosphatase-1	Q12913	PTPRJ, RPTPη
DJ-1	Daisuke and Junko (names of students)	Q99497	Parkinson disease protein-7 (PARK7), MIM:606324
Dos	daughter of sevenless (*Drosophila*, positioned in between sevenless and dras)	Q9VZZ9	
DUSP	dual specificity protein phosphatase (family)		
ERBB2	erythroblastoma protein-2 (EGF receptor)	P04626	

Continued

Abbreviation	Full name/description	SwissProt entry	Other names/OMIM
FcⁿRIIB	low affinity Fc-receptor IIB	P31994	CD32
FYVE	Fab1-YOTB-Vac-EEA1 domain (Zn^{2+}-finger)		
Gab1	Grb2-associated binder 1	Q13480	
GL	protein phosphatase 1 glycogen-targeting subunit in liver	Q86XI6	PPP1R3B
GM	protein phosphatase 1 glycogen-targeting subunit in muscle	Q16821	protein phosphatase-1 regulatory subunit 3A (PPP1RA), PTG
HD-PTP	histidine-rich domain containing protein tyrosine phosphatase	Q9H3S7	PTPN23
HE-PTO	haematopoietic protein tyrosine phosphatase	P35236	PTPN7
HGFR	hepatocyte growth factor receptor	P08581	scatter factor, c-Met
hVH5	human VRH-homologous tyrosine phosphatase-5	Q13202	DUSP8
hYVH1	orthologue of S. cerevisiae YVH1 protein	Q9UNI6	DUSP12
I-1	inhibitory protein-1	Q13522	IPP-1, PPP1R1A
IA2	islet-cell autoantigen-regulated protein-2	Q92932	PTPRN2
IFNAR1	interferon-alpha receptor-1	P17181	
IL-R	interleukin receptors		
ITIM	immunoreceptor tyrosine-based inhibition motif		
JAK2	Janus kinase-2	O60674	
JSP-1	JNK-stimulatory phosphatase-1	Q9NRW4	DUSP22
KAP-1	CDK-associated dual specificity phosphatase	Q16667	CDKN3 (cylin dependent kinase inhibitor-3)
KIR	killer-cell inhibitory receptor (many subtypes)		
Laforin	lafora-type myoclonus epilepsy	O95278	EPM2A (epilepsy myoclonus)

Continued

685

Abbreviation	Full name/description	SwissProt entry	Other names/OMIM
LAR	leukocyte antigen related protein	P10586	PTPRF
LMW-PTP	low molecular weight PTP	P24666	red cell acid phosphatase-1 (ACP1)
Lyn	Lck/Yes-related novel tyrosine kinase	P07948	
MCE	mRNA capping enzyme	O60942	RNGTT (mRNA guanylyltransferase)
MEG2	megakaryocyte protein tyrosine phosphatase-2	P43378	PTPN9, homology with retinaldehyde-binding protein SEC14
Met	methyl-nitro-nitroguanidine-induced oncogene	P08581	scatter factor, hepatocyte growth factor receptor
MKP3	MAPkinase phosphatase-3		DUSP6, Pyst1
MK-Styx	MAPK phosphatase serine/threonine/tyrosine interacting like protein	Q9Y6J8	DUSP24
MTM1	myotubular myopathy	Q13496	myotubularin
MTMR3	MTM-related protein-3	Q13615	ZFYVE10 (Zn finger FYVE domain protein)
MTMR6	MTM-related protein-6	Q9Y217	
MTMR10	MTM-related protein-10	Q6P4Q6	truncated
MTMR13	MTM-related protein-13	Q86WG5	SET-binding factor2 (SBF2)
MYPT1	myosin phosphatase 1 target subunit-1	O14974	PPP1R12A, M110, smooth muscle
MYPT2	myosin phosphatase 1 target subunit-2	O60237	PPP1R12B, skeletal muscle
NFAT2	nuclear factor of activated T cells 2	O95644	NFATc1, NFATc
NIPP1	nuclear inhibitor of protein phosphatase-1	Q12972	PPP1R8
Noonan syndrome		MIM 163950	
NOX1	NADPH-oxidase homologue 1	Q9Y5S8	mitogenic oxidase
NOXA	NOX activator	P19878	p67

Continued

Abbreviation	Full name/description	SwissProt entry	Other names/OMIM
NOXO1	NOX organizer-1	Q8NFA2	p41NOX (PX domain protein)
OB-R	obese receptor	P48357	leptin receptor
PCPTP1	pheochromocytoma-12 (PC12) protein tyrosine phosphatase-1	Q15256	PTPRR
PDP	mitochondrial pyruvate dehydrogenase phosphatase	Q9P0J1	PMM2C
PEP	PEST domain enriched tyrosine phosphatase	Q9Y2R2	PTPN22, Lyp (lymphoid phosphatase)
PIR1	phosphatase interacting with RNA/RNP complex-1	O75319	DUSP11
PP1c	serine threonine phosphatase-1 (alpha) catalytic subunit	P62136	PPP1CA
PP2a	serine threonine phosphatase-2 (alpha) catalytic subunit	P67775	PPP2CA
PP2B	serine threonine phosphatase-2 (beta) catalytic subunit	Q08209	PPP3CA, calcineurin A (CnA)
PP2Cα	serine threonine phoshatase 2C isoform alpha	P35813	PPM1A
PP4	serine threonine phosphatase-4 catalytic subunit	P60510	PPP4C, PP-X
PP4R1	PP4 regulatory subunit-1	Q8TF05	PPP4R1
PP5	serine threonine phosphatase-5 catalytic subunit	P53041	PPP5C
PP6	serine threonine phosphatase-6 catalytic subunit	O00743	PPP6C
PRL1	phosphatase of regenerating liver	Q93096	PTP4A, PTPCAAX
PTEN	phosphatase and tension homology	P60484	mutated in multiple advanced cancers-1 (MMAC1)
PTP	protein tyrosine phosphatase		
PTP1B	protein tyrosine phosphatase 1B	P18031	PTPN1
PTPα	protein tyrosine phosphatase-alpha	P18433	PTPRA
PTP-BAS	protein tyrosine phosphatase basophil	Q12923	PTPN13

Continued

687

Abbreviation	Full name/description	SwissProt entry	Other names/OMIM
PTP-D1	protein tyrosine phoshatase distinct subfamily-1	Q16825	PTPN21
PTP-γ	protein tyrosine phosphatase-gamma	P23470	PTPRG
PTP-H1	protein tyrosine phosphatase in human colon-1	P26045	PTPN3
PTP-μ	protein tyrosine phosphatase-mu	P28827	PTPRM
PTP-PEST	protein tyrosine phosphatase Pro-Glu-Ser-threonine rich sequence	P35831	PTPN12
RP65	regulatory protein 65 kDa (subunit A of PP2a)	P30153	PPP2R1A
SHP-1	Src-homology domain containing tyrosine phosphatase-1	P29350	SH-PTP1, PTPN6, PTP1C
SHP-2	Src-homology domain containing tyrosine phosphatase-2	Q06124	SH-PTP2, PTPN11, PTP2C
SOCS1	suppressor of cytokine signalling-1	O15524	
spinophilin		Q96SB3	PPP1R9B, neurabin-2
Src	sarcoma (soft tissue tumour)	P12931	
SSH1	slingshot-homologue-1	QWYL5	
STAT5	signal transducer and activator of transcription-5	P42229	
Styx	serine/threonine/tyrosine interacting like protein	Q8WUJ0	
tensin-1		Q9HBL0	TNS1
TYK2	non-receptor tyrosine kinase-2	P29597	
VH1	vaccinia virus H1		
VHR	vaccinia virus phosphatase VH1-related	P51452	DUSP3

References

1. Sun H, Tonks NK. The coordinated action of protein tyrosine phosphatases and kinases in cell signaling. *Trends Biochem Sci.* 1994;19:480–485.

2. Tonks NK, Diltz CD, Fischer EH. Purification of the major protein-tyrosine-phosphatases of human placenta. *J Biol Chem.* 1988;263:6722–6730.

3. Charbonneau H, Tonks NK, Kumar S, et al. Human placenta protein-tyrosine-phosphatase: amino acid sequence and relationship to a family of receptor-like proteins. *Proc Natl Acad Sci USA.* 1989;86:5252–5256.

4. Andersen JN, Mortensen OH, Peters GH, et al. Structural and evolutionary relationships among protein tyrosine phosphatase domains. *Mol Cell Biol.* 2001;21:7117–7136.

5. Guan KL, Broyles SS, Dixon JE. A Tyr/Ser protein phosphatase encoded by vaccinia virus. *Nature.* 1991;350:359–362.

6. Gautier J, Solomon MJ, Booher RN, Bazan JF, Kirschner MW. cdc25 is a specific tyrosine phosphatase that directly activates p34cdc2. *Cell.* 1991;67:197–211.

7. Fischer EH, Charbonneau H, Tonks NK. Protein tyrosine phosphatases: a diverse family of intracellular and transmembrane enzymes. *Science.* 1991;253:401–406.

8. Sap J, Jiang YP, Friedlander D, Grumet M, Schlessinger J. Receptor tyrosine phosphatase R-PTP-κ mediates homophilic binding. *Mol Cell Biol.* 1994;14:1–9.

9. Brady-Kalnay SM, Tonks NK. Identification of the homophilic binding site of the receptor protein tyrosine phosphatase PTPμ. *J Biol Chem.* 1994;269:28472–28477.

10. Jia Z, Barford D, Flint AJ, Tonks NK. Structural basis for phosphotyrosine peptide recognition by protein tyrosine phosphatase 1B. *Science.* 1995;268:1754–1758.

11. Barford D, Flint AJ, Tonks NK. Crystal structure of human protein tyrosine phosphatase 1B. *Science.* 1994;263:1397–1404.

12. Tonks NK. PTP1B: from the sidelines to the front lines!. *FEBS Lett.* 2003;546:140–148.

13. Zheng XM, Wang Y, Pallen CJ. Cell transformation and activation of pp60c-src by overexpression of a protein tyrosine phosphatase. *Nature.* 1992;359:336–339.

14. Kenner KA, Hill DE, Olefsky JM, Kusari J. Regulation of protein tyrosine phosphatases by insulin and insulin-like growth factor I. *J Biol Chem.* 1993;268:25455–25462.

15. Ahmad F, Azevedo JL, Cortright R, Dohm GL, Goldstein BJ. Alterations in skeletal muscle protein-tyrosine phosphatase activity and expression in insulin-resistant human obesity and diabetes. *J Clin Invest.* 1997;100:449–458.

16. Tonks NK, Cicirelli MF, Diltz CD, Krebs EG, Fischer EH. Effect of microinjection of a low-Mr human placenta protein tyrosine phosphatase on induction of meiotic cell division in *Xenopus* oocytes. *Mol Cell Biol*. 1990;10:458–463.

17. Cicirelli MF, Tonks NK, Diltz CD, Weiel JE, Fischer EH, Krebs EG. Microinjection of a protein-tyrosine-phosphatase inhibits insulin action in *Xenopus* oocytes. *Proc Natl Acad Sci USA*. 1990;87:5514–5518.

18. Chen H, Wertheimer SJ, Lin CH, et al. Protein-tyrosine phosphatases PTP1B and syp are modulators of insulin-stimulated translocation of GLUT4 in transfected rat adipose cells. *J Biol Chem*. 1997;272:8026–8031.

19. Salmeen A, Andersen JN, Myers MP, Tonks NK, Barford D. Molecular basis for the dephosphorylation of the activation segment of the insulin receptor by protein tyrosine phosphatase 1B. *Mol Cell*. 2000;6:1401–1412.

20. Hubbard SR, Wei L, Ellis L, Hendrickson WA. Crystal structure of the tyrosine kinase domain of the human insulin receptor. *Nature*. 1994;372:746–754.

21. Elchebly M, Payette P, Michaliszyn E, et al. Increased insulin sensitivity and obesity resistance in mice lacking the protein tyrosine phosphatase-1B gene. *Science*. 1999;283:1544–1548.

22. Klaman LD, Boss O, Peroni OD, et al. Increased energy expenditure, decreased adiposity, and tissue-specific insulin sensitivity in protein-tyrosine phosphatase 1B-deficient mice. *Mol Cell Biol*. 2000;20:5479–5489.

23. Dixon JE, Rhodes JD. Expression of a protein tyrosine phosphatase in normal and v-src-transformed mouse 3T3 fibroblasts. *J Cell Biol*. 1992;117:401–414.

24. Bruskin AM, Johnson KA, Hill DE. Effect of protein tyrosine phosphatase 1B expression on transformation by the human neu oncogene. *Cancer Res*. 1992;15:478–482.

25. Carpino N, Wisniewski D, Strife A, et al. p62(dok): a constitutively tyrosine-phosphorylated, GAP-associated protein in chronic myelogenous leukemia progenitor cells. *Cell*. 1997;88:197–204.

26. Dube N, Cheng A, Tremblay ML. The role of protein tyrosine phosphatase 1B in Ras signaling. *Proc Natl Acad Sci USA*. 2004;101:1834–1839.

27. Zinker BA, Rondinone CM, Trevillyan JM, et al. PTP1B antisense oligonucleotide lowers PTP1B protein, normalizes blood glucose, and improves insulin sensitivity in diabetic mice. *Proc Natl Acad Sci USA*. 2002;99:11357–11362.

28. Touw IP, De Koning JP, Ward AC, Hermans MH. Signaling mechanisms of cytokine receptors and their perturbances in disease. *Mol Cell Endocrinol*. 2000;160:1–9.

29. Tartaglia LA. The leptin receptor. *J Biol Chem*. 1997;272:6093–6096.

30. Flint AJ, Tiganis T, Barford D, Tonks NK. Development of 'substrate-trapping' mutants to identify physiological substrates of protein tyrosine phosphatases. *Proc Natl Acad Sci USA*. 1997;94:1680–1685.

31. Zabolotny JM, Bence-Hanulec KK, Stricker-Krongrad A, et al. PTP1B regulates leptin signal transduction in vivo. *Dev Cell*. 2002;2:489–495.

32. Cheng A, Uetani N, Simoncic PD, et al. Attenuation of leptin action and regulation of obesity by protein tyrosine phosphatase 1B. *Dev Cell*. 2002;2:497–503.

33. Ohba M, Shibanuma M, Kuroki T, Nose K. Production of hydrogen peroxide by transforming growth factor-β 1 and its involvement in induction of egr-1 in mouse osteoblastic cells. *J Cell Biol*. 1994;126:1079–1088.

34. Sundaresan M, Yu ZX, Ferrans VJ, Irani K, Finkel T. Requirement for generation of H_2O_2 for platelet-derived growth factor signal transduction. *Science*. 1995;270:296–299.

35. Irani K, Xia Y, Zweier JL, et al. Mitogenic signaling mediated by oxidants in Ras-transformed fibroblasts. *Science*. 1997;275:1649–1652.

36. Chiarugi P, Cirri P. Redox regulation of protein tyrosine phosphatases during receptor tyrosine kinase signal transduction. *Trends Biochem Sci*. 2003;28:509–514.

37. Werner E, Werb Z. Integrins engage mitochondrial function for signal transduction by a mechanism dependent on Rho GTPases. *J Cell Biol*. 2002;158:357–368.

38. Vafa O, Wade M, Kern S, et al. c-Myc can induce DNA damage, increase reactive oxygen species, and mitigate p53 function: a mechanism for oncogene-induced genetic instability. *Mol Cell*. 2002;9:1031–1044.

39. Mukherjee SP, Lane RH, Lynn WS. Endogenous hydrogen peroxide and peroxidative metabolism in adipocytes in response to insulin and sulfhydryl reagents. *Biochem Pharmacol*. 1978;27:2589–2594.

40. May JM, de Haen C. The insulin-like effect of hydrogen peroxide on pathways of lipid synthesis in rat adipocytes. *J Biol Chem*. 1979;254:9017–9021.

41. Baas AS, Berk BC. Differential activation of mitogen-activated protein kinases by H_2O_2 and O_2^- in vascular smooth muscle cells. *Circ Res*. 1995;77:29–36.

42. Seo MS, Kang SW, Baines IC, Tekle E, Chock PB, Rhee SG. Epidermal growth factor (EGF)-induced generation of hydrogen peroxide. Role in EGF receptor-mediated tyrosine phosphorylation. *J Biol Chem*. 1997;272:217–221.

43. Barrett WC, DeGnore JP, Konig S, et al. Regulation of PTP1B via glutathionylation of the active site cysteine 215. *Biochemistry*. 1999;38:6699–6705.

44. Blanchetot C, Tertoolen LG, den Hertog J. Regulation of receptor protein-tyrosine phosphatase α by oxidative stress. *EMBO J*. 2002;21:493–503.

45. Meng TC, Fukada T, Tonks NK. Reversible oxidation and inactivation of protein tyrosine phosphatases in vivo. *Mol Cell*. 2002;9:387–399.

46. Lassegue B, Suh YA, Arnold RS, et al. Cell transformation by the superoxide-generating oxidase Mox1. *Nature*. 1999;401:79–82.

47. Arnold RS, Shi J, Murad E, et al. Hydrogen peroxide mediates the cell growth and transformation caused by the mitogenic oxidase Nox1. *Proc Natl Acad Sci USA*. 2001;98:5550–5555.

48. Lee SR, Kwon KS, Kim SR, Rhee SG. Reversible inactivation of protein-tyrosine phosphatase 1B in A431 cells stimulated with epidermal growth factor. *J Biol Chem*. 1998;273:15366–15372.

49. Denu JM, Tanner KG. Specific and reversible inactivation of protein tyrosine phosphatases by hydrogen peroxide: evidence for a sulfenic acid intermediate and implications for redox regulation. *Biochemistry*. 1998;37:5633–5642.

50. Goldstein BJ, Wu X, Zilbering A, Zhu L, Lawrence JT. Hydrogen peroxide generated during cellular insulin stimulation is integral to activation of the distal insulin signaling cascade in 3T3-L1 adipocytes. *J Biol Chem*. 2001;28:48662–48669.

51. Ravetch JV, Lanier LL. Immune inhibitory receptors. *Science*. 2000;290:84–89.

52. Neel BG, Gu H, Pao L. The 'Shp'ing news: SH2 domain-containing tyrosine phosphatases in cell signaling. *Trends Biochem Sci*. 2003;28:284–293.

53. Hof P, Pluskey S, Dhe-Paganon S, Eck MJ, Shoelson SE. Crystal structure of the tyrosine phosphatase SHP-2. *Cell*. 1998;92:441–450.

54. Yang J, Liang X, Niu T, Meng W, Zhao Z, Zhou GW. Crystal structure of the catalytic domain of protein-tyrosine phosphatase SHP-1. *J Biol Chem*. 1998;273:28199–28207.

55. Yang J, Liu L, He D, et al. Crystal structure of human protein-tyrosine phosphatase SHP-1. *J Biol Chem*. 2003;278:6516–6520.

56. Tsui HW, Siminovitch KA, de Souza L, Tsui FW. Motheaten and viable motheaten mice have mutations in the haematopoietic cell phosphatase gene. *Nature Genet*. 1993;4:124–129.

57. Zhang J, Somani AK, Siminovitch KA. Roles of the SHP-1 tyrosine phosphatase in the negative regulation of cell signaling. *Semin Immunol*. 2000;12:361–378.

58. Daigle I, Yousefi S, Colonna M, Green DR, Simon HU. Death receptors bind SHP-1 and block cytokine-induced anti-apoptotic signaling in neutrophils. *Nat Med*. 2002;8:62–67.

59. Perkins LA, Larsen I, Perrimon N. Corkscrew encodes a putative protein tyrosine phosphatase that functions to transduce the terminal signal from the receptor tyrosine kinase torso. *Cell*. 1992;70:225–236.

60. Raabe T, Riesgo-Escovar J, Liu X, Bausenwein BS, Maroy P, Hafen E. DOS, a novel pleckstrin homology domain-containing protein required for signal transduction between sevenless and Ras1 in *Drosophila*. *Cell*. 1996;85:911–920.

61. Shi ZQ, Yu DH, Park M, Marshall M, Feng GS. Molecular mechanism for the Shp-2 tyrosine phosphatase function in promoting growth factor stimulation of Erk activity. *Mol Cell Biol*. 2006;20:1526–1536.

62. Tang TL, Freeman RM, O'Reilly AM, Neel BG, Sokol SY. The SH2-containing protein-tyrosine phosphatase SH-PTP2 is required upstream of MAP kinase for early *Xenopus* development. *Cell*. 1995;80:473–483.

63. Shi ZQ, Lu W, Feng GS. The Shp-2 tyrosine phosphatase has opposite effects in mediating the activation of extracellular signal-regulated and c-Jun NH2-terminal mitogen-activated protein kinases. *J Biol Chem*. 1998;273:4904–4908.

64. Saxton TM, Ciruna BG, Holmyard D, Kulkarni S, Harpal K, Rossant J, Pawson T. The SH2 tyrosine phosphatase shp2 is required for mammalian limb development. *Nat Genet*. 2000;24:420–423.

65. Tartaglia M, Kalidas K, Shaw A, et al. PTPN11 mutations in Noonan syndrome: molecular spectrum, genotype-phenotype correlation, and phenotypic heterogeneity. *Am J Hum Genet*. 2002;70:1555–1563.

66. Keilhack H, David FS, McGregor M, Cantley LC, Neel BG. Diverse biochemical properties of Shp2 mutants. Implications for disease phenotypes. *J Biol Chem*. 2005;280:30984–30993.

67. Andersen JN, Jansen PG, Mortensen OH, et al. A genomic perspective on protein tyrosine phosphatases: gene structure, pseudogenes, and genetic disease linkage. *FASEB J*. 2004;18:8–30.

68. Noonan JA. Hypertelorism with Turner phenotype. A new syndrome with associated congenital heart disease. *Am J Dis Child*. 1968;116:373–380.

69. Tartaglia M, Mehler EL, Goldberg R, et al. Mutations in PTPN11, encoding the protein tyrosine phosphatase SHP-2, cause Noonan syndrome. *Nat Genet*. 2001;29:465–468.

70. Tartaglia M, Niemeyer CM, Fragale A, et al. Somatic mutations in PTPN11 in juvenile myelomonocytic leukemia, myelodysplastic syndromes and acute myeloid leukemia. *Nat Genet*. 2003;34:146–150.

71. Ostman A, Yang Q, Tonks NK. Expression of DEP-1, a receptor-like protein-tyrosine-phosphatase, is enhanced with increasing cell density. *Proc Natl Acad Sci USA*. 1994;91:9680–9684.

72. Ruivenkamp CA, van Weze IT, Zanon C, et al. Ptprj is a candidate for the mouse colon-cancer susceptibility locus Scc1 and is frequently deleted in human cancers. *Nat Genet*. 2002;31:295–300.

73. Tonks NK, Charbonneau H, Diltz CD, Fischer EH, Walsh KA. Demonstration that the leukocyte common antigen CD45 is a protein tyrosine phosphatase. *Biochemistry*. 1988;27:8695–8701.

74. Thomas ML, Barclay AN, Gagnon J, Williams AF. Evidence from cDNA clones that the rat leukocyte-common antigen (T200) spans the lipid bilayer and contains a cytoplasmic domain of 80,000 Mr. *Cell*. 1985;41: 83–93.

75. Thomas ML. Evidence that the leukocyte-common antigen is required for antigen-induced T lymphocyte proliferation. *Cell*. 1989;58:1055–1065.

76. Byth KF, Conroy LA, Howlett S, et al. CD45-null transgenic mice reveal a positive regulatory role for CD45 in early thymocyte development, in the

selection of CD4+CD8+ thymocytes, and B cell maturation. *J Exp Med*. 1996;183:1707–1718.

77. Yoo LI, Pingel JT, Heikinheimo M, et al. Mutations in the tyrosine phosphatase CD45 gene in a child with severe combined immunodeficiency disease. *Nat Med*. 2000;6:343–345.

78. Majeti R, Xu Z, Parslow TG, Olson JL, Daikh DI, Killeen N, Weiss A. An inactivating point mutation in the inhibitory wedge of CD45 causes lymphoproliferation and autoimmunity. *Cell*. 2000;103:1059–1070.

79. Yamaguchi H, Hendrickson WA. Structural basis for activation of human lymphocyte kinase Lck upon tyrosine phosphorylation. *Nature*. 1996;384:484–489.

80. Jove R, Hanafusa T, Hamaguchi M, Hanafusa H. In vivo phosphorylation states and kinase activities of transforming p60c-src mutants. *Oncogene Res*. 1989;5:49–60.

81. Alexander DR. The CD45 tyrosine phosphatase: a positive and negative regulator of immune cell function. *Semin Immunol*. 2000;12:349–359.

82. Varma R, Campi G, Yokosuka T, et al. T cell receptor-proximal signals are sustained in peripheral microclusters and terminated in the central supramolecular activation cluster. *Immunity*. 2006;25:117–127.

83. Irie-Sasaki J, Sasaki T, Matsumoto W, et al. CD45 is a JAK phosphatase and negatively regulates cytokine receptor signalling. *Nature*. 2001;409: 349–354.

84. Desai DM, Sap J, Schlessinger J, Weiss A. Ligand-mediated negative regulation of a chimeric transmembrane receptor tyrosine phosphatase. *Cell*. 1993;73:541–554.

85. Jiang G, den Hertog J, Su J, Noel J, Sap J, Hunter T. Dimerization inhibits the activity of receptor-like protein-tyrosine phosphatase-α. *Nature*. 1999;401:606–610.

86. Nam HJ, Poy F, Krueger NX, Saito H, Frederick CA. Crystal structure of the tandem phosphatase domains of RPTP LAR. *Cell*. 1999;97:449–457.

87. Yuvaniyama J, Denu JM, Dixon JE, Saper MA. Crystal structure of the dual specificity protein phosphatase VHR. *Science*. 1996;272:1328–1331.

88. Canagarajah BJ, Khokhlatchev A, Cobb MH, Goldsmith EJ. Activation mechanism of the MAP kinase ERK2 by dual phosphorylation. *Cell*. 1997;90:859–869.

89. Stewart AE, Dowd S, Keyse SM, McDonald NQ. Crystal structure of the MAPK phosphatase Pyst1 catalytic domain and implications for regulated activation. *Nat Struct Biol*. 2006;6:174–181.

90. Tonks NK. Protein tyrosine phosphatases: from genes, to function, to disease. *Nat Rev Mol Cell Biol*. 2006;7:833–846.

91. Keyse SM. Protein phosphatases and the regulation of mitogen-activated protein kinase signaling. *Curr Opin Cell Biol*. 2000;12:186–192.

92. Arkinstall S, Nichols A. Dual specificity phosphatases: a gene family for control of MAP kinase function. *FASEB J*. 2000;14:6–16.

93. Alessi DR, Gomez N, Moorhead G, Lewis T, Keyse SM, Cohen P. Inactivation of p42 MAP kinase by protein phosphatase 2A and a protein tyrosine phosphatase, but not CL100, in various cell lines. *Curr Biol*. 1995;55:283–295.

94. Pulido R, Zuniga A, Ullrich A. PTP-SL and STEP protein tyrosine phosphatases regulate the activation of the extracellular signal-regulated kinases ERK1 and ERK2 by association through a kinase interaction motif. *EMBO J*. 1998;17:7337–7350.

95. Farooq A, Zhou MM. Structure and regulation of MAPK phosphatases. *Cell Signal*. 2004;16:769–779.

96. Camps M, Nichols A, Gillieron C, et al. Catalytic activation of the phosphatase MKP-3 by ERK2 mitogen-activated protein kinase. *Science*. 1998;280:1212–1213.

97. Tanoue T, Adachi M, Moriguchi T, Nishida E. A conserved docking motif in MAP kinases common to substrates, activators and regulators. *Nat Cell Biol*. 2000;2:110–116.

98. Charles CH, Abler AS, Lau LF. cDNA sequence of a growth factor-inducible immediate early gene and characterization of its encoded protein. *Oncogene*. 1992;7:187–190.

99. Sun H, Charles CH, Lau LF, Tonks NK. MKP-1 (3CH134), an immediate early gene product, is a dual specificity phosphatase that dephosphorylates MAP kinase in vivo. *Cell*. 1993;75:487–493.

100. Assoian RK. Control of the G1 phase cyclin-dependent kinases by mitogenic growth factors and the extracellular matrix. *Cytokine Growth Factor Rev*. 1997;8:165–170.

101. Brondello JM, McKenzie FR, Sun H, Tonks NK, Pouyssegur J. Constitutive MAP kinase phosphatase (MKP-1) expression blocks G1 specific gene transcription and S-phase entry in fibroblasts. *Oncogene*. 1995;10:1895–1904.

102. Sun H, Tonks NK, Bar SD. Inhibition of Ras-induced DNA synthesis by expression of the phosphatase MKP-1. *Science*. 1994;266:285–288.

103. Brondello JM, Pouyssegur J, McKenzie FR. Reduced MAP kinase phosphatase-1 degradation after p42/p44MAPK-dependent phosphorylation. *Science*. 1999;286:2514–2517.

104. Camps M, Nichols A, Arkinstall S. Dual specificity phosphatases: a gene family for control of MAP kinase function. *FASEB J*. 2000;14:6–16.

105. Dorfman K, Carrasco D, Gruda M, Ryan C, Lira SA, Bravo R. Disruption of the erp/mkp-1 gene does not affect mouse development: normal MAP kinase activity in ERP/MKP-1-deficient fibroblasts. *Oncogene*. 1996;13:925–931.

106. Groom LA, Sneddon AA, Alessi DR, Dowd S, Keyse SM. Differential regulation of the MAP, SAP and RK/p38 kinases by Pyst1, a novel cytosolic dual-specificity phosphatase. *EMBO J*. 2006;15:3621–3632.

107. Sluss HK, Han Z, Barrett T, Davis RJ, Ip YT. A JNK signal transduction pathway that mediates morphogenesis and an immune response in *Drosophila*. *Genes Dev*. 1996;27:45–58.

108. Martin-Blanco E, Gampel A, Ring J, et al. *puckered* encodes a phosphatase that mediates a feedback loop regulating JNK activity during dorsal closure in *Drosophila. Genes Dev*. 1998;12:557–570.

109. Campos-Ortega JA, Hartenstein V. Early neurogenesis in wildtype *Drosophila melanogaster. Roux's Arch Dev Biol*. 1984;193:308–325.

110. Kawakami Y, Rodriguez-Leon J, Koth CM, et al. MKP3 mediates the cellular response to FGF8 signalling in the vertebrate limb. *Nat Cell Biol*. 2003;5:513–519.

111. Li J, Yen C, Liaw D, et al. PTEN, a putative protein tyrosine phosphatase gene mutated in human brain, breast, and prostate cancer. *Science*. 1997;275:1943–1947.

112. Cantley LC, Neel BG. New insights into tumor suppression: PTEN suppresses tumor formation by restraining the phosphoinositide 3-kinase/AKT pathway. *Proc Natl Acad Sci USA*. 1999;96:4240–4245.

113. Maehama T, Dixon JE. The tumor suppressor, PTEN/MMAC1, dephosphorylates the lipid second messenger, phosphatidylinositol 3,4,5-trisphosphate. *J Biol Chem*. 1998;273:13375–13378.

114. Myers MP, Pass I, Batty IH, et al. The lipid phosphatase activity of PTEN is critical for its tumor supressor function. *Proc Natl Acad Sci USA*. 1998;95:13513–13518.

115. Lee JO, Yang H, Georgescu MM, et al. Crystal structure of the PTEN tumor suppressor: implications for its phosphoinositide phosphatase activity and membrane association. *Cell*. 1999;99:323–334.

116. Alessi DR, Kozlowski MT, Weng QP, Morrice N, Avruch J. 3-Phosphoinositide-dependent protein kinase 1 (PDK1) phosphorylates and activates the p70 S6 kinase in vivo and in vitro. *Curr Biol*. 1998;8:69–81.

117. Chow LM, Baker SJ. PTEN function in normal and neoplastic growth. *Cancer Lett*. 2006;20:1–13.

118. Nagata Y, Lan KH, Zhou X, et al. PTEN activation contributes to tumor inhibition by trastuzumab, and loss of PTEN predicts trastuzumab resistance in patients. *Cancer Cell*. 2004;6:117–127.

119. Kim RH, Peters M, Jang Y, et al. DJ-1, a novel regulator of the tumor suppressor PTEN. *Cancer Cell*. 2005;7:263–273.

120. Nagakubo D, Taira T, Kitaura H, et al. DJ-1, a novel oncogene which transforms mouse NIH3T3 cells in cooperation with ras. *Biochem Biophys Res Commun*. 1997;231:509–513.

121. Trotman LC, Wang X, Alimonti A, et al. Ubiquitination regulates PTEN nuclear import and tumor suppression. *Cell*. 2007;128:141–156.

122. Shen WH, Balajee AS, Wang J, et al. Essential role for nuclear PTEN in maintaining chromosomal integrity. *Cell*. 2007;128:157–170.

123. Cori GT, Cori CF. The enzymatic conversion of phosphorylase a to b. *J Biol Chem*. 1945;158:321–332.

124. Stralfors P, Hiraga A, Cohen P. The protein phosphatases involved in cellular regulation. Purification and characterisation of the glycogen-bound form of protein phosphatase-1 from rabbit skeletal muscle. *Eur J Biochem*. 1985;149:295–303.

125. Cohen P. The structure and regulation of protein phosphatases. *Annu Rev Biochem*. 1989;58:453–508.

126. Cohen PT. Novel protein serine/threonine phosphatases: variety is the spice of life. *Trends Biochem Sci*. 1997;22:245–251.

127. Cohen PTW. Nomenclature and chromosomal localization of human protein serine/threonine phosphatase genes. *Adv Prot Phosphatases*. 1994;8:371–376.

128. Ishii K, Yamano H, Yanagida M. Phosphorylation of dis2 protein phosphatase at the C-terminal cdc2 consensus and its potential role in cell cycle regulation. *EMBO J*. 1994;13:5310–5318.

129. Dohadwala M, da Cruz e Silva EF, Hall F, et al. Phosphorylation and inactivation of protein phosphatase 1 by cyclin-dependent kinases. *Proc Natl Acad Sci USA*. 1994;91:6408–6412.

130. Yang J, Roe SM, Cliff MJ, et al. Molecular basis for TPR domain-mediated regulation of protein phosphatase 5. *EMBO J*. 2005;24:1–10.

131. Yamaguchi Y, Katoh H, Mori K, Negishi M. Gα_{12} and Gα_{13} interact with ser/thr protein phosphatase type 5 and stimulate its phosphatase activity. *Curr Biol*. 2002;12:1353–1358.

132. Cohen PT, Schelling DL, da Cruz e Silva OB, Barker HM, Cohen P. The major type-1 protein phosphatase catalytic subunits are the same gene products in rabbit skeletal muscle and rabbit liver. *Biochim Biophys Acta*. 2006;1008:125–128.

133. Terrak M, Kerff F, Langsetmo K, Tao T, Dominguez R. Structural basis of protein phosphatase 1 regulation. *Nature*. 2006;429:780–784.

134. Maynes JT, Bateman KS, Cherney MM, et al. Crystal structure of the tumor-promoter okadaic acid bound to protein phosphatase-1. *J Biol Chem*. 2001;276:44078–44082.

135. MacKintosh C, MacKintosh RW, Inhibitors of protein kinases and phosphatases. *Trends Biochem Sci*. 994;19:444–448.

136. Griffith JP, Kim JL, Kim EE, et al. X-ray structure of calcineurin inhibited by the immunophilin-immunosuppressant FKBP12-FK506 complex. *Cell*. 1995;82:507–522.

137. Coghlan VM, Perrino BA, Howard M, et al. Association of protein kinase A and protein phosphatase 2B with a common anchoring protein. *Science*. 1995;267:108–111.

138. Johnson DF, Moorhead G, Caudwell FB, et al. Identification of protein-phosphatase-1-binding domains on the glycogen and myofibrillar targetting subunits. *Eur J Biochem*. 1996;239:317–326.

139. Hubbard MJ, Cohen P. On target with a new mechanism for the regulation of protein phosphorylation. *Trends Biochem Sci.* 1993;18: 172–177.

140. Chantler PD. Calcium-dependent association of a protein complex with the lymphocyte plasma membrane: probable identity with calmodulin-calcineurin. *J Cell Biol.* 1985;101:207–216.

141. Klee CB, Crouch TH, Krinks MH. Calcineurin: a calcium- and calmodulin-binding protein of the nervous system. *Proc Natl Acad Sci USA.* 1979;76:6270–6273.

142. Borel JF, Feurer C, Gubler HU, Stahelin H. Biological effects of cyclosporin A: a new antilymphocytic agent. *Agents Actions.* 1976;6:468–475.

143. Sihra BS, Kon OM, Durham SR, Walker S, Barnes NC, Kay AB. Effect of cyclosporin A on the allergen-induced late asthmatic reaction. *Thorax.* 1997;52:447–452.

144. McKeon F. When worlds collide: immunosuppressants meet protein phosphatases. *Cell.* 1991;66:823–826.

145. Liu Jr J, Farmer JD, Lane WS, Friedman J, Weissman I, Schreiber SL. Calcineurin is a common target of cyclophilin-cyclosporin A and FKBP-FK506 complexes. *Cell.* 1991;66:807–815.

CHAPTER 22

Notch

Notched wings, Morgan, and the gene theory

Notches in the wings of *Drosophila* were first described in 1916 by T. H. Morgan, who ascribed them to a spontaneous genetic mutation (Figure 22.1). Genes, the elements of heredity postulated by Gregor Mendel some 50 years previously, remained just that: postulates. Morgan appreciated that development and heredity are inextricably linked and that knowledge in this area could not be advanced without an understanding of the nature of the gene. It fell to him to tackle the central question and introduce the science of genetics into embryology, or, as he phrased it, 'approaching a physiological problem with a new and strange discipline to the classical physiology of the schools'. This, and perhaps the fact that they were cheap to work with, explains his involvement with flies.

In writing these paragraphs, we have relied heavily on the work of Scott Gilbert, see http://8e.devbio.com. Details can be found in Gilbert.[1,2]

Thomas Hunt Morgan was awarded the Nobel Prize in Physiology or Medicine in 1933 'for his discoveries concerning the role played by the chromosome in heredity'.

'A defect in the protoplasm often brings about a modified cleavage and also a defective embryo, and this takes place even though the whole of the nuclear material of the unsegmented egg remains present. There seems, therefore no escape from the conclusion that in the protoplasm, and not in the nucleus, lies the differentiating power of the early stages of development'.

'Insofar as it carries a nucleus, every cell, during ontogenesis, carries the totality of all primordia; insofar as it contains a specific cytoplasmic cell body, it is specifically enabled by this to respond to specific effects only ... When nuclear material is activated, then, under its guidance, the cytoplasm of the cell that had first influenced the nucleus is in turn itself changed, and thus the basis is established for a new elementary process, which itself is not only a result but also a cause'.

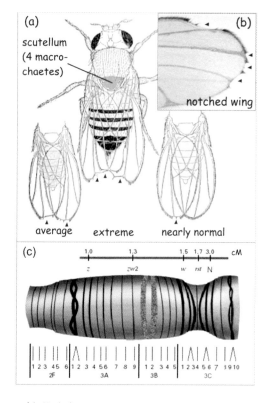

FIG 22.1 Appearance of the Notch phenotype.
(a) Notched wings, first described by Morgan and drawn by Edith Wallace, artist and curator of the *Drosophila* stocks at Columbia University. (b) A photograph of a notched wing. Note the intermittent absence of small sensory bristles (microchaete) along the edge. (c) Localization of the Notch locus on the polytene salivary X chromosome of *Drosophila melanogaster*. z, zeste; zw2, zeste-white2; w, white; N, notch; rst, roughest. Locus distance in centimorgans (cM). 1 cM represents ~300 kilobases.
Image (a) from Morgan.[3] Image (b) from 'The interactive fly', Thomas B. Brody (www.sdbonline.org), section 'Genes involved in tissue and organ development'.

At that time there were two dominant ideas concerning the question of what determines embryonic development. Some declared that the answer must lie in the distribution of the protoplasm. Certain cells contained, the material required to constitute individual organs, while others lacked it. Others had it that all the information resides in the nucleus. Thus, different sets of information would be retained in some cells, not in others, in this way guiding the stages of development.

Initially Morgan denied the evidence for the 'nuclear preformation' theory. This belief was founded, in part, on his own experiments, which revealed the impotence of the nucleus when a segment of cytoplasm was removed. Hans Driesch eventually revealed that the two opposing theories – progressive determination of the cytoplasm by the nucleus versus cytoplasmic control of nuclear function – are like the two sides of a coin. Working with sea urchin eggs, he showed that the nucleus of a cell that is destined originally to become ectoderm can retain the ability to become endoderm or even to generate the entire embryo itself. Driesch correctly concluded that nucleus and cytoplasm interact to guide all the succeeding stages of development.

Compelling evidence for nuclear control of phenotype came from experiments designed to resolve the mechanism of sex determination. Studies of male gametogenesis in the lubber grasshopper, by Sutton and McClung, demonstrated that half of the spermatocytes carry an 'accessory chromosome' that has an effect on sex determination following fertilization. Henking had already noted this accessory chromosome in plants, and not knowing its function, labelled it 'X', henceforth the identifier of the larger of the two universal sex chromosomes. Further evidence came from Nettie Stevens, who studied the chromosomes of mealworm beetles. She reported a

> . . . clear case of sex determination, not by an accessory chromosome but by a definite difference in the character of the elements of one pair of chromosomes of the spermatocyte of the first order, the spermatozoon which contains the small chromosome determining the male sex.

In order to resolve this conflict Morgan commenced an investigation of sex-limited inheritance in *Drosophila*. In 1910, assisted by some white-eyed mutant flies that had entered his stocks by chance, he found that the inheritance of red-eyes is inseparable from the X chromosome. Indeed, the factors that determine eye colour, body colour, wing shape, and sex all segregate together with the X chromosome. This constituted direct evidence for the determination of sex by the chromosomes: he had the provided evidence that genes are heritable elements localized on chromosomes and that their absence or presence affects the phenotype. This finding is known as the *gene theory*.

In 1913, Morgan and A. H. Sturtevant presented the first physical genetic map of the X chromosome of *Drosophila*. From a continuous search of mutants and crossing experiments, genetic maps of other chromosomes came to light. The notched-wing *Drosophila* played a part in all this. By 1933, developmental biology had advanced, especially with the findings of Spemann and others (see page 618). It was understood that environmental factors, operating by cell-to-cell contact or diffusible factors, influence cell fate (the discovery of the 'organizer'). But again the problem of how these interactions change the phenotype of cells remained unsolved.

'From the point of view under consideration, results of this kind are of interest because they bring up once more, in a slightly different form, the problem as to whether the organizer acts first on the protoplasm of the neighboring region with which it comes in contact, and through the protoplasm of the cells on the genes; or whether the influence is more directly on the genes. In either case the problem under discussion remains exactly where it was before. The evidence from the organizer has not as yet helped to solve the more fundamental relation between genes and differentiation, although it certainly marks an important step forward in our understanding of embryonic development'.[4]

We now know that cell environmental factors, acting through receptors, operate signal transduction pathways that directly or indirectly (by changes in protein translation) affect the activity of factors that lead to induction or suppression of transcription. The Notch signal transduction pathway is no exception. It modifies cell fate by silencing some genes and inducing the expression of others, mediating an essential and perhaps universal function in the assignation of cell fates during development.[5] Notch is just one example of how environment, in this case the neighbouring cell, influences cell fate, but its mode of action provides a perfect and even simple answer to the question that Morgan had been asking throughout his career.

Unlike any other system that we have discussed, Notch acts through ligand-induced cleavage of its own intracellular segment that moves straight to the nucleus where it regulates transcription.

One gene, many alleles

The Notched wing *Drosophila* is the consequence of null alleles, abnormal genes that result in loss of function of the gene product. They are the consequence of nonsense mutations, a translocation or an inversion of the chromosome. Null alleles for Notch are lethal in hemizygous males (XY) and only viable in heterozygous females in which they are partially compensated by a functional gene. There are numerous lethal Notch alleles, all indicated as N with superscripts: N^{-40}, N^{NIC}, etc. Females carrying one mutated N allele display the characteristic notched wings, thickened wing veins, and some bristle abnormalities. There are many other, non-lethal, Notch alleles, the consequences of a partial loss of function due to missense mutations, partial deletions, or very low levels of expression. One group, the 'recessive visibles', which fall in three complementation groups, facet (fa), split (spl), and notchoid (Nd), affect either wing or eye morphology. Yet another class comprises the dominant Abruptex (Ax) mutations, which cause missense mutations in the extracellular (ligand-binding) domain of Notch and, surprisingly, enhance the Notch signal.

When Poulson commenced his analysis of flies that had lost the whole or certain bands of the X chromosome, including the band that contains the Notch locus, he discovered a bewildering array of morphological abnormalities affecting nearly all parts of the body. An excess of neural tissue was particularly apparent. This provided the first evidence that Notch has an important role in the early stages of development.[6,7] This was confirmed and extended by the work of Campos-Ortega[8] who demonstrated that the neural hypertrophy is due to a defect that arises very early, roughly 4 h after fertilization. At this stage, precursor cells (neuroblasts) segregate from others that constitute the neurogenic ectoderm. Normally, only the neuroblasts give rise to neural tissue, the remainder of the neurogenic ectoderm becoming epidermis.[9] However, when Notch is absent, nearly all cells proceed towards a neural fate and very few are left to form the hypodermis. This was later explained by a lack of lateral inhibition.[10] We return to this phenomenon below. Here too Notch plays the role of suppressing the default neural fate, thus ensuring that a sufficient number of proneural cells become epidermal, to provide a strong cuticle. Other defects were found in the formation of somites (muscle), the Malpighian tubules (renal function), and the eyes. Using temperature-sensitive mutants (in which the loss of function can be induced by a change of temperature), anomalies were observed in the process of oogenesis, in which lack of Notch activity alters the composition of the egg chamber (excess of polar cells).[11]

Lastly, similar phenotypes occur with mutations at the Delta and Serrato loci that harbour genes coding for the ligands of Notch. Below we describe how Notch is activated and how the signal is transmitted to the nucleus, there to change gene expression and thus cause a switch in cell fate.

Membrane components of the Notch pathway

Notch ligands (DSL proteins)

The notch ligands are single-pass transmembrane proteins that are characterized by an N-terminal DSL motif, named after representatives of the family of ligands present in three different organisms: Delta (human), Serrate (*Drosphila*), and LAG-2 (*Xenopus*) (Figure 22.2). Although this domain is essential for interactions with the receptor, molecular understanding of how ligand and receptor interact is still lacking. The ligands are separated into two classes: Delta (or Delta-like) and Serrate (or Jagged in mammals). The main structural difference between the two is an additional membrane-proximal cysteine-rich region, resembling the von Willebrand factor type C domain, in the Jagged members.[12] Mammals have five genes encoding their ligands, three Delta-like-1,3,4 and two Jagged-1,2. *Drosophila* carries only one Delta and one Serrate whereas *C. elegans* has several Delta homologues (lag-2, apx-1, Arg-1, DSL-1) (see Table 22.1).

Notch receptors

The Notch receptors are also single-pass transmembrane proteins occurring in the plasma membrane as heterodimeric molecules, comprising one extracellular and one transmembrane segment, linked non-covalently (although a disulfide bond has been suggested) (Figures 22.2 and 22.3). The fucosylated Notch precursor is further processed in the trans-Golgi by cleavage of the S1 site by a furin-like convertase.[13] Subsequently the

The many abbreviations used in this chapter are collected together at the end of the chapter.

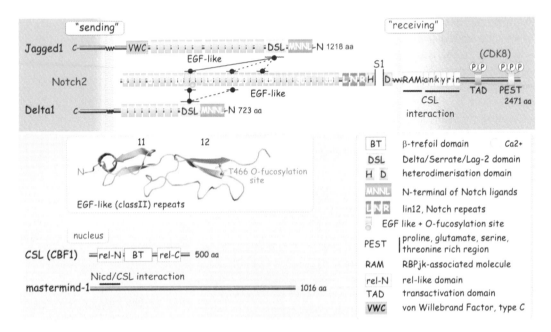

FIG 22.2 Domain architecture of Notch, its ligands and nuclear proteins associated with its intracellular segment.
Both Notch and its ligands contain numerous EGF-like domains. This renders them long and flexible, but the flexibility is restrained at domains that bind Ca^{2+}. The DSL domains interact with Notch at different sites. The mature Notch protein is cleaved in the heterodimerization domain (S1). The intracellular segment contains a RAM domain, approximately 100 residues and loosely defined as a region that commences at the γ-secretase cleavage site (see Figure 22.9) and ends at the first ankyrin repeat. The C-terminal region contains TAD and PEST motifs, both subject to phosphorylation by CDK8. The nuclear protein CSL binds the intracellular fragment of Notch, in which both the BT and C-terminal Rel-like domains take part. Mastermind-1, which binds CSL and the intracellular Notch segment (Nicd) with its N-terminal regions, has no recognizable domains.

fragments are reassembled as a heterodimer. This is essential for Notch activity in mammals, although *Drosophila* appears to get away with uncleaved receptors.[14] The extracellular segment contains numerous (26–36) EGF-like domains and three cysteine-rich repeats (Lin repeats) that mask an essential second extracellular cleavage site (S2) (Figure 22.3). As an indication of its importance, the region covering the Lin repeats is altered in 26% of activating mutations that are associated with T cell acute lymphoblastic leukaemia.[15] The DSL domain of the ligand interacts with EGF-like repeats number 11 and 12, but other regions are not excluded (Figure 22.2). Some of the EGF-like repeats contain Ca^{2+} binding sites, which stiffen the molecule (as in the cadherin adhesion molecules), whereas others are modified by fucosylation on serine or threonine (for instance T466 on EGF-like repeat 12).[16] Mammals have four genes, Notch-1–4, while *Drosophila* carries only one copy and *C. elegans* has two (see Table 22.1).

TABLE 22.1 Components of the Notch signalling pathway

Mammals	*Drosophila*	*C. elegans*	Function
Transmembrane ligands and receptor			
Notch 1-4	Notch	Lin-12, Glp-1	Transmembrane receptor and transcription factor
Delta 1,3,4 Jagged 1,2	Delta, Serrate	APX-1, Lag-2, Arg-1, DSL-1	Transmembrane ligand of Notch receptor
Transcription (co)-factors SCL (CBF1/RBPjk) Mastermind 1-3	Su(H) mastermind	Lag-1 Lag-3	DNA binding transcription factor Transcriptional co-activator
Receptor associated proteins (involved in activity and endocytosis)			
Numb Numbl	Numb Sanpodo	Num-1	Links Notch to α-adaptin (AP-2 component) Membrane localization strictly required for Notch activity (positive), links Notch with Numb (negative)
Glycosylation enzymes of ligands and receptor			
POFUT-1	OFUT-1	OFUT-1	GDP-fucose protein *O*-fucosyltransferase (modifies receptor and ligand, activates signalling)
Lunatic, manic and radical, Fringe	Fringe	no homologue identified	*O*-fucosylpeptide β-1,3-N-acetylglucosaminyltransferase (modifies receptor and ligand, activates signalling)
Regulation of endocytosis			
Dynamin 1, 2 Auxilin, GAK	Shibire Auxilin	Dyn-1 Dnj-25	Required for pinching off vesicle Role in clathrin uncoating (for further processing)
Epsin1,2	Liquid facets (lqf)	Epn-1	Clathrin-associated sorting protein, required for Delta signalling
Transmembrane proteases			
ADAM10, -17	Kuzbanian, Kuzbanian-like, TACE	SUP-17, ADM-4	Metalloproteases, cleaving the S2 site in Notch, prerequisite for Notch signalling
Presenilin 1,2, nicastrin, APH1, PEN2	Presenilin, nicastrin, APH1, PEN2	SEL-12, APH-1, APH-2, PEN2	Proteins of the γ-secretase complex, cleaving the S3 and S4 sites in Notch, activates Notch signalling

Continued

TABLE 22.1 Continued

Mammals	Drosophila	C. elegans	Function
Delta and Serrate E3-ligases (regulation of trafficking)			
Mib1, skeletrophin, neuralized 1,2	Mind bomb 1,2 (Dmib), neuralized	Y47D3A.22	RING-finger type E3 ubiquitin ligases for Delta and Jagged/Serrate, promotes endocytosis, activation of Notch signalling
Notch E3-ligases acting in the cytoplasm (regulation of trafficking)			
Itch, NEDD4	Su(dx), Nedd4	WWP-1	HECT-type E3 ubiquitin ligase, targeting Notch for lysosomal degradation, bind PPSY motif, promotes endocytosis and lysosomal degradation, inhibition of Notch signalling
Deltex 1-4	Deltex	no homologue identified	RING-finger type E3 ubiquitin ligase, binds ankyrin repeats, promotes Notch localization to Rab11-positive vesicles (recycling endosomes), rescue from degradation, activation of Notch signalling
Notch E3-ligase acting in the nucleus			
Fbw7/Sel10	Archipelago	Sel-10	F-box protein, implicated in ubiquitylation of Notch cytosolic fragment, leading to degradation by proteasome, termination of signalling.

Adapted from Fiuza and Arias[17] and Nichols et al.[18]

Functional Notch requires plasma membrane-expression of the tetraspan membrane protein Sanpodo.[19] Little is known, about this protein, which was originally discovered as the homologue of the actin-associated tropomodulin. Sanpodo acts downstream of Delta and upstream of Notch to which it binds. Importantly, loss of its expression at the plasma membrane with subsequent accumulation in endocytic vesicles, renders cells unresponsive to Delta. We return to Sanpodo below (see page 720).

Glycosylation of ligands and receptor

The fucosylation sites on the EGF-like repeats of Notch and its ligands are further modified by addition of N-acetylglucosamines (so-called glycosylation elongation). Lack of the initial fucosylation enzyme, OFUT1, results in a phenotype similar to loss of Notch.[20] It may possibly also act as a chaperone in the rough endoplasmic reticulum, stimulating Notch signalling simply by

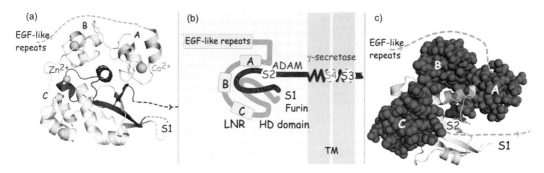

FIG 22.3 Structure of the Notch regulatory region (NRR).

(a) A ribbon representation of the Notch protein. A, B, and C indicate the three Ca^{2+}-binding Lin-12 repeats and the S1 site. Notch is cleaved in the cis-Golgi by a furin-like convertase at the S1-site and arrives at the cell surface as a heterodimer reassembled from the fragments. (b) The Lin-12 repeats mask the S2-cleavage site. The Notch ligand renders this site accessible to the ADAM family of proteases. Cleavage at S1 is ligand-independent, whereas S2 cleaved by ADAM, and S3/S4 cleaved by γ-secretase, are dependent on ligand interaction. LNR, lin-12/Notch repeat; HD, heterodimerization domain. (c) Lin-12 repeats (spheres) covering the S2 site. (2004).

preventing protein misfolding.[21] Fringe, which acts in the Golgi apparatus, generally has the opposite effect, limiting signal output by O-fucose elongation of Notch and its ligands.[22,23] Regulation of OFUT1 and Fringe may contribute to the pattern of Notch activation during development.

Activation of Notch

Signalling through Notch occurs only between cells that are in direct contact with each other.[24] Activation entails proteolytic processing at two sites. First, ligand binding induces cleavage at the S2 site by members of the ADAM family of metalloproteases (Fig 22.3). Then a second cut by the γ-secretase complex, cleaves the S3 site in the transmembrane region.[25,26] There is a further cut by γ-secretase at S4, in the centre of the transmembrane domain,

γ-**Secretase** is a transmembrane multiprotein protease complex. The eight-pass presenilin comprises the catalytic subunit. This is surrounded and stabilized by nicastrin and APH1. PEN2, a two-pass membrane protein, induces endoproteolysis of the catalytic subunit, required for activity.[29] Loss of function of presenilin correlates with accumulation of plaques in Alzheimer patients (extracellular deposition of amyloid-β42 polypeptide-aggregates).[30] It is currently a focus of interest in terms of pharmacological intervention (see Gilbert[2] and Wolfe[29]), in the hope that it may be possible to define new targets in cancers in which mutations of Notch contribute to the pathology (such as T cell acute lymphoblastic leukaemia) or in which Notch acts to prevent progenitor cell differentiation (such as in intestinal adenomas).

Members of **ADAM/ TACE** cleave the juxtamembrane region of transmembrane proteins, causing shedding of the ectodomains (for instance liberation of TNF-α (Figure 16.3, page 490) or EGF Figure 12.25, page 356). The family comprises some 29 members. Of these ADAM10 and ADAM17 (TACE) are able to cleave Notch, whereas ADAM12 cleaves its ligand Delta.

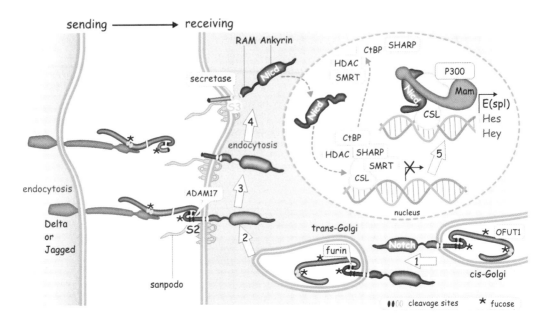

FIG 22.4 Activation of the Notch pathway.

Notch appears at the plasma membrane in a glycosylated (1) and cleaved (2) form. It binds its partner Sanpodo. Binding of ligand (Delta or Jagged) exposes the S2 site, leading to cleavage by ADAM17 (3). This renders Notch susceptible to cleavage at S3 by γ-secretase (4). Endocytosis of the receptor seems a prerequisite for this. The intracellular segment, Nicd, translocates to the nucleus where it binds CSL and causes the recruitment of mastermind (Mam). This leads to a loss of CSL-associated repressors, which are replaced by Mam-associated activators of transcription (5). Important target genes are members of the Enhancer of Split family, E(spl) in *Drosophila*, of which Hes and Hey are the equivalent families in mammals.

but the functional significance of the resulting Notch fragment has not been investigated. We limit our description to the signalling role of the S3 Notch-intracellular domain (Notch-icd or Nicd) (Figure 22.4).

What exactly renders Notch sensitive to proteolytic cleavage following ligand binding remains unclear, but the removal of the inhibitory Lin-repeats, which mask the S2-cleavage site, seems to be a prerequisite. Why the truncated version then becomes susceptible to the action of γ-secretase also remains enigmatic. Curiously, the number of EGF repeats seems to determine the efficiency of proteolysis. Long chains are less efficient, and truncated versions, such as those that occur in the human oncogenes TAN1 (Notch1) and INT3 (Notch4), are constitutively active receptors.[27] Both ADAM and γ-secretase are subject to regulation by post-translational modifications that affect their activity. In which subcellular compartment γ-secretase cuts Notch remains to be resolved. Although clearly evident in early endosomal vesicles, it does not necessarily follow that this is where it acts (see page 713).[28]

Regulation of transcription by the Notch intracellular domain

The untethered intracellular segment of Notch (Nicd) migrates into the nucleus and binds to the transcription factor CSL (Figure 22.4). This interaction first involves the RAM region, which binds the β-trefoil domain of CSL, and then the ankyrin repeats, which bind the C-terminal Rel-homology region (Figure 22.5).[31–33] Together, Nicd and CSL create a binding groove that accommodates the helical structure of Mam, which creates a platform for attachment of further components such as GCNS and the nucleosome acetylation factors CBP or p300. It is also involved in the recruitment of SKIP and, importantly, the cell cycle kinase CDK8 (see page 712).[34] Numerous copies of these complexes form and they are readily visualized as distinct nuclear foci (Figure 22.5).

The Notch transactivation process generally resembles the mechanism of induction of transcription by the Wnt–β-catenin pathway (see page 421). Thus, the Notch and Wnt target genes are normally repressed by the association of corepressors with CSL or with LEF. The Notch transactivation mechanism, however, differs slightly in that translocation of Nicd requires both CSL and MAM in order to initiate transcription on chromatin templates *in vitro*, whereas β-catenin requires only LEF.[34,35] Note that CSL may also repress basal transcription through direct binding to, and inhibition of, TFIIA and TFIID.[36]

Effector genes of Notch signalling

The Nicd-induced transcriptional complex leads to the expression of genes that are members of the *E(spl)* class in *Drosophila* (a complex of at least seven tandem genes) or the *Hes* class in mammals. (*Hes* and *Hey* are the mammalian counterparts of *E(spl)*.) These are transcription factor genes, the so-called Notch-effector genes.[37,38] In mouse and rat, seven *hes* (1–7) and three *hey* (1, 2, and L) genes have been identified. The *Drosophila* e(spl) and the hes and hey gene products belong to a large family of small proteins characterized by a helix–loop–helix (HLH) motif.[39] Well-known members of this family of DNA-binding transcription factors are MyoD (differentiation of muscle tissue) and c-Myc (oncogene, regulation of expression of components of ribosomes).

The N-terminal DNA-binding basic domain of these proteins is contiguous with one of two α-helices separated by a loop (helix–loop–helix) that serves as a dimerization domain and as a platform for additional protein interactions (Figure 22.6). This is followed by two additional α-helical stretches making up the Orange domain, which also serves as an interface for protein interactions, including the formation of homo- and heterodimers of HLH proteins. It also acts as a transcriptional repressor. In Hes proteins, a highly conserved C-terminal tetrapeptide motif -WRPW- recruits the corepressors of the TLE family, of which Groucho (*Drosophila*) is the most familiar (see also page 422). The Hey proteins lack this motif and cannot bind Groucho, but their bHLH

Split and Enhancer of Split

Split is one of the non-lethal alleles of the notch locus. It is a partial loss-of-function mutant that affects formation of the bristles as well as the eyes. The macrochaetae, particularly the dorsocentrals and the scutellars on the mesonotum of the thorax, appear split, doubled, or absent, and so do some of the microchaetae. The eyes are of rough, of unvarying texture, and smaller than the wild type.[40] The partial loss of function is associated with an isoleucine to threonine mutation in the 14th EGF-like repeat.

Enhancer of Split

exaggerates the effects of Split, for the worse. In genetic terms, Split and Enhancer of Split are said to interact.

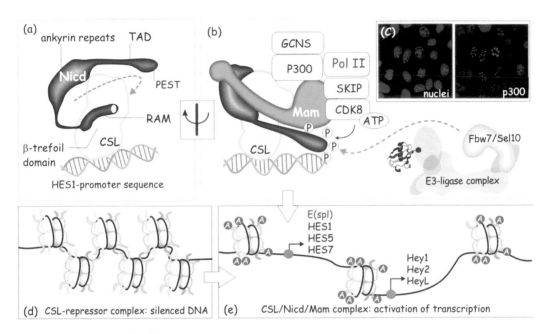

FIG 22.5 The Notch transcriptional complex.

(a) The intracellular Notch segment, Nicd, first binds the β-trefoil domain of CSL and then folds over to attach, by its ankyrin repeats, to the C-terminal domain. (b) This creates a binding site for Mam and a trimeric complex is formed. The C-terminal domain of Mam recruits numerous proteins that cause acetylation of histones, so rendering the DNA amenable for transcription. Among the recruited molecules is CDK8, which later phosphorylates Nicd in both its TAD and PEST regions. Phosphorylated PEST is recognized by the receptor subunit of a nuclear E3-ligase complex, resulting in polyubiquitylation of Nicd, followed by proteasome-mediated destruction. (c) Numerous transcriptional complexes thus formed are readily visible by immunostaining of p300 (d) and (e) Histone acetylation unwinds DNA and transcription follows removal of repressors and recruitment of transcriptional activators. Image (c) from Fryer et al.[34]

(basic helix–loop–helix) motif interacts with yet another histone deacetylase, SIRT1. The corepressors and associated histone deacetylases render the DNA inaccessible to RNA polymerase and thus silence gene expression. In some cases Hes proteins bind CBP and act as transcription factors. Apart from binding DNA and recruiting transcriptional repressors or activators, bHLH proteins have other means of interfering with transcription. For instance, they may sequester other DNA-binding proteins away from the DNA, or they may bind other transcription factors, thus bringing gene silencing repressors to promoter regions where they themselves do not bind. For more information about Hes and Hey proteins, see Fischer and Gessler.[38]

The response to Notch is not fixed by expression of a set of transcription factors that endow certain phenotypes; rather, it acts as a switch with the outcome of its action being determined by the context in which it operates (cell type, developmental stage, other first messengers). The situation is similar to that described for Wnt in Chapter 14. In this way it is applied in cell

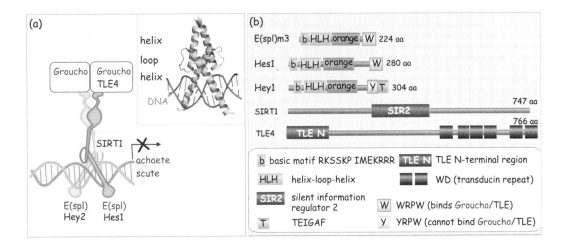

FIG 22.6 Molecular structure and domain architecture of Notch effectors.

(a) Notch target genes are members of the basic helix–loop–helix (bHLH) class of transcription factors (the insert represents the structure of MyoD). These bind DNA as homo- or heterodimers. In *Drosophila*, Notch induces expression of E(spl), which in turn binds Groucho and leads to suppression of expression of achaete and scute, thereby blocking the neural differentiation pathway. In mammalian cells, Notch induces Hes and Hey, members of the same bHLH family. The C-terminal -WRPW- motif of Hes1 interacts with TLE4, whereas Hey2 binds SIRT1 with its bHLH motif. Together they repress gene expression. (PDB file: 1MDY) (b) Domain architecture of Notch effector proteins. SIRT1 and TLE4 are histone deacetylases. (Hey1, 2 and HeyL are also known as Hrt1,2,3, Hesr1,2, Herp 2,1 or Chf2,1.)

fate decisions at many stages of the developing embryo. In *Drosophila*, Notch plays an important role in determining the choice between epidermal and neural development; its activation prevents neurogenesis. More detail on this is given below (see page 720).

Biological functions in humans

In mammals, Hes and Hey have a role in the development of the nervous system, sensory organs (eye, inner ear), pancreas, endocrine cells, and lymphocytes. Hey genes are critical in the cardiovascular system, their dysfunction causing congenital defects, impairments of angiogenic remodelling, and a lack of arterial differentiation.[38]

Gene cassettes and the Enhancer of Split complex

In *Drosophila*, genes are often organized by function on the chromosome. The phenomenon of a group of genes, a *cassette*, having related function and location on the chromosome signifies a so-called *gene complex*. The Enhancer of Split complex, or E(spl) complex, includes eight genes spread over 50 kilobases on the *Drosophila* third chromosome. Other examples of *Drosophila* gene complexes include the antennapedia, bithorax, and achaete/scute (AS) complexes. The hallmark of all these gene complexes is that within any complex the genes are evolutionarily related and jointly regulated. Genes of the E(spl) and the AS complexes regulate neurogenesis and related differentiation pathways. Whereas transcription factors encoded by the AS complex generally activate transcription of other genes, gene products of the E(spl) complex generally repress transcription. Proteins encoded by AS are proneural, initiating neurogenesis, whereas E(spl)-coded proteins prevent neurogenesis. The E(spl) complex contains three additional Notch-responsive, non-bHLH genes: of these, *m4* and *mα* are structurally related, while *m2* encodes a novel protein. For the most part, the genes of the E(spl) complex are redundant; only the nominally defining gene of the complex, ***Enhancer of Split,*** yields a noticeable phenotype when mutated. For further information, see Simpson.[11]

Destruction of the Notch-icd, Nicd

The content of Nicd in the nucleus is vanishingly small because of its rapid degradation. In fact, the assembly of a transcriptionally active protein complex is paralleled by an almost immediate phosphorylation of Nicd and subsequent recognition by a nuclear E3-ubiquitin ligase so that it signals its own demise. Recruitment of CDK8 leads to phosphorylation of Nicd in the transactivation and PEST domain (Figure 22.5). The phosphorylated protein is recognized by the receptor-subunit Fbw7/Sel10 (a WD40 F-box receptor protein) of a nuclear SCF-type E3-ubiquitin ligase complex and the polyubiquitylated Nicd is destroyed by the 26S proteasome.[41,42] If signalling is to be maintained, a continuous supply of Nicd must be provided through ligand-mediated cleavage of the intact membrane protein.

The importance of a continuous supply is well illustrated by the observation that severe suppression of Delta expression in the satellite cells of ageing livers leads to impaired regenerative capacity; due to a failure of Notch signalling, the cells can no longer escape terminal differentiation.[43] Conversely, loss of the PEST domain renders the protein very stable and this contributes to the development of acute lymphoblastic leukaemia (an excess of undifferentiated cells).[15]

Both receptor and ligand trafficking are essential for Notch signalling

One might expect that signal-transmitting cells would exhibit high levels of ligands on their surface so as to activate Notch on adjacent receiving cells. Curiously, in *Drosophila,* most of the Delta is confined to intracellular vesicles and the membrane pool is constantly being removed. Moreover, only ubiquitylated ligands, linked to clathrin-associated sorting proteins, are competent to activate Notch.

Compelling evidence for a role of trafficking came from the finding that shibire (shi), a temperature-sensitive mutant of dynamin, causes a Notch-like phenotype in *Drosophila*. At the restrictive temperature (29°C), Shibire mutants develop an excess of nervous system at the expense of ventral epidermis.[44] Notch signalling is often required to prevent the default neurogenic phenotype (thus to obtain an epidermal cell instead). Therefore, it follows that loss of function of dynamin prevents effective Notch signalling, (even though all the related components, receptors, and ligands, etc., are normally expressed). Dynamin is a GTP-binding protein necessary for the pinching-off (fission) of endocytotic vesicles from the plasma membrane.[45] Cells that express shibire still form clathrin-coated pits but fail to internalize these vesicles. This correlation prompted the question of whether a block in endocytosis could account for the interruption in the communication necessary for normal epidermal and neural cell differentiation? The answer is yes; trafficking of both ligands and receptor affect signalling, though the full picture still remains unclear.

Trafficking of ligand (Delta, Serrate)

Further important evidence for the role of endocytosis in the regulation of Notch signalling came from genetic screens in *Drosophila* and zebrafish (*Danio rerio*). These revealed three key regulators: epsin-1, neur, and mib (Table 22.1). Epsin is a long, almost linear protein that acts as an adaptor for endocytosis, coupling polyubiquitylated protein to α-adaptin (AP-2).[46] Neur and mib are E3-ubiquitin ligases that ubiquitylate Delta and Serrate.[47] Loss of Neur in *Drosophila* and *Xenopus laevis* and of mib in zebrafish results in neurogenic phenotypes resembling those of Notch mutants.[48] Two schemes have been proposed to explain why endocytosis is necessary for proper signalling. The first postulates that uptake of ligand, when bound to the Notch receptor, pulls away the three Lin repeats that normally mask the S2 cleavage site. This 'pulling' renders S2 accessible to ADAM/TACE (see Figures 22.3 and 22.7). The other idea proposes that ligand uptake leads to glycosylation of the EGF-like repeats in the early endosome compartment. This recycling pathway not only renders the ligand competent for receptor binding but may also deliver the ligand to the correct membrane domain in a 'clustered' manner (Figure 22.7). This does not excluded a third possibility, that trafficking of Delta and Serrate, after binding to Notch, is required for signalling within the ligand-presenting sending cell, in a similar way as for Notch in the ligand-receiving cell.[49]

Trafficking of receptor (Notch)

Ubiquitylation and the subsequent endocytosis of Notch has both stimulatory and inhibitory consequences. The inhibitory aspect is rather straightforward; it concerns removal of the unoccupied receptor from the plasma membrane, thereby preventing interaction with ligand. Two ubiquitin E3 ligases stand out in the regulation of endocytosis; NEDD4 and Itch (both HECT-type ligases: see page 469) (Figure 22.8 and Table 22.1). Ubiquitylation conveys Notch into

Epsin is described as a 'natively disordered' polypeptide, meaning that apart from the N-terminal region, it has few structural features.[50] It contains three ubiquitin-interacting motifs (UIM) which, however, bind weakly to ubiquitin. Tight binding therefore requires several interactions, and this may explain why polyubiquitylated proteins are favoured. It connects to PI(4,5)P$_2$ and it carries several AP-2 binding motifs that bind α-adaptin. Epsin also interacts with RhoGAP, suggesting that it has a role in actin dynamics, possibly explaining why it is an important component of the endocytotic machinery (see Figure 15.15, page 471).[46,51,52]

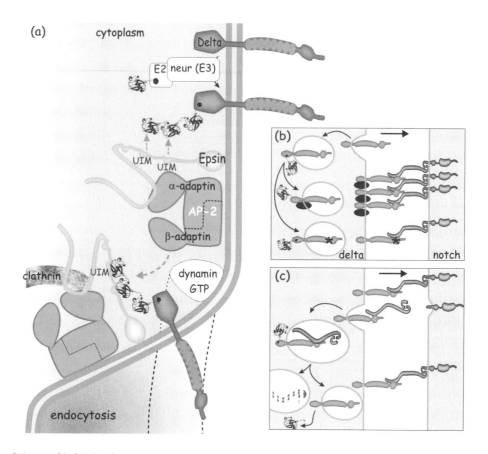

FIG 22.7 Endocytosis of the Delta ligand.

(a) Delta is ubiquitylated by the E3-ligase neur. The polyubiquitylation chain is recognized by the UIM of epsin which links Delta to the adaptor complex AP-2, leading to its endocytosis with the help of the coat protein clathrin. (b) Endocytosis may facilitate Notch signalling in a number of ways. It may be re-expressed at the membrane with important cofactors and in a clustered manner, which may enhance Notch binding and activation. It may also undergo further glycosylation (symbolized as a blue asterisk). (c) Endocytosis may facilitate unmasking of the S2-cleavage site by withdrawing the Lin-12 repeats. Finally, endocytosis may be needed for the effective recycling of ligand after removal of the extracellular Notch segment.

Itch is a gene present in mice that continually scratch their skin. This behaviour is due to aberrant inflammatory responses in several organs, including the skin.[53]

an early endosome, identified by the colocalization of the small GTPase Rab5, and, if no further modifications occur, this is followed by its destruction in the lysosomal compartment (recognized by association with Rab7).

Why Notch is stimulated by ubiquitylation and endocytosis is less clear. One possibility is based on the finding that the catalytic component of γ-secretase (presenilin) predominantly resides in the early and late endosomes, and this might facilitate cleavage of the S2 site[18] (Figure 22.9). Another possibility is that a second round of ubiquitylation, in the early endosomal compartment, rescues

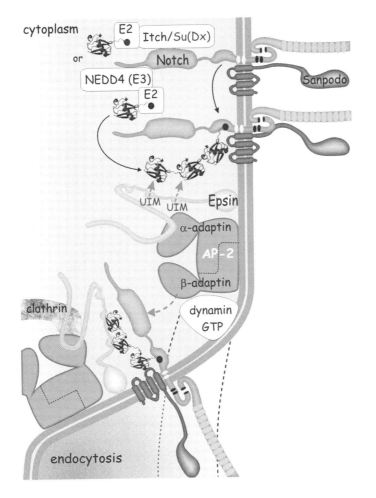

FIG 22.8 Endocytosis of the Notch receptor.

Notch, polyubiquitylated by Itch (Su(Dx) in *Drosophila*) or NEDD4. The ubiquitins are recognized by Epsin and this connects Notch with the adaptor protein complex AP-2. Endocytosis occurs with the help of the coat protein clathrin.

Rab GTPases: These are members of the Ras superfamily of monomeric GTPases involved in intracellular vesicle transport. They regulate the assembly, on the surface of emerging transport vesicles, of (1) coat proteins, necessary for cargo selection and vesicle budding, (2) microtubule binding proteins, necessary for transport, and (3) SNARE and docking proteins, necessary for fusion of the released vesicle with other subcellular membrane compartments.

AP-2: A complex made up of four subunits (two large, α- and $\beta2$-adaptin, and two small, $\mu2$-and $\sigma2$-adaptin), of which μ-adaptin has the role of cargo receptor. The 'ear' domain of α-adaptin interacts with accessory proteins, such as Epsin or numb, and coat proteins, such as clathrin.

Notch from lysosomal destruction, so opening the way for an alternative, non-canonical, signalling pathway. This second round of ubiquitylation occurs through Deltex (a RING-finger type E3 ligase), which binds the ankyrin repeat region.[54] Deltex-mediated ubiquitylation directs Notch towards a recycling endosome, identified by the colocalization of the small GTPase Rab11. From here, Notch either returns to the membrane, possibly modified, and contributes to ligand-mediated signalling or it remains in the cell and signals, by an as yet poorly characterized pathway, in a ligand-independent manner.[55,56] Note that the E3–ubiquitin ligase Cbl, which regulates EGF receptor uptake (see Figure 12.21, page 350) is also associated with Notch receptor uptake and could also play a role in its trafficking. The domain architectures of proteins that take part in these processes are shown in Figure 22.10.

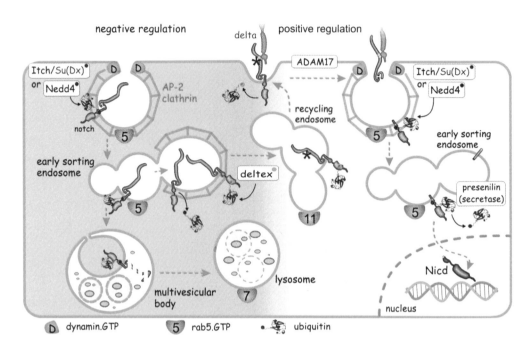

FIG 22.9 Intracellular trafficking of the Notch receptor.

Intracellular trafficking of the Notch receptor has negative and positive effects on its signalling capacity. Receptor endocytosis and transport to early sorting endosomes (Rab5 marker) normally result in partial destruction in multivesicular bodies and then total degradation in the lysosome (Rab7 marker). This effectively removes the receptor from the surface. Alternatively, endocytosis may positively regulate signalling. First, it protects the receptor from destruction and conducting it to recycling endosomes (Rab11 marker) assisted by Deltex-mediated ubiquitylation, which provides a novel sorting signal. From here the receptor returns to the membrane, perhaps in a modified version (indicated by red asterisks) or at a membrane domain more favourable for ligand binding. Secondly, endocytosis also favours proteolysis of the S2-site by bringing the receptor to the compartment that harbours γ-secretase (early and late endosomes). Dynamin (D) and Rabs 5, 7, and 11 are marked by blue numerals. Red spots indicate Itch or NEDD, Blue spots indicate Deltex.

The embryonic life of a fruit fly is brief. It hatches as a first instar larva and then moults successively through second and third instar stages. All three of these instar stages are dedicated to eating. Finally, after a copious intake of food, it pupates. All this takes about 4 days. It then remains as a pupa for another 5 days, at the end of which it emerges as an adult fly.

Notch and sensory progenitor cells of *Drosophila;* the importance of endocytosis

Here we discuss the formation of cuticular patterns, the cellular machinery that generates them, and the genetic circuitry that organizes the machinery. All the events described start at the pupal stage, where the imaginal discs, already visible in third-instar larvae, expand and develop into the body parts of the adult fly. This process is known as *metamorphosis* (**Figure 22.11**). During this process most of the larval structures are destroyed through apoptosis and replaced by tissue derived from the discs, giving rise to thorax, wings, legs, antennae, and genitalia (though not the digestive system). A full-grown

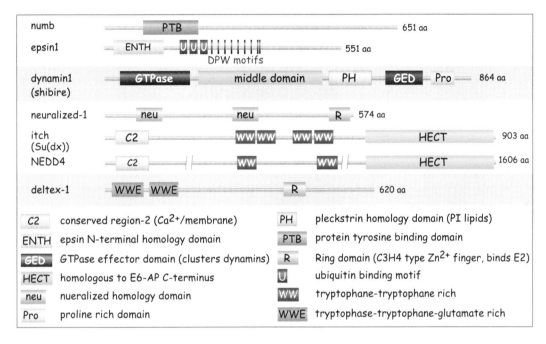

FIG 22.10 Domain architecture of proteins that affect Notch signalling.
Numb and Epsin link Sanpodo and Notch with the endocytotic machinery (AP-2 and clathrin), whereas dynamin is required for fission of the endocytotic vesicle. Neuralized, Itch, NEDD4 and Deltex all are E3–ubiquitin ligases that ubiquitylate the Notch receptor and its ligands, thereby providing sorting signals for the endocytotic machinery.

larva is ~6 mm long, whereas the adult fruit fly is only half that size. When the adults emerge from the pupae, they are fully formed. They become fertile after ~10 h, copulate, the females lay eggs, and the cycle, of roughly 12–14 days, begins again.

Development of mechanoreceptors on thorax and wing

The pair of large wing discs produce both wings and the dorsal site of the thorax (notum) (see Figures 22.11 and 22.12). We focus on the proneural clusters, small patches of cells within these discs. They are detected by expression of neurogenic markers such as the genes *achaete, scute,* and *daughterless* (Figure 22.11b), the products of which give epidermal cells the ability to become sensory organ precursors. The cells surrounding the proneural clusters express high levels of Hairy, a bHLH protein that suppresses the neurogenic genes. Hairy belongs to same family as the previously described

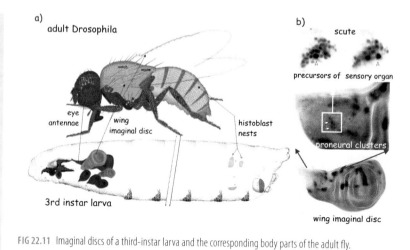

FIG 22.11 Imaginal discs of a third-instar larva and the corresponding body parts of the adult fly.

(a) The imaginal discs harbour cells that form the different body parts of the adult fly during the process of metamorphosis,. Here, only the eye and the wing imaginal discs are indicated (by double-headed arrows). The abdomen arises from histoblast nests. Note that the adult is much smaller than the larva, most of the cells outside the imaginal discs having been removed by apoptosis.

Image adapted with permission from V. Hartenstein, http://flybase.bio.indiana.edu/

(b) The imaginal wing disc contains clusters of cells, proneural clusters, which express the neurogenic genes *achaete/scute*. These clusters later form the mechanosensory organs. In each selected cluster, just one cell that expresses a particularly high level of *achaete/scute* retains the neural fate. The others develop as epidermal cells. Image of immunochemical staining adapted from Cubas et al.[57]

E(spl), Hes, and Hey proteins, but it is regulated differently. These *achaete*-expressing proneural clusters give rise to the bristle bearing sensory organs (mechanosensors) that adorn the dorsal side of the thorax and the edges of the wings (also known as macro- and microchaetes).[58,59] Mechanoreceptors are also abundantly present on the abdomen, legs, and eyes, but these arise from other imaginal discs. (These structures should not be confused with the wing hairs, which are much smaller and arise from epidermal cells, not neuronal cells; see page 421.)

Within the proneural clusters, only the single cell, the sensory organ precursor cell (SOP) has the privilege of developing the organ, providing the neuronal glia, sheath, socket, and bristle cell. The other cells are ruled out through lateral inhibition, and make their contribution in the formation of the epidermis (which produces the cuticular exoskeleton) (Figure 22.12).

Mutations that affect the number or the pattern of sensory organs are thus readily observed by changes in the number or the pattern of the sensory bristles. For example, loss of function mutations in the gene *achaete*, which

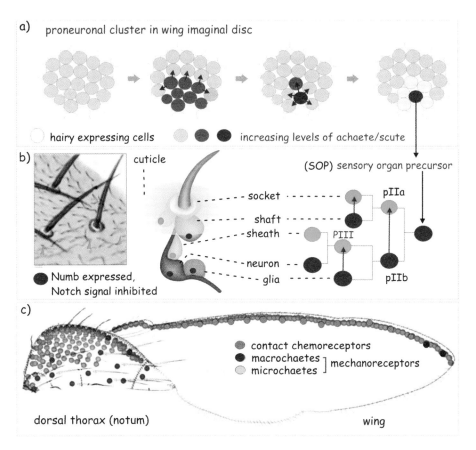

FIG 22.12 Sensory organ development from proneural clusters in the wing imaginal disc.
(a) Proneural clusters in the wing imaginal disc express (among others) the neurogenic genes *achaete* and *scute*. Through lateral inhibition, a progressively smaller group of cells maintains expression of these genes, whereas others lose them and take on an epidermal fate. This process eventually singles out just one cell, designated the sensory organ precursor cell (SOP). (b) The sensory organ precursor cells divide and again through lateral inhibition. Only a single cell maintains the neural phenotype (pIIb), the other (pIIa) losing it. Notch then plays a further role in determining cell lineage (shaft vs socket, sheath/neuron vs glia). The process represents a series of binary switches in which Notch determines the outcome. What we describe here applies for the sensory organs with small bristles but the same principle also operates in development of the organs having large bristles. (c) Schematic presentation of the localization of sensory organs bearing small bristles (microchaetes) and large bristles (macrochaetes), on the thorax and wings of *Drosophila*. Image adapted with permission from V. Hartenstein, http://flybase.bio.indiana.edu/.

codes for an activating bHLH transcription factor that drives differentiation of neurons, lead to flies having a sparse covering of hairs. Conversely, a loss-of-function mutation of the gene *suppressor of hair* results in flies having an excess of sensory bristles (ectopic production).

Scute: from *scutum*, the Roman legionary's shield; knee-pan; scale; piece of bony armour in crocodile, sturgeon, turtle, armadillo, etc.

Scute is a sex-linked gene which controls a stage in the development of a set of bristles, the most characteristic of which are the four on the scutellum (see Figure 22.1), which can be used as an index of activity of the *scute* locus. The number of bristles on the scutellum is what is meant by the 'scute character'. When *scute* mutates, the change in phenotype is extensive and clear-cut. Normally there are four bristles on the scutellum, the number dropping to 0.5 (on average) in males and 1.5 in females in Sc1 mutants.[60]

Achaete: The bristles of the mechanoreceptors are named chaetes. In *Drosophila* there are macrochaetes and microchaetes (see above), present on the edges of the wings and on the dorsal site (notum) of its thorax. Loss of chaetes is referred to as 'achaete'.

Notch and the development of the bristle-containing sensory organ

Comprehensive information regarding the bristle-containing sensory organs can be found in the excellent multimedia resources of Lewis Held, on the website of the Society of Developmental Biology: http://www.sdbonline.org/fly/lewheld/00idheld.htm, section 'the bristle'.

Note that in this example Notch acts predominantly by inhibiting neural differentiation, but this is not always the case. In mouse embryonic stem cells it acts to promote neural fates, a finding that underscores yet again the importance of cellular context in determining the outcome of signalling.[62]

Notch acts at two levels in the development of bristle-containing sensory organs. The first is in the selection of a single sensory precursor cell. Initially all cells of the proneural clusters express both Notch and Delta, but just one cell wins. It possesses a more 'effective' Delta that activates the Notch pathway of its neighbours and, through induction of E(spl), blocks expression of *chaete* and *scute*, thereby shutting down their neurogenic programme (Figure 22.12a). Once the sensory organ progenitor cell has been established, Notch determines the subsequent cell lineage decisions so as to determine the destination of progeny, turning them into a bristle (shaft), socket, glia, neural, or sheath cells (Figure 22.12b). The first (asymmetric) division of the sensory precursor, leading to the progenitor cells, pIIa and pIIb, has been thoroughly investigated.

In the prophase of cell division, just before the condensed chromosomes are aligned by the microtubule cytoskeleton, three proteins of the sensory organ precursor accumulate on one side of the cell. These are Neuralized (E3–ubiquitin ligase), Numb (adaptor), and the AP-2 complex. The daughter cell possessing the higher content of these three proteins loses its ability to receive signals, but has an abundant capacity to transmit, because of very effective trafficking of Delta.[49] Like epsin (see above), Numb is an almost linear (non-globular) protein. It has an N-terminal PTB domain which does not necessarily require a phosphotyrosine in order to bind to other proteins. Numb binds Notch and it binds Sanpodo. Its tail attaches to the ear segment of α-adaptin (AP-2 complex) (Figure 22.13). As a consequence, the daughter cell that expresses Numb loses cell surface expression of Sanpodo, which now concentrates in early endosomal vesicles due to selective uptake by Numb/AP-2.[61] This deprives Notch of Sanpodo and impairs its receptor function. Moreover, Numb binds

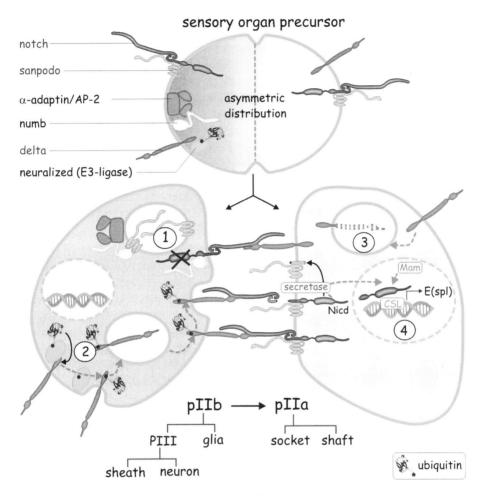

notch

sanpodo

α-adaptin/AP-2

numb

delta

neuralized (E3-ligase)

sensory organ precursor

asymmetric distribution

pIIb ⟶ pIIa

PIII   glia   socket   shaft

sheath   neuron

ubiquitin

FIG 22.13 Asymmetric division of the sensory organ precursor cell determines cell lineage fate of its daughter cells.
During the prophase of the cell cycle, the sensory precursor cells distribute AP-2, Numb, and Neur asymmetrically. The daughter cell having the highest concentration of these components efficiently expresses the ligand Delta. Notch signalling fails due to removal of Sanpodo from the plasma membrane (1) which occurs through Numb/AP-2 mediated uptake of Sanpodo. Neur, on the other hand, stimulates the trafficking of Delta (2), which now activates the Notch exposed on the adjacent cell. This signal-receiving cell seems to have some difficulty expressing its ligand (3). Notch-mediated gene expression (4) ensures that the receiving cell, pIIa, takes on the lineage of socket and shaft, while the descendants of the sending cells, pIIb, provide the neuron, glial, and sheath cells.

the RAM domain of Notch and this may prevent proteolytic processing or, if cleavage still occurs, it may even prevent binding of Nicd to its nuclear target CLS. The presence of Neuralized accounts for very effective ubiquitylation and trafficking of Delta. As shown in Figure 22.7, this may lead to (a) focused cell surface expression of clusters of Delta, (b) association of essential cofactors, or (c) activation of Notch by pulling the inhibitory ankyrin repeats of Notch away from the S2 cleavage site.[48,49]

Notch in the maintenance of an intestinal stem cell compartment

Activation of Notch causes potent inhibition of differentiation in a number of developmental contexts, and it has been associated with amplification of somatic stem cells, such as the neural and haematopoietic stem cells.[63,64] The intestine too has its stem-cell compartments, present in the crypts (see Figure 14.14 page 434). As described in Chapter 14, Wnt signalling plays an important role in driving proliferation of these stem cells (as well as their early progenitors). In the case of the small intestine, they do not differentiate until they are pushed out of the crypt and start moving up the walls of the villi. Once they reach the top, they are shed. Here we indicate how Notch acts in concert with Wnt by preventing differentiation both of the stem cells and of their early progenitors (Figure 22.14). Moreover, Notch signalling is vital to assure a balanced mixture of secretory and absorptive cells.

The role of Notch in the development of intestinal crypts is well indicated by the finding that the intestines of zebrafish that are DeltaD (ligand) mutants have an increased number of secretory goblet cells, at the expense of absorptive epithelial cells (enterocytes).[65] The default for a crypt progenitor cell is differentiation towards goblet cells. To generate a balanced mixture of absorptive and secretory cells, lateral inhibition, mediated by Delta–Notch signalling, is needed. This is also true for mice in which multiple Notch pathway components are expressed. Mice lacking *hes1* have few if any absorptive cells and reveal an excess of secretory and enteroendocrine cells. Induction of a constitutively active form of Notch1 in all the cells of the intestinal epithelium causes a loss of secretory goblet and a reduced number of enteroendocrine cells. Instead, there is an expansion of immature absorptive cells, some of which still divide far beyond their crypts.[66]

Adenomas, in which there is a highly active Wnt pathway, due to loss of function of the adenomatous polyposis coli (APC) protein, also employ the Notch pathway to suppress differentiation. They contain all necessary signalling components and express the *Hes1* effector gene. Strikingly, when APC-deficient mice carrying numerous polyps were fed an inhibitor of γ-secretase, a large number of crypts regained a more or less normal histology, populated by numerous differentiated goblet cells.[67,68]

Conversely, Notch is required for differentiation of stem cells as well as the production of an appropriate fraction of enteroendocrine cells in the stem cell compartment of the midgut in *Drosophila*.[69]

Cross-talk with other signal transduction pathways

We end this chapter with examples that show that Notch interacts with other signal transduction pathways (or the other way round), and that this interaction occurs at different levels.

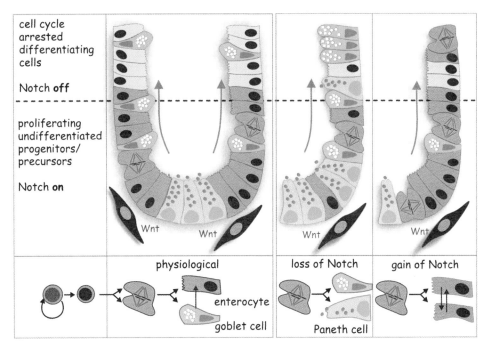

FIG 22.14 Notch regulates the size of the intestinal stem cell compartment and determines cell lineage of progenitor cells. The intestinal crypt, which harbours the stem cell compartment, is under the influence of both Wnt and Notch. Notch prevents differentiation of the stem cell and its early progenitors. It also plays a role, possibly through lateral inhibition, in the lineage decision of the later-stage progenitor cells. Thus, receiving cells maintain an enterocyte (absorptive) phenotype whereas the sending cell turns into either a goblet or a Paneth cell (secretory cell). Loss of Notch therefore gives rise to a minute stem cell compartment and most of the progenitors turn into goblet or Paneth cells. Conversely, an excess of Notch prevents differentiation into secretory cells and leads to an excess of absorptive cells that continue to divide beyond the crypt region.

Cross-talk at the level of the membrane

Notch interacts with the Wnt pathway in *Drosophila*[70] and at least two pathway components are implicated. It binds Dishevelled, an effector protein immediately downstream of the Wnt receptor Frizzled (Fz), and it binds Armadillo (β-catenin), the component of the Wnt pathway that relays the signal into the nucleus (see Chapter 14, in particular Figure 14.8, page 429). Both interactions occur over a broad region of the intracellular domain of Notch. Gain-of-function mutants of Notch antagonize Wnt signalling. The observation that both Dishevelled and Armadillo interact directly with Notch suggests that sequestration of essential signalling components forms the basis of this antagonism. Notch does not blunt the action of Wnt, but it may constitute an important buffer that raises the threshold level for Wnt signalling (Figure 22.15). In other words, Notch acts as a filter that cuts off weak signals (noise) and this could serve to create sharp boundaries between responding and non-responding cells in developing embryos.[71,72]

Cross-talk at the level of gene expression

Notch cooperates with the TGFβ signal pathway at the level of induction of gene expression[75] (Figure 22.16). In myoblast and adult neural stem cells, where addition of Notch and TGFβ prevent differentiation, Nicd and Smad3/Smad4 synergize by forming a protein complex that interacts with CSL. This leads to a higher level of expression of Hes1 than either of the pathways alone can achieve. Hes1 protein, in turn, prevents expression of myogenic factors such as the bHLH protein MyoD.[76]

A similar synergy at the level of transcription occurs between BMP4 and Notch, but the context is different. Here we deal with vascular endothelial cells that show enhanced motility under the influence of BMP4 (member of the TGFβ family of cytokines, see Chapter 20). However, the motility of cultured cells ceases when the density rises to a level where frequent cell-cell contacts occur. During contact, Delta or Jagged1 binds and activates Notch. The subsequently liberated Nicd associates with BMP4-activated Smad1/Smad4 complexes and with CSL. Together they induce high levels of expression of the bHLH protein Hey1 which, in turn, causes destruction of Id1 (Figure 22.16).[73] Expression of Id1, also a bHLH, which sequesters transcription factors away from their favoured promoter elements, is apparently essential for maintaining the motility of endothelial cells. It is known for its role in progression of the G1 phase of the cell cycle. In the context of migration, it suppresses expression of thrombospondin-1, a component of the extracellular matrix that acts to prevent new blood vessel formation (angiogenesis).

Cross-talk at the level of effector genes

In neural precursor cells, both the STAT and Notch pathways are implicated in the promotion of astrocyte differentiation. Recently a surprising connection was found in which the transcription factor Hes1 has the role of gathering STAT3 with its kinase JAK2[74] (in a manner similar to the role of growth factor and cytokine receptors; see Figure 17.9, page 525). In short, Notch induces expression of Hes1 in cortical neuroepithelial cells of mouse embryos. Hes1 then binds either STAT3 or JAK2, most likely in the nucleus. Dimerized forms of Hes1 bring STAT3 together with JAK2. The ensuing phosphorylation causes STAT3 dimerization so that they bind DNA and thus act as transcriptional activators (Figure 22.17). This sequence of events shows that growth factors and Notch act in concert to induce STAT3-mediated cellular responses.

Notch and disease

Defects in Notch2 or Jagged1 are the cause of the Alagille syndrome type 2. This is an autosomal dominant multisystem disorder, defined clinically by hepatic bile duct paucity leading to blockage of bile secretion (cholestasis, children having very pale faeces) in association with cardiovascular (stenosis of lung arteries), skeletal (butterfly-shaped vertebral discs), and ophthalmic manifestations. There

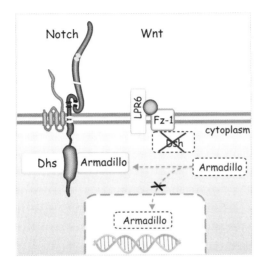

FIG 22.15 Cross-talk between Notch and the Wnt pathway.

In *Drosophila*, Notch binds Dishevelled and Armadillo, components of the Wnt pathway, and thereby increases the threshold of Wnt signalling. This may constitute a mechanism to filter out weak signals (i.e. reduce noise).

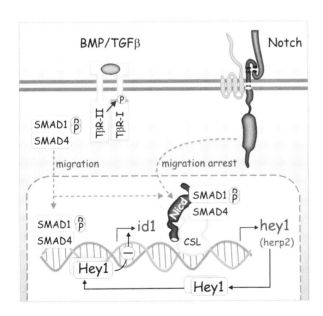

FIG 22.16 Cross-talk between Notch and the TGFβ signal pathway.

The Nicd fragment of Notch interacts with phosphoSmad1/Smad4 and together they bind CSL and drive expression of Hey2. Hey2 dimerizes and binds the promoter of id1, leading to its suppression. With the onset of cell–cell contact, Delta-mediated Notch signalling arrests BMP4-mediated cell migration.

are characteristic facial features (large forehead, deep embedded eyes, and protruding chin) and less frequently, involvement of the renal system.

Reduced Notch-1 and -3 expression is implicated in CADASIL syndrome, a subtype of inherited aortic disease, described as cerebral autosomal dominant arteriopathy with subcortical infarcts and leukoencephalopathy.

Truncation mutations in Notch, leading to increased Notch signalling, are (rarely) involved in T cell acute lymphoblastic leukaemia. Mutations in Delta-like 3 (DLL3) leads to spondylocostal dystosis, with abnormalities including hemivertebrae and blocked vertebrae accompanied by deformities of the ribs which gives rise to breathing problems and opportunistic respiratory infections.[17]

Ras and Notch act in concert to determine cell fate in vulva development of *C. elegans*

As related in Chapter 12, the anchor cells of C. *elegans* present EGF (Lin-3) which binds the EGFR (Let-23) of the p6 cell and that this acts to determine the vulval cell phenotype. The vulval structure develops from precursor cells

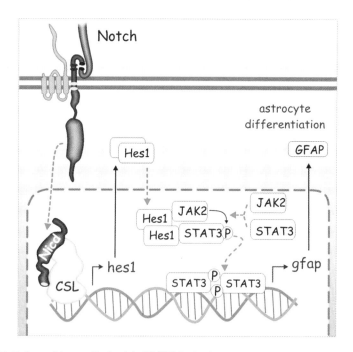

FIG 22.17 Cross-talk between Notch and the JAK/STAT signal pathway.
Notch cooperates with STAT3 in the process of astrocyte differentiation. Nicd induces expression of Hes1, which facilitates STAT3 activation by acting as a platform that gathers both STAT3 and its kinase JAK2. The ensuing phosphorylation renders STAT3 transcriptionally active, through dimerization. This sequence of events gives rise to expression of GFAP, an intermediate filament component and marker for astrocyte differentiation.

that take on three different fates (see Figure 12.11, page 331). The first is reserved for p6, but since the anchor cells may also make weak contact with p5 and p7, or EGF may be shed from the membrane and diffuse in the tissue, it is important that p6 can repress its neighbours' routes of development. This occurs through Ras (Let-60)-mediated expression of Dsl-ligands (see Table 22.1) which in turn activates the Notch (Lin-12) pathway in both p5 and p7. The Notch signal blocks EGFR signalling in a number of ways so enabling the neighbouring p5 and p7 cells to adopt the second fate. Ras also prevents cell membrane expression of the Notch receptor (Lin-12) in p6 cells thus blocking a possible reciprocal action of p5 or p8.

For more information, see http://www.wormbook.org/

List of Abbreviations

Abbreviation	Full name/description	SwissProt entry	Other names/OMIM
achaete	*Drosophila*, loss of macro- and microchaete	P10083	
ADAM10	a disintegrin and metalloprotein domain containing protein-10	O14672	Kuzbanian homologue
ADAM17		P78536	TACE
α-adaptin	AP2 complex α subunit	O95782	AP2A1
Apx-1	*C. elegans*, anterior pharynx in excess protein	P41990	
Arg-1	*C. elegans*	Q17377	homolog of Notch
armadillo		P18824	segment polarity protein, β-catenin homologue
CBF1	C-promotor binding factor-1	see CSL	
CPB	CREB binding protein	Q92793	histone acetyltransferase
CSL	CBF1 Su(H) Lag-1	Q06330	RBPJ
Delta	*Drosophila*	P10041	neurogenic locus protein delta
Delta-like 1		O00548	
Delta-like 2		Q9NYJ7	
Delta-like 3		Q9NR61	spondylocostal dystosis, MIM:277300
Deltex1		Q86Y01	

Continued

Abbreviation	Full name/description	SwissProt entry	Other names/OMIM
Dishevelled-1		O14640	Dvl1
DSL	delta serrate lag-2		
DSL-1	*C. elegans*, DSL domain protein	O45201	
dynamin-1		Q05193	shibire homologue of dynamin
E(spl)M3	*Drosophila*, enhancer of split-M3	Q01068	HLHm3
epsin1		Q9Y6I3	
Fbw7/Sel10	F-box and WD domain containing protein-7/suppressor encancer lin-10	Q969H0	archipelago homologue
fringe		Q24342	*O*-fucosylpeptide 3-β-N-acetylglucosaminanyltransferase
Frizzled		Q9UP38	
GFAP	glial fibrillary acid protein	P14136	intermediate filament component
glp-1	*C. elegans*, germ line proliferation	P13508	
Groucho	*Drosophila*, bushy eyebrows like Groucho Marx	P16371	
Hes1	hairy and enhancer of split related protein1	Q14469	
Hey2	Hes with YRPW motif	Q9UBP5	
Id	inhibitor of DNA binding-1	P41134	
Itch	itchy mice	Q96J02	inflammatory skin disorder
Jagged1		P78504	MIM 118450 Alagille syndrome
Jagged2		Q9Y219	
JAK2	Janus kinase-2	O60674	
LAG-2	*C. elegans*	P45442	longevity-assured gene
lin-12	*C. elegans*, abnormal cell lineage-12	P14585	
Mam	mastermind	Q92585	
Mam	*Drosophila*, mastermind	P21519	neurogenic protein

Continued

Abbreviation	Full name/description	SwissProt entry	Other names/OMIM
MIB1	mind bomb-1	Q86YT6	
MyoD1	myoblast determination protein	P15172	myogenic factor-3
NEDD4	neural precursor cell expressed developmentally down reg. 4	P46934	
neuralized-like		O76050	
Notch	*Drosophila*, notched wing	P07207	
Notch1	notched wing	P46531	Translocation associated notch protein TAN1, MIM:187040
Notch2	notched wing	Q04721	Alagille syndrome, MIM:125310
Notch3	notched wing	Q9UM47	CADASIL syndrome, MIM:25310
Notch4	notched wing	Q99466	INT3 oncogene
Numb	deprived of feeling	P49757	involved in sensory organ development
p300	protein of 300 kDa	Q09472	histone acetyltransferase
POFUT1	peptide *O*-fucosyltransferase-1	Q9H488	
Presenilin1		P49768	PSEN1
Rab11A	Ras-related in brain-11A	P62491	
Rab5A	Ras-related in brain-5A	P20339	
Rab7A	Ras-related in brain-7A	P51149	
RBP-JK	recombination signal sequence binding protein for Jkappa	see CSL	
scute	*Drosophila*, scutellars on mesonotum	P10084	
Serrate	*Drosophila*	P18168	
SIRT1	silent information regulator-like protein-1	Q96EB6	sirtuin family of NAD-dependent deacetylases
SKIP	ski interacting protein	Q13573	SNW1, homologue of *Drosophila* Bx42
Smad1	sma mothers against decapentaplegic-1	Q15797	

Continued

Abbreviation	Full name/description	SwissProt entry	Other names/OMIM
Smad4	sma mothers against decapentaplegic-4	Q13485	
STAT3	signal transducer and activator of transcription-3	P40763	
Su(H)	suppressor of hairless	P28159	
TACE	TNF-alpha converting enzyme	see ADAM17	
TLE4	transducin-like enhancer of split-4	Q04727	

References

1. Gilbert SF. The embryological origins of the gene theory. *J Hist Biol*. 1978;11:307–351.
2. Gilbert SF. In friendly disagreement: Wilson, Morgan, and the embryological origins of the gene theory. *Am Zoologist*. 1987;27:797–806.
3. Morgan TH. *The Physical Basis of Heredity*. Philadelphia: J.B. Lippincott; 1919.
4. Morgan TH. The relation of genetics to physiology and medicine. In: *Nobel Lectures, Physiology or Medicine 1922–1941*. Amsterdam: Elsevier; 1965:313–328.
5. Arias AM. New alleles of Notch draw a blueprint for multifunctionality. *Trends Genet*. 2002;18:168–170.
6. Poulson DF. Chromosomal deficiencies and the embryonic development of *Drosophila melanogaster*. *Proc Natl Acad Sci U S A*. 1937;23:133–137.
7. Poulson DF. Histogenesis, organogenesis and differentiation in the embryo of *Drosophila melanogaster* meigen. In: Demerec M, ed. *Biology of Drosophila*. New York: Wiley; 1950:168–274.
8. Hartenstein V. Genetic analysis of early neurogenesis: dedicated to the scientific contributions of Jose A. Campos-Ortega (1940–2004). *Dev Dyn*. 2006;235:2003–2008.
9. Campos-Ortega JA, Hartenstein V. Early neurogenesis in wildtype *Drosophila melanogaster*. *Roux's Arch Dev Biol*. 1984;193:308–325.
10. Doe CQ, Goodman CS. Early events in insect neurogenesis II. The role of cell interactions and cell lineage in the determination of neuronal precursor cells. *Dev Biol*. 1985;111:206–219.
11. Simpson P. *The Notch Receptors*. Austin TX: R G Landes; 1994.

12. Lissemore JL, Starmer WT. Phylogenetic analysis of vertebrate and invertebrate Delta/Serrate/LAG-2 (DSL) proteins. *Mol Phylogenet Evol*. 1999;11:308–319.

13. Logeat F, Bessia C, Brou C, et al. The Notch1 receptor is cleaved constitutively by a furin-like convertase. *Proc Natl Acad Sci U S A*. 1998;95:8108–8112.

14. Kidd S, Lieber T. Furin cleavage is not a requirement for *Drosophila* Notch function. *Mech Dev*. 2002;115:41–51.

15. Weng AP, Ferrando AA, Lee W, et al. Activating mutations of NOTCH1 in human T cell acute lymphoblastic leukemia. *Science*. 2004;306:269–271.

16. Hambleton S, Valeyev NV, Muranyi A, et al. Structural and functional properties of the human notch-1 ligand binding region. *Structure*. 2004;12:2173–2183.

17. Fiuza UM, Arias AM. Cell and molecular biology of Notch. *J Endocrinol*. 2007;194:459–474.

18. Nichols JT, Miyamoto A, Weinmaster G. Notch signaling – constantly on the move. *Traffic*. 2007;8:959–969.

19. O'Connor-Giles KM, Skeath JB. Numb inhibits membrane localization of Sanpodo, a four-pass transmembrane protein, to promote asymmetric divisions in *Drosophila*. *Dev Cell*. 2003;5:231–243.

20. Okajima T, Irvine KD. Regulation of notch signaling by O-linked fucose. *Cell*. 2002;111:893–904.

21. Okajima T, Xu A, Lei L, Irvine KD. Chaperone activity of protein O-fucosyltransferase 1 promotes notch receptor folding. *Science*. 2005;307:1599–1603.

22. Bruckner K, Perez L, Clausen H, Cohen S. Glycosyltransferase activity of Fringe modulates Notch-Delta interactions. *Nature*. 2000;406:411–415.

23. Okajima T, Xu A, Irvine KD. Modulation of notch-ligand binding by protein O-fucosyltransferase 1 and fringe. *J Biol Chem*. 2003;278:42340–42345.

24. Heitzler P, Simpson P. The choice of cell fate in the epidermis of *Drosophila*. *Cell*. 1991;64:1083–1092.

25. Schroeter EH, Kisslinger JA, Kopan R. Notch-1 signalling requires ligand-induced proteolytic release of intracellular domain. *Nature*. 1998;393:382–386.

26. Mumm JS, Schroeter EH, Saxena MT, et al. A ligand-induced extracellular cleavage regulates γ-secretase-like proteolytic activation of Notch1. *Mol Cell*. 2000;5:197–206.

27. Radtke F, Raj K. The role of Notch in tumorigenesis: oncogene or tumour suppressor? *Nat Rev Cancer*. 2003;3:756–767.

28. Black RA, Rauch CT, Kozlosky CJ, et al. A metalloproteinase disintegrin that releases tumour-necrosis factor-α from cells. *Nature*. 1997;385:729–733.

29. Wolfe MS. The γ-secretase complex: membrane-embedded proteolytic ensemble. *Biochemistry*. 2006;45:7931–7939.

30. Shen J, Kelleher III RJ. The presenilin hypothesis of Alzheimer's disease: evidence for a loss-of-function pathogenic mechanism. *Proc Natl Acad Sci U S A*. 2007;104:403–409.

31. Kovall RA, Hendrickson WA. Crystal structure of the nuclear effector of Notch signaling, CSL, bound to DNA. *EMBO J*. 2004;23:3441–3451.

32. Fortini ME, Artavanis-Tsakonas S. The suppressor of hairless protein participates in notch receptor signaling. *Cell*. 1994;79:273–282.

33. Struhl G, Adachi A. Nuclear access and action of notch in vivo. *Cell*. 1998;93:649–660.

34. Fryer CJ, Lamar E, Turbachova I, Kintner C, Jones KA. Mastermind mediates chromatin-specific transcription and turnover of the Notch enhancer complex. *Genes Dev*. 2002;16:1397–1411.

35. Tutter AV, Fryer CJ, Jones KA. Chromatin-specific regulation of LEF-1-β-catenin transcription activation and inhibition in vitro. *Genes Dev*. 2001;15:3342–3354.

36. Olave I, Reinberg D, Vales LD. The mammalian transcriptional repressor RBP (CBF1) targets TFIID and TFIIA to prevent activated transcription. *Genes Dev*. 1998;12:1621–1637.

37. Lecourtois M, Schweisguth F. The neurogenic suppressor of hairless DNA-binding protein mediates the transcriptional activation of the enhancer of split complex genes triggered by Notch signaling. *Genes Dev*. 1995;9:2598–2608.

38. Fischer A, Gessler M. Delta-Notch – and then? Protein interactions and proposed modes of repression by Hes and Hey bHLH factors. *Nucleic Acids Res*. 2007;35:4583–4596.

39. Massari ME, Murre C. Helix-loop-helix proteins: regulators of transcription in eucaryotic organisms. *Mol Cell Biol*. 2000;20:429–440.

40. Stern C, Tokunaga C. Autonomous pleiotropy in *Drosophila*. *Proc Natl Acad Sci U S A*. 1968;60:1252–1259.

41. Lai EC. Protein degradation: four E3s for the notch pathway. *Curr Biol*. 2002;12:R74–R78.

42. Fryer CJ, White JB, Jones KA. Mastermind recruits CycC:CDK8 to phosphorylate the Notch ICD and coordinate activation with turnover. *Mol Cell*. 2004;16:509–520.

43. Conboy IM, Conboy MJ, Wagers AJ, Girma ER, Weissman IL, Rando TA. Rejuvenation of aged progenitor cells by exposure to a young systemic environment. *Nature*. 2005;433:760–764.

44. Poodry CA, Bryant PJ, Schneiderman HA. The mechanism of pattern reconstruction by dissociated imaginal discs of *Drosophila melanogaster*. *Dev Biol*. 1971;26:464–477.

45. Chen MS, Obar RA, Schroeder CC, Austin TW, Poodry CA, Wadsworth SC, Vallee RB. Multiple forms of dynamin are encoded by shibire, a *Drosophila* gene involved in endocytosis. *Nature*. 1991;351:583–586.

46. Hawryluk MJ, Keyel PA, Mishra SK, Watkins SC, Heuser JE, Traub LM. Epsin 1 is a polyubiquitin-selective clathrin-associated sorting protein. *Traffic*. 2006;7:262–281.

47. Le Borgne R, Remaud S, Hamel S, Schweisguth F. Two distinct E3 ubiquitin ligases have complementary functions in the regulation of delta and serrate signaling in *Drosophila*. *PLoS Biol*. 2005;3(e96).

48. Bray SJ. Notch signalling: a simple pathway becomes complex. *Nat Rev Mol Cell Biol*. 2006;7:678–689.

49. Le Borgne R, Bardin A, Schweisguth F. The roles of receptor and ligand endocytosis in regulating Notch signaling. *Development*. 2005;132:1751–1762.

50. Lobley A, Swindells MB, Orengo CA, Jones DT. Inferring function using patterns of native disorder in proteins. *PLoS Comput Biol*. 2007;3(e162).

51. Aguilar RC, Longhi SA, Shaw JD, et al. Epsin N-terminal homology domains perform an essential function regulating Cdc42 through binding Cdc42 GTPase-activating proteins. *Proc Natl Acad Sci U S A*. 2006;103:4116–4121.

52. Owen DJ, Collins BM, Evans PR. Adaptors for clathrin coats: structure and function. *Annu Rev Cell Dev Biol*. 2004;20:153–191.

53. Perry WL, Hustad CM, Swing DA, O'Sullivan TN, Jenkins NA, Copeland NG. The itchy locus encodes a novel ubiquitin protein ligase that is disrupted in a18H mice. *Nat Genet*. 1998;18:143–146.

54. Matsuno K, Diederich RJ, Go MJ, Blaumueller CM, Artavanis-Tsakonas S. Deltex acts as a positive regulator of Notch signaling through interactions with the Notch ankyrin repeats. *Development*. 1995;121:2633–2644.

55. Ramain P, Khechumian K, Seugnet L, Arbogast N, Ackermann C, Heitzler P. Novel Notch alleles reveal a Deltex-dependent pathway repressing neural fate. *Curr Biol*. 2001;11:1729–1738.

56. Hori K, Fostier M, Ito M, et al. *Drosophila* deltex mediates suppressor of Hairless-independent and late-endosomal activation of Notch signaling. *Development*. 2004;131:5527–5537.

57. Cubas P, de Celis JF, Campuzano S, Modolell J. Proneural clusters of achaete-scute expression and the generation of sensory organs in the *Drosophila* imaginal wing disc. *Genes Dev*. 1991;5:996–1008.

58. Ghysen A. The projection of sensory neurons in the central nervous system of *Drosophila*: choice of the appropriate pathway. *Dev Biol*. 1980;78:521–541.

59. Usui-Ishihara A, Simpson P. Differences in sensory projections between macro- and microchaetes in Drosophilid flies. *Dev Biol*. 2005;277:170–183.

60. Rendel JM. A model relating gene replicas and gene repression to phenotypic expression and variability. *Proc Natl Acad Sci U S A*. 1969;64:578–583.

61. Hutterer A, Knoblich JA. Numb and α-Adaptin regulate Sanpodo endocytosis to specify cell fate in *Drosophila* external sensory organs. *EMBO Rep*. 2005;6:836–842.

62. Lowell S, Benchoua A, Heavey B, Smith AG. Notch promotes neural lineage entry by pluripotent embryonic stem cells. *PLoS Biol.* 2006;4(e121).

63. Shen Q, Goderie SK, Jin L, et al. Endothelial cells stimulate self-renewal and expand neurogenesis of neural stem cells. *Science.* 2004;304:1338–1340.

64. Varnum-Finney B, Xu L, Brashem-Stein C, et al. Pluripotent, cytokine-dependent, hematopoietic stem cells are immortalized by constitutive Notch1 signaling. *Nat Med.* 2000;6:1278–1281.

65. Crosnier C, Vargesson N, Gschmeissner S, et al. Delta-Notch signalling controls commitment to a secretory fate in the zebrafish intestine. *Development.* 2005;132:1093–1104.

66. Fre S, Huyghe M, Mourikis P, Robine S, Louvard D, Artavanis-Tsakonas S. Notch signals control the fate of immature progenitor cells in the intestine. *Nature.* 2005;435:964–968.

67. Milano J, McKay J, Dagenais C, et al. Modulation of notch processing by γ-secretase inhibitors causes intestinal goblet cell metaplasia and induction of genes known to specify gut secretory lineage differentiation. *Toxicol Sci.* 2004;82:341–358.

68. van Es JH, van Gijn ME, Riccio O, et al. Notch/γ-secretase inhibition turns proliferative cells in intestinal crypts and adenomas into goblet cells. *Nature.* 2005;435:959–963.

69. Ohlstein B, Spradling A. The adult *Drosophila* posterior midgut is maintained by pluripotent stem cells. *Nature.* 2006;439:470–474.

70. Axelrod JD, Matsuno K, Artavanis-Tsakonas S, Perrimon N. Interaction between Wingless and Notch signaling pathways mediated by dishevelled. *Science.* 1996;271:1826–1832.

71. Brennan K, Klein T, Wilder E, Arias AM. Wingless modulates the effects of dominant negative notch molecules in the developing wing of *Drosophila. Dev Biol.* 1999;216:210–229.

72. Hayward P, Brennan K, Sanders P, et al. Notch modulates Wnt signalling by associating with Armadillo/β-catenin and regulating its transcriptional activity. *Development.* 2005;132:1819–1830.

73. Itoh F, Itoh S, Goumans MJ, Valdimarsdottir G, et al. Synergy and antagonism between Notch and BMP receptor signaling pathways in endothelial cells. *EMBO J.* 2004;23:541–551.

74. Kamakura S, Oishi K, Yoshimatsu T, Nakafuku M, Masuyama N, Gotoh Y. Hes binding to STAT3 mediates crosstalk between Notch and JAK-STAT signalling. *Nat Cell Biol.* 2004;6:547–554.

75. Klupper M, Wrana JL. Turning it up a Notch: cross-talk between TGFβ and Notch signalling. *Bioessays.* 2005;27:115–118.

76. Blokzijl A, Dahlqvist C, Reissman E, et al. Cross-talk between the Notch and TGF-β signalling pathways mediated by interaction of the Notch intracellular domain with Smad3. *J Cell Biol.* 2003;163:723–728.

Targeting Transduction Pathways for Research and Medical Intervention

The several thousand drugs commonly prescribed in the clinic only involve some 200 molecular entities. Of these, at least half are G-protein-coupled receptors and a large proportion of these are the so-called drug receptors, those that interact with small molecules (adrenaline, acetylcholine, serotonin, histamine, etc.). The lack of viable drug targets can be ascribed in large part to the fact that until about 1980, little was known about most receptors and still less about what they do and how they work.

As we move from the small molecules through peptide (e.g. cytokines) and protein hormones, the challenge of designing appropriate ligands becomes ever more complex. And then there is the challenge of discovering reagents that are selectively agonistic or antagonistic. Protein–protein interactions generally occur over extended and, importantly, flexible surfaces, but it has become apparent that only small areas within these large surfaces,

Medications typically have several names. While in the process of development, they are referred to by a cryptic code. This is partly for reasons of commercial security, and partly because the molecules are often so chemically complex that their full chemical name is too unwieldy to use. Subsequently they acquire a generic name, which (to the knowledgeable) usually gives a clue to their function; and they are generally advertised and sold under a trade name, which may vary from one country to another as well as from one manufacturer to another.

For example, a drug used to treat certain types of cancer is **imatinib**. That's its recommended international non-proprietary name (rINN), which always starts with a lower-case letter. It is currently marketed under the trade names Gleevec (in the USA) or Glivec (in Europe and Australia) – these names always start with a capital letter. Its systematic chemical name is 4-[(4-methylpiperazin-1-yl)methyl]-N-[4-methyl-3-[(4-pyridin-3-ylpyrimidin-2-yl)amino]phenyl]benzamide, and it was coded during development as CGP57148B or STI-571.

so-called *hotspots*, are critical. Targeting these hotspots by small molecules that antagonize protein–protein interactions may thus be feasible, either by binding at critical interfaces or to allosteric sites (that modify the structure of the critical interface).[1]

That such an approach can be successful is demonstrated by the therapeutic success of the fungal metabolites ciclosporin A and tacrolimus (FK506), widely used as immunosuppressants. Both of these act on inhibitory calcineurin-binding proteins (cyclophilin and FKBP12 respectively) and the catalytic subunit of calcineurin (phosphatase PP2B) (see pages 230 and 683). Other examples are the vinca alkaloids, vincristine and vinblastine, which bind β-tubulin at the interface of the α/β-tubulin dimer and inhibit polymerization, or taxanes (paclitaxel, docetaxel), which bind β-tubulin at the inner surface of the microtubule and cause stabilization.

The pharmaceutical industry has responded strongly to the current surge of interest in signal transduction mechanisms, especially in the area of cancer treatment. Here, the need for new drugs is self-evident and this selective approach offers the possibility of developing compounds that may have less severe adverse effects than their cytotoxic precursors.[2]

Numerous highly specific receptor and protein kinase inhibitors have now reached clinical trials. We discuss the development of trastuzumab (Herceptin), gefitinib (Iressa), and imatinib (Glivec/Gleevec) and their application in the treatment of breast cancer, non-small-cell lung cancer, and chronic myeloid leukaemia respectively. We also provide a brief outline of cancer chemotherapy, the rationale behind the choice of the novel signal transduction targets, and resistance to cancer treatment. The concluding message is that fundamental research remains essential in order to understand the success or failure of novel cancer treatments and thus to make advances in this field.

Chemotherapy

The term chemotherapy was coined by Paul Ehrlich at the end of the 19th century to define the treatment of disease by chemical compounds. His idea was that each disease could be treated with a chemical compound specific for that disease. It was the task of the pharmacologist to seek out such compounds by systematic testing of substances that offer therapeutic potential.

Ehrlich himself sought a compound that would cure syphilis and his compound '606' (out of 900 tested) was the arsenical salvarsan (Latin *salvare* meaning save or cure). Arsenic binds to sulfydryl groups and one of the consequences of this is the inhibition of the mitochondrial production of ATP. This is probably the therapeutic site of action of salvarsan. By chance, the organism *Treponema pallidum,* the cause of syphilis, is more sensitive to this arsenical than the human host is. Salvarsan remained the treatment of choice until 1939, when it was superseded by sulfapyrazine.

The term chemotherapy is now generally reserved for the treatment of cancer. The drugs that are current the mainstay of chemotherapy are based on findings and principles of action that date back to around 1917, at the time when armies started to deploy mustard gas as a means to waft enemy soldiers from their trenches. Apart from appalling blistering, the military medics also noticed that soldiers exposed to mustard gas had a greatly reduced white blood cell count, sometimes even losing their bone marrow ('bone marrow aplasia'). Being aware that an excess white cell count equates to leukaemia, they reasoned that mustard gas (or similar compounds) might be effective in the treatment of blood cell cancers.

These compounds cause alkylation of guanine nucleotide bases in DNA, resulting in the inhibition of DNA-dependent DNA polymerase and arrest of cell division (see Figures 23.1 and 23.2). In order for cell division to proceed, the DNA has to be repaired (by excision and replacement of the affected bases). A whole range of compounds (cyclophosphamide, chlorambucil, lomustine, busulfan, etc.) have since been developed, all based on the same general principle.

The platinum-containing drugs cisplatin and carboplatin are variants on this theme (Figure 23.1).

Mustard gas (sulfur mustard) is not related to mustard in any way. It got its name because the impure gas has a yellow-brown colour and its odour resembles mustard and garlic. Mustard gas is best described as a thioether having two reactive Cl atoms. The loss of these in the aqueous environment makes it an electrophilic agent that can form covalent bonds with bases such as the guanine residues (N7) in DNA.

The serendipitous discovery of cisplatin. In the early 1960s, a series of rather way-out experiments in the laboratory of Barnett Rosenberg yielded some peculiar results. An experiment designed to measure the effect of electric currents on cell growth yielded *Escherichia coli* that were 300 times the normal length. The effect was not due to the electrical fields themselves but to a chemical agent that was formed in a reaction between the supposedly inert platinum electrodes and components of the solution. The chemical agent was later identified as cisplatin. Further tests revealed that it prevents cell division but not other growth processes in the bacteria, which continue to elongate. When tested in mice it was found to be effective in eliminating tumours.[3]

Cytotoxic antibiotics and antimetabolites

The finding that metabolites of microorganisms can have cytotoxic effects gave rise to series of anticancer drugs. These compounds, such as the *Streptomyces*-derived doxorubicin, bleomycin, and dactinomycin, emerged in searches for potential antibiotics and are therefore sometimes referred to as cytotoxic antibiotics. They intercalate in the minor groove of DNA and have a number of effects that include chain fragmentation, hindering replication or transcription, or inhibition of topoisomerases (which catalyse the unwinding of DNA during chain replication). More designed are the metabolic inhibitors,

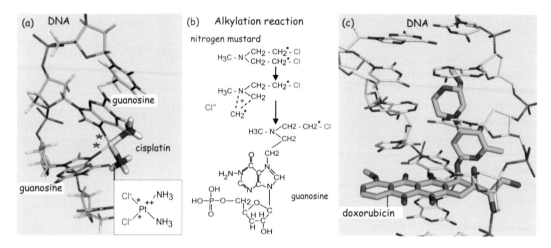

FIG 23.1 Mode of action of alkylating agents and cisplatin.

(a) Binding of cisplatin to two guanine residues in a DNA strand. Binding occurs at the same nitrogen that is bound by the alkylating agents. (b) Nitrogen mustard links to guanosine through methylene groups that have lost a chlorine atom. Two linkages are formed which are either intra- or inter-strand. (c) Doxorubicin, a cytotoxic antibiotic, intercalates between the bases of two DNA strands. In doing so it prevents the action of DNA polymerase (1au5,[4] 2deS[5]).

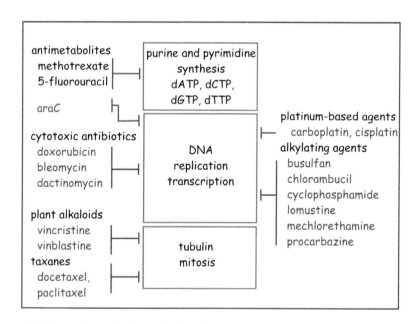

FIG 23.2 Targets of some of the drugs used in chemotherapy.

Three main processes are targeted by the classic chemotherapy drugs. Most affect the making of DNA, both at the level of nucleotide synthesis (antimetabolites) and at the level of the polymerases (cytotoxic antibiotics, platinum-based drugs, and alkylating agents). The third target is the cell division machinery, inhibited by plant alkaloids and taxanes.

so-called antimetabolites, such as araC, an analogue of the pyrimidine bases. This incorporates into forming DNA and prevents the further addition of deoxynucleotides. Other compounds belonging to this group are methotrexate (an analogue of folic acid) and fluoruracil (an analogue of uracil), which inhibit thymidylate synthase and dihydrofolate reductase, so depleting cells of dATP, dTTP, and dGTP. Lastly, there are compounds that interact with the tubulin cytoskeleton that arrest mitosis at the level of metaphase.

The purine pathway to chemotherapy

New drugs come to light in many different ways. Some are compounds extracted from natural sources (bacteria, fungi, sponges, plants). Some are designed. The classic examples are the antihistamine H_2 antagonists, blockers of gastric acid secretion that were synthesized by James Black and his colleagues according to an entirely logical pharmacological rationale.

Many are synthesized randomly and then selected for therapeutic potential in high-throughput screening tests. Others are derivatives of synthetic compounds made for industrial purposes (dyes, for example) but found to have therapeutic activities as well.

The so-called antimetabolites had their origin in screening the therapeutic effects of industrial dyes, but it was design that turned them into useful anticancer and antiviral agents. Gerhard Dogmagk, in the 1930s, discovered that the dye prontosil rubrum had bacteriostatic effects. Its active component was later found to be sulfanilamide. In 1940, Woods and Fildes put forward the antimetabolite theory to explain its action. They suggested that proliferation of bacteria is arrested because of a lack of available nucleotides. This is due to the inhibition of folate synthesis by sulfanilamide, a mimetic of the physiological substrate p-aminobenzoic acid (PABA).

Such findings inspired Gertrude Elion and George Hitchings to design analogues of purines and pyrimidines with the express intention of blocking nucleotide synthesis or nucleotide incorporation into nucleic acids. Such drugs could be useful in diseases that involve proliferation of some sort, such as cancer and bacterial and viral infections. Included in their rich portfolio are 6-mercaptopurine (the first antimetabolite used in the treatment of leukaemia and still applied as an immunosuppressant in its precursor form of azathioprine), trimethoprim (a bacterial dihydrofolate reductase inhibitor), and aciclovir, the first effective antiviral agent.

Good drugs and bad

The doses at which most, if not all, of the substances so far mentioned exert their therapeutic effects overlap those at which they are manifestly toxic. With a therapeutic index of about 1, these might be regarded as distinctly bad

Gerhard Domagk was awarded the Nobel Prize in Physiology or Medicine in 1939 'for the discovery of the antibacterial effects of prontosil', and **Gertrude Elion** and **George Hitchings** shared the Prize in 1988 with **Sir James Black**, 'for their discoveries of important principles for drug treatment'.

drugs. By comparison, penicillin has an index of around 100, indicating that the maximum tolerated dose is 100 times higher than the minimum effective dose. Yet, with respect to cancer chemotherapy, it is the general toxicity of these compounds that bestows therapeutic value.

Unlike cancer cells, normal untransformed cells have the capacity to arrest the cell cycle and then lie in wait until the toxic raid is over. When the 'all clear' sounds, they set about repairing their DNA and only re-commit to proliferation once they are in a fit condition again. Of course, mutations can slip in and so anticancer drugs are themselves also mutagenic and thus also potentially carcinogenic. Cells that fail to repair their DNA commit suicide through the process of apoptosis. Transformed cells have lost some of these qualities (see page 306) and so they are more likely to accumulate alterations that later prove to be fatal. Thus whilst non-transformed cells mainly bounce back after withdrawal of the chemotherapeutic agent, the cancer cells that have been affected tend to regress and the tumour goes into remission.

Because these drugs are as toxic as they are therapeutic, their life-threatening adverse effects dictate their dosage. Each of the drugs has its own dose-limiting effect (see examples in Table 23.1). Hitting the tumour hard without killing the patient depends on knowing precisely how far to push the drug dose.

Combination chemotherapy

The possibility of treating cancer successfully only came into view with the emergence of combination chemotherapy. This is the application of a set of compounds (often four), either in series or sometimes together, interspersed with periods of several weeks to allow sensitive tissues (particularly the

TABLE 23.1 Dose-limiting adverse effects of some anti-cancer drugs

Anticancer drug	Dose-limiting effect
AraC	Cerebral damage (slurred speech, dementia, coma)
Cyclophosphamides	Haemorrhagic cystitis of bladder
Busulphan	Pulmonary fibrosis
Cisplatin	Nephrotoxicity and peripheral neuropathy
Carboplatin	Myelosuppression
Doxorubicin	Cardiotoxicity (myopathy)
Bleomycin	Pulmonary fibrosis
Vincristine	Neurotoxicity

bone marrow and the epithelia) to recover from the assault. The success of combination chemotherapy can be explained in part because a much smaller proportion of the cancer cells escape the treatment. They are unlikely to have resistance genes to all of the compounds employed. The onset of drug resistance is a limiting aspect of all forms of chemotherapy (including, of course, the treatment of infectious diseases with antibiotics and antiviral compounds).

The development of resistance in cancer cells occurs in a number of ways:

- Elimination of the drug by transport proteins (P-glycoprotein).
- Expression of proteins that prevent the inhibitory action of the drug.
- Expression of enhanced amounts of DNA repair enzymes.
- Cells may have mutated target enzymes that are insensitive to the therapeutic agent.
- Cells might modify the drug and render it inoffensive.

Cancer chemotherapy does not in itself make cells resistant. Rather, it selects those pre-existing resistant cells, which, after treatment has ceased, can re-populate the host as drug-resistant clones. This is the phenomenon of relapse. Some cancer cells escape therapy because at the time of treatment they were not engaged in a proliferation cycle, or they were concealed in a hiding place and so eluded the drug. This is yet another reason why repeated treatment is advantageous.

Another aspect of success in cancer treatment is early diagnosis. If successful cancer therapy removes 99.9% of the transformed cells, then the earlier the diagnosis, the fewer surviving cells will be left behind, reducing the chance of relapse. Perhaps these few transformed cells are kept under control by the immune system.[6] Another argument for early diagnosis arises from the high mutability of cancer cells (see page 306). While many of these mutations are silent, others are likely to kill the cell, but there may be a few that propel cell transformation towards a more malignant or drug-resistant phenotype.[7] This chance increases with time and with the proliferation of cells. It is for this reason that screening programmes for the early detection of bowel, breast, prostate, and cervical cancers are heavily promoted.

The current forms of chemotherapy, as well as the application of newer drugs such as those described below, suffer from our insufficient knowledge of the precise identity of most tumours. Ideally, one needs to know which resistance genes and which oncogenes are expressed. There is also a lack of prognostic and predictive markers which provide patient and doctor with an early indication (weeks rather than months) of the expected success or failure of a chemotherapeutic regime. Predictive markers in particular are imperative, because, as already mentioned, time is of the essence if the treatment is to be successful.

Beyond chemotherapy, other weapons in the oncologist's armoury include surgery, radiation, endocrine, and immunotherapy.

Combination chemotherapy can be judged a success or a failure depending on your particular viewpoint. Over the period 1971–2000, the average 10 year survival rate for cancer of the prostate has increased from about 20% to 50%. Other forms of cancer, for example of pancreas and lung, have proved far more obstinate, showing hardly any improvement (England and Wales figures: see http://info.cancerresearchuk.org).

Alternative targets for cancer therapy: towards a scientific rationale

Protein kinases constitute obvious targets for those conditions in which signal transduction pathways have gone awry. In particular, there is much evidence for a decisive role of protein kinases in the development of cancer (see Chapter 11).

By the end of the 1980s, when the search for inhibitors really took off, it was known that the cancer-causing gene (oncogene) of the Rous sarcoma virus encodes a tyrosine protein kinase (v-Src)[8] and that one of the genes responsible for cellular transformation by the avian erythroblastosis virus codes for a truncated version of the EGF receptor.[9] It was recognized that ErbB2, a member of the EGF receptor family, (also known as Her-2 or neu) is associated with rapid growth and metastasis of human breast cancer.[10] In addition, EGF receptors (which are over-expressed in many human tumours[11]), when transfected into cells at high copy numbers, cause constitutive activation of signal transduction, leading to uncontrolled cell division and development of transformed phenotypes.[12-15]

Another early line of evidence linking protein kinases with cancer came from chronic myelogenous leukaemia (CML). This is a relatively rare condition in which 95% of patients show a chromosomal abnormality that gives rise to the expression of a fusion protein Bcr-Abl.[16-19]. c-Abl was already known to be the precursor, or source, of the v-Abl gene carried by the murine Abelson leukaemia virus.[20] The viral oncogene, a fusion product of a cellular Abl and a viral Gag gene, codes for a constitutively activated protein kinase. The fusion protein in CML, just like the viral gene product, is also a deregulated tyrosine kinase and like v-Abl (160 kDa), Bcr-Abl (210 kD) transforms myeloid cells.[21] Mutants that lack the kinase activity are non-transforming. The first unequivocally successful kinase inhibitor to be applied in the treatment of cancer, targets the protein kinase activity of the Bcr-Abl fusion protein.

The abbreviations used in this chapter are listed in at the end of the chapter.

Phorbol esters such as PMA, well established as tumour promoters, bind and activate protein kinase C (PKC).[22] We now recognize that they also bind to many other proteins so the role of PKC in tumour formation remains far from clear (see Chapter 19).

Finally, aberrant proliferation of cells that over-express ErbB2 can be blocked by antibodies raised against that receptor.[23] Also, protein kinase inhibitors, competitors of ATP binding, can block growth factor-induced signalling pathways.[24-26] Collectively, all these lines of evidence have built a strong rationale from which a handful of successful drugs have since been developed.

Inhibiting the EGF family of receptor kinases

EGF receptors undergo various alterations in human tumours. Gene amplification leading to over-expression of ErbB2 is a common occurrence in breast, ovarian, gastric, and salivary gland cancers, and many tumours (or their surrounding stromal cells) generate excessive amounts of EGF-related growth factors. This leads to constitutive activation of the receptors.[27,28] Not surprisingly, in view of its potential as targets for cancer therapy (see Table 23.2), the EGF family of receptors (ErbB-family)is a major focus of attention.

The antibody approach: trastuzumab

Monoclonal antibodies offer the possibility of designing highly specific reagents that can be produced on a large scale and might act as magic

TABLE 23.2 ErbB-targeted therapeutics for clinical practice

Compound	Type	Target	Company
Trastuzumab (Herceptin)	humanized mAb	ERBB2	Genentech/Roche
Pertuzumab (Omnitarg)	humanized mAb	ERBB2	Genentech
Cetuximab (Erbitux)	chimeric mAb	EGFR	ImClone/Merck KGaA/ Bristol-Myers Squibb
Matuzumab	humanized mAb	EGFR	Merck KGaA
Panitumab	humanized mAb	EGFR	Abgenix
Gefitinib (Iressa)	TKI	EGFR	AstraZeneca
Erlotinib (Tarceva)	TKI	EGFR	Genentech/OSI pharmaceuticals
Lapatinib	TKI	EGFR/ERBB2	GlaxoSmithKline
AEE788	TKI	EGFR/ERBB2/ VEGFR	Novartis
CI-1033	TKI (irreversible)	EGFR/ERBB2	Pfizer
EKB-569	TKI (irreversible)	EGFR/ERBB2	Wyeth-Ayerst
EXEL 7647/EXEL 0999	TKI	EGFR/ERBB2/ VEGFR	Exelixis

mAb, monoclonal antibody; TKI, tyrosine kinase inhibitor.
Adapted from Hynes and Lane.[28]

bullets. Their use in research and diagnostics has been widespread and extremely successful. So far, however, they have not lived up to their initial promise as therapeutic agents, particularly in the treatment of cancer. One important limiting factor is the incidence of anti-antibody immune responses.

Inhibitor monoclonal anti-ErbB1 antibodies were raised against partially purified receptors and also A431 cancer cells that over express the receptor.[29] One of the clones, mAb225, formed the basis for the generation of a chimeric antibody named cetuximab (the 'mab' at the end of this made-up generic name tells you that the drug is a monoclonal antibody).[30] This is a potent inhibitor of cancer cells that exhibit autocrine EGF-receptor activation and it has been applied in the treatment of colon cancer, often in combination with a topo-isomerase inhibitor.

A monoclonal antibody, mAb4D5, directed against ErbB2[23] formed the starting point for the generation of trastuzumab, used in the treatment of metastatic breast cancer. Over-expression of ErbB2 occurs in ∼25% of human breast cancers, particularly the more aggressive cases having a poor prognosis. Its structure differs from the other EGF receptors in that it exists permanently in an open configuration[31] and is unable to bind ligands: it is a true orphan receptor. However, it does act as a coreceptor, dimerizing with the other members of this receptor family when they are activated by ligands. The antibody binds the membrane proximal domain of ErbB2 (Figure 23.3).[32] When applied to transformed breast cells it activates the tumour-suppressing $PI(3,4,5)P_3$ phosphatase PTEN.[33] Among many other effects, it also induces expression of the cell cycle inhibitor p27[KIP1] and the retinoblastoma-related protein p130. It diminishes surface expression of ErbB2, restores E-cadherin expression (cell–cell contact) and reduces the release of the pro-angiogenic factor VEGF.[34] Most importantly, it inhibits cell proliferation.

There are several mechanistic explanations that might account for these effects. Thus, the antibody could cross-link the receptor without inducing transphosphorylation. Also, it could stimulate receptor internalization or induce proteolytic cleavage of the extracellular region. The mechanism underlying its clinical efficacy remains unclear but, in view of experience derived from in vitro investigations, this is likely to be multifaceted.

It is important to recognize that antibodies bound to a cell surface have the inherent tendency to recruit immune effector cells that possess Fc-receptors (Fc-γRI, II, or III), such as macrophages and monocytes. These bind cross-linked Fc domains and so in this way antibodies may provoke cell-mediated cytotoxicity. This mechanism appears to be the cause of tumour regression in mouse xenograft models.[34]

Xenografts are fragments of tumour tissue or suspensions of dissociated tumour cells, implanted or injected under the skin of female athymic ('nude') mice. These mice cannot reject the cancer cells and they serve a good in vivo model to study bioavailability, adverse affects, and efficiency of potential therapeutic compounds.

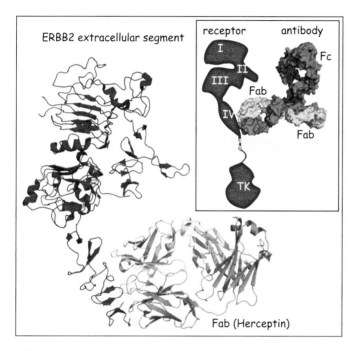

FIG 23.3 The binding of the Fab fragment of trastuzumab to ErbB2.

The main figure shows the extracellular segment of ErbB2 coupled to the monovalent Fab fragment of the monoclonal antibody trastuzumab. The extracellular segment of ErbB2 is composed of four domains (see Figure 12–4, page 321). Domain II (CR1) contains the dimerization finger. Trastuzumab binds to the membrane-proximal region of domain IV (CR2) of ErbB2. Although the size of the antibody is impressive, its attachment appears not to prevent receptor dimerization (1n8z[32]).

The tyrosine kinase inhibitor approach

Most inhibitors of the EGF receptor family of tyrosine kinases are reversible, having high affinity for ATP (of the order of 10^{-9} mol L^{-1}). Some recently developed compounds are irreversible, forming a covalent bond with residue C773 which is unique for the EGF receptors.

Erlotinib

One such compound is erlotinib (Tarceva). It is used as a single-agent treatment for patients having locally advanced or metastatic non-small cell lung cancer, after failure of at least one prior chemotherapy regimen. It has been approved for treatment of metastatic pancreatic cancers, in combination with the standard chemotherapeutic compound gemcitabine (an antimetabolite).

745

Gefitinib

Gefitinib (Iressa) is another EGF-receptor inhibitor. Its structure resembles that of erlotinib, but its future as a drug to fight cancer is less certain. In cell proliferation assays and as an inhibitor of receptor phosphorylation it acts at nanomolar concentrations. Oral administration causes inhibition of tumour growth of mouse xenografts. However, it has a cytostatic, not a cytotoxic effect: withdrawal of the drug after 100 days allowed almost immediate relapse. Initial trials with patients having locally advanced or metastatic non-small-cell lung cancer, which failed to respond to both platinum-based (polymerase inhibitor) and docetaxel (microtubule inhibitor) chemotherapies, appeared promising. Unfortunately, when tested in a large-scale randomized study erlotinib proved to have little effect on survival.

Imatinib: chronic myeloid leukaemia and the Bcr-Abl fusion story

Chronic myeloid leukaemia (CML) is a rare condition in which 96% of the patients who are in the chronic phase carry the same chromosomal abnormality. This is a somatic translocation, t(9:22), between the long arms of chromosome 9 (paternally derived) and chromosome 22 (maternally derived) that gives rise to one very small 'Philadelphia' chromosome (Ph) and one abnormally long chromosome (Figure 23.4). This translocation has been detected in haematopoietic stem (HSC) cells of both myeloid and erythroid lineages, but it is only the myeloid cells that proliferate during the chronic phase of the disease.[35]

The t(9:22) translocation gives rise to a deregulated fusion protein, for instance p210$^{Bcr-Abl}$, that comprises a number of exons of Bcr and all (excepting the first alternatively spliced exons Ia or Ib) of c-Abl (Figure 23.5). The normal c-abl gene codes for a predominantly nuclear tyrosine protein kinase that possesses a number of protein interaction domains (SH3, SH2, DNA, and F-actin binding).

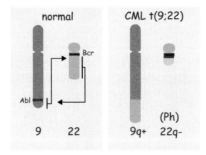

FIG 23.4 Origin of Bcr-Abl.

The fusion protein Bcr-Abl arises by a translocation between chromosomes 9 and 22. The breakage of chromosome 22 occurs at a number of points in the Bcr gene giving rise to different fusion products (190 kDa, 210 kDa, or 230 kDa) associated with various pathologies. Bcr-Abl p210 (see Figure 23.5) is associated with chronic myelogenous leukemia (CML).

The Bcr gene also codes for a multidomain protein having catalytic activities (RhoGAP and RhoGEF and a protein/lipid interaction site (PH). The fusion product is a deregulated tyrosine kinase, localized in the cytosol that causes both the aberrant proliferation and the lack of terminal differentiation of the myeloid cells in chronic myeloid leukaemia. Other forms of Bcr-Abl, resulting from different breakpoints in the Bcr gene, are also encountered in various human leukaemias.

The *Bcr* gene derives its name because it is located at a site of frequent chromosomal breakage, the breakpoint cluster region. *Bcr* can break at different sites, but the major breakpoints are situated after exons 13 or 14. A minor breakage occurs after exons 1 or 19. The 210 kDa chimeric Bcr-Abl protein, present in 95% of CML patients, is the product of breakage either after exon 13 (giving rise to b2a2 fusion product) or 14 (giving rise to b3a2 fusion product).

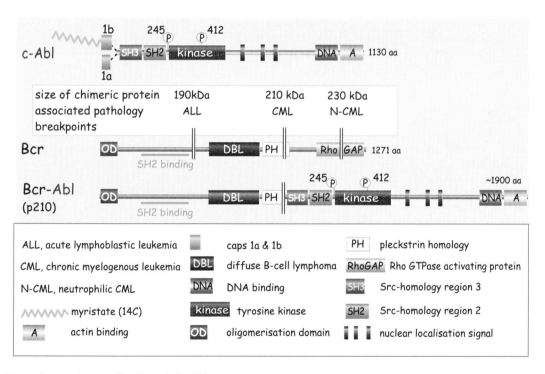

FIG 23.5 Domain architecture of Bcr, Abl and the Bcr-Abl fusion protein.

c-Abl is a cytosolic tyrosine kinase but, unlike Src, it lacks the N-terminal phosphotyrosine. Instead, it has a Cap region (and in the case c-Abl1b, a myristoyl chain), which maintain the kinase in a locked inactive conformation. Bcr tends to break at different points, each associated with a different leukaemia (ALL, CML, or N-CML). The chimeric Bcr-Abl protein comprises the N-terminal segment of Bcr attached to an almost intact kinase, though lacking the Cap region (and the associated myristoyl chain). The consequences of the fusion are described in Figure 23.7.

Similar to the Src-like tyrosine kinases, the SH2 and SH3 domains of the non-transforming c-Abl clamp the catalytic domain in an inactive state, referred to as the 'latched conformation'. However, unlike Src, c-Abl lacks the C-terminal tyrosine that imposes the SH2-SH3 clamp when phosphorylated. In c-Abl, this role is played by the N-terminal Cap region with a myristoyl attachment (figure 23.6).[36] Under physiological conditions, the clamp is relieved by proteins that interact with either the myristoyl attachment, the SH2 domain (amongst others, c-Jun, CAS, Cbl, or EphB2) or the SH3 domain (amongst others, Abi1, SHIP1, Nck, paxillin, or Cbl). Full activation then occurs through phosphorylation in the SH2-kinase linker (Y245, which prevents interaction of SH3 with the linker region) and in the activation segment (Y412, which facilitates access of substrate). These phosphorylations are thought to occur through transphosphorylation (c-Abls phosphorylating each other). Members of the Src-family of protein kinases are other possible candidates that might bestow full catalytic competence on c-Abl.

There are several reasons why the behaviour of the fusion protein Bcr-Abl differs from that of c-Abl (for a review, see Wong and Witte[38]). For instance, Bcr-Abl lacks the Cap domain and the myristoyl attachment, and the Bcr segment of the fusion protein interacts with the SH2 domain of Abl. These all act to prevent the formation of the latched conformation (Figure 23.6 and 23.7). Importantly, unlike c-Abl, Bcr-Abl is a cytosolic protein and must interact with quite different substrates. The presence of the actin-binding domain in

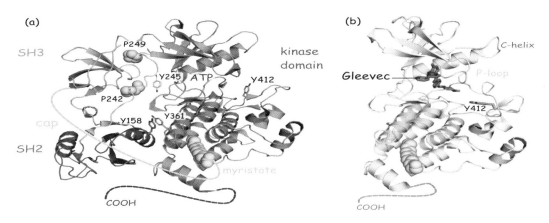

FIG 23.6 Structure of c-Abl.

(a) c-Abl in its closed conformation, with SH3 and SH2 domains clamping the kinase domain in an inactive state. Essential in this is the Cap region, with myristate attached at the C-terminal part of the kinase domain. This interaction creates binding sites between the kinase and SH2 domains. Stacking of pY158 and pY361 is important here. The close apposition of SH3 causes interaction with proline residues in the linker region that bridges the SH2 with the kinase domain (P249 and P242). This interaction is prevented by phosphorylation at Y245. (b) Imatinib/Gleevec occupies the ATP binding pocket and holds the activation segment in a lower position. As a consequence, productive phosphorylation is, of course, out of the question. SH2 and SH3 domains are not shown (1opk,[37] 1opj[37]).

Bcr may also contribute to specific subcellular localizations, thereby offering substrates that c-Abl would never otherwise encounter.

The Bcr-Abl fusion protein has profound effects on cell survival and proliferation. The amount of RasGTP is greatly expanded and the PI 3-kinase pathway is also affected. However, transformation is prevented if dominant negative forms of Ras or PKB (or its inhibitor LY29002) are present. Also, the phosphatidyl inositol 5-phosphatase SHP-1 is down-regulated and the interferon-receptor STAT pathway is inhibited.

The requirement for continuous expression of Bcr-Abl in CML cells has been demonstrated in a mouse model of the disease. Here, expression of Bcr-Abl was controlled by a tetracycline sensitive promoter so that removal of the tetracycline aborted the expression of the fusion protein. When this was done, there was a remission of the leukaemia.[39]

Figure 23.7 outlines modifications that alter the activation state of c-Abl.

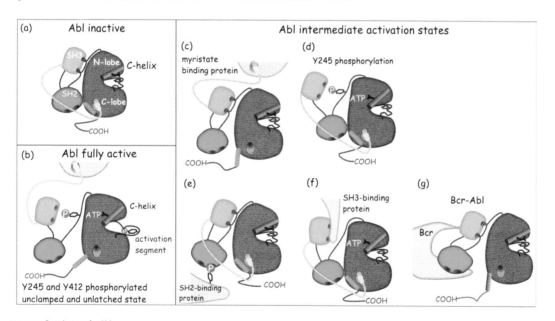

FIG 23.7 Regulation of c-Abl.

Regulation of c-Abl. c-Abl has many different activation states. (a) and (b) represent the inactive and fully active states. Intermediate states, which have limited kinase activity, are represented in panels (c) – (f). (a) Latched conformation, the kinase domain clamped in an inactive state. (b) Displacement of myristate (yellow) from the C-lobe, detachment of the SH3 domain by phosphorylation of Y245 and reorganisation of the activation segment by phosphorylation of Y412 render c-Abl fully competent. (c) Partial activation through displacement of myristate. (d) Partial activation through phosphorylation of the linker region (Y245). (e) Partial activation through displacement of the SH2 domain by a phosphotyrosine-containing protein. (f) Partial activation through displacement of SH3 by a proline rich sequence. (g) The fusion protein Bcr-Abl is also in an intermediate activation state. It lacks the Cap region (together with its myristate chain) and the Bcr moiety binds and holds the SH2 domain away from the kinase domain. Full activation then requires only phosphorylation of Y245 and Y412.

Development of imatinib, inhibitor of c-Abl

Imatinib (Gleevec) emerged from a random screen of 2-phenylaminopyrimidine compounds, known to be inhibitors of PKC-α, as possible inhibitors of PDGF-R tyrosine kinase. The aim was to discover compounds that could suppress the growth of PDGF-R-activated cell lines. The test also included the oncogenic v-Abl as a counter screen, because of its very distant sequence resemblance to PDGF-R. It came as a surprise that the most potent inhibitor of PDGF-R kinase activity, CGP 531716, was also a potent inhibitor of v-Abl. With new potential treatments in mind, it was further optimized for v-Abl kinase inhibition, the compound CGP 57148B, now known as imatinib or by its trade names Gleevec and Glivec, emerging as the most effective (Figure 23.8) (see Druker and Lydon[40] for review). Importantly, imatinib has selective and potent activity in cells possessing the Philadelphia chromosome and yet spares normal cells.[41,42]

Imatinib was first tested in patients who were in the chronic phase of CML and refractory to IFN-α therapy. In one trial, complete normalization of the cellular composition of the blood was obtained after 4 weeks of treatment in 53 out of 54 patients treated in the higher dose range (the drug is seemingly benign: a maximal tolerated dose was not identified and adverse effects were minimal).[43] The amount of detectable Bcr-Abl fused genes in bone marrow cells was considerably reduced and the patients remained free of disease

Despite tight rules and regulations, the reasons why a particular drug might or might not gain approval and why, once approved, it might still be unsuccessful, depends on several factors that are hard to control.

It is possible that adverse effects are not reported to the approval authorities, or that their importance may not be recognized. They may even go unrecognized. Adverse effects may eventually come to light only after prolonged and large-scale application. Drugs may, once firmly established, also be prescribed to patients for whom they are not indicated. In a number of cases they may have unexpected adverse effects or lack therapeutic advantage because patients do not follow the dosage protocols or take over-the-counter, non-prescribed remedies alongside.

We have heard it said that in England during the 1990s, patients spent more on non-prescription remedies than the entire budget of the Medical Research Council. With a rapidly growing and uncontrolled Internet market this can only get worse. Some drugs that have only marginal therapeutic effects gain approval because of the immense pressure applied not only by the pharmaceutical companies to sell their products, but also by patients, newspapers, lawyers, and politicians who demand instant treatment of life-threatening illnesses.

Finally, reasonable or unreasonable litigation may damage the reputation of a company to the extent that drugs have to be taken off the market: the Vioxx drama is a case in point (see http://www.vioxx-recall-lawsuit.com/).

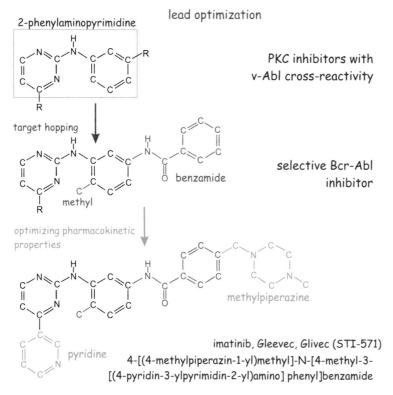

lead optimization

2-phenylaminopyrimidine

target hopping

benzamide

methyl

optimizing pharmacokinetic properties

methylpiperazine

pyridine

PKC inhibitors with v-Abl cross-reactivity

selective Bcr-Abl inhibitor

imatinib, Gleevec, Glivec (STI-571)
4-[(4-methylpiperazin-1-yl)methyl]-N-[4-methyl-3-[(4-pyridin-3-ylpyrimidin-2-yl)amino] phenyl]benzamide

FIG 23.8 Optimization of imatinib as a chemotherapeutic agent.
The discovery that 2-phenylaminopyrimidine inhibitors of PKC also inhibit the unrelated v-Abl oncogene turned attention to its potential use in the treatment of chronic myelogenous leukaemia. Starting with the 2-phenylaminopyrimidine backbone, addition of the benzamidine group increased activity against tyrosine kinases, the methyl group reduced its activity against PKC (so-called 'target hopping'). Addition of a 3'-pyridyl group improved the activity in cellular assays. Subsequent addition of N-methylpiperazine increased water solubility and oral bioavailability, enabling the drug to survive the stomach and to enter the bloodstream.

for over 1 year. Remarkable results were also obtained in subsequent larger-scale trials. Most, >85%, showed a complete cytogenetic response. In assays employing *in situ* hybridization with fluorescent probes, the Bcr-Abl fusion protein became undetectable and the Philadelphia chromosome was absent.

However, even with all this, it does not follow that the tumour cells were completely eliminated, as Bcr-Abl transcripts remained detectable by reverse transcriptase PCR in ~96% of the responding patients. There remains a potential pool of cells from which resistant clones can emerge at a later date. Indeed, after 42 months of follow-up, relapse had occurred in 16% of these patients.[44] The response is much less promising in patients who have reached blast crisis.[43] In some cases this resistance to treatment is due to a very high level of Bcr-Abl expression, in others to mutations of Bcr-Abl.[45,46] With the aim of overcoming drug resistance, new ATP competitors, specific for Bcr-Abl, are currently under development.

Why is treatment of CML successful?

In comparison with other drugs that target EGF receptors, the success of imatinib is not easy to understand, but several arguments can be advanced to explain why targeting Bcr-Abl in CML has been fruitful. The first, despite a long list of reasons predicting that it would work, is simply pure luck. A second argument may be that blood-borne cells differ fundamentally from solid tissue cells. These can switch from being attached to free floating without undergoing the risk of apoptosis, and from this we understand that they have alternative survival pathways (see page 400). Because of their enhanced survival capacity, fewer hits may be required in order to create a leukaemia, but on the other hand, this may have the advantage that there will also be fewer disease genes to target for therapy. A third argument is that CML generally comes to notice in its early chronic benign phase, at a point when many solid tissue tumours would be undetectable. Moreover, because of their very high proliferation rates, somatic mutations in stem cells may be (at least in their initial phase) more likely to give rise to 'monoclonal' tumours. Polyclonality is certainly one of the hurdles that must be overcome in the treatment of solid tissue tumours.

Molecular mechanism of inhibition by imatinib

Imatinib binds in the catalytic cleft of c-Abl, acting as a competitive inhibitor of ATP binding (Figure 23.6). It binds most efficiently to partially activated c-Abl, in which the SH2/SH3 clamp is removed, but the tyrosine (Y412) in the activation segment is not phosphorylated. The drug seeks out most effectively those malignant cells that depend acutely on sustained Bcr-Abl kinase activity for their survival. By precipitating their apoptosis, imatinib greatly reduces the disease burden (1000-fold reduction in cells expressing Bcr-Abl). However, as indicated above, a residue of disease cells remains and some patients never respond at all. This is because leukaemic stem cells, being either quiescent or protected by the bone marrow micro-environment, escape the drug. Alternatively, the resistance may be due to mutations in Bcr-Abl or Bcr-Abl amplification.

Resistance due to mutations at different sites in the Abl moiety of Bcr-Abl

Analysis of resistant cells has revealed a whole range of mutations (Figure 23.9). Some of these act at a distance from the catalytic site and have an allosteric influence on drug binding. They destabilize the autoinhibited conformation of the Abl kinase to which the drug binds, and shift the equilibrium toward the active kinase conformation. This precludes drug binding, but not ATP binding. Other mutations directly involve drug binding. These are P-loop mutations (Y253 or E255), activation segment mutations or mutations that hinder drug occupancy (T315).

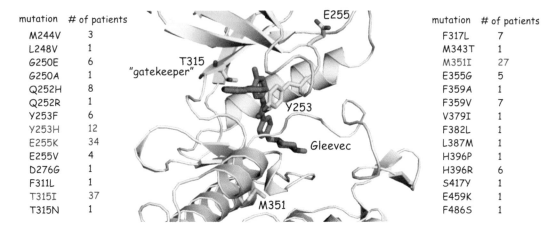

mutation	# of patients		mutation	# of patients
M244V	3		F317L	7
L248V	1		M343T	1
G250E	6		M351I	27
G250A	1		E355G	5
Q252H	8		F359A	1
Q252R	1		F359V	7
Y253F	6		V379I	1
Y253H	12		F382L	1
E255K	34		L387M	1
E255V	4		H396P	1
D276G	1		H396R	6
F311L	1		S417Y	1
T315I	37		E459K	1
T315N	1		F486S	1

FIG 23.9 Mutations in Bcr-Abl that cause insensitivity to imatinib (drug resistance).
Numerous mutations in the Abl moiety have been identified in patients resistant to treatment with imatinib. Only a few of these are direct contact residues; others are remote from the catalytic cleft and their effect is not easily explained. In a survey of 177 patients having such resistance, 28 mutations were identified, 4 of which were predominant (T315I (37), E255K (34), M351I (27), Y253H (12)). T315, the 'gatekeeper' of imatinib, is a key contact residue in the binding pocket. Y253 and E255 regulate the positioning of the P-loop, which acts as a lid on the drug-binding site. How M351 causes resistance is not known. (1opj[37]).
Mutant information from Deininger et al.[48]

More (many more, about 112 resistance mutations in total) were revealed in an ingenious experiment in which the Bcr-Abl gene was randomly mutagenized by insertion into a highly labile strain of *E. coli*. The altered genes were then screened for resistance in BaF3 cells which are normally dependent on the presence of IL-3 for their survival.[49] Cells expressing deregulated forms of Bcr-Abl could grow in the absence of IL3 and of these, some could grow in the presence of imatinib.[50] The mutants identified in this way overlap all the mutants detected in non-responding patients or in patients suffering relapse. We conclude that tyrosine protein kinases are highly plastic entities that can adopt many inactive states so long as they fulfil certain precise structural criteria as they converge towards their active conformation.

A similar conclusion had already been drawn from structural studies of serine/threonine kinases such as MAPkinase, CDK2, and PKA. Their structural conformations differ considerably in their inactive states and also differ with respect to their activation mechanisms. However, their active conformations display clear similarities (for review see Johnson and Lewis.[51])

All this provides a challenge for drug development because those drugs that act on protein kinases in their inactive conformation, while necessarily specific, may more easily suffer from resistance mutations. By contrast, drugs binding to active conformations are likely to be less specific, but may offer the advantage that the kinase is less likely to escape inhibition. The general trend now is to shift attention away from single, highly selective inhibitors towards broader spectrum compounds or to cocktails of specific inhibitors acting in different ways. These approaches open up a range of new therapeutic promise but also the possibility of unacceptable adverse effects.

Other signal transduction components targeted for therapeutic intervention

There are many other components of signal transduction pathways that might be regarded as possible targets for drug therapy. The Ras-MAP kinase pathway has been studied extensively and its components, H-, K- and N-Ras, B-Raf, and MEK1, have been identified as human oncogenes. In particular, it has been suggested that because Ras-mediated cell transformation demands membrane attachment (through polyisoprenoid chains: see page 105), prevention of this post translational modification might stop cancer growth.[52] However, none of the drugs developed for this purpose have shown any benefit in the cancer field, and in many cases mutated Ras escapes their effect. Inhibitors of B-Raf share the same misfortune though the jury is still out for the inhibitors of MEK. Three highly selective MEK inhibitors have recently reached the stage of clinical trials.[53,54] These differ from many other kinase inhibitors in that they do not compete with ATP but bind close to the catalytic cleft, thereby impeding access of the activation segment of ERK (the unique substrate of MEK) (Figure 23.10).

Inhibitors targeting enzymes of the PI 3-kinase pathway (PI 3-kinase, PKB, and mTOR) that controls protein synthesis, cell cycle progression and, importantly, cell survival, also show some promise.[55] Another possible target is Hsp90,[56] involved in protein folding. It is highly expressed in tumours and appears to play a role in the activation and localization of kinases such as Raf, Cdk4/6, v-Src, or PKB. Some of the compounds developed to inhibit protein kinases are listed in Table 23.3.

Towards a different approach in testing cancer drugs?

The general opinion is that with the single exception of imatinib, signal transduction blockers have yet to realize a real breakthrough in cancer treatment. Whereas the limiting factor to advances in this field used to be the definition of drug targets, today it seems to be the design of appropriate and effective clinical trials. We have many compounds, all based on good scientific rationales, all seemingly safe and which offer promising effects in *in vitro* (cell culture) and *in vivo* (mouse xenograft) studies.

If the statins (lovastatin, simvastatin, atorvastatin etc., inhibitors of HMG-CoA reductase) have offered little benefit in the treatment of cancer, they are regarded almost as wonder drugs in the prevention of atherosclerosis and stroke. Polyisoprene (squalene) formation is a key step in the formation of cholesterol and the membrane anchors (farnesyl and geranylgeranyl) that link the Ras GTPases to the plasma membrane (see page 105).

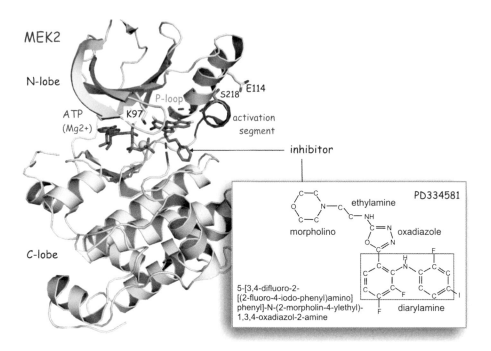

FIG 23.10 Inhibition of MEK2.

PD3334581 inhibits MEK2 by occupying the substrate entry site. It also positions the activation segment in front of the catalytic cleft, making access still more difficult. Moreover, the αC-helix is turned away so that the interaction of E114 with K97, essential for the correct positioning of ATP, is prevented (see also Figure 18.6, page 553 for the activation of PKB). NB: The disordered segment of the protein is represented as red dotted line, with only one of the two 'activation' serines (S218) visible (1s9i[54]).

There are two important obstacles to be faced in the clinical testing of cancer drugs. The first concerns the choice of patients. The old motto 'one drug suits all' is perhaps not appropriate for cancer therapy, not even within carefully defined classes of tumours. Precise predictive markers are needed in order to improve the choice of patients to be included in the trialling of novel compounds. To illustrate this, the two most useful predictors of the responsiveness of breast cancers are the expression of oestrogen receptors (ER) for anti-oestrogen therapy and HER-2 amplification for trastuzumab (Herceptin) therapy. But these only have positive predictive values of 30–50%. At best, only one in two of the selected patients shows a good response to the therapy. Their negative predictive value, at >95%, is more convincing.[57] It is not unthinkable that the compounds currently being rejected have yet to find the right subjects in which to express their therapeutic potential.

The second obstacle concerns the time span of testing. At an early stage it is essential to have good prognostic markers so as to be able to get a realistic impression of the effectiveness of a drug. Once a drug has been accepted

TABLE 23.3 Some compounds targeting signal transduction pathways under scrutiny in clinical trials

Kinome branch	compound	Known targets
TK	AZD-2171	VEGFR
	vandetanib ZD6474	VEGR/EGFR
	vatalanib PTK787	VEGFR, PDGFR, c-Kit
	axitnib AG-013736	VEGFR, PDGFR
	lapatinib (Tykerb) GW572016	EGFR, ERBB2
	CP-751871 (antibody)	IGFR
	dastinib BMS-354825	c-Src, c-Abl
	AMN-107	c-Abl
	dasatinib (Sprycel)	c-Abl, PDFR, c-Kit, c-Src members
	nilotinib (Tasigna)	c-Abl, PDGFR, c-Kit
	sunitinib (Sutent)	VEGFR, c-KIT, PDGFR, Ret, CSFR, Flt3
	sorafenib (Nexavar)	VEGFR2, PDGFR, c-Kit, Flt3, A-Raf, B-Raf
	bevacizumab (Avastin) antibody	VEGFA
	panitumumab (antibody)	EGFR
CMGC (CDK, MAPK, GSK and Cdc-like kinases)	flavopiridol	Pan-CDK
	PD332991	CDK4
STE (MAP2K, MAP3K, MAP4K)	BAY 43-9006	C-Raf
	PD0325901	MEK
	PD184352 (CI-1040)	MEK
	PD318088	MEK1
	PD334581	MEK2
AGC	rapamycin	mTOR

for general clinical application, these markers remain useful in order to verify whether or not tumours are responding.[57] The design of clinical trials should provide the opportunity to assess retrospectively why a particular compound failed to give the expected result in a particular patient.[53] Unfortunately, the pressure exerted both by patients and by the industry to advance new blockbuster drugs limits the scale and value of such retrospective analytical approaches.[58]

For more information about clinical trials we suggest: http://www.cancer.gov/clinicaltrials/learning

List of Abbreviations

Abbreviation	Full name/description	SwissProt entry	Other names, OMIM link
Bcr	breakpoint cluster region	P11274	
B-Raf	B-type rat fibrosarcoma	P15056	BRaf1, MAP3K
c-Abl	cellular homologue of Abelson viral product	P00519	
c-KIT	kitten (oncogene from Hardy-Zuckerman 4 feline sarcoma virus)	P10721	mast/stem cell growth factor receptor
CML	chronic myelogenous leukaemia		MIM:608232
EGFR	epidermal growth factor receptor (ERBB1)	P00533	
ERBB2	erythroblastosis-B2	P04626	Her-2, Neu, MIM:164870
ERK1	extracellular signal-regulated kinase-1	P27361	p44-MAPK, MK03
Flt3	fetal liver tyrosine protein kinase-3	P36888	CD135, stem cell tyrosine kinase (STK1)
GIST	gastrointestinal stromal tumour		MIM;606764
HER2	human EGF receptor-2	P04626	ERBB2, neu, MIM:164870
Hsp90	heat shock protein of 90 kDa	P07900	renal carcinoma-induced antigen
MEK1	MAPK ERK kinase-1	Q02750	MAP2K
neu	oncogene from neuro/glioblastoma cells	P04626	HER-2, ERBB2, MIM:164870
PDGFR-A	platelet-derived growth factor receptor-A	P16234	
Src	sarcoma		
VEGFR	vascular endothelial growth factor receptor type 1	P17948	Fems-like tyrosine kinase-1 (FLT1)

References

1. Arkin MR, Wells JA. Small-molecule inhibitors of protein-protein interactions: progressing towards the dream. *Nat Rev Drug Discov*. 2004;3:301–317.

2. Bridges AJ. Chemical inhibitors of protein kinases. *Chem Rev*. 2001;101:2541–2572.

3. Rosenberg B, VanCamp L. The successful regression of large solid sarcoma 180 tumors by platinum compounds. *Cancer Res*. 1970;30:1799–1802.

4. Yang D, van Boom SS, Reedijk J, van Boom JH, Wang AH. Structure and isomerization of an intrastrand cisplatin-cross-linked octamer DNA duplex by NMR analysis. *Biochemistry*. 1995;34:12912–12920.

5. Cirilli M, Bachechi F, Ughetto G, Colonna FP, Capobianco ML. Interactions between morpholinyl anthracyclines and DNA. The crystal structure of a morpholino doxorubicin bound to d(CGTACG). *J Mol Biol*. 1993;230:878–889.

6. Dunn GP, Bruce AT, Ikeda H, Old LJ, Schreiber RD. Cancer immunoediting: from immunosurveillance to tumor escape. *Nat Immunol*. 2002;3:991–998.

7. Komarova NL, Wodarz D. Drug resistance in cancer: principles of emergence and prevention. *Proc Natl Acad Sci USA*. 2005;102:9714–9719.

8. Collett MS, Purchio FA, Erikson RL. Avian sarcoma virus-transforming protein, pp60src shows protein kinase activity specific for tyrosine. *Nature*. 1980;285:167–169.

9. Downward J, Yarden Y, Mayes E, et al. Close similarity of epidermal growth factor receptor and v-erb-B oncogene protein sequences. *Nature*. 1984;307:521–527.

10. Slamon DJ, Clark GM, Wong SG, Levin WJ, Ullrich A, McGuire WL. Human breast cancer: correlation of relapse and survival with amplification of the HER-2/neu oncogene. *Science*. 1987;235:177–182.

11. Khazaie K, Schirrmacher V, Lichtner RB. EGF receptor in neoplasia and metastasis. *Cancer Metastasis Rev*. 1993;12:255–274.

12. Di Fiore PP, Pierce JH, Fleming TP, et al. Overexpression of the human EGF receptor confers an EGF-dependent transformed phenotype to NIH 3T3 cells. *Cell*. 1987;51:1063–1070.

13. Velu TJ, Beguinot L, Vass WC, et al. Epidermal-growth-factor-dependent transformation by a human EGF receptor proto-oncogene. *Science*. 1987;238:1408–1410.

14. Abelson HT, Rabstein LS. Lymphosarcoma: virus-induced thymic-independent disease in mice. *Cancer Res*. 1970;30:2213–2222.

15. Damm K, Thompson CC, Evans RM. Protein encoded by v-erbA functions as a thyroid-hormone receptor antagonist. *Nature*. 1989;339:593–597.

16. Nowell PC, Hungerford DA. A minute chromosome in human chronic granulocytic leukemia. *Science*. 1960;132:1497.

17. Nowell PC, Hungerford DA. Chromosome studies on normal and leukemic human leukocytes. *J Natl Cancer Inst*. 1960;25:85–109.

18. Rowley JD. A new consistent chromosomal abnormality in chronic myelogenous leukaemia identified by quinacrine fluorescence and Giemsa staining. *Nature*. 1973;243:290–293.

19. Ben-Neriah Y, Daley GQ, Mes-Masson AM, Witte ON, Baltimore D. The chronic myelogenous leukemia-specific P210 protein is the product of the bcr/abl hybrid gene. *Science*. 1986;233:212–214.

20. Heisterkamp N, Groffen J, Stephenson JR. The human v-abl cellular homologue. *J Mol Appl Genet*. 1983;2:57–68.

21. Daley GQ, Van Etten RA, Baltimore D. Induction of chronic myelogenous leukemia in mice by the P210bcr/abl gene of the Philadelphia chromosome. *Science*. 1990;247:824–830.

22. Castagna M, Takai Y, Kaibuchi K, Sano K, Kikkawa U, Nishizuka Y. Direct activation of calcium-activated, phospholipid-dependent protein kinase by tumor-promoting phorbol esters. *J Biol Chem*. 1982;257: 7847–7851.

23. Hudziak RM, Lewis GD, Winget M, Fendly BM, Shepard HM, Ullrich A. p185HER2 monoclonal antibody has antiproliferative effects in vitro and sensitizes human breast tumor cells to tumor necrosis factor. *Mol Cell Biol*. 1989;9:1165–1172.

24. Tomita F, Nomoto H, Takahashi I, Kato Y, Morimoto M. Staurosporine, a potent inhibitor of phospholipid/Ca^{++}-dependent protein kinase. *Biochem Biophys Res Commun*. 1986;135:397–402.

25. Akiyama T, Ishida J, Nakagawa S, et al. Genistein, a specific inhibitor of tyrosine-specific protein kinases. *J Biol Chem*. 1987;262:5592–5595.

26. Markovits J, Linassier C, Fosse P, et al. Inhibitory effects of the tyrosine kinase inhibitor genistein on mammalian DNA topoisomerase II. *Cancer Res*. 1989;49:5111–5117.

27. Salomon DS, Brandt R, Ciardiello F, Normanno N. Epidermal growth factor-related peptides and their receptors in human malignancies. *Crit Rev Oncol Hematol*. 1995;19:183–232.

28. Hynes NE, Lane HA. ERBB receptors and cancer: the complexity of targeted inhibitors. *Nat Rev Cancer*. 2005;5:341–354.

29. Sato JD, Kawamoto T, Le AD, Mendelsohn J, Polikoff J, Sato GH. Biological effects in vitro of monoclonal antibodies to human epidermal growth factor receptors. *Mol Biol Med*. 1983;1:511–529.

30. Goldstein NI, Prewett M, Zuklys K, Rockwell P, Mendelsohn J. Biological efficacy of a chimeric antibody to the epidermal growth factor receptor in a human tumor xenograft model. *Clin Cancer Res*. 1995;1:1311–1318.

31. Garrett TP, McKern NM, Lou M, et al. The crystal structure of a truncated ErbB2 ectodomain reveals an active conformation, poised to interact with other ErbB receptors. *Mol Cell*. 2003;11:495–505.

32. Cho HS, Mason K, Ramyar KX, et al. Structure of the extracellular region of HER2 alone and in complex with the Herceptin Fab. *Nature*. 2003;421:756–760.

33. Nagata Y, Lan KH, Zhou X, et al. PTEN activation contributes to tumor inhibition by trastuzumab, and loss of PTEN predicts trastuzumab resistance in patients. *Cancer Cell*. 2006;6:117–127.

34. Sliwkowski MX, Lofgren JA, Lewis GD, Hotaling TE, Fendly BM, Fox JA. Nonclinical studies addressing the mechanism of action of trastuzumab (Herceptin). *Semin Oncol*. 1999;26:60–70.

35. Fialkow PJ, Gartler SM, Yoshida A. Clonal origin of chronic myelocytic leukemia in man. *Proc Natl Acad Sci USA*. 1967;58:1468–1471.

36. Hantschel O, Nagar B, Guettler S, et al. A myristoyl/phosphotyrosine switch regulates c-Abl. *Cell*. 2003;112:845–857.

37. Nagar B, Hantschel O, Young MA, et al. Structural basis for the autoinhibition of c-Abl tyrosine kinase. *Cell*. 2003;112:859–871.

38. Wong S, Witte ON. The BCR-ABL story: bench to bedside and back. *Annu Rev Immunol*. 2004;22:247–306.

39. Huettner C, Zhang P, Van Etten RA, Tenen DG. Reversibility of acute B-cell leukaemia induced by BCR-ABL1. *Nat Genet*. 2000;24:57–60.

40. Druker BJ, Lydon NB. Lessons learned from the development of an abl tyrosine kinase inhibitor for chronic myelogenous leukemia. *J Clin Invest*. 2000;105:3–7.

41. Druker BJ, Tamura S, Buchdunger E, et al. Effects of a selective inhibitor of the Abl tyrosine kinase on the growth of Bcr-Abl positive cells. *Nat Med*. 1996;2:561–566.

42. Deininger MW, Goldman JM, Lydon N, Melo JV. The tyrosine kinase inhibitor CGP57148B selectively inhibits the growth of BCR-ABL-positive cells. *Blood*. 1997;90:3691–3698.

43. Druker BJ, Talpaz M, Resta DJ, et al. Efficacy and safety of a specific inhibitor of the BCR-ABL tyrosine kinase in chronic myeloid leukemia. *N Engl J Med*. 2001;344:1031–1037.

44. O'Hare T, Corbin AS, Druker BJ. Targeted CML therapy: controlling drug resistance, seeking cure. *Curr Opin Genet Dev*. 2006;16:92–99.

45. Gorre ME, Mohammed M, Ellwood K, et al. Clinical resistance to STI-571 cancer therapy caused by BCR-ABL gene mutation or amplification. *Science*. 2001;293:876–880.

46. Shah NP, Nicoll JM, Nagar B, et al. Multiple BCR-ABL kinase domain mutations confer polyclonal resistance to the tyrosine kinase inhibitor imatinib (STI571) in chronic phase and blast crisis chronic myeloid leukemia. *Cancer Cell*. 2002;2:117–125.

47. Force T, Krause DS, Van Etten RA. Molecular mechanisms of cardiotoxicity of tyrosine kinase inhibition. *Nat Rev Cancer*. 2007;7:332–344.

48. Deininger M, Buchdunger E, Druker BJ. The development of imatinib as a therapeutic agent for chronic myeloid leukemia. *Blood*. 2005;105:2640–2653.

49. Palacios R, Steinmetz M. Il-3-dependent mouse clones that express B-220 surface antigen, contain Ig genes in germ-line configuration, and generate B lymphocytes in vivo. *Cell*. 1985;41:727–734.

50. Azam M, Latek RR, Daley GQ. Mechanisms of autoinhibition and STI-571/imatinib resistance revealed by mutagenesis of BCR-ABL. *Cell*. 2003;112:831–843.

51. Johnson LN, Lewis RJ. Structural basis for control by phosphorylation. *Chem Rev*. 2001;101:2209–2242.

52. Downward J. Targeting RAS signalling pathways in cancer therapy. *Nat Rev Cancer*. 2003;3:11–22.

53. Sebolt-Leopold JS, Herrera R. Targeting the mitogen-activated protein kinase cascade to treat cancer. *Nat Rev Cancer*. 2004;4:937–947.

54. Ohren JF, Chen H, Pavlovsky A, et al. Structures of human MAP kinase kinase 1 (MEK1) and MEK2 describe novel noncompetitive kinase inhibition. *Nat Struct Mol Biol*. 2004;11:1192–1197.

55. Bjornsti MA, Houghton PJ. The TOR pathway: a target for cancer therapy. *Nat Rev Cancer*. 2004;4:335–348.

56. Isaacs JS, Xu W, Neckers L. Heat shock protein 90 as a molecular target for cancer therapeutics. *Cancer Cell*. 2003;3:213–217.

57. Pusztai L. Perspectives and challenges of clinical pharmacogenomics in cancer. *Pharmacogenomics*. 2004;5:451–454.

58. Becker J. Signal transduction inhibitors – a work in progress. *Nat Biotechnol*. 2006;22:15–18.

Protein Domains and Signal Transduction

Modular structure of proteins

Many proteins share regions of similar architecture called *domains*. These are discrete, structurally homologous units that may also behave as functional elements. In previous chapters, we have seen how the domains of signalling proteins can be crucial for their function. We now consider the origin of structural domains and how they have been acquired by proteins throughout evolution, leading to functional diversification. To illustrate how the acquisition or modification of function by domains affects signalling mechanisms, we go on to discuss in more detail a few selected domains that are important in signalling.

Structural domains

The term domain was initially used to describe different spatially organized segments of a protein. These might be defined simply by location. For instance, a transmembrane protein will possess distinct intracellular, extracellular, and transmembrane domains. Alternatively, different regions may have different functions, as in enzymes that have separate catalytic and regulatory domains. Furthermore, similar enzymes may possess corresponding domains that exhibit homologies of sequence and structure.

A wider examination of protein structures reveals a pattern of modularity which extends these notions. That is, there are common, compact structural units which recur in different proteins or which may be repeated within a single polypeptide. These units consist typically of stretches of 40–100 amino acids that fold independently of the main chain as globular, closed structures with a hydrophobic core. These regions are characterized by their structural homology and they are very widespread. Most proteins contain at least two structural domains and many signalling molecules have more. Proteins composed of multiple domains have been termed *mosaic proteins*. The range of 3-dimensional structures that domains adopt seems to be restricted to a small proportion of the theoretically possible folds.

The evolution and shuffling of domains

Some types of domain occur in both archaea and bacteria as well as in eukaryotic organisms, and so it is inferred that they were also present in their common ancestor. Other domains are of recent origin, occurring only in eukaryotic organisms. (Indeed, the incidence of structural domains in microbes is much lower than in eukaryotic cells.) Some of the ancestral domains possess essential metabolic enzymatic properties. Others enable important protein–protein interactions. Details of their origin are lacking, but they may have arisen through the recombination of ancestral short sequences (antecedent domain sequences) to form stable core assemblies. Those with functional importance then persisted.[1]

The commonest mechanism by which biological complexity has developed is through gene duplication. Initially, duplication might generate more mRNA, but it can lead to a change of function. Gene duplication may also be partial (internal duplication), causing a gene to become elongated. An outcome is the repetition of all, or part, of a domain sequence within a protein, and if one of the repeated domains goes on to gather mutations, it may acquire a new function. Another process by which a new domain can be acquired is by the insertion of a sequence from one gene into another by recombination. The N- and C-termini of domains introduced in this way are usually adjacent, so that insertion into the polypeptide chain causes minimal disruption of the fold of the recipient protein. By the same principle, the linear sequence of a structural domain may itself be interrupted by another 'nested' domain. Such insertions occur most readily at loops in the parent structure. For example, the chains that form one of the PH domains of phospholipase C$_\gamma$ are separated by an insert of about 300 residues containing one SH3 and two SH2 domains (see Figure 5.11, page 151).

Although some bacteria have acquired domains from eukaryotic species by horizontal gene transfer, it is through the duplication, insertion, and deletion of genes that proteins principally acquire or lose domains and by which new combinations are generated. The process is termed *domain shuffling*. One way in which this occurs is through the shuffling of exons (again through insertion, duplication, or deletion). For this to work effectively there should be a correspondence between exons and domains; for example, a single exon coding for a single complete domain. Insertion of new exons requires recombination within introns and this has to be achieved without creating frameshifts that would cause misreading of the downstream sequence. To avoid disruption of the reading frame, only symmetrical exons can be inserted into introns with matching phase. Although this might seem unlikely, analysis of the human and other genomes indicates that there has been extensive exon shuffling in the evolution of eukaryotic genomes.[2]

After the emergence of the eukaryotes, development of domains accelerated and their inclusion into proteins by exon shuffling led to a rapid increase in diversity. In addition, a particular feature of multicellular organisms is the need for communication between cells. This became possible with the emergence of molecules having extracellular domains able to take part in intercellular signalling processes. All of this may have contributed to the formation of complex life forms during the period of the metazoan radiation.

In summary, the acquisition of new modules has expanded protein function in a combinatorial fashion, generating diversity without requiring new genes. It may help to account for the relatively small number of genes possessed by such complex species as humans. Indeed, the human genome seems to be the most diverse in terms of domain combinations.

Sequence homology and the acquisition of function

Some types of structural domain show high levels of sequence homology while others are very limited in this respect. Where sequence homology exists, it has tended to preserve basic properties, such as the core structure; where the sequence is variable, as at the loops between secondary structural elements, it can allow variations of function. Thus, differences in sequence may reflect adaptations to meet special requirements or they may just be the result of mutations at non-critical residues. An eroded sequence homology hampers the analysis of genetic history and when structural data are not available, it hinders the recognition of domains.

Intron phase. Introns that lie between complete codons have phase 0. Those that divide codons between the first and second nucleotides are phase 1 and those between the second and third nucleotides are phase 2.

Symmetrical exons are bordered by introns with phases 0–0, 1–1, or 2–2. If frameshifts are to be avoided, these exons can only be introduced into introns of phase 0, 1, or 2, respectively.[3]

The human genome appears to possess 20 000–25 000 genes (www.ensembl.org), far fewer than originally anticipated.

Domain function

We have seen how protein function may be enhanced through the acquisition of structural domains and in preceding chapters we have encountered numerous examples of signalling proteins having domains that are key functional elements. We now consider whether domains with similar structures exhibit similar functions. The domain shuffling mechanisms described above, together with mutation, allow function to be acquired, modified or lost over time. Thus domains of common origin may differ in function and, conversely, structurally dissimilar domains may have similar functions (functional convergence). For example, both SH2 domains and PTB domains bind to phosphotyrosine residues, but they are structurally distinct. By contrast, domains with the same fold can exhibit quite different functions. PTB domains and PH domains share the same basic structure, but PH domains bind to phosphoinositides and to G protein βγ-subunits. To complicate matters further, some domains with so-called characteristic functions may, in some proteins, have lost or never acquired that function, existing simply as structural elements. Thus a particular structural domain in a protein may confer a specific function, but it does not guarantee it.

Catalytic domains

It is not always possible to assign function to a single, spatially distinct region of an enzyme. The residues that form the active sites and that take part in the catalytic process are often from disparate segments of the chain. For this reason, lysozyme was classed as a single domain protein. In other cases, enzyme activity resides in regions that can operate as independent functional units. For example, trypsin-like serine protease domains are widely distributed (trypsin itself, chymotrypsin, choline esterases, elastase, plasminogen, etc.). They are identifiable by their structure, and their presence can be used to predict proteolytic activity (although they are not necessarily active). Other domains having catalytic properties include protein kinase domains (serine/threonine kinase and tyrosine kinase domains), protein phosphatase domains, the X and Y catalytic domains of the phospholipases C, and guanine nucleotide exchange factor (GEF) domains, such as RasGEF and RhoGEF.

Protein interaction domains

A large proportion of known structural domains do not possess any inherent enzyme activity, but do provide sites of interaction with other molecules. Such domains can bring about the association of polypeptide chains with each other, or with lipids, nucleic acids, small molecules, or ions. The binding may be quite specific and may generate inter- or intramolecular interactions. For instance, SH2 domains can recognize and bind phosphotyrosines within specific peptide sequences. Likewise, SH3 domains can bind specifically to proline-rich regions. Alternatively, the binding target may be a membrane lipid; some PH domains bind to phosphoinositide headgroups enabling cytosolic proteins to locate to particular membrane sites. Such domains have been termed *protein interaction domains*.[4] The variety is wide, and some illustrative examples are discussed below. A fuller set of descriptions is provided at http://www.mshri.on.ca/pawson/domains.html.

Protein interaction domains are essential cogs in the machinery of most, if not all cellular signalling pathways. At the simplest level, they may allow two separate components to associate just long enough to ensure the propagation of a downstream signal. Alternatively, they can enable an array of components to assemble into large multimolecular complexes. Importantly, this happens at specific cellular locations determined by interactions of the components with local proteins and phospholipids. Whether simple or complex, these processes are initiated by the touch of a button, often a single event, such as the binding of a ligand to its receptor.

The inventory of domains

Detection

Since the degree of sequence homology among domains of the same type may be limited, their detection using basic sequence alignment tools (e.g. BLAST) can be difficult, particularly when a nested domain interrupts the sequence of another. A better method is to compare a sequence with a statistical profile calculated from a library of known domain sequences. The probability that the test sequence matches the profile is scored, a high score predicting the presence of the domain. The most effective profiles are hidden Markov models, and a number of the publicly accessible protein databases utilize libraries of these models to predict domain architecture (for example Pfam, SMART, TIGRFAM, PANTHER, PIRSF, Gene3D, and SUPERFAMILY). With successive refinements to these tools, previously unperceived domains have been revealed in many proteins. However, differences in definitions and analysis methods mean that domain architectures predicted for a given sequence by the different approaches are not always identical.

Classification

Domains may be classified on the basis of their sequence, structural fold, or function. As already discussed, function and fold are not always correlated, nor are fold and ancestry. Where there are clear sequence similarities, domains have been assigned to families, and where there is structural and functional evidence of descent from a common ancestor, families have been grouped in superfamilies (SCOP) or clans (Pfam[5]). However, because definitions vary, such assignments may differ between databases.

The Interpro database (http://www.ebi.ac.uk/interpro/) integrates information from a range of so-called signature databases, including those mentioned above. Release 12.1 has 12953 protein entries and an inventory of 3585 domain types. However, the number of well characterized families is much smaller, of the order of 50–100.

Examples of domains with roles in signalling

A glance at the protein databases reveals the multidomain structure of many proteins, especially signalling proteins. As more domain types have been identified, it has become clear that the potential combinatorial diversity is huge. As already discussed, signal propagation within cells commonly involves the formation of molecular complexes at particular cellular sites. Such complexes may contain catalytic domains that perform covalent modifications on substrate molecules that are then recognized by domains on other proteins (e.g. SH2 domains binding to phosphorylated tyrosine head groups).

Thus, in response to a ligand binding to its receptor and in a seemingly amorphous and disordered cytoplasm, organized signalling complexes are assembled which recruit yet other molecules to project the signal onwards. Interactions between domains lie at the heart of all this, but although we have accrued extensive information about their incidence in a wide range of proteins, our understanding of the mechanisms in which they take part is limited. Importantly, for the reasons given above, the presence of a particular domain in a protein is no guarantee of function.

The list of domains in signalling proteins is large, and the electronic databases provide the most comprehensive source of information. Table 24.1 summarizes properties of domains that have received mention in previous chapters and the sections that follow provide illustrative examples of some of the most important.

Domains that bind oligopeptide motifs

SH2 domains

SH2 domains are exemplars of protein interaction domains that take part in signalling. They bind to short motifs containing a phosphotyrosine and were first observed in the viral non-receptor protein tyrosine kinase v-fps/fes, in the form of a regulatory region separate and distinct from the kinase domain. The region is conserved among other nrPTKs and is commonly located immediately N-terminal to the catalytic domain, (see Figure 17.1, page 514). It was named SH2 for Src homology region 2.[6] In Src it regulates kinase activity through an intramolecular association with a phosphotyrosine near the

TABLE 24.1 Examples of protein domains important in signalling

Domain	Name/origin	Binds to/associates with	Function	Examples of proteins with the domain
SH2	Src homology 2	pY, some SH3	complex assembly	Src, Btk, RasGAP, PLC-γ, Grb2, PI3K-p85, Vav, Dbl, Shc, STATS,
PTB	Phosphotyrosine binding	pY, phosphoinositides	complex assembly	IRS1, IRS2, Shc
PID	Phosphotyrosine interaction	pY, Y, phosphoinositides	complex assembly	Shc, RGS12
WD40 rpt	Trp-Asp repeats	pS, pT, pS short motifs	complex assembly	Gβ, STE4
SH3	Src homology 3	PXXP, RXXK, tandem Y, α-helical segment	complex assembly	Src, Btk, RasGAP, PLC-γ, Grb2, PI3K-p85, Vav, Nck
PH	pleckstrin homology	phosphoinositide, Gβγ	recruitment to membrane	RasGAP, Sos, PLC-δ1, PKB, Grb2, GRK2/βARK, Vav, Dbl
C2	C2 domain of PKC	phosphoinositide/Ca^{2+}	recruitment to membrane	PKC, RasGAP, PLC-δ1, PLC-γ1, Syt
FYVE Zn finger	Fab1, YOTB/ZK632.12, Vac1, EEA1	phosphoinositide (PI(3)P)	recruitment to membrane	EEA1, SARA
PX	Phox homol.	phosphoinositides	recruitment to membrane	p40^{phox}, p47^{phox}, PLD1
DH	Dbl homol.	Rho GTPase	RhoGEF	βARK, Dbl, Vav, Sos
C1	C1 domain of PKC	DAG	activates PKC	PKC, DAG kinase, Vav
EFh	EF hand	Ca^{2+}	intracellular Ca^{2+} sensor	CaM, troponin C, DAG kinase, calcineurin B, GCAP
cadherin rpt	cadherin repeat	Ca^{2+}	cell adhesion	cadherins
DED	death effector domain	self associates	recruits procaspases	procaspases, FADD
DD	death domain	self associates	regulation of apoptosis	TNF receptors, FADD, TRADD
CARD	caspase recruitment domain	self associates	recruits procaspases	APAF, procaspases
MH1	MAD homology 1	DNA	transcriptional activation	Smads (N-term)

(Continued)

TABLE 24.1 Continued

Domain	Name/origin	Binds to/ associates with	Function	Examples of proteins with the domain
MH2	MAD homology 2	pSXpS (Smads), SARA	transcriptional activation	Smads (C-term)
bromo	brahma protein	acetylated Lys	transcriptional co-activation	Snf-2, pCAF, CBP, p300, TFIID complex
chromo	chromatin organization modifier	methylated Lys	transcriptional repression	Swi-6, HP1
PDZ	post-synaptic density p95, Dlg & ZO1	C-termini of channels/receptors, phosphoinositides	assembly/ membrane recruitment	InaD
RGS	regulator of G protein signalling	$G\alpha$	increases GTPase activity	RGS proteins, GRK2/βARK, RhoGEF
pkinase	protein kinase	Ser, Thr	phosphorylation	PKC, PKA, PKB
Y kinase	protein tyrosine kinase	Tyr	phosphorylation	Insulin receptor, Src, Btk
Zn finger (C2H2)	with 2 Cys and 2 His residues	DNA, RNA, proteins	transcription complex assembly	transcription factors
Zn finger (C4)	with 4 Cys residues	DNA, RNA, proteins	transcription complex assembly	nuclear receptors

See end of chapter for abbreviations.

C-terminus. This keeps the kinase in a compact, inactive conformation. (The mechanism by which this inhibition is lifted is discussed below.) The affinity of SH2 domains for non-phosphorylated tyrosines or for phosphoserines or phosphothreonines is generally negligible.

SH2 domains are present in at least 110 different human proteins, which fall into 11 different functional classes.[7] These may have catalytic activity, such as the receptor and non-receptor tyrosine kinases, RasGAP, phospholipase Cγ, the regulatory subunits of PI3-kinase, and some protein tyrosine phosphatases. In all of these, the ability of the SH2 domain to bind to a phosphotyrosine has important consequences for enzyme activity. SH2 domain-containing proteins that are devoid of a catalytic activity, but carry other protein interaction domains, include adaptor proteins, typified by Grb2.

SH2 domains and protein tyrosine kinases are expressed throughout the animal kingdom, but they are less apparent in other eukaryotic organisms.

There is evidence that they may have coevolved in protozoans that were predecessors of multicellular organisms. For example, the slime mould *Dictyostelium discoideum* encodes 12 proteins having SH2 domains. Some of these are similar to the mammalian STAT proteins that dimerize, through mutual interactions involving SH2 domains and phosphotyrosines (see page 353), to become transcription factors, although this interaction does not appear to be essential for nuclear translocation[8]. In addition, while *Dictyostelium* does not express conventional PTKs, it does encode tyrosine kinase-like proteins.

The credentials of *Dictyostelium* as a transitional species between unicellular and multicellular life are surpassed by choanoflagellate protozoans, such as *Monosiga brevicollis*. This species is also colonial and expresses not only SH2 domains but also PTKs similar to those of mammals.[9] Such organisms may resemble the immediate predecessors of multicellular eukaryotes and it possible that the coordinated evolution of PTKs and SH2 domains contributed to the metazoan radiation.

The structure of the SH2 domain of the nrPTK Lck is depicted in the central panels of Figure 24.1. Like other SH2 domains, it consists of about 100 amino acids and possesses a central antiparallel β-sheet, flanked by two α-helices. The phosphopeptide ligand shown in the figure contains the motif -pYEEI-. Two pockets, on either side of the β-sheet, provide binding sites for the phosphotyrosine and for the isoleucine residue at pY+3, respectively. The combined effect of the two sites is to enable high affinity binding that is phosphorylation-dependent. Within the phosphotyrosine binding pocket, two arginine residues make contacts with the phosphate group (Figure 24.1).

In general, SH2 domains recognize ligand motifs of 4–7 amino acids with an N-terminal pY and, commonly, a hydrophobic residue at pY+3. Different types of SH2 domain have different binding preferences. For example, while the SH2 domain of Src kinase binds optimally to the sequence pYEEI, a mutation of a threonine to a tyrosine in the specificity pocket changes its preference to pYVNV,[11] enabling it to bind Grb2.

SH2 domains have been classified on the basis of their preferred ligand sequences, but an optimal sequence may not be the same as the sequence in an actual target. Ligand selectivity has been investigated *in vitro* using degenerate phosphopeptide libraries and the data indicate considerable binding versatility.[7] For example, Src can bind to pY-[E/D/T]-[E/N/Y]-[I/M/L] and while the C-terminal SH2 domain of phospholipase Cγ prefers pY-[V/I/L]-[E/D]-[P/V/I], it binds to pY-I-I-L-P-D-P in the activated PDGF receptor.

PTB domains

The PTB (phosphotyrosine-binding) domain also recognizes phosphotyrosine residues, but its structure is completely different from that of the SH2 domain. Figure 24.2 shows the PTB domain of IRS-1 (insulin receptor substrate-1). Its

The social slime mould *Dictyostelium discoideum* exists as single cells for most of the time, but under stressful conditions such as starvation, these cells aggregate to form a 'slug' which can migrate as a coherent organism. The cells differentiate, forming a stalk and an independent fruiting body which distributes spores.

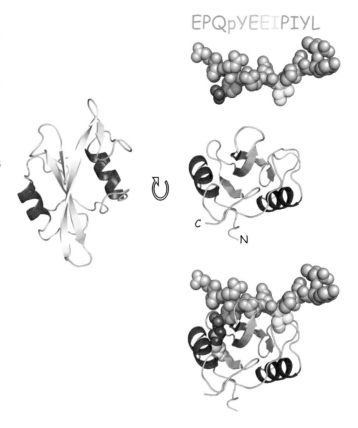

FIG 24.1 SH2 domain of the Src kinase Lck.
The structure on the left shows the central β-sheet and flanking helices. A rotated view is shown in the central panel. A polypeptide containing the -pYEEI- motif is depicted as spheres at the top of the figure. The bottom panel shows the polypeptide bound to the domain. The phosphotyrosine residue (salmon pink) makes a number of contacts with residues that line the pocket, including two arginines, one of which is displayed as grey and blue spheres. The isoleucine residue (yellow) is located in a hydrophobic cleft (1lcj.pdb[10]).

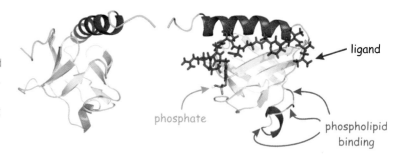

FIG 24.2 The PTB domain of IRS-1. Two views of the domain rotated 90° about a vertical axis. The right-hand structure includes a phosphotyrosyl ligand (LVIAGNPApYRS, shown as blue sticks). Also indicated are three lysine residues (purple) that take part in binding to acidic phospholipid headgroups (1irs[13]).

structure is based on the highly conserved PH domain superfold (see Figure 24.4) adopted by PH, EVH1, and a number of other protein interaction domains. The rigid β-barrel, closed off at one end by a long α-helix, provides a conserved scaffold, the surface of which has been adapted in different ways to bind diverse protein motifs or lipid head groups. Note that PH and EVH1 domains do not bind to phosphotyrosines and, despite their name, nor do a large number of PTB domains. The PTB domains that do, prefer

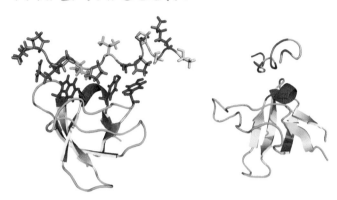

PPRPLPVAPGSSKT

FIG 24.3 The SH3 domain of the Src kinase Fyn complexed with a peptide ligand corresponding to residues 91–104 of the p85 subunit of PI 3-kinase.
The view on the left shows the domain in cartoon format and the peptide as sticks. Three of the aromatic amino acid side chains that define the binding site are also shown as sticks (magenta). The right hand view shows the structure rotated about a vertical axis illustrating the left-handed helical conformation of the ligand (1azg.pdb[15]).

the target motif N-P-X-pY,[12] quite distinct from the sequences recognized by SH2 domains. This is because the binding sites are dissimilar. PTB domains lack binding pockets that bury the target motif, attaching it instead to the side of the β-barrel. The phosphate group remains relatively exposed.

A characteristic that some PTB domains share with PH domains is their ability to bind to acidic phospholipid head groups, such those of the phosphoinositides. This enables them to locate at membrane surfaces where they may then encounter their peptide ligands.[14]

SH3 domains

SH3 (Src homology 3) domains consist of ~60 amino acids arranged to form a compact, twisted β-barrel structure. They bind to proline-rich sequences. These generally consist of 8–10 residues and include at least two prolines in a PxxP motif. A typical SH3 domain is depicted in **Figure 24.3** together with a peptide ligand. Proline-rich sequences are abundant in multicellular species and seem to have coevolved with SH3 domains (and EVH1 and WW domains to which they can also bind). The utility of these sequences stems from the cyclic structure of proline, which limits the range of conformations that it can adopt. It therefore tends to occur at the ends of, or between, regions of secondary structure, particularly at loops and turns. A short sequence having more than one proline adopts a stable, left-handed, extended helical structure (three residues per turn) that can tolerate variations in sequence. Such type II polyproline helices commonly occur on the surfaces of proteins.

The SH3 domain binding surface is hydrophobic[16] with conserved aromatic side chains forming ridges that interact with the grooves of the polyproline helix. About three turns are involved in binding (see Figure 24.3). SH3 domains fall into two main classes that recognize different target sequences. One contains the motif [R/K]xxPxxP (class I), the other PxxPx[R/K] (class II). Proteins with class I motifs will align in the opposite direction to those with class II. Either way, the binding is weak and promiscuous, unless there are additional contributions from flanking residues. This may be the reason why adaptor proteins, such as Grb2 and Nck, possess two or more SH3 domains (see Figure 12.6, page 326). In the case of Grb2, which has the architecture SH3-SH2-SH3, the SH3 domains bind to two neighbouring proline-rich regions on the target, thereby increasing the affinity and specificity of binding. The tandem arrangement of SH2 and SH3 domains found in several signalling proteins provides a conformational mechanism for regulating SH3-dependent interactions through tyrosine phosphorylation.[17]

Note that although the architecture of the other proline recognition domains (EVH1, WW, etc.) is quite distinct from that of SH3 domains, the binding mechanism is very similar.

Phosphoinositide-binding domains

A common feature of signal transduction following receptor activation is the formation of complexes through the recruitment of cytosolic proteins to membrane sites. This may occur through the recognition of membrane phospholipid headgroups by phosphoinositide-binding domains.[18] There are seven main phosphoinositides (see Figure 18.1, page 546) and although they are minority components of cell membranes, they are selectively enriched in different subcellular compartments, such as in the membranes of organelles or in the plasma membrane. For example $PI(4,5)P_2$ is most abundant at particular sites on the inner leaflet of the plasma membrane and on the surface of the Golgi apparatus, while $PI(3)P$ is enriched on endosomal membranes (see below).

$PI(4,5)P_2$ is a substrate for phospholipase C and PI 3-kinase. Whereas $PI(4,5)P_2$ is present in resting cells, the levels of $PI(3,4)P_2$ and $PI(3,4,5)P_3$ are negligible, but rise acutely at the plasma membrane when PI 3-kinase is recruited from the cytosol, as in signalling through the insulin receptor (see page 554). Proteins recruited to the membrane by $PI(3,4)P_2$ and $PI(3,4,5)P_3$ have domains that bind to these 3-phosphoinositides selectively and with high affinity, in contrast to those that bind $PI(4,5)P_2$, which tend to exhibit less preference and much lower affinity. This looser binding allows recruited proteins to explore different areas of membrane, giving them the opportunity to make additional contacts that enhance binding.

Finally, $PI(4,5)P_2$ regulates the actin cytoskeleton. Actin-binding proteins, such as WASP (Wiscott–Aldrich syndrome protein), profilin, cofilin, and gelsolin regulate the assembly and disassembly of filamentous actin. These and other proteins that promote cytoskeletal remodelling adhere to the plasma membrane by binding $PI(4,5)P_2$. Most use short stretches of basic amino acids that bind the acidic head group. These positively charged regions are not structural homology domains; indeed they may be unstructured prior to binding. (Spectrin, on the other hand, binds $PI(4,5)P_2$ through its PH domain.)

Table 24.2 lists the principal phosphoinositide-binding domains with their preferred targets. Before discussing some of these in further detail, it is important to stress that many may also bind peptide ligands.

PH domains

PH or pleckstrin homology domains are among the most common domains in the human proteome.[20] Early studies indicated a general membrane-targeting function enabled by interaction with the head groups of membrane phosphoinositides, but although this is undoubtedly an important property of many PH domains, many bind very weakly and some not at all.[21]

Pleckstrin is the major PKC substrate in platelets.[22,23] Rather unusually, it contains two PH domains, one at each end of the molecule.

TABLE 24.2 Phosphoinositide-binding domains

Domain	Protein	Specificity
PH	PLCδ1	$PI(4,5)P_2$
	βARK	$PI(4,5)P_2$, $PI(3,4,5)P_3$
	GRP1	$PI(3,4,5)P_3$
	PKB/Akt	$PI(3,4)P_2$, $PI(3,4,5)P_3$
	Btk	$PI(3,4,5)P_3$
	Ras GAP1^m	$PI(3,4,5)P_3$
	Ras GAP1^{IP4BP}	$PI(4,5)P_2$, $PI(3,4,5)P_3$
	FAPP1	$PI(4)P$
	TAPP1	$PI(3,4)P_2$
FYVE	EEA1	$PI(3)P$
	Hrs	$PI(3)P$
	Fab1/PIKfyve	$PI(3)P$
	SARA	$PI(3)P$
PX	p40^{phox}	$PI(3)P$
	p47^{phox}	$PI(3,4)P_2$
	SNX1	$PI(3)P$, $PI(3,5)P_2$
	SNX2	$PI(3)P$
	SNX3	$PI(3)P$
ENTH/ANTH	epsin1	$PI(4,5)P_2$
	epsinR	$PI(4)P$
	AP180	$PI(4,5)P_2$

EVH1 domains are involved in cytoskeletal reorganization.

PDZ domains bind to the C-terminal residues of some ion channels and receptors.

Ran-binding domains are present in the nuclear pore complex.

FERM domains are present in cytoskeletal proteins.

PH domains have ~120 amino acid residues. One of the clearest examples of the core structure is provided by the PH domain of GRK2 (G protein receptor kinase/β-adrenergic receptor kinase), depicted in **Figure 24.4a**. It consists of a β-sandwich, formed by seven antiparallel strands. The arrangement resembles a twisted barrel and it is stabilized by a C-terminal α-helix that is packed against one end. Despite low sequence homology among PH domains, this core framework, termed the PH domain superfold, is highly conserved. Moreover, it has been adopted by other domains, none of which have any sequence similarity with PH domains. These include the PTB domains shown in **Figure 24.2**, the EVH1, PDZ, and Ran-binding domains, and also part of the FERM domain. Some of these also bind phosphoinositides, but often at sites that do not correspond to those used by PH domains.

The PH domain of PLCδ1 is unusual in that it has relatively high affinity ($K_D \sim 10^{-6}$ mol L^{-1}) and specificity for PI(4,5)P$_2$.[24] The head-group phosphates bind to basic residues in the loops between the β strands (mostly β1/β2 and β3/β4) as illustrated in **Figure 24.4b**. These help to locate the phospholipase at membrane regions containing PI(4,5)P$_2$, which is also its substrate. However, this is not the whole story. Following this recognition event, hydrophobic contacts, probably involving side chains in the domain's C-terminal region, provide an additional attachment, causing it to insert into the lipid bilayer.[25] Binding to PI(4,5)P$_2$ accelerates this interaction, which otherwise is phosphoinositide-independent.

As mentioned previously, other PH domains that bind PI(4,5)P$_2$ do so only weakly and unselectively, yet many of the proteins that bear them still localize to membranes. This is because these PH domains make additional hydrophobic contacts that do not involve phosphoinositides. Alternatively,

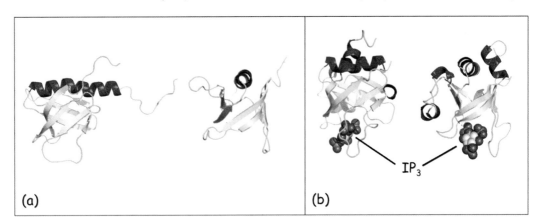

FIG 24.4 PH domains of GRK2 and PLCδ1

(a) Two views of the GRK2 PH domain. The structure on the right is rotated 90° about the y axis. (b) Two similar views of the PLCδ1 PH domain, showing a bound IP$_3$ molecule (spheres) representing the head group of PI(4,5)P$_2$ (2bcj,[26] 1mai[27]).

they may attach to other membrane-associated proteins.[21] Then again, additional contacts may be made by residues outside the PH domain, for instance by other membrane targeting domains. (An example of both of these mechanisms is provided by GRK2.)

The few PH domains that bind specifically to $PI(3,4)P_2$ and $PI(3,4,5)P_3$ have affinities in the submicromolar range. These are present in proteins that can sense the production of these phosphoinositides against a high background of $PI(4,5)P_2$. Such effectors include PKB and PDK1, as well as the guanine nucleotide exchange factors for the GTPase Arf (GRP1, ARNO, and cytohesin) and the non-receptor protein tyrosine kinase Btk. All are recruited to the membrane by 3-phosphorylated polyphosphoinositides.

PH domains may also take part in protein–protein interactions. The best-characterized example is provided by GRK2 that phosphorylates activated β-adrenergic receptors. Its PH domain (Figure 24.4) binds both to $PI(4,5)P_2$ and to the βγ-subunit of the G protein activated by the receptor. Receptor phosphorylation leads to desensitization and down-regulation of signalling. The mechanism is described in Chapter 4 (see page 98).

Other examples of PH domains that participate in protein–protein interactions include those of GRK3, phospholipase C-β2, and the non-receptor protein tyrosine kinase Btk. (Note that the PH domain of PLCβ2 does not bind to phosphoinositides.)

Other phosphoinositide-binding domains

FYVE, PX, and ENTH/ANTH domains bind phosphoinositide headgroups and exhibit preferences that enable the tethering of proteins to specific membranes (Table 24.2). Whereas $PI(4,5)P_2$ is predominant on the inner leaflet of the plasma membrane, other phosphoinositides are enriched on the cytosolic faces of organelles, where they take part in the recruitment of components that form the machinery that controls the vesicular transport of proteins. For example, PI(3)P, $PI(3,5)P_2$, and PI(4)P are markers of early endosomes, late endosomes, and the Golgi apparatus, respectively. In general, the vesicle-mediated trafficking of soluble or transmembrane proteins between compartments is a complex process. Different components control each stage and in many cases these are recruited through the interaction of a phosphoinositide-binding domain with a particular membrane.[28] The initial event involves the induction of curvature at specific points in the membrane of the vesicle-forming compartment. This is followed by the construction of a protein coat to support a vesicle bud, the selection of cargo and then scission of the vesicle and its delivery to the correct destination.

For example, the *early* or *sorting endosome*, the receiving compartment for endocytic vesicles arriving from the plasma membrane, is characterized by the presence of PI(3)P which acts as an anchor for the C-terminal FYVE domain of the protein EEA-1, a marker of early endosomes.[29] The FYVE domain structure (~80 residues, sometimes called a FYVE finger domain) is stabilized by two Zn^{2+} ions, coordinated by the side chains of 8 cysteine residues (**Figure 24.5**; zinc fingers are discussed below). It binds 3-phosphorylated inositides with high specificity,[30] though the interaction is dependent on the presence of the small GTPase Rab5, as well as sequences adjacent to the FYVE domain.[31]

The FYVE domain is also present in PI(3)P-5-kinase (also known as PIKfyve). This lipid kinase is tethered to the membranes of late endosomes, where it phosphorylates PI(3)P to form $PI(3,5)P_2$, a marker of multivesicular bodies and late endosomes.

PX domains (phox homology domains) also bind to 3-phosphorylated inositol lipids. They were originally detected in p40[phox] and p47[phox], subunits of the NADPH oxidase complex responsible for the respiratory burst in neutrophils.[33] p40[phox] binds to PI(3)P on the surface of endosomes. PX domains also occur in the sorting nexins, a large protein family with members that direct the trafficking of membrane proteins, such as receptors, and soluble cargo between cellular compartments. For instance, SNX1 (sorting nexin 1) binds PI(3)P and forms part of a complex that regulates the sorting of specific proteins from the early endosomal compartment to the trans-Golgi network. SNX3 also binds PI(3)P, but directs cargo from the early endosome to the recycling endosome. A key property of SNX1 is the presence of a BAR domain (Bin, amphiphysin, Rvs). These banana-shaped domains act as curvature sensors, binding to highly curved membranes. Furthermore, they are also able to impose curvature upon membranes and thus take part in vesicle formation. This occurs in regions where the nexin has been recruited by PI(3)P. (It remains to be shown if the imposition of curvature is a general property of the sorting nexins.)

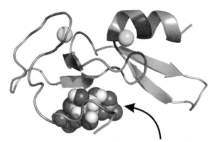

inositol 1,3-bisphosphate

FIG 24.5 The FYVE domain of EEA-1.
The structure shows a bound molecule of inositol 1,3-bisphosphate, corresponding to the head group of PI(3)P (spheres). Also indicated are two Zn^{2+} ions (turquoise spheres) and the residues that coordinate these ions. The two N-terminal cysteines are coloured red and the two C-terminal cysteines are blue. Side chains are not shown (1hyi[32]).

Similar processes occur in clathrin-mediated endocytosis. At the plasma membrane, the components that lead to the formation of endocytic vesicles are recruited to PI(4,5)P$_2$. At an early stage, one of the key proteins that promotes membrane invagination is Epsin,[34] which binds to PI(4,5)P$_2$ through its ENTH domain (Epsin N-terminal homology domain). This interaction causes its N-terminal residues to fold as an amphipathic helix that inserts into the inner leaflet of the bilayer, causing the membrane to bend. (This property is not restricted to Epsin. Other proteins involved in trafficking, including the GTPase ARF and some BAR domain-containing proteins, can also form amphipathic helices.) Epsin also associates with the adaptor AP2 that recruits clathrin monomers to form the coat that supports vesicle budding and mediates cargo selection. A further essential protein involved in the tethering of clathrin is AP180/CALM. This also binds PI(4,5)P$_2$ through its ANTH domain (AP180/CALM N-terminal homology). These domains are about 50% larger than ENTH domains and both are arranged as solenoids of α-helices. Although there is a region that is structurally similar in both, they bind PI(4,5)P$_2$ differently.

Yet another phosphoinositide-binding domain is the C2 domain, discussed in the next section.

For more detail see www.endocytosis.org.

Polypeptide modules that bind Ca^{2+}

Calcium-binding motifs and domains

Many enzymes are sensitive to changes in the local Ca^{2+} concentration. For instance, when the cytosolic [Ca^{2+}] rises above the resting level (40–100 nmol L^{-1}), signalling proteins with Ca^{2+}-binding motifs or domains become activated (see Chapter 8). In addition, cytosolic enzymes that possess Ca^{2+}-binding C2 domains may be recruited to membrane sites through a Ca^{2+}-dependent interaction with membrane phospholipids.

The EF-hand motif

Many Ca^{2+}-binding regions on proteins possess helix–loop–helix structures known as EF-hand motifs. The letters E and F denote helical regions of the muscle protein parvalbumin, in which the motif was first identified. The Ca^{2+} binding site is formed by the loop of $\sim$12 amino acids that links the two α-helices (Figure 24.6). EF-hands generally occur as adjacent pairs. This arrangement allows them to fold compactly as an EF-hand unit. In general, the core of the EF-hand structure is reasonably conserved, but the outer regions vary and the K_D for Ca^{2+} can range from 10^{-5} to 10^{-7} M. The widely expressed, intracellular Ca^{2+}-sensing protein calmodulin possesses two EF-hand units (i.e. four EF-hands). In each motif, the chemical ligands that coordinate Ca^{2+} are oxygen atoms provided by aspartate and glutamate side chains, a peptide bond carbonyl group, and a water molecule.[35,36] Conformational changes in calmodulin that result from Ca^{2+} binding are described in Chapter 8 (see page 225).

FIG 24.6 EF-hand structure. One of the EF-hands of calmodulin showing a bound Ca^{2+} ion (green sphere, almost hidden). The seven oxygen atoms that form the coordination shell are shown as red spheres (1cll.pdb[37]).

C2 domains

Unlike the small EF-hand motifs, C2 domains are substantial structures ($\sim$130 residues), named after the 'second conserved' regulatory domain of protein kinase Cβ (see Figure 9.8, page 254). They exist in a wide range of intracellular proteins and were originally known as Ca^{2+} and lipid-binding domains, but like the EF-hands their affinity for Ca^{2+} is variable. A typical C2 domain is arranged as a rigid, eight-stranded, antiparallel β sandwich. Ca^{2+}-binding is confined to a region defined by three loops on one edge of the structure. In the C2 domain of protein kinase Cβ (**Figure 24.7**), five aspartate residues within two of the loops present oxygen atoms that form a binding pocket which can accommodate up to three Ca^{2+} ions. These bind in a cooperative manner. Note, however, that the coordination sphere around each Ca^{2+} ion is not quite complete. Apart from the anionic residues that bind the Ca^{2+} ions, the side chains of other residues in the loops are mostly positively charged. Thus, when calcium ions enter the pocket, the negative charges are neutralized and the protein can then bind electrostatically to the anionic head group of the membrane lipid phosphatidylserine. This completes the coordination shell around the Ca^{2+} ion and brings about membrane attachment. In a similar fashion, phospholipid binding by the C2 domain of synaptotagmin requires the binding of Ca^{2+}, which acts like an electrostatic switch, changing the surface potential of the Ca^{2+}-binding loops and enabling them to bind to acidic phospholipid head groups.[38]

The tethering of signalling molecules to membranes through C2 domains can be strengthened by interactions involving regions of the domain away from its Ca^{2+}-binding loops. For example the C2 domain of PKC-α possesses additional basic residues in the loop between β-strands 3 and 4, a region on

The binding of one ligand alters the affinity of other site(s). In this case, **positive cooperativity** denotes an enhancement of the affinity at other sites.

Synaptotagmin is a transmembrane protein in secretory vesicles that is important in exocytosis.

FIG 24.7 C2 domain structure.
Bound Ca^{2+} ions are shown (green spheres) with the coordinating oxygen atoms (red spheres). The view on the right shows the structure rotated 90° about a vertical axis (1a25.pdb[41]).

the side of the domain. These residues interact with PI(4,5)P$_2$, targeting PKC-α to areas of the plasma membrane in which this lipid is enriched.[39,40]

While some C2 domains act as Ca^{2+}-dependent membrane-targeting modules, the stoichiometry of binding (number of sites) and the affinity for Ca^{2+} vary widely. Indeed, some C2 domains do not bind Ca^{2+} at all.

Zinc finger domains

Zinc ions play an essential role in maintaining the tertiary structure of a range of proteins, in particular those that bind DNA. To do this, they form stable coordination complexes with cysteine and histidine residues (or less commonly with aspartate or glutamate) present in stretches of some 30–100 amino acids. The bound ions impose tetrahedral geometry on the ligands (N, O, or S) that form the primary coordination sphere. This stabilizes the polypeptide conformation, often in the form of a protrusion or finger. Hence, these Zn^{2+}-binding modules are known as *zinc fingers*. They were first identified in transcription factor IIIA in *Xenopus*, where they occur as repeats of ~30 amino acids.[42] Some 20 zinc finger families have since been characterized.[43] (See also http://www.gene.ucl.ac.uk/nomenclature/genefamily/zincfinger.html.)

These have been classified according to the amino acids that contribute to the coordination shell. The largest family, with several hundred members, encompasses the classical zinc fingers in which Zn^{2+} is coordinated by two cysteines and two histidines (denoted C2H2). An example, the zinc finger domain of ZNF24 is shown in **Figure 24.8**. The coordinating residues are arranged as CxxCy$_{12}$HxxxH (sometimes written as C-2-C-12-H-3-H), where x and y indicate less conserved residues and those in the y region are mainly polar or basic, suitable for nucleic acid binding. However, a single module

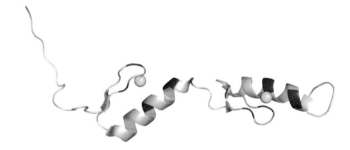

FIG 24.8 Structure of the classical/C2H2 zinc finger domain of ZNF24.
Zinc ions are shown as turquoise spheres. The coordinating residues (side chains not shown) are two cysteines (red) and two histidines (blue) in a $\beta\beta\alpha$ structure. This motif occurs as a tandem repeat. ZNF24 is a transcription factor essential for early embryonic development and cell proliferation[44] (1x6e[45]).

such as this cannot make contact with more than four bases. Recognition of a segment of DNA long enough to represent a transcriptional site generally requires three or more such domains, separated by linkers.

In other families, Zn^{2+} ions are coordinated by sets of four cysteine side chains (C2C2). This occurs in GATA-type zinc fingers (a group that takes its name from the erythroid transcription factor GATA-1, so-called because it binds to the DNA sequence GATA). Again, for this protein to bind with high affinity at a transcription site, more than one C2C2 module is required, in this case a tandem pair separated by a linker. The attachment of C2C2 fingers to DNA is illustrated in Figure 10.8, page 286 for the DNA-binding domain of the glucocorticoid receptor. The tandem C2C2 fingers are separated by a linking sequence, enabling the receptor to bind to the two half-sites on DNA.

A rather different three-dimensional structure is exhibited by the C2C2 finger that occurs in the lipid-binding FYVE domain. In the tandem arrangements of zinc fingers that we have considered so far, the N-terminal C2C2 binds one Zn^{2+} while the C-terminal C2C2 binds the other. In the FYVE domain, alternate pairs of cysteines bind to each Zn^{2+}. This so-called cross-braced conformation is illustrated in Figure 24.5 (compare Figure 24.8).

Another example of the cross-brace occurs in the zinc finger domain that lies within the N-terminal region of conventional PKC. This is the regulatory C1 region that binds the activator, diacylglycerol. It contains two cysteine-rich zinc-binding domains, termed C1A and C1B (see Figure 9.9 page 256). Each of these can bind two Zn^{2+} ions, each complexed by three cysteines and one histidine. Once again, alternate pairs of residues contribute to each site, so stabilizing the tertiary structure.[46,47]

Other zinc finger domains include the RING fingers involved in the ubiquitylation of proteins and in the consequent formation of multiprotein complexes (see figure 15.12 page 468).

Protein kinase domains

Protein kinases share a common domain

The catalytic activity of protein kinases (whether serine/threonine specific or tyrosine specific) is confined to a structurally conserved protein kinase domain. The basic architecture of the kinase domain, as typified by the catalytic subunit of PKA, is illustrated in **Figure 24.9**. The single polypeptide chain folds to form two closely apposed lobes, with an ATP binding site situated in the cleft formed between them. There is also an N-terminal α-helical chain (the A helix) that makes contact with the surface of both lobes.

The N-terminal lobe, the smaller of the two, possesses $\sim$100 amino acids. Although it is myristoylated at its N-terminus, there is no evidence that this group is free to associate with membranes. Instead, it occupies a pocket and

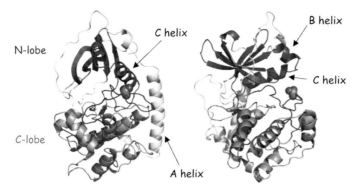

FIG 24.9 Protein kinase domain structure.
Two views of the catalytic subunit of porcine PKA. The structure on the right has been rotated 90° about a vertical axis. The small N-terminal lobe is coloured blue and the large C-terminal lobe magenta. The activation segment is coloured red and the phosphate group on T197 is depicted as spheres. An N-terminal myristoyl group is not shown (1cdk.pdb[49]).

provides structural stability, helping to keep the A-helix in contact with the larger lobe. The principal feature of the small lobe is an antiparallel β-sheet. There are also two α-helical chains, the B- and C-helices. The C-terminal lobe consists of ~200 residues and it possesses mostly α-helical structure, arranged around a stable four-helix bundle. Like other protein kinases, PKA has a central chain of residues called the activation segment or loop (Figure 24.9). The conformation of this segment and in many cases, the phosphorylation of a key residue (T197 in PKA) is critical for catalysis.[48]

Structural elements that regulate kinase activity

The principal components of the catalytic site are shown in **Figure 24.10**. In broad terms, the N-lobe binds ATP, while the C-lobe binds substrate and enables catalysis, though for ATP to bind in the correct orientation in the cleft, it makes contact with residues on both lobes. Two Mg^{2+} ions are also bound (for clarity, these are not shown). The residues on the N-lobe that align the nucleotide include the main chain nitrogens of the glycine-rich loop between β-strands 1 and 2, and the side chain of a lysine (K72) on β-strand 3, that interacts in turn with a glutamate (E91) on the C-helix. The correct location of ATP is very sensitive to the positioning of the C-helix.

In the active conformation of the kinase, T197 is phosphorylated and makes contact with positively charged residues on both the small and large lobes. These interactions effectively seal the cleft and prepare for catalysis by neutralizing the charge on R165 in the catalytic loop, adjacent to D166, the key catalytic residue. With ATP in place, the recognition and binding of a consensus motif on the target protein can proceed. For PKA, the sequence of the motif is Rxx[T/S]h (where h indicates an amino acid with a hydrophobic

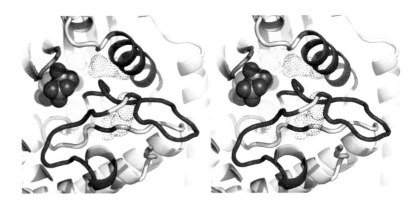

FIG 24.10 Catalytic core of PKA.

In this stereoscopic view (page xxv) of the catalytic site, the key chains are the glycine-rich loop (orange, top left of structure), the C-helix (magenta), the activation segment (red) and the catalytic loop (cyan and blue). Within these, important side chains include those of K72 in β-strand 3 (green dotted spheres, top left), E91 in the C-helix (magenta dotted spheres), and R165 in the catalytic loop (blue dotted spheres, bottom right). Phosphorylated T197 in the activation segment (yellow chain, phosphates not shown) interacts with residues in both lobes, including R165, neutralizing positive charges. D166 (cyan dotted spheres) is the catalytic aspartate. For experimental reasons, the bound nucleotide (spheres) is the ATP analogue, adenylyl imidodiphosphate (1cdk.pdb[49]).

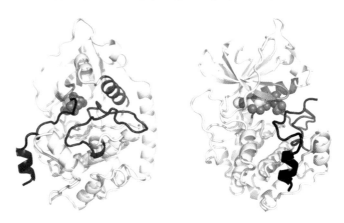

FIG 24.11 Substrate binding at the catalytic site of PKA.

A substrate is mimicked by residues 5–24 of PKI (a PKA inhibitor; black). The key chains in the cleft and the nucleotide are depicted and coloured as in Figure 24.10. The view on the right has been rotated 90° about a vertical axis (1cdk.pdb[49]).

side chain). Binding of substrate in the correct orientation is essential. The serine or threonine hydroxyl group within the motif must be accurately aligned with the terminal phosphate of ATP, as illustrated in **Figure 24.11**. The achievement of this alignment depends critically on the conformation of the activation segment, and in this context the contacts made by pT197 with residues on both lobes are particularly important.

TABLE 24.3 Activation segment phosphorylation

Phosphorylated in the activation segment		Not phosphorylated in the activation segment	
Cyclic AMP-dependent kinase	PKA	Phosphorylase kinase	PHK
Protein kinase B	PKB α, β, γ	Casein kinase I	CKI
Protein kinase C	PKC α, βII	EGF receptor	EGFR
Cyclin-dependent kinase	CDK1 CDK2 CDK7	C-terminal Src kinase	Csk
MAP kinase	ERK1/2	Ca^{2+}/calmodulin kinase	CaMKII
MAP kinase kinase	MEK1	Myosin light chain kinase	MLCK
Raf1 kinase	Raf1		
Ca^{2+}/calmodulin kinase	CaMKI		
Insulin-stimulated kinase	ISPK		
Glycogen synthase kinase	GSK3		
Insulin receptor kinase	IRK		
PDGF receptor	PDGFR		
c-Src family	Src, Yes, Fyn, Fgr, Lyn, Lck, Blk		

Many other kinases are also phosphorylated in their activation segments at positions equivalent to T197 in PKA (see Table 24.3). These include isoforms of PKC, ERKs 1 and 2, MEK1, the cyclin-dependent kinases, CDK2 and CDK7, and the Src family protein tyrosine kinases. Although such phosphorylations are conditional for the activation of these and other kinases, they are not all achieved through autophosphorylation. For example, in PKC, ERK, and the cyclin-dependent kinases, the sequence of the target motif in the activation segment does not match the consensus sequence recognized by the kinase. This implies a requirement for another upstream kinase (for example phosphoinositide-dependent protein kinase 1 for PKC, see page 258).

Other protein kinases have no need for phosphorylation in their activation segments. For instance, on phosphorylase kinase, the side chain carboxyl groups of E182 fulfil a function similar to the phosphate groups of T197 of PKA, neutralizing the positive charge of the catalytic R148 adjacent to D149. (These residues correspond to R165 and D166 in PKA.) Nor does activation segment phosphorylation occur in those kinases in which the catalytic aspartate is preceded by a non-polar residue instead of arginine. Examples are twitchin kinase and MLCK (myosin light chain kinase). In spite of possessing a tyrosine phosphorylation site in its activation segment, the EGF receptor kinase (ErbB1) is not activated by phosphorylation: instead this is achieved through allosteric interactions between the two catalytic domains (see Figure 12.4, page 321).

List of abbreviations

APAF	apoptosis activating factor 1
Akt	see PKB
Arf	GTPase (involved in vesicular transport)
βARK	β-adrenergic receptor kinase (GRK2)
Btk	Bruton's tyrosine kinase
CaM	calmodulin
Caspase	cysteinyl aspartate-specific protease
CBP	histone acetyltransferase, creb-binding protein
DAG	diacylglycerol
Dbl	Rho GEF (diffuse B cell lymphoma oncogene)
EEA1	early endosome antigen 1
FADD	Fas associated protein with death domain
Gα, Gβγ	α and β γ subunits of G proteins
GEF	guanine nucleotide exchange factor
GCAP	guanylate cyclase activating protein
Grb2	SH2-SH3 adaptor (growth factor receptor-bound protein 2)
GRK	G protein coupled receptor kinase (GRK2=βARK)
GRP1	Arf GEF
HP1	non-histone chromosomal protein (heterochromatin protein 1)
InaD	links PKC to PLC in *Drosophila* photoreceptors
IRS1, IRS2	insulin receptor substrate 1 and 2
Nck	SH2-SH3 adaptor
p300	histone acetyltransferase (related to CBP)
p40^{phox}, p47^{phox}	NADPH oxidase subunits
PI3K-p85	P85 subunit of PI 3-kinase

Continued

PKB	protein kinase B (Akt)
PKA, PKC	protein kinase A, protein kinase C
PLC-γ	phospholipase Cγ
PLC-δ	phospholipase Cδ
PLD	phospholipase D
P-rich	proline-rich regions
pS	phosphoserine
pT	phosphothreonine
pY	phosphotyrosine
RasGAP	Ras GTPase activating protein
RGS	regulator of G protein signalling
SARA	Smad anchor for receptor activation
Shc	SH2-containing protein
Smads	transcription factors in TGFβ receptor family pathway
Snf-2	Transcriptional coactivator
Sos	Ras guanine nucleotide exchange factor
Src	Src tyrosine kinase
STATS	signal transducers and activators of transcription
STE4	yeast Gβ of pheromone pathway
Swi-2	transcriptional coactivator
Syt	synaptotagmin
TFIID complex	transcription factor IID complex
TNF	tumour necrosis factor
TRADD	TNF-associated protein with death domain
Vav	adaptor protein and Rho GEF

References

1. Lupas AN, Ponting CP, Russell RB. On the evolution of protein folds: are similar motifs in different protein folds the result of convergence, insertion, or relics of an ancient peptide world?. *J Struct Biol*. 2001;134:191–203.

2. Liu M, Grigoriev A. Protein domains correlate strongly with exons in multiple eukaryotic genomes – evidence of exon shuffling? *Trends Genet* 2004;20:399–403.

3. Graur D, Li W-H. *Fundamentals of Molecular Evolution.* pp. 1–439. Sunderland, MA: Sinauer Associates; 2000.

4. Pawson T, Nash P. Assembly of cell regulatory systems through protein interaction domains. *Science.* 2003;300:445–452.

5. Finn RD, Mistry J, Schuster-Bockler B, et al. Pfam: clans, web tools and services. *Nucleic Acids Res.* 2006;34:D247–D251.

6. Sadowski I, Stone JC, Pawson T. A noncatalytic domain conserved among cytoplasmic protein-tyrosine kinases modifies the kinase function and transforming activity of Fujinami sarcoma virus P130gag-fps. *Mol Cell Biol.* 2006;6:4396–4408.

7. Liu BA, Jablonowski K, Raina M, Arce M, Pawson T, Nash PD. The human and mouse complement of SH2 domain proteins-establishing the boundaries of phosphotyrosine signaling. *Mol Cell.* 2006;22:851–868.

8. Fukuzawa M, Araki T, Adrian I, Williams JG. Tyrosine phosphorylation-independent nuclear translocation of a dictyostelium STAT in response to DIF signaling. *Mol Cell.* 2001;7:779–788.

9. King N, Hittinger CT, Carroll SB. Evolution of key cell signaling and adhesion protein families predates animal origins. *Science.* 2006; 301:361–363.

10. Eck MJ, Shoelson SE, Harrison SC. Recognition of a high-affinity phosphotyrosyl peptide by the Src homology-2 domain of p56lck. *Nature.* 1993;362:87–91.

11. Kimber MS, Nachman J, Cunningham AM, Gish GD, Pawson T, Pai EF. Structural basis for specificity switching of the Src SH2 domain. *Mol Cell.* 2000;5:1043–1049.

12. Songyang Z, Margolis B, Chaudhuri M, Shoelson SE, Cantley LC. The phosphotyrosine interaction domain of SHC recognizes tyrosine-phosphorylated NPXY motif. *J Biol Chem.* 1995;270:14863–14866.

13. Zhou MM, Huang B, Olejniczak ET, et al. Structural basis for IL-4 receptor phosphopeptide recognition by the IRS-1 PTB domain. *Nat Struct Biol.* 1996;3:388–393.

14. Uhlik MT, Temple B, Bencharit S, Kimple AJ, Siderovski DP, Johnson GL. Structural and evolutionary division of phosphotyrosine binding (PTB) domains. *J Mol Biol.* 2006;345:1–20.

15. Renzoni DA, Pugh DJ, Siligardi G, et al. Structural and thermodynamic characterization of the interaction of the SH3 domain from Fyn with the proline-rich binding site on the p85 subunit of PI3-kinase. *Biochemistry.* 1996;35:15646–15653.

16. Yu H, Rosen MK, Shin TB, Seidel Dugan C, Brugge JS, Schreiber SL. Solution structure of the SH3 domain of Src and identification of its ligand-binding site. *Science.* 1992;258:1665–1668.

17. Hu KQ, Settleman J. Tandem SH2 binding sites mediate the RasGAP-RhoGAP interaction: a conformational mechanism for SH3 domain regulation. *EMBO J*. 1997;16:473–483.

18. Lemmon MA. Phosphoinositide recognition domains. *Traffic*. 2003;4:201–213.

19. Takenawa T, Itoh T. Membrane targeting and remodeling through phosphoinositide-binding domains. *IUBMB Life*. 2006;58:296–303.

20. Lander ES, Linton LM, Birren B, et al. Initial sequencing and analysis of the human genome. *Nature*. 2001;409:860–921.

21. Lemmon MA. Pleckstrin homology domains: not just for phosphoinositides. *Biochem Soc Trans*. 2004;32:707–711.

22. Haslam RJ, Koide HB, Hemmings BA. Pleckstrin domain homology. *Nature*. 1993;363:309–310.

23. Imaoka T, Lynham JA, Haslam RJ. Purification and characterization of the 47,000-dalton protein phosphorylated during degranulation of human platelets. *J Biol Chem*. 1983;258:11404–11414.

24. Lemmon MA, Ferguson KM, O'Brien R, Sigler PB, Schlessinger J. Specific and high-affinity binding of inositol phosphates to an isolated pleckstrin homology domain. *Proc Natl Acad Sci USA*. 1995;92:10472–10476.

25. Flesch FM, Yu JW, Lemmon MA, Burger KN. Membrane activity of the phospholipase C-δ1 pleckstrin homology (PH) domain. *Biochem J*. 2005;389:435–441.

26. Tesmer VM, Kawano T, Shankaranarayanan A, Kozasa T, Tesmer JJ. Snapshot of activated G proteins at the membrane: the Gαq-GRK2-G$\beta\gamma$ complex. *Science*. 2005;310:1686–1690.

27. Ferguson KM, Lemmon MA, Schlessinger J, Sigler PB. Structure of the high affinity complex of inositol trisphosphate with a phospholipase C pleckstrin homology domain. *Cell*. 1995;83:1037–1046.

28. Di Paolo G, De Camilli P. Phosphoinositides in cell regulation and membrane dynamics. *Nature*. 2006;443:651–657.

29. Mu FT, Callaghan JM, Steele-Mortimer O, et al. EEA1, an early endosome-associated protein. EEA1 is a conserved α-helical peripheral membrane protein flanked by cysteine 'fingers' and contains a calmodulin-binding IQ motif,. *J Biol Chem*. 1995;270:13503–13511.

30. Gaullier JM, Simonsen A, D'Arrigo A, Bremnes B, Stenmark H, Aasland R. FYVE fingers bind PtdIns(3)P. *Nature*. 1998;394:432–433.

31. Simonsen A, Lippe R, Christoforidis S, et al. EEA1 links PI(3)K function to Rab5 regulation of endosome fusion. *Nature*. 1998;394:494–498.

32. Kutateladze T, Overduin M. Structural mechanism of endosome docking by the FYVE domain. *Science*. 2001;291:1793–1796.

33. Segal AW. How neutrophils kill microbes. *Annu Rev Immunol*. 2005;23:197–223.

34. Chen H, Fre S, Slepnev VI, et al. Epsin is an EH-domain-binding protein implicated in clathrin-mediated endocytosis. *Nature*. 1998;394:793–797.

35. Kretsinger RH. Calcium-binding proteins. *Annu Rev Biochem*. 1976;45: 239–266.

36. Babu YS, Sack JS, Greenhough TJ, Bugg CE, Means AR, Cook WJ. Three-dimensional structure of calmodulin. *Nature*. 1985;315:37–40.

37. Chattopadhyaya R, Meador WE, Means AR, Quiocho FA. Calmodulin structure refined at 1.7 Ångstrom resolution. *J Mol Biol*. 1992;228:1177–1192.

38. Ubach J, Zhang X, Shao X, Südhof TC, Rizo J. Ca^{2+} binding to synaptotagmin: how many Ca^{2+} ions bind to the tip of a C2-domain? *EMBO J*. 1998;17:3921–3930.

39. Verdaguer N, Corbalan-Garcia S, Ochoa WF, Fita I, Gomez-Fernandez JC. Ca^{2+} bridges the C2 membrane-binding domain of protein kinase Cα directly to phosphatidylserine. *EMBO J*. 1999;18:6329–6338.

40. Corbalan-Garcia S, Garcia-Garcia J, Rodriguez-Alfaro JA, Gomez-Fernandez JC. A new phosphatidylinositol 4,5-bisphosphate-binding site located in the C2 domain of protein kinase Cα. *J Biol Chem*. 2003;278:4972–4980.

41. Sutton RB, Sprang SR. Structure of the protein kinase Cβ phospholipid-binding C2 domain complexed with Ca^{2+}. *Structure*. 1998;6:1395–1405.

42. Miller J, McLachlan AD, Klug A. Repetitive zinc-binding domains in the protein transcription factor IIIA from *Xenopus* oocytes. *EMBO J*. 1985;4:1609–1614.

43. Matthews JM, Sunde M. Zinc fingers – folds for many occasions. *IUBMB Life*. 2002;54:351–355.

44. Li J, Chen X, Yang H, et al. The zinc finger transcription factor 191 is required for early embryonic development and cell proliferation. *Exp Cell Res*. 2006;312:3990–3998.

45. Sato M, Tomizawa T, Koshiba S, Inoue M, Kigawa T, Yokoyama S. Solution structures of the C2H2 type zinc finger domain of human Zinc finger protein 24. Deposited RCSB 2005. To be published.

46. Xu RX, Pawelczyk T, Xia TH, Brown SC. NMR structure of a protein kinase C- phorbol-binding domain and study of protein-lipid micelle interactions. *Biochemistry*. 1997;36:10709–10717.

47. Ono Y, Fujii T, Igarashi K, et al. Phorbol ester binding to protein kinase C requires a cysteine-rich zinc finger-like sequence. *Proc Natl Acad Sci USA*. 1989;86:4868–4871.

48. Johnson LN, Noble ME, Owen DJ. Active and inactive protein kinases: structural basis for regulation. *Cell*. 1996;85:149–158.

49. Bossemeyer D, Engh RA, Kinzel V, Ponstingl H, Huber R. Phosphotransferase and substrate binding mechanism of the cAMP-dependent protein kinase catalytic subunit from porcine heart as deduced from the 2.0 Å structure of the complex with Mn^{2+} adenylyl imidodiphosphate and inhibitor peptide PKI(5-24). *EMBO J*. 1993;12:859.

Index

Note: Page numbers in *italics* refer to figures and tables

BMP2 (bone morphogenetic protein-2), 628
Bordetella pertussis, 143, 144
bradykinin, 26
Brat (brain tumor protein), 436
bride-of-sevenless (boss), 329
Brown-Séquard, Charles Edouard, 5
Bruegel, Pieter, 14
Btk Bruton's tyrosine kinase, 786
Buck, Linda, 178
Bungarus multicinctus, 48
busulphan, *740*
Butenandt, Adolf Friedrich Johann, 274

C

c-Abl, 742, 746, *747*, 748, 752
  inhibitor of, 750–752
  regulation of, *749*
  structure of, *748*
c-Cbl, 519
c-Fos, 342, 616
c-Jun, 342, 616
c-myc, 342–343, 614, 709
C-Raf, 334–335
C-TAK1, 346
C1 domains, 256
  as protein–protein interaction domain, 256
C2 domain, 224, 256–257, 780–781
C3 and C4 domains, 257
Ca^{2+}, 519
  free, bound, and trapped, 189–190
  regulation of adenylyl cyclase, 141
  polypeptide modules binding, 779
Ca^{2+} and Mg^{2+}
  approximate levels of, *189*
  distinguishing, 188–189
Ca^{2+}-ATPase (PMCA), *222*
Ca^{2+}-ATPase (SERCA), 190, *222*
Ca^{2+} by cyclic ADP-ribose and NAADP, elevation of, 204–206
Ca^{2+} by sphingosine-1-phosphate (S1P), elevation of, 206–208
Ca^{2+} calmodulin, 619
  pathway, 353
Ca^{2+}-calmodulin-activated protein kinases (CaM-kinases), 226, 228
CaMKI, 228
CaMKII, 228–229

CaMKIII, 229
CaMKIV, 229
Ca^{2+}-calmodulin dependent enzymes
  Ca^{2+}-calmodulin-sensitive adenylyl cyclases and phosphodiesterase, 231–232
  calcineurin, 230–231
  nitric oxide synthase, 232–233
  plasma membrane Ca^{2+} ATPase (PMCA), 229
  Ras guanine nucleotide exchange factor, 229–230
Ca^{2+}-calmodulin-sensitive adenylyl cyclases and phosphodiesterase, 231
Ca^{2+} channels, store-operated, 210–211
Ca^{2+}-induced Ca^{2+}-release, 202
Ca^{2+} influx through plasma membrane channels, 208–209
  receptor-operated channels, 209
  TRPM2 channels, 209
  voltage-operated channels, 208–209
Ca^{2+} ionophores, to rise Ca^{2+}, 191
Ca^{2+} microdomains and global cellular signals, 214
  in electrically excitable cells, 214–215
  in non-excitable cells, 216
Ca^{2+} mobilization and Ca^{2+} entry, 193–194, *194*
Ca^{2+}-promoted Ras inactivator, *see* CAPRI
Ca^{2+}-sensitive photoproteins, 192
Ca^{2+} signals
  decoding, 224
  in electrically excitable cells, 214–215
    cardiac muscle, 215
    nerve cells, 215
    skeletal muscle, 214–215
Caax motif, 105
cachexia, 493
CADASIL, 726
cadherins, 387–392, *387*, *389*, 426
  in central nervous system, 439
  in contact inhibition, 438
  subfamilies, 390–391
cADPR (cyclic ADP-ribose), 205
  structures, *204*

Caenorhabditis elegans, 255, 278, 328, 331–332, 567, 585, *586*, 599, 726
calabar bean, 16
calcineurin, *223*, 230–231, 520, 682
calcineurin B, 519
calcium
  and evolution, 187–188
  as negative regulator, 167–170
calcium-binding by proteins, 221, *222–223*
  Ca^{2+} signals, decoding, 224
  calmodulin, 225–226
    kinases, regulated by, 226–229
  polypeptide modules binding Ca^{2+}, 224
  troponin C, 226
calcium-binding motifs and domains, 779
calcium-dependent enzymes, not regulated by calmodulin, 233
  calpain, 234
  cytoskeletal proteins, 236
  DAG kinase, 234
  neuronal calcium sensors, 233–234
  Ras GEFs and GAPs, 234–236
  synaptotagmin, 234
calcium effectors
  Ca^{2+}-calmodulin dependent enzymes
    Ca^{2+}-calmodulin-sensitive adenylyl cyclases and phosphodiesterase, 231–232
    calcineurin, 230–231
    nitric oxide synthase, 232–233
    plasma membrane Ca^{2+} ATPase (PMCA), 229
    Ras guanine nucleotide exchange factor, 229–230
  calcium-binding by proteins, 221
    Ca^{2+} signals, decoding, 224
    calmodulin, 225–226
    kinases, regulated by calmodulin, 226–229
    polypeptide modules binding Ca^{2+}, 224
    troponin C, 226
  calcium-dependent enzymes, not regulated by calmodulin, 233
    calpain, 234
    cytoskeletal proteins, 236
    DAG kinase, 234

ERK (extracellular signal regulated
        kinase), 291, 349–353, *583*
    activation of, 250–251
ERK1, 349, 592
ERK2, 349
ERK3, 350
ERK4, 350
ERK5, 351
erlotinib, *743*, 745
erysipelas, 488
erythropoietin, *see* EPO
eserine, 16, 43
estradiol, 3, 4, 292
estrogen receptor, *see* ER
Ets (E-26), 342
eubacteria, 132
European Pharmacopoeia, 11
EVH1 domains, 776
EXEL 7647/EXEL 0999, *743*
exocytosis, 234
extracellular matrix, 397, 400
extracellular signal regulated kinase,
        see ERK

F

F-actin, 236
Fab fragment, *745*
FAK (focal adhesion kinase), 400, *401*,
        403, 404–406
farnesyl transferase inhibitors, 105,
        106
FAS ligand, 489
FERM domains, 776
FGF (fibroblast growth factor), 432,
        617, 618
FGFR (fibroblast growth factor
        receptor), 617
fibrinogen, 378, 383
fibroblast growth factor, *see* FGF
fibroblast growth factor receptor, *see*
        FGFR
fibronectin, 395, 418, 625
Finkel–Biskis–Jinkins murine
        osteosarcoma virus, 578
first messengers, 21, *22–23*
    hormones, 25
        common aspects, 28
        cytokines, 26
        growth factors, 25
        lipophilic messengers,
            27–28

neurotransmitters and
        neuropeptides, 27
    vasoactive agents, 26–27
intracellular messengers, 27, 28
    binding affinity measurement,
        30–31
    binding heterogeneity, 30
    cAMP, 34
    down-regulation of receptors,
        33–34
    K_D and EC_{50}, 31–32
    ligands to receptors, binding of,
        29–30
    spare receptors, 32–33
5'-AMP, 246
FK506, 200, 683
FKBP12, 200, 683
fluorescence resonance energy
        transfer, *see* FRET
fluorescent Ca^{2+} indicators, 192–193
fluorophores, 196
fMLP, 494
focal adhesion complexes, 387, 400
focal adhesion kinase, *see* FAK
follicle stimulating hormone, *see* FSH
formylmethionine (fMet), 486
forskolin, 142–143, *143*
Fos, 342, *580*
Fos-c (feline osteosarcoma cellular
        homologue), 592
Foster's Textbook of Physiology, 5
4E-BP1, 566
FRET (fluorescence resonance energy
        transfer), 114–117, *116*, 197
Fringe, 707
Frizzled (Fz), 66, 590
fruit fly, 328–331
FSH (follicle stimulating hormone), *22*,
        25, 249–250
Fura2, 193, *193*
furin, 489
Fus3, 344, 345
FXR, 279, 282, 290
Fyn, 517
FYVE, 777–778

G

G-protein-linked receptors, 354
G protein receptor kinase (GRK2/β
        ARK), 39
G protein receptor kinase family, 98

G proteins, 83, 494
    7TM receptor linkage, 83
    α-subunits, 88
        diversity, 89–92
        evolutionary relationship, *91*
        functions, *90*
        sites interacting with
            membranes/proteins, 92–94
    βγ-Subunits, 94–97
        functions, 97
        negative feedback signals, 98
        signal transmission, 97
    activation without subunit unit
        dissociation, 111–119
    GTPase cycle, 84–86
    heterotrimeric, 83–84
    monomeric, 83
    receptor affinity modulation, 87
    switching off activity, 86–88
GABA (γ-aminobutyric acid), 27, 55
GABA receptor, 45
gag, 101, 742
gallamine, 45
γ-aminobutyric acid, *see* GABA
γ-secretase, 439, 707, 708, 714, 722
GAPs (GTPase activating proteins),
        108, 235
gastrin, 21, *22*
GCAP (guanylyl cyclase activating
        protein), 168
GCN4, 578
gefitinib, *743*, 746
GEFs (guanine nucleotide exchange
        factors), 111, 236
gelsolin, *223*, 224, 236
gene cassettes, 711–712
gene complex, 711–712
gene expression
    activation, 614
    regulation, 422
    repression, 614
gene theory, 701
germline mutations, 668
GFAP (glial fibrillary acidic protein),
        629
GFP (green fluorescent protein), 197,
        276
GHRH (growth hormone releasing
        hormone), 145
Gilman, Alfred G., 1
Gley, Eugene, 544–545
Glivec/Gleevec, 736, 750